The Beginning of the Age of Mammals

The Beginning of

the Age of Mammals

KENNETH D. ROSE

THE JOHNS HOPKINS UNIVERSITY PRESS, *Baltimore*

© 2006 The Johns Hopkins University Press

All rights reserved. Published 2006
Printed in the United States of America on acid-free paper

9 8 7 6 5 4 3 2

The Johns Hopkins University Press
2715 North Charles Street
Baltimore, Maryland 21218-4363
www.press.jhu.edu

Frontispiece: Eocene rodent *Paramys,* reconstruction drawing by
Jay H. Matternes © 1993

Library of Congress Cataloging-in-Publication Data

Rose, Kenneth David, 1949–
    The beginning of the age of mammals / Kenneth D. Rose.
        p.   cm.
    Includes bibliographical references and index.
    ISBN 0-8018-8472-1 (acid-free paper)
    1. Paleontology—Cenozoic.   2. Mammals—History.
3. Mammals—Evolution.   4. Life—Origin.   I. Title.
QE735.R67   2006
569—dc22   2006008096

A catalog record for this book is available from the British
Library.

The last printed pages of the book are an extension of this
copyright page.

For Jennie, Katie, and Chelsea

CONTENTS

Preface xi
Acknowledgments xiii

1 Introduction 1
THE EARLY CENOZOIC MAMMALIAN RADIATION 2
TIMING OF THE CROWN-THERIAN RADIATION 3
MAMMALIAN PHYLOGENY, INTERRELATIONSHIPS,
  AND CLASSIFICATION 5
GEOCHRONOLOGY AND BIOCHRONOLOGY OF THE
  EARLY CENOZOIC 8
PALEOGEOGRAPHIC SETTING DURING THE BEGINNING
  OF THE AGE OF MAMMALS 17
PALEOCENE-EOCENE CLIMATE AND FLORA 20
ORGANIZATION OF THE VOLUME 21

2 Mammalian Skeletal Structure and Adaptations 23
SKULL 24
DENTITION 26
POSTCRANIAL SKELETON 30
SKELETAL ADAPTATIONS 34

3 The Origin of Mammals 41
WHAT IS A MAMMAL? 41
THE EVOLUTIONARY TRANSITION TO MAMMALS 44

4 Synopsis of Mesozoic Mammal Evolution 48
HISTORICAL BACKGROUND 48
THE OLDEST MAMMALS 50
DOCODONTA 55
MULTITUBERCULATA 56
EUTRICONODONTA 61
SYMMETRODONTS 63
EUPANTOTHERES 64
TRIBOSPHENIC MAMMALS 66
MESOZOIC MAMMALS OF UNCERTAIN AFFINITY 70

5 Metatheria: Marsupials and Their Relatives 72
BASAL METATHERIANS 74
PRIMITIVE MARSUPIALS 76

6 Earliest Eutherian Mammals 88

7 Cimolesta 94
DIDELPHODONTA AND OTHER PRIMITIVE CIMOLESTA 94
DIDYMOCONIDAE 97
PANTOLESTA 99
APATOTHERIA 103
TAENIODONTA 105
TILLODONTIA 110
PANTODONTA 114

8 Creodonta and Carnivora 119
CREODONTA 119
CARNIVORA 126

9 Insectivora 138
LEPTICTIDA 140
LIPOTYPHLA 143

10 Archonta: Bats, Dermopterans, Primates, and Tree Shrews 156
CHIROPTERA 157
DERMOPTERA 162
PRIMATES AND PLESIADAPIFORMES 166
SCANDENTIA 197

11 "Edentates:" Xenarthra and Pholidota 198
XENARTHRA 200
PHOLIDOTA 204

12 Archaic Ungulates 211
OLDEST UNGULATE RELATIVES 213
CONDYLARTHRA: ARCHAIC UNGULATES 215
ARCTOSTYLOPIDA 225
MERIDIUNGULATA: ENDEMIC SOUTH AMERICAN UNGULATES 226
DINOCERATA 238

13 Altungulata: Perissodactyls, Hyraxes, and Tethytheres 241
PERISSODACTYLA 244
PAENUNGULATA 257

14 Cete and Artiodactyla 271
CETE AND CETACEA 271
ARTIODACTYLA 285

15 Anagalida: Rodents, Lagomorphs, and Their Relatives 306
PRIMITIVE ASIAN ANAGALIDANS AND POSSIBLE ANAGALIDANS 307
MACROSCELIDEA 310
GLIRES 312

16 Reflections and Speculations on the Beginning of the Age of Mammals 335
EARLY CENOZOIC MAMMAL RECORD 336
SYNOPSIS OF PALEOCENE AND EOCENE MAMMALS 340
A FINAL NOTE 347

Literature Cited 349
Index 399

PREFACE

THIS BOOK IS THE OUTCOME of a decade-long project that began when Robert Harrington, then the science editor for the Johns Hopkins University Press, invited me to write a book on fossil mammals. The need for such a book became apparent from a graduate seminar in mammalian evolution I have taught over the past 20 years at the Johns Hopkins University. While we have witnessed the primary literature in the field increase at an astonishing pace, it became evident that there was a real dearth of general books on the subject. Except for Savage and Long's (1986) *Mammal Evolution* (which is now outdated and gave only a superficial account of many Paleogene groups), there was no available book that synthesized basic data on the extant mammals together with a survey of the rapidly improving mammalian fossil record to provide an overview of mammalian evolution. *The Beginning of the Age of Mammals* is intended to help fill this void by presenting an in-depth account of current knowledge about mammalian evolution in the Early Cenozoic. It is designed to provide both graduate and undergraduate students with a comprehensive summary of the diversity and rich history of mammals, focusing on the early radiations of living clades and their archaic contemporaries. I hope it may serve as a useful reference for professionals as well.

This is a book about fossils. The focus is on the anatomy preserved in the fossil record, and what it implies about relationships, phylogeny, evolution, behavior, paleoecology, and related issues. Other topics, such as geology, paleoflora, climate, and molecular systematics are discussed where they are pertinent, but they are subsidiary to the principal objective, which is to summarize the mammalian fossil record. I have chosen to concentrate on the Early Cenozoic part of that record not just because that is my personal interest, but also because it is the most critical part of the fossil record with regard to the origin and early adaptive radiations of almost all the major clades of extant mammals. Furthermore, substantial recent advances in our knowledge of mammals during this pivotal interval make this summary timely.

I have endeavored to survey the literature through the end of 2004 and have added a few particularly pertinent references that are more recent, in order to furnish a review of all higher taxa of Paleocene and Eocene mammals that is as current as possible. Treatment of different groups is unavoidably uneven, a reflection of multiple factors, including the Early Cenozoic diversity of particular groups, the interest level they have generated, and the intensity at which they have been studied, especially recently. Judgments had to be made as to what was significant enough to be included in a review of this sort and where to include more detail. I hope there have not been serious omissions. I have borrowed liberally from the classification and range data presented by McKenna and Bell (1997, 2002) and have benefited greatly from their vast experience. Although I have not always agreed with their arrangement (and have noted in the text where modifications were necessary), their monumental compilation provided the essential framework, without which this book would have been far more difficult to achieve.

One of the most important aspects of this kind of book is the quality and scope of illustrations. Rather than prepare new figures or redraw existing ones in an attempt at uniformity, I opted to reproduce the best available illustrations of a wide diversity of fossil mammals. The drawback of this approach is that multiple styles of illustration are often combined in the same composite figure. However, I believe the benefit of using original illustrations significantly outweighs the aesthetic of redrawing them all in the same style, with its inherent risk of introducing inaccuracies. For ease of comparison, I have taken liberties in sizing and reversing many images, with apologies to the original artists for anomalies of lighting that may result. I have tried to illustrate at least one member of each Early Cenozoic family (except a few obscure families, and some families of the highly diverse artiodactyls and rodents). Figures were selected to give readers an impression of the diversity of fossil mammals, the state of the evidence, and the most important specimens or taxa.

Throughout the book, my goal has been not just to present current interpretations of the mammalian fossil record but also to highlight the quality of the evidence and analyses on which these inferences are based. I have tried to indicate where the data are particularly sound and convincing, as well as where the evidence is more tenuous or ambiguous. The latter examples should be especially fruitful areas for further research.

I hope that I have been able to impart some of my enthusiasm for mammalian paleontology, and to demonstrate that fossils are not just curiosities but are the key to understanding the extraordinary history of life. George Gaylord Simpson perhaps best captured the allure of paleontology in his classic *Attending Marvels*, recounting his 1930–1931 Scarritt Expedition to Patagonia in search of fossil mammals (Simpson, 1965: 82):

Fossil hunting is far the most fascinating of all sports. I speak for myself, although I do not see how any true sportsman could fail to agree with me if he had tried bone digging. It has some danger, enough to give it zest and probably about as much as in the average modern engineered big-game hunt, and the danger is wholly to the hunter. It has uncertainty and excitement and all the thrills of gambling with none of its vicious features. The hunter never knows what his bag may be, perhaps nothing, perhaps a creature never before seen by human eyes. Over the next hill may lie a great discovery! It requires knowledge, skill, and some degree of hardihood. And its results are so much more important, more worth while, and more enduring than those of any other sport! The fossil hunter does not kill; he resurrects. And the result of his sport is to add to the sum of human pleasure and to the treasures of human knowledge.

ACKNOWLEDGMENTS

AN UNDERTAKING OF THIS SORT could not be accomplished without the support, assistance, and input of many people. First, I thank my colleagues in the Center for Functional Anatomy and Evolution (FAE), Valerie De Leon, Chris Ruff, Mark Teaford, and Dave Weishampel, for their encouragement, illuminating discussions, sharing of knowledge, and numerous other favors. I also thank the FAE graduate students, past and present, who have helped to inspire this book. Many of them have read and corrected chapters, offered helpful insights, or provided information or other assistance, which is much appreciated. Jay Mussell and Mary Silcox deserve special mention for many stimulating discussions and enlightening me on several topics. I am especially grateful to Shawn Zack, who has freely shared his broad knowledge of mammalian fossils and the literature, and who has helped with countless tasks in the preparation of this book. The assistance of Arlene Daniel in the FAE administrative office has been much appreciated during all stages of the project.

I am enormously indebted to Anne Marie Boustani, who undertook the formidable task of scanning, resizing, arranging, and labeling all of the figures in the book, in some cases multiple times, to achieve the best possible result. In so doing she has made a huge and fundamental contribution to this work. She also drafted all the cladograms and a number of original figures. Her dedication, perseverance, and cheerfulness throughout this painstaking process are very much appreciated.

Part of the preparation of this volume was undertaken during my tenure of an Alexander von Humboldt Award in the Institut für Paläontologie at the University of Bonn, Germany, in 2003–2004, for which I thank the A. von Humboldt Foundation. The faculty, staff, and students at the Institut für Paläontologie, especially Prof. Wighart von Koenigswald, created an ideal environment for this work, and I am very grateful for their support.

Throughout preparation of this book, I have consulted with numerous colleagues about their areas of expertise. Their generosity in providing advice, information, casts or images, permission to reproduce illustrations, and other assistance has been overwhelming and has been instrumental in completion of the work. I extend my gratitude to them all. Almost half of those listed sent original photographs, slides, drawings, or electronic images, which often required considerable time and effort on their part. I have attempted to acknowledge here all those who have contributed; nevertheless, as this project has been a decade in development, inadvertent omissions are likely, and I ask the indulgence of anyone overlooked. My appreciation goes to David Archibald, Rob Asher, Chris Beard, Lílian Bergqvist, Jon Bloch, José Bonaparte, Louis de Bonis, Tom Bown, Percy Butler, Rich Cifelli, Russ Ciochon, Bill Clemens, Jean-Yves Crochet, Fuzz Crompton, Demberelyin Dashzeveg, Mary Dawson, Daryl Domning, Stéphane Ducrocq, Bob Emry, Burkart Engesser, Jörg Erfurt, John Fleagle, Ewan Fordyce, Dick Fox, Jens Franzen, Eberhard Frey, Emmanuel Gheerbrant, Philip Gingerich, Marc Godinot, Gabriele Gruber, Gregg Gunnell, Jörg Habersetzer, Gerhard Hahn, Sue Hand, Jean-Louis Hartenberger, Ron Heinrich, Jerry Hooker, Jim Hopson, Yaoming Hu, Bob Hunt, Jean-Jacques Jaeger, Christine Janis, Farish Jenkins, Dany Kalthoff, Zofia Kielan-Jaworowska, Wighart von Koenigswald, Bill Korth, Dave Krause, Conny Kurz, Brigitte Lange-Badré, Chuankuei Li, Jay Lillegraven, Alexey Lopatin, Spencer Lucas, Zhexi Luo, Bruce MacFadden, Thomas Martin, Malcolm McKenna, Jim Mellett, Jin Meng, Michael Morlo, Christian de Muizon, Xijun Ni, Mike Novacek, Rosendo Pascual, Hans-Ulrich Pfretzschner, Don Prothero, Rajendra Rana, Tab Rasmussen, John Rensberger, Guillermo Rougier, Don Russell, Bob Schoch, Erik Seiffert, Bernard Sigé, Denise Sigogneau-Russell, Elwyn Simons, Gerhard Storch, Jean Sudre, Hans-Dieter Sues, Fred Szalay, Hans Thewissen, Suyin Ting, Yuki Tomida, Yongsheng Tong, Bill Turnbull, Mark Uhen, Banyue Wang, Xaioming Wang, John Wible, Jack Wilson, and Shawn Zack.

Wherever possible, illustrators have been acknowledged as well (see the last printed pages of the book). Special thanks are due the following scientific illustrators for allowing reproduction, and in many cases providing images, of their work: Doug Boyer, Bonnie Dalzell, Utako Kikutani, John Klausmeyer, Mark Klingler, Karen Klitz, Jay Matternes, Bonnie Miljour, Mary Parrish, and especially Elaine Kasmer, my illustrator for many years.

I have also benefited from the experience and wisdom of friends and esteemed colleagues who reviewed sections of the text for accuracy, including Rich Cifelli, Mary Dawson, Daryl Domning, John Fleagle, John Flynn, Ewan Fordyce, Jerry Hooker, Christine Janis, Zofia Kielan-Jaworowska, Wighart von Koenigswald, Zhexi Luo, Thierry Smith, Scott Wing, and Shawn Zack. I am grateful to all of them for numerous corrections and clarifications, which improved the text substantially. I am especially indebted to Bill Clemens and Malcolm McKenna, whose critical reading of the entire text and sage advice has been invaluable. Although I have relied on the counsel of these distinguished authorities to avoid errors, omissions, and ambiguities, I did not always follow their suggestions, and any shortcomings that remain are, of course, my own responsibility.

I take this opportunity to acknowledge the encouragement and guidance of several people who fostered my interest in paleontology during my student years (listed more or less chronologically): Dave Stager, Margaret Thomas, Bob Salkin, Don Baird, Nick Hotton, Clayton Ray, Glenn Jepsen, Elwyn Simons, George Gaylord Simpson, Bryan Patterson, and Philip Gingerich. Without their support, particularly at pivotal periods in my life, I would not be a vertebrate paleontologist today.

This project would never have made it to fruition without the able guidance of my editor at the Johns Hopkins University Press, Vincent Burke. To him, as well as to Wendy Harris, Linda Forlifer, Martha Sewall, and Carol Eckhart at the press, and to Peter Strupp, Cyd Westmoreland, and the staff at Princeton Editorial Associates, I extend my sincere thanks for seeing this volume through.

Last but not least, I am most grateful to my family—my wife Jennie and daughters Katie and Chelsea—for their unwavering faith in me and their steadfast encouragement throughout the long gestation of this project, especially when it seemed unachievable. They are to be credited with its completion.

The Beginning of the Age of Mammals

1

Introduction

MAMMALS ARE AMONG THE MOST successful animals on earth. They occupy every major habitat from the equator to the poles, on land, underground, in the trees, in the air, and in both fresh and marine waters. They have invaded diverse locomotor and dietary niches, and range in size from no larger than a bumblebee (the bumblebee bat *Craseonycteris*: body length 3 cm, weight 2 g) to the largest animal that ever evolved (the blue whale *Balaenoptera*: body length 30 m, weight > 100,000 kg). Just over a decade ago, the principal references recognized 4,327 or 4,629 extant mammal species in 21–26 orders (Corbet and Hill, 1991; Wilson and Reeder, 1993), the discrepancy mainly in marsupial orders. The most recent account now recognizes 29 orders of living mammals (the increase mainly reflecting the breakup of Insectivora), with more than 5,400 species in 1,229 genera (Wilson and Reeder, 2005). But many times those numbers of genera and species are extinct. Indeed, McKenna and Bell (1997) recognized more than 4,000 extinct mammal genera, many of which belong to remarkable clades that left no living descendants. The great majority of extinct taxa are from the Cenozoic, the last one-third of mammalian history. What were these extinct forms like? What made them successful, and what led to their eventual demise? How were they related to extant mammals? When, where, and how did the ancestors of modern mammals evolve, and what factors contributed to the survival of their clades?

This book addresses those questions by focusing on the mammalian radiation during the Paleocene and Eocene epochs, essentially the first half of the Cenozoic Era. Although this radiation has attracted far less popular interest than that of dinosaurs, it was a pivotal interval in the history of vertebrates, which set the stage for the present-day mammalian fauna, as well as our own evolution. At its start, the end of the Cretaceous Period, the last nonavian dinosaurs disappeared, leaving a vast, uninhabited ecospace. Mammals quickly moved in, partitioning this landscape in new ways. They were not, however, the first mammals.

Mammals evolved from their synapsid ancestors around the end of the Triassic Period, more than 200 million years ago, and coexisted with dinosaurs, other archosaurs, and various reptiles (among other creatures) for at least 140 million years during the Mesozoic Era. But during that first two-thirds of mammalian history, innovation was seemingly stifled—at least, in comparison to what followed in the early Cenozoic. It is fair to say that mammals *survived* during the Mesozoic but, with a few notable exceptions, rarely flourished. The biggest mammals during that era were little larger than a beaver, and only a few reached that size. Most Mesozoic mammals were relatively generalized compared to the mammals that evolved within the first 10–15 million years of the Cenozoic—although recent discoveries hint at greater diversity than was previously known. Kielan-Jaworowska et al. (2004) present a thorough, current account of mammalian evolution during the Mesozoic.

Like most clades, mammals were severely affected by the terminal Cretaceous mass extinctions. Most Mesozoic mammal radiations became extinct without issue. Indeed, two-thirds of the 35 families of Late Cretaceous mammals listed by McKenna and Bell (1997) disappeared at the end of the Cretaceous. In the northern Western Interior of North America, mammalian extinctions were even more severe, affecting 80–90% of lineages (Clemens, 2002). A small number of clades crossed the Cretaceous/Tertiary (K/T) boundary, most notably, several lineages of multituberculates, eutherians, and marsupials; the latter two groups quickly dominated the vertebrate fauna on land. (Multituberculates are an extinct group of small, herbivorous mammals that were the most successful Mesozoic mammals; see Chapter 4.) Those few lineages that survived the K/T extinctions are the mammals that ultimately gave rise to the diversity of Cenozoic mammals.

It is notable that all three of these groups had existed for at least as long before the K/T boundary as after it, yet the fossil evidence suggests that only the multituberculates radiated widely during the Mesozoic. The Mesozoic was the heyday of multituberculates. They shared the Earth with dinosaurs for 90 million years or more, becoming diverse and abundant in many northern faunas, only to be outcompeted by other mammals before the end of the Eocene. Even those other mammals—metatherians and eutherians (often grouped as therians, or crown therians)—had diverged from a common stem by 125 million years ago. But this divergence occurred well after the multituberculate radiation was under way. Perhaps competition from multituberculates and other archaic mammals—as well as archosaurs—prevented metatherians and eutherians from undergoing major adaptive radiations during the Mesozoic. Whatever the reason, during the Cretaceous, these groups failed to attain anything close to the morphological or taxonomic diversity they would achieve in the first 10–15 million years of the Cenozoic.

THE EARLY CENOZOIC MAMMALIAN RADIATION

The fossil record documents an extensive and rapid—often described as "explosive"—adaptive radiation of mammals during the first third of the Cenozoic, characterized by a dramatic increase in diversity of therian mammals soon after the mass extinctions at the end of the Cretaceous (e.g., McKenna and Bell, 1997; Alroy, 1999; Novacek, 1999; Archibald and Deutschman, 2001). Nearly all of the modern mammal orders, as well as many extinct orders, first appear in the fossil record during this interval (Rose and Archibald, 2005). This era was the "Beginning of the Age of Mammals" alluded to by Simpson (1937c, 1948, 1967).

The adaptive radiation was particularly intense soon after the final extinction of nonavian dinosaurs at the K/T boundary. In the famous Hell Creek section of Montana, for instance, Archibald (1983) found that diversity increased from an average of about 20 mammal species immediately following the K/T boundary to 33 species within the first half-million years, 47 after 1 million years, and 70 after 2–3 million years. For the same intervals, the number of genera rose from about 14 to 30, then 36, and finally 52. Although some of these numbers could be inflated as a result of reworking (discovered subsequent to Archibald's analysis), the overall pattern was upheld in a more recent study by Clemens (2002), who reported that 70% of early Puercan mammals of Montana were alien species new to the northern Western Interior of North America. Similarly, Lillegraven and Eberle (1999) observed a significant mammalian radiation, particularly involving condylarths, at the beginning of the Cenozoic (after the disappearance of nonavian dinosaurs) in the Hanna Basin of southern Wyoming. Only nine mammal species, including just two eutherians, were present in uppermost Cretaceous strata. By contrast, 35 species (75% of them eutherians), almost all presumed immigrants, were recorded from the earliest Paleocene. They further reported that "major experimentations in dental morphology and increasing ranges of body sizes had developed within 400,000 years of the [K/T] boundary" (Lillegraven and Eberle, 1999: 691).

Based on ranges provided by McKenna and Bell (1997), 52 families of mammals are known worldwide from the early Paleocene, but only eight of them continued from the Late Cretaceous—more than 80% were new (Fig. 1.1). Only five therian families are known to have crossed the K/T boundary, two of which are present in late Paleocene or Eocene sediments but have not yet been found in the early Paleocene. On a more local level, Lofgren (1995) reported that the survival rate of mammalian species across the K/T boundary in the Hell Creek area of Montana was only about 10%.

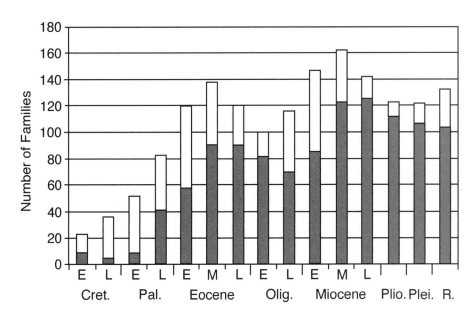

Fig. 1.1. Family diversity of mammals from the Cretaceous to the present. Bars indicate the number of families recorded from each interval; the shaded portion denotes the number of those families also present in the immediately preceding interval. Key: Cret., Cretaceous; E, early; L, late; M, middle; Olig., Oligocene; Pal., Paleocene; Plei., Pleistocene; Plio., Pliocene, R., Recent. (Compiled from McKenna and Bell, 1997, with minor modifications.)

Thus there appears to have been a sharp decline in mammalian diversity at the end of the Cretaceous, followed by a fairly rapid rise in diversity soon after the K/T boundary.

Although the geographic source of many of the newcomers is uncertain, it is important to note that many early Paleocene metatherians and eutherians can plausibly be derived either from other early Paleocene forms or from known Late Cretaceous therian families (including some that did not cross the boundary). For these mammals, it is not necessary to postulate long periods of unrecorded evolution. But it is questionable whether all the diversity that emerged in the Paleocene can be traced to the small number of lineages that we know crossed the K/T boundary. Could the alien species of the northern Western Interior represent clades that were evolving in areas that have not been sampled? And if so, could these clades have existed for a substantial period during the Mesozoic? The answers to these questions are unknown. However, as shown in Fig. 1.1, the fossil record documents that family-level diversity continued to increase through the middle Eocene, then declined somewhat into the early Oligocene, after which it rose again to an all-time high in the middle Miocene (a standing diversity of 162 families). Notably, up to the middle Eocene, the number of new families equaled or exceeded the number that continued from the previous interval.

The present volume is an attempt to summarize current knowledge of the record of this extensive Paleocene-Eocene radiation and the roles of mammals in the world of the Early Cenozoic, which are essential for understanding the structure and composition of present-day ecosystems. This volume focuses on the fossil evidence of these early mammals and what their anatomy indicates about interrelationships, evolution, and ways of life. First it is necessary, however, to touch on several issues that affect the interpretation of that record. These include the timing of the radiation, how phylogenetic relationships are established, the interrelationships and classification of mammals, and the chronologic framework of the Early Cenozoic.

TIMING OF THE CROWN-THERIAN RADIATION

The question of when the therian radiation took place is a contentious issue, whose answer depends on the kind of data employed—paleontological (morphological) or molecular. There are three principal models of the timing of origin and diversification of placental mammals (Archibald and Deutschman, 2001), which also apply generally to the therian radiation (Fig. 1.2):

1. *The explosive model,* in which mammalian orders both originated and diversified in a short period of about 10 million years after the K/T boundary (see also Alroy, 1999; Benton, 1999; Foote et al., 1999);
2. *The long-fuse model,* in which mammalian intraordinal diversification was mostly post-Cretaceous, but interordinal divergence took place in the Cretaceous, when stem taxa of the orders existed (Douady and Douzery, 2003; Springer et al., 2003); and
3. *The short-fuse model,* in which ordinal origin and diversification occurred well back in the Cretaceous (e.g., Springer, 1997; Kumar and Hedges, 1998).

Paleontological evidence generally supports either the explosive model or the long-fuse model, whereas molecular evidence generally supports the short-fuse model.

Let us consider the molecular evidence first. Although this book is about the fossil record, the impact of recent molecular studies on our understanding of mammalian interrelationships and divergence times has been substantial and cannot be ignored. It is chiefly molecular evidence (genetic distance, as measured by differences in nucleotide sequences

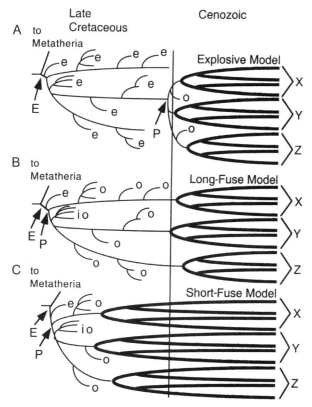

Fig. 1.2. Models of the eutherian mammalian radiation: (A) explosive; (B) long fuse; (C) short fuse. Key: E, Eutheria; e, eutherian stem taxon; io, stem taxon to more than one ordinal crown group; o, ordinal stem taxon; P, Placentalia; X,Y,Z, placental orders. (From Archibald and Deutschman, 2001).

of mitochondrial and nuclear genes) that has been used to suggest that many therian mammal orders originated and diversified during the Cretaceous, some of them more than 100 million years ago (e.g., Hedges et al., 1996; Springer, 1997; Kumar and Hedges, 1998; Easteal, 1999; Adkins et al., 2003). According to this hypothesis, it was the break-up of land masses, not invasion of vacated niches following K/T extinctions, that accounts for the mammalian radiation (Hedges et al., 1996; Eizirik et al., 2001). Other recent molecular studies, however, have produced later divergence times, much closer to the K/T boundary or even early in the Cenozoic, which are more consistent with the fossil record (Table 1.1; Huchon et al., 2002; Springer et al., 2003).

It is often claimed that molecular evidence is more reliable (if not infallible) for assessing divergence times and relationships than is the fossil record, leading some molecular systematists to dismiss fossil evidence entirely. But discordant divergence estimates in different studies—and their variance with the fossil record or with anatomical evidence—raise questions about their dependability. The literature contains many examples of molecular divergence times and phylogenetic conclusions that have subsequently been discredited. Discrepancies in divergence estimates may result from various factors, including the choice of molecular sequences and taxa used, calibration dates, phylogenetic methods applied, and the assumption of a constant rate of molecular change (Bromham et al., 1999; Smith and Peterson, 2002; Springer et al., 2003; Graur and Martin, 2004). It is now known that rates of molecular evolution are heterogeneous both between and within lineages, and at different gene loci (e.g., Ayala, 1997; Smith and Peterson, 2002). Moreover, it appears that molecular clock-based estimates consistently overestimate divergence times (Rodriguez-Trelles et al., 2002). In view of these potential problems, divergence estimates based on molecular data should be viewed with caution.

The fossil record provides the only direct evidence of the occurrence of mammalian orders in the past. But fossils merely indicate the minimum age of a clade, which is likely to be younger than its origin (i.e., its divergence from a sister group or ancestor). Nearly all "modern" orders—those with living representatives—are first seen in the fossil record after the K/T boundary, apparently supporting the explosive model, or possibly the long-fuse model. Indeed, only four extant orders of mammals are potentially known from the Cretaceous, and the ordinal assignments of the relevant fossils are far from secure. They include the monotreme order Platypoda and two living orders of marsupials, Didelphimorphia and Paucituberculata (McKenna and Bell, 1997). Among placental mammals, only a single extant order, Lipotyphla, has so far been tentatively identified in the Late Cretaceous of the northern continents. There is a possible Early Cretaceous record of Lipotyphla from Australia, but it is highly controversial.

Several other Cretaceous fossils might be related to the Cenozoic radiation, but all are too distant morphologically and phylogenetically to be assigned to modern orders. Notable among them are zalambdalestids and zhelestids, the oldest of which are about 85 million years old. Zalambdalestids are considered by some experts to be stem members of the superordinal clade (Anagalida) that includes rodents, lagomorphs, and possibly elephant-shrews (Macroscelidea), whereas zhelestids have been considered to be basal ungulatomorphs (at the base of the ungulate radiation). But recent phylogenetic analyses based on new morphological evidence have challenged these hypotheses. Even if the original assessments were correct, they would at best place a minimum age of 85 million years on some superordinal divergences, which would be consistent with the long-fuse model. Other therians of similar age can be identified as metatherians or eutherians, but they are so primitive that they are not assignable to extant orders or even superordinal clades. It is not until the latest Cretaceous (Maastrichtian or Lancian), the last 5 million years or so before the K/T boundary, that a small number of lineages are present that could represent "modern" clades or stem taxa of extant orders. Thus, taken at face value, the fossil record seems to provide overwhelming evidence that most modern orders did not evolve until the Early Cenozoic.

Robertson et al. (2004) proposed an intriguing scenario that could explain the "explosive" appearance of the early Cenozoic mammalian radiation. They postulated that the terminal Cretaceous bolide impact resulted in a short-term (hours-long) global heat pulse that "would have killed un-

Table 1.1. Estimated age of divergence (in My) of selected placental clades

Taxon	Kumar and Hedges	Divergence estimates	95% credibility interval	Fossils
Placentalia	173	102–131	91–148	125 (85)[a]
Euarchontoglires	>112	85–88	77–94	64
Xenarthra	129 ± 19	66–72	60–79	58
Eulipotyphla	—	73–79	69–84	66[b]
Chiroptera	—	65–66	61–69	52
Primates	—	77–95	70–105	64 (55)[c]
Carnivora	83 ± 4	55–56	50–61	62–64
Cetartiodactyla	83 ± 4	64	62–65	55
Paenungulata	105 ± 7	57–62	54–65	54–55
Perissodactyla	83 ± 4	56	54–58	55
Rodentia	>112 ± 4	70–74	63–81	56
Lagomorpha	91 ± 2	51–71	42–81	48

Notes: Based on molecular sequences of nuclear genes (Kumar and Hedges, 1998) and both nuclear and mitochondrial genes (Springer et al., 2003; middle two columns). The last column shows the approximate age of the oldest known fossils for each clade. Fossil occurrences are discussed in later chapters.

[a] 125 Ma estimate based on *Eomaia,* a basal eutherian; oldest plausible placentals are zalambdalestids and zhelestids from 85 Ma, but even their placental status is controversial.
[b] *Batodon;* could be much older if *Paranyctoides* or *Otlestes* are eulipotyphlans.
[c] Older estimate based on plesiadapiforms; younger estimate based on euprimates.

sheltered organisms directly" (Robertson et al., 2004: 760). They further speculated that a small number of Cretaceous mammal lineages found shelter in subterranean burrows or in the water and survived the heat pulse. In their scenario, it was these lineages that ultimately gave rise to the Cenozoic mammalian radiation. This scenario supports the long-fuse model.

Several other possible explanations for the absence of modern orders in the Cretaceous have been advanced (Foote et al., 1999). Some researchers have claimed that the Cretaceous fossil record is too incomplete to reveal whether the mammalian radiation occurred during the Cretaceous or subsequently (e.g., Easteal, 1999; Smith and Peterson, 2002). Alternatively, it has been argued that Cretaceous fossils of modern orders might actually exist but are unrecognized because they lack any distinguishing characters. In other words, genetic divergence may have preceded morphological divergence (Cooper and Fortey, 1998; Tavaré et al., 2002). Neither argument is very convincing. The possibility that mammals were diversifying somewhere with a poor fossil record, such as Africa or Antarctica (dubbed the "Garden of Eden" hypothesis by Foote et al., 1999), of course cannot be ruled out. Our knowledge of Cretaceous faunas remains limited both geographically and temporally, and the possibility exists that none of the explorations to date has sampled the locations or habitats where the antecedents of modern orders were evolving (see Clemens, 2002, for a recent discussion). Nevertheless, it is also notable that the fossil record of Cretaceous mammals has increased exponentially in recent years, extending into areas and continents where the record was formerly blank; yet no new evidence of the presence of extant orders has materialized. Instead, an array of mostly archaic Mesozoic clades has emerged. Therefore, it is reasonable to conclude that fossils of extant orders have not been discovered in the Cretaceous because they had not yet evolved (Benton, 1999; Foote et al., 1999; Novacek, 1999).

It is also true that if molecular and morphological evolution were decoupled, it might be impossible to recognize early ordinal representatives (in analogy with the genetic but not morphological separation of sibling species). However, no precedent is known for such a lengthy period of significant genetic evolution without concomitant anatomical change, and the fossil record argues against it. Although gaps remain in our knowledge of the origin of many orders, the past decade or so has seen the discovery of many remarkable fossils that appear to document post-Cretaceous transitional stages in the origin of orders, including Rodentia, Lagomorpha, Proboscidea, Sirenia, Cetacea, and Macroscelidea.

Both fossil and molecular evidence are pertinent to resolving the timing of the therian radiation. Better understanding of both are necessary to resolve remaining conflicts. It will also be important to understand the actual effects on the mammalian fauna of physical events, such as the terminal Cretaceous bolide impact.

MAMMALIAN PHYLOGENY, INTERRELATIONSHIPS, AND CLASSIFICATION

There is only one true phylogeny of mammals, and deciphering it is the challenge of mammalian systematics. All phylogenetic studies are works in progress, based on the evidence at hand or, more often, subsets of the available evidence. They should be regarded as hypotheses based on that evidence. Some are better (and presumably more reliable) than others, but none is likely to be the last word on the subject. Each hypothesis is only as good as the evidence it is based on, the characters chosen, how carefully those characters have been examined, and the phylogenetic methods and assumptions employed.

Determining Relationships: The Evidence of Evolution

Two fundamental kinds of evidence are used to determine relationships and phylogeny of mammals and other organisms: anatomical and molecular (genetic). Anatomical evidence usually includes features of the skeleton, dentition, or soft anatomy. Molecular evidence typically consists of sequences of proteins or segments of mitochondrial or nuclear genes. Until the last 25 years or so, mammalian relationships were usually based largely or entirely on anatomical features. The extent of similarity was often the chief criterion, and the distinction between specialized or derived (**apomorphic**) and primitive (**plesiomorphic**) features was often blurred. However, it is now virtually universally accepted that only shared derived features or **synapomorphies** —specialized traits inherited from a common ancestor—are significant for establishing close relationship, whereas shared primitive features (**symplesiomorphies**) do not reflect special relationship.

In practice it is not always self-evident whether a trait is primitive or derived. This distinction, the **polarity** of the trait, is always relative to previous or later conditions, hence its correct determination depends to some extent on the phylogeny we are trying to decipher. It follows that the same character can be derived relative to more primitive taxa and primitive with respect to more advanced taxa. Circularity is avoided by using many independent characters to determine phylogeny; nevertheless, polarity is usually an a priori judgment, based on predetermined ingroup and outgroup taxa. The choice of such taxa (and their character states) ultimately determines the polarity of characters in the ingroup. Thus a change in perceived relationships can result in a change in character polarity. The polarity of some characters is relatively obvious. For example, modification of the forelimbs into wings in bats is an apomorphic condition among mammals, a synapomorphy of all bats, and at the same time a symplesiomorphy of the genera within any family of bats. Less obvious is the polarity of transverse crests or cross-lophs on the upper molars of some basal perissodactyls. This feature has been considered either primitive or derived, depending on the presumed sister-group of perissodactyls. The terms "primitive" or "plesiomorphic" versus "derived" or "apomorphic" are sometimes extended to taxa, to reflect their general morphological condition, but they are more properly restricted to characters.

Of course, not all derived features shared by two animals necessarily reflect close relationship. It is well known that similar anatomical features have independently evolved repeatedly in evolution. Such iterative evolution is often associated with similar function, and it occurs both in groups with no close relationship (**convergence**) and in closely allied lineages with a common ancestor that lacked the derived trait (**parallelism**). Independent evolution of similar traits is called **homoplasy.** The challenge for systematists is distinguishing synapomorphic from homoplastic traits. This problem has long been realized by morphologists, and examples of morphological homoplasy abound. In some cases it is easily recognized by the lack of homology of the similar trait or by significant differences in other characters. For instance, there is ample evidence to demonstrate that the Pleistocene saber-toothed cat *Smilodon* was convergent to the Miocene saber-toothed marsupial *Thylacosmilus,* that creodonts and borhyaenid marsupials were dentally convergent to Carnivora, and that remarkably similar running and gliding adaptations evolved multiple times independently. But whether the specialized three-ossicle middle ear evolved only once in mammals or multiple times convergently is more ambiguous and may require additional evidence (see Chapter 4 for new evidence suggesting multiple origins). Despite widespread assumption to the contrary, molecular sequences are also susceptible to homoplasy, as recent examples demonstrate (e.g., Bull et al., 1997; Pecon Slattery et al., 2000).

Monophyly and Paraphyly

Just as synapomorphic features indicate common ancestry (monophyletic origin), the extent and distinctiveness of synapomorphies reflect proximity of relationship. The term "monophyletic" was long used to indicate descent from a common ancestor, but following Hennig (1966), **monophyly** now usually connotes not just single origin but also inclusion of all descendants from that ancestor (holophyly of Ashlock, 1971). Monophyletic groups or taxa are called **clades.** Groups believed to have evolved from more than one ancestor are referred to as **polyphyletic** and, once demonstrated, are rejected. Such was the case with the original concept of Edentata, which consisted of xenarthrans, pangolins, and aardvarks. Each is now known to constitute a distinct order with a separate origin. However, bats, pinnipeds, rodents, odontocetes, and Mammalia itself have all been claimed to be diphyletic or polyphyletic at some time during the past several decades, but recent analyses once again suggest that all are monophyletic.

The term **paraphyletic** is often applied to groups that are monophyletic in origin but do not include all descendants. Such groups lack unique synapomorphies. Some authors prefer to avoid paraphyletic taxa, or to enclose their names in quotation marks. That convention is not adopted here. Although at first glance elimination of paraphyletic groups would seem to streamline taxonomy, it may instead introduce new problems, including a highly cumbersome hierarchy and taxonomic instability. These problems arise in part because some taxa once thought to be paraphyletic, when better known, are now regarded as monophyletic, and vice versa. Some groups seem to be obviously paraphyletic (e.g., the current conception of condylarths, the stem group of many ungulate orders), but for many others, their status is less clear. For example, phenacodontid condylarths could be either the monophyletic sister taxon of perissodactyls and paenungulates or their paraphyletic stem group. Mesonychia, for the last 30 years regarded as the paraphyletic stem group of Cetacea, is now considered by some to be a monophyletic side branch, as Cetacea appear to be more closely related to artiodactyls. Artiodactyla, long held to be one of the most stable monophyletic groups, could in fact be paraphyletic unless Cetacea are included. These examples highlight the uncertainty of identifying and verifying paraphyly, even in the face of a good fossil record.

Carroll (1988: 13) concluded that as many as half of all species are paraphyletic and that "the existence of paraphyletic groups is an inevitable result of the process of evolution." In fact, it is often the paraphyletic taxa—especially those that gave rise to descendants that diverged sufficiently to be assigned to separate higher taxa—that are of greatest evolutionary interest. Undoubtedly we have only begun to recognize which taxa are paraphyletic. Consequently no attempt is made in this text to eliminate paraphyletic groups. Some, such as Condylarthra, Plesiadapiformes, Miacoidea, and Palaeanodonta, are retained for convenience, and their probable paraphyletic nature noted, pending a better understanding of their relationships.

Phylogeny and Classification

Phylogenetic inferences ideally should be based on all available evidence, but practical considerations restrict most analyses. The majority of studies have been based on either morphological traits or molecular sequences, and usually on only a subset of those data partitions. For example, analyses of fossil taxa are necessarily limited to the anatomy of the hard parts, because soft anatomy and molecular data are not available. In addition, the outcome of phylogenetic analysis may vary depending on such factors as the choice of taxa, outgroups, and characters, the description and scoring of those characters, weighting of characters, and methods used. Consequently there are many reasons not to accept phylogenetic hypotheses uncritically.

Recent attempts to combine morphological and molecular data, optimistically called "total evidence" analysis, suffer from our ignorance of how to analyze such disparate characters meaningfully. How do individual base-pairs in a gene sequence compare with specific anatomical features, and should they be equally weighted in phylogenetic analyses? Total evidence analyses commonly treat individual base-pairs (sometimes even noninformative base-pairs) as equivalent to anatomical characters. Because a single gene segment may consist of hundreds of base-pairs, this practice almost always results in the molecular characters far outnumbering anatomical characters and potentially biasing the outcome.

Another approach to combining data partitions is called "supertree" analysis. This method constructs a phylogeny based on multiple "source trees" drawn from individual phylogenetic analyses of morphological or molecular data (e.g., Sanderson et al., 1998; Liu et al., 2001). It is not clear, however, that this approach is superior to the individual analyses on which it is based. Some of the weaknesses of this approach were summarized by Springer and de Jong (2001).

Phylogenetic analyses typically use such methods as parsimony for morphological data sets and maximum likelihood or Bayesian analysis for molecular data sets. Which method is more likely to yield the most accurate tree is debatable, but it is probable that evolution does not always proceed parsimoniously. The results of these analyses are presented in cladograms that depict hypothetical relationships in branching patterns. The best resolved patterns are dichotomous; unresolved relationships are shown as multiple branches from the same point or node (polytomies).

This text focuses on the morphological evidence for mammalian relationships, although mention is made of contrasting phylogenetic arrangements suggested by molecular analyses. Most chapters include both classifications and cladograms. Although both are based on relationships, their goals are somewhat different. Cladograms place taxa in phylogenetic context by depicting hypotheses of relationship; consequently they are inherently more mutable. A classification provides a systematic framework and should therefore retain stability to the extent possible while remaining "consistent with the relationships used as its basis" (Simpson, 1961: 110; see also Mayr, 1969). Most classifications adopted in individual chapters loosely follow the classification of McKenna and Bell (1997, 2002). Minor modifications, such as changes in rank, are present throughout the book; but where significant departures from that classification are made, they are noted in the text or tables. For ease of reference, families and genera known from the Paleocene or Eocene are shown in boldface in the tables accompanying Chapter 5 and beyond. The cladograms presented reflect either individual conclusions or a consensus of recent studies, and they do not always precisely mirror the classifications.

The taxonomy employed in this volume represents a compromise between cladistic and traditional classifications, while attempting to present a consensus view of interrelationships. Such a compromise is necessary in order to use taxonomic ranks that reflect relationship and indicate roughly equivalent groupings, and at the same time avoid the nomenclatural problems inherent in a nested hierarchy (McKenna and Bell, 1997). The standard Linnaean categories, as modified by McKenna and Bell (1997), remain useful and are employed here, although unranked taxa between named ranks are necessary in a few cases (e.g., Catarrhini and Platyrrhini in the classification of Primates). As pointed out by McKenna and Bell (1997), among others, taxa of the same rank (apart from species) are not commensurate. For example, it is not possible to establish that a family in one order is an equivalent unit to families in other orders (or in the same order, for that matter). Nor are the orders themselves equivalent. Nevertheless, the taxonomic hierarchy does provide a useful relative measure of affinity within groups and of the distance between them.

As recognized in this volume, higher taxa are primarily stem-based. A **stem-based taxon** consists of all taxa that

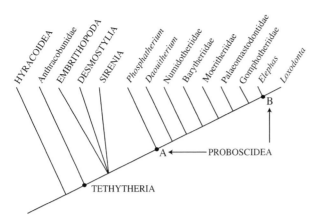

Fig. 1.3. Stem-based versus crown-group definition of taxa, illustrated by the Proboscidea. A crown-group definition limits Proboscidea to node B, equivalent to the extant family Elephantidae. Using a stem-based definition, Proboscidea includes all taxa more closely related to living elephants than to Sirenia or Desmostylia or Embrithopoda, as indicated here at node A. This stem-based definition is adopted in the most recent study of primitive proboscideans (Gheerbrant, Sudre, et al., 2005) and is followed here. See Chapter 13 for details of the proboscidean and tethythere radiations.

share a more recent common ancestor with a specified form than with another taxon (e.g., De Queiroz and Gauthier, 1992). For example, Proboscidea is considered to include all taxa more closely related to extant elephants than to sirenians (Fig. 1.3). Therefore, using a stem-based definition, extinct moeritheriids and gomphotheres are proboscideans. This convention leaves open the possibility that other unknown stem-taxa may exist and could lie phylogenetically outside the known taxa, yet still lie closer to elephants than to any other major clade. Such was the case when the older and more primitive numidotheriids were discovered.

A **node-based taxon** is defined as all descendants of the most recent common ancestor of two specified taxa. In the example above, a node-based Proboscidea could be arbitrarily recognized at the common ancestor of numidotheres and other proboscideans, or of moeritheres and other proboscideans (thus excluding numidotheres). A special category of node-based taxa, which has been applied by some authors to mammalian orders, is the **crown-group.** A crown-group is defined as all descendants of the common ancestor of the living members of a specified taxon (Jefferies, 1979; De Queiroz and Gauthier, 1992). By such a definition, nearly all fossil groups are excluded from Proboscidea, and other well-known basal forms are excluded from higher taxa to which they have long been attributed and with which they share common ancestry and diagnostic anatomical features (Lucas, 1992; McKenna and Bell, 1997). Stem-based taxa are here considered more useful than node-based taxa for reference to the Early Cenozoic mammalian radiation.

The synoptic classification of mammals used in this book is given in Table 1.2. Mammalian relationships based on morphology are shown in Fig. 1.4, and those based on molecular data in Fig. 1.5. Although the discrepancies between morphological and molecular-based phylogenies have garnered considerable attention, it is important to note that there is substantial agreement between most morphological and molecular-based phylogenies (Archibald, 2003). This consensus underscores the significance of the discords that do exist. The two kinds of evidence have been particularly at odds with regard to two conventional orders, Lipotyphla and Artiodactyla, molecular data suggesting that neither is monophyletic. According to molecular analyses, the traditional lipotyphlan families Tenrecidae and Chrysochloridae form a monophyletic group together with Macroscelidea, Tubulidentata, Proboscidea, Sirenia, and Hyracoidea, which has been called Afrotheria. No morphological evidence supporting Afrotheria has been found. Molecular studies also indicate that the order Cetacea is nested within Artiodactyla as the sister group of hippopotamids. These debates are further discussed in the relevant chapters in this volume.

Disagreements also exist at the superordinal level, but the anatomical evidence for higher-level groupings is weak. Thus gene sequences support recognition of four main clades of placental mammals: Afrotheria, Xenarthra, Laurasiatheria (eulipotyphlans, bats, carnivores, pangolins, perissodactyls, artiodactyls, and whales), and Euarchontoglires (primates, tree shrews, flying lemurs, rodents, and lagomorphs), the last two of which form the clade Boreoeutheria (e.g., Eizirik et al., 2001; Madsen et al., 2001; Murphy et al., 2001; Scally et al., 2001; Amrine-Madsen et al., 2003; Nikaido et al., 2003; Springer et al., 2003, 2005). Eizirik et al. (2001) concluded that this superordinal divergence occurred during the Late Cretaceous (about 65–104 Ma) and speculated that it was related to the separation of Africa from South America. These studies further suggest that Afrotheria was the first clade to diverge, followed by Xenarthra (usually considered the most primitive, based on morphology). However, morphological evidence suggests that most of the afrothere groups are nested within the ungulate radiation and are not closely related to tenrecs and chrysochlorids (see Chapters 13 and 15). This inconsistency implies that either the morphological or molecular data must be misleading. Methodological problems that can lead to erroneous phylogenetic conclusions in molecular analyses have been reviewed by Sanderson and Shaffer (2002) and are not further discussed here.

Notwithstanding the substantial contribution molecular systematics has made to our understanding of mammalian relationships, anatomical evidence from fossils plays the predominant role in resolving the phylogenetic positions of extinct taxa and clades for which molecular data are unavailable.

GEOCHRONOLOGY AND BIOCHRONOLOGY OF THE EARLY CENOZOIC

The Paleocene and Eocene epochs make up the first 31 million years of the Tertiary Period of the Cenozoic Era (from 65 Ma to 34 Ma; Fig. 1.6). The chronology of the Paleocene and Eocene used here (Fig. 1.7) is based primarily on that of Berggren et al. (1995b) and McKenna and Bell

Table 1.2. Synoptic higher-level classification of Mammalia used in this book

Class MAMMALIA
        †*Adelobasileus*, †*Hadrocodium*
        †Sinoconodontidae
        †Kuehneotheriidae
     Order †MORGANUCODONTA
     Order †DOCODONTA
     Order †SHUOTHERIDIA
     Order †EUTRICONODONTA
     Order †GONDWANATHERIA
  Subclass AUSTRALOSPHENIDA
     Order †AUSKTRIBOSPHENIDA
     Order MONOTREMATA[1]
  Subclass †ALLOTHERIA
     Order †HARAMIYIDA
     Order †MULTITUBERCULATA
  Subclass TRECHNOTHERIA[2]
    Superorder †SYMMETRODONTA
    Superorder †DRYOLESTOIDEA
     Order †DRYOLESTIDA
     Order †AMPHITHERIIDA
    Superorder ZATHERIA
     Order †PERAMURA
  Subclass BOREOSPHENIDA[3]
     Order †AEGIALODONTIA
  Infraclass METATHERIA
     Order †DELTATHEROIDA
     Order †ASIADELPHIA
   Cohort MARSUPIALIA
    Magnorder AMERIDELPHIA (American marsupials)
     Order DIDELPHIMORPHA (opossums)
     Order PAUCITUBERCULATA (rat opossums, polydolopids, argyrolagids, and kin)
     Order †SPARASSODONTA (borhyaenids)
    Magnorder AUSTRALIDELPHIA (Australian marsupials)
    Superorder MICROBIOTHERIA
    Superorder EOMETATHERIA
     Order NOTORYCTEMORPHIA (marsupial moles)
    Grandorder DASYUROMORPHIA (marsupial mice and cats, numbats, Tasmanian wolf, Tasmanian devil)
    Grandorder SYNDACTYLI
     Order PERAMELIA (bandicoots)
     Order DIPROTODONTIA (kangaroos, phalangers, wombats, koalas, sugar gliders)
  Infraclass EUTHERIA
        †*Eomaia*, †*Montanalestes*, †*Prokennalestes*, †*Murtoilestes*
     Order †ASIORYCTITHERIA
   Cohort PLACENTALIA (placental mammals)
     Order †BIBYMALAGASIA
     Order XENARTHRA (edentates: armadillos, sloths, anteaters)
    Superorder INSECTIVORA
     Order †LEPTICTIDA
     Order LIPOTYPHLA (moles, shrews, hedgehogs, tenrecs, golden moles)
    Superorder †ANAGALIDA[4]
      †Zalambdalestidae[5]
      †Anagalidae
      †Pseudictopidae
     Order MACROSCELIDEA (elephant shrews)
    Grandorder GLIRES
     Mirorder DUPLICIDENTATA
      Order †MIMOTONIDA
      Order LAGOMORPHA (rabbits, hares, pikas)
     Mirorder SIMPLICIDENTATA
        †*Sinomylus*
      Order †MIXODONTIA
      Order RODENTIA (squirrels, beavers, rats, mice, gophers, porcupines, gerbils, guinea pigs, chinchillas, capybaras, etc.)
    Superorder FERAE[4]
     Order †CREODONTA
     Order CARNIVORA (carnivores: cats, dogs, bears, raccoons, hyenas, weasels, otters, badgers, civets, mongooses, seals, walruses)

continued

Table 1.2. Continued

        Mirorder †CIMOLESTA[6]
            †Didymoconidae
            †Wyolestidae
        Order †DIDELPHODONTA
        Order †APATOTHERIA (apatemyids)
        Order †TAENIODONTA
        Order †TILLODONTIA
        Order †PANTODONTA
        Order †PANTOLESTA
        Order PHOLIDOTA (pangolins or scaly-anteaters)
    Superorder ARCHONTA
        Order CHIROPTERA[7] (bats)
        Grandorder EUARCHONTA
            Order DERMOPTERA ("flying lemurs" or colugos)
            Order SCANDENTIA (tree shrews)
            Order PRIMATES (plesiadapiforms, lemurs, lorises, tarsiers, monkeys, apes, humans)
    Superorder UNGULATOMORPHA[8]
        †Zhelestidae[9]
        Grandorder UNGULATA[8] (ungulates: hoofed mammals)
            Order †CONDYLARTHRA[8]
            Order TUBULIDENTATA (aardvarks)
            Order †DINOCERATA (uintatheres)
            Order †ARCTOSTYLOPIDA
            Order ARTIODACTYLA (even-toed ungulates: pigs, hippos, camels, deer, giraffes, antelope, gazelles, sheep, goats, cattle, etc.)
        Mirorder CETE
            Order †MESONYCHIA
            Order CETACEA[10] (whales, dolphins)
        Mirorder †MERIDIUNGULATA[8] (endemic South American ungulates)
            Order †LITOPTERNA
            Order †NOTOUNGULATA
            Order †ASTRAPOTHERIA
            Order †XENUNGULATA
            Order †PYROTHERIA
        Mirorder ALTUNGULATA
            Order PERISSODACTYLA (horses, tapirs, rhinos, †chalicotheres, †titanotheres)
            Order PAENUNGULATA
                Suborder HYRACOIDEA (hyraxes)
                Suborder TETHYTHERIA
                    Infraorder †EMBRITHOPODA
                    Infraorder SIRENIA (sea cows, dugongs)
                    Infraorder PROBOSCIDEA (elephants)

Notes: Classification is modified mainly after McKenna and Bell (1997) and Kielan-Jaworowska et al. (2004). This table and all others presented in this book represent a compromise between traditional and cladistic classifications and are an attempt to provide a consensus view. Ordinal-level and higher taxa are shown in upper case; unassigned taxa immediately below a higher taxon are either plesiomorphic or of uncertain phylogenetic position within that taxon. Many taxa are probably paraphyletic, but no attempt is made in the tables to differentiate them from those believed to be monophyletic; instead these distinctions are discussed in the text. The dagger (†) denotes extinct taxa.

[1] McKenna and Bell (1997) assigned monotremes to the subclass Prototheria and recognized two orders, Platypoda (platypuses) and Tachyglossa (echidnas).
[2] Trechnotheria is essentially equivalent to the concept of Holotheria.
[3] Essentially equivalent to Tribosphenida.
[4] Several taxa considered grandorders by McKenna and Bell (1997) are considered superorders here.
[5] May be basal eutherians.
[6] Monophyly of Cimolesta and interrelationships of its constituents are very uncertain.
[7] Relationship of Chiroptera to other archontans is in dispute.
[8] Monophyly questionable.
[9] Monophyly of Zhelestidae and their relationship to ungulates are controversial.
[10] May be nested in Artiodactyla.

(1997), with modifications as indicated in the following discussion. Geologic periods and epochs may be subdivided into successive stages/ages (chronostratigraphic and geochronologic units) and, in the case of the Cenozoic epochs, land-mammal ages (a biochronologic unit). Land-mammal ages "describe the age and succession of events in mammalian evolution" based on characteristic mammal assemblages, lineage segments, or in some cases first or last appearances (Woodburne, 2004: xiv; see also Walsh, 1998, for an insightful discussion of the definition of land-mammal ages). Although absolute dates have been placed on many of these units using a combination of magnetostratigraphy and radiometric methods, such as high-precision ^{40}Ar/^{39}Ar dating (Berggren et al., 1995b; Gradstein et al., 1995, 2004),

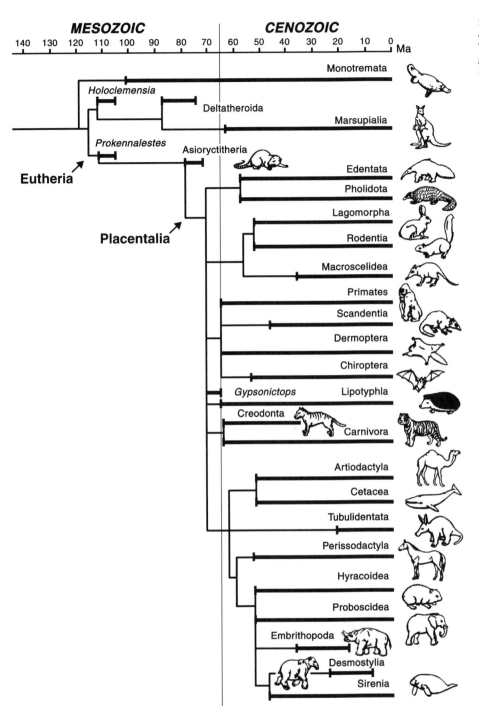

Fig. 1.4. Relationships of higher taxa of mammals based on morphology. Thicker lines indicate documented geologic ranges. (From Wible et al., 2005, based on Novacek, 1999.)

precise dating of some intervals remains tenuous and in some cases controversial.

Geologic time applies worldwide, whereas land-mammal ages are specific to each continent and are relatively well constrained geochronologically only in North America and Europe. Nevertheless, their sequence is reasonably well understood, as is the correlation between North American Land-Mammal Ages (NALMAs) and standard stages/ages (more widely used in Europe than land-mammal ages). For this reason, land-mammal ages and their subdivisions (or stages/ages, particularly in Europe) provide a useful framework for placing fossil mammals in relative chronologic context, and they are employed throughout this volume. As we shall see, the precise age and the correlation of Asian and South American Land-Mammal Ages with those of North America and Europe are more controversial.

Hundreds of radiometric dates are now available for the Mesozoic, permitting a relatively accurate estimate of the age of the oldest known mammals (Gradstein et al., 1995, 2004). Based on these data, mammals first appeared at least 205–210 million years ago, and perhaps as much as 225 million years ago (see Chapter 4). They survived alongside dinosaurs for the first 145 million years of their history, up to the K/T boundary at about 65 million years ago, when the last

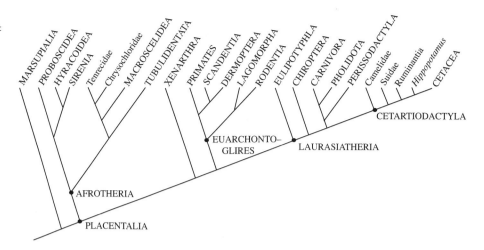

Fig. 1.5. Relationships of higher taxa of mammals based on molecular data. Most higher taxa in these studies are based on only one to five species. Interrelationships among taxa within Afrotheria are particularly unstable. (Modified after Murphy et al., 2001, and Springer et al., 2005.)

nonavian dinosaurs became extinct. The K/T boundary, and thus the base of the Paleocene, is situated near the top of geomagnetic polarity chron C29r and has been dated at 65.0 million years ago (Swisher et al., 1992, 1993; Gradstein et al., 1995) or, most recently, 65.5 million years ago (Gradstein et al., 2004).

Paleocene/Eocene Boundary

The Paleocene/Eocene boundary is situated in the lower part of polarity chron C24r, but its precise position and age have been contentious. Dates range from about 54.8 (Berggren et al., 1995b) to 55.8 million years ago (Gradstein et al., 2004) in various reports over the past decade or so, most centering around 55.0 million years ago. The debate here, as for the Eocene/Oligocene boundary, stems partly from the difficulty of correlating mammal-bearing continental beds with discontiguous marine strata on which much of Cenozoic geochronology is based. As a result, the Paleocene/Eocene boundary has varied relative to the Thanetian/Ypresian Stage/Age boundary in Europe and the Clarkforkian/Wasatchian Land-Mammal Age boundary in North America. For example, different authors have considered the Clarkforkian to be entirely Paleocene, or all or partly of early Eocene age, and the Wasatchian to be entirely Eocene or to have begun during the late Paleocene. In Europe, a stratigraphic gap was found between the Thanetian and the Ypresian, further complicating matters and making precise placement of the boundary uncertain.

This dilemma has been largely resolved by the recent decision to place the beginning of the Eocene at the onset of the isochronous, worldwide Carbon Isotope Excursion (CIE), a major perturbation in the global carbon cycle reflected by a negative excursion in $\delta^{13}C$ (Kennett and Stott, 1991; Dupuis et al., 2003). The ultimate cause of this sudden input of massive amounts of carbon into the atmosphere is controversial (volcanism or comet impact are just two hypotheses; Bralower et al., 1997; Kent et al., 2003), but most authorities agree that it can be traced to the release of methane gas on the ocean floor (Dickens et al., 1995; Katz et al., 1999; Norris and Röhl, 1999; Svensen et al., 2004). The CIE coincided with a brief period of global warming, the Initial Eocene Thermal Maximum (also called the Paleocene-Eocene Thermal Maximum; Sloan and Thomas, 1998; Aubry et al., 2003) and has been recognized in both marine and terrestrial sediments globally. Its onset also coincides with the beginning of the Wasatchian Land-Mammal Age in North America and the beginning of the Ypresian Stage in Europe, which are characterized by substantial faunal turnover, including the abrupt appearance of perissodactyls, artiodactyls, euprimates, and hyaenodontid creodonts. By this convention, the Clarkforkian is entirely of Paleocene age. In northern Europe, cores now fill the former stratigraphic gap and show that the CIE is situated near the base of the "gap," just above the Thanetian (Steurbaut et al., 2003).

Although there is now agreement on exactly where to place the Paleocene/Eocene boundary, controversy persists over its calibration, because no absolute (radiometric) dates are known for this event. Consequently its age has been interpolated based on radiometric dates tied to the geomagnetic polarity time scale, together with data from astronomical cycle stratigraphy. Thus Aubry et al. (2003) dated the start of the CIE at about 55.5 million years ago (but allowed that it could be closer to 55.0 Ma), whereas many other authors place it at 55.0 million years ago (e.g., Bowen et al., 2002; Gingerich, 2003; Koch et al., 2003). However, Röhl et al. (2003: 586) noted that their earlier estimate of 54.98 million years ago (Norris and Röhl, 1999) was "likely to be too young by several 100 k.y." because of inaccuracies in the calibration points used, which suggests that the estimate by Aubry et al. was closer. The most recent time scale placed the Paleocene/Eocene boundary at 55.8 ± 0.2 million years ago (Gradstein et al., 2004).

Aubry et al. (2003) proposed that the name "Sparnacian" be used as a new earliest Eocene stage/age to encompass the time represented by the hiatus between classical Thanetian and Ypresian. The term "Sparnacian" was already applied to early Ypresian faunas by some paleomammalogists (e.g., Savage and Russell, 1983), but the Sparnacian stratotype, as well as some classic Sparnacian assemblages, may not be of

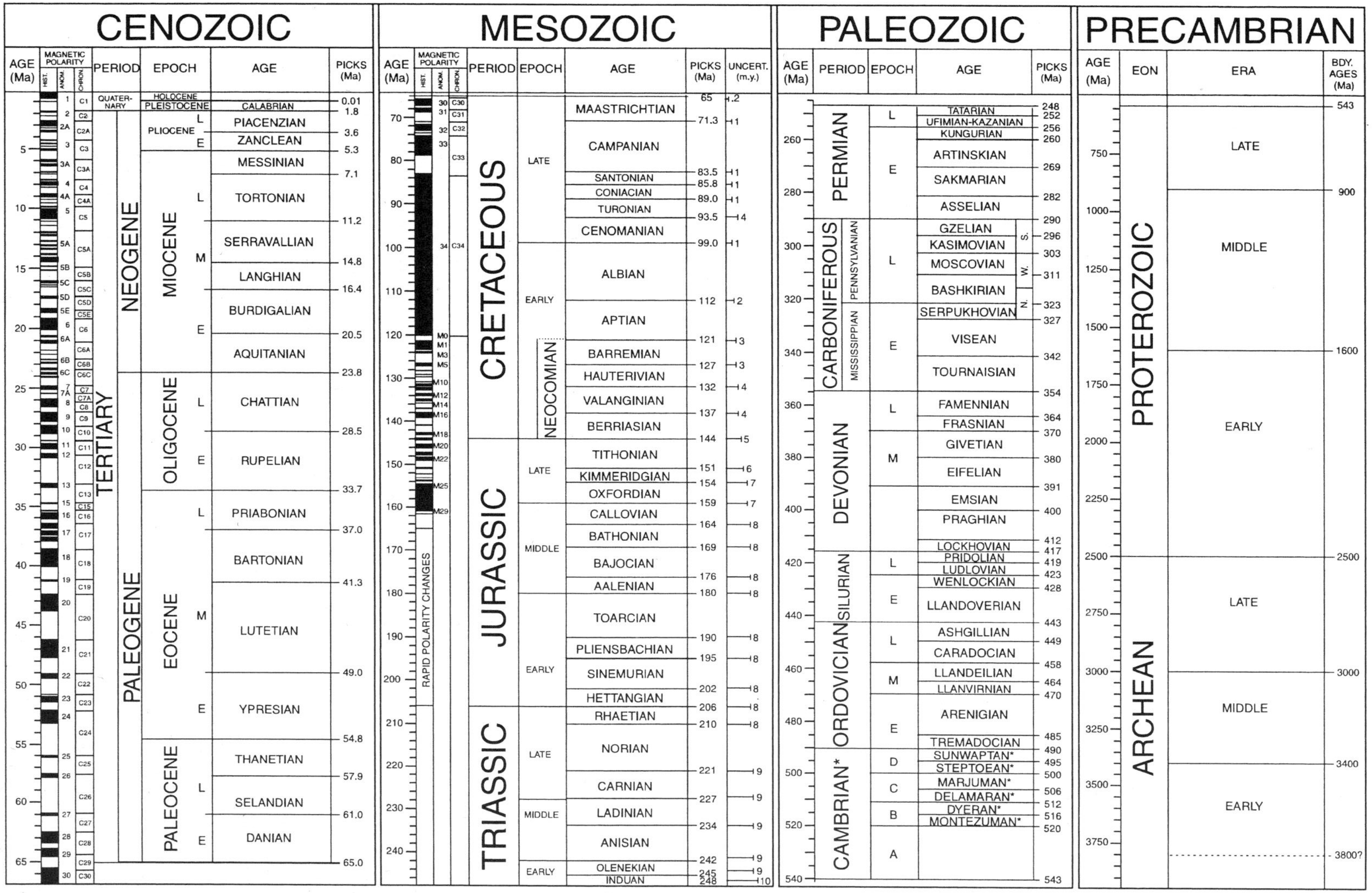

Fig. 1.6. Geologic time scale of the Geological Society of America, 1999. The newly adopted time scale of the International Commission on Stratigraphy (Gradstein et al. 2004) differs in relatively minor ways from this one: epoch and age boundaries during the Paleocene and Eocene are shifted up or down by 0.2–1.0 million years. The largest difference is the Paleocene/Eocene boundary, placed at 55.8 Ma by Gradstein et al. (2004). © Geological Society of America 1999.

EPOCH	Ma	GPTS	EUROPE Stage/Age	N. AMERICA NALMA	ASIA ALMA	S. AMERICA SALMA	Ma
OLIGOCENE		C13n	Rupelian	Orellan	Shandgolian	Tinguirirican	
EOCENE LATE	35	C15n / C16n	Priabonian	Chadronian	Ergilian	Divisaderan	35
EOCENE LATE		C17n	Bartonian	Duchesnean	Sharamurunian	Mustersan	
EOCENE MIDDLE	40	C18n / C19n	Bartonian	Duchesnean	Sharamurunian	Casamayoran	40
EOCENE MIDDLE	45	C20n	Lutetian	Uintan	Irdinmanhan	Casamayoran	45
EOCENE EARLY		C21n	Lutetian	Bridgerian	Arshantan	Casamayoran	
EOCENE EARLY	50	C22n / C23n / C24n	Ypresian	Wasatchian	Bumbanian		50
	55						55
PALEOCENE LATE		C25n	Thanetian	Clarkforkian	Gashatan	Riochican	
PALEOCENE LATE	60	C26n	Selandian	Tiffanian	Nongshanian	Itaboraian	60
PALEOCENE EARLY		C27n	Danian	Torrejonian	Shanghuan	Peligran	
PALEOCENE EARLY		C28n	Danian	Torrejonian	Shanghuan	Tiupampan	
PALEOCENE EARLY	65	C29n	Danian	Puercan			65
CRETACEOUS		C30n	Maastrichtian	Lancian			

early Ypresian age (Hooker, 1998). Consequently, Sparnacian was not accepted in the most recent time scale (Gradstein et al., 2004) and is not used in this volume. It should be remembered that, regardless of the absolute age put on the Paleocene/Eocene boundary, it is coincident with the onset of the CIE.

Eocene/Oligocene Boundary

The Eocene/Oligocene boundary in Europe was long equated with a major episode of faunal turnover called the *"Grande Coupure"* (Stehlin, 1909; Savage and Russell, 1983; Russell and Tobien, 1986; Legendre, 1987; Legendre et al., 1991). In North America the Eocene/Oligocene boundary was believed to correspond to the boundary between the Duchesnean and Chadronian Land-Mammal Ages (Wood et al., 1941). With the advent of high-precision ^{40}Ar/^{39}Ar dating and the correlation of the Eocene/Oligocene boundary in the marine record with the extinction of the planktonic foraminiferan family Hantkeninidae (Hooker et al., 2004), the position of the epoch boundary has been revised on both continents. The boundary is now generally placed within magnetochron C13r at a little less than 34 million years ago (Berggren et al., 1992, 1995b; Prothero and Swisher, 1992; Prothero and Emry, 2004). This time coincides with the boundary between the Priabonian and Rupelian stages (see Fig. 1.7).

The principal faunal turnover at the *Grande Coupure* took place between the Priabonian and Rupelian Stages (Paleogene mammal reference levels MP 20–21; see the section on European Land-Mammal Ages, below), although it is now acknowledged that it was a protracted event. It involved extinction of more than 50% of the indigenous fauna, together with an influx of numerous immigrants from Asia. Disagreement persists over how closely the turnover coincided with the Eocene/Oligocene boundary and whether it was caused by climatic cooling or other factors (see Berggren and Prothero, 1992; Hooker, 1992a; and Legendre and Hartenberger, 1992, for contrasting views). However, the epoch boundary, based on the foram extinction noted above, is now known to be slightly older than the major cooling event that correlates with the *Grande Coupure;* consequently these events are now dated as earliest Oligocene (Hooker et al., 2004). In fact, there were several major phases of faunal turnover in Europe beginning in the middle Eocene and extending into the early Oligocene (Legendre, 1987; Hooker, 1992a; Legendre and Hartenberger, 1992; Franzen, 2003), but none appears to correspond precisely with the Eocene/Oligocene boundary as now recognized. Nevertheless, this revision is so new that most recent accounts continue to place the Eocene/Oligocene boundary at the beginning of the Rupelian Age (Paleogene mammal reference level MP 21) at about 34 million years ago.

In North America, as a result of the revised Eocene/Oligocene boundary, the Chadronian, long considered equivalent to early Oligocene, is now situated in the late Eocene. The Orellan Land-Mammal Age is early Oligocene, and the Chadronian/Orellan boundary coincides with the Eocene/Oligocene boundary (Prothero and Emry, 2004).

European Land-Mammal Ages

As mentioned earlier, standard ages are more widely used for biochronology of European faunas than are the European Land-Mammal Ages (ELMAs), and are therefore used in this text. This preference for the former may have come about because the ELMAs are for the most part equivalent in time to the standard ages ("Dano-Montian" = Danian; Cernaysian = Selandian and Thanetian; Neustrian = most of the Ypresian; Rhenanian = the rest of the Ypresian through the Bartonian; and Headonian = Priabonian; McKenna and Bell, 1997).

For greater resolution than is afforded by the standard ages, faunas are correlated by a series of European reference levels, arranged in sequence by stage of evolution and first and last appearances. In the Paleogene they are numbered from Mammal Paleogene (MP) 1 to MP 30 (Schmidt-Kittler, 1987). Levels MP 1–5 are reserved for lower Paleocene faunas, although only one (Hainin, Belgium) is currently known. MP 6 is used for the late Paleocene site of Cernay, France. When late Paleocene mammals become better known in Europe, more levels will surely be necessary. MP 7–10 are early Eocene (Ypresian), MP 11–16 are middle Eocene (MP 11–13 correspond to Lutetian, MP 14–16 to Bartonian), and MP 17–20 are late Eocene (Priabonian; e.g., Legendre and Hartenberger, 1992). If the *Grande Coupure* actually took place in the earliest Oligocene rather than at the Eocene/Oligocene boundary, as Hooker (1992a) argued, then MP 20 straddles the boundary.

North American Land-Mammal Ages

The sequence of NALMAs initially proposed by Wood et al. (1941) has been widely applied and provides a useful and well-documented biochronology for mammals of North America. Excellent summaries of the NALMAs and their mammal assemblages are found in the two volumes edited by Woodburne (1987, 2004). The NALMAs of interest in

Fig. 1.7. (*opposite*) Early Cenozoic mammalian geochronology and biochronology. Chart shows the time period emphasized in this book (Paleocene-Eocene), approximate age in millions of years (Ma), and correlation with the geomagnetic polarity time scale (GPTS), standard stage/age (commonly used in Europe), and land-mammal ages in North America (NALMA), Asia (ALMA), and South America (SALMA). White bands in GPTS column are intervals of reversed polarity (r), which precede the normal (n, black) interval of the same number. Hatching and dashed lines in ALMA and SALMA denote uncertain boundaries. The position of the boundary between Arshantan and Irdinmanhan ALMAs is unknown. Upper and (especially) lower limits of the Casamayoran SALMA are uncertain. The long span shown reflects this uncertainty, and may overestimate the actual duration of this land-mammal age. (Drafted by W. v. Koenigswald and T. Smith, based on Berggren et al., 1995b; Flynn and Swisher, 1995; McKenna and Bell, 1997; Aubry et al., 2003; Dawson, 2003; Flynn et al., 2003.)

this volume are those of the Paleocene (Puercan, Torrejonian, Tiffanian, and Clarkforkian) and Eocene (Wasatchian, Bridgerian, Uintan, Duchesnean, and Chadronian). These land-mammal ages have been subdivided into sequential biochrons that are variously based on first or last appearances, lineage segments, abundance zones, or assemblage zones. The North American Paleocene-Eocene record is the most nearly continuous in the world, although it is largely concentrated in the region of the Rocky Mountains.

In addition to the details discussed in the preceding sections, the following observations and changes concerning the original concepts may be noted. The Paleocene Puercan and Clarkforkian Land Mammal Ages are the shortest ages, about 1 million years each (Lofgren et al., 2004). Of the Paleocene NALMAs, however, only the Puercan is constrained by radiometric dates, whereas the duration of the others, including the Clarkforkian, is estimated (Clarkforkian was considered to be only half a million years long by Woodburne and Swisher, 1995). The current convention of dividing the Paleocene into only early and late portions (e.g., Berggren et al., 1995a; McKenna and Bell, 1997) results in shifting the Torrejonian NALMA, long considered middle Paleocene, into the early Paleocene. This practice is largely responsible for the apparent temporal range extensions of many mammals discussed later in the volume, although in some cases new evidence has actually extended the range stratigraphically lower into sediments of Puercan age. Land-mammal age occurrences are specified in the text where there might be confusion. The Tiffanian and Clarkforkian together make up the late Paleocene and are believed to account for a little more than half of Paleocene time.

The beginning of the Wasatchian Land-Mammal Age now coincides with the onset of the global CIE, which is also designated as the beginning of the Eocene. Although the exact date of that event is uncertain (but most likely between 55.0 and 55.8 Ma), several $^{40}Ar/^{39}Ar$ dates are now known from tuffs and volcanic ashes of latest Wasatchian age in the Bighorn and Greater Green River basins of Wyoming, ranging from about 50.7 to 52.6 million years ago (Wing et al., 1991; M. E. Smith et al., 2003, 2004). The Wasatchian/Bridgerian boundary appears to be at about 50.6–51.0 million years ago (Smith et al., 2003; Machlus et al., 2004). The Bridgerian, long considered equivalent to the middle Eocene, now straddles the early/middle Eocene boundary; nonetheless, all Bridgerian occurrences were listed as middle Eocene by McKenna and Bell (1997), which could affect some ranges discussed in later chapters. Numerous dates for the Bridgerian range up to slightly younger than 47 million years ago, and the Bridgerian/Uintan boundary is situated in chron C21n at about 46.7 million years ago (Smith et al., 2003).

With the shift of the Eocene/Oligocene boundary to the beginning of the Orellan, the Chadronian (formerly early Oligocene) is now late Eocene; and it is 3 million years long, not 5 million, as previously believed. The Uintan and Duchesnean NALMAs (long considered late Eocene in age) are now correlated with middle Eocene. Several $^{40}Ar/^{39}Ar$ dates on ashes and ignimbrites from Texas and New Mexico indicate that the Duchesnean spanned from 37 to almost 40 million years ago (Prothero, 1996a; Prothero and Lucas, 1996). The Duchesnean/Chadronian boundary is situated near the top of chron C17n.

South American Land-Mammal Ages

The South American mammalian record is relatively incomplete, with discontinuities between all the Paleogene South American Land-Mammal Ages (SALMAs). Nevertheless, a seemingly stable sequence of Cenozoic SALMAs of presumed age has been in use for decades. In the Paleogene, the following sequence has long been recognized: Riochican (late Paleocene), Casamayoran (early Eocene), Mustersan (middle Eocene), Divisaderan (middle or late Eocene), and Deseadan (early Oligocene; Simpson, 1948; Patterson and Pascual, 1968). Relative ages were assigned mainly by stratigraphic position and stage of evolution, as the faunas are entirely endemic. Over the last 30 years or so, however, magnetostratigraphic studies coupled with radioisotopic dates, together with new fossil discoveries, have forced significant revisions in the SALMAs, with particular impact on those of the Paleogene.

Three additional Paleocene land-mammal ages or subages are now recognized that precede the classic late Paleocene Riochican: Itaboraian, Peligran, and Tiupampan (see Fig. 1.7). The Tiupampan fauna was initially thought to come from the El Molina Formation of Late Cretaceous age, but it is now known to come from the overlying Santa Lucía Formation of Paleocene age (Marshall et al., 1995). The Itaboraian is presumed to be earlier late Paleocene (but it derives from fissures, which are difficult to date accurately), whereas the Tiupampan and Peligran are considered successive early Paleocene land-mammal ages (e.g., Flynn and Swisher, 1995). The Riochican appears to correlate with late Paleocene marine strata, but radiometric dates indicate only that it is younger than 63 million years. It may correlate approximately with magnetochron C25n. Low-precision radiometric dates confirm Paleocene age for the three underlying ages as well, but their durations and precise placement within the Paleocene are speculative. Recently, for example, Marshall et al. (1997), using magnetostratigraphy, recalibrated the Paleocene SALMAs and considered all four to be of late Paleocene age, about 55.5–60 million years ago. Furthermore, they concluded that the actual sequence is Peligran-Tiupampan-Itaboraian-Riochican. Because the original Riochican section spanned the entire late Paleocene, they considered all four to be subages of a single late Paleocene Riochican Land-Mammal Age. Most researchers, however, have accepted an early Paleocene age for the Tiupampan and consider it to be the oldest Cenozoic SALMA. This consensus is followed in this volume.

The Peligran Land-Mammal Age is especially problematic. Thought to correlate approximately with the Torrejonian NALMA, it is founded on a new Argentine "fauna" consisting of a few very fragmentary specimens of five mammalian species, together with frogs, turtles, and crocodilians

(Bonaparte et al., 1993). The mammal species include the gondwanathere *Sudamerica* (an enigmatic group whose affinities are very uncertain), the only non-Australian monotreme, and three supposed condylarths, one of which could instead be a dryolestoid. In some respects this assemblage has more of a Mesozoic than Paleocene aspect. Whether this enigmatic fauna proves to be older or younger than Tiupampan, the available fossils are an inadequate basis for establishing a land-mammal age.

Recently there has been even greater change in the concepts of the Eocene SALMAs. New ^{40}Ar/^{39}Ar dates on rocks from the later part of the Casamayoran SALMA (Barrancan subage), conventionally considered early Eocene, yielded the surprising result that they could be as young as late Eocene (35.3–37.6 Ma), almost 20 million years younger than previously thought (Kay et al., 1999). This finding would indicate that the Casamayoran extended much later in time than previously thought and that the Mustersan SALMA is latest Eocene. It also raises the possibility that Riochican could be Eocene, and that there might be an even longer gap in the South American Eocene record than has been acknowledged. But the early Casamayoran fauna (Vacan subage; Cifelli, 1985) is more similar to the Riochican fauna, suggesting that the hiatus is more likely between the Vacan and the Barrancan subages. Flynn et al. (2003) reinterpreted the Casamayoran radioisotopic evidence to indicate a minimum age of 38 million years, and indicated that the lower boundary could be anywhere down to 54 million years ago (see Fig. 1.7), which would equate Casamayoran with most of the early and middle Eocene. This calculation of its duration may be too long, but age constraints are so poor that a more precise estimate is not yet possible.

The revised age estimates for the Casamayoran compress the Mustersan and Divisaderan into a short interval at the end of the Eocene. The relative age and even the validity of the Divisaderan are especially tenuous. Finally, high-precision ^{40}Ar/^{39}Ar dates for the recently proposed Tinguirirican SALMA indicate that it either bridges the Eocene/Oligocene boundary (Flynn and Swisher, 1995) or is of early Oligocene age (Kay et al., 1999). The younger age was upheld by Flynn et al. (2003), who dated the Tinguirirican at 31–32 million years ago but indicated that it might extend back as far as 37.5 million years ago (latest Eocene). Radiometric dates also show that the Deseadan is much younger than long believed, shifting it to late Oligocene (Flynn and Swisher, 1995). Figure 1.7 follows Flynn et al. (2003) for the Eocene SALMAs.

Note, however, that most of these revisions are so recent that they were not known at the time of McKenna and Bell's (1997) compilation, and obviously were unknown to Simpson and other earlier workers. Therefore occurrences and ranges of South American taxa in this text reflect the traditional terminology, namely, that Casamayoran was equivalent to early Eocene, Mustersan and Divisaderan to middle Eocene, and Tinguirirican to the Eocene/Oligocene boundary. Wherever possible, the age of fossils is clarified with the SALMA of origin to avoid confusion.

Asian Land-Mammal Ages

The Asian Land-Mammal Ages (ALMAs) are the most recently named and the most tentative. Several schemes have been proposed over the past two decades or so. The sequence used here follows that of McKenna and Bell (1997), which stems principally from Li and Ting (1983) and Russell and Zhai (1987), although a few of the ages were initially named by Romer (1966). Important modifications were made by Tong et al. (1995) and Ting (1998). A comparison of these reports reveals that there is still no consensus regarding the appropriate name for some of the ALMAs. With a few exceptions, the Asian land-mammal sequence is poorly constrained geochronologically, and the sequence has been based largely on stage of evolution. Therefore further revisions and refinements are to be expected.

There is general agreement that the Shanghuan ALMA is early Paleocene and the Nongshanian ALMA is late Paleocene. Wang et al. (1998), however, suggested that the Nongshanian may overlap with the late early Paleocene, partly based on the first K-Ar date (61.63 ± 0.92 Ma) from the Paleocene of China. Ting (1998) resurrected the Gashatan ALMA, named by Romer (1966), for latest Paleocene faunas that appear to be correlative with the Clarkforkian NALMA. Several names have been used for the first Eocene land-mammal age in Asia, including Ulanbulakian (Romer, 1966) and Lingchan (Li and Ting, 1983; Tong et al., 1995), but Bumbanian, proposed by Russell and Zhai (1987), is now generally accepted. The position of the Paleocene/Eocene boundary relative to the Gashatan and Bumbanian ALMAs has been controversial. However, the discovery that the CIE (and thus the Paleocene/Eocene boundary) is situated between Gashatan and Bumbanian faunas in the Lingcha Formation of China indicates that, at least in that section, Gashatan is entirely late Paleocene and Bumbanian is early Eocene (Bowen et al., 2002). The issue is not fully resolved, however, because it has been suggested that certain other Bumbanian faunas could be older than that of the Lingcha Formation.

Eocene ALMAs following the Bumbanian are very poorly constrained. There is general agreement that three ages can be recognized during the middle Eocene—Arshantan, Irdinmanhan, and Sharamurunian—but their boundaries are very uncertain. The Ergilian ALMA was proposed by Russell and Zhai (1987) as the earliest Oligocene ALMA, but it is now correlated with the late Eocene Priabonian and Chadronian. Consequently, the Shandgolian (Russell and Zhai's middle Oligocene ALMA, equivalent to Ulangochuian of Li and Ting, 1983) is early Oligocene and corresponds to the Rupelian and Orellan Land-Mammal Ages.

PALEOGEOGRAPHIC SETTING DURING THE BEGINNING OF THE AGE OF MAMMALS

The evolution and dispersal of mammals during the early Cenozoic were strongly influenced by the positions of the continental plates, the connections among them, the amount and distribution of subaerial exposure, and the marine barriers separating or dividing continents. The salient aspects of paleogeography at that time summarized here are based primarily on McKenna (1972, 1975b, 1980a, 1983) and Smith et al. (1994).

At the end of the Cretaceous, a wide epicontinental sea extended between the Arctic Ocean and the western Atlantic, dividing North America into eastern and western landmasses (Fig. 1.8A). The western portion was joined to Asia across Beringia (site of the present-day Bering Strait), whereas the eastern part was more closely approximated to Greenland, which was close or joined to northwestern Europe. North America and South America were separated by a wide seaway that connected the Pacific and Atlantic oceans. During the Late Cretaceous and early Paleocene, an epeiric sea apparently divided South America into northern and southern faunal provinces, limiting faunal exchange between the two regions (Pascual et al., 1992; Wilson and Arens, 2001). The southern parts of South America and Australia were close to Antarctica but lacked subaerial connections to that continent. South America and Africa were much closer to each other than they are today, though still separated by a sizable marine barrier. A narrow seaway split northwestern Africa from the rest of that continent, and the Tethys Sea (predecessor of the Mediterranean) came between northern Africa and Europe, which consisted of several islands. The Tethys extended eastward, south of Asia, where it was continuous with the Indian Ocean. India had recently separated from Madagascar and begun its drift northward. The rest of Asia was a large landmass separated from Europe by an epicontinental seaway (the Obik Sea to the north and the Turgai Straits at the southern end), which joined the Arctic Ocean to the Tethys Sea. This was the paleogeographic setting at the beginning of the Age of Mammals. Interchange of land mammals between any of the landmasses separated by marine barriers could only have occurred by Simpson's sweepstakes dispersal (Simpson, 1953; McKenna, 1973).

By the end of the early Paleocene a major lowering of sea level was under way, exposing more extensive land areas. North America was now a single landmass, as the epicontinental sea had diminished to a narrow extension from the Caribbean northward to the middle of the continent. Land bridges joined North America to northern Europe and to Asia, allowing faunal exchange. The Eurasian epicontinental sea also receded, exposing land bridges or islands between Europe and western Asia (Iakovleva et al., 2001). India was almost halfway to its junction with Asia.

The brief interval of global warming at the beginning of the Eocene (the Initial Eocene Thermal Maximum) resulted in increased continental temperatures as well as surface warming of high-latitude oceans (Sloan and Thomas, 1998). These changes turned the high-latitude North Atlantic land bridge (and, to a lesser extent, the North Pacific Bering bridge) into a hospitable corridor for mammalian dispersal. Geophysical evidence in fact suggests the presence of two North Atlantic land bridges during the late Paleocene–early Eocene: the northern De Geer Route and the southern Thulean Route (Fig. 1.8B, numbers 2 and 3). The De Geer Route—which was probably farther south in the early Tertiary, near the present-day Arctic Circle—joined northern Scandinavia, Svalbard (including Spitsbergen), northern Greenland, and northern Canada in the region of Ellesmere Island, and could have served as a direct passage between northwestern Europe and the Western Interior of North America. The Thulean bridge would have connected the British Isles to Greenland via the Faeroe Islands and Iceland, a geothermal "hot spot" in the early Cenozoic (Knox, 1998).

Although little fossil evidence is known from along these proposed land bridges, Simpson (1947: 633) long ago established that the extent of exchange between Europe and North America indicated that these land masses were "zoogeographically essentially a single region at this time." About 50–60% of early Eocene mammal genera from northwestern Europe are shared with western North America (Savage, 1971; McKenna, 1975b; Smith, 2000). In contrast, only one-third of earliest Eocene genera were shared by northern and southern Europe, suggesting that the continent was sporadically divided by some kind of barrier during the Paleogene, but whether it was geographic or climatic is unknown (Marandat, 1997). Ellesmere Island, which was within the Arctic Circle and at about the same latitude in the Eocene as it is today, has produced early-to-middle Eocene mammals and reptiles (crocodilians) that indicate a warm climate (Dawson et al., 1976; West et al., 1977; McKenna, 1980a). Several of the mammalian taxa are similar at the generic or family level to those found on both continents and suggest dispersal across Ellesmere in both directions (Eberle and McKenna, 2002). The effect of highly variable periods of daylight (and seasonal darkness) on the biota at such high latitudes remains problematic. By the middle Eocene (Lutetian), faunal disparities indicate that the opening of the North Atlantic by sea floor spreading had already interrupted the Euramerican land bridges.

The Bering land bridge (Beringia, Fig. 1.8B, number 1) seems to have been emergent throughout most of the Cenozoic (Marincovich and Gladenkov, 1999). However, it was evidently at even higher latitude (about 75° N) during the late Paleocene and early Eocene than it is today and consequently may have acted as a filter rather than a corridor (McKenna, 2003). Nonetheless, similar taxa found on both continents at that time (e.g., arctostylopids, uintatheres, carpolestids, omomyids), many of which are unknown from Europe, indicate faunal exchange. A more southern bridge across the Aleutian area may have existed as well, but probably not before the middle Eocene (McKenna, 1983).

Europe continued to be separated from Asia for part of the early Cenozoic by a marine barrier consisting of the Obik Sea and, at the southern end, the Turgai Strait. Current evidence suggests, however, that occasional subaerial connections may have been present at the northern and southern ends (Fig. 1.8B, numbers 4 and 5), particularly around the Paleocene/Eocene boundary (Iakovleva et al.,

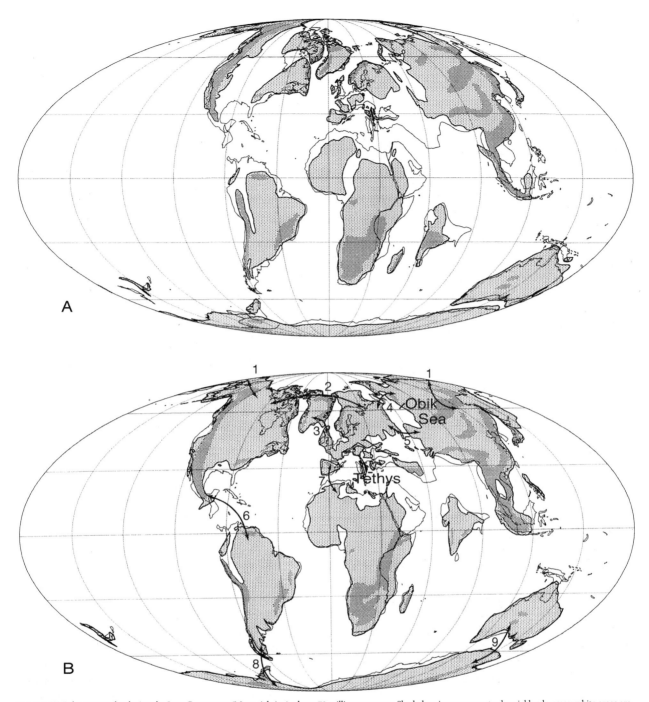

Fig. 1.8. (A) Paleogeography during the Late Cretaceous (Maastrichtian), about 70 million years ago. Shaded regions represent subaerial landmasses; white areas are oceans; lines show present-day coastlines. (B) Paleogeography during the early Eocene, about 53 million years ago. Numbered arrows indicate hypothesized dispersal routes during the Early Paleogene: 1, between Asia and North America via Bering land bridge; 2, De Geer route; 3, Thulean route; 4, between Asia and Europe at the northern end of the Obik Sea; 5, across the Turgai Strait; 6, probable sweepstakes dispersal between North and South America via Central America or perhaps a Caribbean archipelago; 7, between southern Europe and north Africa; 8, between South America and Antarctica; 9, between Antarctica and Australia. Some routes shown as marine barriers in this reconstruction might have been intermittently subaerial during the Early Cenozoic. (Modified from Smith et al., 1994.)

2001). A marine recession at the Eocene/Oligocene boundary finally exposed significant land bridges across the former seaway, allowing the immigrations from Asia that characterized the *Grande Coupure*.

It is now generally thought that the Indian Plate began to collide with Asia in the late Paleocene. Beck et al. (1998) even hypothesized that this collision could have precipitated the CIE (by triggering the release of organic carbon from the northern continental shelf of India) and the concomitant climatic and biotic changes that took place at the Paleocene/Eocene boundary. Fossil evidence regarding the time of collision is equivocal. Frogs and crocodilians of Laurasian affinity and the mammal *Deccanolestes* (see Chapter 10) have been cited as evidence of limited contact with Asia as early as the Late Cretaceous (e.g., Jaeger et al., 1989; Sahni and Bajpai, 1991; Prasad et al., 1994), but other records (fishes, turtles, and dinosaurs) imply that some animals dispersed from Madagascar or Africa to India in the Late Cretaceous (about 80 Ma; Sahni, 1984).

South America was isolated from other continents through much of the Cenozoic, and most of its endemic early Cenozoic mammal fauna seems to be derived from at least two sweepstakes dispersal events, an earlier one (no later than early Paleocene) from North America, and a later event (late Eocene) from Africa. Close proximity or a possible land connection between Patagonia and the Antarctic Peninsula is implied by the discovery in Antarctica (Seymour Island) of a small number of typically Patagonian taxa. The late middle Eocene age of the assemblage (Bartonian, or in the gap between the early and late Casamayoran) suggests that these were relict taxa that were isolated from the early Paleogene Patagonian fauna (Reguero et al., 2002). Nevertheless, the presence in Antarctica of marsupials believed to lie near the base of the Australian radiation supports the hypothesis that therian mammals reached Australia through Antarctica by the early Eocene, and probably before then (Woodburne and Case, 1996).

Known Early Cenozoic faunas from Africa are largely confined to a few areas of the northern Sahara, with an important exception from the middle Eocene of Tanzania (see Chapter 10). Although many groups appear to be endemic, there are hints of affinities with European faunas, which might have dispersed between present-day Spain and Morocco.

PALEOCENE-EOCENE CLIMATE AND FLORA

The world of the Paleocene and Eocene was very different from that of today. It was much warmer and more equable during most of that interval than at any other time during the Cenozoic (Wing and Greenwood, 1993). Temperatures varied little seasonally or latitudinally, mid-latitudes were largely frost-free, and there were no polar ice caps. Conditions were generally wet or humid. A paleotemperature curve reconstructed from deep-sea oxygen isotope records (Zachos et al., 2001) shows that early Paleocene temperatures continued as high as, or higher than, those at the end of the Cretaceous. Following a slight decline at the start of the late Paleocene (59–61 Ma), ocean temperatures increased steadily through the rest of the late Paleocene and the early Eocene (52–59 Ma) and peaked in the late early Eocene, about 50–52 million years ago (the Early Eocene Climatic Optimum, the warmest interval of the past 65 My). Thereafter, temperatures deteriorated more or less continuously to the end of the Eocene, when an abrupt, substantially cooler interval corresponded approximately with the Eocene/Oligocene boundary (or more accurately, the earliest Oligocene). This interval also corresponds with the appearance of permanent ice sheets in Antarctica for the first time in the Cenozoic, and possibly Northern Hemisphere glaciation as well (e.g., Coxall et al., 2005). Antarctic glaciation probably resulted in part from changes in ocean circulation following the isolation of Antarctica. The only significant interruption in these overall trends was the Initial Eocene Thermal Maximum, the short-term global warming alluded to earlier, which further raised temperatures for about 100,000 years at the beginning of the Eocene (Sloan and Thomas, 1998). A few other episodes of elevated temperature during the early Eocene have been identified recently, but they are of lesser magnitude (e.g., Lourens et al., 2005). The relatively high temperatures of the Paleocene and Eocene have led to the characterization of this interval as a "greenhouse," compared to the "ice house" of the post-Eocene.

Deep ocean temperatures during the Paleocene and early Eocene, deduced from oxygen isotope ratios in benthic foraminifera, ranged from 8 to 12° C (Zachos et al., 2001). Continental temperatures have been estimated from the proportion of leaves with entire (untoothed) margins, which has been shown to be higher in warmer climates (Wolfe, 1979; Wilf, 1997), from multivariate analysis of leaf physiognomy (Wolfe, 1993, 1994), and from oxygen isotope composition analyzed from paleosols and fossil teeth (Fricke et al., 1998; Koch et al., 2003). Although estimates based on these different methods do not always agree, the overall pattern is consistent. For western North America, leaf-margin analysis (supported by oxygen isotope data from foraminifera) documents an increase in mean annual temperature (MAT) from 10 to 15–18° C during the last 0.5 million years of the Cretaceous, followed by an abrupt drop to about 11° C just before the K/T boundary (Wilf et al., 2003). MAT remained at about 11° C through at least the first half of the Puercan, except for a brief, small increase immediately after the K/T boundary (probably of about 3° C, according to Wilf et al., 2003, rather than the 10° C increment reported by Wolfe, 1990). Nevertheless, these early Paleocene floras contain palms. Somewhat later in the early Paleocene (about early Torrejonian) temperatures rose again, and tropical rainforest was present in Colorado (Johnson and Ellis, 2002).

Leaf-margin analyses indicate that MAT in western North America increased from about 13 to more than 15° C during the last 2 million years of the Paleocene, and from about 18° C near the beginning of the Eocene to more than 22° C during the late early Eocene (the Early Eocene Climatic Optimum), with a possible brief cooler interval (dipping to about 11° C) in the middle of the early Eocene (Hickey, 1977; Wing, 1998b; Wing et al., 1999). For comparison, present-day MAT in Wyoming is about 6° C, with a much greater annual range than during the Early Cenozoic. Oxygen isotope analyses indicate that MAT during the Initial Eocene Thermal Maximum was 3–7° C higher than just before and just after that interval (Fricke et al., 1998; Koch et al., 2003). Wolfe (1985) estimated that latest Paleocene MAT was as high as 22–23° C in the northern High Plains. He later estimated early Eocene temperatures to have been at least 27° C at paleolatitude 45° N, and 19° C at 70° N in North America (Wolfe, 1994). Even the lower temperature estimates for the late Paleocene and early Eocene are within the range for present-day subtropical and paratropical rainforests (Hickey, 1977). The annual temperature range was small in the early Eocene, but increased substantially as the climate cooled toward the end of the Eocene.

Based on his higher temperature estimates, Wolfe (1985) inferred that tropical rainforest covered broad areas of the continents to latitude 50° during the latest Paleocene and early Eocene (the warmest interval of the Cenozoic), with paratropical rainforest extending to latitude 60–65°. Broad-leaved evergreen forest and palms extended to 70°. Farther poleward (e.g., on Ellesmere Island) were low-diversity forests of deciduous broad-leaved trees and deciduous conifers, such as *Glyptostrobus* (bald cypress) and *Metasequoia* (dawn redwood), which apparently were tolerant of seasonal darkness. One effect of a relatively frost-free climate at high latitudes—or, at least, a climate without persistent frost—was that forests of these deciduous angiosperms and conifers spread between Europe and North America, and even across Beringia (Manchester, 1999; Tiffney, 2000). Undoubtedly this situation made it easier for mammals also to disperse along these routes.

Like vertebrates, plants suffered major extinctions across the K/T boundary (e.g., Wolfe and Upchurch, 1986). Floras from immediately above the K/T boundary in North America tend to be dominated by ferns, which are among the first plants to reappear after major environmental disruption, such as the K/T boundary bolide impact (Wing, 1998a). Thereafter, floral diversity increased slowly, and recovery of angiosperms—which were decimated by the bolide impact—took hundreds of thousands of years. Paleocene floras of western North America are typically characterized by a low diversity of deciduous broad-leaved trees, and many of the taxa had very broad ranges (Wing, 1998a; Manchester, 1999). There are more deciduous taxa than are usually present in tropical or subtropical floras. This relative abundance could be a result of terminal Cretaceous extinctions of evergreens, or it may indicate that continental interiors were somewhat cooler than has been inferred. In the late Paleocene and early Eocene, floras consisted of mixed deciduous and evergreen broad-leaved trees. During the climatic optimum of the late early Eocene, there was a higher proportion of evergreen species. Later Eocene cooling led to greater floristic zonation, which in turn may have stimulated a general dietary shift among mammals (e.g., rodents, perissodactyls) from mainly frugivory to more specialized browsing and folivory (Collinson and Hooker, 1987). Broad-leaved evergreen vegetation was mostly restricted to below latitude 50°, whereas farther poleward there was mixed conifer forest (Wolfe, 1985). Latitudinal variation in temperature was still relatively low, however, so that rainfall had a stronger influence on vegetation patterns (Wing, 1998a). Following the dramatic cool episode at the end of the Eocene, temperate deciduous and conifer forests prevailed in the mid-latitudes.

The principal constituents of North American Paleocene and Eocene floras are summarized here based on Brown (1962), Hickey (1977), Upchurch and Wolfe (1987), Wing (1998a,b, 2001), and Manchester (1999). Common elements of the Paleocene flora of the Western Interior were walnuts and hickories (Juglandaceae), birches (Betulaceae), witch hazels (Hamamelidaceae), elms (Ulmaceae), dogwoods (Cornaceae), ginkgos (Ginkgoaceae), oaks (*Quercus*), sycamores (*Platanus*), katsuras (*Cercidiphyllum*), and the genera *Averrhoites* (Oxalidaceae?) and *Meliosma* (Sabiaceae). *Glyptostrobus* and *Metasequoia* (Taxodiaceae) predominated in backswamps. Several of these, including *Glyptostrobus, Metasequoia, Platanus,* and *Palaeocarpinus* (Betulaceae), were present during the Paleocene on all three northern continents (Manchester, 1999). Ground cover consisted of ferns, horsetails (*Equisetum*), and other low herbaceous plants, for grasses did not dominate in open habitats until the latest Oligocene or earliest Miocene (Strömberg, 2005). Palms were essentially limited to the southern half of the continent. Early Eocene floras included many of the same taxa, but also more subtropical taxa. Poplars, ginkgos, and hazelnuts were present; relatives of laurels (Lauraceae), citrus (Rutaceae), and sumac, mango, and cashew (Anacardiaceae) helped to form the canopy. Still abundant in swamp forests were the widespread conifers *Glyptostrobus* and *Metasequoia*. Other common swamp plants during the warm early Eocene include palms, palmettos, cycads, tree ferns, ginger, magnolia, laurel, hibiscus, and the floating fern *Salvinia*. Many of these plants are similar to the largely tropical or subtropical flora present in the early and middle Eocene of England (Collinson and Hooker, 1987).

Wing and Tiffney (1987) proposed that the interaction between land vertebrates and angiosperms during the Late Cretaceous and Early Cenozoic had profound effects on both floras and faunas. The extinction of dinosaurs at the end of the Cretaceous altered selective pressures on the plant community by eliminating large herbivores. This change in pressure, in turn, may have led to denser vegetation, intensified competition among plants, and selection for larger seeds—floral changes that would have stimulated the radiation of arboreal frugivores, but might have stifled diversification of larger terrestrial herbivores. Although such a model is consistent with many Paleocene quarry assemblages from the northern Western Interior, it is less consistent with assemblages from the San Juan Basin, New Mexico, which are dominated by larger terrestrial herbivores. The relationship between floras and faunas is complex and not yet well understood. For example, mammalian diversity is not always correlated with floral diversity (e.g., Wilf et al., 1998), nor are major changes in the structure of mammal and plant communities necessarily closely associated (Wing and Harrington, 2001).

ORGANIZATION OF THE VOLUME

Chapter 2 provides an overview of mammalian skeletal anatomy and the principal features of the skeleton and dentition that are used to interpret diet, locomotion, and other aspects of behavior in fossil mammals. A review of the origin of mammals follows in Chapter 3, and a synopsis of mammalian evolution during the Mesozoic in Chapter 4, as the background to the Early Cenozoic radiation that is the principal focus of the book. The Multituberculata, a Mesozoic clade that survived into the Early Cenozoic and was a significant constituent of many Paleocene faunas, is covered in the latter chapter. In Chapter 5 the fossil record of Metatheria from the Cretaceous through the Eocene is presented. Basal eutherians of the Cretaceous, the primitive antecedents of the Cenozoic placental radiation, are highlighted in Chapter 6.

Chapters 7 through 15 summarize the Paleocene-Eocene fossil record of eutherian mammals. In some cases pertinent early Oligocene groups are discussed as well. Chapters generally group taxa that are, or have been, thought to be monophyletic; but for some taxa the evidence for monophyly is weak at best, and the association is really more one of convenience. Cladograms and classification tables are included in Chapters 4 through 15 to help readers place taxonomic groups in phylogenetic context. In the tables a dagger symbol (†) is used to indicate extinct taxa, and families and genera known from the Paleocene or Eocene are shown in boldface. Unless otherwise indicated, most classifications used in the book are modified after McKenna and Bell (1997, 2002). All Paleocene-Eocene higher taxa are listed, but complete listings of all later Cenozoic and Recent taxa are omitted for some of the most diverse orders.

Chapter 7 covers the primitive cimolestan "insectivores" as well as several clades that have been associated with them or are thought to be their descendants, including didymoconids, pantolestans, apatotheres, taeniodonts, tillodonts, and pantodonts. In Chapter 8 the creodonts and carnivorans are reviewed. Insectivora, including leptictids and lipotyphlans are the subject of Chapter 9. The early fossil record of the Archonta, including bats, dermopterans, tree shrews, and primates, is detailed in Chapter 10. Chapter 11 concerns the xenarthrans, pangolins, and palaeanodonts—mammals loosely grouped as "edentates," although there is little convincing evidence for relationship of the xenarthrans to the others. Under the heading of archaic ungulates, the subject of Chapter 12, are grouped condylarths as well as an assortment of other primitive ungulates, including uintatheres, arctostylopids, and the extinct South American ungulates (litopterns, notoungulates, pyrotheres, astrapotheres, and xenungulates). This grouping, too, is one of convenience and does not imply any special relationship. Chapter 13 describes the Altungulata, which comprises perissodactyls, hyracoids, and tethytheres (sirenians, proboscideans, and arsinoitheres). Cetacea, archaic mesonychians, and artiodactyls are discussed in Chapter 14. Chapter 15 summarizes the fossil record of Anagalida: the rodents, lagomorphs, and possible relatives, including elephant shrews and several fossil clades. The final chapter provides a retrospective on mammalian evolution during the beginning of the Age of Mammals.

2

Mammalian Skeletal Structure and Adaptations

THE MAMMALIAN SKELETON HAS BEEN evolving for more than 200 million years, since it originated from that of nonmammalian cynodonts, resulting in variations as different in size and adaptation as those of bats, moles, horses, elephants, and whales. Therefore, to assume that there is a living species that displays the "typical" mammalian skeleton would be naive and misleading. Nevertheless, all mammalian skeletons represent variations on a fundamental theme, and in terms of the addition or loss of skeletal elements, mammals have, in general, remained rather conservative. The objective of this chapter is to review the skeleton of generalized mammals as a foundation for the discussion of mammalian dentition and osteology throughout this book, and to briefly survey some of the variations on this theme.

Compared to the skeletons of lower tetrapods, those of mammals are simpler (with fewer elements, because of fusion or loss of bones) and better ossified (with more bone and less cartilage in adults). Both conditions probably contribute to greater mobility and speed of movement. One of the most important consequences of thorough ossification is more precisely fitting limb joints. The articular ends of reptile limbs are covered in cartilage. Because reptile bones grow in length throughout life by gradual ossification of this cartilage, a distinct articular surface never forms. By contrast, the articular ends, or epiphyses, of mammalian limb bones (and certain bony features associated with muscle attachment, such as the femoral trochanters) develop from separate centers of ossification from the one that forms the shaft, or diaphysis. Growth in length occurs at the cartilaginous plates between the shaft and the epiphyses, thus allowing the formation of well-defined articular surfaces, even in animals that are still growing.

SKULL

The human skull consists of 28 bones (including the three middle-ear ossicles), most of which are very tightly articulated or fused together. This constitutes a reduction of the cranial arrangement characteristic of primitive mammals (e.g., several elements fuse to form the human temporal or occipital bones). Nevertheless, mammals as a rule have fewer skull bones than do reptiles. There are several excellent general accounts of the anatomy of the mammalian skull (e.g., Romer and Parsons, 1977; Barghusen and Hopson, 1979; Novacek, 1993) that supply detailed information. The summary presented here is based partly on these accounts.

The braincase of mammals is generally much larger than that of reptiles. Besides protecting the brain, it provides a surface of origin for the temporalis muscles (used in mastication) laterally, and for neck muscles on the posterior surface, or occiput. Additional attachment area for these muscles is provided by the midline sagittal crest and the transversely oriented lambdoidal (=nuchal or occipital) crest at the top of the occiput. Typically the braincase consists of paired frontals (fused in humans) and parietals dorsally, an occipital at the back, and paired squamosals laterally, below the frontals and parietals (Fig. 2.1). An interparietal may be present between the parietals and the occipital. Developmentally the occipital bone consists of several elements, including the supraoccipital, the basioccipital, and the paired exoccipitals (which include the occipital condyles and surround the foramen magnum, through which the spinal cord passes), but these are usually fused in adults. The squamosal bones contain the glenoid fossa posteriorly, for articulation with the condylar process of the dentary, and form the back of the zygomatic arch. Anterior to the basioccipital on the ventral surface are the basisphenoid and presphenoid; paired alisphenoids and orbitosphenoids extend laterally from the basisphenoid and presphenoid to form part of the lateral wall of the braincase. Commonly these bones fuse to form the complexly shaped sphenoid bone (as in humans). A delicate midline bone, the ethmoid, forms part of the floor of the braincase and extends into the upper nasal cavity.

Situated more or less between the squamosal and the basioccipital is the auditory region of the basicranium, which contains the tympanic or middle-ear cavity, with its three tiny auditory ossicles (malleus, incus, and stapes), and the inner ear, with its bony labyrinth enclosing the cochlea and semicircular canals. These are the organs of hearing and balance. The ossicular chain extends from the tympanic membrane (eardrum) laterally, to the fenestra ovalis (oval window) on the ventrolateral wall of the petrosal, which receives the footplate of the stapes. The eardrum is usually supported by a ringlike or tubular ectotympanic (=tympanic) bone. Within the dense petrosal bone, or otic capsule, is the inner ear. The cochlear canal in living mammals (except monotremes) is coiled, resembling a snail shell. There are always at least $1\frac{1}{2}$ turns (about $2\frac{1}{2}$ in humans). A coiled cochlea can accommodate a longer basilar membrane—which supports the spiral organ for hearing, within the cochlea—in a smaller space, and is therefore usually a good indicator of auditory acuity.

In many mammals, as in humans, the petrosal, ectotympanic, and squamosal bones synostose to form the temporal bone. The tympanic cavity and otic capsule in mammals are typically surrounded and protected by a bubble-like bony structure, the auditory bulla, behind which the mastoid portion of the petrosal is often exposed. The bulla, which forms the floor of the tympanic region, is a mammalian innovation. When present in marsupials it usually forms from the alisphenoid, whereas in placentals it is variously constructed of the ectotympanic, entotympanic, petrosal, or a combination of these or other elements. The bony anatomy of the auditory region, particularly bullar composition, and the pattern of vascular grooves on the ventral surface of the petrosal created by branches of the internal carotid artery (which usually flows through this region en route to the brain) are important considerations in mammalian systematics.

The basicranium is also important because of its numerous foramina that transmit the 12 pairs of cranial nerves and various vessels to and from the brain. The nerves and vessels do not actually penetrate the basicranium; instead, during development the bone ossifies around them. Cranial nerves (CNs) serve many vital functions: they are responsible for the special senses, control muscles and supply sensory innervation to the head and neck, and provide parasympathetic autonomic innervation to thoracic and abdominal viscera as well as glands and smooth muscle in the head. They are numbered from front to back as they emerge from the base of the brain. The configuration of the basicranial foramina through which the nerves enter or leave the cranial cavity also weighs heavily in phylogenetic interpretations. The foramina and nerves may be summarized as follows:

Cribriform plate of the ethmoid bone—A perforated bone at the anterior floor of the braincase, through which nerve bundles of CN I, the olfactory nerve (which provides the sense of smell), pass from the roof of the nasal cavity to reach the olfactory bulbs of the brain.

Optic foramen—The opening that transmits CN II, the optic nerve (the nerve of vision), through the orbitosphenoid to the orbit and the eyeball.

Sphenorbital foramen (=anterior lacerate foramen, superior orbital fissure)—An opening between the orbitosphenoid and alisphenoid through which CNs III, IV, V^1, and VI reach the orbit. CN III (oculomotor nerve), IV (trochlear nerve), and VI (abducent nerve) supply muscles that move the eye; V^1 (the first division of the trigeminal nerve, called the ophthalmic nerve) is sensory to the eye, orbit, and forehead.

Foramen rotundum—A hole in the alisphenoid that is the usual pathway of CN V^2 (the second or maxillary division of the trigeminal nerve) to the floor of the orbit, where the nerve gives off sensory branches to the maxillary sinus and upper teeth. Its termination passes

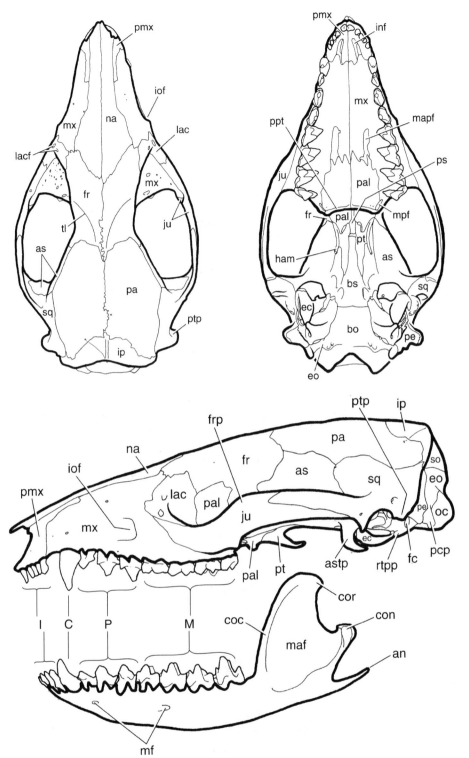

Fig. 2.1. Anatomy of the skull of a primitive mammal (*Monodelphis*). The dental formula exhibited by this marsupial is 5.1.3.4/4.1.3.4. Key: an, angular process; as, alisphenoid; astp, alisphenoid tympanic process; bo, basioccipital; bs, basisphenoid; C, canines; coc, coronoid crest; con, mandibular condyle; cor, coronoid process; ec, ectotympanic; eo, exoccipital; fc, fenestra cochleae; fr, frontal; frp, frontal process of the jugal; ham, hamulus; I, incisors; inf, incisive foramen; iof, infraorbital foramen; ip, interparietal; ju, jugal; lac, lacrimal; lacf, lacrimal foramen; M, molars; maf, masseteric fossa; mapf, major palatine foramen; mf, mental foramina; mpf, minor palatine foramen; mx, maxilla; na, nasal; oc, occipital condyle; P, premolars; pa, parietal; pal, palatine; pcp, paracondylar process of the exoccipital; pe, petrosal; pmx, premaxilla; ppt, postpalatine torus; ps, presphenoid; pt, pterygoid; ptp, posttympanic process; rtpp, rostral tympanic process of the petrosal; so, supraoccipital; sq, squamosal; tl, temporal line. (Modified from Wible, 2003.)

through the infraorbital foramen in the maxilla, emerging onto the snout or face to provide sensory innervation to this area.

Foramen ovale—The opening that transmits CN V^3 (third or mandibular division of the trigeminal nerve) through the alisphenoid to the mouth, where it supplies the masticatory muscles and is sensory to the cheek and most of the tongue. One branch enters the dentary posteromedially through the mandibular foramen to supply the lower teeth and gums, and emerges through one or more mental foramina (anterolaterally on the dentary) to provide sensation to the chin area.

Internal acoustic meatus—Visible only from inside the cranial cavity, this opening in the petrosal bone transmits CN VII (facial nerve) and CN VIII (vestibulocochlear nerve) into the ear region, where the latter runs to ganglia in the inner ear associated with hearing and balance. After giving off several branches within the petrosal (including two special sensory branches involved with taste), CN VII emerges through the stylomastoid foramen at the back of the basicranium to supply the muscles of the snout or face.

Posterior lacerate foramen (=jugular foramen)—This large foramen between the basioccipital and the otic capsule transmits CNs IX (glossopharyngeal nerve), X (vagus nerve), and XI (accessory nerve), as well as the internal jugular vein. CN IX innervates aspects of the tongue, pharynx, and middle ear, and CN X innervates the pharynx, larynx, most palatal muscles, and thoracic and abdominal viscera. CN XI supplies two muscles of the neck and back (sternocleidomastoid and trapezius).

Hypoglossal canal (=anterior condyloid foramen)—Paired or multiple openings within the foramen magnum and just anterior to the occipital condyles, which transmit CN XII (hypoglossal nerve) to the tongue muscles.

A few other cranial openings are of note. The carotid canal carries the internal carotid artery into the cranial cavity to supply the brain. Upon entering the basicranium, usually near the posterior lacerate foramen, the artery traverses the carotid canal in the alisphenoid and emerges into the cranial cavity immediately above the middle lacerate foramen (=lacerate foramen of human anatomy) at the front of the alisphenoid. An opening at the front of the auditory bulla joins the middle-ear cavity to the back of the throat via the cartilaginous part of the auditory (eustachian) tube.

The facial skeleton or snout includes the bones around the orbit (except the frontal), the nose, and the mouth. The paired maxillae hold most of the upper teeth and make up a large part of the secondary (hard) palate, a mammalian characteristic, which separates the oral and nasal cavities. They also usually form the front of the zygomatic arches and often contribute to the floor or anterior rim of the orbit. Enclosed within each maxilla is a large cavity, or sinus, which adjoins the nasal cavity. Similar sinuses are found in the sphenoid, ethmoid, and occasionally the frontal and various basicranial bones. The pneumatization created by these sinuses reduces the weight of the skull, contributes to vocal resonance, and at the same time provides advantageous muscle attachments. The premaxillae contain the incisors and form the front of the palate and the anterolateral wall of the nasal cavity. The back of the palate consists of the palatine bones, which also define the lower margin of the choanae, or internal nares. The bony palate has paired incisive foramina in front and palatine foramina at the back, which carry nerves and vessels. In some mammals, the incisive foramina lead to Jacobson's organ (=vomeronasal or accessory nasal organ). Marsupials often have additional openings in the palate, called vacuities. Behind and above the palatines, and anterior to the presphenoid, is a small midline element, the vomer, which, together with the ethmoid, divides the two nasal cavities. Attached to the lateral walls of the nasal cavity are the turbinals: delicate, scroll-like structures of cartilage or bone that expand the surface area of the nasal cavity. The lower one on each side is a separate element, the inferior turbinate (=maxilloturbinal or inferior nasal concha). Situated medial to the maxillae and premaxillae and behind the external nares are the nasals. At the anteromedial margin of the orbit are the lacrimal bones, pierced by the nasolacrimal (=lacrimal) canal, which contains a duct that drains lacrimal fluid from the eye into the nose. The paired zygomatic or jugal (=malar) bones are positioned between the maxilla and squamosal on each side. The zygomatic thus forms the middle of the zygomatic arch, a bony bar on the outside of the orbitotemporal fossa, which protects the eye and provides attachment area for the masseter muscle.

The mandible or lower jaw in mammals, which contains all the lower teeth, consists of a pair of dentaries, in contrast to the multi-element lower jaw of reptiles and nonmammalian therapsids. The two dentaries are either joined by ligaments or co-ossified at the front (the mandibular symphysis). Behind the toothrow is the ascending ramus, with a coronoid process for attachment of the temporalis muscle, and a condyle that articulates with the squamosal. The medial side of the condylar neck is also the site of insertion of the lateral pterygoid muscle. At the posteroinferior margin of the jaw is the angular process, for attachment of the masseter (laterally) and medial pterygoid muscles (medially). These are all chewing muscles, supplied by CN V^3.

DENTITION

Probably more than any other part of the skeleton, the dentition of fossil mammals plays a critical role in taxonomy, assessment of phylogenetic position, and interpretation of behavior (primarily diet, but also such activities as grooming, gnawing, or even digging). In part, this reflects the durability of teeth (enamel, the hard outer layer of most mammal teeth, is the hardest substance in the body), which accounts for why they are generally more common than other skeletal remains. But it is also because the dentition usually exhibits species-specific differences, not so readily distinguished in other parts of the skeleton, that can often be detected even in individual teeth. Especially useful general accounts of the dentition in vertebrates generally, and mammals in particular, include Gregory (1922), Peyer (1968), and Hillson (1986).

One of the characteristics of mammals (inherited from their nonmammalian cynodont ancestors) is the regional differentiation of the dentition into incisors (I), canines (C), premolars (P), and molars (M), known as heterodonty (Fig. 2.1). The postcanine teeth are collectively called cheek teeth. Incisors are typically involved in procuring and ingesting food. Canines usually function for stabbing or holding prey, for aggression, or for display. Premolars hold or prepare food for the molars, which shear, crush, and grind the food. In most mammals, the antemolar teeth are replaced once during life, a diagnostic mammalian condition called diphyodonty. The first set of teeth, the deciduous or milk teeth (indicated by "d," such as dP_4), erupts more or less in sequence from front to back, followed by the molars, which are actually part of the first generation of teeth. Most of the antemolar teeth are sequentially replaced by permanent teeth after some or all of the molars are in place.

The number of teeth present in each part of the dentition varies among mammals and is an important taxonomic characteristic. It is expressed in shorthand by the **dental formula**, I.C.P.M/I.C.P.M, which specifies the number of teeth in each quadrant, that is, on each side, above and below. Thus the dental formula of primitive extant placentals is $3.1.4.3/3.1.4.3 \times 2 = 44$; this translates to three incisors, one canine, four premolars, and three molars in each upper and each lower quadrant, for a total of 44 teeth. These teeth are conventionally identified as $I^{1-3}C^1P^{1-4}M^{1-3}/I_{1-3}C_1P_{1-4}M_{1-3}$. The dental formula of primitive marsupials (e.g., the opossum *Didelphis*) is $5.1.3.4/4.1.3.4 \times 2 = 50$. Generalized marsupials typically differ from primitive living placentals, then, in having more incisors (and more of them in the upper jaw than in the lower), one more molar, and one less premolar. The postcanine teeth are conventionally identified as P1–3, M1–4 (in both upper and lower jaws), although some accounts use a different numbering system. Obviously, the number of teeth has varied considerably among mammals. Some primitive Mesozoic types had more premolars and/or molars than do most modern species, whereas some living mammals have many homodont (similar) teeth (e.g., porpoises) and others have no teeth at all (e.g., anteaters). As a general rule, however, no mammal has more than one canine, and living marsupials and placentals rarely increase the number of premolars and molars beyond the primitive state.

Fossil evidence suggests, however, that the primitive eutherian dental formula was 5.1.5.3/4.1.5.3. To achieve the dental formula common in the most generalized living placentals, it is probable that incisors were lost from the back of the series and a premolar was lost from the middle (P3), very early in the history of placentals (McKenna, 1975a; Novacek, 1986b). This hypothesis suggests that the four premolars present in most primitive extant placentals could be dP1.P2.P4.P5, although the last two are conventionally identified as P3 and P4. This convention has been adopted because there is little direct evidence of how the reduction to four premolars took place, and whether it represents a single event in eutherians or occurred multiple times. Whatever position was lost, there is general agreement that the remaining teeth are probably homologous, and they continue to be almost universally identified as (d)P1.P2.P3.P4. This practice is also followed here, acknowledging that it is an assumption. Nearly all Cenozoic placentals have no more than three incisors and four premolars, hence a dental formula of 3.1.4.3/3.1.4.3 may be considered the primitive condition among Paleocene and Eocene mammals.

Although the dental formula is an important characteristic of mammals, equally or more important are the homologies of the teeth. For example, an enlarged central incisor evolved in many clades of mammals, but the tooth involved is not always homologous. In some cases it is I1, whereas in others it is I2 or a retained deciduous I2. When all the incisors are present, homologies are easily determined, but deciphering true homologies when the number of incisors is reduced to one or two requires developmental or evolutionary evidence. Unusually specialized premolars have also arisen independently in various lineages, as demonstrated by their occurrence at different tooth loci in different clades.

Several positional and other descriptive terms are commonly used when describing teeth. **Buccal** refers to the external or lateral surface, which faces the cheek (=**labial**, facing the lips, especially near the front of the jaw); **lingual** denotes the internal or medial surface, toward the tongue. The anterior end of the toothrow is also called **mesial**, the posterior end **distal**. Tooth length is measured mesiodistally, whereas width is measured transversely (buccolingually). Teeth are implanted in the **alveoli** (sockets) of the jaw by the **root**(s); the **neck** is approximately at the gum line, and most of what is exposed is the **crown**, usually covered by enamel. **Enamel** is an extremely hard, largely inorganic substance composed of hydroxyapatite crystallites. The underlying **dentine** is an avascular tissue consisting of hydroxyapatite, collagen, and water, and is softer than the enamel. **Cementum** is a bonelike tissue usually found covering the roots of teeth, but it is also found in the crowns of the teeth of many herbivores. Teeth with relatively low crowns are characterized as **brachydont**, whereas those with high crowns (higher than the roots, or higher than the length or width; Simpson, 1970c) are **hypsodont**. Teeth that grow continuously throughout life and never form roots are called **hypselodont** (e.g., Simpson, 1970c) or **euhypsodont** (Mones, 1982); these are essentially equivalent terms. The most obvious examples of hypselodont teeth are the incisors of rodents, but the condition has evolved independently in multiple lineages, and at different tooth loci.

Incisors may be small to very large and ever-growing, and the crowns vary from pointed to broad and spatulate, chisel-like, bilobed, or multicuspate; upper incisors tend to be larger than lowers. Canines are usually relatively large, conical teeth, but in some forms they are reduced or lost, whereas in others they are huge and saberlike or gliriform (like rodent incisors) and may be ever-growing. Both incisors and canines are almost always single-rooted. Premolars may be simple with one main cusp, or more complex, sometimes closely resembling molars. They usually increase in size and complexity posteriorly. In some mammals the posterior premolars are greatly enlarged, and in this case they may be swollen or bladelike. Despite these interesting variations in antemolar teeth, which are sometimes diagnostic of particular taxa, the crown morphology of molars is particularly distinctive and almost always carries substantial weight for taxonomy, phylogenetic assessment, and dietary inference.

Extant mammals, as well as most of the fossil groups dealt with in this book, have molars derived from the basic tribosphenic condition that evolved in the Mesozoic ancestors

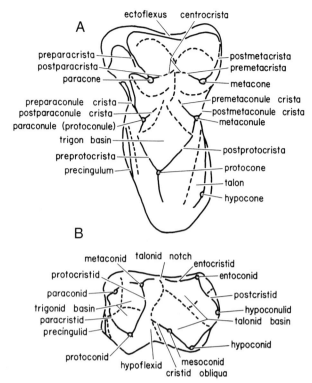

Fig. 2.2. Structure of tribosphenic molars (anterior to the left): (A) left upper; (B) left lower. (From Bown and Kraus, 1979.)

of marsupials and placentals. Some more primitive groups, discussed in the chapter on Mesozoic radiations (Chapter 4), were not yet tribosphenic. Here I focus on the structure of tribosphenic molars (Fig. 2.2) and defer a discussion of how tribosphenic molars evolved until Chapter 4. In general, tribosphenic molars have divided roots, two for each lower molar, located below the trigonid and talonid, and three for each upper, under each of the three main cusps. Generalized tribosphenic upper molars are transversely wider than they are long, and the three main cusps form a triangle with two cusps arranged buccally and one lingually. On the buccal side, the **paracone** is anterior, the **metacone** posterior; the lingual cusp is the **protocone.** Between the paracone-metacone and the buccal margin of the tooth is the **stylar shelf,** on which smaller cusps may be present, such as the parastyle, stylocone, mesostyle, and metastyle. Conules (**paraconule** and **metaconule**) are commonly present between the paracone or metacone and the protocone. A **hypocone** is frequently developed posterolingually, especially in herbivores, and may result in a quadrate upper molar. It is generally assumed that these cusps on adjacent teeth of an individual are serially homologous.

Mammalian cusp nomenclature is largely topographic: although it is probable that the three main cusps are almost always homologous across species, this is not true for the hypocone, mesostyle, and various other smaller cusps, which have demonstrably arisen multiple times independently (Van Valen, 1994a; Hunter and Jernvall, 1995). Indeed, developmental evidence has shown that relatively small changes during tooth formation can result in substantial changes in the size or number of small cusps (Jernvall, 2000). Although this instability helps to explain the frequent appearance of new cusps in different clades, it also means that variations in small cusps may have little phylogenetic significance, which should be remembered when using minor variations in cusp pattern as evidence for or against relationship.

Cusps are often joined by crests, and in some teeth crests predominate. The centrocrista is the crest between the paracone and metacone in generalized molars. When this crest is better developed, or when it links the centrocrista to the parastyle, metastyle, or mesostyle, it is called the ectoloph. Other crests are usually named with respect to the cusps they join. For instance, the preprotocrista and postprotocrista run anteriorly and posteriorly to the protocone, from the paracone or paraconule, and metacone or metaconule, respectively. Parallel transverse crests joining the paracone to the protocone and the metacone to the hypocone are the protoloph and metaloph, respectively. They are particularly well developed in herbivorous forms. A low (basal) shelf on any margin of the tooth is a cingulum.

Tribosphenic lower molars are longer than wide and consist of a trigonid anteriorly and a talonid posteriorly. As its name implies, the **trigonid** consists of three cusps, but in tribosphenic molars these cusps are arranged in a triangle that is inverted compared to that of the upper molars. Lower molar features end in the suffix -id; the two lingual cusps of the trigonid are the **paraconid** and **metaconid**, and the buccal cusp is the **protoconid**. The trigonid is almost always taller than the talonid. When it first evolved, the **talonid** was little more than a short "heel" with a single cusp, but in tribosphenic molars it usually has two or three cusps, the **entoconid** lingually, **hypoconid** buccally, and the **hypoconulid** in between. As in the upper molars, crests commonly join various cusps: the paracristid (or paralophid) between paraconid and protoconid, the protocristid (sometimes called the metacristid or metalophid in certain mammals) between protoconid and metaconid, and the postcristid (=hypolophid) between hypoconid, hypoconulid, and entoconid. The cristid obliqua is a crest that runs from the hypoconid anteriorly to the back of the trigonid, often oriented obliquely toward the metaconid. An entocristid may be present mesial to the entoconid. The crests of the talonid usually encircle a depression of variable size, forming a **talonid basin** that occludes against the protocone. Additional talonid cusps are sometimes present, including a metastylid behind the metaconid (really an accessory trigonid cusp), an entoconulid anterior to the entoconid, or a mesoconid on the cristid obliqua. A basal cingulum (cingulid) is often present on the buccal side, sometimes extending to the mesial or distal ends but almost never lingually.

Molars with sharp or bladelike cusps or crests are described as **secodont** or **sectorial**; specialized sectorial teeth called **carnassials** are characteristic of carnivorous mammals. Teeth with low, rounded cusps are **bunodont**. An occlusal pattern dominated by crescentic crests with a mesiodistal long axis is **selenodont,** whereas a pattern characterized by transverse ridges is **lophodont**. These and other

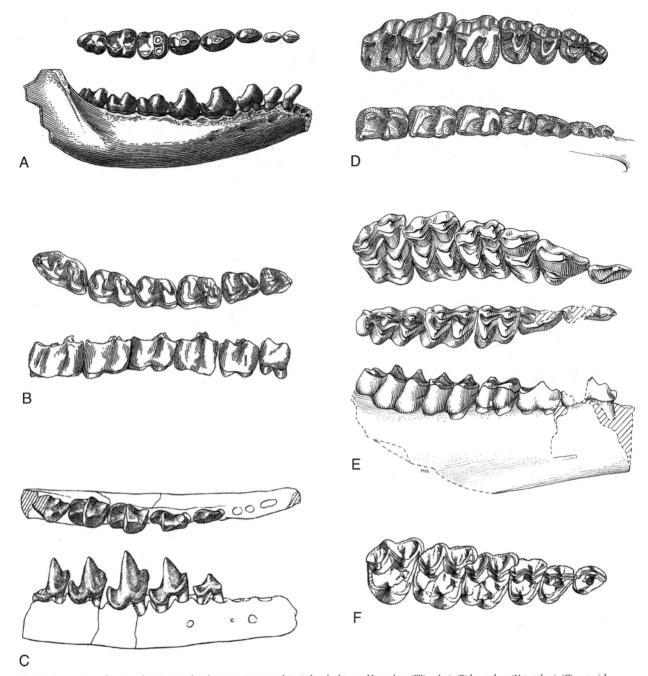

Fig. 2.3. Comparison of various dentitions and molar types (not to scale): (A) brachydont and bunodont (*Ellipsodon*); (B) hypsodont (*Notostylops*); (C) sectorial (*Batodonoides*); (D) lophodont (*Triplopus*); (E) selenodont (*Poabromylus*); (F) bunoselenodont and dilambdodont (*Eotitanops*). (A from Matthew, 1937; B from Simpson, 1948; C from Bloch et al., 1998; D from Radinsky, 1967a; E from Wilson, 1974; F from Osborn, 1929.)

modifications of the primitive tribosphenic pattern have enabled mammals to adapt for diverse diets (Fig. 2.3) and are one of the keys to their success.

The microscopic structure of the enamel also provides information relevant to phylogeny and function (e.g., Koenigswald and Clemens, 1992; Koenigswald, 1997a,b). Enamel is composed of long, needlelike crystallites of carbonate hydroxyapatite. In the most primitive Mesozoic mammals, the crystallites are parallel and radiate from the enamel-dentine junction to the surface. This relatively simple type of enamel is called aprismatic or nonprismatic enamel. In most mammals, however, the crystallites combine into bundles called prisms, each of which is surrounded by a prism sheath, also composed of crystallites. Although there is considerable variation in the morphology of the prisms and their sheaths, the significance of this variation is unknown. Groups of prisms are often arranged in the same orientation. In some cases all the prisms are oriented similarly and are either arranged radially from the enamel-dentine junction (radial enamel) or bend together (tangential enamel). In most eutherian mammals that weigh more than a few kilograms, the enamel consists of decussating groups of prisms that change orientation together, known as Hunter-Schreger bands (HSB; see Fig. 15.14). This specialized arrangement of prisms is thought to help strengthen the enamel, but the functional significance of different types of HSB is poorly understood. Although some patterns of HSB appear to be phylogenetically significant, the extent of homoplasy can make it difficult to distinguish them from functionally related patterns. Enamel microstructure is particularly important in rodents and is further discussed in Chapter 15.

POSTCRANIAL SKELETON

Although dental and cranial anatomy have generally received more attention in mammalian paleontology than has the postcranial skeleton, the skeleton is a critical source of information on body size, locomotion, habitat preference, and many other aspects of paleobiology. Postcranial characters are also playing an increasingly significant role in phylogenetic analyses, as it becomes more accepted that these features are no more subject to homoplasy than are dental or cranial features (Sánchez-Villagra and Williams, 1998). The skeleton (Fig. 2.4) can be divided into axial and appendicular parts. The axial skeleton comprises the skull and trunk, including the vertebral column, sternum, and ribs. The appendicular skeleton encompasses the limbs and limb girdles.

The segmented vertebral column provides support and flexibility and protects the spinal cord. It is also closely associated with locomotion. In mammals it is differentiated into five regions, each with its own distinctions: cervical, thoracic, lumbar, sacral, and caudal, abbreviated as C, T, L, S, and Ca, respectively (Fig. 2.5). Nearly all mammals have seven cervical vertebrae, a remarkable conservatism probably resulting from developmental constraints (Galis, 1999); the only exceptions are found among sloths (six to nine) and manatees (six). Other regions are much more variable.

Among extant mammals, thoracic vertebrae may number 9–25 (usually 12–15), lumbars 2–21 (usually 4–7), sacrals 3–13 (usually 3–5), and caudals 3–50 (Flower, 1885; Lessertisseur and Saban, 1967a; Wake, 1979). As a rule, individual vertebrae consist of a body (centrum) and a vertebral (neural) arch bearing a median dorsal spinous process and two pairs of articular processes (zygapophyses). The anterior or prezygapophyses face more or less dorsally or medially, whereas the postzygapophyses face ventrally or laterally. From the side of the arches or centra extend the transverse processes. Both neural arch and transverse processes tend to be much reduced in most of the tail. In the thoracic and lumbar regions, an additional process, the metapophysis (mamillary process), may project from the prezygapophysis, and an anapophysis (accessory process) may extend caudally below the postzygapophysis.

Cervical vertebrae are distinguished by having a very large vertebral foramen (for passage of the spinal cord) and foramina in the transverse processes (except C7) through which the vertebral arteries pass en route to the cranial cavity. In most mammals the cervical centra tend to be short, but in some mammals, such as the giraffe, they are very long. The first two cervicals, called the atlas (C1) and axis (C2), are diagnostic of mammals. The ringlike atlas, which lacks a centrum, articulates with the occipital condyles and allows flexion and extension of the head. The axis has an anterior projection, the dens (odontoid process), which is a neomorphic addition to the atlas centrum rather than its homologue (Jenkins, 1969a). The dens is held by ligaments against the ventral arch of the atlas, serving as a pivot for rotation of the head-atlas complex. The neural spine of the axis tends to be very prominent. Cervical ribs are present in monotremes and some primitive fossil mammals.

Thoracic vertebrae are readily distinguished because they articulate with ribs. The head (capitulum) of each rib artic-

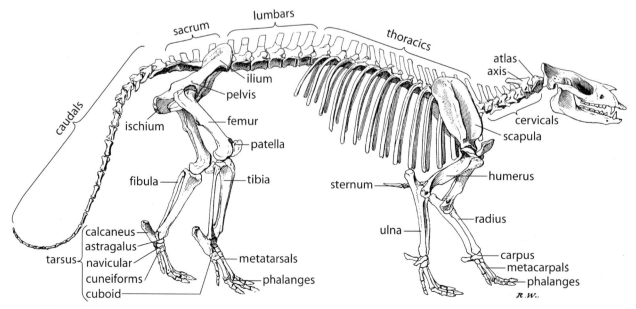

Fig. 2.4. Skeleton of a generalized mammal, Eocene *Phenacodus*. (Modified from Osborn, 1898a.)

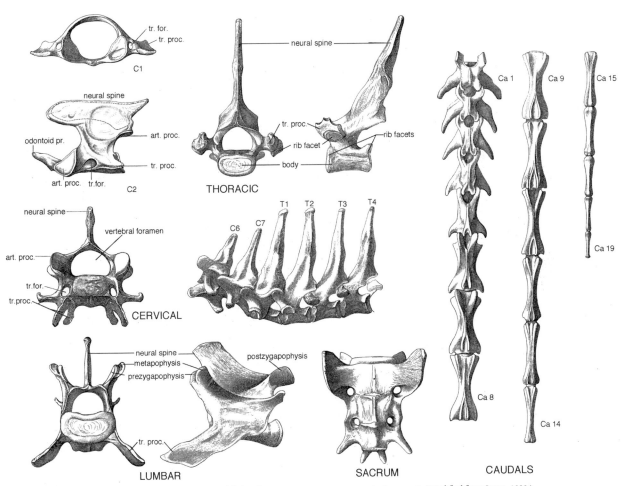

Fig. 2.5 Mammalian vertebrae. Key: art., articular; Ca, caudal; for., foramen; pr. or proc., process; tr., transverse. (Modified from Jayne, 1898.)

ulates at the junction of two centra, and the tubercle (tuberculum) of the rib articulates with the transverse process. Thoracic vertebrae have progressively larger centra and smaller vertebral foramina, moving caudally in the series. The spinous processes of the anterior thoracics tend to be high and posteriorly inclined. At the caudal end of the series, the orientation changes, becoming somewhat anteriorly directed. Near the end of the series is a transitional vertebra with a vertical spine, called the anticlinal vertebra.

The lumbar vertebrae typically have the largest and longest centra. The transverse and spinous processes are well developed and anteriorly directed. Metapophyses and anapophyses tend to be most prominent in this region.

The sacrum is the only part of the vertebral column in which the individual elements are typically fused. The number of fused vertebrae varies considerably among taxa and sometimes involves "sacralization" of adjacent caudal or lumbar vertebrae. The sacrum articulates with the ilia at a tight-fitting synovial joint mainly involving the first sacral.

The tail is a particularly variable part of the vertebral column, which can differ dramatically in both vertebral number and size. The caudal centra tend to be shorter and more robust proximally, and cylindrical and elongate distally. Proximal caudals usually have neural arches, transverse processes, and zygapophyses, which are greatly reduced or lost distally (see, e.g., Youlatos, 2003). Haemal arches, or chevron bones, project ventrally from between the centra in some mammals.

The ribs of extant mammals consist of a bony portion, which articulates with the vertebral column, and a costal cartilage (sternal rib), between the ventral end of the rib and the sternum. The sternal ribs are normally ossified in some primitive mammals, such as monotremes and xenarthrans. As already mentioned, most ribs have two articular surfaces for the vertebrae, the capitulum (which meets the demifacets on adjacent vertebral centra) and the tuberculum (which articulates with the transverse process). The tubercles decrease in size caudally so that only a capitulum remains on some posterior ribs. Posterior ribs may join preceding ribs rather than having a separate sternal attachment, or may be "floating," with no attachment to the sternum.

The sternum is a segmented, midline bony structure, which articulates with the shoulder girdle at its cranial end and with the ribs between successive sternebrae. The first sternebra, or manubrium, is commonly enlarged; the last is the xiphisternum.

The limbs of mammals have diversified for a wide variety of locomotor and other functions, an appreciation of which requires an understanding of comparative anatomy.

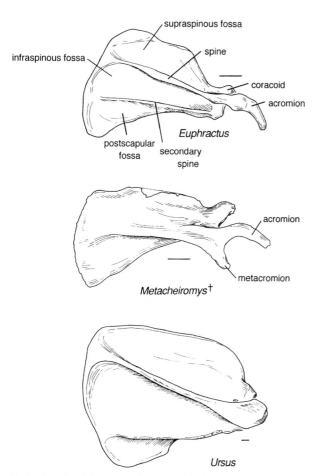

Fig. 2.6. Scapulae of three mammals. Dagger (†) indicates extinct taxon. Scale bars = 1 cm. (Modified from Rose and Emry, 1993.)

An excellent comparative account of the limb skeleton of diverse mammals is presented by Lessertisseur and Saban (1967b). Each limb consists of the limb girdle and proximal, intermediate, and distal segments. The shoulder (pectoral) girdle is simplified in most mammals compared with other tetrapods, consisting of only the scapula (Fig. 2.6) and clavicle; the (posterior) coracoid, formerly a separate element, is incorporated as a process of the scapula. The clavicle forms a strut between the sternum and the scapula. The scapulae, however, have no direct bony connection to the trunk, but are suspended by muscles on the sides of the anterior thoracic region. In living mammals, except monotremes, the scapular spine divides the outer surface into supraspinous and infraspinous fossae, and an acromion process projects from the distal end of the spine. Monotremes are primitive, however, in retaining separate anterior and posterior coracoids and an interclavicle, as in some therapsids. Moreover, they have no scapular spine and no distinct supraspinous fossa; the anterior margin of the scapula is homologous with the spine of other mammals. In all mammals the scapula articulates with the head of the humerus at the glenoid fossa.

Distal to the shoulder girdle the forelimb skeleton consists of the humerus, the radius and ulna, and the manus (Figs. 2.7–2.10). Many surface features of the humerus are related to muscle attachments (e.g., greater and lesser tuberosities [tubercles]; deltoid and pectoral crests or a combined deltopectoral crest; teres tubercle; medial and lateral epicondyles; supinator crest, also called the lateral supracondylar ridge or brachialis flange). So, too, are the ulnar

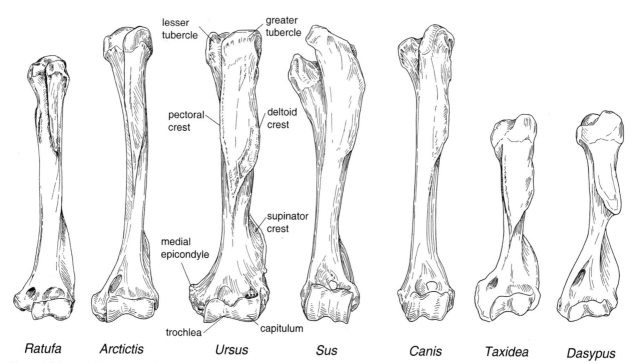

Fig. 2.7. Left humeri of extant mammals (not to scale). *Ratufa* and *Arctictis* are arboreal, *Ursus* is generalized, *Sus* and *Canis* are cursorial, and *Taxidea* and *Dasypus* are fossorial.

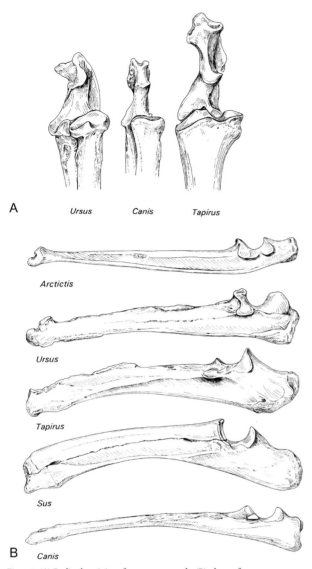

Fig. 2.8. (A) Radioulnar joint of extant mammals; (B) ulnae of extant mammals. Taxa same as in Fig. 2.7 except *Tapirus* (cursorial). (From O'Leary and Rose, 1995.)

olecranon process and the bicipital tuberosity and certain crests of the radius. The elbow is a complex joint involving three articulations: between the humeral trochlea and the ulna (a hinge), the humeral capitulum and the radial head (often a pivot), and the proximal radius and ulna (a potential gliding joint; Fig. 2.8). In higher primates, as well as some other arboreal mammals, the radius has substantial freedom to rotate on its long axis, allowing pronation (in which the distal radius crosses over the ulna so the palm faces downward or backward) and supination (in which the radius and ulna are parallel and the palm faces upward or forward). In most mammals the forearm and manus are normally held in the pronated position, and in some the elbow joint is modified to restrict or prevent supination.

The manus consists of the carpus, metacarpus, and phalanges (Fig. 2.10). Primitively the carpus comprises nine elements—arranged essentially in proximal and distal rows—some of which have been lost or fused in some mammals.

From medial to lateral (in the typically pronated manus of quadrupeds), the proximal row consists of scaphoid, lunate (lunar), cuneiform (triquetrum of human anatomy), and pisiform. Distally the radius articulates with the scaphoid and lunate, which are fused in some mammals, such as carnivorans, whereas the ulna usually articulates with the cuneiform and pisiform. Composing the distal carpal row (medial to lateral) are the trapezium, trapezoid, magnum (capitate in humans), and unciform (hamate of humans). In many primitive mammals a centrale is present as a separate element, usually between the scaphoid-lunate and trapezoid-magnum. Typically the trapezium articulates with metacarpal I, the trapezoid with metacarpal II, the magnum with metacarpal III, and the unciform with metacarpals IV and V.

Distal to the carpus are the digits, generally five in number, each of which has a metacarpal, and either two phalanges (in digit I, the pollex or thumb) or three (all others), resulting in a phalangeal formula of 2-3-3-3-3. The terminal or ungual phalanges are modified to bear claws, hoofs, or nails, and they vary considerably in form in relation to both phylogeny and function (Fig. 2.11; see also Fig. 2.17). The form and number of metacarpals and their phalanges also vary considerably among mammals. The forelimbs of many mammals have become modified in connection with other behaviors besides locomotion.

The pelvic girdle consists of the ilium, ischium, and pubis on each side, fused together to form a single innominate (hip) bone; the innominates articulate with the vertebral column at the sacroiliac joints, and with each other at the pubic symphysis (Fig. 2.12). All three pelvic elements meet and fuse within the acetabulum, which forms a socket for the femoral head. Primitive mammals, including extant monotremes and marsupials, also have epipubic ("marsupial") bones. Although epipubic bones have generally been assumed to be related to pouch support, a recent study indicates that they also (or alternatively) function as levers between abdominal muscles and the femur during locomotion (Reilly and White, 2003). Also associated with the pelvis is the baculum (os penis) found in many mammals, and used as a taxonomic character in rodents and carnivorans.

The femur, tibia and fibula, and pes (foot) make up the hind limb skeleton (Figs. 2.13, 2.14). The femur is generally the longest of the limb elements. It typically has three muscular processes on the proximal half, the greater, lesser, and third trochanters. Distally the femoral condyles articulate with the proximal tibia. The fibula may be strong and free from the tibia (joined at each end by synovial or fibrous joints), co-ossified at one or both ends, or reduced and virtually lost. There are seven tarsal bones: the astragalus (talus), calcaneus, navicular, cuboid, and three cuneiforms (Fig. 2.14). The astragalus (Fig. 2.15), which is supported by the calcaneus, articulates with the tibia proximally and the navicular distally. The navicular articulates with the three cuneiforms (ento-, meso-, and ectocuneiform), which in turn articulate with metatarsals I–III, respectively. The calcaneus articulates distally with the cuboid, which usually articulates

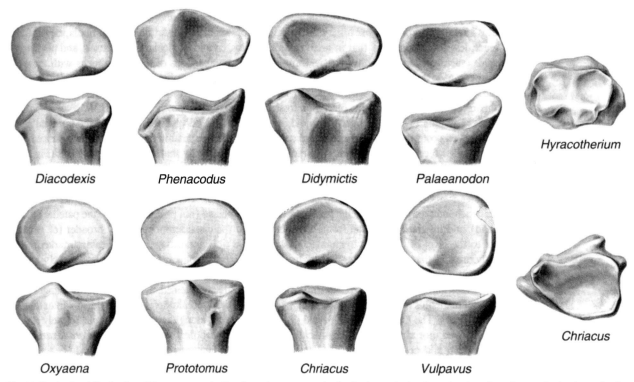

Fig. 2.9. Proximal and distal radius of Eocene mammals. First four columns are proximal radius in proximal and anterior views; last column is distal radius in distal view. Differences in shape affect mobility of the radius and reflect locomotor diversity. (From Rose, 1990, and O'Leary and Rose, 1995.)

with metatarsals IV and V. Perhaps more than any other part of the postcranial skeleton, the anatomy of the tarsals, particularly the astragalus and calcaneus, has played an important role in the determination of phylogenetic relationships of mammals (e.g., Matthew, 1937; Szalay, 1977, 1994). As in the manus, there are five metapodials (called metatarsals in the foot), and the same complement of phalanges as in the manus. The first pedal digit is the hallux, and it is often somewhat divergent from the other phalanges.

Several sesamoid bones are also associated with the limb skeleton. These are generally small, nodular elements encased within muscle tendons and located near joints. The best known is, of course, the patella (knee cap). Additional sesamoids are associated with various digital flexor tendons in both the manus and pes of many mammals. They usually serve to enhance leverage of the muscle in which they are contained.

SKELETAL ADAPTATIONS

From small, probably terrestrial, carnivorous or insectivorous Mesozoic ancestors, mammals have diversified to occupy almost every major environment throughout the world. They have evolved a remarkable diversity of skeletal adaptations for life in the air, in trees, on land, under ground, and in water. Due in part to the versatile tribosphenic molar, mammalian dentitions have become modified for almost every conceivable diet, including leaves, grass, roots and tubers, seeds, fruits, sap, nectar, bark, meat, fish, mollusks, krill, insects and other invertebrates, and even bones.

Some mammals have relatively generalized teeth that can handle a diet of mixed plant and animal items; they are omnivores.

The generalized anatomy described in the preceding section is often modified in similar ways in different clades, in association with similar diets, habitats, and lifeways. This tendency leads to the phenomena of convergence and parallelism—the independent acquisition of similar morphology in distantly related and closely related organisms, respectively. The resulting resemblances are known as homoplasy. Here I review some of the characteristics of the dentition and skeleton in mammals adapted for different lifestyles.

The dentitions of many extant insectivores, like those of primitive Paleogene mammals, are relatively little changed from those of their Cretaceous ancestors. They have secodont teeth with high, sharp cusps, often joined by sharp, bladelike crests. Their incisors are often enlarged and procumbent, and the trigonids of the lower molars tend to be much higher, and commonly larger, than the talonids. Both the ectoloph of the upper molars and the occluding lower molar crests may be arranged in the form of a W, a condition termed **dilambdodont** (e.g., shrews and moles; Fig. 2.16A). Dilambdodonty promotes more efficient cutting (Butler, 1996) and also occurs in some herbivorous lineages (see Fig. 2.3F). In some insectivores (e.g., golden moles, tenrecs, *Solenodon*) the upper molar paracone and metacone are connate (closely appressed and joined at the base) and set well in from the buccal margin, so the ectoloph forms a V-shape, and the protocone is reduced; this configuration is

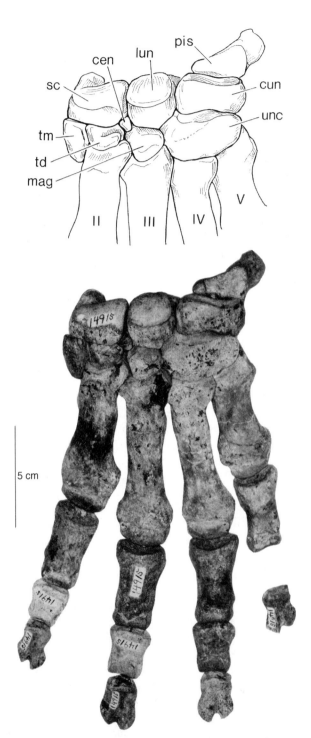

Fig. 2.10. Mammalian carpus and manus, exemplified by Eocene *Pachyaena*. Roman numerals indicate metacarpals. Key: cen, centrale; cun, cuneiform; lun, lunate; mag, magnum; pis, pisiform; sc, scaphoid; td, trapezoid; tm, trapezium; unc, unciform. (From Rose and O'Leary, 1995.)

flexid), and shearing occurs predominantly between the anterior crest of the upper molar (preparacrista) and the back of the trigonid (protocristid; Asher et al., 2002). Bats and some small primates, such as tarsiers, also have insectivorous dentitions.

Myrmecophagous mammals, which specialize on a diet of ants and termites, include members of several orders. The most extreme forms (echidna, anteaters, and pangolins) have lost all the teeth and have shallow, delicate mandibles. Those that retain teeth (numbat, some armadillos, aardvark, and aardwolf; Fig. 2.16E) tend to have small, homodont teeth, sometimes reduced in number and lacking enamel, but some have more than the usual number of simple teeth. The skull is often elongate and tubular, in association with a long, protrusile tongue. Many myrmecophagous mammals have evolved fossorial skeletons (see below) that enable them to tear into ant and termite nests.

Carnivorous (meat-eating) mammals typically have small incisors, large canines, and one or more pairs of upper and lower cheek teeth specialized into cutting blades called carnassials (Fig. 2.16C). In Carnivora the carnassials are P^4 and M_1, but other teeth are modified into carnassials in the extinct Creodonta. The most strictly carnivorous forms, such as cats, are termed hypercarnivores. They have long, sharp carnassial blades and have reduced or lost the molars behind the carnassials. Some carnivorans have evolved away from the original carnivorous diet of their ancestors. Omnivorous and frugivorous carnivorans, such as bears, raccoons, and palm civets, have broad, bunodont teeth and lack specialized carnassials. Carnivores tend to have well-developed temporalis muscles. Consequently, the skull generally has a prominent sagittal crest (reduced in frugivorous forms), and the coronoid process of the dentary is large. The mandibular condyle is situated at about the level of the toothrow, which maximizes power at the carnassials.

Certain specialized faunivorous diets are associated with particularly unusual dentitions. (Faunivorous is a general term for a diet consisting of animals of any kind.) Piscivorous (fish-eating) mammals, such as seals and dolphins, often have simple, homodont, conical teeth, in some cases greatly exceeding the normal number. Walruses and some seals and otters (Carnivora) eat mollusks and sea urchins, using teeth that are either peglike or broad and flat, for crushing hard objects. Mysticete whales, which filter-feed on plankton, have lost the teeth and replaced them with keratinous, straining baleen plates suspended from the maxilla.

Herbivores (plant-eating mammals, including ungulates and some rodents and primates) can usually be recognized by their broad grinding molars, and (in ungulates) a tendency toward molariform premolars. Beyond these general similarities, however, herbivores have achieved considerable dental diversity. They may be brachydont or hypsodont. Some are bunodont, but more often their molariform teeth have multiple shearing crests; lophodonty or selenodonty is common. The incisors of herbivores often form a cropping apparatus that is separated from the cheek teeth by a gap, or diastema. In some forms the upper incisors are absent

described as **zalambdodont**. Zalambdodonty, or a close approximation to it, occurs in various noninsectivoran clades as well. In highly zalambdodont forms, the metacone may be lost and the paracone may be near the center of the tooth; the lower molars tend to have very tall trigonids and greatly reduced talonids (Fig. 2.16B). The paracone occludes in the notch between trigonid and talonid (called the hypo-

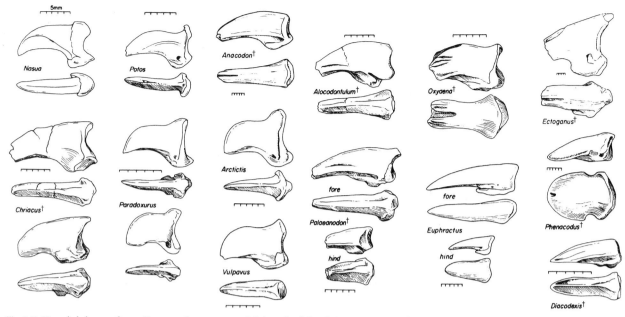

Fig. 2.11. Ungual phalanges of some Eocene and extant mammals in lateral and dorsal views. Compare with Fig. 2.17. Scale bars = 5 mm. (From Rose, 1990.)

and the lowers work against a corneous pad covering the premaxilla. The enamel of specialized herbivores shows complex infolding with dentine windows and cementum.

There are several specialized kinds of herbivory. **Frugivores** (herbivores that feed primarily on fruit; e.g., fruit bats, some monkeys and apes, kinkajou) tend to have brachydont, bunodont teeth, with minimal development of shearing crests. As noted above, frugivorous carnivorans lost their shearing teeth through evolution. Some small marsupials and primates feed on tree gum and sap, for which they have evolved large, procumbent incisors used to gouge through bark. As might be expected, their molars are generally very low crowned, with indistinct surface features. **Nectivorous** forms (nectar and pollen feeders), including certain bats and marsupials, also reduce the cheek teeth, in the most extreme case to just a few vestigial pegs (the honey possum, *Tarsipes*).

Folivores are herbivores specialized for feeding on leaves. They typically have lophodont or bilophodont (with two transverse ridges) cheek teeth. Examples include tapirs, langurs and colobus monkeys, and the koala. In some folivores the enamel is crenulated and multiple shearing blades are present (Fig. 2.16D). As a result of their heritage, tree sloths and their extinct relatives differ from other folivores in having simple cylindrical teeth. The most specialized folivores are grazers. Grazers have evolved various mechanisms to cope with a diet of grass, which contains a high component of abrasive silica phytoliths. Commonly the teeth of grazers are hypsodont, with multiple lophs (selenodont, as in ruminant artiodactyls) or complex enamel patterns (as in horses; Fig. 2.16F). In the most specialized forms the cheek teeth grow continuously throughout most of the life of the animal.

The skulls of more specialized herbivores are often elongate, to accommodate molarized premolars. Ungulate skulls are often adorned with horns, antlers, or bony protuberances. The herbivore mandible is deep in back, with a large angular process where the well-developed masseter and medial pterygoid muscles attach. The latter, particularly, are related to transverse movement of the jaw during chewing, which is especially important in herbivores. In contrast to

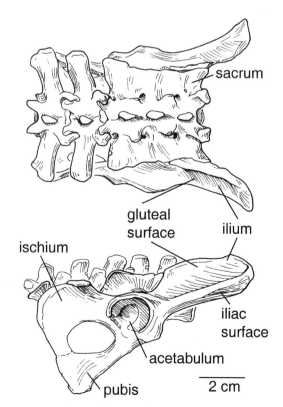

Fig. 2.12. Mammalian pelvis (innominate) and sacrum, exemplified by late Eocene *Patriomanis*. (Modified from Rose and Emry, 1993.)

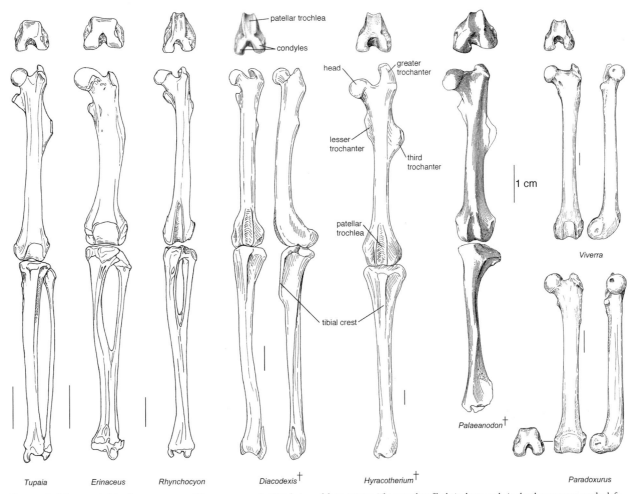

Fig. 2.13. Left femur and tibia of some extant and Eocene mammals. Distal view of femur at top. The complete fibula is shown only in the three genera on the left; the distal fibula is indicated for *Diacodexis*. *Tupaia* is scansorial, *Erinaceus* is generalized terrestrial, *Rhynchocyon* is cursorial, *Diacodexis* and *Hyracotherium* were cursorial, *Palaeanodon* was fossorial, *Viverra* is generalized terrestrial, and *Paradoxurus* is arboreal. Dagger (†) indicates an extinct Eocene taxon. Scale bars = 1 cm.

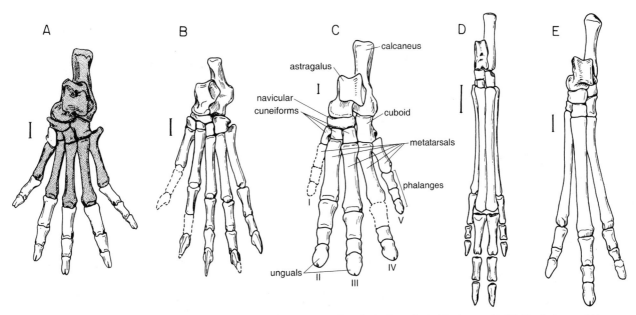

Fig. 2.14. Feet of Eocene mammals: (A) *Oxyaena*, generalized terrestrial; (B) *Chriacus*, arboreal; (C) *Phenacodus*, incipiently cursorial; (D) *Diacodexis*, cursorial/saltatorial; (E) *Hyracotherium*, cursorial. Scale bars = 1 cm. (Modified from Rose, 1990.)

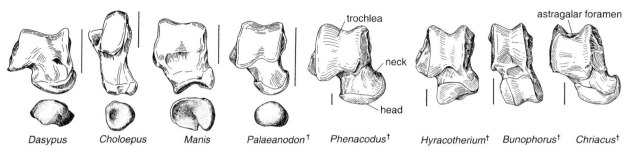

Fig. 2.15. Right astragali of some extant and Eocene mammals. Distal view of the astragalar head is shown for the first four genera. Dagger (†) indicates an extinct Eocene taxon. Left scale bars = 1 mm; right scale bars = 5 mm.

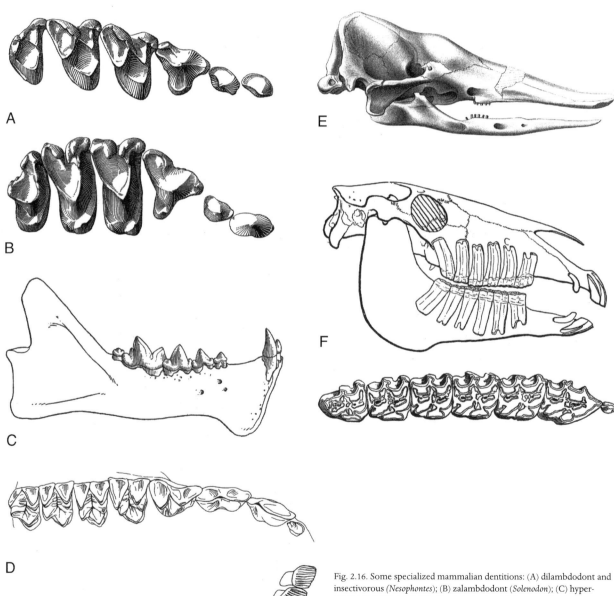

Fig. 2.16. Some specialized mammalian dentitions: (A) dilambdodont and insectivorous (*Nesophontes*); (B) zalambdodont (*Solenodon*); (C) hypercarnivorous (*Dinictis*); (D) dilambdodont and folivorous/frugivorous (*Cynocephalus*); (E) myrmecophagous (*Stegotherium*); (F) hypsodont grazer (*Equus*), skull and crown view of upper teeth. (A–B from McDowell, 1958; C from Matthew, 1910b; D from MacPhee et al., 1989; E from Scott, 1903–1904; F from Gregory, 1951.)

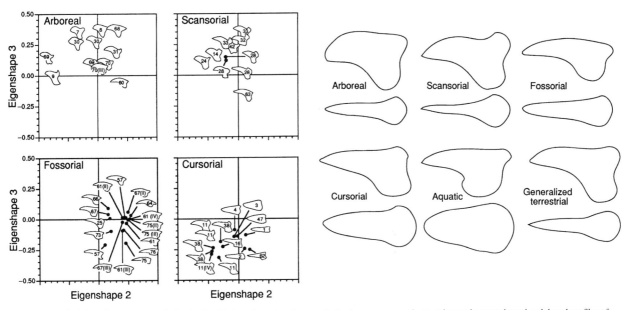

Fig. 2.17. Ungual phalanx shape in various behavioral guilds based on eigenshape analysis of extant mammals. At right are the mean lateral and dorsal profiles of each locomotor group. (From MacLeod and Rose, 1993.)

the situation in carnivores, the coronoid process is typically reduced, and the mandibular condyle is positioned well above the toothrow in herbivores.

The postcranial skeletons of mammals also have distinctive modifications that reflect their habitat, locomotion, or lifestyle. Particularly useful accounts of the skeletal characteristics of different locomotor groups can be found in Gambaryan (1974), Hildebrand et al. (1985), Van Valkenburgh (1987), and Hildebrand (1995). The primitive mammalian skeleton, from which more specialized skeletal adaptations evolved, was presumably a rather generalized one that enabled progression on uneven substrates and was, therefore, conducive to both terrestrial and arboreal environments. In part, this versatility resulted because most basal mammals were very small, and for them there was probably little difference between the varied substrates of the forest floor and those of brush, vines, tree trunks, and branches (Jenkins, 1974). Climbing was therefore very likely part of their locomotor repertoire. However, some of the most ancient mammals for which skeletons are known were already specialized for particular lifestyles, hence the primitive state for mammals remains uncertain. Among living mammals, several locomotor categories are recognized, which also reflect habitat (see, e.g., Eisenberg, 1981).

Many living mammals are adept climbers and spend considerable time in the trees. Those that forage and shelter in trees are considered **arboreal**, whereas able climbers that also spend much of their time on the ground are **scansorial**. Mammals in both of these categories have similar skeletal specializations for maximizing mobility at the shoulder, elbow, wrist, hip, and ankle, although these modifications tend to be more extreme in arboreal forms (Figs. 2.7, 2.13, 2.14). They can rotate the radius to supinate the forearm. The manus and pes are typically plantigrade (with palms and soles in contact with the substrate) and adapted for grasping, often with abducted or opposable pollex and hallux. Some specialized arboreal mammals—including sciurids and procyonids—have evolved anatomical modifications that allow them to hyperinvert or "reverse" the hind feet, thus enabling them to descend from trees headfirst or to hang upside down (Jenkins and McClearn, 1984). The digits of most arboreal mammals bear sharp, curved, laterally flattened claws (formed of keratin), which are supported by bony ungual phalanges of similar shape (Fig. 2.17). In arboreal primates and hyracoids, however, the unguals bear nails. The tail of arboreal mammals is usually long and may be prehensile.

Many primates, carnivores, xenarthrans, and marsupials are arboreal or scansorial. Some highly arboreal mammals (e.g., dermopterans, phalangers, flying squirrels) have evolved a skin membrane, or patagium, which enables them to glide between tree branches. These **glissant** forms tend to have delicate, elongate limb elements and specializations in the manus and pes associated with attachment and control of the patagium. In bats, the only **volant** (flying) mammals, the forelimbs are modified to support active wings. The skeleton is very lightly built and delicate. The forelimb bones in particular are very long and slender, with elongate digits that support the wing membrane. Mobility at the shoulder, elbow, and wrist is greatly restricted. The hind limbs are small and very thin.

Terrestrial mammals spend most or all of their time on the ground. Although some are able, if infrequent, climbers, others are incapable of climbing trees. Generalized terrestrial mammals (e.g., hedgehogs, tenrecs, civets, some bears) lack clear modifications for specific locomotor specializations. They may show some restriction of mobility at limb joints, but not to the extent seen in cursorial forms. Their foot posture ranges from plantigrade to digitigrade (supported by the digits, with palm and heel off the ground). The claws are usually longer, not as curved, and broader ventrally than in scansorial or arboreal forms.

There are several specialized categories of terrestrial mammals, described in this and the following paragraphs. **Cursorial** mammals are adapted for running, and their skeletons show modifications that increase stride length and rate, which results in greater speed (Hildebrand, 1995). They have elongate limbs, with the intermediate and distal segments especially long and slender (Figs. 2.13, 2.14). Muscle masses tend to be concentrated in the proximal part of the limb to reduce the weight of the distal portion. The limb joints are modified to restrict motion to a parasagittal plane. The bony crests and processes for muscle attachment tend to be reduced compared to those of climbers and diggers, and are situated closer to the joints they affect, an adaptation for speed. The clavicle is usually absent, and the ulna and the fibula are often reduced or fused to the radius and tibia, respectively. Runners usually have long to very long metapodials, the lateral ones sometimes reduced or lost. Fusion of some of the carpals or tarsals is common. Hoofs are often present. When claws are retained, the terminal phalanges supporting them are longer, less curved, and broader than in climbing mammals (Fig. 2.17). The stance of cursors is typically either digitigrade (standing on the digits, with the metapodials raised off the ground, as in various carnivores) or unguligrade (standing on the terminal phalanx or hoof, as in most ungulates).

Saltatorial mammals are specialized for jumping, and are usually propelled by the hind limbs (e.g., rabbits). The skeleton generally resembles that of cursors, with similar limitations on joint mobility, but the hind limbs are usually much longer than the forelimbs (see, e.g., Fig. 9.4). When the hind limbs are used together for bipedal jumping, as in kangaroos, jerboas, and kangaroo rats, the gait is called ricochetal. The intermembral index ([length of humerus + radius]/[length of femur + tibia] $\times$ 100) of ricochetal mammals is less than 50, compared to an average index of 75 in generalized mammals (Howell, 1944). The tibia and fibula are usually fused at one end or both ends for stability, and the metatarsals may be exceptionally long. Some bounding mammals have fused cervical vertebrae to provide neck stability. Such primates as tarsiers, galagos, and some lemurs are arboreal saltators.

Very heavy terrestrial mammals are described as **graviportal** (with limbs adapted for supporting heavy weight; see Figs. 7.25, 12.29B, 13.12, 13.23). Most graviportal mammals are large ungulates (e.g., elephant, hippopotamus, rhinoceros) and, therefore, presumably evolved from somewhat cursorial antecedents. They stand with straight, columnar limbs, an adaptation to minimize the stresses imposed on the limb bones. Unlike typical cursors, the intermediate limb segments (radius and tibia) are shorter than the proximal segments. The manus and pes have robust, spreading digits with short, broad phalanges and hoof-bearing unguals.

Mammals adapted for digging are **fossorial** (e.g., golden moles, armadillos, badgers, pocket-gophers, various squirrels, other rodents). The most specialized fossorial mammals (moles, marsupial mole) are subterranean, seeking food and shelter underground and rarely coming to the surface. The term "fossorial" is sometimes restricted to just these subterranean dwellers, the term "semi-fossorial" being used for diggers that live above ground. In this text the broader usage is applied. Fossorial mammals typically have robust skeletons with strong limb girdles and short, heavily built limb bones (particularly the forelimb) that have prominent crests and processes for muscle attachment (Figs. 2.7, 2.13). The ulnar olecranon process tends to be very prominent and long, but the functional length of the intermediate segment of the forelimb is much less than that of the proximal segment. The elements of the manus are short and stout, except for the claws (especially of the middle digit), which may be greatly enlarged. Claws of diggers tend to be longer, shallower, less curved, and ventrally wider than those of climbers (MacLeod and Rose, 1993). Fossorial mammals that also use the head and teeth for digging have a wedge-shaped skull, with a broad lambdoid crest for attachment of neck muscles. In some diggers several cervical vertebrae are fused. The tail is generally reduced in subterranean forms.

Terrestrial mammals adapted for swimming (e.g., otters, beavers, muskrats, capybaras) are termed **semi-aquatic**. Their limb bones are usually short and stout, with prominent crests and processes for muscle attachment, similar to those of fossorial mammals. The humerus may have a slightly S-shaped profile. Manus and pes tend to have short, spreading digits, which are often webbed. The tail may be long and muscular, and the hind limbs are often specialized for propulsion.

Some mammals have become more committed to life in the water, and are described as **aquatic** or natatorial (swimming). Most aquatic mammals are marine, but some frequent freshwater. The body of aquatic mammals is often long and streamlined. The neck is commonly very short, and cervical vertebrae may be fused. The forelimbs are short and modified into paddles or flippers with elongate digits, and, in whales, extra phalanges. The hind limbs may be modified like the forelimbs (as in seals), reduced, or vestigial (as in manatees, whales, and dolphins). Limb joint mobility is often severely restricted.

The anatomical adaptations described in the preceding paragraphs have known functional associations in extant mammals. Applying this knowledge to fossils enables educated inferences on the lifeways of extinct mammals.

3

The Origin of Mammals

WHAT IS A MAMMAL?

Living mammals are easily recognized by a suite of characteristics that distinguish them from all other vertebrates. Most obvious are an external covering of hair (except in certain highly specialized types) and nourishment of the young by milk produced in the mother's mammary glands. The heart has four chambers, allowing separation of blood flow to the lungs (for reoxygenation) from circulation to the rest of the body. There is a muscular diaphragm, related to increased oxygen consumption. Mammals are endothermic and, consequently, generally have higher metabolic rates and higher activity levels than are found in other vertebrates except birds. Vision, hearing, and olfaction tend to be highly developed, and the brain (especially the cerebrum) is relatively larger and more complex than in other vertebrates. Most of these features, however, are rarely (or never) preserved in fossils.

Fortunately, many skeletal features diagnostic of extant mammals are often preserved in fossils. These include a single lower jaw bone, the dentary; a dentary-squamosal articulation between the lower jaw and the skull; three middle ear ossicles; diphyodonty (two sets of teeth with sequential replacement in all except some primitive Mesozoic forms, but molars not replaced); heterodont dentition, typically with complex molar crowns and multiple molar roots, and associated with precise occlusion; a secondary bony palate; a single bony nasal opening; paired occipital condyles; five regionally differentiated sections of the vertebral column, the first two vertebrae at the cranial end modified to allow rotation; ribs usually limited to the thoracic region; modification of the shoulder girdle (including further reduction of the coracoid); reorganization of the pelvic girdle (elongation of the ilium and separation

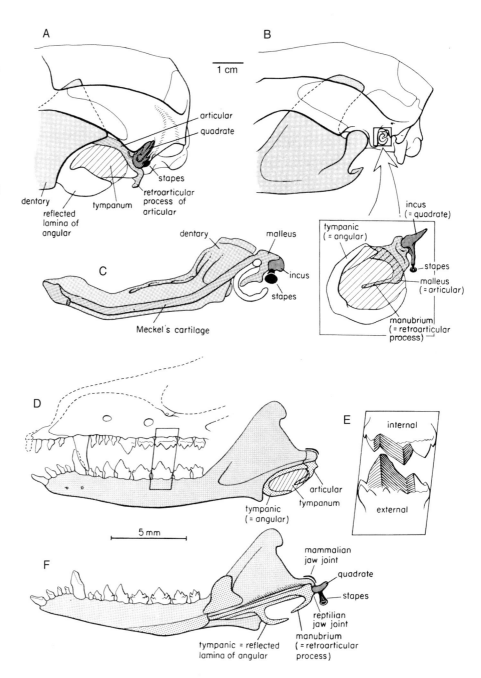

Fig. 3.1. Evolution of mammalian characters: transformation of jaw joint and origin of middle-ear ossicles. (A) The cynodont *Thrinaxodon*, in which the quadrate and articular functioned both as the jaw joint and part of the hearing apparatus; (B) the extant marsupial *Didelphis*, in which the jaw joint is between the dentary and squamosal, and the auditory ossicles (modified from the quadrate and articular of cynodonts; enlarged in inset) are located behind the jaw joint; (C) lower jaw of fetal mammal, showing the developmental similarity to the phylogenetic origin of jaw and auditory features; (D) reconstruction of *Morganucodon* (compare with A); (E) occlusal relationships in *Morganucodon*; (F) medial view of the lower jaw of *Morganucodon*, showing relationship of dentary-squamosal jaw joint to postdentary bones and auditory ossicles. (From Crompton and Jenkins, 1979.)

of its gluteal and iliac surfaces, reorientation of the ischium and pubis posterior to the acetabulum); and separate centers of ossification for the shaft (diaphysis) and ends (epiphyses) of long bones, which result in better-defined joints and determinate growth (Figs. 3.1, 3.2; see also Figs. 2.1, 2.4). The skull and postcranial skeleton of mammals generally comprise fewer elements than in nonmammalian tetrapods, as a result of both fusion and loss of bones.

Although there is little difficulty in distinguishing mammals from other vertebrates in present-day faunas, it has long been recognized that the distinction breaks down when one considers the fossil record. The transition between mammalian forerunners (cynodont therapsids, discussed in the next section) and the earliest mammals now includes many known intermediate stages (discussed in the next section and in Chapter 4) that document the mosaic evolution of "mammalian" traits. Consequently, how to recognize the first mammal has become controversial: which character(s) should be considered most important for recognizing a mammal? Even if a node-based definition of Mammalia is applied, practical identification of mammals (or any other taxon) in the fossil record is ultimately based on anatomical characters. The acquisition of a well-developed dentary-squamosal joint as the only jaw articulation has traditionally been considered to be the most important indication that the mammalian boundary has been crossed, but even here transitional forms are known that possess this articulation in combination with a joint between the articular and the

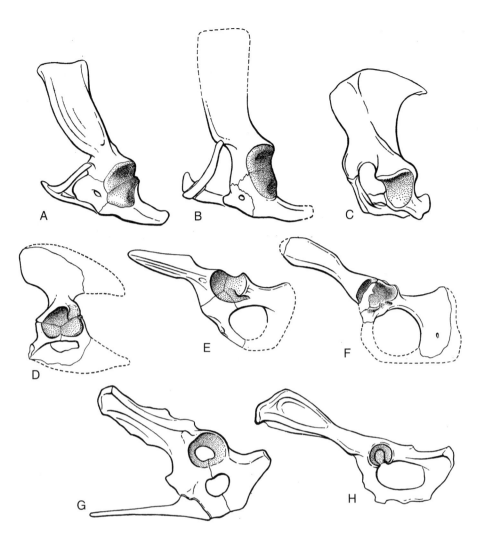

Fig. 3.2. Evolution of mammalian characters. (A–C) Shoulder girdles and (D–H) pelvic girdles of cynodonts and primitive mammals (not to scale): (A, D) cynodont; (B, F) *Morganucodon*; (C, G) echidna (extant monotreme); (E) *Oligokyphus* (tritylodont); (H) *Tupaia* (tree shrew, an extant placental). (From Jenkins and Parrington, 1976.)

quadrate or between the surangular and the squamosal. Other important mammalian synapomorphies include postcanine teeth with two or more roots, diphyodont rather than continuous alternate tooth replacement, a petrosal promontorium (the bony swelling enclosing the cochlea and forming the medial wall of the middle-ear cavity), and a bony floor of the cavum epiptericum (the fossa in the braincase that houses the sensory ganglion of the trigeminal nerve, cranial nerve V; Miao, 1991; Kielan-Jaworowska, 1992; Novacek, 1993; Luo et al., 2002).

As the fossil record of the therapsid-mammal transition improves, it has also become apparent that many "mammalian" characters arose independently multiple times by convergence, making recognition of Mammalia even more problematic. (For example, critical features that may have evolved more than once among mammals include the three ossicles of the middle ear and the dentary-squamosal joint.) Rowe (1988) therefore proposed that a distinction be made between the definition of the group, based on ancestry and taxonomic content, and its diagnosis, based on morphology.

Adopting the "crown-group" concept (that higher taxa should be restricted to descendants of the most recent common ancestor of two or more extant lineages), Rowe defined Mammalia as all taxa stemming from the last common ancestor of monotremes and therian mammals. Although this approach may seem to provide a neat solution to the ambiguity of what constitutes a mammal, such a restrictive definition excludes many fossil groups long accepted as mammals on anatomical grounds. At the same time, it necessitates the creation of several new higher taxa (Mammaliamorpha, Mammaliaformes) to encompass successive outgroups to Mammalia. Perhaps most objectionable is the volatile composition of Mammalia that results from the instability of the position of monotremes (e.g., Lucas, 1992). A crown-group definition of Mammalia is no more biologically real (or less arbitrary) than any other definition, and in this case conflicts with widely held morphological definitions. In agreement with most students of Mesozoic mammals, a more inclusive stem-based definition of Mammalia is employed in this volume (see also Luo et al., 2002; Kielan-Jaworowska et al., 2004; Kemp, 2005), essentially equivalent to Rowe's Mammaliaformes. Thus Mammalia as used here includes all taxa more closely related to monotremes and therians than to tritheledonts or tritylodonts (see Fig. 4.2).

THE EVOLUTIONARY TRANSITION TO MAMMALS

The ancestors of mammals, Synapsida, diverged from basal amniotes—protothyrid captorhinomorphs—at least 300 million years ago, in the Pennsylvanian Period. As the oldest and most primitive amniotes, protothyrids were also ultimately ancestral to reptiles (including lizards, snakes, and turtles) and archosaurs (crocodilians, dinosaurs, and birds). Synapsids include two successive radiations, the Pennsylvanian-Permian Pelycosauria, and the largely Permo-Triassic Therapsida (see Carroll, 1988, for an excellent summary). Although synapsids were long classified as reptiles, it is now accepted that they shared a more recent ancestry with mammals. Therapsids arose in the Permian from sphenacodontid pelycosaurs (which include the carnivorous "sail-backed" *Dimetrodon* from Texas). The Cynodontia of the late Permian-Triassic were the most mammal-like therapsids.

Note that cladistically, mammals are, therefore, successively nested within synapsids, pelycosaurs, therapsids, and cynodonts. These names were long applied (in what is now regarded as a paraphyletic sense) only to nonmammalian Paleozoic and early Mesozoic representatives, excluding mammals (e.g., Romer, 1966; Carroll, 1988). For convenience, the names are used here in that sense, rather than modifying them with the term "nonmammalian" each time they are mentioned.

Through the Permian and Triassic, a succession of cynodonts progressively acquired mammal-like anatomy, including heterodont dentition, postcanine teeth with three longitudinally aligned cusps, a pair of occipital condyles, a secondary palate, differentiation in the vertebral series, confinement of ribs mainly to the thoracic region, modified limb girdles, better-defined limb joints, and less sprawling posture (Figs. 3.3, 3.4). Particularly important was progressive enlargement of the dentary at the expense of the postdentary bones. Many of these features were already evident in the well-known Early Triassic cynodont *Thrinaxodon* (Jenkins, 1971). As the dentary enlarged in some advanced cynodonts, it approached or came in contact posteriorly with the squamosal bone, creating a secondary jaw joint beside the old "reptilian" articular-quadrate joint. In *Probainognathus* this secondary jaw joint was between the surangular and the squamosal (Figs. 3.4, 3.5), whereas in *Diarthrognathus* it was between the dentary and the squamosal. The bones of the "reptilian" jaw joint (articular and quadrate) were probably also involved in transmitting sound to the stapes and would eventually become the malleus and incus of the mammalian middle ear. Concomitantly the cheek teeth became more complex, and further modifications of the jaw and skull permitted reorientation of the jaw muscles. These changes led to more precise occlusion. The accumulation of these mammal-like features in cynodonts leaves little question that they were the progenitors of mammals. However, it has become increasingly clear that mammal-like specializations arose repeatedly among cynodonts, making the precise ancestry of Mammalia difficult to decipher.

The iterative evolution of mammalian characters in multiple lines of cynodonts led to the prevailing view during much of the twentieth century that mammals constitute a polyphyletic grade rather than a clade, a view strongly influenced by the work of George Gaylord Simpson, Everett C. Olson, and Bryan Patterson (see Luo et al., 2002). However, most authorities since about 1970 have concluded, as did Gregory (1910), that Mammalia (=Mammaliaformes of Rowe, 1988, and McKenna and Bell, 1997) is monophyletic

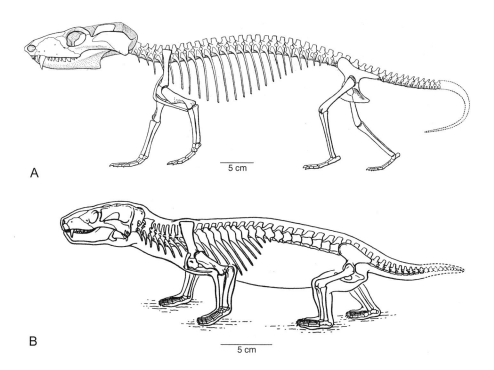

Fig. 3.3. Skeletons of advanced cynodonts: (A) *Probelesodon*; (B) *Thrinaxodon*. (A from Romer and Lewis, 1973; B from Jenkins, 1984.)

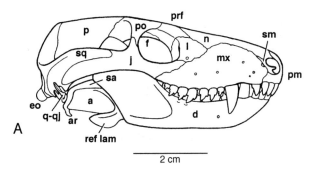
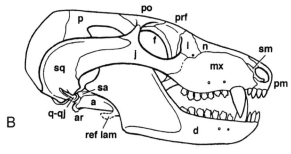
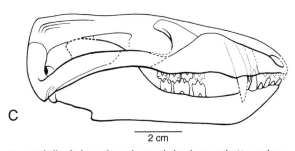

Fig. 3.4. Skulls of advanced cynodonts and a basal mammal: (A) cynodont *Thrinaxodon*; (B) cynodont *Probainognathus*; (C) *Sinoconodon*, a primitive mammal. Key: a, angular; ar, articular; d, dentary; eo, exoccipital; f, frontal; j, jugal; l, lacrimal; mx, maxilla; n, nasal; p, parietal; pm, premaxilla; po, postorbital; prf, prefrontal; q-qj, quadrate-quadratojugal; ref lam, reflected lamina; sa, surangular; sm, septomaxilla; sq, squamosal. (A, B from Hopson and Kitching, 2001; C from Crompton and Sun, 1985.)

(e.g., Hopson and Crompton, 1969; Crompton and Jenkins, 1973, 1979; Rougier et al., 1996a; Luo et al., 2002; Kielan-Jaworowska et al., 2004). But as late as the 1990s some distinguished researchers still hinted at the possibility that Mammalia as it is widely conceived could be polyphyletic (Lillegraven and Krusat, 1991; Kielan-Jaworowska, 1992).

Some advanced cynodonts, called gomphodonts, evolved broad, complex teeth—somewhat reminiscent of some mammalian teeth—in association with a herbivorous diet. Despite this apparent approach toward a mammalian dentition, gomphodonts were not particularly closely related to mammals. Most early mammals were very small and had sharp teeth indicative of an insectivorous habit, making it much more likely that they descended from carnivorous/insectivorous cynodonts. Furthermore, some experts now are persuaded that the two families of gomphodonts (Traversodontidae and Diademodontidae) achieved their herbivorously adapted dentitions in parallel.

Many authorities accept that the late Triassic–early Jurassic Tritheledontidae (also called ictidosaurs), including *Diarthrognathus* and *Pachygenelus* (Fig. 3.6), are the cynodonts most closely related to mammals (e.g., Hopson and Barghusen, 1986; Shubin et al., 1991; Crompton and Luo, 1993; Luo, 1994; Hopson and Kitching, 2001; Kielan-Jaworowska et al., 2004). Although this hypothesis was initially based primarily on the dentition, a recent comprehensive analysis including cranial and postcranial skeletal characters as well as the dentition also supports this interpretation (Luo et al., 2002). Tritheledonts were small cynodonts, some with skulls only a few centimeters long. The teeth of some types, such as *Pachygenelus*, are similar in size and morphology to those of morganucodontids (basal mammals; see Chapter 4) and, like the latter, have prismatic enamel (Gow, 1980). However, the dental formula and details of the dental anatomy and enamel microstructure suggest that known tritheledonts cannot be directly ancestral to mammals. According to Bonaparte and Barberena (2001), postcranial and dental features suggest that the cynodonts *Therioherpeton* and *Prozostrodon*, both from the Upper Triassic of Brazil, are also closely related to mammals, although not as closely as tritheledonts.

Alternatively, Tritylodontidae, once considered mammals because of dental and general cranial resemblances to multituberculates, have also been championed as the sister-group of mammals (Fig. 3.7). Although some authorities (e.g., Sues, 1985) have argued that they are more closely related to gomphodont cynodonts, numerous synapomorphies seem to support a close alliance between tritylodonts and mammals (Kemp, 1983; Wible, 1991; Rowe, 1993; Martinez et al., 1996). These include such features as cheek teeth with multiple roots, absence of prefrontal and postorbital bones, a partially floored cavum epiptericum (the fossa for the trigeminal nerve ganglion), postdentary bones that are similar to the auditory ossicles of primitive mammals, and many other cranial characters (e.g., Sues, 1986), as well as an odontoid process (dens) on the axis vertebra, details of shoulder and pelvic structure, and the presence of an astragalar canal. As in tritheledonts and basal mammals, the postdentary jaw bones are reduced relative to their state in other cynodonts. However, tritylodonts have a primitive quadrate-articular jaw joint and enlarged incisors separated by diastemata from the complex cheek teeth—a specialized, rodentlike pattern. These features exclude known forms from direct mammalian ancestry and raise the possibility that some of the mammalian traits of tritylodonts arose independently.

Hahn et al. (1994) proposed that the Upper Triassic Dromatheriidae (in which they included the South American *Therioherpeton* and several other genera whose phylogenetic positions previously were ambiguous) were even closer to mammals, suggesting that these animals occupied a transitional zone between cynodonts and mammals. Like tritheledonts, tritylodonts, and mammals, dromatheriids (where known) lack prefrontal and postorbital bones in the skull. The teeth have a single row of laterally compressed

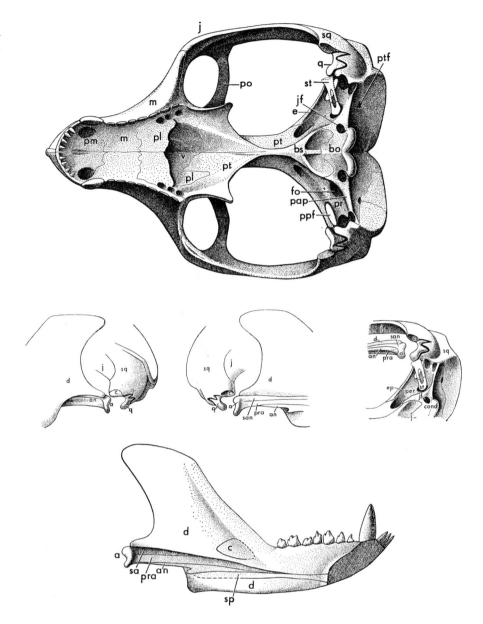

Fig. 3.5. Skull of *Probainognathus*, showing enlarged dentary (d) approaching the squamosal (sq) and bringing the surangular (sa) into contact with the squamosal. A quadrate (q)-articular (a) jaw joint was also present. (From Romer, 1970.)

cusps (typically three principal cusps), recalling those of tritheledonts and morganucodontids (Fig. 3.8). Premolar and molar morphologies can be distinguished, but the teeth lack cingula, and their roots are incompletely divided. Most of the genera are represented only by isolated teeth; hence their precise phylogenetic position (and even whether they are closely related to each other) is in dispute (e.g., Sues, 2001). Nonetheless, the current consensus is that they are not particularly closely related to mammals.

Bonaparte et al. (2003) recently described two new genera of small, advanced cynodonts (*Brasilodon* and *Brasilitherium*) from the Late Triassic of Brazil that may be closer to the ancestry of mammals than any other forms yet found. Both are known from skulls, which lack the prefrontal and postorbital bones, and *Brasilitherium* has morganucodontid-like lower teeth. Their phylogenetic analysis placed these genera closer to *Morganucodon* than are either tritheledonts or tritylodonts.

Kemp (2005) recently provided an excellent summary of the evidence for a relationship between various cynodonts and mammals. He postulated that the choice (and probable lack of independence) of anatomical characters used in various phylogenetic analyses may explain why a consensus on the sister-group of mammals has eluded researchers.

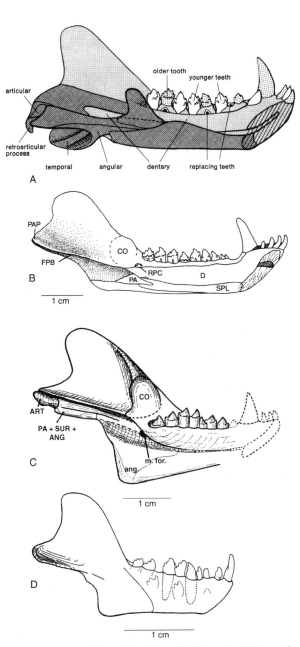

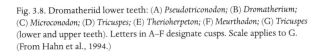

Fig. 3.6. Lower jaws of advanced cynodonts: (A) *Thrinaxodon*; (B) *Prozostrodon*; (C) *Diarthrognathus*; (D) *Pachygenelus*. Key: ANG, angular; ang., angle of dentary; ART, articular; CO, coronoid; D, dentary; FPB, fossa for postdentary bones; m. for., mandibular foramen; PA, prearticular; PAP, prearticular process; RPC, replacing postcanine; SPL, splenial; SUR, surangular. (A from Crompton and Parker, 1978; B from Bonaparte and Barberena, 2001; C from Crompton, 1963; D from Crompton and Luo, 1993.)

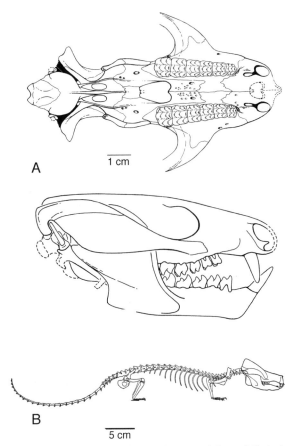

Fig. 3.7. Tritylodontids: (A) skull of *Kayentatherium*; (B) skeleton of *Oligokyphus*. (A from Sues, 1983; B from Kühne, 1956.)

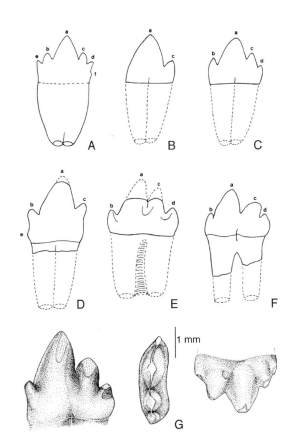

Fig. 3.8. Dromatheriid lower teeth: (A) *Pseudotriconodon*; (B) *Dromatherium*; (C) *Microconodon*; (D) *Tricuspes*; (E) *Therioherpeton*; (F) *Meurthodon*; (G) *Tricuspes* (lower and upper teeth). Letters in A–F designate cusps. Scale applies to G. (From Hahn et al., 1994.)

4

Synopsis of Mesozoic Mammal Evolution

AFTER MAMMALS EMERGED FROM CYNODONTS, they coexisted with dinosaurs for about 150 million years during the Mesozoic. The Mesozoic radiation of mammals consisted largely of groups that became extinct by the end of that era, without direct descendants; but some of them have been identified as structural, if not actual, stages in the evolution of the therian mammals prevalent today. Therefore a review of mammalian evolution during the Mesozoic will help to place Early Cenozoic mammals in perspective. All three major groups of living mammals—monotremes, metatherians, and eutherians—had evolved by the end of the Early Cretaceous, but they were not yet diverse or abundant. The only other group from the Mesozoic that unquestionably survived for a significant period into the Cenozoic (but is now extinct) is the Multituberculata.

Until recently Mesozoic mammals were quite rare, and our knowledge of most species (with a few notable exceptions) was restricted to the dentition. Over the past decade or so, however, new data have been accumulating at an astonishing rate, promoting tremendous strides in our knowledge of this early phase of mammalian evolution. Nevertheless, many relationships remain contentious, and in the last several years the field has been in a constant state of flux (see Cifelli, 2001, for a particularly useful recent review, and Kielan-Jaworowska et al., 2004, for a comprehensive and authoritative account). This chapter provides a brief summary of the current state of knowledge. A current classification of Mesozoic mammals is shown in Table 4.1.

HISTORICAL BACKGROUND

Not so long ago, Mesozoic mammals were assigned to a relatively small number of higher taxa, whose relationships seemed more or less understood. According to this

Table 4.1. Synoptic classification of Mesozoic mammals (excluding Metatheria and Eutheria)

Class MAMMALIA
   †*Adelobasileus*, †*Hadrocodium*[1]
   †Sinoconodontidae
   †Kuehneotheriidae
  Order †MORGANUCODONTA
   †Morganucodontidae
   †Megazostrodontidae
  Order †DOCODONTA
  Order †SHUOTHERIDIA
  Order †EUTRICONODONTA
   †Amphilestidae
   †Triconodontidae
   †Austrotriconodontidae
  Order †GONDWANATHERIA
 Subclass AUSTRALOSPHENIDA
  Order †AUSKTRIBOSPHENIDA
  Order MONOTREMATA
 Subclass †ALLOTHERIA
   †Theroteinidae
   †Eleutherodontidae
  Order †HARAMIYIDA[2]
   †Haramiyidae
  Order †MULTITUBERCULATA
   Superfamily †Plagiaulacoidea
  Suborder †CIMOLODONTA
   Superfamily †Ptilodontoidea
   Superfamily †Taeniolabidoidea
   Superfamily †Djadochtatherioidea
 Subclass TRECHNOTHERIA
  Superorder †SYMMETRODONTA
   †Amphidontidae
   †Tinodontidae
   †Spalacotheriidae
  Superorder †DRYOLESTOIDEA
   †Vincelestidae[3]
  Order †DRYOLESTIDA
   †Dryolestidae
   †Paurodontidae
  Order †AMPHITHERIIDA
   †Amphitheriidae
  Superorder ZATHERIA
  Order †PERAMURA
   †Peramuridae
   †Arguitheriidae
   †Arguimuridae
 Subclass BOREOSPHENIDA
  Order †AEGIALODONTIA
   †Aegialodontidae
 Infraclass METATHERIA
 Infraclass EUTHERIA

Notes: Modified after Kielan-Jaworowska et al. (2004). The dagger (†) denotes extinct taxa.

[1] *Hadrocodium* appears to be closer to crown-group Mammalia than are morganucodonts or sinoconodontids, but its precise position is uncertain.
[2] Haramiyids were considered to be a possible sister group of tritylodontid cynodonts by Luo et al. (2002), as shown in Fig. 4.2, but are now generally considered to be closer to mammals. Theroteinidae and Eleutherodontidae were included in the paraphyletic Haramiyida by Butler (2000) and Butler and Hooker (2005).
[3] Phylogenetic position uncertain, probably a dryolestoid or a zatherian.

view (e.g., Crompton and Jenkins, 1973, 1979), a group called morganucodonts lay at or near the base of a dichotomy between nontherian (or prototherian) and therian mammals (Fig. 4.1). Nontherians included the living monotremes, whereas therians comprised all other extant mammals. On the nontherian side, morganucodonts (then regarded as basal triconodonts) were believed to have given rise to other triconodonts, as well as to the docodonts, and questionably to the haramiyids, which were considered possible ancestors of the Multituberculata. Although their ancestry was unknown, monotremes were unambiguously grouped with nontherians. Therian mammals were seen as evolving from *Kuehneotherium*, itself derived from a morganucodont or sharing a common ancestor with morganucodonts. From *Kuehneotherium*, which was considered a basal symmetrodont, evolved the other symmetrodonts on the one hand, and eupantotheres on the other. Eupantotheres were considered ancestral to the therians—the marsupials and placentals (Fig. 4.1B).

While parts of this appealing scenario remain essentially valid, recent discoveries and an explosion of interest in this early episode of mammalian history have led to a great expansion of known forms, and with it, the realization that the Mesozoic radiations of mammals were far more complex than previously imagined. Supposed differences in braincase construction that were the basis of the dichotomy between nontherian and therian mammals are now known to be inaccurate, and this bipartite division of Mammalia has been largely abandoned (Kielan-Jaworowska, 1992). Consequently, there is considerable disagreement among experts concerning the sequence of divergence of the various early clades and even the definition of Mammalia itself. Especially volatile and controversial are the relationships of monotremes and multituberculates to each other and to other mammals, which vary depending on the anatomical system analyzed. Here it is important to realize that the position of monotremes directly affects the content of crown-group Mammalia (Rowe, 1988). Part of the controversy stems from the difficulty in determining which mammalian traits are synapomorphous and which ones may have evolved multiple times independently, and on this there is little agreement. This situation led Lillegraven and Krusat (1991: 43) to conclude that "parallel development of similar features was an all-pervasive phenomenon within early evolution of the Mammalia, making the unravelling of phylogenetic relationships among its basal groups a daunting, yet highly interesting, task." A current view of Mesozoic mammalian relationships is shown in Figure 4.2.

Among the most important characters that evolved early in mammalian evolution and contributed to their success are increasing brain capacity, the tribosphenic molar and associated changes in mastication, a single jaw joint between the dentary and the squamosal bones, a middle ear with three ossicles (see Fig. 3.1), and changes in limb posture related to increased activity. In the following sections, the evolution of these key mammalian features is emphasized.

At this point it should be noted that the term "therian" has been used both formally and informally with various

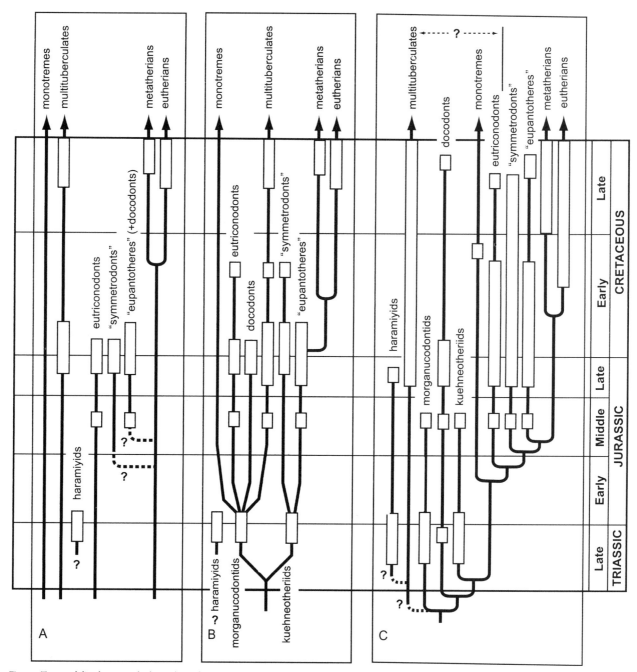

Fig. 4.1. Temporal distribution and relationships of Mesozoic mammals. (A) View of a polyphyletic Mammalia, widely held in first half of the twentieth century and based mainly on Simpson (1928); (B) monophyletic Mammalia, with a dichotomy between nontherian and therian clades, based on Hopson and Crompton (1969); (C) current view of relationships, based on Luo, Crompton, and Sun (2001) and Luo et al. (2002). (From Cifelli, 2001.)

connotations. Although the basic dichotomy between Prototheria (nontherians) and Theria has been largely abandoned, the name Theria is still generally used to refer to the crown-group of metatherians and eutherians and their close relatives (e.g., Rowe, 1988; Hopson, 1994; McKenna and Bell, 1997). Usage here follows this convention.

THE OLDEST MAMMALS

Arguably the oldest and most primitive known mammal is *Adelobasileus,* based on the back half of a skull from the Late Triassic (late Carnian, about 225 Ma) of Texas (Lucas and Luo, 1993). This unique fossil predates the next oldest mammals by at least 10 million years, and shares with later mammals several derived features, including configuration of certain cranial foramina, morphology of the occipital condyles, and presence of a bony floor of the cavum epiptericum. The incipient development of a promontorium to house the cochlea is anatomically intermediate between the conditions in cynodonts and early mammals. Unfortunately, the fossil lacks teeth, a lower jaw, and other parts of the skeleton that might corroborate its mammalian status.

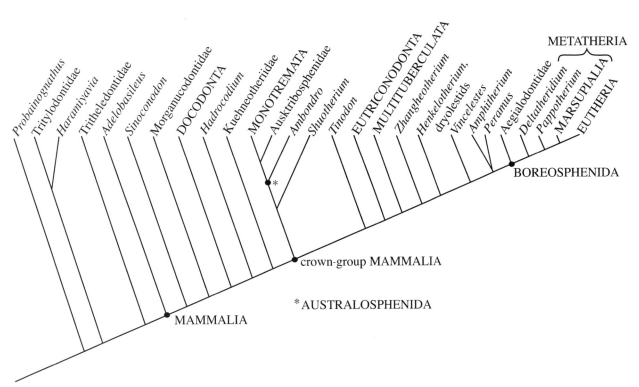

Fig. 4.2. Relationships of Mesozoic mammals. The current consensus places haramiyids (including *Haramiyavia*) and Multituberculata as sister taxa within Mammalia. (Modified after Luo et al., 2002.)

Until these are found, *Adelobasileus* will remain a taxon of problematic relationships.

Haramiyids and Possible Relatives

Two other families that appeared in the Late Triassic (?Norian-Rhaetic), Haramiyidae and Theroteinidae, could be the oldest known mammals, but are also problematic. Both have been suggested to have possible affinities with multituberculates because of dental resemblance. Theroteinidae are known only from isolated teeth from France that have complex crowns and preprismatic enamel (Sigogneau-Russell et al., 1986; Hahn et al., 1989). They could belong to primitive mammals or to advanced cynodonts, but similarities to teeth of haramiyids and primitive multituberculates suggest that they are probably mammals. The Middle Jurassic eleutherodontids (*Eleutherodon*), based on isolated teeth from England and China (Kermack et al., 1998; Butler and Hooker, 2005; Maisch et al., 2005), represent a third family perhaps related to theroteinids or haramiyids; but it is difficult to reach definitive conclusions based on these isolated teeth. Citing similarities in their molariform teeth, Butler (2000) united these three families in the order Haramiyida. If multituberculates originated from within this group, Haramiyida would be a paraphyletic assemblage. The grouping of Haramiyida and Multituberculata is called Allotheria.

Until recently, haramiyids were also known only from isolated teeth. They resemble multituberculate molars in being relatively low crowned and having two parallel, peripheral rows of cusps, longitudinally arranged and separated by a median furrow. These teeth were presumed to be molars, but their orientation and position in the toothrow were uncertain. Their mammalian status did not appear to be in question, however, for they have multiple roots, preprismatic enamel, and wear facets indicating precise interlocking occlusion, which in turn suggests diphyodont tooth replacement (Sigogneau-Russell, 1989). Wear patterns even seem to indicate a palinal (longitudinal and backward) chewing stroke, as in multituberculates (Butler and MacIntyre, 1994).

Much more complete haramiyid fossils discovered in Greenland (Jenkins et al., 1997) confirm that the isolated teeth were indeed molars and clarify their orientation. These fossils, named *Haramiyavia* (Fig. 4.3), also show that the postcranial skeleton was generally similar to that of morganucodonts, the most primitive mammals for which the skeleton is known. The lower dentition further resembles that of multituberculates in having procumbent incisors separated from the cheek teeth by a conspicuous diastema. According to Jenkins et al. (1997), the occlusal relationships of the teeth contradict those observed in other haramiyids and indicate a predominantly orthal (vertical) chewing stroke. This observation might suggest that *Haramiyavia*, at least, was not as closely related to multituberculates as had been supposed. However, Butler (2000) suggested that limited palinal movement probably did occur during the power stroke of chewing in *Haramiyavia*, which would support its position as a transitional form leading to multituberculates. Butler (2000) placed *Haramiyavia* in its own family, Haramiyaviidae.

Haramiyavia retained larger postdentary bones than did morganucodontids, suggesting that haramiyids could be an

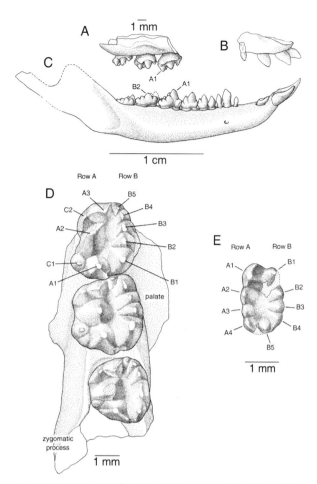

Fig. 4.3. *Haramiyavia*, jaw and dentition: (A, B) upper dentition; (C) mandible; (D) right maxilla with M^{1-3}; (E) right M_3. Anterior at top in D and E. Labels designate cusps. (From Jenkins et al., 1997.)

even earlier offshoot of the mammalian stem. Unfortunately, the jaw joint is not preserved in the fossils of *Haramiyavia*, but the presence of larger postdentary bones suggests that a quadrate-articular joint was still functional. Potentially more significant is the evidence for orthal and palinal jaw movements during chewing in Haramiyida and multituberculates. This mode of occlusion, which differs fundamentally from that in other mammals, led Butler (2000; see also Butler and Hooker, 2005) to hypothesize that allotheres diverged from other mammals before they evolved unilateral shearing and transverse jaw movements—which would be very early indeed. Primitive haramiyidans document transitional stages in the development of palinal occlusion. Whether haramiyidans are related to multituberculates or represent a separate branch of primitive mammals, or even cynodonts, remains unsettled (e.g., Butler and MacIntyre, 1994; Butler, 2000; Luo et al., 2002).

Morganucodonts, *Sinoconodon*, and Kuehneotheriids

Prior to these discoveries, the oldest mammals were long held to be those from fissure-fillings in Wales that have usually been considered of Late Triassic (Rhaetic) age. However, their uncertain age was indicated by the label "Rhaeto-Liassic" often applied to these fossils, and the age of the fissures now appears to be younger rather than older (late Rhaetic-Liassic, or Sinemurian: latest Triassic–Early Jurassic; Kermack et al., 1981; Clemens, 1986; Kielan-Jaworowska, 1992). They are usually assigned to the families Morganucodontidae and Kuehneotheriidae, which have essentially triconodont-like cheek teeth. Morganucodontidae (formerly considered to be basal "triconodonts") are known from teeth and skulls, whereas Kuehneotheriidae (formerly basal "symmetrodonts") are represented only by jaws and teeth. Chinese deposits of similar age (Liassic) have produced additional skulls of *Morganucodon* and of another primitive form, *Sinoconodon* (Fig. 4.4A). These taxa, though perhaps not the oldest, are widely considered to be the most primitive mammals (Crompton and Luo, 1993). *Eozostrodon*, based on two isolated teeth, has often been considered a synonym of *Morganucodon*, but it is possibly a distinct morganucodontid (Kielan-Jaworowska et al., 2004).

Morganucodonts (now considered to include the families Morganucodontidae and Megazostrodontidae) were small shrew- to mouse-sized animals (10–30 g; Jenkins and Crompton, 1979) that were widely distributed during the latest Triassic(?)–Early Jurassic, occurring in Europe, Asia, Africa, and North America (Fig. 4.4). The anatomy of morganucodonts indicates that they occupy a central position at the base of the mammalian radiation. Kermack et al. (1973, 1981) and Jenkins and Parrington (1976) detailed the anatomy of morganucodonts, which combines derived mammalian traits with primitive cynodont features. The dental formula of *Morganucodon* varies within and among species: 3–5.1.4–5.3–4/4–5.1.4–5.3–5 (Kielan-Jaworowska et al., 2004). As in eutriconodonts, the premolars are simple, with one main cusp, and the molar cusps are linearly arranged (Fig. 4.5). Based on this morphology, morganucodonts were previously regarded as primitive triconodonts; however, the similarities are now generally considered to be plesiomorphic. Current consensus separates morganucodonts from eutriconodonts and places them at the base of mammals, whereas eutriconodonts are thought to be closer to therian mammals (e.g., Luo et al., 2002).

The dentary of morganucodonts is mammal-like in having a large coronoid process and a well-developed condylar process that articulated with the squamosal. From the lateral side, this appears to be the only jaw joint, but medial to it a functional quadrate-articular jaw joint was also still present. In *Morganucodon* the angular process is situated well anterior to that of more derived mammals, which led Jenkins et al. (1983) to identify it as a pseudangular process. The skull of *Morganucodon* lacks prefrontal and postorbital bones, as in other mammals and advanced cynodonts, but primitively retains a septomaxilla. Although first reported to retain tabular bones, as in cynodonts (Kermack et al., 1981), subsequent studies have concluded that tabulars are absent in *Morganucodon* (e.g., Luo et al., 2002).

The vertebral column of morganucodonts is more regionally differentiated than in cynodonts and shows other

mals and tritylodontids, in having a long ilium with separate gluteal and iliac surfaces and a large obturator foramen. The femur has a spherical head and trochanters arranged as in mammals (and tritylodontids, but not other cynodonts). The ankle, however, shows few mammalian specializations except for the presence of an astragalar foramen. Morphology of the phalanges and a probably abducted hallux suggest grasping ability. Together these features suggest that morganucodonts had rather generalized skeletons that enabled them to climb as well as scramble on the ground.

Sinoconodon has triconodont-like cheek teeth, which differ from those of morganucodonts in lacking well-developed cingula (Fig. 4.4A). As in morganucodonts, there is a bony separation between the orbits, and the jaw joint is between the squamosal and the dentary, but the postdentary bones are more reduced than in morganucodonts (Crompton and Sun, 1985)—a presumably derived condition. In other ways, however, *Sinoconodon* seems to be more primitive than *Morganucodon*. Cochlear structure was more primitive (Luo et al., 1995), and the absence of consistent wear facets indicates that *Sinoconodon* lacked precise molar occlusion. It resembles cynodonts in retaining a large septomaxillary bone and multiple replacement of the incisors and canines; in addition, the posterior molars were replaced once (Zhang et al., 1998). These features suggest that *Sinoconodon* diverged from the mammalian stem earlier than morganucodontids and could be the sister group of all other mammals (Wible, 1991; Crompton and Luo, 1993; Luo et al., 2002).

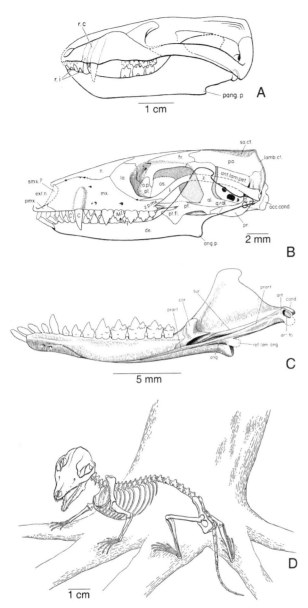

Fig. 4.4. Basal mammals: (A) *Sinoconodon* skull; (B, C) *Morganucodon* reconstructed skull and lower jaw (medial view); (D) *Megazostrodon* skeleton. (A from Crompton and Sun, 1985; B, C from Kermack et al., 1973 and 1981; D from Jenkins and Parrington, 1976.)

modifications associated with mammal-like movements, including an essentially mammalian atlas-axis complex and an enlarged cervical canal (reflecting expansion of the spinal cord in the region of the brachial plexus, which in turn suggests more complex neural control of the forelimbs). At the same time, the pectoral girdle is distinctly cynodont-like. The scapula lacks a supraspinous fossa, and there are two coracoids, although only the posterior one contributes to the glenoid fossa. The humerus is rather therian-like at the proximal end, with a hemispherical head and a pair of tuberosities; but distally it has an ulnar condyle as in cynodonts, rather than an ulnar trochlea as in advanced mammals. The proximal and distal articulations of the humerus are twisted relative to each other. These humeral features suggest a sprawling stance. The pelvis is derived, as in mam-

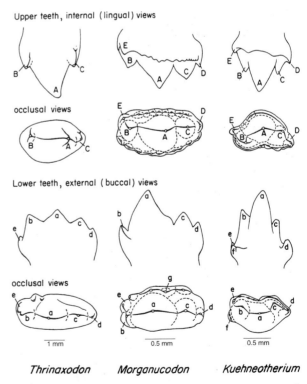

Fig. 4.5. Comparison of *Thrinaxodon*, *Morganucodon*, and *Kuehneotherium* teeth, based on Crompton (1963) and Hopson and Crompton (1969). Letters designate cusps. (From Jenkins, 1984.)

Kuehneotherium (Fig. 4.5) was formerly regarded as a basal symmetrodont belonging either to the family Tinodontidae or to a separate family, Kuehneotheriidae. The current view is that kuehneotheriids occupy a position near the base of mammals (Cifelli, 2001; Luo et al., 2002). The lower dental formula of *Kuehneotherium* is 4?.1.6.4 or 5 (Gill, 1974). As in morganucodonts and eutriconodonts, the cheek teeth have three principal cusps, with the central one tallest. The main cusp of the upper molars is probably homologous with the paracone of therians, whereas that of the lowers is thought to be homologous with the protoconid. In contrast to morganucodonts and eutriconodonts, however, the cusps are not directly aligned, but form an obtuse angle, with the front and back cusps rotated slightly lingually on the lower teeth and buccally on the uppers, foreshadowing the arrangement in tribosphenic therians. The upper cusps probably represent the stylocone, paracone, and metacone of therian molars, whereas the lowers are probably the three trigonid cusps (followed by an incipient talonid cusp; Patterson, 1956). Whether this cusp rotation is homologous with that in true symmetrodonts and therians or evolved independently is controversial. The relative position of upper and lower cusps during occlusion also differs from that in *Morganucodon*, being shifted so that each tooth opposed parts of two others. These progressive dental features suggest that *Kuehneotherium* could be closer to the stem of the therian radiation than any other Rhaeto-Liassic taxon (Crompton and Jenkins, 1979). Luo et al. (2002), however, recently questioned the supposed close relationship between *Kuehneotherium* and extant mammals. *Kuehneotherium* primitively retained much reduced postdentary bones and a double jaw articulation, but the dentary-squamosal joint was predominant. In true symmetrodonts only the dentary-squamosal joint was present.

Woutersia is a possible relative of *Kuehneotherium* known from isolated teeth from Rhaetian deposits in France. It differs from *Kuehneotherium* in having cusps on the lingual cingulum (one on upper molars, two on lowers), thus broadening the teeth, a possible early adaptation for crushing (Sigogneau-Russell and Hahn, 1995). Butler (1997) be-

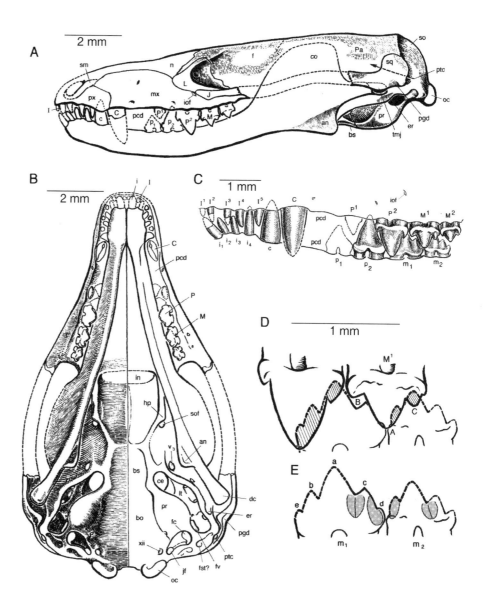

Fig. 4.6. *Hadrocodium* skull and teeth: (A, B) restored skull; (C) lateral view of restored dentition; (D, E) occlusion and wear facets. (From Luo, Crompton, and Sun, 2001.)

lieves that it may represent a transitional form leading to docodonts.

Hadrocodium

Luo, Crompton, and Sun (2001) described a slightly younger animal with triconodont-like teeth, *Hadrocodium*, based on a shrew-sized skull from the Sinemurian (Early Jurassic, 195 Ma) of China (Fig. 4.6). It was one of the smallest known mammals, estimated to have weighed only 2 g. *Hadrocodium* is significant in showing several unexpectedly progressive features for such an ancient mammal. Its skull is wide posteriorly and the braincase is relatively large. There is a single jaw articulation, between the dentary and the squamosal bones; and there is no postdentary groove, implying that the middle-ear ossicles were already separate from the lower jaw and attached to the skull. These features, together with wear on the molar teeth, suggest that the single known specimen represents an adult or subadult (hence its small size is not attributable to being a juvenile). *Hadrocodium* had a primitive incisor count of 5/4 but had a reduced number of premolars and molars (dental formula 5.1.2.2/4.1.2.2). The derived features of *Hadrocodium* indicate that it is more closely related to crown-group Mammalia than is either *Sinoconodon* or *Morganucodon*, but the reduced number of cheek teeth make *Hadrocodium* too specialized to be on the direct line to therian mammals.

DOCODONTA

Although unknown before the Middle Jurassic, and therefore not among the oldest known mammals, docodonts are considered to be one of the most archaic mammalian groups. Their remains were first discovered more than a century ago in the Late Jurassic Morrison Formation of Wyoming and Colorado, where they are found together with the bones of giant sauropod dinosaurs. They have subsequently been discovered at several sites in Europe and Asia. A purported docodont (*Reigitherium*) has been reported from the Late Cretaceous of Patagonia in South America (Pascual et al., 2000), but this attribution is questionable (Kielan-Jaworowska et al., 2004). Rougier, Novacek, et al. (2003) reported new specimens of *Reigitherium* that show the absence of postdentary bones as well as dental features that suggest that it is a dryolestoid, as originally proposed by Bonaparte (1990).

Most docodonts are known solely from the dentition, which includes complex, broad cheek teeth. In *Docodon*, the lower molars are rectangular and the buccal cusps are higher than the lingual cusps; the upper molars are hourglass shaped and transversely wider than long (Fig. 4.7B). The teeth of docodonts have been cited as evidence that the group evolved from morganucodonts (e.g., Crompton and Jenkins, 1979). According to this hypothesis, the wide molars of docodonts evolved by expansion of the lingual cingula of typical morganucodont molars. Cusps on the cingula eventually enlarged, and transverse crests formed, joining them to the original (lateral) cusps. The morganucodont *Megazostrodon*, which has a well-developed lingual cingulum and cingular cusps on the lower molars, represents a plausible morphologic stage from which docodonts might have evolved (Crompton, 1974). As noted above, it is also possible that docodonts evolved in a similar manner from kuehneotheriids. Docodonts also evolved precise molar occlusion in association with their complex molar crowns. These derived conditions are superficially similar to those characterizing therians, but they are different enough to indicate that they arose independently.

Insight on the phylogenetic position of docodonts is afforded by the best-known docodont, *Haldanodon*, from the Late Jurassic Guimarota lignites (swamp deposits) of Portugal (Lillegraven and Krusat, 1991; Martin and Krebs, 2000). *Haldanodon* is represented by dozens of jaws, several skulls (Fig. 4.7A), and a skeleton. Based on its robust limb skeleton—especially the scapula with a postscapular fossa, the broad humerus with a prominent deltopectoral crest, an elongate ulnar olecranon process, and short and robust

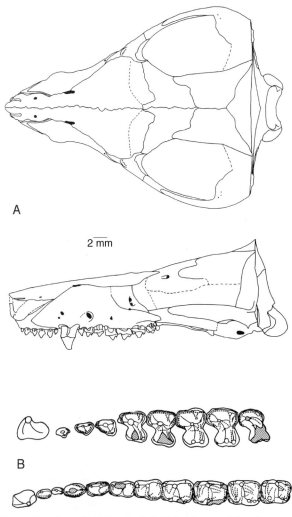

Fig. 4.7. Docodonts: (A) *Haldanodon* skull; (B) *Docodon* upper left and lower right dentitions, anterior to left, buccal at top. (A from Lillegraven and Krusat, 1991; B from Jenkins, 1969b.)

phalanges—*Haldanodon* appears to have been fossorial (Krusat, 1991; Martin, 2005). However, its occurrence in lignite deposits suggests that it may also have been semiaquatic, similar to extant desman moles. *Haldanodon* resembles cynodonts in several plesiomorphous cranial features that are present in more derived states in morganucodontids. The presence in *Haldanodon* of a large septomaxilla in the nasal region, retention of larger accessory ("postdentary") jaw bones and a larger stapes than in morganucodontids, and several other cynodont-like features could indicate that this genus was more primitive than *Morganucodon* and diverged even earlier from the mammalian stem. At the same time, several other cranial characters of *Haldanodon* are derived, like those of other early mammals. Lillegraven and Krusat suggested that *Haldanodon* could have acquired many of its "mammalian" traits earlier than, and independently from, morganucodontids, which would suggest that Mammalia is polyphyletic. Subsequent phylogenetic analyses (e.g., Rougier et al., 1996a; Luo et al., 2002), however, support a monophyletic Mammalia that includes *Haldanodon*.

A new docodont recently reported from the Middle Jurassic of China provides additional evidence on the relationships and behavior of these archaic mammals (Ji et al., 2006). Based on a partial skeleton, *Castorocauda* is the largest known docodont, almost half a meter long from the snout to the end of the tail. Its skeleton is adapted for swimming and burrowing, supporting the interpretation that docodonts were semiaquatic. Phylogenetic analysis confirmed that docodonts are a primitive mammalian clade more derived than morganucodonts but less derived than *Hadrocodium*. *Castorocauda* is the oldest mammal preserving evidence of fur.

MULTITUBERCULATA

Multituberculates were the longest-lived order of mammals except for monotremes, recorded with certainty from Upper Jurassic through upper Eocene sediments, a time-span of more than 100 million years (from about 155 to 40 Ma). They have no living descendants. Isolated upper second molars from the Middle Jurassic (Bathonian) of England, assigned to the new genera *Kermackodon* and *Hahnotherium*, have recently been identified as multituberculate based on wear and inferred occlusal relationships (Butler and Hooker, 2005). This finding extends the range of the group back another 10 million years. As noted previously, multituberculates may be related to Late Triassic and Jurassic haramiyidans, and the two groups are sometimes united in the Allotheria to reflect this relationship. Multituberculates were abundant in many Mesozoic and Early Cenozoic faunas of the northern continents. Recent discoveries have extended their Mesozoic range into northern Africa and South America, although they are still very rare from those regions (and the African teeth may instead belong to haramiyidans, according to Butler and Hooker, 2005). All were small, mostly shrew- to rat-sized, the largest reaching the size of a beaver.

The anatomy of multituberculates has been reviewed by Hahn (1978), Clemens and Kielan-Jaworowska (1979), Krause and Jenkins (1983), and Kielan-Jaworowska et al. (2004), among others. Multituberculates are distinguished by their unique dental complex, which includes in the lower jaw a single enlarged, somewhat rodentlike incisor separated from the cheek teeth by a diastema, no canine, one to four bladelike lower premolars with oblique ridges joined to apical serrations, and molars with multiple low cusps arranged in two longitudinal rows (Figs. 4.8, 4.9). In some multituberculates, including taeniolabidoids and djadochtatheres, the lower incisor is very large and has enamel essentially restricted to the labial half of the tooth. The upper series has one to three incisors, I^2 enlarged, usually no canine, and premolars and molars with two or three longitudinally arranged rows of cusps separated by longitudinal grooves. M^2 is medially offset relative to M^1 in all but the most primitive forms. Consequently the central groove of M^2 occludes with the lingual cusp row of M_2, whereas the groove of M^1 occludes with the buccal cusp row of M_1 (Butler and Hooker, 2005). All known multituberculates are dentally so derived that it has not been possible to determine the homologies of the cusps with those of therians. The incisors and the anterior premolars were diphyodont (with deciduous precursors), but only the most primitive multituberculates (Paulchoffatiidae) are known to have had both deciduous and permanent P_4. In all others, the bladelike, deciduous P_4 seems to have been retained throughout life, and erupted in a unique way, by rotating anterodorsally about 90° into position (Greenwald, 1988).

Analysis of tooth morphology and microwear indicates that most multituberculates had a unique two-stroke masticatory cycle (Krause, 1982). First, food held in place by the last upper premolar was sliced by the bladelike lower premolar(s) as the dentary moved orthally (upward). Then the lower jaw moved palinally (backward), grinding the food between the molar cusp rows. Molar occlusion usually occurred bilaterally, although the unfused symphysis probably allowed occasional unilateral occlusion (Wall and Krause, 1992; Gambaryan and Kielan-Jaworowska, 1995). Unlike in most therians, there was no transverse component in the grinding stroke. The second part of the chewing cycle is superficially like the grinding phase in rodents (which is also often bilateral), but in the latter the chewing stroke is propalinal (forward). Although multituberculates have often been considered herbivorous analogues of rodents, considerations of body size and microwear suggest that they were omnivores that consumed a variety of items, including seeds, nuts, and small invertebrates.

The lower jaw of multituberculates is typically short and deep. It is derived compared to that of many other Mesozoic mammals in consisting entirely of the dentary, which articulates with the squamosal; there are no postdentary bones, except for a vestigial coronoid bone in one of the earliest forms, the plagiaulacid *Kuehneodon*. The symphysis is unfused, which allowed the dentaries to move independently from each other to a considerable extent. The dentary lacks an angular process.

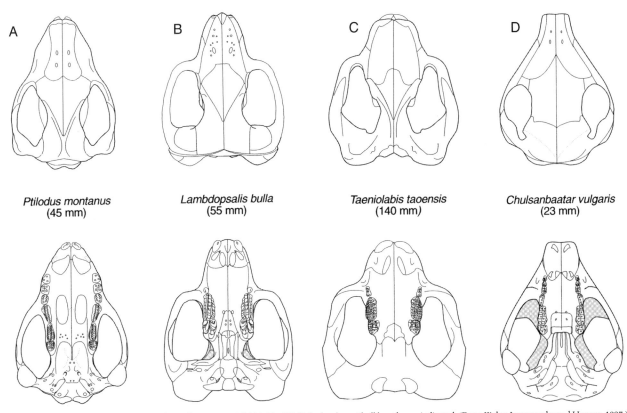

Fig. 4.8. Multituberculate skulls: (A) ptilodontoid; (B, C) taeniolabidoids; (D) djadochtathere. Skull lengths are indicated. (From Kielan-Jaworowska and Hurum, 1997.)

The skull is typically low and broad, with a short, rather wide snout and well-developed zygomatic arches, which consist mainly of the maxilla and squamosal. The jugal, previously thought to be absent, has been identified on the medial side of the arch in several multituberculates (Hopson et al., 1989). The orbit lacked a bony floor, and the eyes were directed laterally. Elements earlier believed to be plesiomorphic tabular and ectopterygoid bones have since been shown to be parts of the mastoid and alisphenoid, respectively (Kielan-Jaworowska et al., 1986; Hurum, 1998). In fact, no mammal has been shown to have tabular bones. Endocranial casts show that the olfactory bulbs of multituberculates were relatively very large. The slightly curved cochlea is also primitive; it resembles that of *Morganucodon* more than the bent cochlea of monotremes, and differs markedly from the coiled cochlea of marsupials and placentals (Kielan-Jaworowska and Hurum, 2001). The lateral wall of the braincase, however, composed of a reduced alisphenoid and enlarged anterior lamina, is a derived condition shared with monotremes (Hopson and Rougier, 1993). Moreover, multituberculates are seemingly advanced in having three middle-ear ossicles arranged like those in living mammals (Miao and Lillegraven, 1986; Hurum et al., 1996; Rougier et al., 1996b). Whether they are homologous with those of other mammals or evolved independently, however, is a matter of contention.

The postcranial skeleton, known in only a few forms, indicates that multituberculates had epipubic ("marsupial") bones and that the limbs were abducted, probably resulting

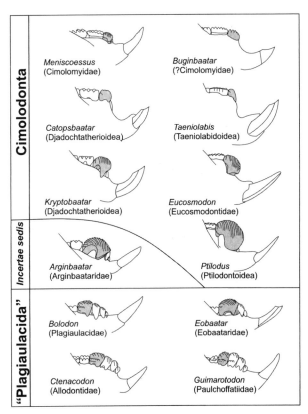

Fig. 4.9. Multituberculate lower jaws. "Plagiaulacida" is equivalent to Plagiaulacoidea as used in this chapter. (From Kielan-Jaworowska and Hurum, 2001.)

in a sprawling stance (Kielan-Jaworowska and Gambaryan, 1994; Gambaryan and Kielan-Jaworowska, 1997). At least one multituberculate, the Late Cretaceous djadochtathere *Bulganbaatar,* may have had a mobile pectoral girdle and more advanced, parasagittal forelimb posture like that of higher therians, based on a specimen with articulated forelimbs (Sereno and McKenna, 1995). But Gambaryan and Kielan-Jaworowska (1997) challenged this interpretation, which appears to conflict with other evidence. As further discussed below, the few known skeletons display considerable diversity: some multituberculates were specialized for arboreal life, whereas others were terrestrial, and still others were fossorial.

Multituberculates have traditionally been classified in three suborders or superfamilies: Plagiaulacoidea (essentially a primitive grade of Mesozoic multituberculates), and the probably monophyletic Ptilodontoidea and Taeniolabidoidea (Clemens and Kielan-Jaworowska, 1979; Simmons, 1993). A fourth monophyletic subdivision was recently recognized: Djadochtatherioidea (=Djadochtatheria, hereafter called djadochtatheres), which includes nearly all Late Cretaceous multituberculates of Mongolia (Kielan-Jaworowska and Hurum, 1997, 2001). The last three groups comprise the monophyletic Cimolodonta (McKenna and Bell, 1997; Kielan-Jaworowska and Hurum, 2001). Although plagiaulacoids are grouped essentially by primitive characters, cimolodonts are united by several synapomorphies, including the loss of I^1, loss of P_{1-2}, and great reduction or loss of P_3 (Kielan-Jaworowska and Hurum, 2001). All cimolodonts have prismatic enamel, whereas most plagiaulacoids lacked prisms. However, the prismatic enamel of cimolodonts almost surely evolved independently from that of therian mammals.

The classification of multituberculates has been especially mercurial in recent years (see the historical review by Kielan-Jaworowska and Hurum, 2001), probably owing chiefly to two factors—the rapid accumulation of new discoveries and the exploration of relationships using cladistic methods. Several new classifications have been proposed in the past decade, but there is currently no generally accepted arrangement. Although most classifications are more or less consistent with the groups listed above, the placement of both genera and higher taxa varies (see, for example, Kielan-Jaworowska et al., 2004).

Plagiaulacoidea

Plagiaulacoids include the oldest and most primitive multituberculates—families Paulchoffatiidae, Allodontidae, Plagiaulacidae, and a few others (Kielan-Jaworowska and Hurum, 2001; Kielan-Jaworowska et al., 2004). They are best known from Late Jurassic and Early Cretaceous sites in Europe and North America. In addition, several Early Cretaceous forms from Asia and one tooth from northern Africa have been described in recent years. Plagiaulacoidea is a paraphyletic assemblage (including successive sister taxa of later multituberculates), hence it is not surprising that its composition is particularly unstable—for example, regarding the proper familial allocation of various genera and even which families belong here. The dental formula is 3.0–1.4–5.2 / 1.0.3–4.2. I^2 is multicusped and the largest of the upper incisors, I^3 may also be large and multicusped, and several genera retain the upper canine (Hahn, 1977, 1993). The three or four lower premolars form a bladelike cutting edge, with the anteriormost premolar smallest and lowest. In paulchoffatiids, unlike later multituberculates, the premolars usually were not markedly larger than the molars. Heavy apical wear in some of them suggests that these teeth were used for grinding. A row of basal cuspules is developed on the buccal surface of the posterior lower premolars. A reduced jugal bone was still present, as was a vestigial coronoid, which is absent in later multituberculates. All the cranial nerves entering the orbit apparently passed through a single large sphenorbital fissure. These features suggest that paulchoffatiids are the most primitive multituberculates. However, allodontids are plesiomorphic in having a small I^3, smooth enamel, a premolar formula of 5/4, and well-separated cusps on the lower molars. Nevertheless, their lower premolars are more derived than those of paulchoffatiids (Kielan-Jaworowska et al., 2004). One bizarre plagiaulacoid, Early Cretaceous *Arginbaatar* from Asia, had an unusually large P_4 that rotated anteriorly over the two more anterior premolars, gradually pushing them out of the jaw as the animal aged (Kielan-Jaworowska et al., 1987b).

Djadochtatherioidea

The Asian djadochtatheres are united by several apomorphic cranial features relating mainly to the anatomy of the frontal and lacrimal bones. Unlike many other multituberculates, most of the dozen djadochtathere genera are known from skulls and often postcranial skeletons as well (Fig. 4.10). The skull is short and broad and, in several genera, triangular in superior view. The cheek teeth have fewer cusps and P_4 has fewer ridges than in ptilodontoids and taeniolabidoids (Kielan-Jaworowska and Hurum, 1997). Some forms retain cervical ribs. According to Kielan-Jaworowska and Gambaryan (1994), both the forelimbs and the hind limbs of djadochtatheres were held in a primitive abducted, sprawling posture, and the feet were abducted about 30° from the sagittal plane. These authors also identified an incipient supraspinous fossa in djadochtatheres, but it is much less developed than in therians. Long spinous processes of the lumbar vertebrae imply well-developed erector spinae muscles, as are found in mammals with jumping ability. Djadochtatheres were terrestrial animals that evidently progressed by an asymmetrical gait punctuated by occasional jumps.

Ptilodontoidea

Ptilodontoids are first known from the Late Cretaceous and survived until the late Eocene. They were the most diverse early Cenozoic multituberculates, with 15 of the

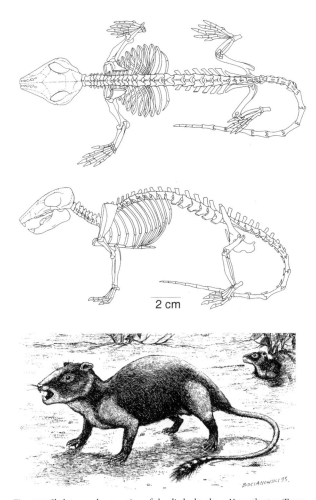

Fig. 4.10. Skeleton and restoration of the djadochtathere *Nemegtbaatar*. (From Kielan-Jaworowska and Gambaryan, 1994.)

16 genera known from Paleocene or early Eocene strata. They ranged in size from a small mouse to a squirrel (*Sciurus*). The two or three families are known from North America, Europe, and Asia. Ptilodontoids had a longer, more slender lower incisor than in plagiaulacoids (Fig. 4.11). There are at most four upper premolars and only one or two lower premolars. P_3, when present, is reduced to a single-rooted peg, whereas P^4 is elongate, and P_4 is large and bladelike, its crown usually extending well above the molars and bearing 8 to 16 serrations. The upper molars often have cingula or an extra row of cusps compared to plagiaulacoids. The first molars are longer than the second molars.

The postcranial skeleton of *Ptilodus* (Fig. 4.12) displays numerous arboreal adaptations, including a divergent hallux, a long and probably prehensile tail, and tarsal modifications that facilitated hindfoot abduction and reversal, thus allowing the animal to descend trees headfirst (Plate 1.1; Jenkins and Krause, 1983). The mouse-sized neoplagiaulacids account for two-thirds of ptilodontoid genera and existed from the Late Cretaceous through the Eocene. A neoplagiaulacid from the Chadronian (latest Eocene, approximately 35 million years ago), usually identified as *Ectypodus*, was the last occurring multituberculate (Krishtalka et al., 1982).

Multituberculates, particularly ptilodontoids, experienced something of a resurgence during the Paleocene in western North America. In the richest Torrejonian through Clarkforkian quarry assemblages from Wyoming and Montana, multituberculates typically account for 15–20% of all mammal species and from 12–25% of the individuals represented (Rose, 1981; Krause, 1986). At Swain Quarry in southern Wyoming—perhaps the richest known Paleocene site—43% of the 28,000 mammal teeth collected belong to multituberculates, mainly ptilodontoids (Rigby, 1980). Ptilodontoids were generally rare after the Paleocene, although they were moderately common in the early Eocene Four Mile fauna of Colorado (McKenna, 1960a) and abundant in one early Eocene quarry sample from the Bighorn Basin of Wyoming (26% of individuals; Silcox and Rose, 2001).

Taeniolabidoidea

Taeniolabidoids (Fig. 4.8B–C) are known from the Late Cretaceous through early Eocene of North America, Asia, and Europe. Two of the three known families persisted into the Early Tertiary: Taeniolabididae are known from the Paleocene of North America and Asia, while Eucosmodontidae lived into the early Eocene in North America and Europe. (Survival of eucosmodontids into the Eocene, however, is based on the genera *Neoliotomus* and *Microcosmodon*, both of which were excluded from this family by Kielan-

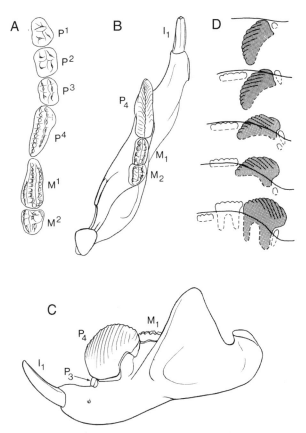

Fig. 4.11. (A–C) *Ptilodus* left upper and lower dentitions; (D) P_4 replacement in multituberculates; anterior to the right. (A–C from Krause, 1982; D from Greenwald, 1988.)

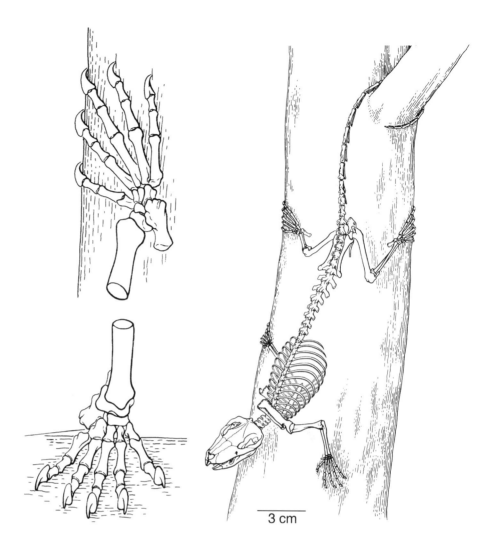

Fig. 4.12. *Ptilodus* skeleton and foot. Reconstructions at left show the right hind foot in normal terrestrial stance (below) and reversed for headfirst descent from trees (above). (From Jenkins and Krause, 1983.)

Jaworowska and Hurum, 2001.) In these multituberculates I_1 is hypsodont and in some forms possibly ever-growing, and the enamel is limited to a ventrolateral band. There are two upper incisors (I^{2-3}). From one to four upper premolars are present, and only one or two lower premolars. P_4 is typically bladelike, but in Taeniolabididae it is reduced to a small tooth with only a few apical cusps. In Eucosmodontidae, by contrast, P_4 can have up to 15 serrations. The molars of taeniolabidoids are often as large as or larger than the premolars. Early Paleocene *Taeniolabis* was the largest known multituberculate, reaching the size of a large beaver. Some recent studies suggest that eucosmodontids are not so closely related to taeniolabidids as long believed and should be excluded from the Taeniolabidoidea (e.g., Kielan-Jaworowska and Hurum, 2001).

Skeletal remains are known for several taeniolabidoid taxa, and they indicate diverse habits. The late Paleocene taeniolabidid *Lambdopsalis* from Asia was fossorial. This interpretation is based on many features, including fused neck vertebrae (C2–3), a robust humerus, and a thick, keeled manubrium sterni (Kielan-Jaworowska, 1989; Kielan-Jaworowska and Qi, 1990). The incisors of *Lambdopsalis* had pigmented enamel, presumably indicating a hard, iron-bearing outer layer of enamel (Akersten et al., 2002), which would have been useful if the teeth were used for digging. The skull of *Lambdopsalis* was wedge-shaped and had very large, inflated petrosals, superficially resembling bullae, that housed an expanded vestibular apparatus and an uncoiled cochlea. These features, as well as the structure of the auditory ossicles, suggest that *Lambdopsalis* was adapted for low-frequency sound reception, another indication of fossorial habits (Miao, 1988; Meng and Wyss, 1995). The morphology of the middle ear of *Lambdopsalis* is particularly similar to that of living monotremes, which is taken by some authors as evidence of a special relationship between them (e.g., Meng and Wyss, 1995). Other taxa traditionally considered taeniolabidoids, such as *Eucosmodon*, show arboreal adaptations similar to those of *Ptilodus* (Krause and Jenkins, 1983).

Relationships and Extinction of Multituberculates

Despite substantial knowledge of their anatomy, the ancestry of multituberculates remains enigmatic. Their highly apomorphic dentition and the possibility that haramiyids or theroteinids might be related to multituberculates suggest that they are a very ancient clade that could have originated independently from other mammals. However, the mammalian jaw joint, virtual absence of postdentary bones, diphyodont tooth replacement, and presence of mammal-like middle-ear ossicles (Miao and Lillegraven, 1986; Meng and Wyss, 1995; Hurum et al., 1996; Rougier et al., 1996b) support their inclusion in a monophyletic Mammalia, as does the unequivocal presence of hair (Meng and Wyss, 1997). Various cranial features (e.g., cribriform plate, ossified ethmoid plate) are shared with monotremes and therians (Hurum, 1994). However, there remains a wide gulf between multituberculates and therians in many other aspects of their anatomy. This paradoxical association of very primitive traits with autapomorphic and derived therian-like features has made it very difficult to decipher the phyletic position of multituberculates relative to other mammals. In recent years, they have been considered a primitive offshoot of the mammalian stem (e.g., Kielan-Jaworowska, 1992; Miao, 1993; McKenna and Bell, 1997), the sister taxon of monotremes (e.g., Kemp, 1983; Wible and Hopson, 1993), the sister taxon of Theria (Rowe, 1988), or somewhere in between (e.g., Rougier et al., 1996a; Luo et al., 2002).

Why such a diverse and successful group as multituberculates became extinct remains a conundrum. Perhaps they were competitively inferior to placentals. The brains of multituberculates were relatively large among Mesozoic mammals (encephalization quotient, or EQ, between 0.37 and 0.71) but relatively much smaller than in average modern mammals (Krause and Kielan-Jaworowska, 1993; Kielan-Jaworowska and Lancaster, 2004). Although competition with condylarths, rodents, plesiadapiforms, and early euprimates was probably a factor in their disappearance (Van Valen and Sloan, 1966; Krause, 1986), it seems unlikely to be the full explanation, as multituberculates coexisted successfully with one or more of these placental groups in several Early Tertiary faunas. Reproductive biology may also have contributed. Kielan-Jaworowska (1979) suggested that the pelvic outlet of the djadochtathere *Kryptobaatar* was too small to allow eggs to pass through, which may indicate that it gave birth to tiny, altricial young, as do living marsupials.

Whatever the reason for their demise, multituberculates were the most successful Mesozoic mammalian group, dispersing through most of the world. More than 40 Cretaceous genera in 12 families were listed by McKenna and Bell (1997), and at least nine more Cretaceous genera have been described since then (Kielan-Jaworowska et al., 2004). But multituberculates had declined sharply by the end of the Cretaceous: only five of the families and just four genera continued into the Paleocene. Remarkably, multituberculates radiated again in the Early Cenozoic to become abundant constituents of northern Paleocene faunas. About 30 Paleocene genera are recognized. Multituberculates diminished quickly after the Paleocene, only two families persisting into the Eocene, and just one genus beyond the early Eocene. By the end of the Eocene, the last of the multituberculates had disappeared.

EUTRICONODONTA

Until recently, triconodonts were viewed as including three families: Morganucodontidae (then including Megazostrodontidae), Triconodontidae, and Amphilestidae (Jenkins and Crompton, 1979). A fourth family, Austrotriconodontidae, was based on very fragmentary fossils from the Late Cretaceous of South America (Bonaparte, 1994). Two additional genera, *Dinnetherium* and *Jeholodens*, significant because of their excellent state of preservation, have proven difficult to accommodate within these four families. *Dinnetherium* has been considered an amphilestid (Jenkins and Schaff, 1988), a morganucodontid (Luo, 1994; Rougier et al., 1996a), and most recently a megazostrodontid (Kielan-Jaworowska et al., 2004).

Recent studies indicate that this traditional concept of triconodonts represents a grade of primitive mammals rather than a monophyletic group. Morganucodonts, as noted earlier, are now widely considered to be basal mammals, whereas amphilestids, triconodontids, and *Jeholodens* (together comprising the Eutriconodonta) share a more recent common ancestry with advanced therians and appear to be monophyletic (Rougier et al., 1996a; Luo et al., 2002). Eutriconodonts are derived compared to morganucodonts in having a pterygoid fossa on the medial side of the dentary and in lacking an angular process and a postdentary trough (which is associated with retention of postdentary bones; Kielan-Jaworowska et al., 2004).

Eutriconodonts (Fig. 4.13) were a very successful Mesozoic group, being known from Jurassic and Cretaceous strata and existing on all continents except Australia and Antarctica, but they left no Cenozoic descendants. Most taxa are known only from isolated teeth or jaw fragments, although several important skulls and skeletons have substantially improved our understanding of eutriconodonts over the past 20 years.

The triconodonts derive their name from their narrow molars with three principal longitudinally-aligned cusps, an arrangement similar to that in morganucodonts as well as the presumed cynodont ancestors of morganucodonts (and therefore considered primitive). A much smaller cingular cusp is present distally on both upper and lower molars. In most types the central cusp (designated "A" on the upper teeth, "a" on the lowers; Jenkins and Crompton, 1979) is most prominent, but in triconodontids the three cusps are of about equal height, giving the molar series a saw-tooth appearance. Both eutriconodonts and morganucodonts are further distinguished by having precise molar occlusion, as reflected by consistently developed shearing facets. The way the teeth occlude varies, however. In morganucodontids and triconodontids upper and lower molars occlude essentially one on one, whereas in megazostrodontids and amphilestids the main upper cusp occludes between the high cusps of two adjacent lower molars (Jenkins and Crompton, 1979). Where known, eutriconodonts have a straight cochlea, as in docodonts, multituberculates, and their cynodont ancestors, unlike the bent or coiled cochlea of higher therians (Rougier et al., 1996a).

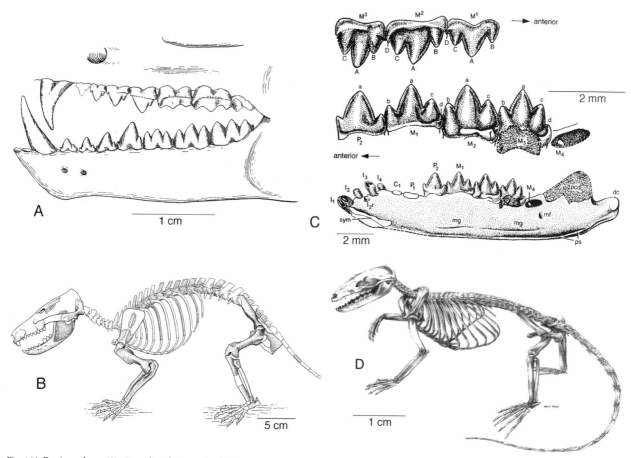

Fig. 4.13. Eutriconodonts: (A) triconodontid *Trioracodon*, left dentition (lateral view); (B) amphilestid *Gobiconodon*, skeleton; (C, D) *Jeholodens* right dentition and skeleton. (A from Simpson, 1928; B from Jenkins and Schaff, 1988; C, D from Ji et al., 1999.)

Triconodontids are known from the Upper Jurassic through Upper Cretaceous. Where known, the incisors are reduced in number compared to morganucodonts and there are either three or four premolars and from three to five molars, depending on the genus (Jenkins and Crompton, 1979). Austrotriconodontids differ in having a larger central cusp on the lower molars, and a bladelike arrangement on the uppers, with the highest cusp in front and the next three cusps successively lower (Bonaparte, 1994).

Amphilestids have been recorded from Middle Jurassic through Lower Cretaceous strata. The dentition is best known in the lower jaws of such genera as *Phascolotherium* and *Amphilestes*, in which the lower dental formula is 3 or 4.1.4.5 (Jenkins and Crompton, 1979). In these forms the cheek teeth are characterized by a large central cusp flanked by smaller cusps in front and back.

The most completely known amphilestid is *Gobiconodon* (Fig. 4.13B; now sometimes assigned to its own family Gobiconodontidae). It is among the most wide-ranging Cretaceous mammals, being known from the Early Cretaceous of Asia, western Europe, north Africa, and Montana (Jenkins and Schaff, 1988; Kielan-Jaworowska and Dashzeveg, 1998; Cifelli, 2000; Li et al., 2003; Sigogneau-Russell, 2003b). *Gobiconodon* is unusual in having enlarged, caninelike incisors, reduced canines, and replacement of the molars. The scapula is distinctly therian-like, with a large supraspinous fossa, and the humerus has a grooved trochlea for the ulna, unlike the ulnar condyle of morganucodonts. The forelimb skeleton of *Gobiconodon* is relatively much more robust than that of morganucodonts. The humerus has a prominent deltopectoral crest and is very broad distally, and as in morganucodonts it displays torsion (i.e., the articular ends are twisted relative to each other). The phalanges are relatively stout and the terminal phalanges are especially large, with prominent extensor and flexor processes. When found in therians these forelimb traits are typically associated with digging habits.

Gobiconodon and its close relative *Repenomamus* (Early Cretaceous of China) were large mammals for the Mesozoic, reaching at least the size of the opossum *Didelphis*. One species of *Repenomamus* had a skull more than 15 cm long and was about the size of the wolverine *Gulo*. It is the largest known Mesozoic mammal—apparently large enough to consume small dinosaurs, based on one individual, whose stomach contents consisted of a juvenile ceratopsian *Psittacosaurus* (Hu et al., 2005). Y. Wang et al. (2001) reported the presence of an ossified Meckel's cartilage in the jaws of *Repenomamus* and *Gobiconodon*, which they interpreted as an intermediate stage in the evolution of the definitive (i.e., fully) mammalian middle ear.

A virtually complete skeleton of a previously unknown eutriconodont, *Jeholodens*, was recently reported from Early Cretaceous beds of China (Fig. 4.13C,D; Ji et al., 1999). It is autapomorphic in having a reduced number of premolars compared to other eutriconodonts (dental formula 4.1.2.3/4.1.2.4). *Jeholodens* had generalized body proportions and a primitive sprawling posture. Surprisingly, however, it also had numerous derived anatomical features typical of therian mammals (some of which also occur in *Gobiconodon*). For instance, the scapula has a large supraspinous fossa, the coracoid is fused to the scapula, the humeral epicondyles are reduced, and there is an incipient trochlea for the ulna. As the dentition of eutriconodonts would seem to preclude them from direct ancestry to therians, some or all of these therian-like traits could be homoplasies.

The conflicting characters of eutriconodonts have led to instability of their phylogenetic position with respect to other mammals. Although they are now usually placed within crown-group Mammalia (e.g., Luo et al., 2002), this position is far from certain.

SYMMETRODONTS

Symmetrodonts were small shrew- to mouse-sized animals, known mainly from teeth and jaw fragments. They are considered to lie at or near the base of the therian radiation because they are the first mammals to show a nearly symmetrical triangular arrangement of the three main cusps on the upper and lower molariform teeth. The pattern varies from obtuse angled in primitive and some derived types to acute angled in the most derived. The resulting reversed triangles of the upper and lower molars form a series of oblique shearing edges. This configuration was a marked advance compared to the condition in eutriconodonts, and is regarded as an important step in the evolution of the tribosphenic molar. Mammals that possess a reversed-triangle molar pattern—symmetrodonts, eupantotheres, and therian mammals—are sometimes united in the higher taxon Holotheria. The concept of Holotheria is problematic, however, because its contents are controversial (variously including or excluding *Kuehneotherium*, monotremes, and eutriconodonts) and because of recent arguments that its defining feature, the reversed-triangle molar pattern, has evolved more than once. For these reasons, the name Holotheria was rejected in the most recent treatise on Mesozoic mammals (Kielan-Jaworowska et al., 2004).

Where known, the lower incisors and canine are small, and there are 7–11 postcanines (Fig. 4.14). The premolariform teeth are simple and have one main cusp with accessory cusps in front and behind. The lower molars primitively had a small talonid, which was lost in more derived forms. The dentary is long and slender and has no angular process, and the articular condyle is above the level of the toothrow. Postdentary bones were present in the most primitive forms (kuehneotheriids), but are absent in more derived types (Cassiliano and Clemens, 1979). Symmetrodonts are recorded from the latest Triassic (if kuehneotheriids are included; otherwise Early Jurassic) to Late Cretaceous, and are known from all continents except Australia and Antarctica.

The 20 or so known genera of symmetrodonts are classified in at least three and as many as seven families. The disparity has arisen because many genera are known only from isolated teeth, the affinities of which are often unclear. Like the traditional concept of triconodonts, symmetrodonts appear to be a paraphyletic assemblage. Late Triassic or earliest Jurassic *Kuehneotherium* (see Fig. 4.5) has been regarded as the most primitive symmetrodont, but recent studies suggest that it is actually not very closely related to later symmetrodonts (Rougier et al., 1996a; Luo et al., 2002; Kielan-Jaworowska et al., 2004).

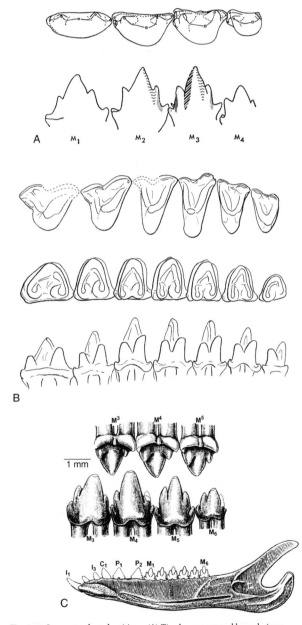

Fig. 4.14. Symmetrodont dentitions: (A) *Tinodon*, crown and buccal views of left lower molars; (B) *Spalacolestes*, left upper and right lower cheek teeth (crown and lingual views); (C) *Zhangheotherium*, left upper and lower molars and right lower jaw in medial view. (A from Crompton and Jenkins, 1967; B from Cifelli and Madsen, 1999; C from Hu et al., 1997.)

Most later symmetrodonts are classified in the families Amphidontidae, Tinodontidae, and Spalacotheriidae. The first two had obtuse-angled molars, whereas spalacotheriids had acute-angled molars in which the three main cusps form a tight triangle (superficially similar to the trigonid of tribosphenic therians) and the talonid has been greatly reduced or lost. Although these families have long been grouped in the Symmetrodonta based on their rotated cusps, Luo et al. (2002) recently suggested that obtuse-angled forms represent a grade of primitive mammals that bear no special relationship with acute angled forms. Most spalacotheriids had well-developed shearing surfaces on the front and back of the trigonid and on opposing surfaces of the upper molars (Cifelli and Madsen, 1999). Spalacotheriids had fewer premolars and more molars than *Kuehneotherium;* the dental formula was ?.1.3.6–7/3+.1.3.6–7; Cassiliano and Clemens, 1979; Sigogneau-Russell and Ensom, 1998). The specialized dentition of spalacotheriids suggests that they were not directly ancestral to advanced therians, although they are believed to be more closely related to them than are the obtuse-angled symmetrodonts.

Although most symmetrodonts are known only from teeth and jaws, a nearly complete skeleton of a new symmetrodont, *Zhangheotherium* (Fig. 4.14C), was reported from the same Early Cretaceous site in China that yielded the eutriconodont *Jeholodens* (Hu et al., 1997). *Zhangheotherium* had fewer postcanines than in most other symmetrodonts: the dental formula is 3.1.2.5/3.1.2.6. It also had an uncoiled cochlea, in contrast to therians. The postcranial skeleton shows many features intermediate between those of multituberculates or monotremes and those of therians, particularly in the shoulder girdle (e.g., presence of a supraspinous fossa, retention of a smaller interclavicle than in monotremes), elbow joint (incipient trochlea for ulna), pelvis, and femur. At the same time, these features indicate that *Zhangheotherium* had an abducted, sprawling forelimb posture more like that of monotremes than like the parasagittal posture of advanced therians. Cifelli and Madsen (1999) consider *Zhangheotherium* to be the most primitive known spalacotheriid symmetrodont. A closely allied new genus, *Maotherium,* was recently proposed, based on a skeleton preserving fur impressions from Jurassic/Cretaceous boundary strata in China (Rougier, Ji, and Novacek, 2003).

A variety of new symmetrodonts and eupantotheres has been found recently in the Late Cretaceous of Argentina, indicating both greater diversity and broader distribution of these groups than previously suspected (Bonaparte, 1990, 1994). At the same time, they have blurred the distinction between the two groups. Unfortunately, most are known only from isolated teeth or fragmentary dentitions, making interpretation tenuous. Other new genera of symmetrodonts have been reported from North Africa and several parts of Asia, but their precise relationships with other symmetrodonts are also uncertain (Sigogneau-Russell and Ensom, 1998).

EUPANTOTHERES

The eupantotheres (Dryolestoidea and Peramura of McKenna and Bell, 1997; Figs. 4.15, 4.16) occupy a structural and phylogenetic position essentially between symmetrodonts, on the one hand, and aegialodonts + therian mammals on the other. Among eupantotheres, dryolestoids (dryolestids and paurodontids), amphitheriids, and peramurans are successively more closely related to crown therians. This conclusion is founded on the anatomy of the teeth and jaws, which constitute almost all known fossils. Eupantotheres are derived compared to symmetrodonts in having wider upper than lower teeth (although they still lack the protocone of tribosphenic forms), larger talonids on the lower molars, and a well-developed angular process on the dentary (Figs. 4.16, 4.17; Simpson, 1928; Kraus, 1979). The coronoid process was high. The dentary was long and slender in most types, but usually shorter and deeper in paurodontids. The trigonid cusps of the lower molars and, to a lesser extent, the cusps of the upper molars are arranged in reversed acute triangles. Where the dental formula is known, most forms have four incisors, a canine, four simple premolariform teeth, and four to nine molariform teeth. Eupantotheres had a dentary-squamosal jaw joint. Although small accessory (postdentary) bones were still present in the Upper Jurassic paurodontid *Henkelotherium,* they did not participate in the jaw articulation (Krebs, 1991). A vestigial coronoid bone was present in the lower jaw of primitive dryolestids (Martin, 1999b), but postdentary bones were probably absent in the Early Cretaceous dryolestid *Crusafontia* (Kraus, 1979; Krebs, 1993).

Eupantotheres have been found on all continents except Antarctica and range from the Middle Jurassic through Late Cretaceous and possibly early Paleocene. Again, some of the best preserved eupantothere fossils come from the Late Jurassic Guimarota Mine in Portugal (Martin and Krebs, 2000). Most eupantotheres were small shrew- to mouse-sized mammals, although the South American *Mesungulatum* and *Vincelestes* were somewhat larger. Body size and dental morphology indicate that eupantotheres were insectivorous or carnivorous.

Eupantotheres have traditionally been assigned to the families Amphitheriidae, Dryolestidae, Paurodontidae, and Peramuridae (Kraus, 1979). Like triconodonts and symmetrodonts, eupantotheres now appear to be a paraphyletic structural grade. Middle Jurassic *Amphitherium* is the oldest well-known eupantothere (but still known only from the lower dentition). Although once considered the most primitive eupantothere, *Amphitherium* is now thought to be more derived than dryolestids in having larger talonids. It had 11 postcanines, interpreted as five premolars and six molars (Butler and Clemens, 2001). The molars had a well-formed trigonid followed by a much lower, bladelike talonid with one cusp. Amphitheriids are an important structural stage in the evolution of tribosphenic molars (Crompton, 1971). The Middle Jurassic amphitheriid *Palaeoxonodon* has

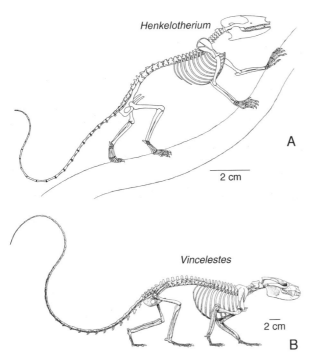

Fig. 4.15. Eupantothere skeletons: (A) *Henkelotherium;* (B) *Vincelestes.* (A from Krebs, 1991; B from Rougier, 1993.)

America, however, but from the Guimarota Mine in Portugal (the same site that produced the docodont *Haldanodon*). *Henkelotherium* is represented by a nearly complete skeleton (the only one known for eupantotheres), which has a more advanced pectoral girdle than in monotremes and morganucodonts (Fig. 4.15A). As in therians, only the scapula and clavicle are present, and there is a large supraspinous fossa. The pelvic girdle has a long iliac blade and retains epipubic bones. A long tail and sharp, curved claws suggest arboreal habits or at least climbing capability. *Henkelotherium* differs from the Morrison paurodontids in having more postcanines (dental formula is 4–5.1.4.5/4.1.4.7; Krebs, 1991).

Several new monotypic "eupantothere" families thought to be closely related to therian mammals have been described from the Cretaceous of South America, Asia, and Africa over the past decade or so, indicating that mammals dentally approaching therians were nearly cosmopolitan and rather diverse. Among the most important of these is *Vincelestes* (Fig. 4.15B) from the Early Cretaceous of Argentina. It has been considered to be related to tribosphenic therians because its upper molars have a lingual expansion

a somewhat better developed talonid on the lower molars. Nonetheless, a protocone is not yet present on the upper molars; they are triangular, with lingual paracone, less lingual metacone, and buccal parastyle, stylocone, and metastyle (Sigogneau-Russell, 2003a). *Palaeoxonodon* is sometimes considered a peramuran.

Dryolestidae were the most diverse eupantotheres, and have been found in the Late Jurassic of North America and Europe, the Early Cretaceous of Europe, and the Late Cretaceous and possibly early Paleocene of South America. The dryolestid *Leonardus* and several taxa assigned to closely allied separate families from the Los Alamitos Formation of Argentina are the latest known eupantotheres (Bonaparte, 1990, 1994), unless the early Paleocene *Peligrotherium* is actually a dryolestoid, as recently suggested by Gelfo and Pascual (2001) and Rougier, Novacek, et al. (2003). Dryolestids, such as *Krebsotherium* from Guimarota, usually have 12 postcanines in both upper and lower jaws, four premolars and eight molars (Fig. 4.17). The trigonids are tall and anteroposteriorly compressed ("closed"), with the metaconid almost as high as the protoconid, and the talonids are smaller than in other eupantotheres. The upper molars are transversely very wide. Dryolestids replaced all of their antemolar teeth, like placental mammals but unlike marsupials, which replace only dP3 (Martin, 1999b). Martin believes this trait implies that their reproduction was unlike that of marsupials.

Paurodontidae are known primarily from the Late Jurassic Morrison Formation of North America. Most have a short, robust mandible containing eight postcanines, and the molars have a shorter talonid than in *Amphitherium*. The best-known paurodontid, *Henkelotherium*, is not from North

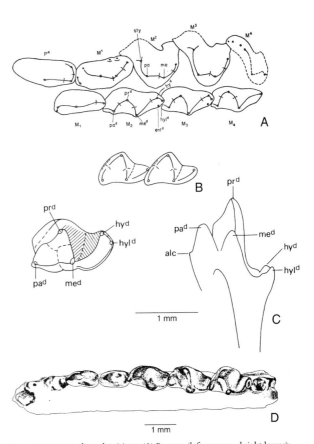

Fig. 4.16. Eupantothere dentitions: (A) *Peramus* (left upper and right lower); (B) *Amphitherium;* (C) *Kielantherium* (right molar in crown and lingual views); (D) *Arguimus* (right lower teeth). Key: alc, anterolingual cuspule; ent^d, entoconid; hy^d, hypoconid; hyl^d, hypoconulid; me, metacone; me^d, metaconid; pa, paracone; pa^d, paraconid; pr^d, protoconid; sty, stylocone. (A from Clemens, 1971; B from Clemens, 1970; C from Crompton and Kielan-Jaworowska, 1978; D from Dashzeveg, 1994.)

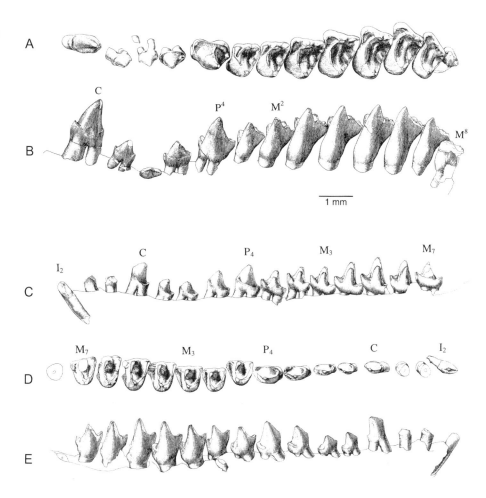

Fig. 4.17. Dentition of the dryolestid *Krebsotherium*: upper teeth in (A) crown and (B) lingual views; lower teeth in (C) lingual, (D) crown, and (E) buccal views. (From Martin, 1999b.)

where a protocone would be expected to form, and it has a partly coiled cochlea (about 270°) and several other derived conditions of the ear region found in marsupials and placentals (Hopson and Rougier, 1993; Rowe, 1993; Rougier et al., 1996a). But there is disagreement over whether *Vincelestes* really had an incipient protocone (Sigogneau-Russell, 1999). Moreover, it is dentally too derived (dental formula 4.1.2.3/2.1.2.3, with a small talonid present only on M_2, and the first premolar and last molar reduced), and probably too late in time (Neocomian), to have been directly ancestral to marsupials and placentals. *Vincelestes* may be a basal member of Zatheria, the higher taxon that also includes peramurid eupantotheres (McKenna and Bell, 1997; Martin, 2002).

Peramuridae are best known from the Late Jurassic *Peramus* of England (Sigogneau-Russell, 1999). *Peramus* is generally similar to *Amphitherium* (Fig. 4.16A,B), but differs in several respects that suggest that it represents a more derived stage in the evolution of tribosphenic molars. *Peramus* has only eight postcanines—both lower and upper—variously interpreted as four premolars and four molars or, now, usually as five premolars and three molars. Significantly, the talonids of M_{1-2} (assuming three molars) have an incipient basin bordered by a second cusp in addition to the one present in *Amphitherium*. The last premolars are higher crowned than the adjacent teeth. The upper molariform teeth are dominated by a large paracone followed by a much lower metacone, flanked by stylar cusps. The lingual border is somewhat inflated but there is still no protocone (Clemens and Mills, 1971; Sigogneau-Russell, 1999). Although these conditions suggest an approach toward the tribosphenic molars of marsupials and placentals, Dashzeveg and Kielan-Jaworowska (1984) concluded that peramurids may already be too derived to be directly ancestral to modern therians. Sigogneau-Russell (1999), however, considered permaurids to be structurally intermediate between symmetrodonts and tribosphenic mammals.

The Arguitheriidae and Arguimuridae, from the Early Cretaceous of Mongolia (Dashzeveg, 1994), have been considered to have more progressive molars than those of amphitheriids and peramurids. *Arguitherium* has a relatively open trigonid and an incipient talonid basin, whereas *Arguimus* (Fig. 4.16D) has well-developed trigonids and unbasined talonids bearing three cusps. Together with the recently described *Nanolestes* from the Late Jurassic–Early Cretaceous of Portugal, these forms are perhaps best regarded as peramurid-grade stem zatherians (Sigogneau-Russell, 1999; Martin, 2002).

TRIBOSPHENIC MAMMALS

Evolution of tribosphenic molars—uppers with three principal cusps arranged in a triangle (the trigon), the buccal paracone and metacone and a lingual protocone; and lowers with a three-cusped triangular trigonid and a basined talonid for occlusion with the protocone (see Fig. 2.2)—was one of the most important anatomical innovations in mammalian history. It laid the stage for the great diversity in dentitions of therian mammals. Tribosphenic molars can grind as well as shear food. With this dental structure, mammals were able to expand widely into omnivorous, herbivorous, and other specialized dietary niches.

Until quite recently it was assumed that tribosphenic molars evolved only once in mammals, first appearing in the Cretaceous Northern Hemisphere aegialodontids and eventually leading to marsupials and placentals. This established dogma has been challenged by the discovery of several apparent tribosphenic mammals from southern continents, which do not seem to fit the widely held model. In contrast to the conventional view, which assumes a monophyletic northern origin of all tribosphenic mammals, two competing hypotheses have been advanced: that tribosphenic mammals evolved first in the Southern Hemisphere, much earlier than previously thought (e.g., Flynn et al., 1999; Sigogneau-Russell et al., 2001; Woodburne et al., 2003), or that tribosphenic mammals evolved independently in the Northern and Southern Hemispheres (e.g., Luo, Cifelli, and Kielan-Jaworowska, 2001; Luo et al., 2002; Rauhut et al., 2002; Kielan-Jaworowska et al., 2004). According to the latter hypothesis, marsupials and placentals derive from the northern radiation, whereas monotremes are the only extant remnants of the southern radiation. It is not yet clear which, if either, hypothesis is correct.

Southern Tribosphenic Mammals: Australosphenida (Monotremes and Extinct Relatives)

Mesozoic Australosphenidans

Several very ancient tribosphenic mammals, all based on lower dentitions, have recently been reported from the Southern Hemisphere, complicating what had seemed to be a relatively straightforward record of the origin of tribosphenic mammals on northern continents. They are regarded as tribosphenic because their lower molars have three-cusped trigonids and the talonids have three peripheral cusps bordering a talonid basin, implying the presence of an upper molar protocone. There is currently no consensus on whether the tribosphenic condition in these mammals is homologous with that of Holarctic therians. Luo and colleagues (Luo, Cifelli, and Kielan-Jaworowska, 2001; Luo et al., 2002) offered the intriguing hypothesis that these animals—*Ambondro, Ausktribosphenos, Steropodon* (Fig. 4.18), and related forms including extant monotremes—belong to an independent, southern radiation of tribosphenic mammals, which they have called Australosphenida.

Ausktribosphenos is based on a lower jaw from the Early Cretaceous of Australia that shows a precociously hedgehoglike molar pattern (Rich et al., 1997), leading its describers

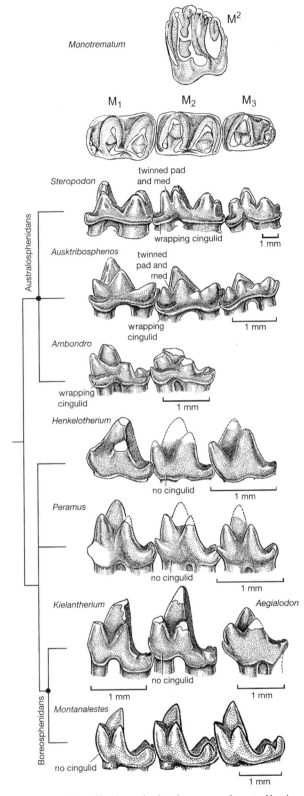

Fig. 4.18. Dentitions of basal australosphenidans, eupantotheres, and basal boreosphenidans. (From Luo, Cifelli, and Kielan-Jaworowska, 2001; Luo et al., 2002.)

to suggest that it could be one of the oldest known placentals. Others contend that *Ausktribosphenos* is not a placental because of the position of the mandibular foramen (in contact with the Meckelian groove), presence of a posteromedial depression presumably for postdentary bones, and different structure of the angular process than in therians—features suggesting that *Ausktribosphenos* might have been derived independently from symmetrodonts (Kielan-Jaworowska et al., 1998). The shape of the dentary and the last premolar are symmetrodont-like, but the molars are rather different from those of symmetrodonts. Subsequent discovery of a second ausktribosphenid jaw showed that an angular process is, in fact, present at the back of the dentary, as in therians (Rich et al., 1999). These authors continue to argue that, based on that feature and the presence of five premolars and three molars, including a submolariform last premolar and molars with a talonid basin and low trigonid, *Ausktribosphenos* was, after all, a primitive placental (Rich et al., 1999, 2002; Woodburne et al., 2003). One of the problems with this interpretation is that *Ausktribosphenos* does not resemble known Cretaceous eutherians; rather, it bears a (probably superficial) resemblance to middle and later Tertiary erinaceids. Recently, Rich, Flannery, et al. (2001) named another ausktribosphenid, *Bishops,* from the late Early Cretaceous of Australia. They classified it, too, as a placental, even though it has six premolars, in contrast to any known placental. The affinities of ausktribosphenids are puzzling, but few paleontologists have accepted placental ties. A relationship to monotremes seems more likely than to placentals (e.g., Sigogneau-Russell et al., 2001; Luo et al., 2002).

Even more unexpected was the discovery of a jaw with three tribosphenic molars from the Middle Jurassic (about 167 Ma) of Madagascar (Flynn et al., 1999). Named *Ambondro mahabo,* it is 25 million years older than *Tribotherium,* the oldest tribosphenic mammal known from the African continent. It is primitive in having an open molar trigonid, but already has a fully developed talonid basin.

The first South American australosphenidan was recently described from late Middle Jurassic strata of Argentina (Rauhut et al., 2002; Martin and Rauhut, 2005). *Asfaltomylos* is based on a mandibular fragment with several teeth, including three tribosphenic molars. The dentary is primitive in having an anteriorly placed mandibular foramen at the front of a postdentary trough, implying the presence of small postdentary bones. *Asfaltomylos* appears to be closely allied with, but slightly more primitive than, *Ambondro* and is therefore the most basal australosphenidan. The discovery of this South American form reinforces the notion of a widespread Mesozoic radiation of australosphenidans.

Steropodon, based on a jaw containing three molar teeth from the late Early Cretaceous Lightning Ridge local fauna of Australia, has been interpreted as a monotreme (Archer et al., 1985). Its molars have a mesiodistally compressed trigonid separated by a deep notch from a short lophlike talonid. The configuration is reminiscent of the tribosphenic cusp pattern of marsupials and placentals, but differs in the mode of wear, which suggests that *Steropodon* may not have had a protocone on the upper molars (which are unknown). Based on this resemblance, *Steropodon* and monotremes were considered aberrant therian mammals, representing a lineage separate from that leading to marsupials + placentals (Kielan-Jaworowska et al., 1987a).

Upon restudy of these fossils, however, Luo, Cifelli, and Kielan-Jaworowska (2001) found that *Steropodon* shares derived dental features (an anterolingual cingulid, a low trigonid, and a twinned paraconid and metaconid) with *Ausktribosphenos* and *Ambondro.* On this basis they postulated that these taxa, together with other monotremes, represent an endemic Gondwanan radiation of mammals, Australosphenida, that evolved tribosphenic molars independently from the Holarctic Boreosphenida. According to this hypothesis, the tribosphenic molar arose at least twice and apparently evolved earlier in the Southern Hemisphere than in Laurasia. It also suggests that monotremes are the sole survivors of the australosphenidan radiation, whereas marsupials and placentals represent the boreosphenidans.

This interpretation has been questioned by some authors. Sigogneau-Russell et al. (2001) suggested that the presence of an anterolingual cingulid is not restricted to australosphenidans, but rather is a widespread primitive feature in early mammals. In their view, *Ambondro* could be as closely related to boreosphenidans as it is to australosphenidans. They therefore postulated that the tribosphenic molar had a Gondwanan origin, spreading from there to other parts of Gondwana as well as to Laurasia (and diversifying differently in these two regions). Finally, Woodburne et al. (2003) considered *Ambondro, Asfaltomylos,* and *Bishops,* as well as *Ausktribosphenos,* to be eutherians, and suggested a single, presumably southern, origin of tribosphenic mammals. Additional fossils obviously will be important in testing these novel hypotheses.

Monotremes

The living platypus (Ornithorhynchidae) and echidnas (Tachyglossidae) of Australia and New Guinea are all that remain of the Monotremata, a group that diverged from other mammals during the Mesozoic. Because they are endothermic, have hair, suckle their young, and have several other mammalian synapomorphies, they are undisputed mammals. Nonetheless, their primitive nature and long separation from other living mammals are underscored by their unique characteristic of laying eggs. Monotremes are more primitive than other extant mammals (and resemble morganucodonts) in possessing cervical ribs, a therapsid-like shoulder girdle with an interclavicle and both coracoids, and a sprawling forelimb posture. The scapula lacks a spine and a supraspinous fossa. In addition, the skull retains a septomaxilla, as in cynodonts and morganucodonts, and the cochlea is only bent (not coiled as in marsupials and placentals) and lacks the bony laminae within the cochlear canal found in marsupials and placentals. The pelvic girdle is more therian-like than the shoulder girdle, and there are large epipubic bones in both sexes.

Superimposed on this basically primitive structure are several specialized (autapomorphous) features. The skull is usually described as birdlike, the rostrum being either narrow and beaklike (in Tachyglossidae) or shaped like a duck's bill (in Ornithorhynchidae). Adult monotremes are edentulous, but juvenile platypus retains milk "molars." The jugal bones are small or absent, and the zygomae are formed from processes of the maxilla and squamous temporal.

The primitive nature of monotremes indicates that they must have diverged from the mammalian stem quite early in the history of mammals, yet the fossil record of monotremes is extremely poor and is almost entirely limited to Australia. Possible derivation from peramurid or dryolestoid eupantotheres has been suggested, but no transitional fossils have been found. As discussed above, it is possible that monotremes instead are part of a Gondwanan tribosphenic radiation; but we are just getting our first glimpses of this supposed australosphenidan clade, and its origin remains obscure. Recently Woodburne (2003) advanced the hypothesis that monotremes are relicts of a Mesozoic, mainly Gondwanan radiation of *pre*tribosphenic mammals, unrelated to australosphenidans.

Pleistocene remains of both living families are known, as is a large Miocene echidna (*Zaglossus*), but they are similar to living forms and contribute little to understanding either the relationships or evolution of Monotremata. A Miocene ornithorhynchid, *Obdurodon*, was initially described from isolated teeth that resemble the transient teeth of juvenile platypus (Woodburne and Tedford, 1975). Both *Obdurodon* and juvenile *Ornithorhynchus* have vaguely bilophodont "molar" teeth comprised of two parts separated by a transverse valley. Each part consists of a high internal cusp joined by transverse crests to one or two labial cusps. Subsequent discoveries of *Obdurodon*, including a complete skull, confirm that it is a fossil platypus (Archer, Murray, et al., 1993). The realization that these Miocene teeth belonged to fossil ornithorhynchids proved critical to identification of still older monotremes—all of which are known only from teeth or jaws.

Since the mid-1980s several intriguing, much older specimens thought to be monotremes have come to light. Three of them are based on lower jaw fragments from the late Early Cretaceous (Aptian) of Australia. They have been placed in two different families. *Steropodon*, whose molar structure resembles that of *Obdurodon*, was discussed in the preceding section. A second *Steropodon*-like form, *Teinolophos*, was recently reported from the late Early Cretaceous Flat Rocks locality in Australia (Rich et al., 1999; Rich, Vickers-Rich, et al., 2001). It was initially interpreted as a eupantothere but is now considered to be a monotreme. Significantly, *Teinolophos* retains a trough on the medial surface of the mandible, a primitive feature reflecting the presence of accessory (postdentary) bones. The retention of postdentary bones, in turn, implies that the middle-ear ossicles were not yet separate from the lower jaw and, therefore, that the three-ossicle chain evolved independently at least twice in mammals (Rich et al., 2005; Martin and Luo, 2005).

Kollikodon, from Lightning Ridge, Australia (the same locality as *Steropodon*), was initially based on a lower jaw fragment with three teeth. The name, meaning "bun tooth," alludes to its peculiar, quadrate lower molars with four rounded and inflated cusps (perhaps adapted to feeding on crustaceans)—which reminded its describers (Flannery et al., 1995) of hot cross buns! The upper teeth are also multicusped and very bunodont (Kielan-Jaworowska et al., 2004). This dental morphology is strikingly different from that of other monotremes, and its monotreme affinity remains to be demonstrated. If it is indeed a monotreme, it indicates that two quite divergent lineages already existed in the Early Cretaceous.

The only known Early Tertiary monotreme, *Monotrematum* (Fig. 4.18), comes from the early Paleocene ("Peligran") of southern Argentina (Pascual et al., 1992, 2002). It is also the only monotreme known from outside the Australian region. Although only two teeth have been found, it can be securely identified as a platypus based on its close resemblance to *Obdurodon*. Presence of a monotreme in South America indicates that the present-day restriction of the group to Australia is but a remnant of a once much wider geographic distribution.

Northern Tribosphenic Mammals: Boreosphenida (Metatherians, Eutherians, and Related Therians)

Aegialodontia

From the late Early Cretaceous come two closely allied (perhaps synonymous) genera that have been widely considered to lie close to the stem of Holarctic tribosphenic mammals, or Boreosphenida (essentially =Tribosphenida of McKenna and Bell, 1997, or Theria + Aegialodontia of general usage), or at least to represent a critical stage in the origin of tribosphenic molars (Crompton, 1971). Included here are *Aegialodon* (Lower Cretaceous Wealden beds of England) and *Kielantherium* (late Early or early Late Cretaceous from Khovboor, Mongolia). *Kielantherium* (Fig. 4.16C) is advanced beyond such eupantotheres as *Arguimus* (Fig. 4.16D) in having a talonid basin and an entocristid. Recently Sigogneau-Russell et al. (2001) described a third, even older aegialodont genus, *Tribactonodon*, from the earliest Cretaceous (Berriasian) Purbeck Group of England. *Tribactonodon* is based on a lower molar with a tall trigonid and a long, narrow talonid with three cusps.

Although aegialodont upper teeth are unknown, the structure and wear pattern of their lower molars indicate that a protocone was present on the upper molars, making aegialodonts the most primitive known truly tribosphenic mammals from Laurasia. The lower jaw held four or possibly five premolars and four molars (Dashzeveg and Kielan-Jaworowska, 1984). Disagreements persist over the correct homologies of the postcanines and, therefore, the dental formula of Aegialodontia and Peramura. Kielan-Jaworowska (1992) argued that Aegialodontia was ancestral to Metatheria but not Eutheria, but subsequently aegialodonts have been depicted as the sister taxon of Theria (Metatheria + Eutheria; e.g., Luo et al., 2002).

"Tribotheres"

Tribotheria was proposed to encompass primitive tribosphenic forms whose teeth lack diagnostic traits of Metatheria or Eutheria (Butler, 1978; Sigogneau-Russell, 1991). Consequently "Tribotheria" is an evolutionary grade rather than a clade. Nonetheless, the term "tribothere" remains a useful informal name for these fossils—most known only from isolated teeth or fragmentary jaws—which are often difficult to assign and are otherwise given the cumbersome moniker "therians of metatherian-eutherian grade." Most are probably basal members of Boreosphenida; however, it may eventually be possible to assign many of them to either Metatheria or Eutheria when they are more completely known. A variety of "tribotheres" have been described from the Cretaceous of North America, including *Comanchea, Picopsis, Potamotelses, Kermackia, Pappotherium*, and *Holoclemensia* (Fig. 4.19), the last two originally considered to be basal eutherian and metatherian, respectively. Substantially older tribotheres are now known from Africa, and their antiquity has raised significant questions regarding the time of origin of tribosphenic molars. *Tribotherium*, based on an upper molar from the earliest Cretaceous of Morocco, pushed back the first occurrence of mammals with tribosphenic molars by 30 million years (Sigogneau-Russell, 1991).

Although the three groups of living mammals (monotremes, marsupials, and placentals) are differentiated primarily by reproductive strategy, it has long been held that differences in dental pattern and morphology also distinguish primitive marsupials and placentals and can be used to assign fossils to one group or the other. With the discovery of an increasing number and variety of Cretaceous mammalian remains, however, it has become clear that the "defining" dental characters were acquired in mosaic fashion. Consequently, it is often not possible to identify a fossil taxon as definitively metatherian or eutherian based on teeth alone, and it is even possible that some fossils represent tribosphenic clades that diverged before the metatherian-eutherian dichotomy. Once the dichotomy occurred, probably during or before the Early Cretaceous, nearly all higher therians seem to lie closer to either Metatheria or Eutheria; but this inference has often been based on characters that are not demonstrably derived and may well be primitive. Experts still disagree on whether certain early forms represent true marsupials or other metatherian clades and on how to recognize the oldest true placentals.

MESOZOIC MAMMALS OF UNCERTAIN AFFINITY

A few Mesozoic mammals are so unusual that they cannot be placed with confidence in any of the established higher taxa, and their broader relationships remain uncertain. One

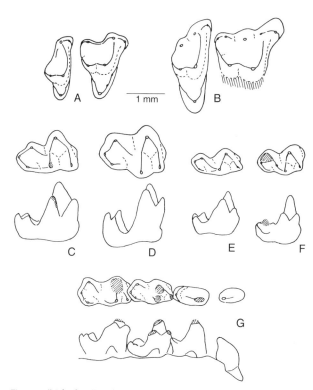

Fig. 4.19. "Tribothere" teeth: upper molars of (A) *Pappotherium;* (B) *Holoclemensia*. Left lower molars of (C) *Pappotherium;* (D) *Holoclemensia;* (E) *Kermackia;* (F) *Trinititherium*. (G) Lower teeth of *Slaughteria*. C–G in crown and lingual views. (From Butler, 1978.)

such form is *Shuotherium*, from the Middle and Upper Jurassic of China and England, whose teeth are superficially similar to the tribosphenic teeth of therians. Its lower molars have a trigonid and a "talonid," but the talonid-like structure is attached to the *front* of the trigonid (Chow and Rich, 1982). Such a "pseudotalonid" is present also in the derived docodont *Simpsonodon* (Kermack et al., 1987). Upper molars from the Middle Jurassic of England and the Late Jurassic of China that possibly represent *Shuotherium* are tricuspid, superficially like those of tribosphenic mammals. But they differ in wear pattern and crown morphology (the cusps are paracone, metacone, and a lingual "pseudoprotocone") from those of advanced tribosphenic therians (Sigogneau-Russell, 1998; Wang et al., 1998). Furthermore, they differ from each other in overall shape as well as size of the stylar shelf, suggesting that both may not belong to the same genus. Various authors have proposed that *Shuotherium* is a highly specialized docodont, an aberrant symmetrodont, or (most recently) the sister taxon of the Australosphenida (Kermack et al., 1987; Kielan-Jaworowska, 1992; Luo et al., 2002). Kielan-Jaworowska et al. (2002) postulated that the anterolingual cingulid characteristic of australosphenidans was the precursor of the pseudotalonid in *Shuotherium*.

Even stranger are the Gondwanatheria, a rare and bizarre group apparently restricted to the Southern Hemisphere. Gondwanatheres are known from the Upper Cretaceous–Paleocene of South America, the middle Eocene of Antarctica, and the Upper Cretaceous of Madagascar and India

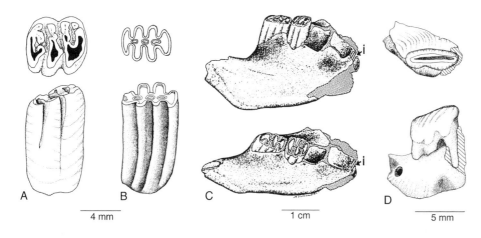

Fig. 4.20. Gondwanathere dentitions: (A) *Gondwanatherium* molar; (B) *Sudamerica* molar; (C) *Sudamerica* jaw (i indicates posterior end of evergrowing incisor); (D) jaw initially referred to *Ferugliotherium*, now considered to belong to an undetermined multituberculate. (A and B from Bonaparte, 1990; C from Pascual et al., 1999; D from Kielan-Jaworowska and Bonaparte, 1996.)

(Bonaparte, 1990; Krause et al., 1992, 1997; Krause and Bonaparte, 1993; Reguero et al., 2002). Four genera have been described: *Gondwanatherium* and *Ferugliotherium* (Upper Cretaceous, Patagonia), *Lavanify* (Upper Cretaceous, Madagascar), and *Sudamerica* (Paleocene, Patagonia; Fig. 4.20). A dentary with a large, procumbent incisor and five single-rooted, hypsodont cheek teeth from the Cretaceous of Tanzania might represent the first African gondwanathere (Krause et al., 2003). Gondwanatheres include the oldest known hypsodont mammals, and at least one, *Sudamerica*, had an ever-growing incisor. This trait suggests an abrasive diet, and perhaps fossorial or semiaquatic habits like those of beavers (Koenigswald et al., 1999). The recent discovery of silicified phytoliths representing several kinds of grasses in the Late Cretaceous of India suggests the possibility that gondwanatheres were the earliest mammalian grazers (Prasad et al., 2005).

Gondwanatheres were first identified from isolated molars that are so different from those of contemporaneous mammals that their broader attribution was (and is) uncertain. The molars are weakly bi- or trilobate, with three transverse ridges separated by furrows—a pattern superficially suggestive of some derived rodent teeth. Additional isolated teeth and a dentary fragment containing P_4 (questionably belonging to *Ferugliotherium*; Fig. 4.19D) suggested affinity with multituberculates (Krause et al., 1992; Kielan-Jaworowska and Bonaparte, 1996): the incisors, like those of multituberculates, have a limited band of enamel, and P_4 is bladelike, with oblique ridges, as in multituberculates. In addition, microwear on the occlusal surface of the molars indicates that the lower jaw moved palinally (posteriorly), as in multituberculates generally. *Ferugliotherium* has brachydont molars, whereas other known gondwanatheres have very hypsodont molars. The enamel microstructure, consisting mainly of radial enamel and interprismatic matrix, provides little insight on the relationships of gondwanatheres (Koenigswald et al., 1999). *Sudamerica*, from the early Paleocene ("Peligran") of Argentina, is the only gondwanathere known to have survived into the Cenozoic. A newly found lower jaw of *Sudamerica* (Fig. 4.19C), however, has four hypsodont molariform teeth but no bladelike tooth, which casts doubt on the multituberculate affinities of gondwanatheres and the attribution of the supposed *Ferugliotherium* P_4 (Pascual et al., 1999). Indeed, Kielan-Jaworowska et al. (2004) now regard the dentary with the bladelike P_4 to be a multituberculate of uncertain affinity rather than a gondwanathere. A possible sudamericid has recently been reported from the middle Eocene La Meseta Formation of Seymour Island, Antarctica (Reguero et al., 2002). The unusual southern distribution of gondwanatheres suggests that they represent a very ancient mammalian clade. Additional evidence that will resolve gondwanathere relationships is eagerly awaited.

Finally, there is the newly described *Fruitafossor*, a mouse-sized mammal from the Late Jurassic Morrison Formation of Colorado (Luo and Wible, 2005). *Fruitafossor* is unique among Mesozoic mammals in having simple, tubular teeth, the molars apparently open-rooted (ever-growing), thus superficially resembling the dentition of armadillos. The dental formula is ?.1.3.3/3.1.3.3. The skeleton is robust, the forelimbs converging toward those of living monotremes and moles (Talpidae). The humerus is short and very wide, the olecranon is elongate and medially inflected, and the manus is broad, with four digits bearing wide, flat terminal phalanges. *Fruitafossor* is reported to have xenarthrous articulations between the lumbar vertebrae, a specialization otherwise known only in the placental order Xenarthra; but this trait would clearly have evolved convergently, as there is otherwise no evidence of relationship to Xenarthra. The anatomy of *Fruitafossor* suggests that it was a specialized fossorial mammal that fed on insects and other small invertebrates, but its relationships remain uncertain. Although comparable dental reduction in extant mammals is often associated with myrmecophagy, ants and termites have not been reported from before the Cretaceous, strongly implying that social insects were not the diet of this Jurassic mammal. *Fruitafossor* offers a glimpse of even greater diversity than we have come to expect among Mesozoic mammals.

5

Metatheria
Marsupials and Their Relatives

THE TERM METATHERIA IS USED to unite marsupials and their presumed extinct relatives, including the Deltatheroida and the Asiadelphia. Although generally overshadowed by placental mammals, marsupials persist today, and during the Cenozoic they underwent diverse radiations in South America and Australia, where they still predominate. The oldest known metatherians are from the Cretaceous. Deltatheroidans are a largely Asian clade restricted to the Cretaceous. They are generally considered to be the sister group of Marsupialia, or of a clade of marsupials and other primitive metatherians. Alternatively they could be the sister taxon of all living therians (i.e., eutherians + marsupials; Luo et al., 2002). Asiadelphians, based primarily on the Asian genus *Asiatherium,* are variously considered to be another branch of metatherians, the sister group of Marsupialia, or a primitive clade of marsupials. All of these groups primitively share the postcanine dental formula of three premolars and four molars and have upper molars with a wide stylar shelf and one or more stylar cusps, but no hypocone. Some later metatherians, however, especially Australian clades, evolved a cusp in the position of a hypocone (probably a displaced metaconule).

The fossil record indicates that metatherian and eutherian mammals had already diverged by early in the Cretaceous (Cifelli, 1993a; Eaton, 1993). Some molecular studies suggest an even earlier split, in the Jurassic. Most of the Mesozoic metatherian clades became extinct by the end of the Cretaceous. A number of recent discoveries have greatly expanded our knowledge of these primitive metatherians.

Based on some molecular studies, it has been suggested that marsupials are the sister taxon of monotremes (e.g., Janke et al., 1997, 2002). These authors resurrected

W. K. Gregory's abandoned term Marsupionta (subsequently shown to be based on shared primitive characters) for this supposed clade. However, as detailed by Luo et al. (2002), substantial anatomical evidence and even most molecular data indicate that metatherians are more closely related to eutherians than to any other mammals.

Extant metatherians (marsupials) differ from eutherians in many ways, the most obvious of which concern reproductive anatomy and development: the female reproductive tract is bifid; the gestation period is very short (8–42 days; Moeller, 1990); the young are altricial, and organ systems and limbs may be only partly developed at birth; most species have a pouch or marsupium, where the young complete their development (this external "womb" explains Linnaeus's name for the opossum *Didelphis,* "double womb," although it might also be a reference to the double uterus); and epipubic bones project forward from the pubic bones (associated with locomotion and/or support of the pouch or of the developing young). Since, apart from epipubic bones (which are also known to be present in monotremes, multituberculates, and basal eutherians), these traits are not preserved in fossils, how can metatherians be recognized in the fossil record? Fortunately, there are several characters of the skull and teeth that separate most early marsupials from placentals. They include an auditory bulla composed primarily of the alisphenoid; large openings, or vacuities, in the palate; an inflected angular process of the dentary (possibly a retained primitive trait); more upper than lower incisors; simple tribosphenic upper molars lacking a hypocone, but with a wide stylar shelf bearing multiple cusps; three simple premolars followed by four molars, the last premolar being the only tooth replaced; and lower molars often with an unreduced paraconid and twinned hypoconulid and entoconid (see Figs. 2.1, 5.1). But, as in other evolutionary transitions, anatomical features were acquired sequentially, and the most primitive forms, transitional between "tribotheres" and metatherians, lack some of these "diagnostic" traits. For example, some Cretaceous teeth that are otherwise marsupial-like have poorly developed stylar cusps or lack twinning of the hypoconulid and entoconid.

Dental anatomy plays a prominent role in identifying the oldest metatherians. The disposition of stylar cusps on the upper molars has been considered to be particularly significant. The stylar cusps of metatherians are typically designated by the letters A through E, starting at the front of the molar (Fig. 5.1). As in eutherians, cusp A is identified as the parastyle, cusp C the mesostyle, and cusp E the metastyle. Cusp B, which is joined to the paracone by the paracrista, is also called the stylocone, whereas cusp D has no other designation. It should be realized, however, that there is little evidence that these cusps are homologous with the stylar cusps of eutherians; hence, these designations are primarily topographic. Indeed, some of these cusps probably arose multiple times within marsupials. Cusps B and D are often larger than the others (e.g., in didelphoids), and not all cusps are present in most forms.

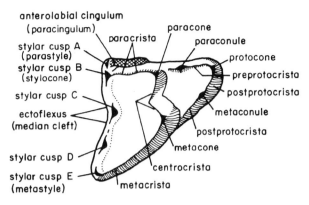

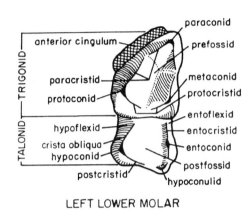

Fig. 5.1. Terminology for metatherian molars. (From Marshall, 1987.)

Recent discoveries suggest that postcranial features, particularly the carpus and tarsus, may also be important in distinguishing the earliest metatherians and eutherians (Luo et al., 2003).

Many different classifications of marsupials have been proposed over the past 20 years or so, but none seems to have achieved the level of consensus. (The classification used here is presented in Table 5.1.) There is even substantial disagreement as to the higher-level assignments of many non-Australian families. But most current authorities agree that marsupials comprise multiple ordinal-level taxa, in contrast to the single order Marsupialia widely used a generation ago. In addition, there is agreement that, except for some Cretaceous forms (sometimes placed in a separate clade, Alphadelphia), there is a basic dichotomy between New World and Australian marsupials (cohorts or magnorders Ameridelphia vs. Australidelphia; Szalay, 1982, 1994). Figure 5.2 depicts one interpretation of relationships that illustrates this dichotomy, but it differs in many details from the classification used in this chapter. It is also evident that Australian marsupials evolved from an American form, most likely a microbiothere. A modified version of McKenna and Bell's (1997) classification is used as a framework here. In this scheme, Ameridelphia includes didelphid opossums and other didelphimorphs, as well as paucituberculates and

Table 5.1. Classification of Metatheria

Infraclass METATHERIA
    †*Holoclemensia,* †*Sinodelphys*
    Order †DELTATHEROIDA
    Order †ASIADELPHIA
    Cohort MARSUPIALIA
        †*Kokopellia,*[1] †*Iugomortiferum,*[1]
        †*Anchistodelphys*[1]
    Magnorder AMERIDELPHIA
        Order DIDELPHIMORPHIA
            †**Peradectidae**[2]
            Didelphidae
            †Sparassocynidae
        ?Order DIDELPHIMORPHIA
            †**Pediomyidae**[2]
            †**Stagodontidae**[2]
            †**Protodidelphinae**
        Order PAUCITUBERCULATA
            Superfamily †Caroloameghinioidea
                †Glasbiidae
                †**Caroloameghiniidae**
            Superfamily Caenolestoidea
                †**Sternbergiidae**
                Caenolestidae
                †Palaeothentidae
                †Abderitidae
            Superfamily †Polydolopoidea
                †**Sillustaniidae**
                †**Polydolopidae**
                †**Prepidolopidae**
                †**Bonapartheriidae**
            Superfamily †Argyrolagoidea
                †Argyrolagidae
                †**Groeberiidae**
                †Patagoniidae
        Order †SPARASSODONTA
            †**Mayulestidae**
            †**Hathliacynidae**
            †**Borhyaenidae**
            †**Proborhyaenidae**
            †Hondadelphidae
    Magnorder AUSTRALIDELPHIA
        Djarthia
        Superorder MICROBIOTHERIA
            Microbiotheriidae
        Superorder EOMETATHERIA
            Order NOTORYCTEMORPHIA
            Grandorder DASYUROMORPHIA
            Grandorder SYNDACTYLI
                Order PERAMELIA
                Order DIPROTODONTIA

Notes: Modified after McKenna and Bell (1997). The dagger (†) denotes extinct taxa. Families and genera in boldface in this table are known from the Paleocene or Eocene.

[1] Could lie just outside Marsupialia.
[2] May lie outside Ameridelphia; Case et al. (2004) separated these families together with Caroloameghiniidae in the Alphadelphia, comparable in rank to the magnorders used here. McKenna and Bell (1997) included Peradectidae within Didelphidae.

the extinct carnivorous borhyaenids and their relatives; Australidelphia comprises South American microbiotheres and all Australian marsupials. Case et al. (2004), mostly following Marshall et al. (1990), restricted Paucituberculata to the caenolestoids, assigning all the other paucituberculate taxa in Table 5.1 to a separate order Polydolopimorphia.

BASAL METATHERIANS

Little over a decade ago, almost nothing was known of the earliest metatherians. Since then, a flurry of new fossil discoveries in North and South America and Asia has expanded our knowledge so rapidly that our notions of primitive metatherians and their phylogeny have been in a state of flux. Several of the recently found Cretaceous forms seem to occupy important roles in the evolution of Metatheria.

One of the most intriguing new fossils is mouse-sized *Sinodelphys* (Fig. 5.3), the oldest known mammal that is probably closely related to Metatheria. It is based on a skeleton recently found in the 125-million-year-old Lower Cretaceous Yixian Formation of Liaoning Province, China (Luo et al., 2003), the same deposits that produced skeletons of *Jeholodens* and *Zhangheotherium*. *Sinodelphys* differs from other metatherians in dental formula (4.1.4.4/4.1.4.3)—having an equal number of upper and lower incisors, one more premolar, and one less lower molar—and in lacking an inflected mandibular angle. But it has several derived features characteristic of marsupials, including the shape of the upper incisors, closely approximated (but not twinned) hypoconulid and entoconid, and modifications of the wrist (enlarged scaphoid, hamate, and triquetrum) and ankle (broad navicular, oblique calcaneocuboid joint) that are indicative of arboreal or scansorial habits. Based on this combination of features, Luo et al. (2003) considered *Sinodelphys* to be a basal metatherian more primitive than *Deltatheridium;* but its precise phylogenetic position relative to other metatherians is uncertain.

Also of note (though not new) is *Holoclemensia* (see Fig. 4.20), which was established on isolated teeth from the Lower Cretaceous (Albian) Trinity Formation of Texas. Although initially considered a basal metatherian, it was subsequently classified as a "tribothere" (Butler, 1978; McKenna and Bell, 1997), but was once again placed at the base of Metatheria by Luo et al. (2003). Its upper molars have a larger paracone than metacone and a wide stylar shelf with several stylar cusps, including an enlarged cusp C. The lower molars have a tall protoconid, somewhat reduced paraconid, and closely approximated hypoconulid and entoconid (Slaughter, 1971). Cifelli (1993b) considered *Holoclemensia* to represent a structural stage leading to marsupials (see Fig. 5.7).

Deltatheroida

Deltatheroidans were primitive, tribosphenic marsupial-like therians, mainly from the Late Cretaceous of central Asia. There are two families, Deltatheridiidae and Deltatheroididae. Besides having a marsupial-like postcanine formula (three premolars and four molars), *Deltatheridium* (Fig. 5.4), the best-known deltatheroidan, resembles marsupials in replacing only the last premolar (P^3_3), and in having an inflected mandibular angle (Rougier et al., 1998). The upper molars have a very wide stylar shelf with several cusps. Unlike primitive marsupials, however, different stylar cusps are

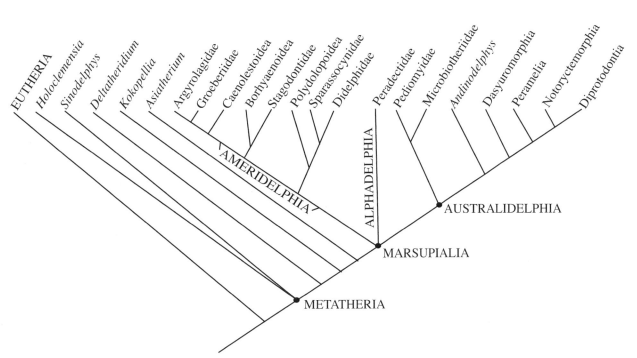

Fig. 5.2. Cladogram of metatherian and marsupial relationships, modified after Marshall et al. (1990) and Luo et al. (2002). This arrangement differs in several ways from the one used in this chapter. In particular, in this chapter, Peradectidae and Pediomyidae are included in Ameridelphia and Stagodontidae is not considered to be closely related to borhyaenoids (see Table 5.1).

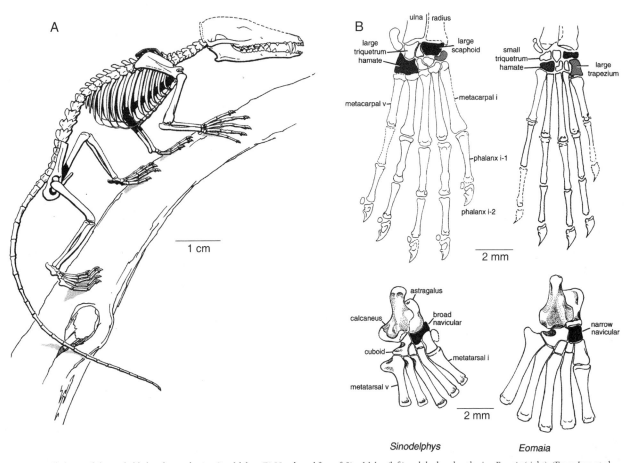

Fig. 5.3. (A) Skeleton of the probable basal metatherian *Sinodelphys*. (B) Hands and feet of *Sinodelphys* (left) and the basal eutherian *Eomaia* (right). (From Luo et al., 2003.)

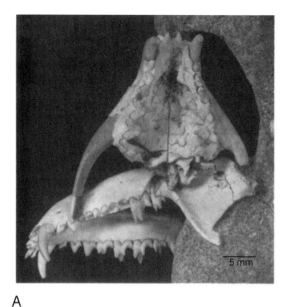
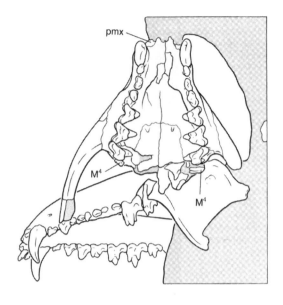
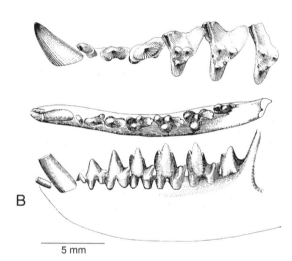

Fig. 5.4. Primitive metatherian *Deltatheridium:* (A) skull and mandible; (B) left upper and lower dentition. Key: pmx, premaxilla. (A from Rougier et al., 1998; B from Kielan-Jaworowska, 1975b.)

emphasized (A, B, B_1, and E; Kielan-Jaworowska, 1975b), the molar talonids are narrow, the hypoconulid and entoconid are not twinned, and the last molar (M^4_4) is vestigial. In addition, there is one less incisor above and below (four incisors over three incisors) than in primitive marsupials. These features indicate that *Deltatheridium* is not a marsupial in the strict sense. Deltatheroidans are usually considered to represent either an independent clade of metatherians (together with Asiadelphia, composing Holarctidelphia; Szalay, 1994), or alternatively to lie just outside of crown therians (Metatheria + Eutheria; e.g., Luo et al., 2002).

Asiadelphia

This higher taxon is based principally on *Asiatherium* (Fig. 5.5), a mouse-sized animal known from a skull and articulated skeleton from the Late Cretaceous of the Gobi Desert, Mongolia (Szalay and Trofimov, 1996). Like deltatheroidans, *Asiatherium* resembles marsupials in having three premolars and four molars and a somewhat inflected mandibular angle. It is more marsupial-like than deltatheroidans in having paraconids lower than metaconids and twinned hypoconulid-entoconid cusps on the lower molars. However, the stylar shelf of the upper molars is narrower and the stylar cusps much weaker than in primitive marsupials. The upper molars further differ from those of marsupials in having expanded precingula and postcingula, the latter resembling an incipient hypocone shelf. Thus, *Asiatherium* seems to be related to marsupials, but, like deltatheroidans, it probably belongs to a separate clade of metatherians. The skeleton of *Asiatherium* is similar to that of generalized terrestrial therians. Epipubic bones are present.

PRIMITIVE MARSUPIALS

From the late Early Cretaceous (Albian, about 98 Ma) of Utah comes another primitive metatherian, *Kokopellia* (Fig. 5.6A), known from a well-preserved lower jaw (Cifelli, 1993a; Cifelli and Muizon, 1997). It resembles marsupials in dental formula (three premolars, four molars) and various

Metatheria: Marsupials and Their Relatives 77

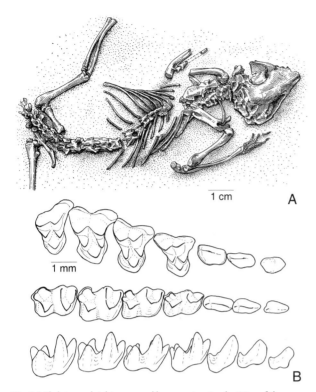

Fig. 5.5. Skeleton and right upper and lower postcanine dentition of the primitive metatherian *Asiatherium*. (From Szalay and Trofimov, 1996.)

dental traits (simple premolars, P_3 tall, first molar smaller than the others) but lacks a twinned hypoconulid-entoconid, often regarded as a key synapomorphy of marsupials. Here again, as in the transition to Mammalia, anatomical traits that have been widely used to identify a clade—in this case, Metatheria—evidently were acquired in mosaic fashion. Ironically, this staggered acquisition of defining characters has resulted in a blurring of the taxonomic boundary as the record has improved. The somewhat younger (Campanian) *Anchistodelphys* (Fig. 5.7) and *Iugomortiferum* are also very similar to marsupials but differ slightly in stylar cusp development (weaker cusps, with cusp D absent or variable; Cifelli, 1990b). These taxa either represent stem marsupials or the closest sister taxa to Marsupialia.

A diversity of mouse-sized Late Cretaceous forms that appear to be true marsupials, at least based on teeth, has been described since the late 1980s. These dental taxa, most known only from isolated teeth, have diagnostic marsupial traits, such as twinned entoconid-hypoconulid and several stylar cusps including the consistent presence of stylar cusp D (Cifelli, 1993a; Eaton, 1993). Among them, *Aenigmadelphys*, *Protalphadon*, and especially *Iqualadelphis* (Fig. 5.7), dating from the Aquilan and Judithian land-mammal ages (=Campanian), have been suggested to be the most primitive known marsupial genera, based partly on the absence or variable presence of stylar cusp C (e.g., Clemens, 1979; Cifelli, 1990a; Marshall et al., 1990; Cifelli and Johanson, 1994). *Protalphadon* (Fig. 5.8D) is based on species previously included in *Alphadon* but later separated from it on the basis of the absence or weakness of stylar cusp C. Other rela-

tively large species formerly assigned to *Alphadon* are now placed in *Turgidodon*. These Late Cretaceous genera are variously allocated to Didelphidae, Peradectidae, or basal positions in Didelphimorphia or Marsupialia.

Despite the multiplicity of generic names and the instability of their higher taxonomic positions, all these early marsupial genera are dentally very similar and are characterized by primitive "didelphoid" or "didelphimorph" molars: the uppers are triangular with the paracone usually larger than the metacone, distinct conules, and a wide stylar shelf with several stylar cusps (usually A, B, and D, or A–D); the lowers have tall protoconids, paraconids only slightly lower than metaconids, and basined talonids with closely approximated hypoconulid and entoconid. Generic distinctions are based on minor variations in expression of the stylar cusps, presence or absence of cusp C, relative height of the trigonid and width of the talonid, and overall dental proportions.

Alphadon or a closely allied form (Figs. 5.7, 5.8B,C) has been considered to be the source of all post-Cretaceous marsupials (e.g., Clemens, 1966, 1979; Fox, 1987; Marshall

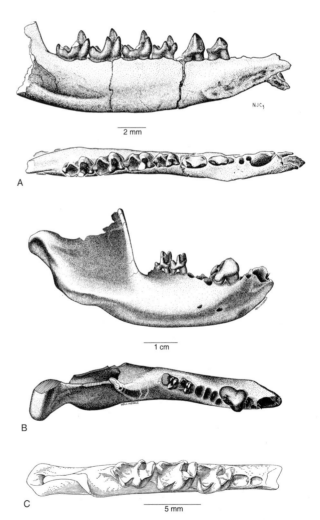

Fig. 5.6. Cretaceous basal metatherian dentitions: (A) *Kokopellia*, left lower jaw; (B) stagodontid *Didelphodon*, right lower jaw; (C) *Pediomys*, right lower jaw with M_{2-4}. (A from Cifelli, 1993a; B from Fox and Naylor, 1986; C from Clemens, 1966.)

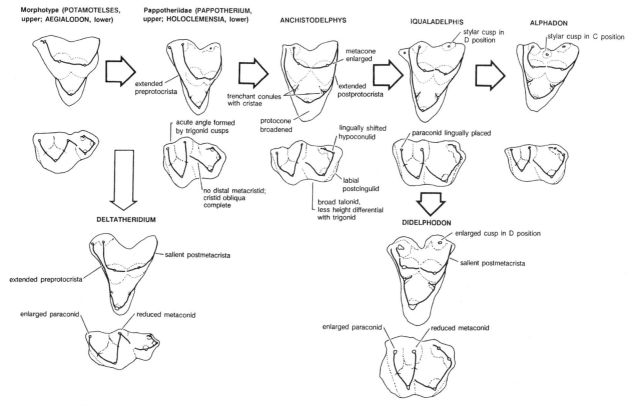

Fig. 5.7. Evolution of metatherian molars. (From Cifelli, 1993b.)

et al., 1990). It has conventionally been allocated to the Didelphidae (e.g., Clemens, 1966; Fox, 1979a, 1987; McKenna and Bell, 1997), an attribution that reinforced the view of didelphids as the most primitive marsupials and the ultimate source of all later forms. The classification of *Alphadon* has been controversial—other authors assign it to the Peradectidae (e.g., Cifelli, 1990a; Marshall et al., 1990) or to an expanded Pediomyidae (Szalay, 1994)—but there is little question that it represents one of the most primitive marsupials. Although forms with slightly more primitive teeth are now known, *Alphadon* is still a good model for the ancestral molar structure of most post-Cretaceous marsupials. Consequently, Peradectidae (or Didelphidae) is probably paraphyletic. During the Early Tertiary, marsupials potentially derived from these Cretaceous forms dispersed to all continents, including Antarctica.

Early descendants of these Cretaceous peradectids (or didelphids) include the mainly North American Cretaceous families Stagodontidae and Pediomyidae (sensu stricto), both characterized by somewhat derived teeth relative to other Cretaceous marsupials (e.g., Clemens, 1979; Fox, 1987). Pediomyids (Figs. 5.6C, 5.8A) were small marsupials characterized by upper molars with a narrower stylar shelf than in *Alphadon* and related forms and reduction or loss of stylar cusps B (stylocone) and C (Clemens, 1966, 1979; Fox, 1979b). Based on these dental features, Marshall et al. (1990) grouped pediomyids with Australidelphia (specifically, microbiotheres), a relationship that remains to be corroborated by other evidence.

Stagodontids were the first marsupials to modify the dentition for carnivory. They include some of the largest Cretaceous mammals, *Didelphodon* (Fig. 5.6B), reaching the size of the opossum *Didelphis*. Like eutherian carnivores and borhyaenoid marsupials, stagodontids evolved shearing between the paracristid (prevallid) of the lower molars and the posterior crest (postvallum) of occluding upper molars. They also had large posterior premolars that may have been used to crack shellfish, bones, or other hard materials. Palatal vacuities, a distinctive feature of marsupial skulls, are present in both stagodontids and *Alphadon*, suggesting that they were primitively present in marsupials (Fox and Naylor, 1995). The putative stagodontid *Pariadens*, based on a jaw from the earliest Late Cretaceous (Cenomanian) of Utah, appears to be the oldest known "definitive" marsupial (Cifelli and Eaton, 1987). It was initially referred questionably to the Stagodontidae, based on its relatively large size compared to other Cretaceous marsupials, molars that increase in size posteriorly, and larger paraconid than metaconid; but subsequent discoveries call into question its referral to Stagodontidae (Fox and Naylor, 1995).

Presumably because of their carnivorous specializations, Stagodontidae were classified by Marshall et al. (1990) as basal borhyaenoids in the order Sparassodonta (extinct South American carnivorous marsupials). However, other authors variously assign them to Didelphimorphia (or equivalent) or separate from other marsupials, either in the higher taxon Archimetatheria (which unites Stagodontidae and Pediomyidae; Szalay, 1994) or in Alphadelphia (Case et al.,

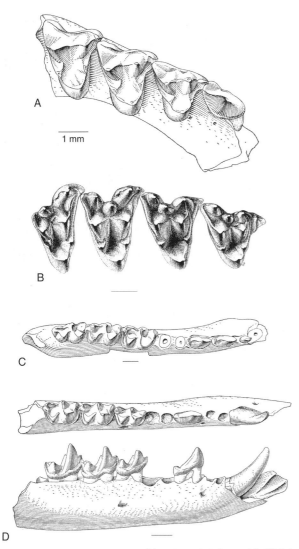

Fig. 5.8. Primitive Cretaceous marsupial dentitions: (A) *Pediomys*, right P³M¹⁻³; (B) *Albertatherium* (alphadontine), right upper molars; (C) *Alphadon*, right lower dentition; (D) *Protalphadon*, right lower dentition. All scale bars = 1 mm. (A, C, and D from Clemens, 1966; B from Fox, 1987.)

2004). The weight of the evidence indicates that borhyaenoids are more closely related to didelphoids than to stagodontids (see below).

Further diversity in Cretaceous marsupials is indicated by *Glasbius* (see Fig. 5.12A), a mouse-sized form with low, bunodont cusps; upper molars with enlarged cusps B and D (but all five stylar cusps present); and broad lower molars with a strong ectocingulid variably bearing distinct cuspules (Clemens, 1966). *Glasbius* is generally thought to be related to the South American early Tertiary Caroloameghiniidae (which is variously assigned to Peradectidae, Pediomyidae, or Paucituberculata); but Reig et al. (1987) thought *Glasbius* was more closely allied with microbiotheres, and Szalay (1994) grouped it with *Peradectes* and *Alphadon* in his expanded Pediomyidae. It remains possible that resemblances between *Glasbius* and either of the South American clades are convergent.

Krause (2001) attributed a partial molar from the Late Cretaceous of Madagascar to an indeterminate marsupial, but Averianov et al. (2003) suggested that it might instead belong to a eutherian, specifically, a zhelestid. Pending more conclusive remains, the presence of marsupials in Madagascar should be regarded as doubtful.

Although nearly all remains of Cretaceous marsupials consist of teeth and jaws, a few skull elements have been reported. Isolated Late Cretaceous petrosals, thought to belong to primitive marsupials, lack grooves for branches of the internal carotid artery but already had a fully coiled cochlea, which presumably allowed for a longer basilar membrane and relatively acute high-frequency hearing (Meng and Fox, 1995).

Paleontologists have long debated the place of origin of marsupials. Revised age estimates of key South American localities have shifted the balance in favor of North America, where the record of early marsupials is both older and more diversified than anywhere else (Cifelli, 1993a, 2000). However, the presence of a diversity of basal metatherians in Asia, at least one of which is even older than those from North America, has rekindled the debate about where metatherians and marsupials first originated and diversified. In any case, the Cretaceous radiation of marsupials appears to have been predominantly (or exclusively) Holarctic. The major radiations of post-Cretaceous marsupials, however, took place in South America and, later, Australia.

Ameridelphia

Peradectidae and Didelphidae

Although some of the Cretaceous marsupials in the foregoing discussion (those referred variously to Peradectidae or Didelphidae, as well as stagodontids and pediomyids) are sometimes included in Ameridelphia, they may well belong to a more plesiomorphic paraphyletic assemblage that includes the ancestors or sister taxa of other ameridelphians as well as australidelphians. By the Paleocene and Eocene, however, more than two dozen genera usually accepted as early members of the extant family Didelphidae (opossums: *sensu lato*, as used by McKenna and Bell, 1997) are known, mostly from North America and Europe, and especially South America. A few are now known from Asia and northern Africa as well, but they were clearly very rare on those continents and never diversified there. These animals were dentally conservative, showing only minor variations in dental anatomy, and they are central to most concepts of Ameridelphia.

Peradectids in the strict sense (Peradectinae of some authors) are a group of small marsupials known mainly from the Paleocene–Eocene of North and South America and Europe, and presumably derived from an *Alphadon*-like form. There is no dispute that they were didelphid-like, but whether they should be subsumed in Didelphidae is a contentious issue. Mouse-sized *Peradectes* was present on all three continents. Peradectid skeletons from the middle Eocene of Messel, Germany, show specializations for arboreal life, including a long, prehensile tail—preserved in coiled position—

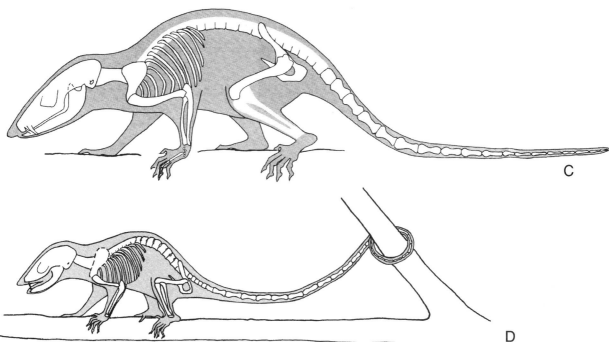

Fig. 5.9. Marsupial skeletons from the middle Eocene of Messel, Germany: (A, C), *Amphiperatherium;* (B, D) *Peradectes.* Scale bar = 2 cm for (A), 1 cm for (B). (A and B courtesy of C. Kurz and the Hessisches Landesmuseum Darmstadt; C and D from Koenigswald and Storch, 1992.)

and greatly reduced lumbar transverse processes, which allowed more flexibility of the lower spine (Fig. 5.9B,D; Koenigswald and Storch, 1992; Kurz, 2001).

Uncontested didelphids were diverse and widespread in the Early Tertiary, being represented by herpetotheriines in North America, Europe, northern Africa, and Asia, and three subfamilies (including Didelphinae) from the Paleocene (and possibly the latest Cretaceous) of South America. Although they persisted into the Miocene in North America and parts of the Old World, no other marsupials subsequently populated those regions (until the Pleistocene in North America). *Amphiperatherium* (Fig. 5.9A,C) and a tiny mouse-sized herpetotheriine are known from skeletons from Messel, which have shorter tails than peradectids and well-developed lumbar transverse processes. These features suggest that they inhabited the forest floor, or at least were less committed tree dwellers than were peradectids (Koenigswald and Storch, 1992; Kurz, 2001). Interestingly, all the Messel marsupials have an opposable hallux, which apparently bore a nail. The other ungual phalanges in both terrestrial and arboreal types are short and dorsoventrally deep, similar to those of arboreal mammals generally (MacLeod and Rose, 1993), which suggests that the presumed terrestrial forms evolved from arboreal marsupials and retained the ability to climb.

The oldest marsupials from South America come from the early Paleocene Santa Lucía Formation of Tiupampa, Bolivia. Despite the antiquity of this site, even conservative estimates identify at least five families of marsupials already present there (Muizon, 1998), demonstrating that once they were established in South America, marsupials diversified rapidly. This cladistic diversity is inferred from relatively minor differences, however, and the dentitions of these early marsupials are generally quite similar and can be characterized as "didelphoid."

Didelphids were especially diverse in the Paleocene of South America, where at least a dozen genera representing this family are recognized from three main sites: the early Paleocene at Tiupampa, Bolivia; the late Paleocene of Itaboraí, Brazil (Fig. 5.10A–D; see e.g., Paula Couto, 1952b, 1962, 1970a; Marshall and Muizon, 1988); and the late Paleocene of Laguna Umayo, Peru (formerly thought to be late Cretaceous or early Paleocene, but redated as late Paleocene or earliest Eocene; Sigé et al., 2004). Marshall (1987) recognized 16 genera of didelphids and didelphid-like forms from Itaboraí. Based on the anatomy of isolated postcranial bones from Itaboraí, both terrestrial and arboreal marsupials were present in the fauna (Szalay and Sargis, 2001).

The Tiupampan didelphid *Pucadelphys* (Fig. 5.11A–C, Plate 1.2) is one of the best-known marsupials, being represented by virtually complete skulls and articulated skeletons with a long, nonprehensile tail (Marshall et al., 1995). It is apparently primitive in lacking an alisphenoid bulla (Muizon, 1994), the presence of which is a hallmark of marsupials; but it must be noted that very few early marsupials are known from skulls preserved well enough to assess this feature. *Pucadelphys* was an agile generalist, probably mainly terrestrial and capable of digging (as suggested by the occurrence of the skeletons in burrows), but also capable of climbing, perhaps resembling present-day dasyurids more than didelphids in locomotor habits (Muizon, 1998; Argot, 2001; Muizon and Argot, 2003). The Tiupampan didelphoids provide a structural link between North American Cretaceous didelphoids and later South American didelphoids, such as those from Itaboraí (Muizon and Cifelli, 2001).

Besides didelphids, at least two other clades of ameridelphian marsupials thrived in South America during the Ceno-

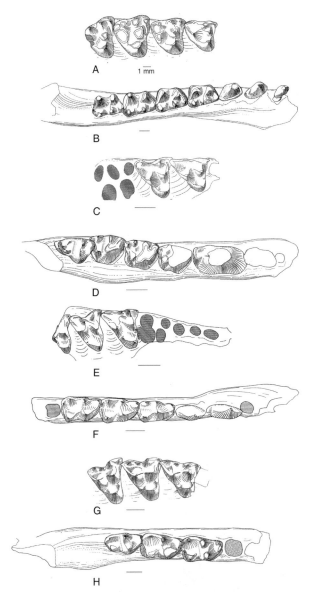

Fig. 5.10. Dentitions of South American Paleocene marsupials: (A) Protodidelphine *Protodidelphis*, right M^{1-4}; (B) protodidelphine *Guggenheimia*, left P_1–M_4; (C) didelphine *Marmosopsis*, right M^{1-2}; (D) eobrasiliine *Gaylordia*, right P_3–M_4; (E, F) caenolestoid *Carolopaulacoutoia*, right upper and lower dentitions; (G, H) microbiothere *Mirandatherium*, right upper and lower molars. A–D are didelphids. All scale bars = 1 mm. (From Marshall, 1987.)

zoic: Paucituberculata and Sparassodonta. Both are believed to have originated from didelphoid marsupials, and the oldest known members of both groups are also found at Tiupampa. It should be realized that assignment of many of these Paleocene marsupials to nondidelphoid groups is generally based on hindsight, citing rather subtle features. All of the Tiupampan marsupials are anatomically very similar.

Paucituberculata

This order is often used to encompass caenolestoids (including the living rat opossum *Caenolestes* and its extinct relatives) and the extinct argyrolagoids, caroloameghinioids, and polydolopoids (e.g., Aplin and Archer, 1987; McKenna

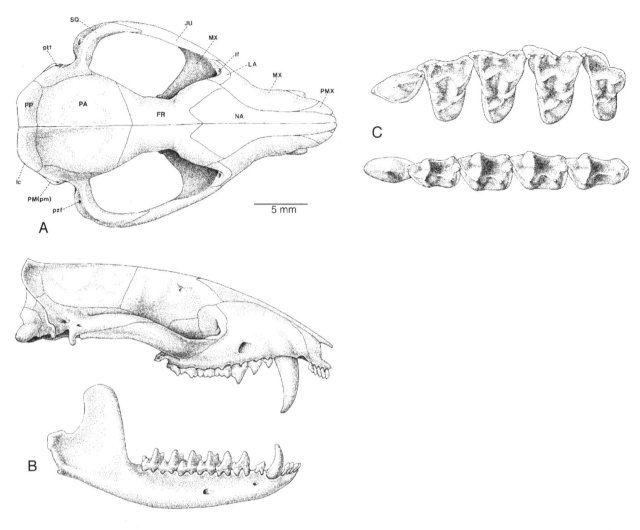
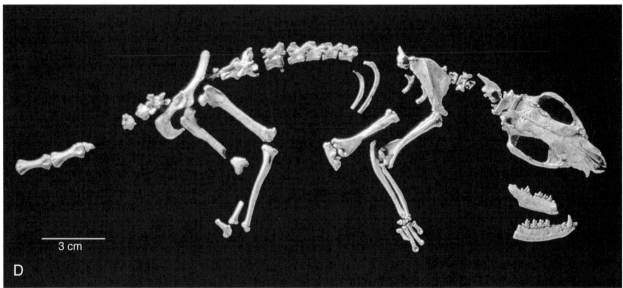

Fig. 5.11. Marsupials from the early Paleocene of Tiupampa, Bolivia: (A–C) *Pucadelphys* skull and left dentition; (D) *Mayulestes* skeleton. Key: FR, frontal; JU, jugal; LA, lacrimal; lc, lambdoid crest; lf, lacrimal foramen; MX, maxilla; NA, nasal; PA, parietal; PM (pm), mastoid part of petrosal; PMX, premaxilla; PP, postparietal; ptf, post-temporal foramen; pzf, postzygomatic foramen; SQ, squamosal. (A–C from Marshall et al., 1995; D courtesy of Christian de Muizon.)

and Bell, 1997). Marshall (1987) united the same assemblage under the name Polydolopimorphia. The precise relationships of these clades to each other, as well as to other marsupials, is problematic, and it is far from certain that all belong to a monophyletic group. For example, Marshall et al. (1990) and Szalay (1994) classified caenolestoids and argyrolagoids in the same clade (Marshall's Paucituberculata, Szalay's Glirimetatheria), but allocated polydolopoids to a separate order, Polydolopimorphia, and caroloameghinioids to either Peradectidae (Marshall et al., 1990) or suborder Sudameridelphia (together with polydolopoids and borhyaenoids; Szalay, 1994). Molecular studies have reached different conclusions about the relationships of extant forms (e.g., Kirsch et al., 1997) but, of course, cannot place the extinct groups.

The most primitive marsupials included in Paucituberculata are the caroloameghinioids, a group of small mammals characterized by low-crowned, bunodont, "didelphoid" teeth. North American Cretaceous *Glasbius* (Fig. 5.12A) is usually considered the oldest known caroloameghinioid, whereas *Roberthoffstetteria,* from the early Paleocene at Tiupampa, Bolivia, is the oldest caroloameghiniid (Fig. 5.12B–D). *Roberthoffstetteria* has broad molars with swollen, rounded cusps (including very prominent stylar cusps B, C, and D), suggesting a more omnivorous diet than in other contemporary marsupials.

Goin et al. (2003) suggested that *Roberthoffstetteria* is the sister taxon of polydolopoids, rather than a caroloameghiniid, based on such features as a thick dentary; thick enamel; and a very large, lingual metaconule positioned like a hypocone. On this basis, they proposed that *Roberthoffstetteria* represents a structural stage in the evolution of polydolopoid molars from a form like *Glasbius*. If this hypothesis is corroborated, it would constitute another lineage in addition to *Alphadon* that had Paleocene descendants.

Szalay (1994) used an expanded concept of Caroloameghiniidae, which included several taxa that other authors usually include in the Didelphidae (highlighting the primitive dental structure in this group), as well as the Protodidelphinae, an assemblage of primitive Paleocene genera whose affinities are uncertain. Protodidelphines have been considered to be basal members of Ameridelphia, Didelphimorphia, or Polydolopimorphia by other authors.

Polydolopoids include four families and a dozen genera of dentally derived marsupials primarily from the Paleocene and Eocene of South America. At least two genera are known from Antarctica. They are generally characterized by "plagiaulacoid" lower dentition (Simpson, 1933), a complex including enlarged incisors, an anterior diastema (or greatly reduced anterior teeth behind the incisor) followed by an enlarged shearing tooth (in this case, P_3^3), and small, broad, brachydont molars. This dental pattern is already present in the oldest polydolopid, Itaboraian *Epidolops* (Paula Couto, 1952a). *Epidolops* (Fig. 5.13) retains four lower molars, whereas Casamayoran forms typically have lost M_4. *Prepidolops* is the sole representative of the family Prepidolopidae. Although known so far only from the Eocene (Casamayoran-

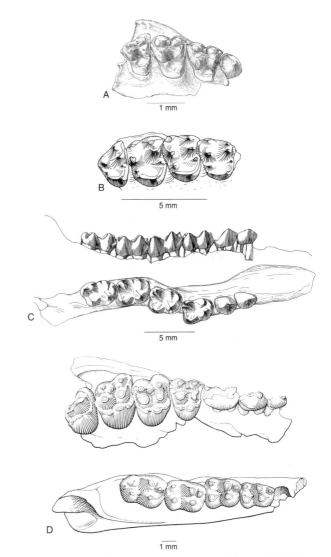

Fig. 5.12. Marsupial dentitions: (A) Late Cretaceous *Glasbius,* right P^3–M^3; (B–C) early Eocene *Caroloameghinia,* right M^{1-4} and P_2–M_4; (D) early Paleocene *Roberthoffstetteria,* right P^1–M^4 and M_{1-4}. (A from Clemens, 1966; B and C from Reig et al., 1987; D from Marshall et al., 1983.)

Mustersan), it seems to represent the most plesiomorphic branch of polydolopoids, or perhaps a transitional stage between didelphoids and polydolopoids (Pascual, 1980a,b). Its combination of didelphoid-like molars, tall and pointed P_3, and compressed anterior teeth result in a striking convergence toward certain omomyid primates. Casamayoran *Bonapartherium* (Bonapartheriidae) has enlarged premolars and unusual quadrate, bunoselenodont molars, the uppers lacking a stylar shelf (Pascual, 1980a). *Sillustania,* a putative polydolopoid based on isolated teeth from Chulpas, Peru, was initially thought to date from Cretaceous/Tertiary (K/T) boundary strata (Crochet and Sigé, 1996); but it is now believed to be of late Paleocene or earliest Eocene age (Sigé et al., 2004).

Caenolestoids are not well known before the Deseadan SALMA (late Oligocene). Nevertheless, one Paleocene form, the Itaboraian sternbergiid *Carolopaulacoutoia,* is assigned to this clade and is its oldest known representative (Paula Couto,

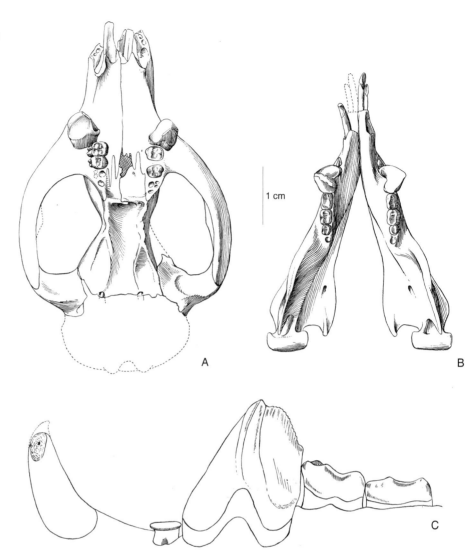

Fig. 5.13. Late Paleocene polydolopoid *Epidolops:* (A) skull; (B) mandible; (C) left lower dentition. (From Paula Couto, 1952a.)

1970a; McKenna and Bell, 1997). Marshall (1980) suggested that caenolestoids evolved from didelphoid ancestors, which is consistent with the dentition of *Carolopaulacoutoia* (Fig. 5.10E,F).

Argyrolagids, best known from the Neogene, were highly specialized jerboa-like, bipedal marsupials with enormous orbits (Simpson, 1970a). Were it not for possession of several diagnostic marsupial traits—presence of four molars, a medially inflected mandibular angle, palatal vacuities, and an alisphenoid bulla (Sánchez-Villagra and Kay, 1997)—argyrolagids might not be recognized as marsupials at all. Their origin and relationships remain uncertain, although they are usually allied with groeberiids, based mainly on the enlarged incisors. A mandibular fragment from the Eocene/Oligocene boundary strata (Tinguirirican) of Chile could represent the only pre-Deseadan record of the family (Wyss et al., 1994), but it has since been tentatively referred to the groeberiid genus *Klohnia* (Flynn and Wyss, 1999).

Groeberia (Fig. 5.14), from the middle Eocene (Mustersan? and Divisaderan SALMAs) of Argentina and Chile, is the oldest groeberiid and the oldest argyrolagoid (if groeberiids are indeed related to argyrolagids). Its affinities have long been enigmatic. It resembles rodents in having a pair of enlarged, ever-growing incisors with enamel limited to the anterior surface (Simpson, 1970c). The enamel has two layers, as in rodents, but unlike rodents, the inner layer is much thinner and consists of tangential enamel that lacks Hunter-Schreger bands (Koenigswald and Pascual, 1990). Moreover, the skull has palatal openings, an inflected mandibular angle, and four molars. Based on these last three features, most current authorities classify groeberiids as marsupials (Pascual et al., 1994; Flynn and Wyss, 1999). Most recent accounts suggest that groeberiids are closely related to Argyrolagidae, but the evidence is not conclusive, and some authors believe their similarities are convergent.

Sparassodonta (Borhyaenoids)

Borhyaenoids are an extinct clade of South American carnivorous marsupials that existed from the Paleocene through the Pliocene (Marshall, 1978). They were dentally convergent toward placental creodonts and carnivorans, and are the only South American marsupials that evolved Hunter-Schreger bands in the enamel (Koenigswald and Goin, 2000).

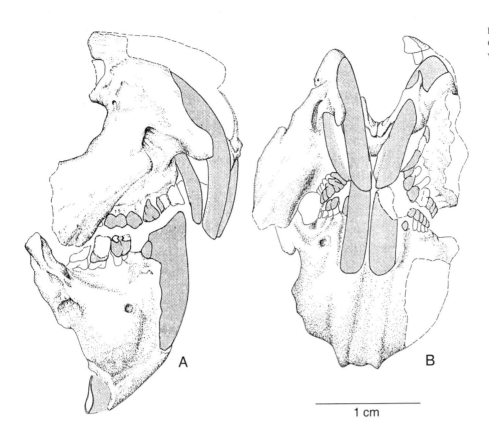

Fig. 5.14. Snout and mandible of Eocene *Groeberia*. (A) lateral view; (B) anterior view. (From Pascual et al., 1994.)

Mayulestes (Fig. 5.11D), a hedgehog-sized animal known from a skull and skeleton from the early Paleocene at Tiupampa, Bolivia, is considered to be the oldest borhyaenoid (Muizon, 1994, 1998). Compared to didelphids, its upper molars have a reduced paracone, and its lower molars have a paracristid with a carnassial notch and narrow and elongate talonids with a reduced entoconid—features suggestive of incipient carnivorous adaptation. The postcranial skeleton was adapted for an agile, arboreal life, as inferred from the prehensile tail, reduced humeral tuberosities and prominent supinator crest, contour of the ulna, shape of the radial head, and many other features (Muizon, 1998; Argot, 2001). Like *Pucadelphys*, *Mayulestes* appears to be primitive in lacking a tympanic process of the alisphenoid, which is usually considered diagnostic of marsupials (Muizon, 1994).

Andinodelphys, the largest didelphid from Tiupampa, has a reduced paracone and entoconid, suggesting that it could represent a transitional stage between didelphids and borhyaenoids (Muizon et al., 1997). If so, it implies that resemblances between stagodontids and borhyaenoids are convergent. Marshall et al. (1990), however, placed *Andinodelphys* as the sister taxon to Australian australidelphians. In a recent study, Muizon and Argot (2003) found that *Pucadelphys*, *Andinodelphys*, and *Mayulestes* show a gradient of increasing size and arboreal adaptation, although all were agile forms adept both on the ground and in the trees.

In addition to Mayulestidae, borhyaenoids comprise four or five families (sometimes regarded as subfamilies of Borhyaenidae), three of which are present in the early Tertiary. Hathliacynidae, exemplified by Itaboraian-Casamayoran *Patene* (Fig. 5.15A), were small to medium-sized borhyaenoids (Marshall, 1981). Early Paleocene *Allqokirus*, based on isolated teeth from Tiupampa, has been considered to be either the oldest hathliacynid or a mayulestid. Although dentally more primitive than other borhyaenoids, these genera show the first stages of carnivorous specialization: incipient reduction of the upper molar protocone and the lower molar metaconid and talonid, together with elongation of the postmetacone crista. The auditory bulla was large and was composed of the alisphenoid and periotic. Borhyaenids (sensu stricto) include a number of more specialized, dog-sized Eocene genera (e.g., *Angelocabrerus* [Fig. 5.15D], *Plesiofelis*) with reduced protocones and paracones on the upper molars, and sectorial lower molars with enormous roots and reduced metaconids and talonids (Simpson 1948, 1970b; Marshall, 1978).

The most carnivorously adapted early Tertiary borhyaenoids, referable to the Proborhyaenidae, were also the largest. Casamayoran *Callistoe* (Fig. 5.15C), known from a nearly complete skull and most of the skeleton, was a terrestrial, wolf-sized animal with massive canines and molars developed as bladelike carnassials (Babot et al., 2002). The last premolars (P^3_3) were also enlarged, as might be expected in a bone-crushing form. In *Callistoe* and all other proborhyaenids, the canines were ever-growing. Casamayoran *Arminiheringia* (Fig. 5.15B) was dentally one of the most specialized borhyaenoids (Marshall, 1978). It was even larger than *Callistoe*, with a more robust skull with widely flaring zygomae and a long, fused mandibular symphysis (Simpson, 1948). The dentition of these large proborhyaenids

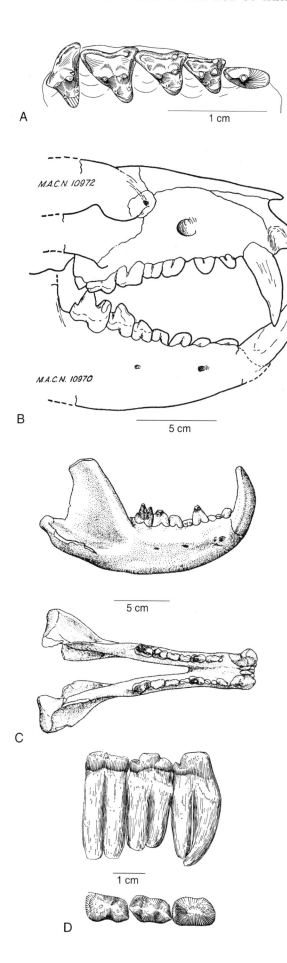

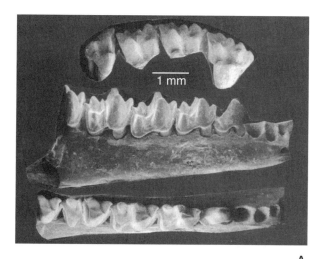

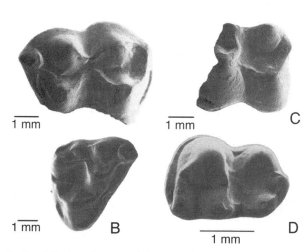

Fig. 5.15. Borhyaenoids: (A) *Patene*, right P³–M⁴ (B) *Arminiheringia;* (C) *Callistoe* mandible in lateral and dorsal views; (D) *Angelocabrerus*, left P₃–M₂ in lingual and crown views. (A from Marshall, 1981; B from Simpson, 1948; C from Babot et al., 2002; D from Simpson, 1970b.)

bears a superficial resemblance to that in some placental carnivores, such as hyaenids and oxyaenid creodonts.

Australidelphia

Current evidence suggests that the Australian radiation of marsupials emerged from South American Microbiotheriidae, a relict of which, the monito del monte (*Dromiciops australis*), survives in southern Chile and Argentina to the present day. Together, these marsupials are classified as the clade Australidelphia, which is supported by tarsal specializations indicative of a grasping hind foot typically associated with arboreal habits (Szalay, 1982). Molecular data also generally support the monophyly of microbiotheres

Fig. 5.16. Early Eocene(?) marsupials from Australia: (A) *Djarthia*, right upper and lower dentition; (B–C) *Thylacotinga*, fragmentary upper and lower teeth; (D) *Tingamarra*, right lower molar. (A from Godthelp et al., 1999; B–C from Archer et al., 1993a; D from Godthelp et al., 1992.)

and Australian marsupials (e.g., Burk et al., 1999). Microbiotheres were small, opossum-like marsupials represented by several early Tertiary genera, including Tiupampan *Khasia* and Itaboraian *Mirandatherium* (Fig. 5.10G,H). The Paleogene genera differ from didelphids and resemble later microbiotheres in several details of dental morphology, including reduction of the stylar shelf and stylar cusps on the upper molars, similar P_3 structure, talonids wider than trigonids on lower molars, and a reduced last lower molar (Marshall, 1987).

Teeth of several marsupials, including a microbiothere and opossum-like didelphimorphs and polydolopids, have recently been found in the middle Eocene La Meseta Formation of Seymour Island, Antarctica (Goin and Carlini, 1995; Goin et al., 1999). Although their strongest affinities are with Itaboraian and Patagonian faunas, their presence in Antarctica—the presumed route for dispersal to Australia—suggests that the Australian marsupial fauna could have reached that continent earlier than previously thought (perhaps by the K/T boundary, according to Goin et al., 1999).

The oldest and most primitive Australian marsupial is the recently named *Djarthia murgonensis* (Fig. 5.16A), known from upper and lower jaw fragments from the Tingamarra Local Fauna. The formation has been dated at 54.6 Ma, just above the Paleocene/Eocene boundary (Godthelp et al., 1992, 1999). Nevertheless, Woodburne and Case (1996) suggested that the fossil could actually be as young as Oligocene. The cheek teeth of *Djarthia* are very reminiscent of those of New World didelphids; they have a wider stylar shelf and larger stylar cusps than in microbiotheres. *Djarthia* is suitably ancient and unspecialized to lie near the ancestry of all other Australian marsupials and suggests an ultimate didelphoid origin for Australidelphia. A few other early Eocene marsupial teeth were previously described from the Tingamarra Local Fauna (*Thylacotinga*, Fig. 5.16B,C), but they were too incomplete to indicate clear affinities (Archer, Godthelp, and Hand, 1993). Woodburne and Case (1996) postulated that *Tingamarra* (Fig. 5.16D), a genus based on a lower molar originally attributed to a placental (condylarth), is more favorably interpreted as a marsupial close to protodidelphines.

Earliest Eutherian Mammals

EUTHERIA ARE DIFFERENTIATED FROM METATHERIA on the basis of reproductive anatomy and biology and the presence of a trophoblast during development (an extra-embryonic layer that surrounds the inner cell mass of the embryo; Novacek, 1986a; Lillegraven et al., 1987). Such a characterization, however, cannot be applied to fossils. Consequently, morphologists and paleontologists have sought reliable features of the hard tissues that can be used to recognize each group. As detailed in Chapter 5, there are several dental and skeletal traits that differ more or less consistently between the two groups.

Especially distinctive is the dental formula. The primitive eutherian dental formula was long considered to be 3.1.4.3/3.1.4.3, in contrast to the primitive metatherian formula 5.1.3.4/4.1.3.4. Evidence that many archaic Cretaceous eutherians had more than three incisors and that some had five premolars (McKenna, 1975a) led to reconsideration of the primitive eutherian dental formula (Novacek, 1986b), but the problem is not yet resolved. Some experts believe the "extra" premolar is a retained milk tooth (dP$_2$: Luckett, 1993), an interpretation supported by the apparent retention of dP$_2$ in dryolestids (Martin, 1997b). The presence of five premolars in sirenians, previously thought to support a primitive premolar number of five, has more recently been interpreted as a synapomorphy of Sirenia (Domning, 1994). Discoveries of several well-preserved basal eutherians in the past decade or so have contributed new information on the primitive eutherian dental formula.

Surprisingly, the recently improved fossil record of the most primitive eutherians has contributed little to the understanding of placental interrelationships, because new discoveries have led to the realization that there is little convincing evidence for assigning them to modern orders. Nonetheless, some taxa appear to be distinctly

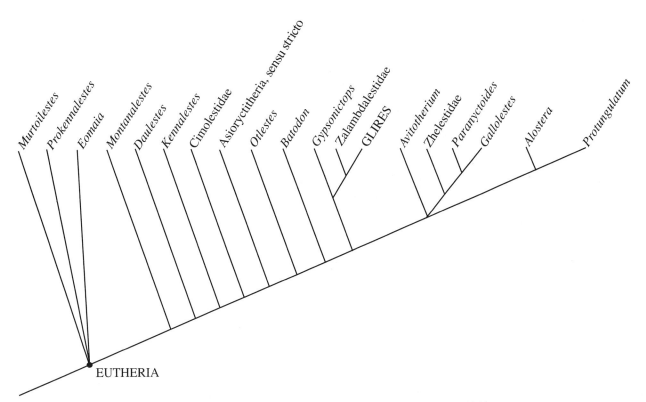

Fig. 6.1. Proposed relationships of the most primitive known eutherians. (Simplified after Archibald et al., 2001, and Archibald, 2003.)

more primitive or more derived than others, permitting a first-order approximation of relationships among these primitive eutherians, which is shown in Figure 6.1.

The terms Eutheria and Placentalia are sometimes used interchangeably, but there is a growing convention to restrict Placentalia to the last common ancestor of the extant orders and all its descendants, whereas the broader Eutheria includes additional stem taxa that are closer to placentals than to marsupials. In practice, it is not always clear which primitive eutherian taxa should be excluded from Placentalia, and in this book all eutherians except the Cretaceous forms discussed in this chapter are considered to be placentals. Indeed, some of these Cretaceous taxa could also turn out to be placentals.

Fossils that depart from the metatherian pattern and display features and dental formulae like those of later placentals are first known from the Early Cretaceous of Asia and western North America. Until recently these earliest eutherians were known only from dentitions, the oldest of which dated from the Aptian-Albian boundary on both continents, thus raising doubt about where they originated.

The recent discovery of a complete skeleton of a probable eutherian, older than any known previously, provides evidence of the anatomy of the earliest eutherians and suggests that they arose in Asia. *Eomaia scansoria* (Fig. 6.2) comes from the Early Cretaceous Yixian Formation of China and is dated to about 125 million years ago (Ji et al., 2002). It was a shrew-sized mammal, estimated to have weighed only 20–25 g—less than an ounce. *Eomaia* is more primitive than other eutherians in having a slightly inflected mandibular angle and retaining Meckel's groove on the mandible. It also retains epipubic (marsupial) bones. Unlike metatherians, however, the dental formula is 5.1.5.3/4.1.5.3, the premolars are simple and not molarized, and the molar hypoconulid and entoconid are not twinned (Fig. 6.3A). The skeleton was adapted for scansorial or arboreal habits, which is particularly apparent in the anatomy of the phalanges.

About 10 million years younger is *Prokennalestes,* another shrew-sized form with five premolars and three molars, from the ?Aptian or Albian of Khovboor, Mongolia (Kielan-Jaworowska and Dashzeveg, 1989). It is sometimes assigned to the leptictidan family Gypsonictopidae (McKenna and Bell, 1997). As in metatherians, the upper molars of *Prokennalestes* have a moderately wide stylar shelf with cuspules and lack a hypocone, but the paracone is higher than the metacone and the lower molars have a low paraconid and a centrally placed hypoconulid rather than a twinned entoconid-hypoconulid, characteristics of eutherians. A petrosal recently referred to *Prokennalestes* shows a combination of primitive mammalian and derived therian traits, the latter including a cochlea showing one full coil, which is the oldest known occurrence of this feature (Wible et al., 2001). *Murtoilestes* is a closely related form, based on a few isolated teeth from Russia that may be a little older than *Prokennalestes* but younger than *Eomaia* (Averianov and Skutschas, 2001).

North American *Montanalestes* (Fig. 6.3B), from the Albian (about 110 Ma) of Montana, is the oldest North American eutherian-like form. It is known from a lower jaw with four or five premolars (the precise number is uncertain) and three molars. The last premolar has a metaconid and an

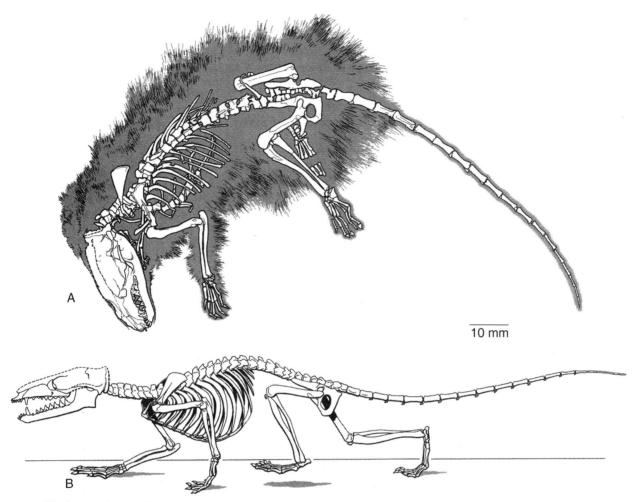

Fig. 6.2. Oldest known eutherian, Early Cretaceous *Eomaia* from China. (A) Skeleton with fur impression; (B) skeletal reconstruction. (From Ji et al., 2002.)

anterobasal cusp, indicating a gradation from simple premolars to complex molars, a eutherian hallmark (Cifelli, 1999).

Whether these Early Cretaceous mammals are the earliest eutherians or are better considered as "proto-eutherian" will be decided by further evidence. It is notable that the traits that distinguish the earliest eutherians from metatherians are, for the most part, primitive features.

Apart from *Eomaia*, the oldest well-preserved eutherian mammals come from the Late Cretaceous of central Asia (Figs. 6.3, 6.4; Plates 1.3, 1.4). *Kennalestes, Asioryctes, Ukhaatherium, Zalambdalestes,* and *Barunlestes,* from ?Campanian strata (about 75 Ma) of the Gobi Desert of Mongolia, and *Daulestes,* from the Coniacian of Uzbekistan, are each known from skulls, and all but *Daulestes* and *Kennalestes* from partial or complete skeletons as well. The first three genera are now united in the higher taxon Asioryctitheria (Novacek et al., 1997), based on details of cranial anatomy, although *Kennalestes* was earlier placed in Leptictidae or Gypsonictopidae. *Daulestes* is probably also an asioryctithere. It has a cochlea with one full turn—more than in multituberculates or monotremes, but less than in other known placentals (or marsupials) except for *Prokennalestes* and *Zalambdalestes* (McKenna et al., 2000). This and other primitive features of the ear region suggest that marsupials and placentals convergently acquired many of their derived auditory traits (Wible et al., 2001).

Not unexpectedly, asioryctitheres have many plesiomorphic traits, including more incisors than in other eutherians (5/4 in *Asioryctes* and *Ukhaatherium,* as in primitive marsupials; 4/3 in *Kennalestes;* the incisor count in *Daulestes* is uncertain); primitive skull structure, including such features as a large facial exposure of the lacrimal bone; generalized limbs with separate radius-ulna and tibia-fibula; a primitive ankle with an ungrooved astragalar trochlea and ventrally curved calcaneal tuber; and epipubic bones (Novacek et al., 1997; Horovitz, 2000, 2003). The four premolars are simple except the last one (semimolariform), the lower molars have very tall, mesiodistally compressed trigonids with reduced paraconids and low, basined talonids, and the upper molars are transversely wide, with well-developed parastyle and/or metastyle and no hypocone. These features suggest that asioryctitheres lie near the base of the eutherian radiation.

Zalambdalestes and *Barunlestes,* assigned to the family Zalambdalestidae, are considerably more specialized (Kielan-Jaworowska, 1978, 1984). The skull (Fig. 6.4D) is larger than

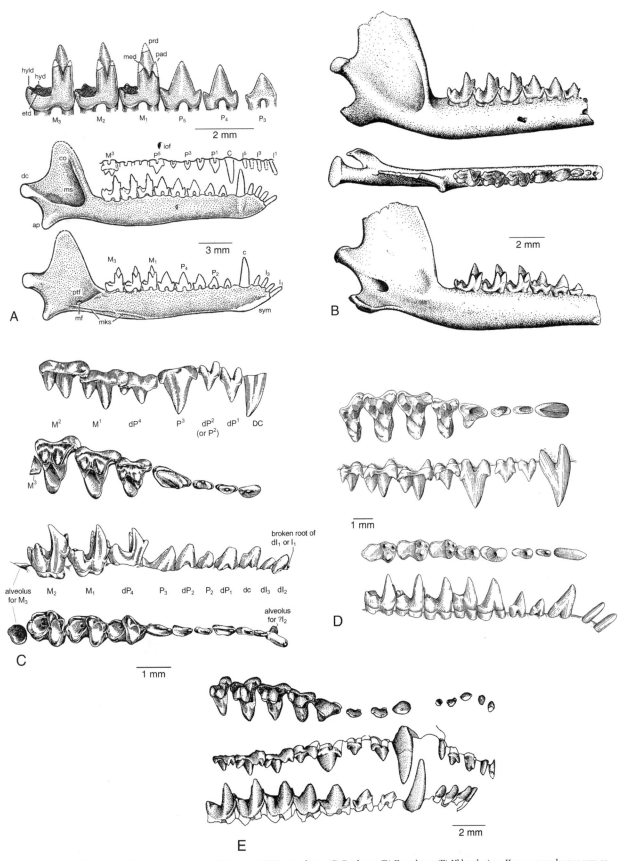

Fig. 6.3. Dentitions of primitive Cretaceous eutherians: (A) *Eomaia;* (B) *Montanalestes;* (C) *Daulestes;* (D) *Kennalestes;* (E) *Ukhaatherium.* Key: ap, angular process; co, coronoid process; dc, dentary condyle; etd, entoconid; hyd, hypoconid; hyld, hypoconulid; iof, infraorbital foramen; med, metaconid; mf, mandibular foramen; mks, meckelian sulcus; ms, masseteric fossa; pad, paraconid; prd, protoconid; ptf, pterygoid fossa; sym, symphysis. (A from Ji et al., 2002; B from Cifelli, 1999; C from McKenna et al., 2000; D from Kielan-Jaworowska, 1969; E from Novacek et al., 1997.)

bald et al., 2001; see Chapter 15 on Anagalida). Nonetheless, zalambdalestids, like asioryctitheres, are primitive in retaining epipubic bones. The coexistence in the Late Cretaceous of such differently adapted animals as zalambdalestids and asioryctitheres is evidence that the eutherian radiation was already well under way. But the presence of epipubic bones—perhaps related to either a pouch or external support of altricial young—suggests that these Cretaceous eutherians had not yet achieved the prolonged gestation characteristic of placentals (Novacek et al., 1997).

From the same strata in Uzbekistan that produced *Kulbeckia* come several genera with relatively broad, brachydont molars that seem to foreshadow conditions later developed in the hoofed mammals. Called zhelestids, these archaic eutherians have been interpreted as the stem group of the grandorder Ungulata (Archibald, 1996; Nessov et al., 1998; see Chapter 12).

The oldest uncontested eutherians from North America are Late Cretaceous (Campanian) *Paranyctoides* and the slightly younger *Gallolestes* and *Gypsonictops* (Lillegraven and McKenna, 1986; Cifelli, 2000; Fig. 6.5). *Paranyctoides* was recently reported from Asia (Uzbekistan) as well (Archibald and Averianov, 2001). Compared to other Cretaceous "insectivores" they have relatively low molar trigonids and broad talonid basins. The last premolar in all three is sub-

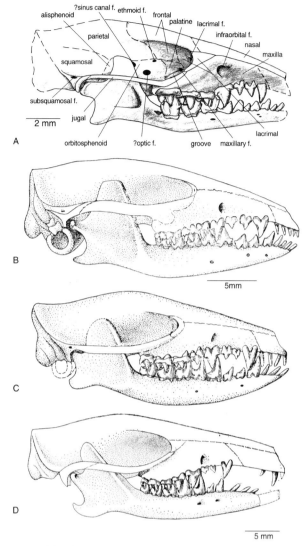

Fig. 6.4. Skulls of primitive Cretaceous eutherians: (A) *Daulestes*; (B) *Asioryctes*; (C) *Kennalestes*; (D) *Zalambdalestes*. Key: f., foramen. (A from McKenna et al., 2000; B–D from Kielan-Jaworowska, 1975a.)

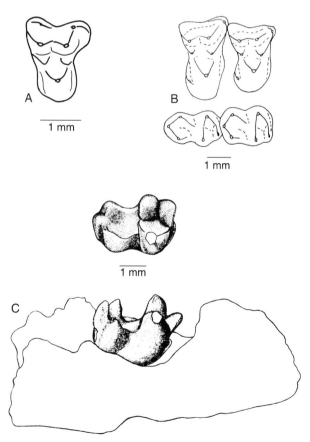

Fig. 6.5. Cretaceous eutherian teeth: (A) *Paranyctoides*, right upper molar; (B) *Gypsonictops*, right M^{1-2} and left M_{2-3}; (C) *Gallolestes*, right lower molar. (A from Butler, 1990; B from Butler, 1977; C from Clemens, 1980.)

in other Cretaceous eutherians and has an elongated snout and a long diastema between the incisors and canine. There are only three incisors in each quadrant, the anterior ones enlarged and strongly procumbent. The last lower premolar (P_4) is more molariform, the molar trigonids are lower and further compressed, the molar talonids are broader, and the upper molar stylar shelves are more reduced than in asioryctitheres. *Kulbeckia*, from deposits in Uzbekistan that are probably about 10 million years older than the Gobi Desert zalambdalestids, is the most primitive zalambdalestid (Archibald et al., 2001; Archibald and Averianov, 2003). It had at least three and possibly four or five upper incisors and lacked a significant diastema. The limbs of zalambdalestids (see Fig. 15.2) are long, the tibia and fibula fused, and the joints specialized for running and hopping, including a deeply grooved astragalus. In all these features zalambdalestids resemble rabbits and elephant shrews, and some recent studies suggest they could be the stem group of Glires (e.g., Archi-

molariform, and both *Paranyctoides* (at least one species) and *Gypsonictops* are known to have had five lower premolars. (These genera are further discussed in Chapter 9.) *Paranyctoides* is usually considered a lipotyphlan insectivore, variously assigned to Nyctitheriidae or basal Soricomorpha (shrews and related forms), whereas *Gallolestes* is usually assigned to Leptictida (McKenna and Bell, 1997); but Archibald and Averianov (2001) argue that they are basal eutherians of uncertain ordinal affinity, possibly related to zhelestids. *Gypsonictops* is a leptictidan.

7

Cimolesta

SEVERAL CLADES OF PRIMITIVE EUTHERIANS are included in the Cimolesta, proposed as an order of the grandorder Ferae by McKenna (1975a). Some of the higher taxa that are discussed here were formerly assigned to the Insectivora *sensu lato* or to the Proteutheria, a "wastebasket taxon" that has been used to encompass various primitive placental groups of uncertain affinity (e.g., Didelphodonta, Pantolesta, Apatotheria; see, e.g., Romer, 1966). However, it is very unlikely that either Proteutheria or Insectivora is a natural group when these primitive clades are included. Other taxa included in McKenna's Cimolesta have traditionally been accorded ordinal rank by most researchers (e.g., Taeniodonta, Tillodontia, Pantodonta). Unfortunately, the evidence (mainly dental) that they constitute a monophyletic assemblage descended from or sharing a common ancestry with Cimolestidae is not much stronger and has not been tested by phylogenetic analysis. Hence their unification in Cimolesta should be regarded as tentative and is used here largely for convenience. Ordinal status for most of these groups is maintained here, and accordingly Cimolesta is considered a mirorder (Table 7.1).

DIDELPHODONTA AND OTHER PRIMITIVE CIMOLESTA

The most plesiomorphic group of the Cimolesta, Didelphodonta (as employed by McKenna and Bell, 1997) comprises the single family Cimolestidae, which is widely considered on dental criteria to include the ancestors or sister taxa of most or all of the other groups these authors assigned to Cimolesta. Didelphodonts also seem to be related to the early Tertiary families Palaeoryctidae and Wyolestidae, as well as to the origin of creodonts and carnivores. These inferred relationships are reflected by

Table 7.1. Synoptic classification of Cimolesta

Superorder FERAE
  Mirorder †CIMOLESTA
        †**Didymoconidae**[1]
        †**Wyolestidae**[1]
    Order †DIDELPHODONTA
        †**Palaeoryctidae**[1]
        †**Cimolestidae**
    Order †APATOTHERIA
        †**Apatemyidae**
    Order †TAENIODONTA
        †**Stylinodontidae**
    Order †TILLODONTIA
        †**Esthonychidae** (=†Tillotheriidae)
    Order †PANTODONTA
        †**Harpyodidae**
        †**Bemalambdidae**
        †**Pastoralodontidae**
        †**Pantolambdidae**
        †**Titanoideidae**
        †**Barylambdidae**
        †**Cyriacotheriidae**
        †**Pantolambdodontidae**
        †**Coryphodontidae**
    Order †PANTOLESTA
        †**Pantolestidae**
        †**Pentacodontidae**
        †**Paroxyclaenidae**
        †**Ptolemaiidae**
    Order PHOLIDOTA
        †**Eomanidae**
        †**Patriomanidae**
        Manidae
    ?Order PHOLIDOTA
      Suborder †PALAEANODONTA
        †**Escavadodontidae**
        †**Epoicotheriidae**
        †**Metacheiromyidae**
      Suborder †ERNANODONTA
        †**Ernanodontidae**

Notes: Modified after McKenna and Bell (1997). McKenna and Bell consider Cimolesta to be an order, and the orders listed here to be suborders. The dagger (†) denotes extinct taxa. Families in boldface are known from the Paleocene or Eocene.

[1] Relationships uncertain.

grouping all these taxa in the grandorder Ferae (McKenna and Bell, 1997).

Traditionally the genera now assigned to Cimolestidae were included in a broadly conceived family Palaeoryctidae, but current usage usually restricts the latter family to the most specialized zalambdodont forms. Members of both families, sometimes loosely called "palaeoryctoids," are generally characterized by cheek teeth with high, sharp cusps; transversely wide upper molars with a broad stylar shelf but lacking pre- and postcingula and hypocones; lower molars with tall trigonids and reduced talonids; and an emphasis on transverse shearing of the front and back of the trigonid against the posterior and anterior crests, respectively, of successive upper molars (Novacek, 1986a; Figs. 7.1, 7.2). Cimolestids tend to have a primitive dental formula of 3.1.4.3/3.1.4.3, whereas palaeoryctids lack the first premolar, so the dental formula is typically 3.1.3.3/3.1.3.3. Late Cretaceous

Asioryctes, which was originally assigned to the Palaeoryctidae, resembles palaeoryctoids in some dental and cranial features, but most of these features are plesiomorphic. *Asioryctes* is currently placed in the basal eutherian taxon Asioryctitheria.

Cimolestids are thought to retain one of the most plesiomorphic dentitions among placentals. Thus it is reasonable to place them in a basal position to many other groups; nevertheless, few synapomorphic features link them specifically with any later taxa. Most cimolestids were approximately the size of living hedgehogs. Unfortunately, little is known about them beyond the dentition. *Cimolestes*, best known from the latest Cretaceous and early Paleocene of western North America, is generally considered the most primitive and is the only Mesozoic cimolestid genus. Its

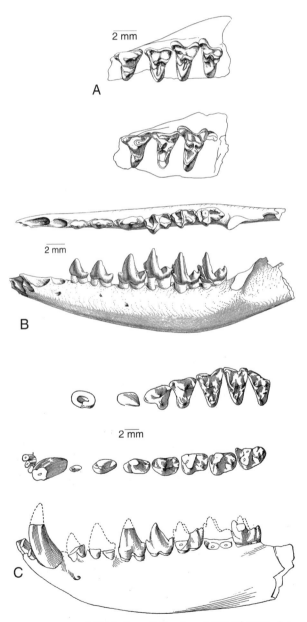

Fig. 7.1. Dentitions of didelphodonts: (A) *Procerberus*, left P^4–M^3; (B) *Cimolestes*, left P^4–M^2 and P_2–M_3; (C) *Didelphodus*, left upper and lower dentitions. (A from Lillegraven, 1969; B from Clemens, 1973; C from Matthew, 1918.)

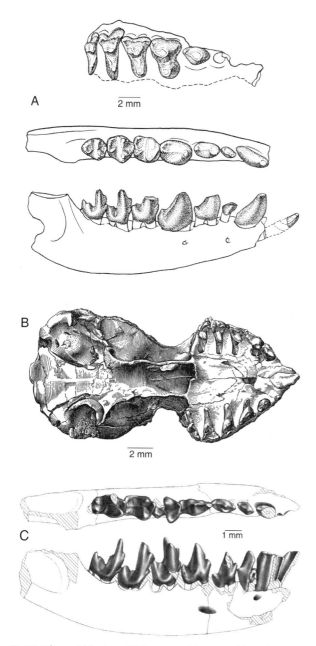

Fig. 7.2. Palaeoryctid dentitions: (A) *Aaptoryctes*, right upper and lower dentitions; (B) *Palaeoryctes* skull (ventral view); (C) *Palaeoryctes* right lower dentition. (A from Gingerich, 1982; B from McDowell, 1958; C from Bloch et al., 2004.)

dentition is characterized by simple premolars, the lower ones dominated by one main cusp, and molars with a generalized tribosphenic pattern. The lower molars have tall trigonids and small, shallowly basined talonids with three cusps. The uppers have three main cusps, small conules, no hypocone, and a broad stylar shelf with few cusps. Based on these dental traits, *Cimolestes* has been considered to be the sister group and possible ancestor of both creodonts and carnivorans (see Chapter 8). Therefore it is almost certainly paraphyletic, and the variation encompassed by its several species might be more appropriately separated into more than one genus.

Several other cimolestid genera are known from the Paleocene and Eocene of North America and Europe, as well as the late Paleocene of Africa and the early Eocene of Asia. Early Paleocene *Procerberus* of western North America is currently considered a cimolestid because of various dental features shared with *Cimolestes* (Lillegraven, 1969; McKenna and Bell, 1997). But it also resembles leptictids in having submolariform last premolars and in its molar wear pattern, hence its family assignment is debatable (Novacek, 1986a). One of the best-known cimolestids is *Didelphodus* (early to middle Eocene of North America and Europe), which differs in only minor ways from *Cimolestes* (e.g., having somewhat lower trigonids and broader stylar shelves).

Palaeoryctidae, known from the Paleocene and Eocene of western North America and possibly northern Africa and Eurasia, were tiny shrew- to mouse-sized insectivores with zalambdodont or nearly zalambdodont molars (Figs. 7.2, 7.3). Current usage restricts the family to *Palaeoryctes, Eoryctes, Aaptoryctes, Ottoryctes,* and perhaps one or two other genera (MacPhee and Novacek, 1993; McKenna and Bell, 1997; Bloch et al., 2004). *Palaeoryctes minimus* from the late Paleocene of Morocco is among the smallest known mammals, with molars less than 1 mm long (Gheerbrant, 1992). The premolars of palaeoryctids are typically simple, tall, and sharp, except in Tiffanian *Aaptoryctes*, in which the last premolars are enlarged and swollen and have blunt cusps for crushing (Gingerich, 1982). Palaeoryctid upper molars are strongly transverse, with connate (twinned) paracone and metacone usually set well in from the buccal margin; a high, constricted protocone; and often a prominent parastyle. There are deep embrasures between the upper cheek teeth, which accommodate the very high trigonids of the lower molars; the talonids are typically small or vestigial. These features were presumably associated with a diet of insects and other small invertebrates.

Several well-preserved skulls of palaeoryctids are known, which show that palaeoryctids had rather short snouts, a bony auditory bulla, and poorly developed zygomatic arches or none at all. The squamosal in *Eoryctes* and *Ottoryctes* is expanded on each side into a broad lambdoidal plate similar to that in apternodontids (Bloch et al., 2004; see the section on Soricomorpha in Chapter 9). The bulla consisted of the ectotympanic in *Palaeoryctes* (Butler, 1988) but was possibly derived from the petrosal and/or the entotympanic in Wasatchian *Eoryctes* (Thewissen and Gingerich, 1989). In either case it differs from lipotyphlans, which typically have a basisphenoid bulla or lack an ossified bulla (Novacek, 1986a). In *Eoryctes* and *Ottoryctes*, the promontorial and stapedial branches of the internal carotid artery (inside the auditory bulla) were enclosed in bony tubes, whereas in *Palaeoryctes* grooves (but no tubes) were variably present (McDowell, 1958; Van Valen, 1966; Thewissen and Gingerich, 1989; Bloch et al., 2004). The significance of these arterial tubes, however, is unknown.

Palaeoryctids have been considered to be related either to genera now classified as cimolestids (Van Valen, 1967; Butler, 1988; McKenna and Bell, 1997) or to lipotyphlans (Lillegraven et al., 1981), specifically, soricomorphs (McKenna et al., 1984). Association with soricomorphs is based partly

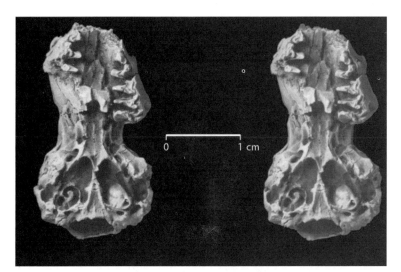

Fig. 7.3. *Eoryctes* skull in ventral view. (Stereophotograph from Thewissen and Gingerich, 1989.)

on mutual presence of a piriform fenestra in the skull (an opening in the bony roof of the middle ear), but this feature is of questionable significance (Butler, 1988; MacPhee and Novacek, 1993). Palaeoryctids share certain other cranial features with various lipotyphlans but differ from them in other ways (as noted above), leaving their relationship to Lipotyphla uncertain: they are either related or evolved convergently, probably from a *Cimolestes*-like form (Butler, 1988). Although palaeoryctids are assigned here to Cimolesta, following McKenna and Bell (1997), a relationship to Lipotyphla seems equally likely, based on current understanding.

Sometimes associated with palaeoryctids are the Early Tertiary Micropternodontidae. They are alternatively grouped with soricomorph lipotyphlans, and in the present account they are covered under that heading in Chapter 9.

Knowledge of cimolestids and palaeoryctids is largely limited to teeth and a few partial skulls; only very fragmentary or isolated postcrania (often attributed without dental association) are known. Unfortunately, this paucity of material severely limits our ability to reconstruct their behavior. A fragmentary humerus associated with *Palaeoryctes* suggests that it was an efficient digger (Van Valen, 1966). Isolated ankle bones attributed to *Cimolestes* and *Procerberus* (Fig. 7.4) share derived features with leptictids that imply abilities for plantar flexion and eversion typical of terrestrial mammals (Szalay and Decker, 1974). These features include an oblique and inclined posterior calcaneoastragalar facet and a distally placed peroneal tubercle on the calcaneus, an astragalus with a moderately long neck, grooved trochlea, isolated sustentacular facet, and no dorsal astragalar foramen. However, this interpretation should be viewed cautiously until confirmed by associated skeletons.

DIDYMOCONIDAE

Didymoconids (Fig. 7.5) are an enigmatic group of carnivorous Early Tertiary mammals probably endemic to Asia. The half dozen or so genera, most of which are known only from teeth and jaws, are characterized by loss of P^1_1 and M^3_3, and presence of simple premolars; transverse tritubercular upper molars with connate paracone and metacone and a somewhat reduced stylar shelf, and usually lacking a hypocone or postcingulum; and lower molars with tall trigonids and narrow talonids (Gingerich, 1981b). The oldest known didymoconid skulls, belonging to latest Paleocene *Archaeoryctes* from Mongolia and late early Eocene ?*Hunanictis* from China, are posteriorly broad, with small orbits, frontal-maxillary contact in the orbit, a reduced jugal, an ossified auditory bulla composed mainly of the entotympanic (?*Hunanictis*), and various other specialized cranial features (Meng, Ting, and Schiebout, 1994; Lopatin, 2001). Oligocene *Didymoconus* had a short, broad skull, with flaring zygomae, a broad occiput, and a prominent sagittal crest (Fig. 7.5B). The mandible was deep and the canines large. Postcrania of didymoconids are rare and poorly known. Best represented is *Didymoconus*, which had moderately robust forelimb elements, including short metacarpals and phalanges suggestive of fossorial habits (X. Wang et al., 2001). The humerus of middle Eocene *Ardynictis* was short and very robust with a prominent supinator crest, as in diggers, but the deltopectoral crest (typically well developed in diggers) seems to have been low and weakly developed (Lopatin, 2003b).

The phylogenetic position of didymoconids is uncertain. The dentition is consistent with relationship to Cimolesta or Leptictida, but this observation seems to be based primarily on primitive similarity. Didymoconidae most recently has been placed in Leptictida (McKenna and Bell, 1997) or Insectivora (Meng, Ting, and Schiebout, 1994; Lopatin, 2001; X. Wang et al., 2001), the latter used in the broadest sense. Cranial features have been cited as evidence of insectivoran affinity (Meng, Ting, and Schiebout, 1994), but the entotympanic bulla indicates that didymoconids are not lipotyphlans.

Possibly related to the didymoconids are the Wyolestidae, comprising three early to middle Eocene genera, two from

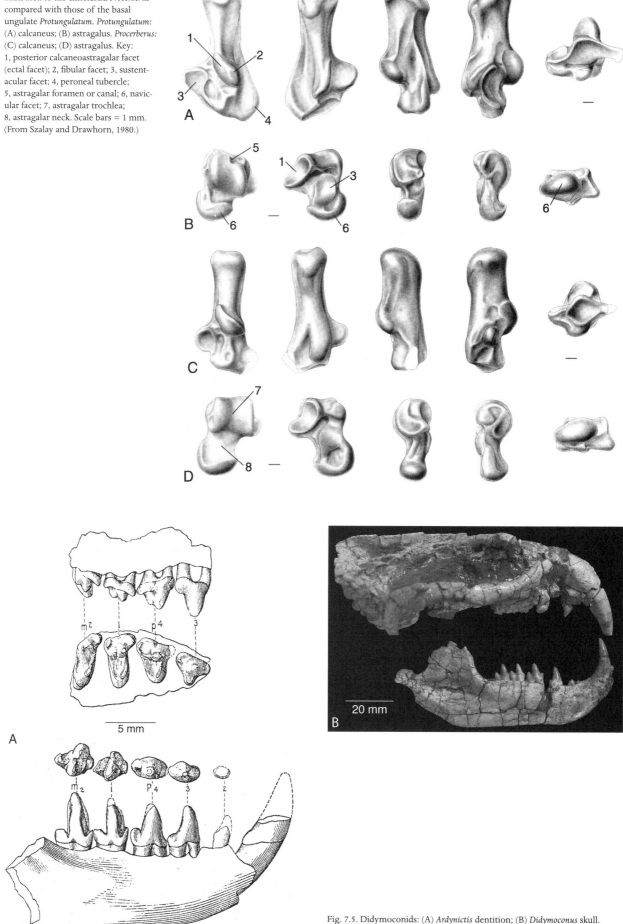

Fig. 7.4. Calcaneus and astragalus attributed to the cimolestid *Procerberus* compared with those of the basal ungulate *Protungulatum*. *Protungulatum*: (A) calcaneus; (B) astragalus. *Procerberus*: (C) calcaneus; (D) astragalus. Key: 1, posterior calcaneoastragalar facet (ectal facet); 2, fibular facet; 3, sustentacular facet; 4, peroneal tubercle; 5, astragalar foramen or canal; 6, navicular facet; 7, astragalar trochlea; 8, astragalar neck. Scale bars = 1 mm. (From Szalay and Drawhorn, 1980.)

Fig. 7.5. Didymoconids: (A) *Ardynictis* dentition; (B) *Didymoconus* skull. (A from Matthew and Granger, 1925b; B from Wang, Downs, et al., 2001.)

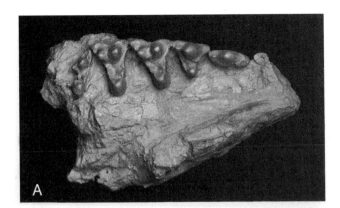
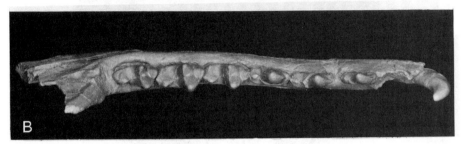
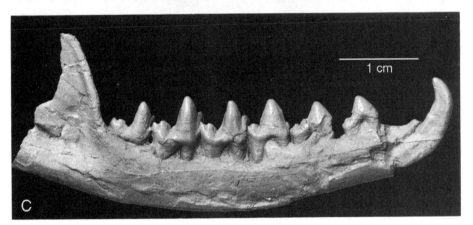

Fig. 7.6. Right dentition of early Eocene *Wyolestes:* (A) right P^3–M^3; (B, C) right dentary with C_1–M_3 in crown and buccal views. (From Novacek et al., 1991.)

Asia and one—*Wyolestes* (Fig. 7.6), the best known—from western North America (Gingerich, 1981b; Novacek et al., 1991). Wyolestids are known from relatively complete dentitions, which are very generalized and similar to those of didymoconids (again mainly in primitive features), except for retaining the first premolar and last molar. The upper molars are essentially tritubercular and triangular with more separated paracone and metacone than in didymoconids, and the lowers have larger talonids. Relationship to hyaenodontid creodonts has also been suggested, but the wyolestids have not yet been linked with any other group based on demonstrably synapomorphic traits. Gingerich (1981b) assigned wyolestids to the Didymoconidae, which he allied with Mesonychia. Subsequent authors have left Wyolestidae unassigned (Eutheria *incertae sedis:* Novacek et al., 1991) or placed them in the Cimolesta (McKenna and Bell, 1997).

The primitive anatomy and diverse opinions about didymoconids and wyolestids suggest that they may represent early offshoots of the eutherian stem.

PANTOLESTA

This order includes several Paleogene families—Pantolestidae, Pentacodontidae, Paroxyclaenidae, and Ptolemaiidae—that appear to be related based on dental similarities (Fig. 7.7). Although the dentition in most forms retains the same generalized pattern seen in didelphodonts, most pantolestans are more derived in having lower-crowned molars with more rounded cusps; uppers with a narrower stylar shelf and (in pantolestids and pentacodontids) a wide posterolingual cingulum, often bearing a distinct hypocone cusp; lowers with relatively lower trigonids and broader, basined talonids; and larger premolars. This suite of features suggests greater emphasis on crushing and grinding than in cimolestids and adaptation to a hard (durophagous) diet. The dental formula is generally 3.1.4.3/3.1.4.3, as in many primitive eutherians. Skulls are known for at least one member of each family of pantolestans, and in most cases they are moderately robust, with large canines; wide snouts; and broad, well-developed occipital regions,

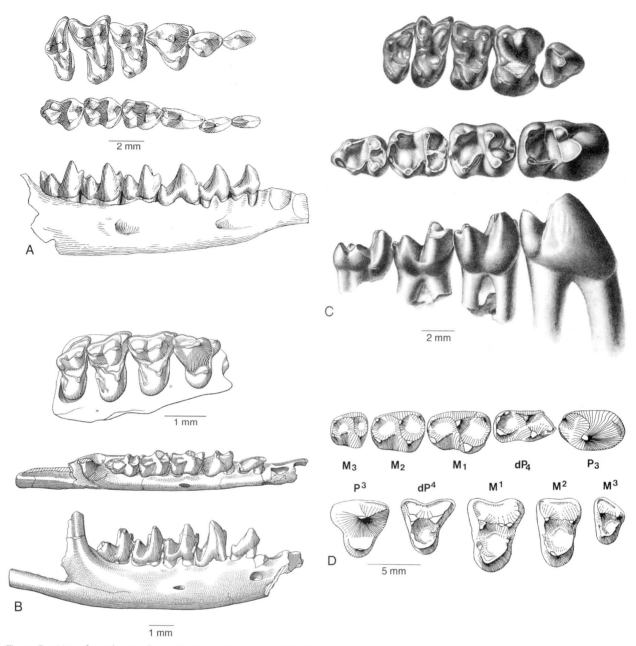

Fig. 7.7. Dentitions of pantolestans (all are right side except bottom row of D): (A) Paleocene pantolestid *Propalaeosinopa*; (B) late Paleocene todralestid *Todralestes*; (C) Paleocene pentacodontid *Aphronorus*; (D) middle Eocene paroxyclaenid *Kopidodon*. (A from Simpson, 1936; B from Gheerbrant, 1994; C from Gazin, 1959, 1969; D from Koenigswald, 1983.)

thus superficially resembling those of some of the less specialized extant carnivorans. There is no postorbital bar, so the orbital and temporal fossae are confluent, but a postorbital process exists in some forms. The infraorbital foramen was large, reflecting a large maxillary nerve that innervated a sensitive snout with tactile vibrissae.

In the few pantolestans in which it is known, the skeleton is comparatively robust with a long, well-developed tail. Some appear to have been terrestrial, with digging ability, and semiaquatic, whereas others were almost certainly arboreal. The fossorial skeleton of some primitive pantolestans shares particular similarities with that of palaeanodonts (see Chapter 11), suggesting a close relationship between these two clades.

The oldest known pantolestan is the pantolestid *Propalaeosinopa* (Fig. 7.7A), first known from the early Paleocene of western North America (possibly late Puercan [Cifelli et al., 1995], but common by the Torrejonian). Pentacodontids were also well differentiated by Torrejonian time, indicating that Pantolesta must have originated very early in the Paleocene or perhaps before.

Pantolestids are known from Paleocene to Oligocene strata of North America and Europe and have also been found in Asia and Africa. In unworn specimens, the molar cusps are moderately sharp; typically, heavy wear resulted in rounded or flattened cusps and exposure of dentine. This type of wear suggests a hard diet, such as mollusks or hard seeds, or perhaps the incidental ingestion of considerable grit. Dental morphology is consistent with omnivory. The dentary has a high and broad coronoid process that is well excavated laterally, and a prominent, hooklike angular process, suggesting that all the jaw muscles were well developed (Matthew, 1909). The skulls of middle Eocene *Pantolestes* and *Buxolestes* had wide snouts, an elongate temporal region, and a prominent, broad lambdoid crest for attachment of neck muscles that supported and extended the head.

Incomplete skeletal remains of *Pantolestes* from the middle Eocene of Wyoming have long been interpreted to indicate semiaquatic habits (Matthew, 1909). This conclusion was partly based on the humerus, which resembles that of otters in being short, robust, and slightly S-shaped in lateral profile, with prominent deltopectoral and supinator crests. This interpretation has been corroborated by exceptional skeletons of middle Eocene *Buxolestes* from Messel, Germany, and late early Eocene *Palaeosinopa* from the Green River Formation of Wyoming (Fig. 7.8, Plate 2.1). They also have an otterlike humerus, as well as a long tail that is particularly robust proximally, as in otters (*Lutra*), which employ the tail for propulsion when swimming (Koenigswald, 1980, 1987; Pfretzschner, 1993; Rose and Koenigswald, 2005). The hind limbs were also well developed and probably assisted in swimming. The prominent, broad nuchal crest on the occipital region of the skull, and the large spinous process on the axis vertebra, reflect the presence of powerful neck muscles characteristic of aquatic mammals. The terminal phalanges are long, slightly curved, and somewhat broadened, reminiscent of those in the beaver *Castor*. Fish remains and other bones have been found in the stomach region of both *Buxolestes* and *Palaeosinopa*. Most of the skeletal characteristics of pantolestids are also quite similar to those of fossorial mammals, and it is probable that pantolestids were able diggers as well.

Todralestes (Fig. 7.7B), represented by abundant dental remains from the late Paleocene Ouarzazate Basin in Morocco, appears to be closely related to pantolestids (Gheerbrant, 1994). It is variously placed in its own family, Todralestidae, or included in Pantolestidae. *Todralestes* was smaller than other pantolestans.

Pentacodontidae (sometimes considered a subfamily of Pantolestidae) are distinguished by their unusual swollen fourth premolars, which are often much larger than the molars, and a tendency to wear the cheek teeth very heavily (Fig. 7.7C). The molars have relatively low, blunt cusps and a very narrow stylar shelf. In some forms they decrease in size posteriorly. The single known skull, belonging to late Paleocene *Aphronorus*, is similar in size and form to that of the hedgehog *Erinaceus* (Gingerich, Houde, and Krause, 1983). The auditory bulla is not preserved and apparently was not ossified. The teeth in this individual and some other pentacodontid specimens are so heavily worn that little remains of the original crown morphology, indicating a particularly durophagous diet, or one that incorporated much grit. Recently reported postcrania of *Aphronorus*—the first known for any pentacodontid—are robust and support a close relationship with pantolestids (Boyer and Bloch, 2003). All known pentacodontids come from western North America, mainly from Paleocene deposits, although one genus (*Amaramnis*) lived during the early Eocene.

Evidence was recently reported that the pentacodontid *Bisonalveus* might have been venomous (Fox and Scott, 2005). If so, it is the oldest known poisonous mammal. The upper canine is deeply grooved on its anterior surface, comparable to the grooved lower second incisor of the venomous extant insectivoran *Solenodon*. In the latter, the groove helps direct venom from a modified submandibular gland to the prey. In *Bisonalveus*, some other specialized gland (possibly derived from the parotid) would have been involved.

Paroxyclaenidae are known principally from the Eocene of Europe; one genus has been reported from Asia. Although most taxa are represented only by teeth and jaws, the raccoon-sized *Kopidodon* (Fig. 7.9, Plate 2.2) is known from several skeletons preserved in the middle Eocene oil shale from Messel, Germany. They provide substantial evidence concerning its anatomy and paleobiology (Koenigswald, 1983; Clemens and Koenigswald, 1993). As in other pantolestans, the posterior premolars of *Kopidodon* and other paroxyclaenids are moderately robust, and the molars broad and low crowned (Fig. 7.7D). Compared to pantolestids, however, the talonids of *Kopidodon* are narrower and the stylar shelf wider. The molars of paroxyclaenids decrease in size posteriorly, and the uppers lack a hypocone (Russell and Godinot, 1988). The skull of *Kopidodon* is rather short, with a broad muzzle and distinct sagittal and nuchal crests; a bony auditory bulla was apparently lacking. The skeleton is generally similar in proportions to those of extant procyonids but more robust. As in pantolestids, all the limb bones are stout, and the humerus bears prominent deltopectoral and supinator crests. The mobile elbow joint, however, allowed substantial supination of the forearm, unlike pantolestids. The ankle was also very flexible. The feet were plantigrade, with divergent first digits, and the claw-bearing terminal phalanges were short, deep, and laterally compressed, as in extant arboreal carnivores (Rose, 1988; Clemens and Koenigswald, 1993; MacLeod and Rose, 1993). The long caudal series, coupled with the outline of soft tissues preserved in one of the Messel skeletons, show that *Kopidodon* had a long, bushy tail, which must have resembled that of the binturong *Arctictis* and some squirrels. These features indicate that *Kopidodon* was an arboreal animal with a probably omnivorous diet, perhaps filling an ecological niche similar to that of the smaller arctocyonids in the Paleogene of North America and some procyonids and viverrids today.

Ptolemaiidae (Fig. 7.10) were large pantolestans that lived in Egypt during the late Eocene and early Oligocene. The skull was about the size of that of the coyote *Canis latrans*. Unlike other pantolestans, their lower teeth had hypsodont crowns, and both uppers and lowers lacked cingula (Bown

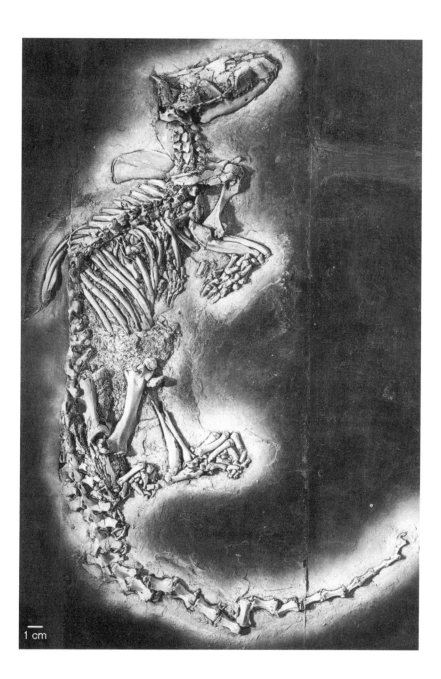
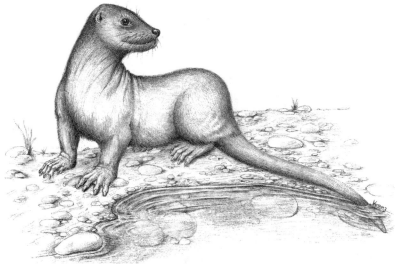

Fig. 7.8. *Buxolestes,* a pantolestid from the middle Eocene of Messel. (From Pfretzschner, 1993.)

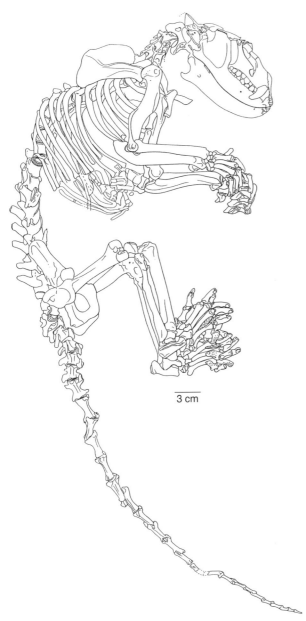

Fig. 7.9. *Kopidodon*, a paroxyclaenid from the middle Eocene of Messel. (From Clemens and Koenigswald, 1993.)

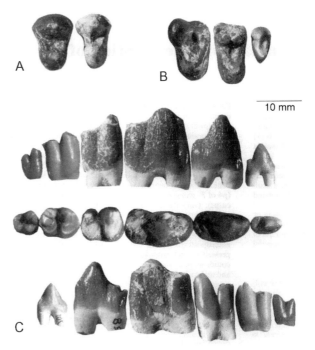

Fig. 7.10. Dentition of *Ptolemaia* from the late Eocene–early Oligocene of Egypt: (A) right P^{3-4}; (B) left M^{1-3}; (C) right P_2–M_3 in buccal, crown, and lingual views. (From Simons and Bown, 1995.)

and Simons, 1987; Simons and Bown, 1995). When unworn, the crowns bear low, rounded to moderately sharp cusps arranged essentially as in pantolestids, although the upper molars lack hypocones. But the cusps quickly wore away, exposing wide areas of dentine, which suggests thin enamel, heavy use, or both. The premolars were robust, P_{3-4} being larger than the molars. The lower molars present a columnar appearance from the buccal side, and the trigonids are barely taller than the talonids. The molars decrease in size posteriorly. The single known skull of *Ptolemaia* appears to be generally similar to those of other pantolestans, but is too crushed to provide many details. Relationship to paroxyclaenids has been suggested, and many aspects of the dentition also resemble those of pentacodontids.

Pantolesta were never abundant or very diverse, but they include some of the most successful semiaquatic mammals of the Early Tertiary. They were also among the first placentals to exploit the hard-object feeding niche. The latest known pantolestans disappeared in the Oligocene, leaving no descendants.

APATOTHERIA

This small but highly distinctive group is known from the Paleocene and Eocene of Europe and North America; a single rare genus, *Sinclairella*, survived into the Oligocene in North America. The half-dozen known genera are all included in the family Apatemyidae. Based on dental morphology, their relationships have been variously considered to lie with "proteutherians," Plesiadapiformes, or Cimolesta, but none of these proposals has been convincingly demonstrated. Apatemyids were small to medium-sized animals, mostly mouse- to squirrel-sized, the largest species reaching the size of a beaver (*Castor*). The full range of sizes is known among species of the European genus *Heterohyus* (Koenigswald, 1990).

Apatemyids are known primarily from jaws and teeth (Fig. 7.11), which are characterized by an enlarged front tooth in each quadrant, a reduced number of mostly small teeth between this front tooth and the first molar, and rather low-crowned molars (West, 1973a,b; Gingerich and Rose, 1982). The large front tooth is usually considered to be the first incisor; I_1 is procumbent, whereas I^1 is more nearly vertical. The dental formula is interpreted as 2.0.2.3/1.0.3.3 in most Paleocene members (*Jepsenella* and *Labidolemur*).

Torrejonian-Tiffanian *Unuchinia*, however, is more primitive than other apatemyids in retaining two lower incisors (Gunnell, 1988). Later apatemyids (*Apatemys*, *Heterohyus*, and *Sinclairella*) have only two lower premolars. The lower premolars of apatemyids contrast sharply with those of other early eutherians. The anterior lower premolar (P_2) enlarges, becoming an elongate, bladelike tooth that overhangs the back of the large incisor. The middle lower premolar (P_3) of early apatemyids was a small, one-rooted tooth and is therefore presumed to be the one that was lost in later forms. These later apatemyids retain a large, bladelike P_2, but P_4 is reduced to a one-rooted vestige. Unlike in many cimolestans, the lower molar trigonids are only a little higher than the talonids. The unique addition of a fourth, anterobuccal cusp makes the trigonid trapezoidal instead of triangular, as in other placentals. The talonids are broad, basined, and rounded in back, but usually narrower than the trigonids except on M_1. In some of the more derived apatemyids M_3 is elongate, as in some plesiadapiforms. The upper molars are relatively narrow transversely and have three main cusps (conules very small or absent), together with a small hypocone and a moderate stylar shelf with a pronounced parastyle.

Details of cranial anatomy of apatemyids are best seen in the unique skull of *Sinclairella* (Scott and Jepsen, 1936; Fig. 7.12), unfortunately now lost. Like many insectivores, it apparently lacked an ossified auditory bulla. The upper incisors of *Sinclairella* were enormously enlarged and caniniform; together with the relatively short face and narrow rostrum they give the skull a superficial resemblance to those of rodents, the phalangeroid marsupial *Dactylopsila* (striped possum), and the lemur *Daubentonia* (aye-aye), all of which also have enlarged incisors used in food procurement.

Several complete articulated skeletons of the European genus *Heterohyus* are known from the middle Eocene Messel site in Germany (Koenigswald and Schierning, 1987; Koenigswald, 1990; Kalthoff et al., 2004; Fig. 7.13, Plate 3.1). In addition, a partial skeleton of *Labidolemur* is known from the Clarkforkian of Wyoming (Bloch and Boyer, 2001), and a complete, articulated skeleton of *Apatemys* (Plate 3.2) was recently described from the late early Eocene Green River Formation of Wyoming (Koenigswald et al., 2005). These skeletons provide an unusually detailed view of apatemyid anatomy and insight into their paleobiology. They have numerous arboreal specializations, such as flexible ankles; terminal phalanges that are short, deep, and laterally compressed; and a long, bushy tail (an outline of which is preserved in one of the Messel specimens). Most remarkable, however, is the modification of the second and third digits of the hand into elongate probes, comparable to the third and fourth digits of *Daubentonia* and the fourth digit of *Dactylopsila* (Fig. 7.14). In analogy with these two extant mammals, it is inferred that apatemyids used their enlarged incisors to gouge into bark and rotten wood in search of wood-boring grubs and other insects, which they retrieved with their long, slender fingers. They may even have used these fingers, like *Daubentonia*, to tap on the wood to deter-

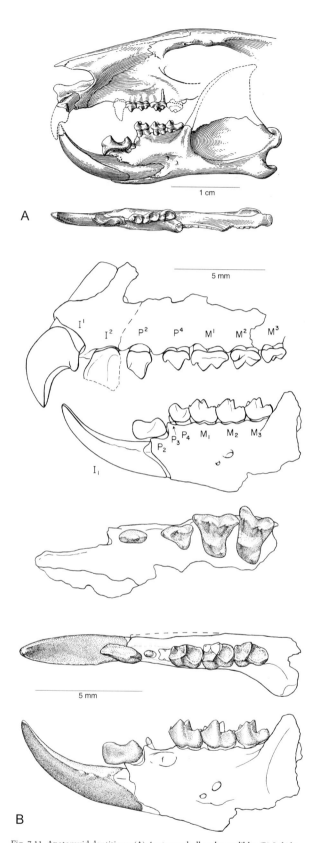

Fig. 7.11. Apatemyid dentitions: (A) *Apatemys* skull and mandible; (B) *Labidolemur*. (A from Matthew, 1921; B from Gingerich and Rose, 1982.)

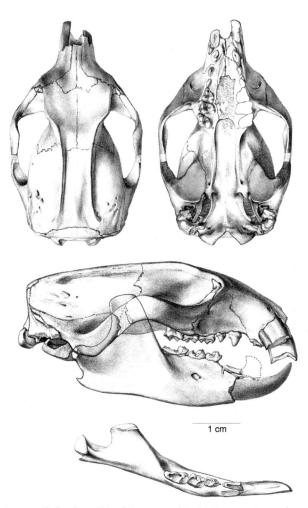

Fig. 7.12. Skull and mandible of the apatemyid *Sinclairella*. (From Scott and Jepsen, 1936.)

mine where to excavate. Like species of *Dactylopsila* today (Flannery, 1994), the apatemyid genera all share this specialized adaptation but differ in proportions of the fingers. Thus apatemyids exhibit remarkable convergence toward these two unrelated extant mammals and were the first of the three to adopt one of the most bizarre foraging behaviors known in mammals.

TAENIODONTA

The Taeniodonta—from Greek roots meaning "ribbon tooth," in allusion to the restricted band of enamel on the canines and cheek teeth of specialized forms—constitute a small and peculiarly specialized order of mammals whose broader affinities appear to lie with Cimolesta. The bizarre morphology of specialized taeniodonts has long obscured their relationships, but the most primitive representatives are said to share tarsal and/or dental traits with cimolestids and pantolestids (Szalay, 1977; Schoch, 1986), suggesting affinities with the Cimolesta. This hypothesis was strengthened when Eberle (1999) proposed that the large, bunodont cimolestid *Alveugena* (Fig. 7.15A), from the early Paleocene (Puercan) of Wyoming, could be the closest relative and possible ancestor of taeniodonts. Unfortunately, it is known only from skull fragments and the upper dentition, making comprehensive comparison impossible. The recent discovery of a possible Late Cretaceous taeniodont (discussed below) further complicates the issue of taeniodont origins and casts doubt on a direct relationship between *Alveugena* and taeniodonts. Earlier suggestions of resemblance and relationship to edentates or tillodonts, however, are now considered superficial and convergent.

Taeniodonts are known principally from the early Paleocene through middle Eocene of North America. In addition, two records from the early Eocene of Europe have been considered to represent taeniodonts; they are reviewed below. The North American forms comprise ten genera in two families (or two subfamilies of a single family), Conoryctidae and Stylinodontidae.

Taeniodonts have rather simple, tritubercular molars with low, bunodont cusps that were quickly obliterated by heavy wear (Fig. 7.15) and hence are preserved only in young individuals. The upper molars are transversely narrower than those of many other cimolestans. They have a reduced stylar shelf, no pre- or postcingula, and no hypocone. The paracone and metacone are separate and very buccally situated, and the protocone and conules are low and very lingual. The lower molars also lack cingula, and the trigonids and talonids are typically more or less equal in size and height. The molars decrease in size posteriorly. The skeleton of taeniodonts primitively was relatively robust, and became progressively more so during their evolution.

Several morphologic trends occurred in taeniodont evolution, leading to the distinctive or unique specializations of Eocene stylinodontids (Patterson, 1949; Schoch, 1986; Lucas et al., 1998). There was increasing extension of the enamel on the buccal side of lower cheek teeth and on the lingual side of uppers, a condition known as crown hypsodonty. This condition led to elongated roots and, ultimately, hypselodonty (ever-growing teeth with open roots). The canine teeth in particular underwent progressive hypertrophy and hypselodonty as well. These trends were presumably a response to an increasingly abrasive diet, perhaps associated with an unusual foraging behavior. The skull and mandible of taeniodonts (Fig. 7.16) became shorter and deeper through time, and the skeleton became conspicuously more robust. At the same time, taeniodonts were increasing in body size. Several of these trends occurred in parallel in the two families.

The most primitive taeniodonts are assigned to the paraphyletic family Conoryctidae (Schoch, 1986; Lucas et al., 1998). They were small to medium-sized (5–15 kg) generalized omnivores restricted to the early Paleocene; *Conoryctes*, the latest occurring representative, became extinct at the end of the Torrejonian (Lofgren et al., 2004). In the most primitive conoryctid, *Onychodectes* (Figs. 7.15B,C, 7.16C), the skull was rather long and narrow, with confluent orbital and temporal fossae, and a shallow dentary. The molars already show heavy wear, indicating a proclivity for an abrasive diet. The canine was prominent, the incisors small, and

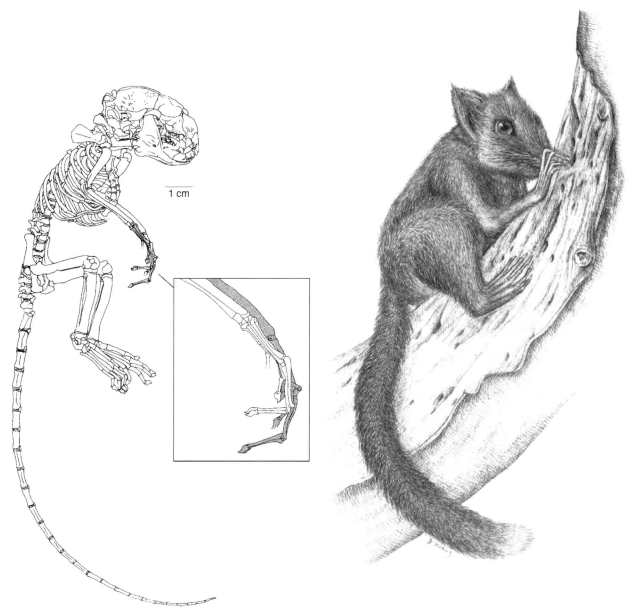

Fig. 7.13. *Heterohyus,* an apatemyid from the middle Eocene of Messel. (From Koenigswald, 1990.)

the premolars simple. The dental formula, 3.1.4.3/3.1.4.3, remained the same in most taeniodonts except for the tendency to reduce the number of incisors. *Onychodectes* and *Conoryctella* retained three lower incisors, whereas *Conoryctes* may have had only two.

The limbs of *Onychodectes* (Fig. 7.17) were generalized in proportion, but somewhat robust, suggesting climbing or digging capabilities. The humerus is broad distally and has a prominent deltopectoral crest, and the ulnar olecranon is pronounced. The manus and pes are pentadactyl, with small claws. There was a long and well-developed tail. Apart from being a little larger and more robust, *Onychodectes* was similar in body form, and perhaps habits, to the opossum *Didelphis* (Schoch, 1986).

Stylinodontidae were already divergent from Conoryctidae in the early Paleocene (middle Puercan) and possibly

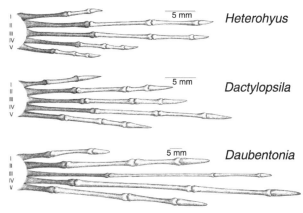

Fig. 7.14. Convergence of the manus in the apatemyid *Heterohyus,* the phalangeroid marsupial *Dactylopsila,* and the lemuroid primate *Daubentonia.* (From Koenigswald, 1990.)

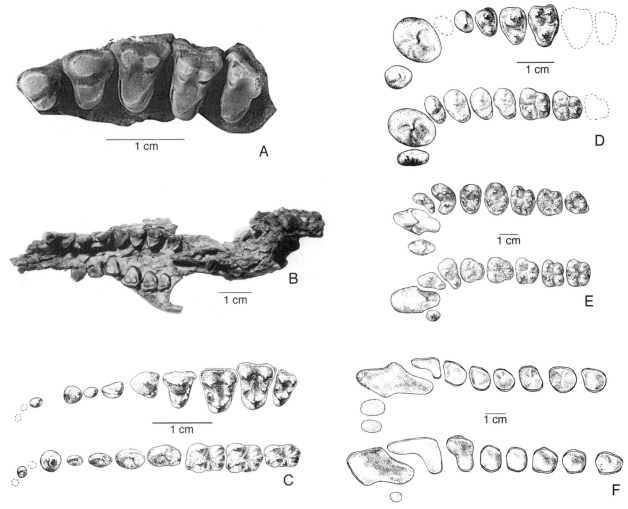

Fig. 7.15. Dentitions of taeniodonts and a possibly related cimolestid (anterior to left): (A) left upper teeth of the Puercan cimolestid *Alveugena*; (B) skull of early Paleocene *Onychodectes* (teeth typically worn); (C) *Onychodectes*, left upper and right lower teeth (unworn); (D) early Paleocene *Wortmania*, left upper and right lower teeth; (E) late Paleocene–early Eocene *Ectoganus*, left upper and right lower teeth; (F) Eocene *Stylinodon*, left upper and right lower teeth. (A from Eberle, 1999; B–F from Schoch, 1986.)

before (see below). Most were larger and more robust than conoryctids (10–110 kg; Schoch, 1986; Lucas et al., 1998). Early Paleocene *Wortmania* (Figs. 7.15D, 7.16B), the oldest unequivocal member of the family, is more derived than conoryctids in having a somewhat deeper skull and mandible, only one incisor in each quadrant, larger canines, and transversely oriented lower premolars—all hallmarks of the family. The coronoid process of the mandible was noticeably larger than in conoryctids, suggesting larger temporalis muscles. The skeleton was markedly more robust, and there were large, recurved claws on the manus.

These traits were further accentuated in other stylinodontids. The skull and mandible became shorter and deeper, especially the symphysis, which was solidly fused. The sagittal crest increased in prominence, and the occiput broadened. In Paleocene stylinodontids such as *Psittacotherium* (Fig. 7.16A), the canines were very large, but not yet evergrowing. In Eocene *Ectoganus* and *Stylinodon* (Fig. 7.15E,F), however, they became huge, ever-growing, and gliriform, with enamel restricted to the labial surface, superficially resembling the enlarged incisors of rodents and trogosine tillodonts. In similar fashion to rodents, this morphology maintained a sharp, chisel-like cutting edge on the canines as the softer dentine wore down behind the hard enamel border. At the same time, the crowns of the cheek teeth became progressively more hypsodont. In the most derived taeniodont, middle Eocene *Stylinodon*, all teeth were evergrowing, and the posterior premolars and molars were reduced by wear to cylindrical dentine pegs with buccal and lingual bands of enamel. Unworn molars, however, show a bilophodont crown pattern.

Stylinodontids had massive skeletons (Fig. 7.18). The limb elements are short and stout: the humerus is very broad distally and has enormous supinator and deltopectoral crests, the ulnar olecranon is very prominent, and the radius is much shorter than the humerus. The bones of the manus and pes are also short and stout, and include numerous large flexor sesamoids in the manus (Turnbull, 2004). The digits of the manus bore particularly large, curved claws, which were long presumed to have been used for

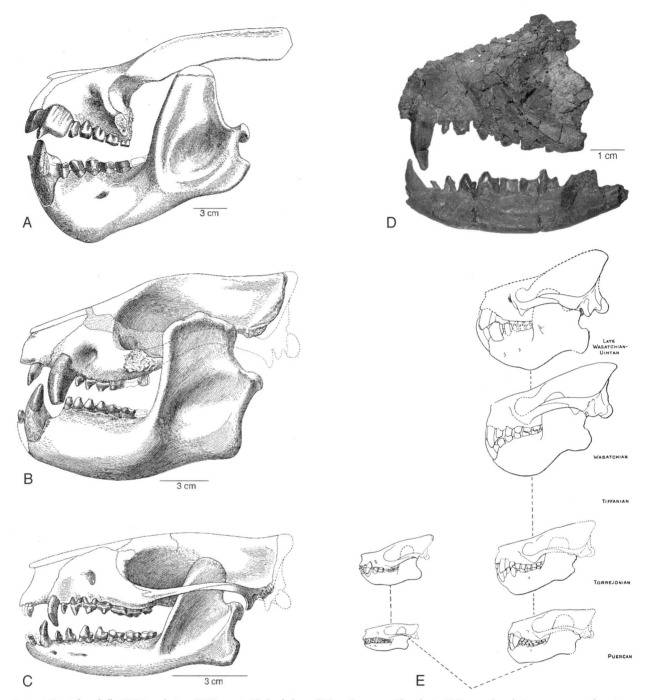

Fig. 7.16. Taeniodont skulls: (A) *Psittacotherium;* (B) *Wortmania;* (C) *Onychodectes;* (D) Late Cretaceous *Schowalteria;* (E) Patterson's evolutionary sequence of taeniodonts, illustrated by skulls. From the bottom, left side shows *Onychodectes* and *Conoryctes;* right side shows *Wortmania, Psittacotherium, Ectoganus,* and *Stylinodon.* (A–C from Matthew, 1937; D courtesy of R. C. Fox; E from Patterson, 1949.)

burrowing and digging up subterranean food items. Thus Schoch (1986: 1) concluded that derived stylinodonts were proficient rooters and grubbers, analogous to "an aardvark, with the head of a pig." More recently, however, Turnbull (2004) observed that *Stylinodon* lacks the abrasive scratches to be expected in a rooter. Instead he hypothesized that it used its powerful forelimbs to hook vines and branches and pull them through the teeth to strip off leaves and fruit. Whether such a diet could explain the hyselodonty in *Stylinodon,* not to mention its extremely robust limb elements, is conjectural.

An analysis of taeniodont tooth wear by scanning electron microscopy might help to resolve this controversy.

A new North American taeniodont, *Schowalteria* (Fig. 7.16D), was recently described from the latest Cretaceous (Lancian) of Alberta, Canada (Fox and Naylor, 2003). It is based on a snout and dentary, similar in size to its contemporaries *Cimolestes magnus* and *Didelphodon vorax* and is thus smaller than other taeniodonts. The cheek teeth are very heavily worn, obscuring most surface details, but a narrow stylar shelf and shallow ectoflexus (indentation of the buc-

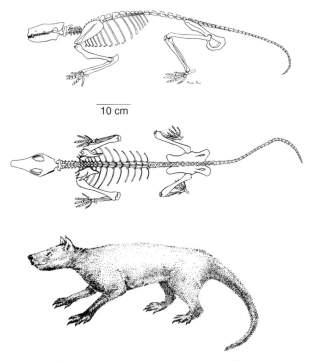

Fig. 7.17. Early Paleocene taeniodont *Onychodectes*. (From Schoch, 1986.)

cal margin) are evident. According to Fox and Naylor, several features—including the robust canines with restricted enamel; enlarged central incisor (interpreted as I_2, I_1 being absent); short, deep snout; and structure of the zygomatic arch—suggest that *Schowalteria* is related not just to taeniodonts, but specifically to stylinodontids. If their interpretation is correct, it challenges the hypothesis that *Alveugena* is the sister taxon of Taeniodonta. Nonetheless, no group other than cimolestids has been advanced as a likely relative. Thus the existence of taeniodonts in the Late Cretaceous might be indirect evidence of unknown older cimolestids. Although *Schowalteria* appears to be more plesiomorphic than other taeniodonts in nearly all dental features, it is already more derived than primitive taeniodonts in lacking I_1. This feature raises the possibility that its taeniodont-like features could be convergent.

Taeniodonts may well have been restricted to North America. Their basic interrelationships are shown in Fig. 7.19B. The only non-North American taxa that have been assigned to the order are early Eocene *Eurodon* and *Lessnessina* from Europe, which have been interpreted as conoryctids (Estravis and Russell, 1992). They are represented so far only by a few dental specimens, which are low crowned, bunodont, and very small. Although these specimens share certain features with primitive conoryctids such as *Onychodectes*, they contrast with taeniodonts in other features (the presence of strong cingula on upper molars, expanded talonid on M_3) that are reminiscent of condylarths, such as

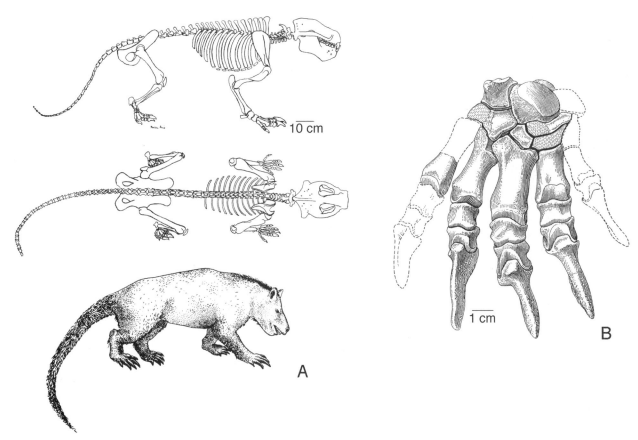

Fig. 7.18. Stylinodont taeniodonts: (A) Eocene *Stylinodon*; (B) right forefoot of Paleocene *Psittacotherium*. (A from Schoch, 1986; B from Matthew, 1937.)

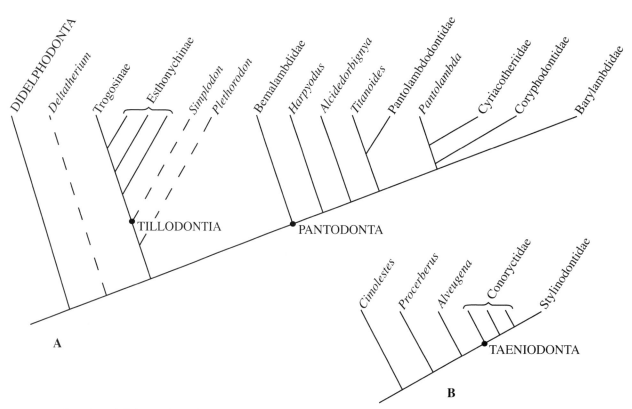

Fig. 7.19. (A) Relationships of tillodonts and pantodonts; dashed lines indicate tentative positions. (B) Relationships of taeniodonts. The position of Late Cretaceous *Schowalteria* (not shown) is uncertain; it is more primitive in some features than the basal conoryctid *Onychodectes*, but in other features it appears to be a basal stylinodontid (Fox and Naylor, 2003). (A based mainly on Muizon and Marshall, 1992, and Lucas, 1993, 1998; B mainly after Schoch, 1986, and Eberle, 1999.)

hyopsodontids. *Lessnessina*, in fact, was initially identified as a periptychid condylarth (Hooker, 1979) and most recently was considered to be a hyopsodontid (Hooker and Dashzeveg, 2003). The relationship of these mammals to taeniodonts should be considered questionable until more definitive evidence is known.

TILLODONTIA

Tillodontia (Fig. 7.19A) is a relatively small and highly distinctive group of archaic mammals that was widespread across the Northern Hemisphere during the Paleogene. The fifteen or so known genera have a wide range in size, but are dentally relatively conservative. All are assigned to the same family, and most experts have recognized the group as a distinct mammalian order. The family name Tillotheriidae technically has priority, but Esthonychidae has been almost universally used since Gazin's (1953) revision half a century ago and is therefore used here.

The dentition of tillodonts (Fig. 7.20) is characterized by rodentlike incisors with restricted enamel, the second pair enlarged and in more advanced types ever-growing. These gliriform incisors (especially in the most derived forms) bear a superficial resemblance to the hypertrophied canines of taeniodonts, but they are not homologous. In most forms the dental formula is 2.1.3.3/3.1.3.3. However, a small P1 was present in several of the most primitive genera (Paleocene *Benaius*, *Lofochaius*, and *Yuesthonyx*, and Eocene *Basalina*,

all from Asia), and I_1 and I_3 were lost in middle Eocene *Tillodon*. The teeth between the enlarged incisors and P_3 are typically reduced and may be separated by small gaps, creating a functional diastema resembling that of rodents. The cheek teeth are broad, with the trigonids only a little higher than the talonids, and they often show heavy wear, exposing broad dentinal areas. The fourth premolars above and below are submolariform. The lower cheek teeth are moderately high crowned and columnar (without cingula) on the buccal side, and lower crowned lingually. The crowns of the lower molars are incipiently selenodont and bear a metastylid cusp behind the metaconid. M_3 has an extended hypoconulid lobe. The upper molars have a moderate stylar shelf, small conules, and a hypocone cusp or lobe that makes M^{1-2} more or less quadrate. A few genera developed a mesostyle. These features, together with aspects of the postcranial anatomy, suggest that tillodonts were herbivorous or omnivorous, feeding on tough vegetation, fruits, or roots and tubers. Gingerich and Gunnell (1979) observed grooves on the incisors, which may indicate that they were used to pull up roots or coarse vegetation (the name "tillodont" comes from Greek roots meaning "teeth that pull out"). They speculated that a diet that inadvertently incorporated considerable grit might explain the dental anatomy and wear.

Esthonychidae is usually considered to include two subfamilies, Esthonychinae and Trogosinae. The middle to late Eocene trogosines are distinguished principally by their

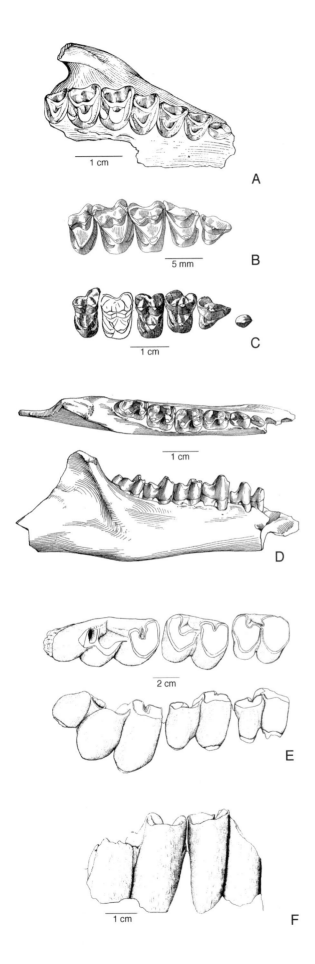

extremely hypertrophied, gliriform, and continuously growing second incisors, associated with progressive hypsodonty of cheek teeth and variable reduction or loss of premolars. There is no serious challenge to the monophyly of Trogosinae. Esthonychines, however, are characterized by their less enlarged, rooted I_2 (a primitive trait). They almost certainly include the ancestors of Trogosinae and are thereby paraphyletic. For this reason, McKenna and Bell (1997) placed all these tillodonts in a single subfamily.

The skull and skeleton of tillodonts are known from only a few specimens. The skull is best known in trogosines (Fig. 7.21A), which have well-defined temporal fossae, a moderate sagittal crest, flaring zygomae, and a narrow but elongate snout associated with the enlarged incisors. Esthonychines, known from skulls that are less well preserved, appear to be basically similar. The mandibular symphysis is fused and elongate in all but the most primitive tillodonts (e.g., *Azygonyx*, *Interogale*).

The postcranial skeleton is best known in the dentally derived Bridgerian genus *Trogosus*, in which it is robust but otherwise rather generalized. The limb bones are heavy and bear strong crests, suggesting powerful musculature that is consistent with digging and perhaps also with climbing. In *Esthonyx*, *Trogosus*, and presumably other tillodonts, the carpus still has all nine unfused carpals (the primitive eutherian condition) and the manus is pentadactyl and tipped by ungual phalanges that were curved, laterally compressed, and unfissured (Gazin, 1953). The manus bears a close overall resemblance to that of the Paleocene arctocyonid *Claenodon* (quite possibly a primitive resemblance), particularly in the form and arrangement of the carpals and the terminal phalanges. The other phalanges are relatively shorter and broader than in *Claenodon*. The tibia is noticeably shorter than the femur, which is a tendency of clawed herbivores that load the hind limbs more heavily than the forelimbs (Coombs, 1983). The pes is poorly known.

Only fragmentary postcrania are known for the more primitive esthonychines *Azygonyx* and *Esthonyx*. What little exists is not very different from the larger trogosine postcrania, but there is clearer indication of scansorial tendencies, especially in *Azygonyx*, as reflected by a round radial head; shallow astragalar trochlea; and curved, laterally compressed claws (Gingerich, 1989). *Esthonyx* was more terrestrially adapted than was *Azygonyx*, as inferred from the more ovoid proximal radius and well-defined patellar groove on the femur, but it was probably also a capable climber (Rose, 2001a).

Azygonyx, which includes some species formerly referred to *Esthonyx* (Gingerich and Gunnell, 1979), is the oldest North American tillodont and one of the most primitive members of the family. It first appears at the beginning of

Fig. 7.20. Tillodont dentitions (A–C right upper teeth and D–F right lower teeth): (A) *Plethorodon*; (B) *Simplodon*; (C, D) *Esthonyx*; (E) *Trogosus*; (F) *Higotherium*. (A from Huang and Zheng, 1987; B from Huang and Zheng, 2003; C from Gazin, 1953; D from Simpson, 1937b; E, F from Miyata and Tomida, 1998.)

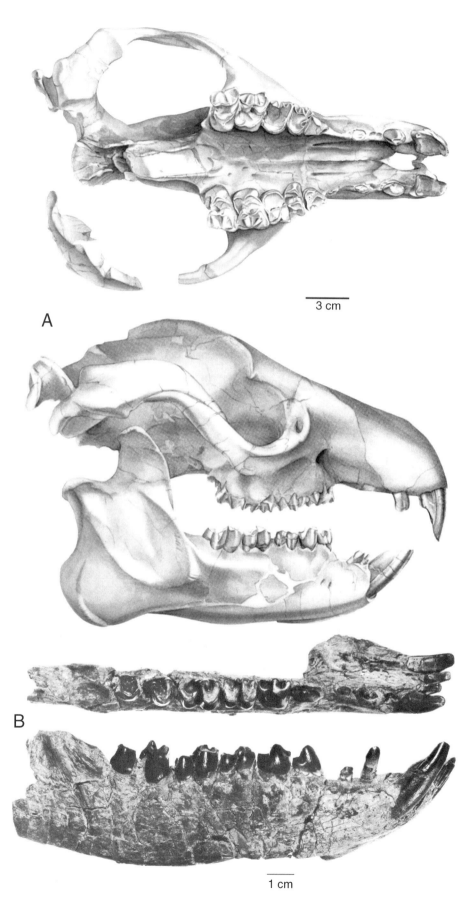

Fig. 7.21. Tillodonts: (A) Bridgerian *Trogosus* skull and mandible; (B) Wasatchian *Megalesthonyx* partial mandible. (A from Gazin, 1953; B from Rose, 1972.)

the Clarkforkian and is characteristic of this land-mammal age. *Azygonyx* survived into the early Wasatchian, where it coexisted with the most common tillodont, *Esthonyx* (Fig. 7.20C,D). The latter, however, persisted through the Wasatchian. The European genus *Plesiesthonyx* appears to be closely allied with *Azygonyx* (Baudry, 1992). These and other esthonychines were about the size of a raccoon or smaller.

Older or more primitive forms assigned to Tillodontia are now known from Europe and Asia. These include *Franchaius,* from the lower Eocene of Europe (Baudry, 1992); *Benaius, Lofochaius, Meiostylodon,* and *Huananius,* from the early Paleocene of China (e.g., Huang and Zheng, 1999; Wang and Jin, 2004); and *Yuesthonyx* from the late Paleocene of China (Tong et al., 2003). Late Paleocene *Interogale* and the poorly known ?middle Paleocene *Anchilestes* from China, both initially assigned to the order Anagalida, could also be primitive tillodonts (Ting and Zheng, 1989). Like tillodonts, *Interogale* has an enlarged I_2, reduced I_3, absent P_1, and elongated hypoconulid lobe on M_3. However, the incisors lack restricted enamel, a derived characteristic of all other tillodonts, indicating that it is more primitive in this regard than other tillodonts. *Anchilestes* resembles tillodonts in having a shelflike ectocingulum on upper teeth, a well-developed hypocone, and relatively low trigonid, but these characters also occur in other mammalian groups. All these Chinese Paleocene forms are much smaller than North American tillodonts, and some are more primitive in having wider stylar shelves and lacking a distinct hypocone on upper molars. It may be that some are not tillodonts at all; indeed several authors have noted the difficulty of distinguishing them from pantodonts (e.g., Ting and Zheng, 1989; Wang and Jin, 2004)—which suggests that the two groups are closely allied.

Trogosines almost certainly arose from an esthonychine close to *Megalesthonyx* (Fig. 7.21B). The latter is a rare late Wasatchian form from Wyoming that is essentially intermediate between other esthonychines and trogosines in size, most dental features, and age (Rose, 1972). *Megalesthonyx,* like *Esthonyx,* had rooted second incisors, unlike the ever-growing incisors of trogosines. The presence of a mesostyle on the upper molars associates *Megalesthonyx* with the Asian genera *Adapidium* and *Yuesthonyx* (though the latter is more primitive) and precludes the single known species from being directly ancestral to trogosines.

Trogosines (Fig. 7.21A) were relatively large mammals for the Eocene, reaching the size of small bears and weighing at least 150 kg (Lucas and Schoch, 1998a). The most derived forms were the middle Eocene genera *Tillodon* from North America, and *Higotherium* (Fig. 7.20F) and *Chungchienia* from Asia. In addition to the rootless I_2 (which in one specimen of *Chungchienia* is 26 cm long!), the Asian genera also had remarkable hypsodont to hypselodont (evergrowing) cheek teeth (Chow et al., 1996; Miyata and Tomida, 1998). The enamel of the cheek teeth of *Chungchienia* is limited to the buccal surfaces only and extends all the way to the open roots. The columnar, very hypsodont condition of these teeth gives them a superficial resemblance to horse (*Equus*) teeth in lateral view. *Chungchienia* seems to have been gnawing and ingesting very abrasive material, but precisely what led to the evolution of such an unusual dentition is unknown. When known from only a few fragments, *Chungchienia* was believed to be either an aberrant edentate or a taeniodont. Its true affinities only became apparent after the discovery of more complete material.

Based on dental and postcranial anatomy it was long thought that tillodonts were related to, or possibly derived from, the condylarth family Arctocyonidae (e.g., Gregory, 1910; Gazin, 1953; Van Valen, 1963; Rose, 1972). This view was undoubtedly influenced by the relative abundance, diversity, and antiquity of North American tillodonts, as well as the presence on this continent of diverse arctocyonids in older strata. However, the discovery of older and more primitive tillodonts in Asia, with dental resemblances to pantodonts, increasingly suggests that these two groups are more closely related (Gazin, 1953; Chow and Wang, 1979; Wang and Jin, 2004). Lucas (1993), while considering many of the resemblances between those two groups to be convergent, nevertheless regarded them as sister taxa. Other authors have emphasized the hypothesized link with pantodonts by including Tillodontia as a suborder of Pantodonta (Chow and Wang, 1979; Marshall and Muizon, 1988). Neither of these phylogenetic hypotheses has been rigorously tested, however; and it remains possible that dental similarities between tillodonts and pantodonts, such as dilambdodonty, arose independently (e.g., Gazin, 1953; Ting and Zheng, 1989; Lucas, 1993), perhaps from a common ancestor among the Cimolestidae. This view is consistent with recognition of both Tillodontia and Pantodonta as separate orders or suborders of the Cimolesta, as proposed by McKenna and Bell (1997).

Early Paleocene (Torrejonian) *Deltatherium* has been considered a pivotal genus variously assigned to arctocyonid condylarths, Pantodonta, or Tillodontia. Its phylogenetic position is still very much in dispute, largely because it exhibits certain derived features of these groups superimposed on a very primitive dental pattern. The upper molars retain a plesiomorphic trituberular pattern with a moderately wide stylar shelf; in addition there are small conules and a prominent lingual cingulum with a small hypocone (presumably derived). The lowers have sharp cusps and moderately high trigonids (primitive) but are relatively broad and low crowned (derived). The ambiguity in its allocation suggests that it could occupy a phylogenetic position near the divergence of Tillodontia, Pantodonta, or both, linking them to Arctocyonia (Van Valen, 1988); however, absence of the first premolar would seem to preclude a position ancestral to either tillodonts or pantodonts. Lucas (1993, 1998) similarly regarded *Deltatherium* as a sister taxon of both Tillodontia and Pantodonta, perhaps closer to tillodonts than to pantodonts, but concluded that all of them evolved from a "didelphodontine" (i.e., cimolestid) rather than an arctocyonid ancestor.

Also potentially pertinent to the origin of Tillodontia is early Paleocene *Plethorodon* (Fig. 7.20A), based on a snout with complete postcanine dentition from China. Although initially referred tentatively to the Pantodonta (Huang and Zheng, 1987), *Plethorodon* has less transverse upper cheek teeth, a shallower ectoflexus (buccal indentation), and better developed pre- and especially postcingula (hypocone shelf)—all of which are characteristics of tillodonts. *Plethorodon* could be a primitive tillodont (e.g., McKenna and Bell, 1997) or another offshoot of a possible pantodont-tillodont common ancestor. The recently described *Simplodon* (Fig. 7.20B), based on a maxillary dentition from the middle Paleocene of China, may also be relevant to the origin of tillodonts (Huang and Zheng, 2003). Its molars lack conules and a hypocone but have a well-developed posterolingual shelf.

Ting and Zheng (1989) proposed that Anagalida (in their view then, including Zalambdalestidae) is the closest relative of Tillodontia, but this hypothesis appears unlikely. The higher-level relationships of Tillodontia therefore remain unsettled, but existing evidence favors a relationship to pantodonts and an Asian origin from cimolestids, probably very early in the Paleocene.

PANTODONTA

The Pantodonta were a group of heavily built omnivorous and herbivorous mammals that were moderately common and diverse in the Northern Hemisphere during the Paleocene and Eocene. One genus is known from South America. About two dozen genera in 10 families are currently recognized (McKenna and Bell, 1997). Most of the families are restricted to the Paleocene and are known from either North America or Asia, but not both. Pantolambdodontidae and Coryphodontidae extended into the Eocene as well, and the latter family is the only group of pantodonts to disperse across the three Holarctic continents. Pantodonts were long grouped with ungulates, either together with uintatheres as amblypods, as paenungulates (Simpson, 1945), or linked with arctocyonids via *Deltatherium* (e.g., Van Valen, 1988). In recent years there has been a growing consensus that they are closely allied with tillodonts and derived from Cimolestidae (e.g., Muizon and Marshall, 1992; Lucas, 1993; McKenna and Bell, 1997). Pantodonts tended toward ponderous size, some probably exceeding 500 kg and having skulls half a meter or more in length. They include some of the largest mammals of their time. Nevertheless, as in many clades, some of the most primitive members weighed less than 10 kg.

Pantodonts are characterized mainly by dental traits (Simons, 1960; Lucas, 1998; Figs. 7.22, 7.23). Their most important synapomorphy is the distinctive V-shaped ectoloph of the posterior upper premolars (P^{3-4}). The upper molars are usually dilambdodont (with a W-shaped ectoloph), although the latter tendency was poorly developed in the most primitive members. In Asian families, the paracone and metacone tend to be closer together and more lingually situated than in the North American families (Ting et al., 1982). Nearly all pantodonts lack a hypocone, and the conules are typically small. P^3–M^3 usually have a moderate to deep indentation, or ectoflexus, of the buccal margin. The lower cheek teeth are also dilambdodont and have broad, high metalophids with tall metaconids and much lower paracristids with reduced paraconids. The dental formula is primitive, 3.1.4.3/3.1.4.3, with no significant diastemata. The incisors are small and the canines large and sometimes saberlike.

The postcranial skeleton is known in several genera and is plesiomorphic (with no elements lost or fused) and robust. The feet are pentadactyl and usually hoofed. In addition to the presence of hoofs in most genera, evidence for possible ungulate ties has come from the overall similarity of the foot skeleton, especially the tarsus, to that of arctocyonid "condylarths" such as *Protungulatum* (Szalay, 1977), but these similarities are quite possibly primitive (Lucas, 1993). At the same time, Szalay (1977) has pointed out that basal pantodonts lack the tarsal specializations present in cimolestids, an argument against their relationship. Proposed pantodont-cimolestid relationship is based primarily on dental resemblances between the oldest pantodonts and Wasatchian *Didelphodus* (Muizon and Marshall, 1992; Lucas, 1993).

The classification and interrelationships of pantodonts (Fig. 7.19A) are controversial, but there is general agreement that the most primitive pantodonts are the Paleocene genera *Harpyodus* and *Alcidedorbignya* and the family Bemalambdidae.

Bemalambdidae (Figs. 7.22B, 7.23A; see also Fig. 7.26) are represented by several small to medium-sized early Paleocene species from China (Zhou et al., 1977). They have the hallmark upper premolars of pantodonts, but the upper molars are very transverse, almost zalambdodont (not at all dilambdodont), with closely appressed or connate paracone and metacone. The stylar shelf of P^3–M^3 is very wide, and the ectoflexus is deeply incised. The lower cheek teeth have the typical pantodont morphology. Bemalambdids had low, rather short skulls with a broad snout, flaring zygomatic processes, and a very small braincase. Deep temporal fossae, a prominent sagittal crest, and a high mandibular coronoid process suggest relatively better-developed temporal musculature than in later pantodonts. The postcranial skeleton was robust. One species had a particularly massive humerus, suggesting a propensity for digging.

Harpyodus (Fig. 7.22A) was a very small pantodont from the early and late Paleocene of China (Wang, 1979). Like *Bemalambda*, it has typical pantodont premolars and upper molars with closely appressed or connate paracone and metacone and a very wide stylar shelf. It differs from *Bemalambda* and most other pantodonts, however, in having a distinct hypocone, presumably an autapomorphy of this genus.

Alcidedorbignya (Figs. 7.22C, 7.23B), from the early Paleocene Tiupampa local fauna of Bolivia, may be older than the pantodonts from China (Muizon and Marshall, 1992). It is small, about the same size as *Harpyodus*, but differs from the Chinese forms in having the paracone and metacone of the upper molars separate rather than connate as in *Harpyodus* and *Bemalambda*. Although this feature has been interpreted as a derived trait of *Alcidedorbignya,* its polarity is uncertain. The stylar shelf is wide and there is a deep notch, or ectoflexus, in the buccal margin; there is no mesostyle. *Alcidedorbignya* has a shelflike postcingulum, sometimes bearing a small hypocone. The upper molars of all three genera lack the inverted V-shaped centrocrista and mesostyle, which contribute to the W-shaped ectoloph characteristic of other pantodonts. It is not clear which of these three genera has the most primitive upper molars—that is, whether separate or connate paracone-metacone is the plesiomorphic condition in pantodonts.

The Paleocene Pantolambdidae are usually considered the most primitive North American pantodonts. *Pantolambda* (Figs. 7.22–7.26), from the early Paleocene (Torrejonian), was typical. It was a medium-sized animal, about the size of the wolverine *Gulo,* characterized by typical pantodont upper premolars, dilambdodont upper and lower molars, and a robust skeleton (Matthew, 1937). Several late Paleocene

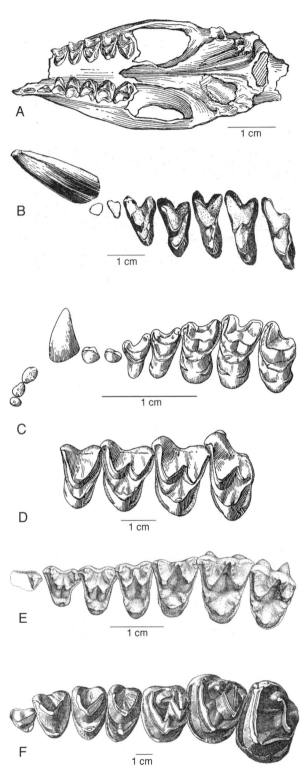

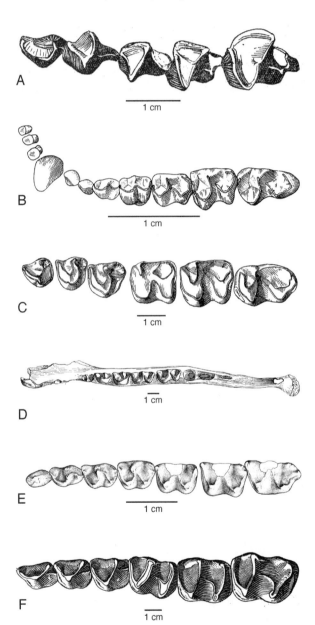

Fig. 7.22. Upper dentition of pantodonts (anterior to left): (A) *Harpyodus* skull; (B) *Bemalambda;* (C) *Alcidedorbignya;* (D) *Pantolambda;* (E) *Cyriacotherium;* (F) *Coryphodon.* (B)–(F) show left dentition. (A from Wang, 1979; B from Zhou et al., 1977; C from Muizon and Marshall, 1992; D from Matthew, 1937; E from Rose and Krause, 1982; F modified from Osborn and Granger, 1931.)

Fig. 7.23. Left lower dentition of pantodonts: (A) *Bemalambda;* (B) *Alcidedorbignya;* (C) *Pantolambda;* (D) *Pantolambdodon;* (E) *Cyriacotherium;* (F) *Coryphodon.* (A from Zhou et al., 1977; B from Muizon and Marshall, 1992; C from Matthew, 1937; D from Granger and Gregory, 1934; E from Rose and Krause, 1982; F modified from Osborn and Granger, 1931.)

Fig. 7.24. Skulls of North American pantodonts. (A) *Coryphodon;* (B) *Barylambda;* (C) *Titanoides;* (D) *Haplolambda;* (E) *Caenolambda;* (F, G) *Pantolambda.* (From Simons, 1960.)

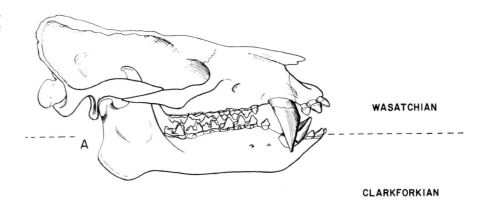
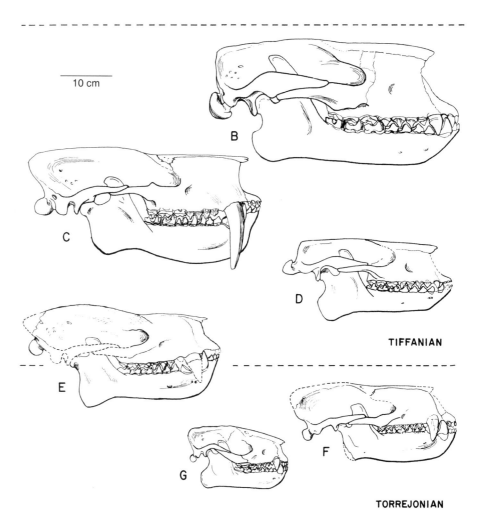

pantodonts seem to be closely related to, or perhaps derived from, pantolambdids, based on dental similarities. *Barylambda* (Figs. 7.24–7.26), best known of the Barylambdidae, was a large (650 kg), lumbering pantodont from the late Paleocene of Wyoming and Colorado. It had a small head, a graviportal pelvis and hindlimb skeleton, and an especially robust tail, suggesting an ability for bipedal browsing, as in extinct giant ground sloths (e.g., Simons, 1960; Coombs, 1983; Lucas, 1998). Barylambdids were apparently supplanted by *Coryphodon* in the latest Paleocene (Clarkforkian) of western North America (Gingerich and Childress, 1983), though *Coryphodon* was surely quadrupedal. Other probable relatives of pantolambdids were the families Titanoideidae and Cyriacotheriidae.

Late Paleocene *Titanoides* (Figs. 7.24, 7.25B, 7.26) of western North America, the sole genus of Titanoideidae, differs from other North American pantodonts in having claws rather than hoofs. Based on the widespread presence of hooflike terminal phalanges in all other pantodonts, the presence of claws clearly must be a derived condition in

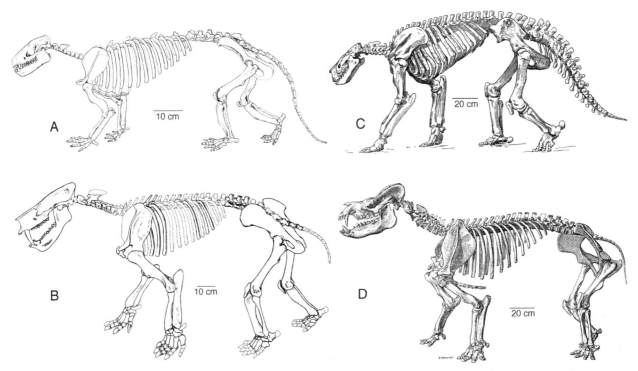

Fig. 7.25. Skeletons of pantodonts: (A) *Pantolambda;* (B) *Titanoides;* (C) *Barylambda;* (D) *Coryphodon.* (A, B from Simons, 1960; C from Lavocat, 1955, after Patterson; D from Osborn, 1898b.)

Titanoides. The cheek teeth of *Titanoides* are basically similar to those of pantolambdids. *Titanoides* had large, saberlike upper canines, typically unworn, whose function is therefore difficult to decipher (Coombs, 1983). Apart from the claws, the skeleton is similar to that in other pantodonts. *Titanoides* was evidently somewhat adapted for digging but, unlike *Barylambda,* had no adaptations for bipedal browsing (Coombs, 1983).

Based partly on the presence of claws, Lucas (1993, 1998) allied *Titanoides* with the Asian pantolambdodontid *Archaeolambda,* the only other pantodont known to have had claws (Huang, 1977; Lucas, 1982). But *Archaeolambda* was much smaller (about 7 kg). Perhaps it was the combination of small body size and a relatively gracile, clawed skeleton that led Lucas (1998) to suggest that *Archaeolambda* was arboreal. No detailed analysis of the skeleton has been presented, however, so this novel hypothesis cannot yet be evaluated.

Pantolambdodontids (Fig. 7.23D) were mainly late Paleocene–Eocene pantodonts further characterized by shallow jaws, small canines and premolars, lower molar trigonids that are larger than the talonids and bear high paraconids, and especially pronounced W-shaped ectolophs on M^{1-2} (Lucas, 1993; Huang, 1995). One species of *Pantolambdodon* had an elongate, low rostrum with a large, high nasal opening, indicating the presence of a tapirlike proboscis (Ding et al., 1987). Another Asian group, Pastoralodontidae, had a greatly expanded mandibular angle, resembling that of a hippopotamus (Chow and Qi, 1978). Huang (1995) relegated pastoralodonts to a subfamily of Pantolambdodontidae.

Cyriacotherium ("Sunday beast") differs from all other pantodonts in having molariform premolars (Figs. 7.22E, 7.23E). P^{3-4} have W-shaped ectolophs rather than the simpler V-shape characteristic of other pantodonts, and the lower premolars have larger talonids. This striking contrast with other pantodonts was the basis for assigning *Cyriacotherium* to its own family, Cyriacotheriidae (Rose and Krause, 1982). Some subsequent researchers, however, cited this feature as grounds for excluding *Cyriacotherium* from the Pantodonta and instead proposed that it may be a plagiomenid or a mixodectoid (Van Valen, 1988; Muizon and Marshall, 1992; Lucas, 1993). The matter has not been resolved, and the phylogenetic position of *Cyriacotherium* remains controversial; but either way there has been remarkable dental convergence to one group. Cyriacotheriids resemble pantolambdids in several derived features, including lower molar dilambdodonty, lingually placed hypoconulids, and reduced or absent entoconids, as well as in having small incisors that enlarge slightly from I_1 to I_3. They lack certain dental synapomorphies of plagiomenids, such as accessory stylar cusps, skewed cheek teeth, and an enlarged I_1. *Cyriacotherium* was somewhat smaller than *Pantolambda* but larger than *Harpyodus* and is known only from the late Paleocene of the northern Rocky Mountain region.

Coryphodontidae were large, derived pantodonts of the late Paleocene and Eocene, probably also related to pantolambdids. However, intermediate stages leading to the unusual upper molars of *Coryphodon* are unknown. Named in 1845 by Richard Owen, *Coryphodon* has been well known from Europe and North America for more than a century. It first appeared in the late Paleocene (Clarkforkian) of North America and is particularly abundant in early Eocene (Wasatchian) strata; it is also known from early

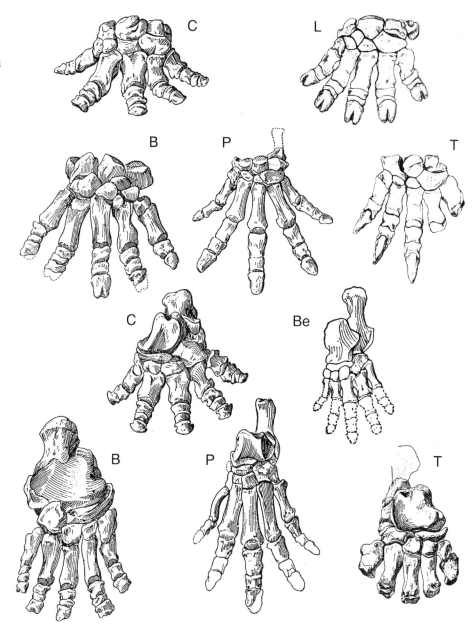

Fig. 7.26. Forefeet (top two rows) and hind feet of pantodonts. Key: B, *Barylambda;* Be, *Bemalambda;* C, *Coryphodon;* L, *Leptolambda;* P, *Pantolambda;* T, *Titanoides*. (From Lavocat, 1955; Simons, 1960; Zhou et al., 1977.)

Eocene beds of Europe and Asia. Several other genera of coryphodontids are found in middle and upper Eocene deposits of Asia.

Coryphodon is known from numerous skulls and skeletons (Figs. 7.24–7.26, Plate 3.3). It was graviportal, with a massive skeleton and rather short limbs. As in other pantodonts, the radius and ulna are separate, as are the tibia and fibula. The distal limb segments are distinctly shortened and the feet are broad and spreading and bore short, wide hoofs. The lophodont teeth and large canines are reminiscent of the arrangement in the hippopotamus and, together with the skeleton, suggest that *Coryphodon* was semiaquatic, feeding largely on aquatic vegetation. Simons (1960) described grooves on the lower canines that may have been caused by pulling up tough vegetation.

Coryphodon was sexually dimorphic in body size and canine size. Large individuals approximated the size of a steer. Body mass estimates are highly variable, depending on whether they are based on teeth, particular postcranial elements, or overall body length, and which regression is employed. Estimates range from 90 to 800 kg for various species (Uhen and Gingerich, 1995). Even assuming it only reached the middle of this range, *Coryphodon* was without doubt one of the biggest early Eocene mammals. Its brain, however, was relatively among the smallest known for an animal of its size, about 90 g in an animal of 500 kg (Savage and Long, 1986)!

The antiquity, diversity, and plesiomorphic state of Asian members of both Pantodonta and Tillodontia suggest that continent as a center of origin of these clades.

Creodonta and Carnivora

THE ORDER CARNIVORA AND THE extinct order Creodonta include the most carnivorously adapted placental mammals, and they have been widely considered to be sister taxa, within the superorder Ferae (Fig. 8.1, Table 8.1). Except for a general dental similarity between them, however, there is surprisingly little evidence that they constitute a monophyletic group (e.g., Flynn et al., 1988). For example, although both groups have teeth adapted for carnivory, the carnassial teeth are not homologous in creodonts and carnivorans (Fig. 8.2). Possible synapomorphies of the two orders include the presence of an ossified tentorium (a bony shelflike projection separating parts of the brain) and a few basicranial and tarsal similarities, but they are not very compelling (Flynn et al., 1988; Wyss and Flynn, 1993). There is also a close correspondence in postcranial osteology between various primitive carnivorans (miacoids) and creodonts, but as yet, none of the similarities has been shown to be synapomorphic, and most are thought to be primitive. Consequently the case for monophyly of a creodont-carnivoran clade is weak. Nevertheless, no preferable phylogenetic arrangement is obvious. Although possible relationships between Carnivora and other major clades (Pholidota, Lipotyphla, Primates, Chiroptera, and various ungulates) have been suggested, largely based on molecular evidence, none of these (except possibly Pholidota) is particularly persuasive either (Flynn and Wesley-Hunt, 2005).

CREODONTA

Also known as Pseudocreodi or "archaic carnivores," creodonts are a group of extinct carnivorous mammals that thrived during the Eocene and Oligocene in North

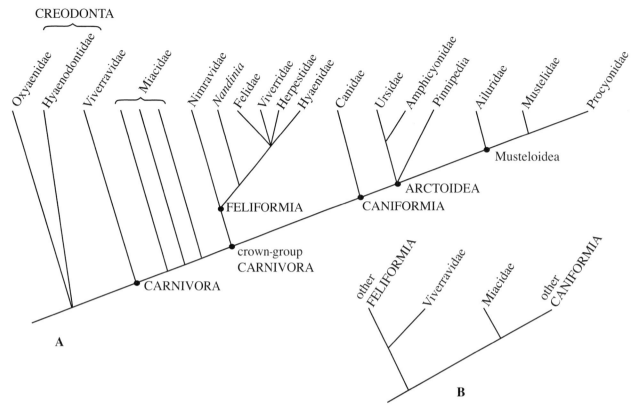

Fig. 8.1. (A) Relationships of creodonts and carnivorans. Alternatively, Amphicyonidae could be the most basal branch of Caniformia, and Pinnipedia could be the sister taxon of Musteloidea. Canidae constitute the Cynoidea. (B) Alternative view of basal carnivoran relationships. (A based on Flynn and Wesley-Hunt, 2005; B modified from McKenna and Bell, 1997, and Flynn and Galiano, 1982.)

America and from the Eocene into the Miocene in the Old World. Indeed, they were the most abundant carnivorous mammals in the Old World during the Paleogene (Muizon and Lange-Badré, 1997). They coexisted with true carnivores, order Carnivora, during their entire duration, but were eventually replaced by members of the "modern" order. The reason for this replacement remains a mystery. The available evidence does not particularly support competitive displacement, as creodonts were declining in diversity through most of the Eocene, and carnivorans did not diversify morphologically to fill the vacant niches (Van Valkenburgh, 1999; Wesley-Hunt, 2005). Because they were similarly adapted, creodonts are usually considered together with Carnivora (in the higher taxon Ferae) and could be their sister group, although as noted above, the evidence for this alliance is weak.

Some popular and general accounts still include a diversity of primitive carnivorous mammals, such as mesonychids and arctocyonids, under the heading of creodonts. But virtually all current paleomammalogists have abandoned this antiquated view and restrict the name Creodonta to two families, Oxyaenidae and Hyaenodontidae. These two families are united by the presence of more posterior carnassials (specialized shearing teeth) than in the Carnivora, as well as by various primitive aspects of the skeleton. The teeth modified as carnassials—often two or even three teeth in each jaw, although the more posterior tooth is typically the principal carnassial—usually differ in the two families, however; and there are few, if any, skeletal synapomorphies. These factors suggest that Creodonta, although a convenient and widely used term, is not a natural group (Polly, 1996), and that the two families may have emerged independently from cimolestan ancestors. *Cimolestes* (see Chapter 7) is a plausible morphotype for both families of creodonts (Lillegraven, 1969).

The dental formula of creodonts is primitively 3.1.4.3/3.1.4.3, but many forms have reduced the number of molars (oxyaenids, most limnocyonine hyaenodontids), incisors, or premolars (Denison, 1938). The canines are always large. Creodont premolars are simple, with one primary cusp and variable anterior and posterior accessory cusps; a low protocone is sometimes present on the posterior upper premolars. The molars are primitively tribosphenic. The lower molars have tall trigonids and narrow talonids, the trigonids often wider than the talonids, and the talonids often reduced, especially on the more posterior molars. The lower carnassial typically has a reduced metaconid and a sectorial paracristid. The upper molars are triangular, with connate (closely joined) paracone and metacone, small conules, and a well-separated and often reduced protocone. The metastyle of the upper carnassials is large and joined to the metacone by a bladelike crest that occludes against the paracristid. The last upper molar is usually reduced, but the last lower molar ranges anywhere from very large to very small in different taxa.

Table 8.1. Classification of Creodonta and Carnivora

Superorder FERAE
   Order †CREODONTA
         †Hyaenodontidae
         †Oxyaenidae
   Order CARNIVORA
         †Viverravidae[1]
         †Miacidae[2]
    Suborder FELIFORMIA
         †Nimravidae
         Felidae
         Viverridae
         Nandiniidae
         Herpestidae
         Hyaenidae
    Suborder CANIFORMIA
      Infraorder CYNOIDEA
         Canidae
      Infraorder ARCTOIDEA
        Parvorder URSIDA
          Superfamily Ursoidea
           Ursidae
           †Amphicyonidae[3]
          Superfamily Phocoidea (=Pinnipedia)
           †*Enaliarctos*
           Otariidae
           Phocidae
           Odobenidae
        Parvorder MUSTELIDA
         Mustelidae
         Procyonidae

Notes: Modified after McKenna and Bell (1997). The dagger (†) denotes extinct taxa. Families in boldface are known from the Paleocene or Eocene.

[1] Previously considered stem feliforms; now considered stem carnivorans.
[2] Previously considered stem caniforms; now considered paraphyletic stem carnivorans.
[3] Sometimes placed in a separate superfamily †Amphicyonoidea; relationship to Ursidae is uncertain; could be the sister taxon of other caniforms (Wesley-Hunt and Flynn, 2005).

Many of these dental features, as well as some of the postcranial traits mentioned in the next paragraph, are also characteristic of the most primitive carnivorans (miacoids). The overall dental morphology of creodonts is also similar—strikingly so in some cases—to that of various borhyaenid and dasyuroid marsupials, but in this case the resemblances are certainly convergent. As observed by Muizon and Lange-Badré (1997), the recurrence of this complex of dental features in multiple mammalian groups, many of which are unequivocally only distantly related, constitutes strong evidence of its frequent homoplasy. This observation is further cause to be suspicious of a close relationship between creodonts and Carnivora.

As in many mammals, the enamel of creodont teeth exhibits a pattern of decussating prisms called Hunter-Schreger bands (HSB). The HSB have an unusual zigzag arrangement in taxa whose gross dental morphology suggests bone-crushing habits (Stefen, 1997), but the significance of this pattern is uncertain, as a similar pattern occurs in some herbivorous Paleogene mammals (Koenigswald and Rose, 2005).

Many creodonts had disproportionately large heads. Primitive characteristics of creodonts compared to Carnivora include strong sagittal and lambdoidal crests on the skull, an unossified auditory bulla, relatively short and generalized limbs (in early representatives), carpus with a centrale and separate scaphoid and lunate bones, and pentadactyl feet. The terminal phalanges are fissured at the tip and tend to be broader in oxyaenids than in hyaenodontids. Most creodonts were generalized terrestrial animals, but some early representatives were scansorial, and some later hyaenodontids were specialized cursors. Despite their primitive features, however, creodonts were not subordinate to carnivorans in dental adaptation for carnivory.

Creodonts were widespread and successful during the Early-Middle Tertiary, so why did they become extinct? Many were larger than contemporaneous carnivorans, and equalled or surpassed them in carnivorous adaptation. Although they were once thought to have had smaller brains than did contemporary carnivorans, it is now known that creodonts and carnivorans generally had similar-sized brains, and that both groups showed comparable brain expansion during the Early Tertiary (Radinsky, 1977). Thus competitive exclusion does not appear to be the explanation (Van Valkenburgh, 1999). Whatever the cause, oxyaenids became extinct by the late Eocene, whereas hyaenodontids persisted almost to the Pliocene (in parts of the Old World; only through the Oligocene in North America). Possibly contributing to their demise were their generally more conservative postcranial skeletons, but the full explanation is unknown.

Oxyaenidae

Oxyaenids were a mainly North American group of creodonts with relatively short, broad skulls and deep, robust dentaries. The primary carnassial teeth are M^1/M_2 (often with assistance from P^4 and M_1), and the third molars are absent, a derived feature relative to most hyaenodontids (Figs. 8.2, 8.3). All show clear specializations for meat eating, although the carnassials are only weakly developed in the earliest forms, and some seem to have been better adapted for crushing bones than cutting flesh. Certain later types even achieved catlike or hyena-like hypercarnivorous dental adaptations.

Most oxyaenids had rather long bodies and short, robust limbs with relatively short middle and distal segments (Fig. 8.4D). The weakly grooved astragalus articulates distally with both the navicular and the cuboid. The stance was probably plantigrade. Oxyaenid locomotor capabilities have proven difficult to assess, in part because they are so generalized, and in part because there are no obvious modern analogues. Oxyaenids are often described as wolverine-like, "terrestrial ambulatory" predators (e.g., Matthew, 1909; Denison, 1938; Gunnell and Gingerich, 1991). Prominent crests on the humerus and a well-developed olecranon process on the ulna suggest digging ability as well. Some features of the skeleton (e.g., divergent first digit, robust humerus, nearly flat astragalar trochlea indicating a mobile upper ankle joint) suggest an arboreal ancestry, and it is even possible that some early oxyaenids were scansorial.

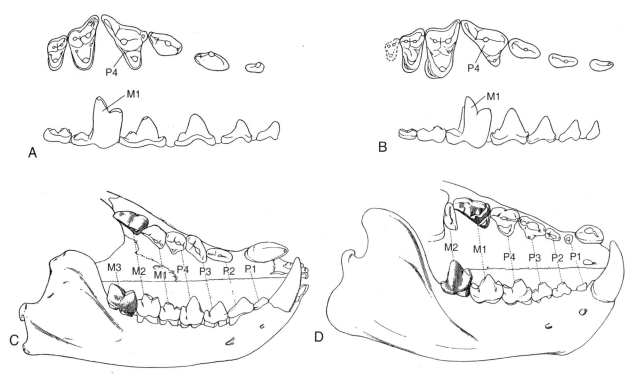

Fig. 8.2. Carnassial positions in carnivorans (A, B) and creodonts (C, D): (A) viverravid *Didymictis*; (B) miacid *Miacis*; (C) hyaenodontid *Hyaenodon*; (D) oxyaenid *Oxyaena*. Carnassials are P^4/M_1 in carnivorans, M^2/M_3 in hyaenodontids, and M^1/M_2 in oxyaenids. (Modified after Matthew, 1909.)

Oxyaenids were less diverse and of much shorter duration than hyaenodontids. About ten genera are recognized. The earliest genus, *Tytthaena* (Fig. 8.3A), appeared in North America in the late Paleocene (middle Tiffanian; Gunnell and Gingerich, 1991)—before hyaenodontids but after the oldest Carnivora—perhaps as an immigrant from Asia. *Tytthaena* was a rare, housecat-sized oxyaenid known only from teeth found in Wyoming. By the end of the Paleocene (Clarkforkian) several oxyaenid genera representing three different subfamilies were present, but their pedigree is uncertain. Late Paleocene *Dipsalodon*, *Palaeonictis*, and *Dipsalidictis* were the first big carnivorous mammals of the Early Cenozoic except for mesonychids, some species attaining the size of small bears. The latter two genera survived into the Eocene, but were much less common than their relative *Oxyaena* (Figs. 8.3B, 8.4A,C), which was one of the most prevalent carnivorous mammals in western North America during the Wasatchian. It, too, reached the size of a small bear—much larger than any contemporary true carnivoran. Both *Oxyaena* and *Palaeonictis* dispersed to Europe during the early Eocene. They were evidently terrestrial animals, whereas *Dipsalidictis* had more gracile limbs and more flexible elbow and ankle joints, suggesting scansorial habits (Gunnell and Gingerich, 1991).

One lineage of oxyaenids (Ambloctoninae), typified by *Palaeonictis* (Fig. 8.3E), widened the premolars, deemphasized the carnassials, and reduced M_2, perhaps reflecting a more omnivorous or durophagous diet. A relative of *Palaeonictis*, Wasatchian *Ambloctonus*, evolved in a somewhat different direction, achieving catlike shearing by reducing M_2 to a bladelike paracristid, although this molar was smaller than M_1. Hypercarnivory was achieved by the bear-sized middle Eocene oxyaenines *Patriofelis* (from North America and Europe; Fig. 8.4B,D) and *Sarkastodon* (Asia; Fig. 8.3D), which were probable descendants of *Oxyaena*. They had very robust premolars and independently evolved a bladelike M_2 consisting only of the paracristid, but here it is the largest cheek tooth, giving them a distinctly hyena-like dentition. With such slicing molars and crushing premolars they could have consumed both meat and bones (Gazin, 1957; Gunnell, 1998).

Carnivory reached its peak in the late Wasatchian-Bridgerian *Machaeroides* and *Apataelurus* (Machaeroidinae; Fig. 8.3C), which not only evolved a huge bladelike M_2 consisting of a hypertrophied paracristid, as in *Patriofelis* and *Sarkastodon*, but also had long upper canines protected by a prominent ventral flange at the front of the jaw (Denison, 1938; Scott, 1938; Dawson et al., 1986). These specializations made machaeroidines the oldest known saber-toothed predators (not to be confused with machairodontines, true saber-toothed cats, which did not appear until the Miocene). Their affinities remain problematic; they have been variously associated with oxyaenids (Dawson et al., 1986; Gunnell, 1998) or with limnocyonine hyaenodontids (Denison, 1938; McKenna and Bell, 1997).

Hyaenodontidae

Hyaenodontids (Figs. 8.5–8.8) were much more diverse than oxyaenids, consisting of about 50 genera found in Eocene-Oligocene deposits of North America and Eocene-

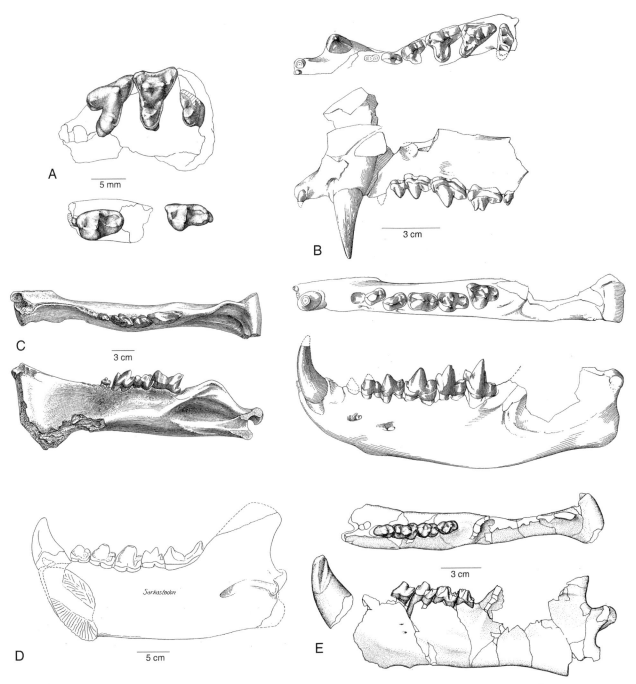

Fig. 8.3. Dentitions of oxyaenids (all left side except D): (A) *Tytthaena;* (B) *Oxyaena;* (C) *Apataelurus;* (D) *Sarkastodon* (right mandible, medial view); (E) *Palaeonictis.* (A from Gingerich, 1980; B from Matthew 1915a; C from Scott, 1938; D from Denison, 1938; E from Rose, 1981.)

Miocene beds of Eurasia and Africa (Lange-Badré, 1979; Gingerich and Deutsch, 1989; McKenna and Bell, 1997). They are variously placed in two to four subfamilies, whose relationships are not well understood. The Proviverrinae is a paraphyletic assemblage of the most primitive genera, whereas more derived genera are assigned to Hyaenodontinae, Limnocyoninae, or Pterodontinae (Polly, 1996). Proviverrinae and Pterodontinae are sometimes subsumed in Hyaenodontinae (e.g., McKenna and Bell, 1997). Hyaenodontids were particularly successful in the early and middle Eocene of North America and Europe, which had several genera in common. In Europe oxyaenids disappeared after the early Eocene, and hyaenodontids filled many of the vacant niches. Hyaenodontids are one of several higher taxa that generally herald the beginning of the Eocene across Laurasia (Gingerich, 1989; Smith and Smith, 2001). An exception is *Prolimnocyon,* the earliest member of the Limnocyoninae, which was evidently present, but rare, in the latest Paleocene of Asia (Meng et al., 1998). This occurrence hints at an Asian source for the group.

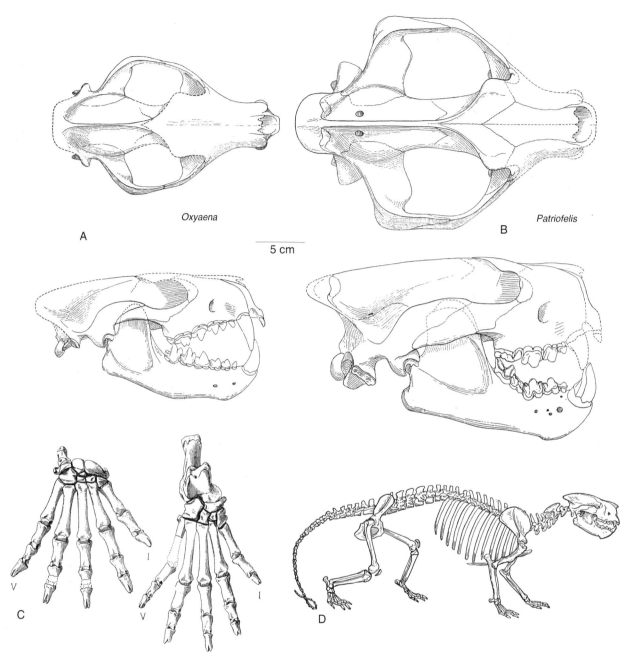

Fig. 8.4. Oxyaenids: (A) skull of *Oxyaena*; (B) skull of *Patriofelis*; (C) right forefoot and hind foot of *Oxyaena*; (D) skeleton of *Patriofelis*. (A–C from Denison, 1938; D from Gregory, 1951.)

If Hyaenodontidae form a clade with Oxyaenidae, as is commonly assumed, it implies the existence of Hyaenodontidae by no later than the Tiffanian, perhaps in some as-yet unsampled region. Furthermore, the derived dental state of oxyaenids (loss of third molars) relative to hyaenodontids, suggests that stem creodonts, which probably would have resembled primitive hyaenodontids, were present well before the Tiffanian.

Hyaenodontids generally had longer, narrower skulls and shallower jaws than those of oxyaenids (Figs. 8.5, 8.6), which are probably primitive conditions. They differ further from oxyaenids in primitively retaining three molars, with the principal shearing usually taking place between M^2 and M_3, though more anterior molars often were also involved. In fact, these characteristics apply only to the subfamilies Proviverrinae, Hyaenodontinae, and Pterodontinae. In contrast, limnocyonine hyaenodontids greatly reduced or lost the third molars; hence their carnassials were always developed at M^1/M_2, as in oxyaenids. In addition, many limnocyonines had shorter, broader skulls, similar to those of oxyaenids. Other aspects of the dentition and postcranial anatomy, however, imply closer relationship to proviverrines than to oxyaenids (Denison, 1938; Polly, 1996; Gunnell, 1998). Their early dental reduction implies a ghost lineage of proviverrines well back into the Paleocene.

Compared to oxyaenids, hyaenodontids tended to have longer, more gracile limbs, with more mediolaterally compressed but typically fissured terminal phalanges. At least some species were digitigrade (Fig. 8.7). Several early Eocene genera (the limnocyonine *Prolimnocyon* and the proviverrines *Prototomus, Tritemnodon,* and *Pyrocyon*) show varying degrees of adaptation for scansorial locomotion (Gebo and Rose, 1993; Rose, 2001a). Their contemporary, *Arfia* (also a proviverrine), has characteristics more typical of incipient cursors: a humerus with a prominent greater tuberosity and a supratrochlear foramen, and a femur with a high greater trochanter and a well-defined patellar groove. Paradoxically, *Arfia* seems to have had a very flexible ankle that allowed limited hind foot reversal (Gingerich and Deutsch, 1989), a specialization found in arboreal mammals that descend trees headfirst. Presumably this flexibility would have made the ankle less stable on the ground. Late early Eocene *Gazinocyon* was more clearly cursorial, based on ankle modifications that kept movement in the parasagittal plane (Polly, 1996). Bridgerian *Sinopa* shows many of the same features and was probably incipiently cursorial (Matthew, 1906). These two genera foreshadow the specializations that are better developed in *Hyaenodon*. The mixture of scansorial and terrestrial features found in many early hyaenodontids, however, suggests that they were generalists that spent significant amounts of time both on the ground and in the trees, not being restricted to either habitat. Many of these early members were not as big as typical oxyaenids, being fox-sized or smaller (ranging from 1 to 15 kg; Egi, 2001), and their dental modifications for carnivory were less well developed.

Quercitherium (Fig. 8.5C) from the middle and late Eocene of Europe was one of several genera that evolved swollen premolars probably used for crushing hard food items, such as bones or mollusks (Lange-Badré, 1979). Middle Eocene *Lesmesodon* (Fig. 8.8) from Messel is known from complete skeletons with soft-tissue impressions, revealing a bushy tail like that of a squirrel or a fox (Morlo and Habersetzer, 1999). It is also distinctive among hyaenodontids in having unfissured terminal phalanges. The skeleton indicates that *Lesmesodon* was a generalized terrestrial animal.

Later genera, such as *Hyaenodon* (later Eocene–early Oligocene of North America and Eurasia), included species with a head ranging in size from that of a housecat to that of a lion (Fig. 8.6B; Mellett, 1977). As mentioned earlier, however, the head was out of proportion with the body, so most species of *Hyaenodon* probably weighed only 10–40 kg, while the largest species of the genus weighed about 120 kg (Egi, 2001). *Hyaenodon* was widespread in the northern continents during the middle Tertiary, evidently occupying the vacant ecological niche left by the extinction of oxyaenids. Although small species of *Hyaenodon* may have been scansorial, larger species were relatively long-legged for creodonts and had many cursorial specializations (Fig. 8.7D).

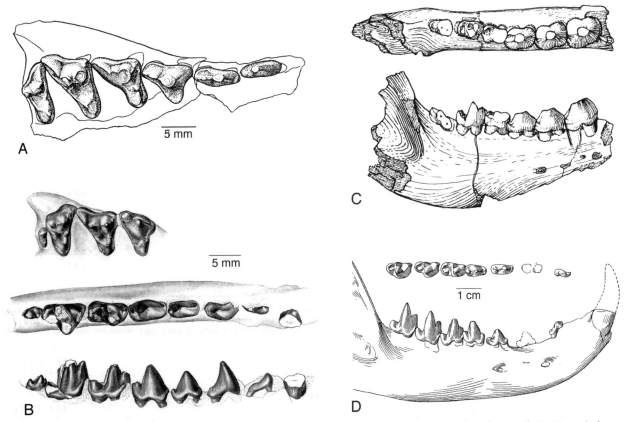

Fig. 8.5. Right dentitions of hyaenodontids: (A) *Gazinocyon* upper teeth; (B) *Prolimnocyon* upper and lower teeth; (C) *Quercitherium* lower teeth; (D) *Tritemnodon* lower teeth. (A from Gingerich and Deutsch, 1989; B from Gebo and Rose, 1993; C from Lange-Badré, 1975; D from Matthew, 1915a.)

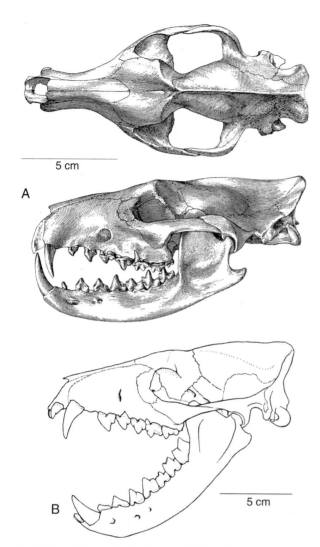

Fig. 8.6. Skulls of hyaenodontids: (A) *Sinopa*; (B) *Hyaenodon*. (A from Matthew, 1906; B from Mellett, 1977.)

They were digitigrade, with a reduced deltopectoral crest on the humerus; perforated olecranon fossa; deep humeral trochlea articulating with a broad proximal radius (which limited forearm supination); a long, deep patellar groove on the femur; a compact tarsus with a relatively deeply grooved astragalar trochlea; and closely appressed metatarsals (Scott and Jepsen, 1936; Mellett, 1977).

Hyaenodon had robust premolars and specialized bladelike carnassials (M^2/M_3) analogous to those of hyenas (which did not appear until the Miocene). The bladelike structure of the molar trigonids, formed by the paracristid, was enhanced by the loss of the metaconid. Late Eocene *Pterodon* achieved hypercarnivorous molars in the same way, which was long considered an indication of close relationship to *Hyaenodon*. Polly (1996) argued, however, that the loss of the metaconid occurred independently in the two genera.

Mellett (1977) documented a chain of events that led to changes in the skull and dentition of *Hyaenodon*, eventually resulting in larger body size, greater gape, and more efficient shearing. He postulated that lengthening of the principal shearing blades, the metacrista (=postmetacrista) of the upper molars and the paracristid of the lowers, and reorientation of these crests from oblique to almost mesiodistal enabled *Hyaenodon* to exploit larger prey and therefore to increase in size. Efficient function of the deciduous carnassial (dP^4) necessitated an occluding tooth below and led to selection for precocial eruption of a lower carnassial, M_1. This tooth is typically heavily worn or prematurely lost in *Hyaenodon*. Subsequent loss of M^3 allowed the posterior expansion of the M^2/M_3 carnassials so characteristic of *Hyaenodon*, and eventually caused further changes in the jaw joint and chewing muscles. These traits surely made *Hyaenodon* one of the most formidable predators of the middle Tertiary.

Yet even *Hyaenodon* was dwarfed by the later Eocene *Hemipsalodon*, the largest North American hyaenodontid at more than 400 kg (Egi, 2001), and the early Miocene hyaenodontid *Megistotherium*, one of the largest carnivorous mammals ever, with a skull about 66 cm long—nearly twice as big as that of a bear or a lion. *Megistotherium* may have weighed as much as 800 kg (Savage, 1973).

As already noted, limnocyonines differ from other hyaenodontids in skull shape, reduction or loss of third molars, and development of more anterior carnassials. For these reasons they are sometimes placed in their own family (e.g., Gunnell, 1998). Wasatchian *Prolimnocyon*, the oldest limnocyonine, seems to have been at least partly scansorial, whereas Bridgerian *Thinocyon* was similar in size and postcranial anatomy to a mink (*Mustela vison*)—about 1–1.5 kg. The larger (8–16 kg) middle Eocene *Limnocyon* was a generalized, perhaps semifossorial, terrestrial form (Matthew, 1909; Gebo and Rose, 1993; Egi, 2001).

CARNIVORA

Carnivorans have been the principal group of predaceous mammals throughout much of the Cenozoic, although in the Early Cenozoic creodonts were equally or more successful. Many extant carnivorans remain primarily meat eaters, but some lineages have evolved away from that regimen toward omnivory, frugivory, myrmecophagy, piscivory, and other specialized diets. Carnivorans occupy arboreal, scansorial, cursorial, and fossorial niches on land and have also invaded both freshwater and marine environments. They are naturally occurring throughout the world and its oceans, although terrestrial carnivorans never reached Australia (or Antarctica) without human intervention.

Many different classifications of Carnivora are in use (the one followed here is shown in Table 8.1). The order is usually divided into two large clades, the Feliformia (=Aeluroidea)—felids, viverrids, herpestids, hyaenids, and nimravids (cats, civets, mongooses, hyenas, and false sabertooths, respectively)—and the Caniformia—canids, ursids, amphicyonids, pinnipeds, procyonids, and mustelids (dogs, bears, bear-dogs, seals and walruses, raccoons, and weasels, respectively). All of the caniforms except canids are often united in a presumed monophyletic group Arctoidea. The oldest and most primitive known carnivorans are the Paleocene and Eocene

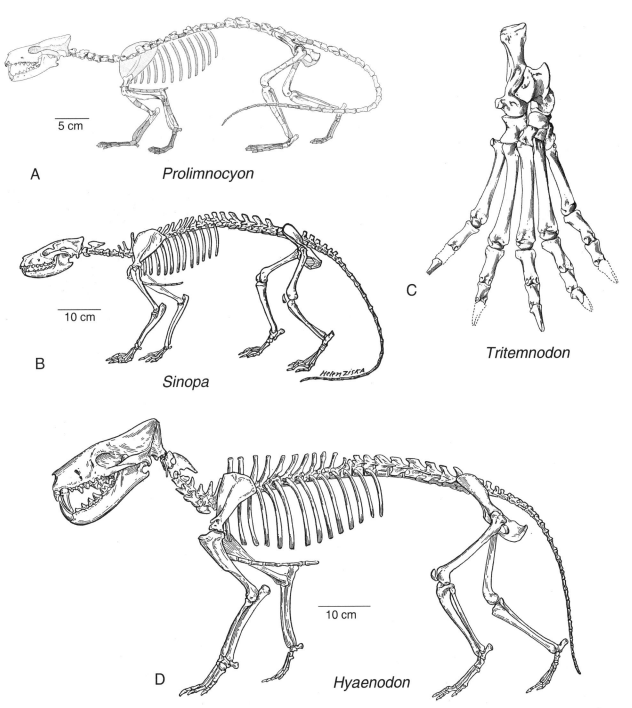

Fig. 8.7. Hyaenodontids: (A) *Prolimnocyon;* (B) *Sinopa;* (C) right foot of *Tritemnodon;* (D) *Hyaenodon.* (A from Gebo and Rose, 1993; B, D from Gregory, 1951; C from Matthew, 1909.)

miacoids, comprising the families Viverravidae and Miacidae. Miacoids are a paraphyletic assemblage whose two families were previously considered to be basal members or sister taxa of Feliformia (Viverravidae) and Caniformia (Miacidae; Flynn and Galiano, 1982; Flynn et al., 1988; Hunt and Tedford, 1993; McKenna and Bell, 1997; Flynn, 1998). In more recent analyses miacoids are found to lie outside of the two crown clades (Wyss and Flynn, 1993; Flynn and Wesley-Hunt, 2005; Wesley-Hunt and Flynn, 2005; see Fig. 8.1). Those authors prefer to restrict Carnivora to the crown clades and use the term Carnivoramorpha for the larger group that includes miacoids.

The principal distinguishing characteristics of Carnivora concern the dentition and the basicranium (especially auditory structures). Like creodonts, carnivorans primitively retain a plesiomorphic placental dentition of 3.1.4.3/3.1.4.3, but reduction in this number is common. The hallmark of the Carnivora is the specialization of P^4/M_1 as bladelike carnassials (Fig. 8.2), a modification usually assumed to have occurred only once at these tooth loci (but see below).

Fig. 8.8. *Lesmesodon*, a hyaenodontid from the middle Eocene of Messel. (Restoration from Morlo and Habersetzer, 1999; skeleton courtesy of Hessisches Landesmuseum Darmstadt.)

Associated with this adaptation, the lower jaw moves mainly orthally (up and down), with little transverse motion possible; this restriction is further ensured by a tight-fitting cylindrical temporomandibular joint. Carnassials are present in most living and extinct carnivorans but have been lost or greatly modified in procyonids, ursids, and pinnipeds. When specialized for shearing, P^4 is triangular, with the paracone centrally located on the buccal side of the tooth, the protocone shifted anterolingually (well anterior to the paracone), and a long, bladelike metastylar crest separated from the paracone by a carnassial notch. M_1 is larger than the other molars and has a tall trigonid with a well-developed paracristid blade enhanced by a carnassial notch between the protoconid and paraconid. The second and third molars are reduced in size or lost.

In extant carnivorans the auditory bulla, which surrounds the middle-ear cavity, consists of three elements: the ecto-tympanic (sometimes simply called the tympanic) and two entotympanics, rostral and caudal (Hunt, 1974b). There has been a widespread tendency among carnivorans to enlarge the middle-ear cavity, usually by expanding the bulla, which enhances hearing ability. Variations in the anatomy of the three bullar elements, particularly the caudal entotympanic, as well as in petrosal anatomy, have played an important role in deciphering carnivoran relationships (e.g., Hunt, 1974b, 1987, 1989, 1991, 1998c, 2001; Fig. 8.9). For example, the ectotympanic is the largest bullar element in arctoids, whereas the caudal entotympanic tends to be larger in canids and feliforms. In addition, the internal carotid artery (ICA), which supplies the brain in many mammals, is usually reduced in carnivorans, many of which get their primary blood supply to the brain through the external carotid.

Caniforms and feliforms differ fundamentally in construction of the auditory bulla and in the carotid circulation

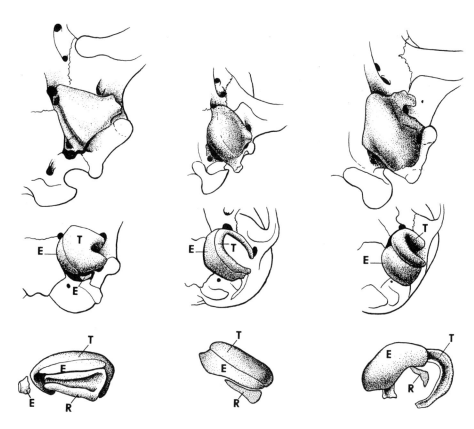

Fig. 8.9. Auditory structure in carnivores. Elements forming the auditory bulla in (left to right) a bear (arctoid), a dog (cynoid), and a cat (feloid). Top row shows ventral view of adult bulla, middle row shows ventral view of neonatal bulla. Bottom row shows isolated neonatal bulla in medial view, to reveal the rostral entotympanic, which is not exposed ventrally. Key: E, caudal entotympanic; R, rostral entotympanic; T, ectotympanic. (From Hunt and Tedford, 1993.)

that supplies the brain. In most feliforms (except nimravids and the extant *Nandinia*) the bulla is divided into two chambers by a bony septum, the anterior chamber composed of ectotympanic and rostral entotympanic, and the posterior chamber made by the caudal entotympanic (Hunt and Tedford, 1993). The ICA is reduced or absent, the primary endocranial blood supply instead coming through the external carotid via a pair of arterial networks (or retia; Hunt, 1974b). Caniforms have a bulla with a single chamber and no septum (except in canids, which have a partial septum), and the blood supply to the brain comes through the ICA. Unfortunately, the bulla of miacoids is unknown, hence we lack this important criterion for establishing relationship to feliforms or caniforms. However, indentations in the basicranium of some recently described skulls of Bridgerian and later miacids suggest the presence of a loosely attached compound bulla (either ossified or cartilaginous) consisting of ectotympanic and entotympanic elements (Wesley-Hunt and Flynn, 2005).

Also distinctive of extant carnivorans is a large braincase, with the coronal (frontal-parietal) suture situated well behind the postorbital constriction, owing to cerebral expansion (Wyss and Flynn, 1993). Miacoids differ in having relatively smaller brains, and a more anterior coronal suture. The skeleton of terrestrial carnivorans is usually relatively generalized, but is sometimes overprinted with specializations for climbing, running, or digging. The feet tend to be conservative, typically remaining pentadactyl, and the posture is plantigrade or digitigrade. A fused scapholunate in the carpus is a diagnostic trait of extant carnivorans, but the two elements remain separate in most miacoids. Similarly, extant carnivorans lack a third trochanter on the femur, but it is present in miacoids. Marine carnivorans (pinnipeds) show major limb modification or reduction.

Following the work of Lillegraven (1969) there has been general agreement that the teeth of both carnivorans and creodonts can be plausibly derived from those of Cretaceous Cimolestidae, such as *Cimolestes* (Fig. 8.10). Hunt and Tedford (1993) suggested that *Cimolestes* is more closely related to Carnivora than to Creodonta and that different lineages of the genus may have given rise to the two families of miacoids (Viverravidae and Miacidae). This hypothesis raises the possibility that the order Carnivora is diphyletic and that the classic synapomorphy of the order, P^4/M_1 carnassials, arose more than once (in fact, it is also present in hedgehogs; see Chapter 9). According to Hunt and Tedford, viverravids such as Torrejonian *Simpsonictis* might have evolved from a Late Cretaceous species of *Cimolestes* that had lost its third molars prior to the development of carnassial teeth, whereas miacids could have evolved later from a separate species of *Cimolestes* (which retained third molars) and evolved carnassials independently. Fox and Youzwyshyn (1994) disagreed, however, and postulated a more primitive eutherian ancestry of Carnivora involving neither Creodonta nor Cimolestidae. In view of these uncertainties, the precise timing of the origin of Carnivora is unknown. Unfortunately, the available fossil evidence from the critical interval (Late Cretaceous–early Paleocene) is unable to resolve the matter conclusively.

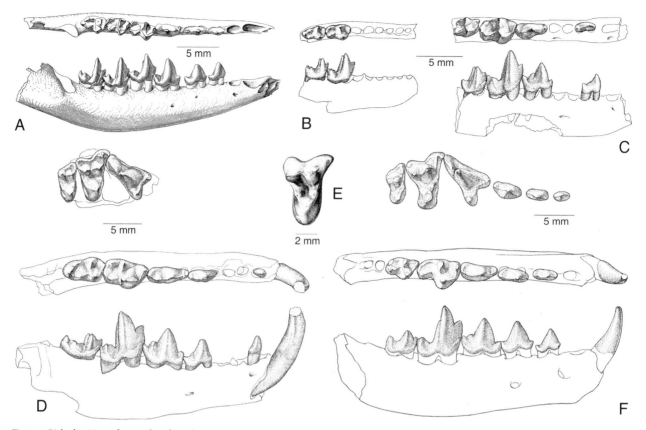

Fig. 8.10. Right dentitions of miacoids and *Cimolestes*: (A) *Cimolestes* lower teeth; (B, C) lower jaws of two different species of *Simpsonictis*; (D) *Viverravus* upper and lower teeth; (E) *Ravenictis* upper molar; (F) *Uintacyon* upper and lower teeth. (A from Clemens, 1973; B–D from Gingerich and Winkler, 1985; E from Fox and Youzwyshyn, 1994; F from Gingerich, 1983a.)

Miacoids

Unequivocal Carnivora, as indicated by the presence of the P^4/M_1 carnassial pair, are first known from the Paleocene of North America. These earliest carnivorans belong to two primitive families, Viverravidae and Miacidae, often grouped as the paraphyletic Miacoidea (but formerly considered subfamilies of a stem family Miacidae). Several members of both families were also present in Europe during the Eocene, whereas only a couple of miacoids are known from the Paleocene-Eocene of Asia. By the end of the Eocene miacoids had disappeared everywhere and were quickly replaced by more modern carnivorans. Most miacoids ranged from weasel-sized to a little larger than a fox, or roughly 100 g to 10 kg.

Miacidae, in the strict sense, are characterized by retention of third molars (a primitive trait) together with reduction or loss of the parastyle on P^4 and loss of calcaneofibular contact (derived traits). They share these features with caniforms. In Viverravidae the third molars are absent, the parastyle on P^4 strong, and the fibula articulates with the calcaneus—features in common with feliforms. Based on these criteria, the two families have been considered to be the earliest representatives of the two major clades of extant Carnivora (Flynn and Galiano 1982; Flynn, 1998), although definitive evidence from the basicranium is lacking.

In addition, most (but not all) miacoids have separate scaphoid and lunate bones in the carpus; fusion of these elements is often considered a diagnostic trait of Carnivora. Consequently, Viverravidae and Miacidae are currently considered to be stem taxa that lie outside the two crown clades of Carnivora (Wyss and Flynn, 1993; Flynn and Wesley-Hunt, 2005; Wesley-Hunt and Flynn, 2005). According to this view, Viverravidae is the sister group of all other Carnivora (making the loss of third molars in this family an autapomorphy), whereas the paraphyletic Miacidae are probably closer to the crown clade.

The oldest securely dated carnivoran, *Ravenictis* (Fig. 8.10E), comes from the early Paleocene of Saskatchewan, but it is represented only by an isolated upper molar, which is not diagnostic at the family level. Consequently it provides little information beyond extending the geologic range of the order. *Ictidopappus* (North America) and *Pappictidops* (China) are nearly as old and also known only from dentitions. They are variously regarded as primitive viverravids or as basal carnivorans of uncertain affinity. By the late early Paleocene (Torrejonian), however, several genera of undoubted viverravids, including *Protictis* and *Simpsonictis*, were present in western North America (Gingerich and Winkler, 1985; Flynn, 1998). The earliest record of Miacidae, despite their more primitive dental formula, is not until the latest Paleocene (Clarkforkian) of North American (e.g.,

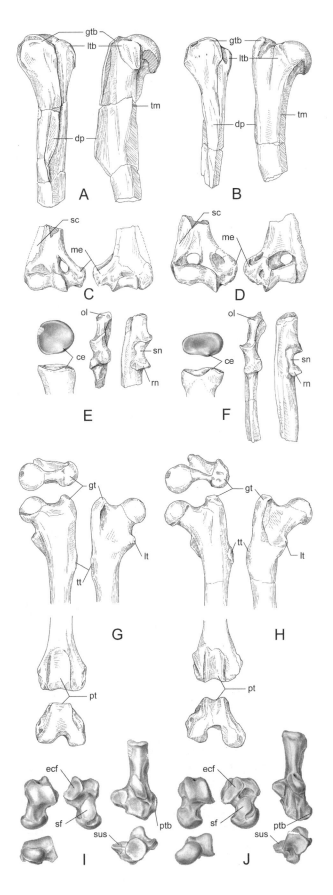

Uintacyon; Fig. 8.10F). However, if Carnivora is monophyletic, they must have existed much earlier (at least as early as the oldest viverravids). This early origin would presumably hold true even if Carnivora is not monophyletic and the two families arose independently from *Cimolestes* (which is known principally from the Cretaceous and early Paleocene). At present, however, there is little fossil evidence to favor this interpretation over a common origin of miacoids.

Besides their dichotomy in dental formulae, viverravids and miacids also differed in locomotor adaptation, as reflected in their appendicular skeletons (Fig. 8.11). Many features in the limbs of viverravids, such as early Eocene *Didymictis,* indicate that they were terrestrial and probably incipiently cursorial, although they probably retained the ability to climb, not unlike extant *Viverra* (Heinrich and Rose, 1997). These features include a prominent greater tuberosity, reduced deltopectoral crest, supratrochlear foramen, and wide radial head in the forelimb, and a posteriorly directed lesser trochanter, well-defined patellar trochlea, moderately grooved astragalar trochlea, narrow and more elongate calcaneus, smaller and more distal peroneal tubercle on the calcaneus, and several other tarsal characteristics in the hind limb. Miacids, however, were adapted for scansorial and arboreal habitats. *Vulpavus* (Fig. 8.12) and *Miacis* resemble living palm civets and coatimundis in having a sharp deltopectoral crest, shallow humeral trochlea and olecranon fossa, round proximal radius, medially directed lesser trochanter, shallow patellar groove, and nearly flat astragalar trochlea (Heinrich and Rose, 1995, 1997). Most of these features are associated with increased joint mobility, as would be expected in arboreal animals. Both miacids and viverravids had relatively short, laterally compressed terminal phalanges.

Most authorities agree that miacoids were the source group for more advanced feliforms and caniforms. However, transitional taxa or plausible ancestors for most of the modern families have not been identified. Canidae, which can be derived from *Miacis* or a closely allied form, is an exception, as discussed below.

Feliformia

Not until the latest Eocene and earliest Oligocene do unequivocal feliforms appear in the fossil record. The early Oligocene Phosphorites of Quercy, France, have produced the most diverse assemblage of primitive feliforms, including skulls of several genera that seem to be close to the base

Fig. 8.11. Comparison of limb elements of miacids (left column: A, E, G, I, *Vulpavus;* C, *Uintacyon*) and viverravids (right column: B, D, F, H, J, *Didymictis*): (A–D) right humerus, proximal and distal ends; (E–F) right radius and ulna (proximal); (G–H) left femur, proximal and distal ends; (I–J) left astragalus and calcaneus. Key: ce, capitular eminence; dp, deltopectoral crest; ecf, ectal facet; gt, greater trochanter; gtb, greater tuberosity; lt, lesser trochanter; ltb, lesser tuberosity; me, medial epicondyle; ol, olecranon process; pt, patellar trochlea; ptb, peroneal tubercle; rn, radial notch; sc, supinator crest; sf, sustentacular facet; sn, semilunar notch; sus, sustentaculum tali; tm, teres major tubercle; tt, third trochanter. (Figure prepared by R. E. Heinrich; modified from Heinrich and Rose, 1997.)

Fig. 8.12. Eocene miacid *Vulpavus*. (Restorations by J. Matternes.)

of viverrids and felids, as shown by their possession of a two-chambered auditory bulla (Hunt, 1998c). Although they have been variously referred to these modern families, they differ relatively little from each other in dental or basicranial anatomy, which suggests that the Quercy fauna samples the beginning of the modern feliform radiation (Hunt, 1989, 2001). Their sudden appearance in Europe just after the Grande Coupure (Remy et al., 1987) indicates that they are immigrants, perhaps from Asia. The other extant feliform families, Hyaenidae and Herpestidae, did not appear until the Miocene. *Nandinia,* the extant African palm civet, is the most primitive living feliform. It was long considered to be a viverrid but is now usually placed in its own family.

Palaeoprionodon, best known from Quercy, is the oldest feliform with viverrid ear structure (Hunt, 1989, 1998c). Viverrids, an Old World family that includes the extant civets and Asian palm civets, are generally considered to be primitive feliforms. *Stenoplesictis* (Fig. 8.13A), from Quercy and probable late Eocene deposits of Alag Tsab, Mongolia, has been regarded as the oldest viverrid, based on dental resemblance (e.g., Dashzeveg, 1996). However, its auditory region, although clearly two-chambered and therefore feliform, differs from that of both viverrids and felids; hence *Stenoplesictis* has been considered a stem feliform (Hunt, 1991, 1998c; Peigné and Bonis, 1999). The oldest viverrid-like skeleton is that of *Asiavorator* (Fig. 8.13B) from the early Oligocene of Mongolia. It closely resembles that of extant civets and genets and was primarily terrestrial but probably retained the ability to climb trees (Hunt, 1998c).

The earliest felids also come from Quercy. *Proailurus* and *Stenogale* (Fig. 8.13C,D), known from jaws at Quercy (but no recognized ear regions), can be confidently identified as felids, based on the derived petrosal anatomy of early Miocene representatives (Hunt, 1991, 1998c). Although *Proailurus* has long been recognized to be a felid, prior to Hunt's study *Stenogale* was usually identified as a viverrid or a basal feliform. In these basal felids M_1 has a well-developed shearing blade formed by the tall paraconid and protoconid and intervening carnassial notch; the metaconid is already reduced or lost. The early radiation of felids took place in the Old World; they did not reach North America until well into the Miocene. The close resemblance among these early feliforms indicates that felids and viverrids are sister taxa.

The remaining feliform family, Nimravidae, was contemporaneous with the oldest feliforms discussed above, appearing in the late Eocene of North America and Eurasia (Martin, 1998). Nimravids were the earliest saber-toothed carnivorans (Fig. 8.14). They were once thought to be felids, which they resemble in having a short face, hypercarnivorous dentition, and retractile claws, but analysis of dental characters led Flynn and Galiano (1982) to unite nimravids with caniforms. These catlike late Eocene to Miocene "paleofelids," or false saber-tooths, are now placed in a separate family whose relationships remain unsettled (Flynn et al.,

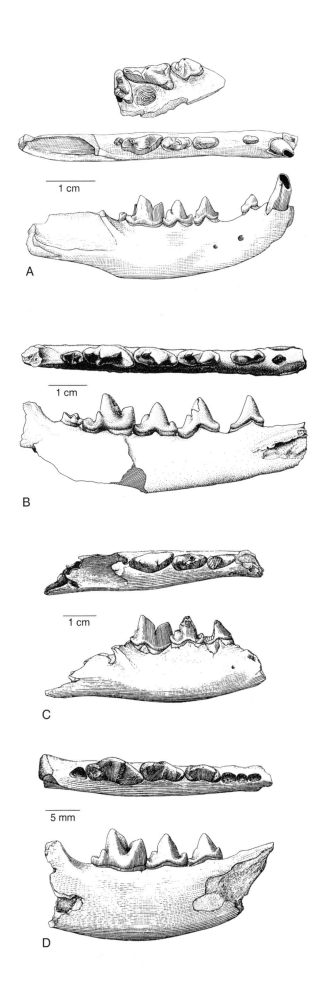

1988; Bryant, 1991). Most recent studies ally them with feloids, based on the reduction of posterior molars and possession of hooded terminal phalanges that bore retractile claws (e.g., Hunt, 1987; Bryant, 1991; Wyss and Flynn, 1993; Martin, 1998; Flynn and Wesley-Hunt, 2005). Nevertheless, nimravids differ from felids and most other feliforms in several cranial details, including having an essentially single-chambered auditory bulla with a uniquely formed anterior septum, a caudal entotympanic that is only partially ossified, and a different conformation of basicranial foramina (Hunt, 1987). Their origin remains obscure.

The earliest nimravids, *Dinictis* and *Hoplophoneus* of western North America, were already saber-toothed, with large, serrated, and laterally compressed upper canine teeth and a protective bony flange on the mandible. Some species reached the size of cougars or jaguars (about 100 kg). *Hoplophoneus* was short-legged and more like an ambush predator, whereas *Dinictis* had longer limbs and was more cursorially adapted, like living pursuit predators (Martin, 1998). It is likely that they were also able to climb trees. These early nimravids (subfamily Nimravinae) became extinct by the beginning of the Miocene, perhaps partly as a result of the spread of grasslands (Bryant, 1996). They were succeeded in the late Miocene by barbourofeline nimravids and saber-toothed felids.

Late Eocene and Oligocene *Palaeogale* (Fig. 8.15) may also be mentioned here. Long considered a primitive mustelid, this widespread Holarctic taxon is now thought to be a basal feliform (Baskin, 1998) or possibly even a viverravid (Hunt, 1989). Like other feliforms, it has a bladelike trigonid on M_1, but it differs from feliforms in having a single-chambered bulla. The third molars are lost and the second molars are very small or absent. *Palaeogale* could be a pivotal form in the early radiation of modern carnivorans.

Caniformia

Caniforms can be divided into two clades, Cynoidea (canids) and Arctoidea (all other caniforms; see Fig. 8.1). Arctoids are united by two synapomorphies, a suprameatal fossa (a hollow in the dorsolateral wall of the middle-ear cavity) and the loss of M^3 (Wolsan, 1993; Wolsan and Lange-Badré, 1996); each subgroup of arctoids has its own distinctive morphology of the suprameatal fossa. Most have a single-chambered auditory bulla composed mainly of the ectotympanic (Hunt, 1974b). Whereas early arctoids were common and diverse in Europe (particularly at Quercy) but sparse in North America, early canids were common in North America but did not reach the Old World until the late Miocene (Hunt, 1998a). Current evidence indicates that canids (dogs) originated in North America, whereas arctoids

Fig. 8.13. Feliform dentitions: (A) *Stenoplesictis*, right upper and lower teeth; (B) *Asiavorator*, left P_2–M_2; (C) *Proailurus*, right P_3–M_1; (D) *Stenogale*, left P_3–M_1. *Proailurus* and *Stenogale* are considered the oldest felids. (A from Peigné and Bonis, 1999; B from Hunt, 1998c; C, D from Bonis et al., 1999.)

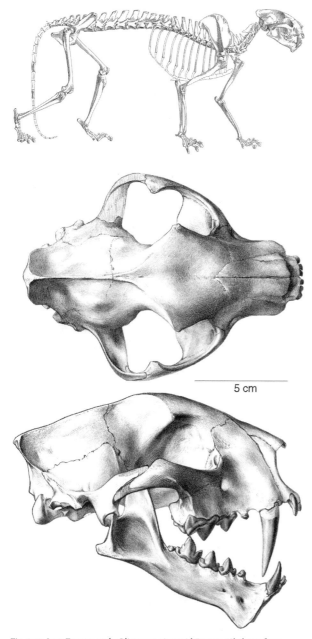

Fig. 8.14. Late Eocene–early Oligocene nimravid *Dinictis*. (Skeleton from Matthew, 1901; skull from Scott and Jepsen, 1936.)

North America for all of their early history, not reaching the Old World until the latest Miocene.

The oldest and most primitive canids are the hesperocyonines of the late middle Eocene (Duchesnean) to the middle Miocene (Barstovian). They resemble certain species of *Miacis* or *Procynodictis* closely enough—especially dentally—to suggest an ancestral or sister-group relationship (Wang and Tedford, 1994; Munthe, 1998; Fig. 8.16), making Canidae the only modern carnivoran family that can be linked to a specific miacid.

The earliest and best-known form is *Hesperocyon,* a common element of faunas from the White River Group characterized by a trenchant talonid on M_1 (primitive for canids). Unlike that of miacids, its auditory bulla is fully ossified and composed mainly of the caudal entotympanic, with contributions from the ectotympanic and rostral entotympanic; the caudal entotympanic forms a partial septum within the middle-ear cavity. The bullar anatomy is thus similar in detail to that in modern canids (Hunt 1974a,b; Wang and Tedford, 1994). In addition, the internal carotid artery, rather than crossing the promontorium as in miacids, is situated medial to the promontorium and outside the bulla on its medial surface, and the stapedial branch is absent. *Hesperocyon* was about the size of a small fox, but was proportioned more like extant civets and mongooses. The limb skeleton of *Hesperocyon* is intermediate between that of its arboreal, plantigrade, miacid progenitors and that of cursorial, fully digitigrade later canids (X. Wang 1993, 1994). A vestigial clavicle was still present. The terminal phalanges were short, deep, and laterally compressed, and may have been retractile. Based on these features, Wang (1993) concluded that *Hesperocyon* was a plantigrade animal, mainly scansorial in

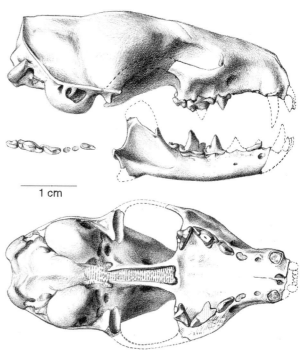

Fig. 8.15. Skull of the basal feliform *Palaeogale*. (From Scott and Jepsen, 1936.)

evolved in the Old World and dispersed multiple times to North America.

Canidae

Canids comprise dogs, foxes, wolves, coyotes, and jackals, which are cursorial, relatively omnivorous carnivorans. Canids have a primitive placental dental formula except for loss of M^3. Although they have well-developed carnassials, their molars also have basins for crushing. The limbs tend to be slender and moderately elongate, and the feet are digitigrade and functionally four-toed. As in many cursorial mammals, the clavicle is lost in extant members. Canids are virtually cosmopolitan today, but they were restricted to

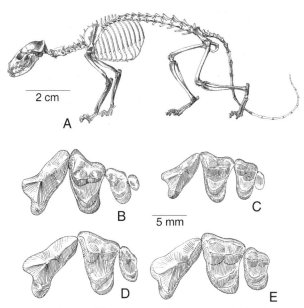

Fig. 8.16. (A) Skeleton of the primitive canid *Hesperocyon*. Left P^4 and upper molars: (B, C) *Miacis*; (D) basal canid *Prohesperocyon*; (E) basal canid *Hesperocyon*. (A from Matthew, 1901; B–E from Wang and Tedford, 1994.)

habit but incipiently cursorial as well. *Hesperocyon* appears to be broadly ancestral to most later lineages of canids. These include a diversity of closely related Oligocene hesperocyonines as well as the borophagines, or "hyaenoid dogs," whose earliest representatives (e.g., early Oligocene *Oxetocyon*) had bunodont, hypocarnivorous teeth similar to those of procyonids (Munthe, 1998).

Chadronian *Prohesperocyon* lacks the partial intrabullar septum characteristic of *Hesperocyon* and all other canids. Its dentition approaches that of the miacid *Procynodictis* more closely than that of *Hesperocyon* or any other canid. For these reasons *Prohesperocyon* is considered the most primitive known canid and the sister taxon of all other canids (X. Wang, 1994).

Ursoidea

The bears (Ursidae) and bear-dogs (Amphicyonidae) are carnivorans that are sometimes united in the Ursoidea, which first appear in the Duchesnean and Chadronian of North America. Both families are usually regarded as arctoids, but the phylogenetic position of amphicyonids remains ambiguous. *Parictis* (Fig. 8.17A), a rare, primitive arctoid from the Chadronian and Orellan of North America, is usually considered to be the oldest known ursid (Hunt, 1998a). Although bears today are among the largest terrestrial carnivores, *Parictis* was small (2 kg). Like ursids, it had a primitive dental formula except for the loss of M^3, robust premolars, and broad, relatively low-crowned molars with large basins. In Europe, the closely related *Amphicynodon* (Fig. 8.17B) from Quercy (not to be confused with amphicyonids; see below) had similar dental features and also seems to occupy a phylogenetic position near the beginning of ursids. The relationships of these taxa continue to be problematic, however. *Parictis* has been considered to be a canid (Scott and Jepsen, 1936), a member of a new ursoid family Subparictidae (Baskin and Tedford, 1996), and even a basal pinniped (Phocoidea; McKenna and Bell, 1997). Cirot and Bonis (1992) regarded *Amphicynodon* as a stem arctoid, possibly near the origin of both ursids and musteloids.

Identification of *Parictis* as a primitive pinniped is not as surprising as it may seem, because the basicranial and carnassial morphologies suggest that pinnipeds evolved from an ursid (Hunt and Barnes, 1994). Pinnipeds are otherwise unknown until the Miocene, however, and their precise origin is uncertain. A recent molecular analysis placed pinnipeds as the sister group of musteloids (Flynn et al., 2005).

Aside from amphicynodonts, the ursid radiation took place primarily later in the Cenozoic (Miocene and thereafter). The only exception is *Cephalogale*, the oldest member of the hemicyonine ursids, which first appeared in the late Eocene of Asia and the early Oligocene of Europe (McKenna and Bell, 1997). Unlike other bears, which modified their dentition for omnivory, hemicyonines retained well-developed carnassials and did not elongate their molars (Hunt, 1998a).

Amphicyonids, or bear-dogs, first appear in the late middle Eocene (Duchesnean) of North America and slightly later in Europe. The oldest North American form is *Daphoenus* (Figs. 8.17D, 8.18), best known from the White River Group of the mid-continent; *Cynodictis* from the latest Eocene of Quercy is the oldest European form. Amphicyonids rapidly became widely distributed. Several lineages, representing three subfamilies, evolved from these two genera by the end of the Eocene (Hunt, 1998b). These early representatives were small (<5 kg, no bigger than a small fox), but some later amphicyonids reached 200 kg.

Amphicyonids exhibit a mixture of bearlike and doglike features, which has caused confusion about their affinities. In some features they appear to be closely related to canids, whereas others suggest they are close to the base of the arctoid radiation (Wolsan, 1993; Wang and Tedford, 1994). A recent phylogenetic analysis found amphicyonids to be the sister taxon of all other caniforms (Wesley-Hunt and Flynn, 2005), making the name bear-dog truly appropriate. They had a short snout and a single-chambered bulla like that of ursids, in which the ectotympanic is the main element. In early amphicyonids the bulla was incompletely ossified and consisted of a crescent-shaped ectotympanic (Hunt, 1974b). In addition, the basioccipital bone of amphicyonids is excavated to house an enlarged inferior petrosal sinus, which in analogy with bears probably contained a loop of the internal carotid artery for cooling blood en route to the brain (Hunt, 1977). The dental formula was primitively 3.1.4.3/3.1.4.3, although M^3 was lost and the premolars reduced in some lines. Like canids (but unlike most ursids), they retained shearing carnassials and had a triangular (not quadrate) M^1 with three main cusps—primitive features that misled early workers to ally them with canids. *Daphoenictis* converged on felids in having a large, bladelike lower carnassial. Most amphicyonids had relatively generalized skeletons. *Daphoenus* was a subdigitigrade to digitigrade cursor whose feet retained

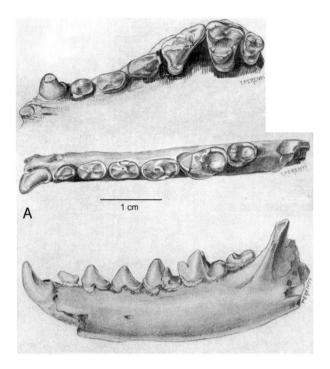
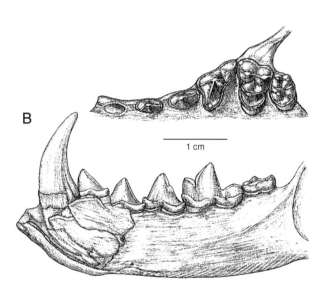
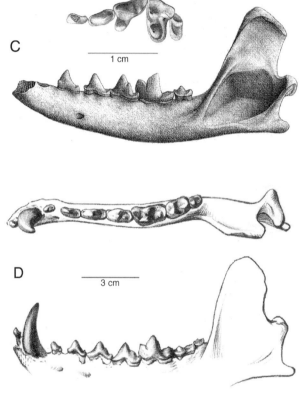

Fig. 8.17. Left dentitions of primitive arctoids: (A) basal ursid *Parictis*; (B) *Amphicynodon*; (C) basal musteloid *Mustelictis*; (D) amphicyonid *Daphoenus*. (A from Clark and Guensburg, 1972; B from Cirot and Bonis, 1992; C from Bonis, 1997; D from Scott and Jepsen, 1936.)

the flexibility expected in a climber (Hunt, 1996). It is believed that this skeletal form gave rise to both more cursorial types and more robust, bearlike forms (Hunt, 1998b).

Musteloidea (=Mustelida)

Under this heading are included mustelids (the most diverse extant carnivorans, including weasels, skunks, and otters) and procyonids (raccoons and coatis). The beginnings of the musteloid radiation are found in the late Eocene and early Oligocene of western North America and Europe, but details remain to be resolved. The oldest musteloids are *Mustelavus* (late Chadronian-Orellan, North America) and *Mustelictis* (early Oligocene, Europe; Fig. 8.17C). Basal members at this stage are very similar, and it is uncertain whether these genera should be allocated to either family or are better considered stem taxa. Their musteloid status is affirmed by a low trigonid on M_1, absence of both upper and lower third molars, and presence of a suprameatal fossa (Bonis, 1997). Additional dental characters (single-rooted first premolars, reduced second molars, and reduced metaconule and postprotocrista on M^1) suggest that they are primitive mustelids (Wolsan, 1993; Bonis, 1997; Baskin, 1998). McKenna and Bell (1997), how-

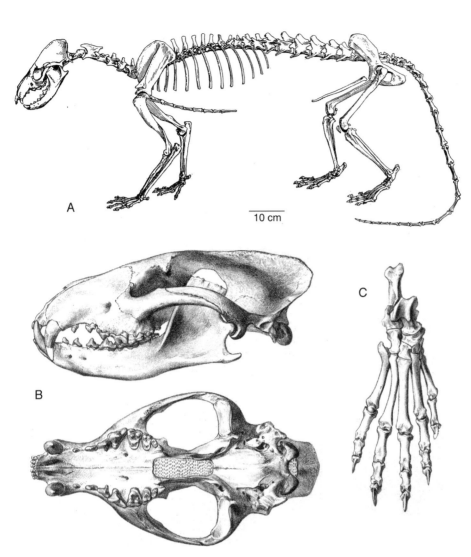

Fig. 8.18. Amphicyonid *Daphoenus*: (A) skeleton; (B) skull; (C) right foot. Scale applies to A. (From Scott and Jepsen, 1936.)

ever, consider *Mustelavus* and *Mustelictis* to be synonyms of *Pseudobassaris*, the oldest known procyonid (Wolsan, 1993; Wolsan and Lange-Badré, 1996). Skulls of the latter from Quercy, however, have an inflated, single-chambered auditory bulla and a deep suprameatal fossa like that of procyonids, whereas the suprameatal fossa of *Mustelictis* is shallow. Consequently, *Mustelictis*, at least, appears to be distinct from *Pseudobassaris*.

9

Insectivora

ALTHOUGH THE TERM INSECTIVORA and the vernacular forms insectivoran and "insectivore" are widespread in both popular and scientific literature, they have been used to refer to very different associations of eutherians. To mammalogists, Insectivora has usually been considered to include the living hedgehogs, moles, shrews, solenodons, tenrecs, golden moles, and their immediate fossil relatives, which are alternatively (and preferably) united as Lipotyphla (Fig. 9.1A). Anatomy provides weak support (a small number of characters) for a monophyletic Lipotyphla (e.g., Asher et al., 2003; Mussell, 2005). Molecular evidence, however, conflicts with the traditional concept of Lipotyphla and suggests that Lipotyphla is polyphyletic (Fig. 9.1B), as further discussed below. Lipotyphlans are typically viewed as very primitive eutherians because they retain many plesiomorphic features, often including a basic tribosphenic molar pattern, but this pattern has been modified, sometimes substantially, in some families. Until fairly recently, tree shrews (Tupaiidae) and elephant shrews (Macroscelididae)—once grouped as Menotyphla—were also often included in the Insectivora (e.g., Romer, 1966; Vaughan, 1978), but they are now assigned to separate ordinal-level groups, Scandentia and Macroscelidea, respectively, following Butler (1972).

Many paleontological accounts have employed a broader concept of Insectivora that includes not just lipotyphlans but also some or all of the following so-called "archaic insectivores" (many of which Romer, 1966, included in his Proteutheria): leptictids, palaeoryctids, apatemyids, pantolestids, pentacodontids, mixodectids, and a few other families. Most of these families have proven difficult to place phylogenetically, and they have little in common except relatively unmodified dentitions. Indeed, with the possible exception of the first two families, there is no good evidence

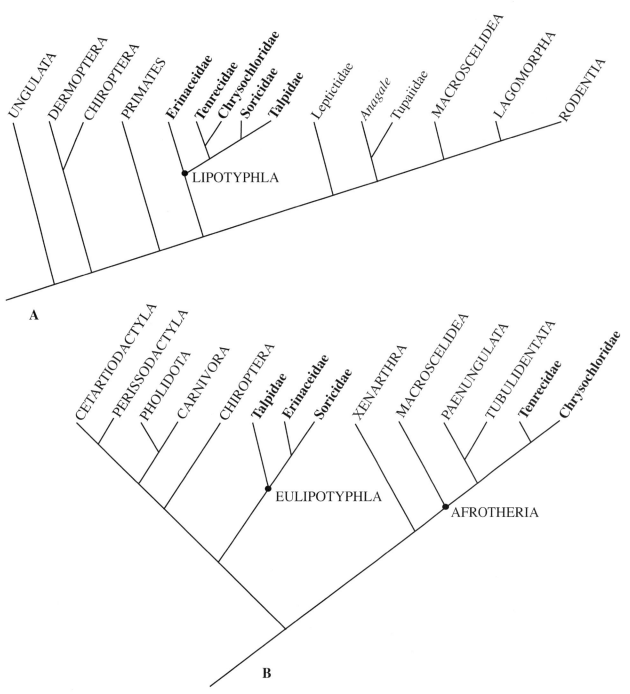

Fig. 9.1. Lipotyphlan relationships: (A) based on morphology; (B) based on gene sequences. (Positions of taxa within Afrotheria vary in different molecular studies.) Conventional lipotyphlan families are shown in bold. Solenodontidae were not included in the analyses on which these trees are based; they appear to be soricomorphs closest to Soricidae among extant families. (A simplified after Asher et al., 2003; B modified after Douady et al., 2002, and Asher et al., 2003.)

that they compose a monophyletic group with Lipotyphla. Consequently, an Insectivora of such broad composition is a true taxonomic wastebasket that cannot be defined or characterized except by retention of primitive eutherian features. Fortunately, most of these families are now assigned to other higher taxa, as they are in this book (see Chapters 7 and 10).

Based on a comprehensive analysis of cranial anatomy across the major groups of eutherians, Novacek (1986a) concluded that Leptictida is the sister group of Lipotyphla and proposed that the superorder Insectivora be used to encompass these two orders. This morphologically based arrangement is adopted here (Table 9.1), although the relationship remains to be compellingly demonstrated. It has also been argued that palaeoryctids are closely related to lipotyphlans, specifically, soricomorphs or tenrecoids (e.g., Lillegraven et al., 1981; McKenna et al., 1984; Thewissen and Gingerich, 1989; MacPhee and Novacek, 1993), but this hypothesis is also weakly based. Butler (1988) rejected soricomorph, and presumably lipotyphlan, affinities for palaeoryctids. Here

Table 9.1. Classification of Insectivora

```
Superorder INSECTIVORA
   Order †LEPTICTIDA
            †Gypsonictops
         †Leptictidae
         †Pseudorhyncocyonidae¹
   Order LIPOTYPHLA
         †Adapisoriculidae
      Suborder ERINACEOMORPHA
         Erinaceidae
         †Sespedectidae
         †Scenopagidae²
         †Amphilemuridae³
         †Adapisoricidae
         †Creotarsidae
         †Chambilestidae⁴
      Suborder SORICOMORPHA
         †Geolabididae
         Superfamily Soricoidea
            †Nyctitheriidae⁵
            Soricidae
            †Plesiosoricidae
            †Nesophontidae
            Solenodontidae
            †Micropternodontidae
            †Apternodontidae
         Superfamily Tenrecoidea⁶
            Tenrecidae
         Superfamily Talpoidea⁵
            †Proscalopidae
            Talpidae
            †Dimylidae
      Suborder SORICOMORPHA?
            †Otlestes,⁷ †Batodon,⁸ †Paranyctoides
      Suborder CHRYSOCHLOROMORPHA
         Chrysochloridae
```

Notes: Modified mostly after Novacek, 1986a; MacPhee and Novacek, 1993. The dagger (†) denotes extinct taxa. Families in boldface in this table are known from the Paleocene or Eocene.

[1] Named as a subfamily of Leptictidae.
[2] Sometimes considered a subfamily of Sespedectidae.
[3] Probably includes Dormaaliidae.
[4] May belong in Soricomorpha.
[5] Sometimes assigned to Erinaceomorpha.
[6] This taxon has also been used to unite Tenrecidae and Chrysochloridae.
[7] Probable synonym of *Bobolestes*; better considered a basal eutherian or possibly a zalambdalestoid (Archibald and Averianov, 2001; Averianov and Archibald, 2005).
[8] Assigned to Geolabididae by McKenna and Bell (1997).

palaeoryctids are included under Cimolesta, following McKenna and Bell (1997).

LEPTICTIDA

Leptictida is an ordinal-level group that accommodates the Early Tertiary Leptictidae and the closely allied Late Cretaceous *Gypsonictops* (Novacek, 1986a). McKenna and Bell (1997) included two additional families in their superorder Leptictida: Kulbeckiidae (now considered a junior synonym of Zalambdalestidae) and Didymoconidae (here tentatively included in Cimolesta). They also united *Gypsonictops* with a half-dozen other Cretaceous genera, including *Prokennalestes*, *Kennalestes*, and *Zhelestes*, in the family Gypsonictopidae. This association, however, is mainly based on primitive characters. *Prokennalestes* is indeed a very primitive eutherian and its precise position is uncertain (see Chapter 6), whereas the last two genera are now generally transferred to Asioryctitheria and Ungulatomorpha (Zhelestidae), respectively. Consequently, Leptictida is used here in the more restricted sense of Novacek (1986a).

Late Cretaceous *Gypsonictops*, known from jaws and teeth from western North America, is the oldest known leptictidan and the only pre-Cenozoic representative. It is the sister taxon of Leptictidae (Novacek, 1977, 1986a). Like early Tertiary leptictids (but unlike many other insectivorans), *Gypsonictops* (Fig. 9.2A) has molariform posterior premolars; the lower molariform teeth have less elevated, more mesiodistally compressed trigonids and relatively larger, broader talonids than those of its contemporary *Cimolestes* (Clemens, 1973). The upper molars are less transverse than in *Cimolestes*, with a reduced stylar shelf and well-developed postcingula. *Gypsonictops* differs from leptictids and most other eutherians in having five premolars (the primitive state for eutherians), the last one slightly less molariform than in leptictids, and in some other details of tooth structure.

Leptictidae includes 10 genera, mostly from North America. All were small, roughly the size of hedgehogs. The teeth, such as in Paleocene–early Eocene *Prodiacodon* (Fig. 9.2B), are similar to those of *Gypsonictops* except that there are only four premolars (presumably the central premolar, dP_3, has been lost), the last lower premolar has an elongate trigonid with a stronger paraconid, and the upper molars have a small hypocone (Novacek, 1977). These trends were accentuated in Chadronian-Orellan *Leptictis* (=*Ictops*; Figs. 9.2C, 9.3, Plate 4.1), for which exquisitely preserved skulls and skeletons are known. The skull has been described in detail by Butler (1956) and Novacek (1986a). It has a long, tapered rostrum, a deep antorbital fossa for snout muscles, paired parasagittal (temporal) crests, an entotympanic bulla, and a reduced lacrimal bone restricted to the orbit (Fig. 9.3). The olfactory region is extensive and well developed, between and anterior to the orbits. The postcranial skeleton of leptictids has been described by Cavigelli (1997) and Rose (1999a, in press). The forelimbs are relatively short and moderately robust, suggesting they were adapted for digging. The hind limbs are exceptionally long and more slender, the femur with a narrow and elevated patellar groove, and the tibia and fibula extensively fused distally. The astragalar trochlea is deeply grooved; the tarsals and metatarsals moderately elongated. Overall, the skeleton suggests that leptictids were terrestrial animals, capable of occasional bouts of running or jumping and digging with the front legs. Less well-preserved skulls and skeletal remains of Paleocene and early Eocene leptictids, such as *Palaeictops*, do not differ appreciably from *Leptictis*, except that tibiofibular fusion is limited to the distal end in Torrejonian *Prodiacodon*.

Leptictids were long held to be closely related to lipotyphlans, because of alleged craniodental similarities to

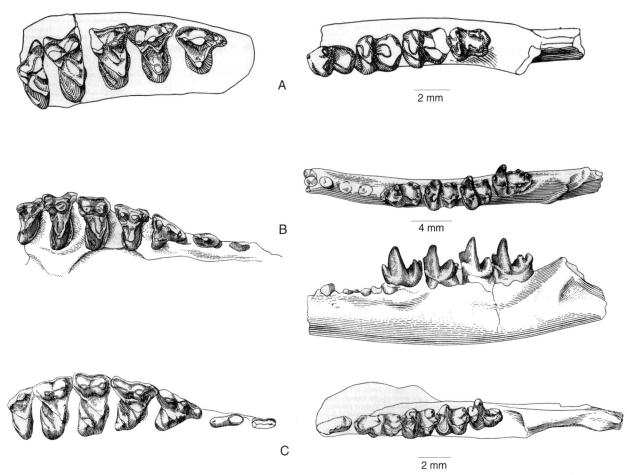

Fig. 9.2. Leptictidan dentitions, right upper teeth (left column), left lower teeth (right): (A) Late Cretaceous *Gypsonictops;* (B) Paleocene *Prodiacodon;* (C) late Eocene–early Oligocene *Leptictis.* (A, C from Lillegraven, 1969; B from Novacek, 1977.)

hedgehogs (e.g., Gregory, 1910; Butler, 1956). This alliance was questioned by later authors (McDowell, 1958; McKenna, 1975a) and was also subsequently rejected by Butler (1972), because of differences in the anatomy of the orbit and the presence of an entotympanic bulla in leptictids, unlike the basisphenoid or unossified bulla of lipotyphlans. Following detailed cranial analysis, however, Novacek (1986a; MacPhee and Novacek, 1993) revived the hypothesis of leptictid-lipotyphlan ties, based on numerous shared derived cranial traits. Nevertheless, McKenna and Bell (1997) continued to separate these two groups at the superordinal level (Leptictida vs. Preptotheria for Lipotyphla and most other eutherians). Alternatively, it is possible that leptictids are more closely related to macroscelideans and Glires than to lipotyphlans (Rose, 1999a; Asher et al., 2003). Thus, in spite of detailed knowledge of the anatomy of leptictids, their relationships remain controversial.

Leptictids have always been construed as very primitive eutherians. Consequently, ancestral-descendant relationships between leptictids and various other eutherians, including Primates, were considered likely through the 1960s and 1970s (e.g., Van Valen, 1965; McKenna, 1966; Bown and Gingerich, 1973; Clemens, 1973). The current consensus is that leptictids are among the most primitive known eutherians and offer important information on plesiomorphic eutherian features, but close relationship to other eutherians remains to be convincingly demonstrated.

Probably closely related to leptictids are two European genera with the ungainly family name Pseudorhyncocyonidae—an allusion to their superficial resemblance to the living elephant shrews, particularly the genus *Rhynchocyon*. *Leptictidium* (Fig. 9.4, Plate 4.2) is the better-known genus, being represented by several complete skeletons from the middle Eocene of Messel, Germany (Storch and Lister, 1985; Maier et al., 1986; Koenigswald et al., 1992a). They were the largest leptictidans, ranging in length from just over a half meter to a little less than a meter. *Leptictidium* had a long, slender snout with antorbital muscle fossae (as in leptictids), suggesting a mobile snout, as in elephant shrews. The molars are generally similar to those of leptictids, although the upper molars are less transverse, and the fourth premolars are molariform. *Leptictidium* is more derived than leptictids in having wide diastemata separating the anterior teeth from the cheek teeth. Stomach contents, preserved in several

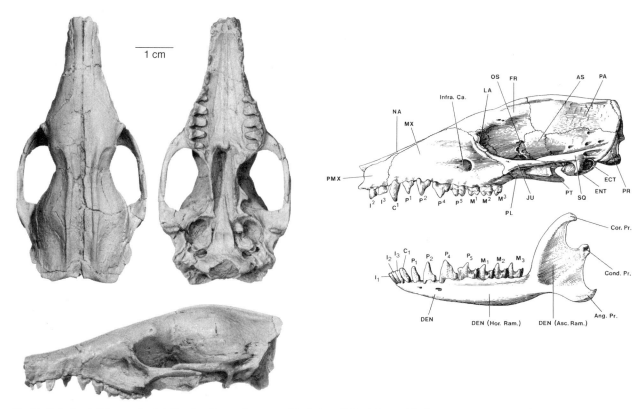

Fig. 9.3. Skull of early Oligocene *Leptictis*. Key: Ang. Pr., angular process; AS, alisphenoid; Cond. Pr., condyloid process; Cor. Pr., coronoid process; DEN, dentary; DEN (Asc. Ram.), ascending ramus; DEN (Hor. Ram.), horizontal ramus; ECT, ectotympanic; ENT, entotympanic; FR, frontal; Infra. Ca., infraorbital canal; JU, jugal; LA, lacrimal; MX, maxilla; NA, nasal; OS, orbitosphenoid; PA, parietal; PL, palatine; PMX, premaxilla; PR, petromastoid; PT, pterygoid; SQ, squamosal. (From Novacek, 1986a.)

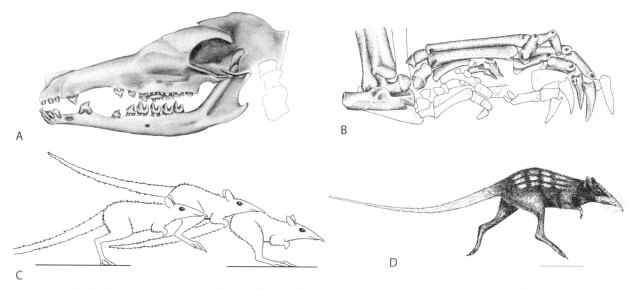

Fig. 9.4. *Leptictidium* from the middle Eocene of Messel: (A) skull; (B) right hind foot. Two hypotheses of locomotion in *Leptictidium*: (C) bipedal saltation; (D) bipedal running. (A, B from Storch and Lister, 1985; C from Frey et al., 1993; D from Maier et al., 1986.)

Messel skeletons, reveal (as the teeth suggest) that *Leptictidium* ate insects and small vertebrates.

The forelimbs of *Leptictidium* are extremely short and the hind limbs elongate. The tibia and fibula are separate, although apparently closely appressed and perhaps joined by strong ligaments for much of their length (Frey et al., 1993). The tail was twice as long as the body. These unusual body proportions indicate that *Leptictidium* must have been functionally bipedal, but its mode of progression is controversial. According to one hypothesis (Maier et al., 1986), it was a bipedal cursor but did not jump or hop. However, Frey et al. (1993) suggested that *Leptictidium* was more likely saltatorial, like extant mammals with similar limb proportions. The latter interpretation was supported by a recent biomechanical analysis, which cited the flexible lumbar region and long, rodlike ilia of *Leptictidium* as particular characteristics of bipedal hoppers (Christian, 1999).

LIPOTYPHLA

The Lipotyphla, as traditionally defined, consists of six extant families—Erinaceidae (hedgehogs), Soricidae (shrews), Talpidae (moles), Solenodontidae (solenodons), Tenrecidae (tenrecs), and Chrysochloridae (golden moles)—and a dozen or so extinct families believed to be closely related to extant forms. Two principal clades have been widely recognized, based on anatomical evidence: Erinaceomorpha, for hedgehogs and their relatives, and Soricomorpha, comprising the others (Butler, 1988). Subsequent accounts place the golden moles in a separate clade, Chrysochloromorpha, of equivalent rank (MacPhee and Novacek, 1993; McKenna and Bell, 1997). Because they are restricted to the Miocene to Recent of Africa, they are not discussed here. Several authors include moles and their close relatives in Erinaceomorpha rather than Soricomorpha (McDowell, 1958; Van Valen, 1967; McKenna and Bell, 1997), but others consider moles to be closely related to shrews (e.g., Butler, 1988). Molecular data conflict with the conventional composition of Lipotyphla, however, as further discussed below.

Lipotyphlans resemble one another largely in primitive eutherian traits. Derived features that would demonstrate monophyly of the group—especially osteological characters—have proven difficult to establish. Butler (1988) proposed several derived features of Lipotyphla, including a reduced pubic symphysis, reduced or absent jugal, and an expanded maxilla that displaces the palatine in the orbital wall. He listed three additional traits that relate to soft tissues and cannot be judged in fossils (a hemochorial placenta, mobile proboscis, and the absence of the cecum). Various other features Butler noted, including several in the auditory region, are arguably primitive. MacPhee and Novacek (1993), reevaluating Butler's list, were able to confirm only two or three of his traits as probable lipotyphlan synapomorphies: those pertaining to the cecum and the pubic symphysis, and possibly the maxillary character. Subsequently Whidden (2002) found that similar snout musculature was associated with the mobile proboscis in all six extant lipotyphlan families and added this as a lipotyphlan synapomorphy. Recently two more potential lipotyphlan synapomorphies were identified: a small optic canal and a distally placed peroneal tubercle on the calcaneus (Mussell, 2005). Most of these features are rarely or never preserved in fossils, however, and in practice, the extinct lipotyphlan families are assigned to the order primarily because of dental similarity (in many cases the only available evidence), which may be substantially plesiomorphic. At present, therefore, the anatomical evidence for lipotyphlan monophyly is weak.

Hence it is not surprising that recent analyses of genetic sequences in modern mammals indicate that Lipotyphla is not monophyletic. Instead, most of these studies find that only a subset of lipotyphlans—talpids, soricids, and erinaceids, plus or minus *Solenodon*—constitute a monophyletic group, which has been called Eulipotyphla (Douady et al., 2002; Fig. 9.1B). Most molecular studies indicate that talpids are the most primitive eulipotyphlans, with erinaceids and soricids more derived sister taxa. This conclusion is at variance with the fossil record of these groups, as talpids are the last of these clades to appear (late Eocene). Unexpectedly, molecular evidence unites tenrecs and chrysochlorids with tethytheres, hyracoids, tubulidentates, and macroscelidids in a clade called Afrotheria (e.g., Stanhope et al., 1996, 1998; Madsen et al., 2001; Murphy et al., 2001; Douady et al., 2002; Malia et al., 2002). The grouping of tenrecs and golden moles has been called Afrosoricida, which is equivalent to Tenrecoidea as used by McDowell (1958), but not the Tenrecoidea of earlier authors.

So far, the proposed order Afrotheria has been recognized, albeit strongly, solely from molecular evidence; no compelling anatomical evidence supporting this supposed clade has been identified. A recent combined evidence analysis (morphological and molecular data) supported Afrotheria, but analysis of the expanded morphological data set alone weakly favored Lipotyphla rather than Afrotheria (Asher et al., 2003). It is noteworthy that, despite their numerous unusual specializations, chrysochlorids have the three traits listed above that appear to be synapomorphous for Lipotyphla (MacPhee and Novacek, 1993). Moreover, Whidden (2002) found that both chrysochlorids and tenrecs possess the same distinctive and possibly unique snout musculature found in other lipotyphlans, in contrast to the pattern seen in afrotheres. Phylogenetic analyses that combine molecular and morphological data, however, invariably support Eulipotyphla and place Tenrecoidea within or as the sister group of Afrotheria (Asher et al., 2003; Mussell, 2005), probably because the molecular "characters" far outnumber and overwhelm the morphological ones.

Consequently the composition of Lipotyphla remains equivocal, although it seems increasingly probable that the group, as long conceived, is not a natural one. Pending conclusive evidence regarding the phylogenetic position of tenrecs and chrysochlorids, the conventional composition of Lipotyphla is adopted here. This assumption has little effect on the ensuing discussion, however, as neither tenrecs nor chrysochlorids are known from the early Tertiary. Most of the remainder of this chapter thus centers on eulipotyphlans.

In addition to the features listed earlier, characteristics of extant lipotyphlans include a well-developed olfactory sense, a tendency toward small eyes, small external ears, presence of a ringlike ectotympanic and a relatively horizontal tympanic membrane, absence of the entotympanic, an auditory bulla (when present) that is composed of the basisphenoid, a zygomatic arch that is often incomplete (associated with the reduced jugal), and a foramen ovale surrounded by the alisphenoid (McDowell, 1958; Novacek, 1986a; Butler, 1988). A basisphenoid bulla might be another synapomorphy of Lipotyphla, but its absence in multiple lineages, including shrews, *Solenodon,* and some erinaceomorphs, complicates this interpretation (MacPhee et al.,

1988), for experts disagree as to whether its absence reflects a derived condition (its loss) or a primitive state. The molars of lipotyphlans, though usually preserving the basic tribosphenic pattern, nonetheless vary considerably, from relatively bunodont in erinaceids, to sharp-cusped and dilambdodont in shrews and moles, to high-crowned and zalambdodont in golden moles and tenrecs. In primitive Paleogene forms that are generally considered lipotyphlans, the molar morphology is much closer to the plesiomorphic eutherian condition. Most living lipotyphlans have crests on the bases of the molars, between or extending from the roots, that fit into grooves in the alveolar bone; they are incipiently developed in some Paleogene forms (Butler, 1988). The anterior incisors in extant forms are often enlarged (especially in shrews), and the canines are reduced. The premolars are usually simple, but in basal lipotyphlans the last premolar is typically semimolariform. Most lipotyphlans have generalized skeletons and are terrestrial, but many are specialized for fossorial life, and a few are semiaquatic.

Because many of the extinct families of lipotyphlans are known only or primarily from dentitions, there remains substantial ambiguity about their interrelationships, particularly about whether specific families belong to Erinaceomorpha or Soricomorpha. This ambiguity might imply that these animals have not diverged far from the lipotyphlan (or eulipotyphlan) stem, at least in their dental morphology. Indeed, as discussed below, many of these families have rather similar dental characteristics, and the distinctions often come down to subtle details, about which even experts do not agree. Consequently, the assignments adopted here could well change as our understanding of these animals improves.

Adapisoriculidae

This family of tiny, primitive eutherians from the Paleocene and early Eocene of Europe and North Africa are probably, but not certainly, lipotyphlans. Only fragmentary dentitions and isolated teeth are known. Late Paleocene *Afrodon* from Morocco and *Bustylus* from France (Fig. 9.5A,C) are the most primitive known adapisoriculids (Gheerbrant and Russell, 1991; Gheerbrant, 1995). These shrew-sized animals had simple trituburcular upper molars with a wide stylar shelf and no hypocone. The lower molars had large, open trigonids slightly wider than the talonids. More derived members, such as *Adapisoriculus,* showed a tendency toward dilambdodonty, with a mesostyle and several marsupial-like stylar cusps on the upper molars, but unlike marsupials, their lower molars have reduced paraconids. According to Gheerbrant (1995), *Garatherium* (Fig. 9.5B), a dilambdodont form originally described as a didelphoid marsupial from Africa, is an adapisoriculid.

Erinaceomorpha

The Erinaceomorpha includes the hedgehogs (Erinaceidae) and at least five Early Tertiary families, as well as several genera unassigned to families (McKenna and Bell, 1997). The

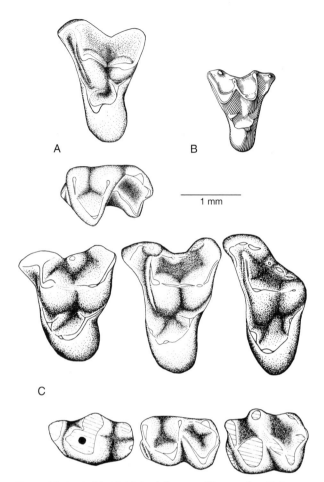

Fig. 9.5. Adapisoriculids: (A) *Afrodon,* left upper and lower molars; (B) *Garatherium,* right upper molar; (C) *Bustylus,* left upper molars and left P_4–M_2. (A from Gheerbrant, 1995; B from Crochet, 1984; C modified from Gheerbrant and Russell, 1991.)

only living erinaceomorphs are erinaceids, small insectivorans restricted to the Old World today but present in North America throughout much of the Tertiary. Living erinaceids range from mouse-sized to a little larger than the muskrat *Ondatra* (2 kg). The largest erinaceid, Miocene *Deinogalerix* (Freudenthal, 1972), was considerably larger, with a body the size of the badger *Taxidea* and a skull as big as in the coyote *Canis latrans*. Erinaceids have relatively large eyes for lipotyphlans and are more primitive than some other lipotyphlans in retaining a complete zygomatic arch. The skeleton is generalized, with moderately robust long bones, the distal half of the tibia and fibula co-ossified (as in soricids and talpids), pentadactyl feet, and a plantigrade stance.

The most primitive erinaceids have an essentially primitive eutherian dental formula (i.e., primitive for the Cenozoic: 3.1.4.3/3.1.4.3; see Chapter 6), but some derived species lose an upper incisor or one or more premolars, and some fossil taxa reduced the dentition even more. Erinaceids tend to have more bunodont cheek teeth than other lipotyphlans, reflecting a more omnivorous diet. M^{1-2} are quadrate with a narrow stylar shelf and a large hypocone. The lower molars are rectangular with subequal trigonids and talonids. Erinaceids emphasize shearing between P^4 and M_1 (like the carnassials of carnivorans), so P^4 has a long metacristid and M_1 has a large open trigonid with an extended paracristid (Butler, 1988). Associated with this modification is a distinct gradient in molar size: M1 > M2 > M3. Fossil genera that exhibit these features have usually been assigned to the Erinaceomorpha.

Primitive Erinaceomorphs

While there is general agreement about which fossils represent primitive erinaceomorphs based on their dental morphology, there is considerably less agreement about their interrelationships. The two dozen or so genera of Early Tertiary erinaceomorphs, most of them known only from dentitions, have been placed in several families, including the extant Erinaceidae, but the composition of these families (Adapisoricidae, Amphilemuridae, Dormaaliidae, Sespedectidae, Scenopagidae, Creotarsidae, and Chambilestidae), and even their names, are very unstable. This is not surprising, considering that analysis of dental attributes in extant hedgehogs suggests that teeth alone may be inadequate for deducing erinaceomorph relationships (Gould, 2001). Many of these early erinaceomorphs have been confused with primates or hyopsodontid condylarths because of their low-crowned, bunodont molars with reduced paraconids and broad talonid basins, and it is not entirely certain that all are really erinaceomorphs. Percy Butler, one of the most distinguished authorities on insectivores, has even suggested that several taxa widely accepted as primitive erinaceomorphs, such as *Macrocranion* and *Pholidocercus,* may not be lipotyphlans at all (Butler, 1988). Until a more probable relationship is convincingly demonstrated, however, they are included here as erinaceomorph lipotyphlans.

Where known, these early Tertiary erinaceomorphs retain a dental formula of 3.1.4.3/3.1.4.3. Like erinaceids, the members of the various erinaceomorph families listed above have relatively low-crowned, bunodont molars with moderately to well-developed hypocones on M^{1-2}. The fourth premolars are premolariform to submolariform (with a molariform trigonid and variably developed talonid). Most are more primitive than erinaceids, however, in having the first two molars subequal in size and the third molar only slightly smaller. Moreover, the trigonids are mesiodistally compressed, so that the paracristid of M_1 is not extended and the P^4/M_1 shearing function is poorly developed (Butler, 1988). Otherwise, the various genera differ in relatively minor details of dental anatomy, and all were probably omnivorous. Some were very bunodont (e.g., *Adapisorex, Pholidocercus, Sespedectes*), suggesting that they consumed more fruit and other plant matter than other early insectivorans. This hypothesis is confirmed by skeletons of *Pholidocercus* from the middle Eocene of Messel, Germany, which preserve stomach contents of leaves or fruit in addition to insects (Koenigswald et al., 1992a).

Several Paleocene and early Eocene genera have been identified as the most primitive known erinaceomorphs (Fig. 9.6). The oldest well-known genus is *Adunator* (=*Mckennatherium*), first recorded from the Torrejonian in North America and the Thanetian in Europe (Krishtalka, 1976a; Bown and Schankler, 1982; Butler, 1988). Its molars are very primitive, with relatively acute cusps, the lowers with reduced paraconids and the uppers with variably developed lingual hypocones. The first two molars are about the same size, whereas the third is slightly smaller. The fourth premolars are semimolariform to molariform, and the other premolars are simple. *Adunator* is lower crowned and more bunodont than leptictids and cimolestids, but closely resembles primitive nyctitheriid soricomorphs (see below), differing primarily in minor details of molar morphology.

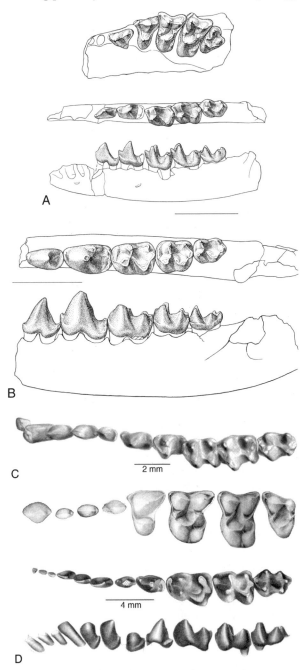

Fig. 9.6. Left dentition of primitive erinaceomorphs: (A) *Diacocherus* upper and lower teeth; (B) *Litocherus,* P_3–M_3; (C) *Ankylodon,* lower teeth in crown view; (D) *Macrocranion,* upper and lower teeth. (A and B from Gingerich, 1983b; C from Lillegraven et al., 1981; D from Tobien, 1962.)

Late Paleocene *Diacocherus* is closely allied or perhaps synonymous (Gingerich, 1983b). Another very primitive North American erinaceomorph is late Paleocene *Litocherus*. *Litocherus* is more derived than *Adunator* in having more bunodont molars that decrease in size posteriorly, more closely approaching erinaceids in these features (Gingerich, 1983b; Novacek et al., 1985). *Chambilestes,* based on a maxillary dentition from the early Eocene of Tunisia, could be the oldest (and only pre-Miocene) African erinaceomorph (Gheerbrant and Hartenberger, 1999). It has low-crowned, transverse upper molars with a small hypocone, resembling the North American erinaceomorph *Scenopagus*. However, *Chambilestes* also resembles the soricomorph *Centetodon* in the same features, leaving its affinities ambiguous. The source of these basal erinaceomorphs has not been established.

Skeletons of Eocene erinaceomorphs from Messel are among the most complete lipotyphlan fossils known. One of these is *Macrocranion* (usually referred to Amphilemuridae or Dormaaliidae; Fig. 9.7), which is present in the Eocene of both Europe and North America. The name is a misnomer, referring to the seemingly disproportionately large head in the holotype, which now appears to be an artifact of poor preservation (Maier, 1979). The Messel skeletons of *Macrocranion* reveal many details about its anatomy and lifestyle (Maier, 1979; Koenigswald et al., 1992a). *Macrocranion* had a long mobile snout, short forelimbs and long hind limbs with extensively fused tibia-fibula and elongate metatarsals, and a moderately long tail. The terminal phalanges were short and stout and seem to have supported hooflike claws, unlike the longer and more curved claws of diggers or the deep, laterally compressed claws of climbers. *Macrocranion* was a small terrestrial runner, and some species were probably saltatorial (Storch, 1996). Small orbits, large auditory bullae, and details of the nasal region indicate that *Macrocranion* relied on its auditory, olfactory, and tactile senses more than on vision. Stomach contents of the Messel specimens reveal a diet including fish, insects, and plant matter. Although the erinaceomorph affinities of *Macrocranion* are widely accepted, some features of its skeleton suggest that it may not even be a lipotyphlan (see also Butler, 1988).

The closely related *Pholidocercus* (Amphilemuridae; Fig. 9.8) has more generalized body proportions than those of *Macrocranion,* but the Messel skeletons reveal several very distinctive features (Koenigswald et al., 1992a). The bones of the forehead are sculptured with vascular impressions, indicating the presence of a horny pad, and the tail is encased in imbricating bony scales along its entire length. Several specimens preserve the outlines of coarse, bristly fur on the back. The significance of these features is uncertain, but they may have been defense mechanisms. The forelimbs of *Pholidocercus* are moderately robust and the terminal phalanges deeply fissured, perhaps associated with digging.

Erinaceidae

Erinaceids are the most derived erinaceomorphs. The oldest fossils often identified as erinaceids are *Litolestes,*

Fig. 9.7. *Macrocranion* from the middle Eocene of Messel. (Top panel courtesy of G. Storch; lower panels from Storch, 1993a.)

Leipsanolestes, and *Cedrocherus* from the late Paleocene of Wyoming and Montana (Krishtalka, 1976a; Novacek et al., 1985; McKenna and Bell, 1997; Fig. 9.9A,B). Several more erinaceid genera have been described from the Eocene of North America and Asia. Most of them are known only from dentitions, which are morphologically similar to other erinaceomorphs but specifically resemble modern erinaceids in exhibiting a clear decrease in size from M1 to M3. However, most of them still lack the P^4/M_1 shear characteristic

Soricomorpha

This lipotyphlan group has been considered to comprise the shrews (Soricidae), moles and their relatives (Talpoidea), *Solenodon* (Solenodontidae), and the tenrecs (Tenrecidae), as well as about half a dozen extinct families (Butler, 1972, 1988). Although they all have various features in common with shrews, attribution of all of them to a monophyletic Soricomorpha is by no means firmly established (and is contradicted by molecular evidence). Some authorities consider talpoids and the extinct family Nyctitheriidae to be erinaceomorphs, and others have challenged the lipotyphlan status of some other extinct families (Apternodontidae and Micropternodontidae). Thus Soricomorpha as commonly conceived is a rather eclectic assemblage that includes dilambdodont and zalambdodont forms that range from generalized terrestrial quadrupeds to the most specialized living diggers. Of the extant families, only shrews and moles have an Early Tertiary fossil record.

Soricomorphs have smaller eyes than those of erinaceomorphs, lack a postorbital constriction of the skull, tend to have a medially expanded mandibular condyle and a relatively shorter infraorbital canal, and are said to have a more

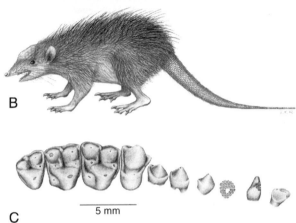

Fig. 9.8. *Pholidocercus*, an erinaceomorph from the middle Eocene of Messel: (A) skeleton; (B) restoration; (C) right upper teeth. (A courtesy of W. von Koenigswald; B and C from Koenigswald and Storch, 1983.)

of later erinaceids, a fact that led Butler (1988) to exclude the above-named genera from this family. *Litolestes* had bilobed lower incisors (Schwartz and Krishtalka, 1976). With one exception noted below, none of these early erinaceids are assignable to living subfamilies. Regardless of whether these taxa belong to Erinaceidae or to some more primitive erinaceomorph lines, they are probably closer to the erinaceid stem than any other known forms.

The oldest known representative of an extant erinaceid subfamily is *Eochenus*, from the middle Eocene of China (Fig. 9.9C). Based on numerous dental similarities, Wang and Li (1990) assigned it to the subfamily Galericinae, which includes present-day moon rats or gymnures. One distinctive feature is the presence of procumbent, bilobed incisors, as in *Litolestes*. A second middle Eocene galericine (*Eogalericius*) was recently described based on fragmentary dentitions from Mongolia (Lopatin, 2004). *Neurogymnurus* and *Tetracus*, galericines known from the Phosphorites of Quercy, France, are the oldest European hedgehogs (Crochet, 1974). They first appear immediately after the Grande Coupure (Remy et al., 1987). Other hedgehogs followed in the early Oligocene of Eurasia and North America (e.g., *Amphechinus*, *Galerix*, *Proterix*), but the group is unknown in Africa until the Miocene (McKenna and Bell, 1997).

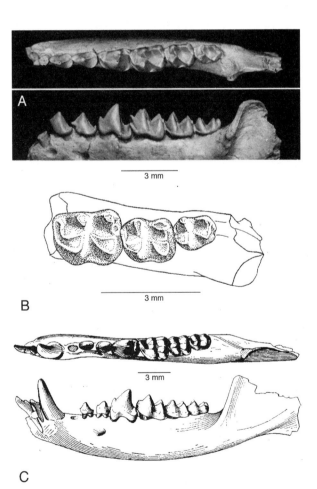

Fig. 9.9. Primitive erinaceid dentitions: (A) *Litolestes*, left P_2–M_3; (B) *Cedrocherus*, left M_{1-3}; (C) *Eochenus*, left lower jaw. (A from Novacek et al., 1985; B from Gingerich, 1983b; C from Wang and Li, 1990.)

specialized alisphenoid canal (Butler, 1988). But variation in these features raises questions about their utility for defining the Soricomorpha, and unequivocal synapomorphies characterizing this group are difficult to identify (MacPhee and Novacek, 1993). Butler (1988) listed several features shared by shrews and moles, including dilambdodont molars and details of muscular and cranial anatomy, which suggest that these two extant families are sister groups. This interpretation is supported by some recent molecular studies (e.g., Stanhope et al., 1998) but not by others.

Various groupings of zalambdodont insectivorans have occasionally been recognized (e.g., Tenrecoidea for tenrecs, *Solenodon*, and various extinct families; e.g., Simpson, 1945), but most are no longer accepted. However, increasing evidence suggests that tenrecs and golden moles (Chrysochloridae) may be closely allied. Both zalambdodonty and dilambdodonty are believed to have arisen multiple times among lipotyphlans (Butler, 1988).

Notable in this context is *Widanelfarasia*, a possible lipotyphlan based on lower jaws from the late Eocene of Egypt. This tiny "insectivore" is characterized by relatively tall trigonids and narrow talonids on the lower molars, features suggestive of incipient zalambdodonty and potential relationship to tenrecoids (Seiffert and Simons, 2000). These features, however, are also reminiscent of didelphodont cimolestans. To determine whether *Widanelfarasia* could represent a stem tenrecoid will require more complete dentitions as well as additional intermediate forms that would link it to Miocene tenrecoids. Molecular evidence, however, suggests that tenrecoids separated from their closest relatives (macroscelidids, according to molecular data) at the end of the Cretaceous and that the diversification within tenrecs was already under way at the time *Widanelfarasia* existed (Douady and Douzery, 2003).

Primitive Soricomorphs

The oldest putative soricomorphs are very diminutive forms from the Late Cretaceous that have been ascribed to three different families (McKenna and Bell, 1997). All are known only from dentitions, which are so primitive that attribution to Soricomorpha, and even Lipotyphla, must be regarded as very tentative. Nonetheless, they merit consideration here, because if even one of them is properly assigned to Lipotyphla, it would represent the only demonstrable Mesozoic record of a "modern" (extant) ordinal-level clade.

The oldest is *Otlestes* (Fig. 9.10A), from the Cenomanian of Uzbekistan. It is very primitive in having eight two-rooted postcanines, interpreted as five premolars and three molars (Nessov et al., 1994). It is also said to retain a remnant of the coronoid bone. The lower molars have tall trigonids with high protoconids and low talonids, and the last premolar (P_5) is semimolariform. The simple upper molars lack pre- or postcingula and have a reduced stylar shelf compared to that of basal eutherians. *Otlestes* is so plesiomorphic that it was grouped with *Prokennalestes* in the family Otlestidae by Kielan-Jaworowska and Dashzeveg (1989); but it is more derived than *Prokennalestes* in having less transverse upper molars with a narrower stylar shelf and larger lower molar talonids. These features are not, of course, found only in soricomorphs. Consequently, Archibald and Averianov (2001) assigned *Otlestes* to Eutheria incertae sedis. Recently, however, they considered *Otlestes* to be a junior synonym of *Bobolestes* and assigned the taxon tentatively to Zalambdalestoidea, based on relatively weak evidence (Averianov and Archibald, 2005).

Batodon (Fig. 9.10B), from the latest Cretaceous (Lancian) of western North America, is the smallest known Cretaceous mammal at a body mass of about 5 g (Wood and Clemens, 2001). It has simple, transverse upper molars with a narrow precingulum, a wider posterolingual cingulum, and a reduced stylar shelf compared to cimolestids and palaeoryctids. The lower molars have tall trigonids and relatively long talonids, which are narrower than the trigonids. *Batodon* has been interpreted as a palaeoryctid *sensu lato* (Lillegraven, 1969) or the most primitive geolabidid soricomorph (Krishtalka and West, 1979; McKenna and Bell, 1997; Bloch et al., 1998), but most recently it was deemed a primitive eutherian of uncertain relationship (Wood and Clemens, 2001).

Paranyctoides (see Fig. 6.5A) is known from the Campanian of western North America and the Turonian-Santonian equivalent strata of Asia (Fox, 1984a; McKenna and Bell, 1997; Archibald and Averianov, 2001). It is estimated to have been somewhat larger than *Batodon* (about 10–15 g; Wood and Clemens, 2001). At least some specimens of the Asian species had five lower premolars, the last of which is submolariform. The lower molars are lower crowned than those of most other Cretaceous eutherians and have open trigonids, a low paraconid, and large, wide talonid basins. The upper molars are less transverse than in other Cretaceous eutherians, but are primitive in retaining a moderate stylar shelf with stylar cusps. Derived traits include the presence of small conules, pre- and postcingula, and in some specimens a tiny hypocone. These features resemble both lipotyphlans (specifically, nyctitheriids) and very primitive ungulates, and they suggest the possibility of a relationship between these two groups deep in the Cretaceous (Fox, 1984a; Archibald and Averianov, 2001). *Paranyctoides* has been variously identified as a primitive nyctitheriid soricomorph (Fox, 1984a), a basal soricomorph (McKenna and Bell, 1997), or a possible relative of zhelestid ungulatomorphs (Archibald and Averianov, 2001).

Geolabididae

Geolabidids are small soricomorphs from the Early Tertiary of North America (Fig. 9.10C,D). Although originally described as a subfamily of Erinaceidae (McKenna, 1960b), features of the mandible support affinity with soricoids. Apart from Cretaceous *Batodon*, geolabidids are first known from the early Eocene and survived into the early Miocene. The best-known genus, *Centetodon*, existed throughout this

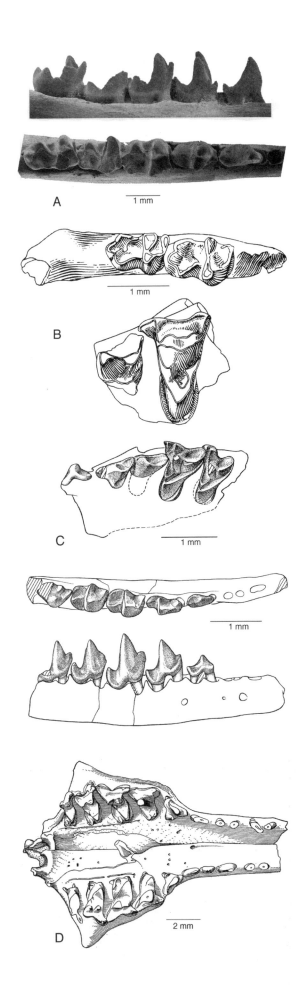

entire interval and was geographically widespread (Lillegraven et al., 1981). Where known, the skull retains a complete zygomatic arch, but modifications of the angular process of the jaw and the mandibular condyle (an incipient double articulation in *Centetodon*) foreshadow specializations characteristic of soricids. The dental formula was 3.1.4.3/3.1.4.3.

Geolabidids have higher-crowned lower cheek teeth than in erinaceomorphs, with tall trigonids (P_4–M_3) that are slightly wider than the talonids. P_4 has a large metaconid and narrow talonid but is less molarized than in nyctitheres. In *Centetodon*, short diastemata between the premolars contribute to a long snout, which tapers abruptly in front of P^3. The lower incisors of *Centetodon* are bilobed, not unlike those of some primitive erinaceids. The upper molars of geolabidids may be characterized as incipiently zalambdodont; they are transverse and have a relatively wide stylar shelf, connate paracone and metacone, small conules, and well-developed pre- and postcingula, the latter often with a small hypocone. Geolabidids were probably more strictly insectivorous than were erinaceomorphs. This family includes the smallest known mammal, early Eocene *Batodonoides vanhouteni*, whose teeth indicate a body weight of less than 2 g, smaller than any living mammal (Bloch et al., 1998).

Nyctitheriiidae

Nyctitheres ("night beast") were shrew-sized, presumably insectivorous mammals, known primarily from dentitions from the late Paleocene and Eocene of North America and Europe (Fig. 9.11). One genus, *Saturninia*, persisted into the Oligocene of Europe, and another European genus, *Darbonetus*, is present only in the early Oligocene. A few genera of nyctitheriids, based on very fragmentary dental remains, have been reported from the late Paleocene and early Eocene of central Asia.

Like many other primitive lipotyphlans, nyctitheres have a lower dental formula of 3.1.4.3. The dentition of nyctitheres is characterized by multilobed lower incisors; elongate but simple lower premolars except for P_4, which is submolariform; and lower molars with a low paraconid, tall protoconid and metaconid, and a broad talonid basin with a tall, acute hypoconid (Robinson, 1968; Krishtalka, 1976b; Sigé, 1976). The lower molar trigonids tend to be slightly more elevated and the cusps somewhat more acute than in typical primitive erinaceomorphs. The upper molars of such primitive nyctitheres as *Leptacodon* (McKenna, 1968; Fig. 9.11A)—the most primitive genus if *Paranyctoides* is excluded—have a rather generalized tribosphenic pattern with three main cusps, small conules, and a posterolingual cingulum with a small hypocone. The stylar shelf is reduced and the buccal margin indented.

Fig. 9.10. Primitive soricomorph dentitions: (A) *Otlestes*, left P_3–M_3 (soricomorph affinities questionable); (B) *Batodon*, right M_{2-3} and left upper molars; (C) *Batodonoides*, left P^2–M^2 and right P_3–M_3; (D) *Centetodon*, snout with upper dentition. Note extremely small size of B and C. (A from Nessov et al., 1994; B from Lillegraven, 1969; C from Bloch et al., 1998; D from McKenna, 1960b.)

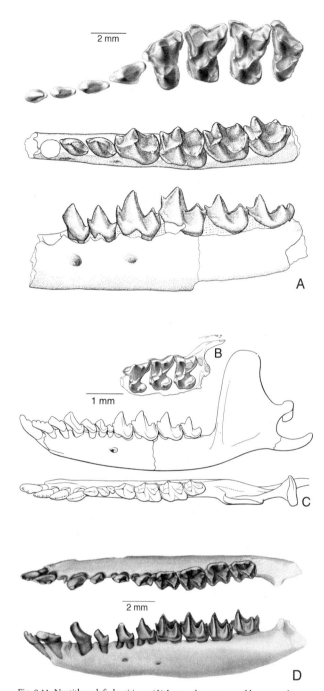

Fig. 9.11. Nyctithere left dentitions: (A) *Leptacodon*, upper and lower teeth; (B) *Saturninia*, P⁴–M²; (C) *Amphidozotherium*, left lower jaw; (D) *Ceutholestes*, left lower jaw. (A from McKenna, 1968, and Rose, 1981; B, C from Sigé, 1976; D from Rose and Gingerich, 1987.)

Nyctitheres have been variously allied with erinaceomorphs (e.g., Van Valen, 1967; Robinson, 1968; Sigé, 1976) or, more recently, soricomorphs (e.g., Butler, 1988; McKenna and Bell, 1997). They do, in fact, resemble shrews and moles in several details of dental and mandibular morphology, and they have been suggested as the ancestral stock from which shrews and moles evolved (Dawson and Krishtalka, 1984). Despite the disagreement over subordinal affinities, there has been no dispute that nyctitheres are lipotyphlans—until recently. According to Hooker (2001), isolated tarsal bones attributed to nyctitheres resemble those of scandentians and plesiadapiformes more than those of lipotyphlans, suggesting that nyctitheres might actually be archontans instead of lipotyphlans, contrary to dental evidence. This claim should be reexamined when more complete (associated) fossils are available. These bones indicate that the foot could be inverted, an ability typical of scansorial mammals.

Micropternodontidae

This small Early Tertiary family (eight genera) is variously allied with soricomorph insectivorans (McKenna et al., 1984; McKenna and Bell, 1997) or with palaeoryctids (Van Valen, 1967; Butler, 1988). All three of these groups have an opening in the roof of the middle ear, the piriform fenestra, which some experts have considered to indicate special relationship.

Micropternodontids are primarily an Asian group, the oldest of which are the early Paleocene genera *Prosarcodon* and, questionably, *Carnilestes* (McKenna et al., 1984; Wang and Zhai, 1995; Fig. 9.12A). North American *Micropternodus* from the late Eocene and Oligocene was the last surviving member (Russell, 1960; Stirton and Rensberger, 1964). Curiously the older genera have lost the third molars (dental formula 3.1.4.2/3.1.4.2), whereas the latest occurring genera, *Micropternodus* and middle Eocene *Sinosinopa*, retain M_3^3. This discrepancy indicates that either there is a long ghost lineage for these forms or micropternodontids are not monophyletic.

The sharp-cusped lower molars of micropternodontids have tall, wide trigonids and narrow talonids that are relatively elongate in early forms but mesiodistally shortened in *Micropternodus* (Fig. 9.12B). P_4 is submolariform. The upper molars of early forms are very transverse (less so in *Carnilestes*), with closely appressed paracone and metacone; a small, very lingual hypocone on M¹; and a relatively narrow stylar shelf but prominent styles. The hypocone is larger in late Paleocene *Sarcodon* and relatively enormous on P⁴–M² of *Micropternodus*. The latter has an almost zalambdodont P⁴, together with dilambdodont upper molars with a relatively wide stylar shelf. These upper molars resemble those of some shrews rather closely, but the lowers are much less similar. If micropternodontids are monophyletic, the resemblance to shrews must have been acquired independently.

Few postcrania of micropternodontids have been identified. Isolated humeri described as *Cryptoryctes*, from late

More derived genera, such as North American *Nyctitherium* and European *Saturninia* and *Amphidozotherium* (Fig. 9.11B,C), have a large posterolingual lobe bearing a hypocone. Other derived features present in some nyctitheres are a fully molarized P_4 (*Ceutholestes*; Fig. 9.11D) and a W-shaped ectoloph on upper molars (*Pontifactor*). Based partly on these features, *Ceutholestes* was placed in its own family (Rose and Gingerich, 1987) and *Pontifactor* was considered to be possibly a bat (Gingerich, 1987; Butler, 1988); but both genera are probably better interpreted as specialized nyctitheres.

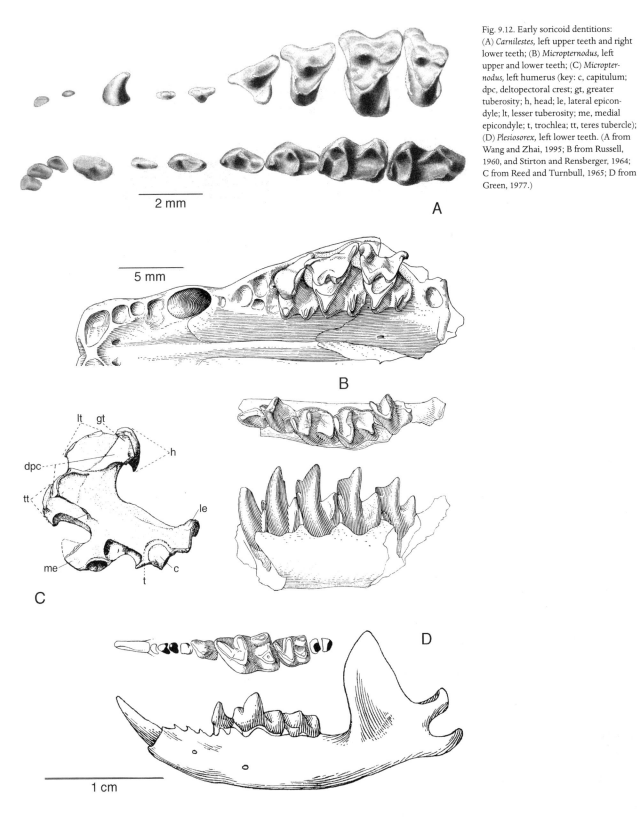

Fig. 9.12. Early soricoid dentitions: (A) *Carnilestes*, left upper teeth and right lower teeth; (B) *Micropternodus*, left upper and lower teeth; (C) *Micropternodus*, left humerus (key: c, capitulum; dpc, deltopectoral crest; gt, greater tuberosity; h, head; le, lateral epicondyle; lt, lesser tuberosity; me, medial epicondyle; t, trochlea; tt, teres tubercle); (D) *Plesiosorex*, left lower teeth. (A from Wang and Zhai, 1995; B from Russell, 1960, and Stirton and Rensberger, 1964; C from Reed and Turnbull, 1965; D from Green, 1977.)

Eocene sites, are now thought to belong to *Micropternodus* (Fig. 9.12C). As in moles, the humerus is as wide as it is long, the tubercle for insertion of the teres major muscle is extraordinarily large, and the deltopectoral crest and bicipital groove are almost perpendicular to the long axis of the humerus. If properly attributed, these humeri indicate that micropternodontids were highly fossorial animals adapted for humeral-rotation digging, as in extant moles (Reed and Turnbull, 1965).

Apternodontidae

Apternodonts were extraordinary lipotyphlans. They include four North American genera from Eocene–early

Oligocene rocks, an undescribed form from the late Paleocene, a newly described middle Eocene genus from Mongolia, and a possible apternodont from the middle Eocene of China (McKenna and Bell, 1997; Tong, 1997; Robinson and Kron, 1998; Lopatin, 2003a). All are strongly zalambdodont, resembling the living tenrecs and *Solenodon* in dental morphology. The upper molars are dominated by a large and tall lingual cusp (apparently the displaced paracone) and a broad, V-shaped stylar shelf. Lingually there is a low cingulum that may bear a small protocone, but no metacone or hypocone is present. The lower molars have high trigonids and greatly reduced talonids. The anterior incisors are enlarged and procumbent. The medial side of the mandibular coronoid process is excavated for jaw muscle attachment, somewhat as in shrews. An ossified auditory bulla is lacking in the two genera for which the skull is known (*Apternodus* and *Oligoryctes*; Asher et al., 2002).

Duchesnean-Orellan *Apternodus* (Figs. 9.13, 9.14) is the best-known genus. The various species range in size from shrew- to hedgehog-sized and have a dental formula of 2.1.3.3/3.1.3.3. The posterior premolars of *Apternodus* are often relatively large, and P^4_4 are submolariform, P_4 with a well-developed trigonid. Late Eocene *Apternodus baladontus* has peculiarly swollen and rounded canines and adjacent teeth, which were presumably used for crushing hard-shelled arthropods or crustaceans. The skull of *Apternodus* is bizarrely expanded at the back, forming a pair of dorsoventrally oriented lambdoid plates ("lateral auditory plates" of McDowell, 1958) composed of the squamosal, petromastoid, and occipital bones (Asher et al., 2002). Posteriorly the plates contribute to a wide nuchal crest, and anteriorly they are hollow, containing an extension of the tympanic cavity called the epitympanic recess. The lateral plates are reminiscent of similar structures on the skull of the extant golden mole *Chrysospalax*, in which they also enclose the epitympanic recess. In chrysochlorids the recess is occupied by a greatly enlarged malleus, which is involved in low-frequency sound transmission (Cooper, 1928; see Rose and Emry, 1983: fig. 4). Like *Apternodus*, golden moles are zalambdodont. A close relationship between them is not likely, but the analogy suggests similar lifestyles. Golden moles are fossorial, and the adaptive features they share with *Apternodus* are related to fossorial habits. However, fragmentary postcrania of *Apternodus* illustrated by Asher et al. (2002) are only moderately robust and do not show specializations for digging; instead they indicate a more generalized terrestrial mammal.

Bridgerian to Orellan *Oligoryctes* is a shrew-sized form that has been interpreted as a close relative of *Apternodus* (Hough, 1956). It had a more complete dental formula, 3.1.3.3/3.1.3–4.3. *Oligoryctes* lacks the lambdoid plates found at the back of the braincase in *Apternodus*, but is more derived in having a medially "pocketed" mandibular coronoid process, approaching the condition in shrews. (Asher et al., 2002). Early Eocene *Parapternodus* and *Koniaryctes* are the oldest described apternodonts and are known only from

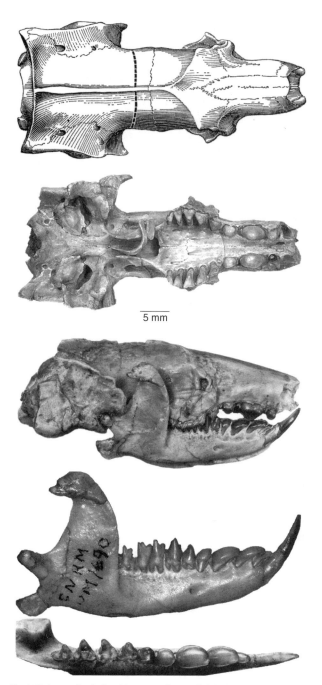

Fig. 9.13. *Apternodus* skull and jaw. (Dorsal view of skull from McDowell, 1958; photographs from Asher et al., 2002.)

a few jaw fragments. *Parapternodus* resembles *Oligoryctes* in having a pocketed coronoid.

Some authors have associated *Apternodus* with extant *Solenodon*, but this grouping seems to be based largely on zalambdodonty (a condition known to have arisen multiple times) and perhaps the occurrence of both genera in the New World (in contrast to the Old World zalambdodonts, the tenrecs and golden moles). McDowell (1958) enumerated many differences between them, and concluded that *Apternodus* was probably not even lipotyphlan. Subsequent authors (e.g., McKenna, 1975a; Butler, 1988), however, re-

Fig. 9.14. Restoration of *Apternodus*. (From Asher et al., 2002.)

affirmed the lipotyphlan affinities of *Apternodus* and noted that it shares many features with soricomorphs, including pigmented teeth as in soricids. Also as in soricids, the zygomatic arch is apparently absent and the coronoid process is vertically oriented, but these traits are not restricted to soricids. In a recent comprehensive analysis of apternodonts, Asher et al. (2002) found that Apternodontidae as generally conceived is a paraphyletic group, with Soricidae nested within it and *Solenodon* as the sister taxon. Consequently they recognized three families for North American apternodonts, Apternodontidae (*Apternodus* only), Oligoryctidae (*Oligoryctes*), and Parapternodontidae (*Parapternodus* and *Koniaryctes*).

Plesiosoricidae

A handful of poorly known genera from the Eocene to Miocene of Eurasia and the Miocene of North America are placed in this family (McKenna and Bell, 1997), which has been considered the sister group of soricids (Wang and Li, 1990). The best-known forms, *Butselia* and *Plesiosorex* (Fig. 9.12D), are shrewlike in having an enlarged lower incisor (but smaller than in shrews) and reduced canine and premolars, a posterolingual expansion with large hypocone on P^4, loss of the zygomatic arch, a vertically oriented coronoid process, P^4/M_1 shear, and reduction in molar size from M_1 to M_3 (Butler, 1988). The last two traits also characterize erinaceids. Unlike that of shrews, however, P_4 of plesiosoricids is often submolariform, with a prominent metaconid, and the upper molars have well-developed stylar cusps, particularly cusps B (the stylocone) and D. Also in contrast to shrews, *Plesiosorex* primitively retained a full lower dental formula of 3.1.4.3 (Green, 1977). *Butselia* is incipiently zalambdodont, with a reduced metacone and a wide stylar shelf on the uppers and high trigonids on the lowers. *Pakilestes*, from the early or middle Eocene of Pakistan, and *Ernosorex*, from the middle Eocene of China, are considered the oldest known plesiosoricids by McKenna and Bell (1997). Both are known from only one or two teeth or jaw fragments, making their relationship to this family very tenuous. Indeed, Wang and Li (1990) placed *Ernosorex* in the Soricidae, an assignment followed here.

Soricidae

By far the most diverse and successful lipotyphlans, shrews account for three-quarters of all living insectivoran species (about 300) and are nearly cosmopolitan. Ranging in body size from 2 g to more than 100 g, they include the smallest extant mammals except for the bumblebee bat (*Craseonycteris*). All shrews have a unique double temporomandibular joint, consisting of an upper joint surface homologous with that of other mammals and a secondary lower joint surface that is a neomorph. Other characteristics of shrews include a long, pointed angular process; absence of an auditory bulla; and loss of the zygomatic arch. In addition, all shrews except those of the most primitive subfamily, Heterosoricinae (sometimes considered a separate family), have a deep fossa that opens into the medial side of the coronoid process of the mandible. This fossa accommodates the internal temporal muscle, which replaces the masseter in these animals (Repenning, 1967).

Dental features of shrews include enlarged anterior incisors, the lower one procumbent and often with a serrated or multilobed crown; reduction in size and number of other antemolar teeth, which are one-rooted and often crowded and overlapping; P_4 primitively unicuspid and triangular; P^4/M_1 shear developed (as in erinaceids); and dilambdodont molars. In contrast to talpids, the entoconid is usually large and separate from the posterior crest, or hypolophid, and may be joined to the metaconid by a crest. Members of the subfamily Soricinae have enamel pigmented by the deposition of iron (goethite crystallites) in a superficial aprismatic layer external to the typical white hydroxyapatite enamel (Koenigswald, 1997a; Akersten et al., 2002). According to Akersten et al., this specialized enamel is harder than normal enamel and may be an adaptation to resist abrasion.

Most of these features are already present in the oldest unequivocal soricids (Fig. 9.15B,C), the heterosoricines *Domnina* from the middle Eocene (Uintan) to early Miocene of North America (Repenning, 1967; Krishtalka and Setoguchi, 1977) and *Quercysorex* from the early Oligocene of Europe (Crochet, 1974; Engesser, 1975; Rzebik-Kowalska, 2003). Although lacking the double jaw joint of later shrews, these genera have a medially expanded mandibular condyle with an incipiently bipartite articular surface (Repenning, 1967; Engesser, 1975). *Domnina* already has pigmented enamel, which suggests that pigmentation was a primitive feature of shrews that has been lost in some extant members (Crocidurinae). Even at this early stage, the dentition of the most primitive species is reduced to 1.5.3/1.5.3 (an enlarged incisor, five intervening teeth, and three molars; Repenning, 1967).

Two recently named Asian forms could be slightly older and more primitive shrews than *Domnina*. *Ernosorex* (Fig. 9.15A), from the middle Eocene of China, resembles soricids

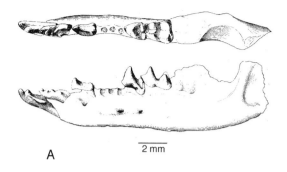

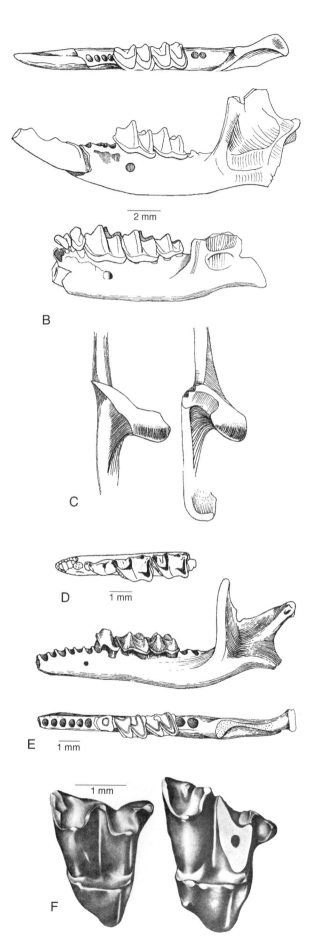

Fig. 9.15. Primitive shrews and moles: (A) *Ernosorex;* (B) *Quercysorex;* (C) mandibular condyles of *Domnina* (left) and *Quercysorex;* (D) *Soricolestes;* (E) *Myxomygale;* (F) *Eotalpa*, right upper molars. (A from Wang and Li, 1990; B from Crochet, 1974, and Engesser, 1975; C from Engesser, 1975; D from Lopatin, 2002; E from Crochet, 1974; F from Sigé et al., 1977.)

in having large procumbent incisors with lobate crowns; anteriorly inclined and overlapping antemolar teeth; a triangular P_4; and M_1 with a narrow, open trigonid and much wider talonid. It is more primitive than shrews in several features, including molar talonid structure and the retention of three lower incisors, which are remarkably similar to those of nyctitheriids (bilobed I_1, four-lobed I_2). At the same time, I_2 is similar to the enlarged incisor of many shrews (the homology of which is uncertain). It is possible that *Ernosorex* represents a transitional stage between nyctitheres and soricids, a relationship proposed by Dawson and Krishtalka (1984). *Soricolestes* (Fig. 9.15D), from the middle Eocene of Mongolia, has shrewlike molars but is more primitive than other soricids in having one more anterior tooth (1.6.3) and a semimolariform P_4 (Lopatin, 2002). Unlike all but heterosoricines, it has a deep masseteric fossa on the outside of the mandible and little indication of a temporal fossa on the medial aspect. In neither of the Asian genera is the structure of the jaw joint known. For now, the soricid status of both genera should be considered tentative. Lopatin (2002) interpreted *Soricolestes* to be dentally intermediate between nyctitheriids and soricids. Apart from these Eocene records, the oldest Asian shrews date from the early Oligocene (Rzebik-Kowalska, 2003). The supposed Late Cretaceous shrew *Cretasorex* from Uzbekistan may well be a soricine, but it is most likely a late Cenozoic contaminant (Nessov et al., 1994).

Although a number of shrew genera were present in Europe and North America by the Oligocene, it was not until the Miocene that the family became truly diverse and widespread.

Talpoidea

This group, composed of Talpidae (moles) and two related mid-Tertiary families, the North American Proscalopidae and the mainly European Dimylidae, is variously included in Erinaceomorpha or Soricomorpha. Talpids appear to be more generalized than soricids in lacking a transversely broad mandibular condyle and retaining a slender zygomatic arch, but it has been suggested that both of these are actually secondary traits that evolved from a more soricid-like condition (McKenna, 1975a; Butler, 1988). Talpids are apparently more derived in having a basisphenoid auditory bulla (which is absent in soricids), although it is possible that this trait is a primitive condition. They are also more derived in having vestigial eyes and numerous other fossorial specializations, particularly in the forelimbs. In addition to the typical shoulder joint between the scapula and humerus, moles have a secondary articulation between the clavicle and greater tuberosity. Moles have evolved their own unique method of digging in which the humerus rotates around its long axis (Reed, 1951, 1954; Yalden, 1966). However, the most primitive living moles are postcranially less specialized and more similar to shrews (Reed, 1951).

The cheek teeth of primitive moles are quite similar to those of shrews, but the incisors are smaller and less specialized and more anterior teeth are retained. In addition, moles tend to lack a hypocone on the upper molars (but not always; see below), in contrast to the well-developed hypocone of shrews. The entoconid of the lower molars is usually joined by a crest to the hypoconid, and there is usually a small accessory cusp immediately posterior to the entoconid. Talpid molars are dilambdodont and specialized for M^1/M_2 shear, rather than the P^4/M_1 shear characteristic of erinaceids and soricids (Butler, 1988). Butler (1988) listed several features, in addition to dental and postcranial similarities, shared by talpids and soricids; he regarded these two families to be sister groups. This pairing is consistent with possible derivation of both families from the Nyctitheriidae (Dawson and Krishtalka, 1984).

The oldest known moles come from the late Eocene of Europe. *Eotalpa* (Fig. 9.15F), based on isolated upper molars from England, is considered to be the earliest talpid (Sigé et al., 1977). It is dilambdodont, and its upper molars lack a hypocone, suggesting that this cusp was primitively absent in moles. However, the presence of a hypocone in many primitive lipotyphlans and in the plesiomorphic extant mole *Uropsilus* makes the primitive state for moles ambiguous. By the early Oligocene (just after the *Grande Coupure*) several talpid genera existed, including *Myxomygale* and *Mygatalpa*, the oldest known desman (Crochet, 1974; Remy et al., 1987; Fig. 9.15E). Their lower molars differ only in minor details from those of contemporary shrews, but the antemolar dentition is more conservative (lower dental formula 3.1.3–4.3). Talpids apparently did not reach North America until the late Oligocene.

During the Oligocene and Miocene, relatives of talpids called proscalopids occupied the mole ecological niche in North America. The dentition of proscalopids is generally talpidlike, including dilambdodont molars. Like moles, proscalopids have a highly fossorial skeleton, but the forelimb functioned differently than in any living mammal, and the head was also used for digging, as in chrysochlorids (Barnosky, 1981). Although proscalopids are first known with certainty from the latest Eocene (Chadronian) of western North America, a humerus from the Paleocene (Torrejonian) of Montana is similar enough to suggest that the family might already have existed at that time (Simpson, 1937a; Reed and Turnbull, 1965).

Dimylids are another presumed offshoot of the talpoid stem, known from the Oligocene and Miocene of Europe (Hürzeler, 1944). Their teeth were lower crowned and more bunodont than those of other talpoids. Dimylids are generally characterized by having enlarged P^4–M^1 and P_4–M_1, whose crowns often overhang the lateral side of the dentary, a condition known as exodaenodonty. The third molars are typically lost, but a greatly reduced M_3 is retained in the oldest member, early Oligocene *Exodaenodus*.

10

Archonta
Bats, Dermopterans, Primates, and Tree Shrews

MORPHOLOGICAL EVIDENCE FROM RECENT and fossil mammals, including a suite of derived arboreal adaptations, suggests that primates, bats, dermopterans, and tree shrews compose a monophyletic superordinal group, the Archonta (Table 10.1). The notion of Archonta dates back to Gregory (1910) but has become popular again among morphologists in recent years (Szalay, 1977; Novacek and Wyss, 1986; Novacek, 1992a,b; Szalay and Lucas, 1993, 1996; McKenna and Bell, 1997; Silcox et al., 2005). Most morphological studies favor sister-group relationships of Scandentia + Primates and Chiroptera + Dermoptera, although there is some evidence from fossils reported over the past 15 years or so to suggest a special dermopteran-primate relationship (Primatomorpha; e.g., Beard, 1993a,b). These relationships are depicted in Figure 10.1. Some morphological studies, however, have questioned the basis for a monophyletic Archonta (e.g., MacPhee, 1993).

Molecular studies are equivocal on the concept of Archonta, most supporting an alliance of primates, tree shrews, and flying lemurs (called Euarchonta) but excluding bats, which appear to be more distantly related (e.g., Adkins and Honeycutt, 1991; Allard et al., 1996; Madsen et al., 2001; Murphy et al., 2001). These studies conclude that euarchontans are most closely related to Glires (rodents and lagomorphs), the two composing one of four major placental clades, Euarchontoglires. According to these studies, bats are part of another major clade, Laurasiatheria, which includes carnivorans, modern ungulates, and lipotyphlan insectivores (see Fig. 1.5).

The current morphological concept of Archonta, including the addition of the extinct primatelike Plesiadapiformes, is used here for convenience; but it should be understood that a special relationship among these four extant orders (especially of bats to the others) remains controversial. No stem fossils that would clearly link these

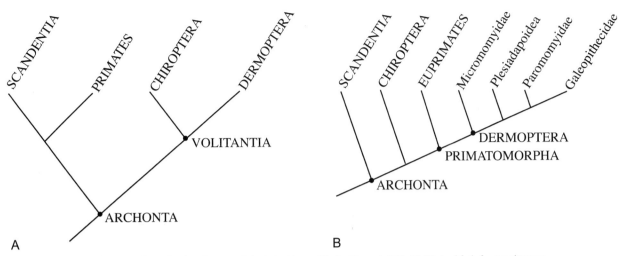

Fig. 10.1. Two views of archontan relationships based on morphological evidence: (A) after Novacek (1992a,b); (B) simplified after Beard (1993a).

orders have been found. Late Cretaceous *Deccanolestes* (Fig. 10.2) from India has been postulated to be a basal archontan or archontan ancestor, based on referred isolated tarsal bones (Prasad and Godinot, 1994; Hooker, 2001). But it is otherwise known only from isolated teeth, which are primitive and palaeoryctoid-like rather than showing any derived archontan features (Prasad et al., 1994; Rana and Wilson, 2003). More complete, associated fossils are needed to test this hypothesis.

CHIROPTERA

Bats, which constitute the order Chiroptera, are among the most easily recognized mammals. They are the only mammals that have achieved powered flight. Recent classifications recognize 22 to 24 families of bats (see Table 10.1), most still extant, with about 200 living genera and more than 1,100 living species—about 20% of all extant mammal species (e.g., Corbet and Hill, 1991; McKenna and Bell, 1997; Simmons and Geisler, 1998; Simmons, 2005; Wilson and Reeder, 2005). Conventionally, two suborders of bats have been widely distinguished, Megachiroptera and Microchiroptera, which are strongly supported by anatomical evidence. Megachiropterans account for about a quarter of all extant bat genera and almost 20% of species, but comprise only a single family, Pteropodidae, the fruit bats or "flying foxes." They include the largest bats, which attain wingspans of 1.7 m but rarely weigh more than 1.5 kg; some nectar-feeding fruit bats are much smaller (15 g; Nowak, 1999). Megabats are restricted to the Old World, whereas Microchiroptera are cosmopolitan in distribution.

This long-accepted classification has recently been challenged by analyses of molecular sequence data, which suggest that Microchiroptera is not monophyletic. These studies conclude that rhinolophids and several related families are more closely allied with Pteropodidae than with the other microchiropteran families (e.g., Hutcheon et al., 1998; Teeling et al., 2002, 2005). Such an arrangement, however, requires that the kind of echolocation typical of rhinolophids evolved multiple times independently but was lost in pteropodids. Pending corroboration of this seemingly unlikely scenario, the conventional classification is used here.

The living bats are characterized by their delicate skeletons, especially the forelimbs, which support a patagium, or wing membrane. Distinctive chiropteran skeletal features, nearly all of which are associated with flight, include (summarized from Koopman, 1984; Simmons and Geisler, 1998; Vaughan et al., 2000):

The sternal elements are fused and the sternum usually has a prominent ventral keel associated with the origin of wing muscles;
The forelimb bones are slender and very elongate, especially the radius and the second through fifth fingers, to which the wing membrane attaches;
Muscular crests and processes are poorly developed, except for the humeral tubercles, which are usually prominent; in microchiropterans the greater tubercle (called the trochiter in bats) often forms an accessory articulation with the scapula;
The elbow and wrist joints are modified to restrict mobility to flexion-extension;
The ulna and fibula are usually greatly reduced;
The number of manual phalanges is commonly less than the primitive formula of 2-3-3-3-3;
In the manus, claws are present only on the pollex and (in Megachiroptera) digit II, but all of the hind digits bear claws;
The acetabulum is reoriented, and the ankle joint substantially modified, both in connection with use of the hind limbs for suspension rather than locomotion; and
A bony or cartilaginous rod, called the calcar, extends from the ankle to support the hind end of the patagium.

The dentitions of recent bats are diverse, those of insectivorous species typically being sectorial, with dilambdodont

Table 10.1. Classification of Archonta

Superorder ARCHONTA
    †*Deccanolestes*
  Order CHIROPTERA[1]
    Suborder MEGACHIROPTERA
      †*Archaeopteropus*
      Pteropodidae[2]
    Suborder MICROCHIROPTERA
      †*Eppsinycteris,* †*Australonycteris*
      †Icaronycteridae
      †Archaeonycteridae
      †Hassianycteridae
      †Palaeochiropterygidae
      Infraorder YINOCHIROPTERA
        Rhinopomatidae
        Craseonycteridae
        Nycteridae
        Megadermatidae
        Hipposideridae[3]
        Rhinolophidae
      Infraorder YANGOCHIROPTERA
        Emballonuridae[4]
        †**Philisidae**
        Vespertilionidae
        Molossidae
        Mystacinidae
        Noctilionidae
        Mormoopidae
        Phyllostomidae
        Myzopodidae
        Furipteridae
        Thyropteridae
        Natalidae
  Grandorder EUARCHONTA
    †**Mixodectidae**[5]
    Order DERMOPTERA
      †**Plagiomenidae**[6]
      Galeopithecidae
    Order SCANDENTIA
      Tupaiidae
    Order PRIMATES

Notes: Compiled and modified mainly after McKenna and Bell, 1997, and Simmons and Geisler, 1998. See Table 10.2 for classification of Primates. The dagger (†) denotes extinct taxa. Families and genera in boldface are known from the Paleocene or Eocene.

[1] Molecular evidence suggests Chiroptera are not closely related to other archontans; but morphology suggests they could be the sister taxon of Dermoptera (these two higher taxa are sometimes united in the Volitantia).
[2] Molecular data place Pteropodidae as the sister group of Yinochiroptera.
[3] Sometimes included in Rhinolophidae.
[4] Considered the sister group of all other extant Microchiroptera by Simmons and Geisler (1998).
[5] Probably closely allied with Plagiomenidae.
[6] Dermopteran status questionable.

upper molars (presumably the primitive state); those of frugivores usually low-crowned and bunodont; and those of nectivores degenerate. In most insectivorous forms the molar hypoconulid has shifted toward the lingual border of the tooth, behind the entoconid, rather than being more centrally situated between the hypoconid and entoconid. There is typically a well-developed postcristid joining the hypoconid with either the entoconid or the hypoconulid. The auditory bulla consists of the ectotympanic or a combination of entotympanic and ectotympanic bones (Novacek, 1986a). In microchiropterans the premaxillae and postorbital processes tend to be reduced.

Microchiropterans are one of the few mammalian groups that have evolved the ability to use high-frequency sound for echolocation. Ultrasonic pulses, which may vary in frequency or duration, are generated by the larynx and emitted through the mouth or nose, in some cases being directed by the elaborate external nasal apparatus. The auditory regions of microchiropterans are specialized to receive the high-frequency echoes. The cochlea is unusually large, often compressing the basioccipital, and the malleus and stylohyal are also distinctively modified. Microchiropterans use echolocation both for orientation when maneuvering and for tracking moving prey. Although some megachiropterans (which are frugivorous) use a primitive kind of echolocation to detect large, stationary objects, it involves audible tongue clicks and is believed to have evolved independently from the sophisticated echolocation of microchiropterans (Simmons and Geisler, 1998). Indeed, the fossil record of bats suggests that microchiropteran-like echolocation evolved very early in the history of bats and may have been lost in the ancestor of megachiropterans.

Most mammalogists have long assumed that wings and powered flight evolved only once in mammals, but a serious challenge to the view that Chiroptera is monophyletic was advanced in the 1980s (e.g., Pettigrew, 1986; Pettigrew et al., 1989). It was argued that megabats share complex visual neural pathways with primates, and are therefore more closely related to primates than to microchiropterans. This unorthodox view stimulated considerable research, with the result that there is now substantial morphological and molecular evidence that strongly corroborates bat monophyly (e.g., Wible and Novacek, 1988; Adkins and Honeycutt, 1991; Simmons, 1994, 1995; Miyamoto, 1996). The resemblance in pteropodid and primate visual pathways therefore seems to be convergent.

The sister taxon of Chiroptera, however, is still unresolved. Morphological data have long suggested Dermoptera as the sister group, the resulting higher taxon designated as Volitantia (see Simmons, 1994, for a summary). Several morphological studies have supported a monophyletic Volitantia (e.g., Novacek and Wyss, 1986; Thewissen and Babcock, 1992; Szalay and Lucas, 1993, 1996; Simmons, 1995; Silcox, 2001). Molecular data, however, fail to support either Volitantia or a monophyletic Archonta; instead they consistently separate bats from other archontans.

It is often supposed that an arboreal insectivore gave rise to bats, but in general there is little direct evidence for this assumption, and it has not been supported by phylogenetic studies. An exception is the Early Tertiary lipotyphlan family Nyctitheriidae, whose members are dentally very similar to early Eocene bats. The notion of a possible link between this family and Chiroptera can be traced back at least to Matthew (1918), who based the alliance on long, slender limb bone fragments found with teeth of *Nyctitherium* (but see Hooker, 1996, for a contrary opinion). This hypothesis

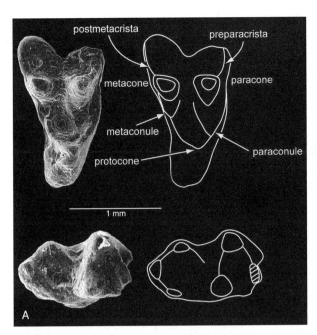
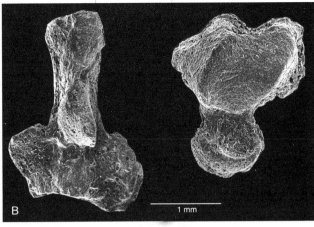

Fig. 10.2. Cretaceous *Deccanolestes* from India, a possible basal archontan: (A) upper right and lower left molars; (B) left calcaneus and right astragalus. (A from Rana and Wilson, 2003; B from Prasad and Godinot, 1994.)

would seem to conflict with other morphological data suggesting that bats are archontans, but if Hooker (2001) is right that nyctitheres are in fact closer to archontans than to lipotyphlans, the possibility of a nyctithere-bat-archontan relationship should be reexamined. The problems of the origin and higher relationships of Chiroptera promise to be challenging issues for some time to come.

Early Eocene Bats

The oldest known bats date from the early Eocene. Nine genera, perhaps representing as many as six families, are reported from this interval, from deposits in North America (*Icaronycteris, Honrovits*), Europe (*Icaronycteris?, Archaeonycteris, Palaeochiropteryx, Eppsinycteris, Ageina,* and *Hassianycteris*), Asia (*Icaronycteris?* and *Archaeonycteris?*; Rana et al., 2005), Africa (*Dizzya*: Philisidae), and Australia (*Australonycteris*; Simmons and Geisler, 1998; Figs. 10.3, 10.4). All are generally regarded as microchiropterans (see below). The diversity and nearly cosmopolitan distribution of early bats indicate that the order must have originated much earlier.

Most early Eocene bat fossils consist of teeth or jaws only and belong to extinct families. For some of these genera there are middle Eocene skeletons that allow confirmation that the older teeth belong to bats. For others, however, in the absence of evidence concerning the key anatomical innovation (wings), we are faced with the difficulty of recognizing the order from dental characteristics alone—for example, dilambdodont upper molars with a deep buccal indentation and no hypocone, and sectorial lower molars with a lingual hypoconulid. Such features suggest that *Honrovits* (Fig. 10.4C) may be related to the extant family Natalidae (Beard et al., 1992), and that *Eppsinycteris* (Fig. 10.4B), from the basal Eocene, could be an emballonurid (Hooker, 1996), making these potentially the oldest records of extant bat families. Because primitive bat teeth are similar to those of certain early Tertiary lipotyphlan insectivores and didelphoid marsupials, however, it can be difficult to verify the chiropteran affinities of isolated teeth or fragmentary dentitions. Indeed, *Eppsinycteris* was initially described as an erinaceomorph lipotyphlan, and its identity as a bat has been challenged by some bat experts (Storch et al., 2002).

Wyonycteris (Fig. 10.3C), founded on dentitions from the late Paleocene–early Eocene (Clarkforkian-Wasatchian) of Wyoming, could be the oldest known bat (Gingerich, 1987). Like many bats, it is very small and has dilambdodont upper molars and a lingual hypoconulid on the lowers. Chiropteran affinities are problematic, however, because it lacks a buccal cingulum on the lower molars and certain other chiropteran synapomorphies (Hand et al., 1994). Hooker (1996) has suggested that it may be a lipotyphlan insectivore. In fact, *Wyonycteris* is also very similar in upper molar form to *Pontifactor*, a nyctitheriid lipotyphlan. It seems more likely that *Wyonycteris* is a lipotyphlan dentally convergent on chiropterans.

Early Eocene *Icaronycteris* (Fig. 10.5, Plate 5.1) stands apart from other early Eocene bats, however, in being known from several exquisitely preserved skeletons from the Green River Formation of Wyoming (Jepsen, 1970; Novacek, 1987; Simmons and Geisler, 1998). They demonstrate that this early bat had fully developed wings and differed only in relatively small details from recent bats. *Icaronycteris* was a small bat, about 30 cm in wingspan and weighing about 15 g. It is more primitive than all other bats in having a relatively shorter radius and a complete phalangeal formula (2-3-3-3-3) in the wing (Jepsen, 1966; Novacek, 1987; Simmons and Geisler, 1998). In addition, it has a longer tail than most, if not all, other bats and a claw on its second manual digit as well as on the pollex (as in *Archaeonycteris* and recent megabats). The dental formula (2.1.3.3/3.1.3.3) is the most primitive

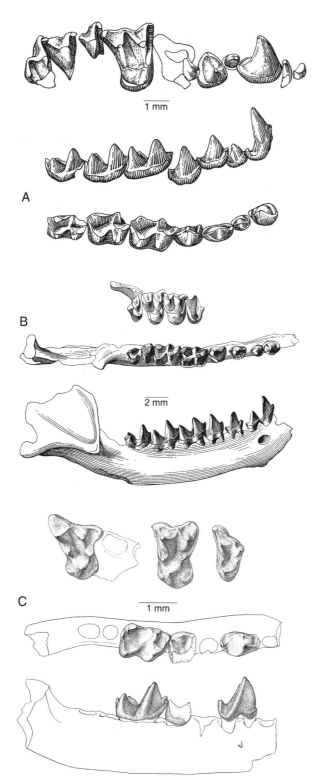

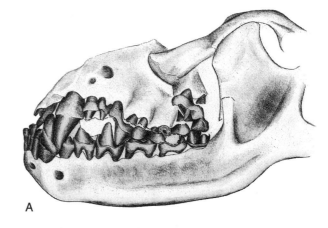

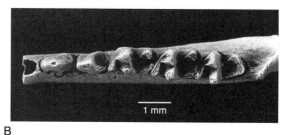

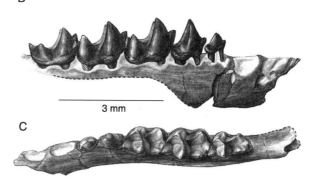

Fig. 10.4. Eocene bats: (A) *Archaeonycteris*, skull; (B) *Eppsinycteris*, right P_3–M_3; (C) *Honrovits*, right P_3–M_3. (A from Smith and Storch, 1981; B from Hooker, 1996; C from Beard et al., 1992.)

Fig. 10.3. Primitive bat dentitions: (A) middle Eocene *Palaeochiropteryx*, right upper and lower teeth; (B) middle Eocene *Lapichiropteryx*, right P^4–M^3 and C_1–M_3; (C) latest Paleocene–early Eocene *Wyonycteris*, a possible bat or nyctitheriid lipotyphlan, right and left upper molars and right lower dentition. (A from Russell and Sigé, 1970; B from Tong, 1997; C from Gingerich, 1987.)

known for bats and is shared by *Archaeonycteris*, *Palaeochiropteryx*, and *Hassianycteris*. Teeth possibly representing *Icaronycteris* have been reported from the late Paleocene (Clarkforkian) of Wyoming (Gingerich, 1987). If confirmed, they would extend the temporal range of definitive Chiroptera into the Paleocene.

Middle and Late Eocene Bats

An extraordinary assemblage of middle Eocene (Lutetian) bats has been collected from the famous Messel site in Germany (e.g., Sigé and Russell, 1980; Habersetzer and Storch, 1987, 1989; Habersetzer et al., 1992, 1994). Hundreds of superbly preserved bat skeletons, some even indicating the shape of the wing membrane, have been found, a situation unparalleled in the fossil record. Most are assigned to seven species of the genera *Archaeonycteris* (Plate 5.2), *Palaeochiropteryx*, and *Hassianycteris* (Plate 5.3), each of which represents a separate family. *Palaeochiropteryx* was a small

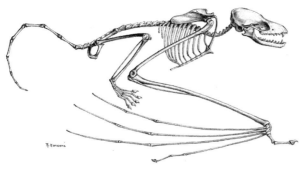

Fig. 10.5. Restoration of *Icaronycteris*. (From Jepsen, 1970.)

bat about the same size as *Icaronycteris*. *Archaeonycteris* was slightly larger, and *Hassianycteris* was the largest known Eocene bat, reaching a wingspan of a half-meter and body mass of 65 g. Like *Icaronycteris*, *Archaeonycteris* (Fig. 10.4A) had primitive tribosphenic molars lacking the high postcristid of other bats, a claw on the second digit of the hand, and a well-developed fibula. Both genera apparently lacked a calcar. In this combination of characters they are more primitive than all other bats. The cochleae of *Icaronycteris* and *Archaeonycteris* were only moderately enlarged, at the small end of the size range for microbats but at the upper end of the range for megabats. The cochleae of *Palaeochiropteryx* and *Hassianycteris* were of comparable size to those of living microchiropterans with relatively small cochleae (Habersetzer and Storch, 1993). Thus it may be inferred that sophisticated ultrasonic echolocation had already evolved in these Eocene bats, although it was probably less highly developed than in living microbats.

Because they are so well preserved, the Messel fossils provide a rare glimpse into the paleoecology of bats. By reconstructing wing shapes, Habersetzer and Storch (1989; Habersetzer et al., 1992, 1994) were able to compare flight dynamics among the Messel bats (Fig. 10.6). They found that *Hassianycteris* had a wide wingspan and relatively narrow wings craniocaudally. This configuration translates into high wing loading (body mass / wing area) and a high aspect ratio (wing span^2 / wing area), factors that suggest rapid flight above the forest canopy. *Palaeochiropteryx*, by contrast, had broad wings with low wing loading and a low aspect ratio, characteristic of slow, maneuverable, low flyers. *Archaeonycteris* (and *Icaronycteris*) had less specialized broad wings with a smaller wingspan (high wing loading and low aspect ratio) and probably occupied mid-level open spaces. The Messel bats thus provide early evidence of niche-partitioning, each foraging at a different level. Together, the wing shape, cochlear anatomy, and stomach contents suggest that *Hassianycteris* foraged by catching beetles, cockroaches, and other insects in flight (aerial hawking); *Archaeonycteris* and *Icaronycteris* were probably perch hunters that took beetles and other insects from surfaces and plants closer to the ground; and *Palaeochiropteryx* used an intermediate foraging strategy to catch moths and caddisflies close to the ground (Norberg, 1989; Simmons and Geisler, 1998). Fossilized scales of nocturnal moths—presumably slow, low-flying insects—

have been found in the gut regions of many *Palaeochiropteryx* specimens. *Palaeochiropteryx* is abundant at Messel, accounting for three-fourths of all bat specimens. As a slow, low-flying bat, it may have succumbed more readily to toxic gases that are believed to have accumulated above the ancient Messel lake. *Cecilionycteris* is a closely allied palaeochiropterygid known from slightly younger strata at Geiseltal, Germany.

Middle Eocene bats have been reported from Asia and Africa as well. *Lapichiropteryx* (Fig. 10.3B), a close relative of *Palaeochiropteryx*, is based on jaws and teeth from China (Tong, 1997); and isolated teeth of a bat similar to *Icaronycteris* or *Palaeochiropteryx* have been found in Pakistan (Russell and Gingerich, 1981). *Tanzanycteris*, known from a partial skeleton from Tanzania, is the oldest African bat (Gunnell et al., 2003). It has an enlarged cochlea indicative of echolocation typical of microchiropterans. Unfortunately, no teeth were found, leaving its relationships to other microchiropterans uncertain.

Archaeonycteris, *Palaeochiropteryx*, and *Icaronycteris*, together with a few other genera known mainly from teeth (e.g., *Cecilionycteris*), are sometimes grouped as archaic bats and have been assigned by some authors to a separate suborder, Eochiroptera (Van Valen, 1979; Sigé, 1991), regarded as the possible stem group for both Microchiroptera and Megachiroptera. The concept of Eochiroptera has not gained general acceptance, however, because it was based on primitive resemblances. Moreover, subsequent studies indicate that all of these bats share derived features with Microchiroptera, to the exclusion of Megachiroptera. In particular, all have an enlarged cochlea, together with specialized expansions of the malleus and stylohyal, features related to echolocation and diagnostic of microchiropterans (Novacek, 1987; Habersetzer and Storch, 1993; Simmons and Geisler, 1998).

At least 18 more genera of bats, all microchiropterans, are recorded from the middle or late Eocene (Simmons, 2005). They include the first unequivocal representatives of four extant families: Rhinolophidae (Old World leaf-nosed bats, including Hipposideridae), Emballonuridae (sac-winged or ghost bats), Molossidae (mastiff or free-tailed bats), and Megadermatidae (false vampire bats), as well as genera believed to be related to Vespertilionidae and Natalidae. Most are known from fragmentary remains, sometimes including postcrania. Five genera of rhinolophids are recorded from Europe during this interval, including *Hipposideros* and *Rhinolophus*, both of which survive to the present. These are the oldest records for any extant bat genus, and probably the oldest for any still-living mammalian genus (with the possible exception of the primate *Tarsius*; see below). Of course, how reliable these fossils are as records of the existence of extant genera in the Eocene is debatable.

Tachypteron (Plate 5.4), known from well-preserved skeletons from the middle Eocene of Messel, and *Vespertiliavus*, from the middle and late Eocene Quercy Phosphorites in France, are the oldest uncontested emballonurids (Sigé, 1988; Storch et al., 2002). *Tachypteron* had a narrow wing,

Fig. 10.6. Proposed niche-partitioning by flight level in Eocene bats from Messel, Germany, based on wing shape. (From Habersetzer and Storch, 1989.)

suggesting adaptation for continuous, rapid flight. Storch et al. (2002) found that its postcranial skeleton, wing membrane outline, and cochlear size are remarkably close to those of the extant emballonurid *Taphozous*. Upper teeth ascribed to *Wallia*, a middle Eocene (Uintan) taxon from Saskatchewan that was initially described as a lipotyphlan insectivore, were subsequently reinterpreted as the oldest known molossid bat (Legendre, 1985). By the late Eocene, molossids were also known in Europe (*Cuvierimops*). Late Eocene *Necromantis* is considered the oldest known megadermatid. Middle Eocene–Oligocene *Stehlinia* is variously considered the oldest known vespertilionid or a natalid relative (Sigé, 1974).

Definitive megachiropterans are unknown in the fossil record before the Oligocene, when *Archaeopteropus* appeared in Italy. A megachiropteran was recently reported from the late Eocene of Asia (Ducrocq et al., 1993), but it is based on a single premolar, and secure identification will require more complete specimens. The existence of microchiropterans in the early Eocene (or possibly before) indicates that megachiropterans were probably already in existence, although it is also possible that they evolved somewhat later from a microchiropteran. Unfortunately, the only known skeleton of *Archaeopteropus* was destroyed in World War II (Simmons and Geisler, 1998), leaving many questions about the origin of megabats unanswerable until new specimens are discovered.

DERMOPTERA

The colugos or flying lemurs make up the order Dermoptera, whose name (literally, "skin wing") refers to the fur-covered membrane, or patagium, which stretches from the neck to the hands, between the limbs, and from the feet to the tail. The patagium enables colugos to glide over 100 meters with little loss in elevation (Nowak, 1999). Only two species of living dermopterans survive today in the forests of southeast Asia. Both are usually included in the genus *Cynocephalus* of the family Galeopithecidae (McKenna and Bell, 1997; Wilson and Reeder, 2005, use the family name Cynocephalidae), although some experts assign them to separate genera. Roughly the size of large squirrels (1–2 kg), they are nocturnal and strictly arboreal, making behavioral study challenging. Living dermopterans have elongate, very slender limb elements, a keeled sternum, and broad ribs (Fig. 10.7A). The antebrachium (forearm) is longer than the brachium (arm), and the ulna is reduced and fused with the radius. Carpal and tarsal modifications allow the animal to adjust foot positions easily to the uneven substrate of the arboreal realm. The digits (longer in the manus than in the pes) are tipped by short, deep, recurved, and laterally compressed claws, which are characteristic of arboreal mammals. The skull is broad and relatively flat, with the orbits almost fully encircled by bone. The flattened auditory bulla is made up almost entirely of the ectotympanic bone (Hunt and Korth, 1980).

Colugos consume mainly leaves, buds, flowers, and fruit, and the dentition is specialized for an herbivorous diet (Fig. 10.7B). The premolars and molars are relatively low-crowned and sharp-cusped, with well-developed shearing crests, particularly a prominent W-shaped ectoloph (outer crest) on the upper molars. The posterior premolars are molariform, and the enamel of the molariform teeth is crenulated. When worn, this crenulated enamel results in sharp-edged crevices that enhance shearing function. Diagnostic of living dermopterans are the unique, comblike, procumbent lower incisors, each of which has 5–20 tines. They occlude against a horny pad in the wide diastema that separates the upper incisors and are apparently used for grooming as well as feeding (Aimi and Inagaki, 1988).

The fossil record of dermopterans is poor. Only a single fossil, for which the name *Dermotherium* (Fig. 10.7C) was proposed, has been referred to the extant family. It is a poorly preserved lower jaw fragment with two molars, from the late Eocene of Krabi, Thailand (Ducrocq et al., 1992). The teeth closely resemble those of *Cynocephalus*, to the extent that they can be compared. Unfortunately, this fossil indicates little more than the likelihood that dermopterans had diverged from other mammals by the end of the Eocene.

With so few living representatives and such a meager fossil record, it would be reasonable to conclude that dermopterans are too obscure to be of interest to paleobiologists. However, in addition to their possible relationship to bats discussed above, the controversial association of two Early Tertiary groups—Plagiomenidae and Plesiadapiformes—with Dermoptera makes consideration of this group particularly interesting.

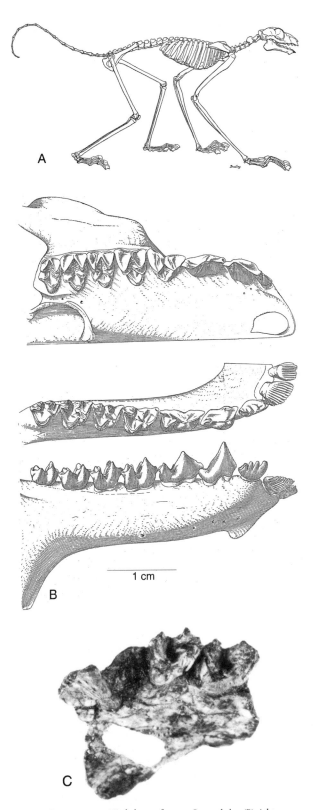

Fig. 10.7. Dermopterans: (A) skeleton of extant *Cynocephalus*; (B) right upper and lower dentition of *Cynocephalus*; (C) right lower jaw with M_{1-2} of late Eocene *Dermotherium*. (A from Gregory, 1951, after Blainville; B courtesy of M.C. McKenna; C courtesy of S. Ducrocq.)

Plagiomenidae and Mixodectidae

The North American Paleocene–Eocene family Plagiomenidae, known primarily from jaws and teeth, has generally been allocated to Dermoptera ever since the initial description of early Eocene *Plagiomene* almost 90 years ago (Matthew, 1918). Plagiomenids evidently preferred high latitudes, being found in the northern Rocky Mountain region (Wyoming north to Alberta) and within the Arctic Circle (Ellesmere and Axel Heiberg islands, Canada). In the latter region they were among the most common and speciose placentals during the Eocene.

The posterior premolars of plagiomenids are molariform, and the molariform teeth bear a particular resemblance to those of extant *Cynocephalus*. Plagiomenid lower molars are broad, with exceptionally wide talonids and comparatively low trigonids. In both *Plagiomene* (Fig. 10.8C) and *Cynocephalus* the upper molars are divided essentially in half by a deep transverse valley that runs from the buccal edge between the paracone and the metacone to the protocone; a similar transverse valley separates the trigonid from the talonid in the lower molariforms (Rose and Simons, 1977). The hypocone is small or absent, the lower molariform teeth have much higher lingual than labial cusps, and the enamel is crenulated. The lower incisors of *Plagiomene* have bifid crowns, interpreted as an incipient stage leading to the comblike incisors of modern colugos (Rose, 1973). Plagiomenids are further characterized by the development of multiple stylar cusps and strong conules on the upper molariform teeth and a mesial shifting of lingual cusps relative to buccal ones, which gives the teeth a distinctive skewed appearance. Plagiomenids primitively had the dental formula, 3?.1.4.3/3.1.4.3, but more derived members of the family lost from one to three anterior teeth; galeopithecids have only two upper incisors and two premolars above and below.

The basicranium of plagiomenids, known only from one poorly preserved skull of *Plagiomene*, is very derived and unlike that of *Cynocephalus* (MacPhee et al., 1989). The tympanic cavity and adjacent parts of the basicranium in *Plagiomene* are highly pneumatized, and the auditory bulla is very complex, consisting of as many as seven elements. Evolution of such a complex bulla would be very improbable in a close relative of dermopterans and suggests that plagiomenids had diverged significantly from any common stem with galeopithecids. Nevertheless, the dentition remains strongly suggestive of dermopteran ties, and no more likely basal stock for galeopithecids has been identified. Plagiomenidae are retained in Dermoptera by McKenna and Bell (1997). A recent phylogenetic analysis based on morphological evidence once again allies plagiomenids with extant Dermoptera (Silcox, 2001).

The oldest plagiomenid, *Elpidophorus* (Fig. 10.8B), appeared at the end of the early Paleocene (Torrejonian). *Elpidophorus* was for many years assigned to the family Mixodectidae (see below), but derived dental resemblances indicate closer relationship with plagiomenids (Rose, 1975b). Its morphology supports a relationship between the two families. Latest Paleocene–early Eocene *Plagiomene* from Wyoming probably evolved from a form like *Elpidophorus*. Both genera had moderately high, sharp molar cusps. Closely allied is *Ellesmene*, one of at least six species of plagiomenids known from Ellesmere Island (Dawson et al., 1993). The presence of diverse plagiomenids in the Canadian Arctic is remarkable, because this area was well north of the Arctic Circle even in the Eocene.

A second lineage of smaller, more bunodont plagiomenids, assigned to the subfamily Worlandiinae, coexisted with *Plagiomene*. *Worlandia* (Fig. 10.8D) had only two lower incisors, the first one relatively larger than in *Plagiomene*, and a somewhat enlarged fourth premolar (Rose, 1982a). Another lineage of plagiomenids, represented by *Tarka* and *Tarkadectes* (McKenna, 1990), is known from the middle and late Eocene (Uintan-Chadronian). In contrast to other plagiomenids, they have a premolariform P_3 and a short P_4 with a molariform trigonid and almost no talonid, features suggesting that they had been evolving for a considerable period independently from other members of the family. The precise relationship of tarkadectines to other plagiomenids is not well understood.

The oldest plagiomenids are dentally similar to Torrejonian Mixodectidae, a small and poorly known family, also primarily from the Rocky Mountain region. Mixodectidae is generally thought to be the sister group or ancestor of Plagiomenidae. Mixodectids, such as *Eudaemonema* (Fig. 10.8A), have low-crowned lower molars with very broad talonid basins and tall metaconid and entoconid cusps; their upper molars have large, lingual hypocones and a W-shaped ectoloph with a prominent mesostyle. The last premolars are premolariform to submolariform, and there is an enlarged, procumbent lower incisor (Szalay, 1969a; Gunnell, 1989). Several of these features foreshadow the more derived conditions of plagiomenids. Mixodectids have sometimes been considered "proteutherians" (see Chapter 7) or relatives of Early Tertiary Microsyopidae (see Plesiadapiformes, this chapter), but relationship to Plagiomenidae seems most secure. The postcranial skeleton of *Mixodectes* shows arboreal specializations similar to those of both plesiadapiforms and extant dermopterans (Szalay and Lucas, 1996), supporting their inclusion within Euarchonta.

Szalay and Schrenk (2001) identified a partial skeleton from the middle Eocene of Messel, Germany, as a possible plagiomenid. It has obvious significance as potentially the first postcranial evidence known for the family. Unfortunately, the specimen, which consists of only the hind limbs and tail, is badly crushed, making many details impossible to decipher. The most important feature is the presence of short, very deep, and laterally compressed terminal phalanges with prominent flexor tubercles. They closely resemble those of arboreal mammals and bear particular resemblance to those of extant colugos, though they are much smaller. The specimen is close in size and morphology to the arboreal apatemyid *Heterohyus*, also known from Messel. It was initially thought to differ from *Heterohyus* in having a separate fibula and in some details of the feet, but

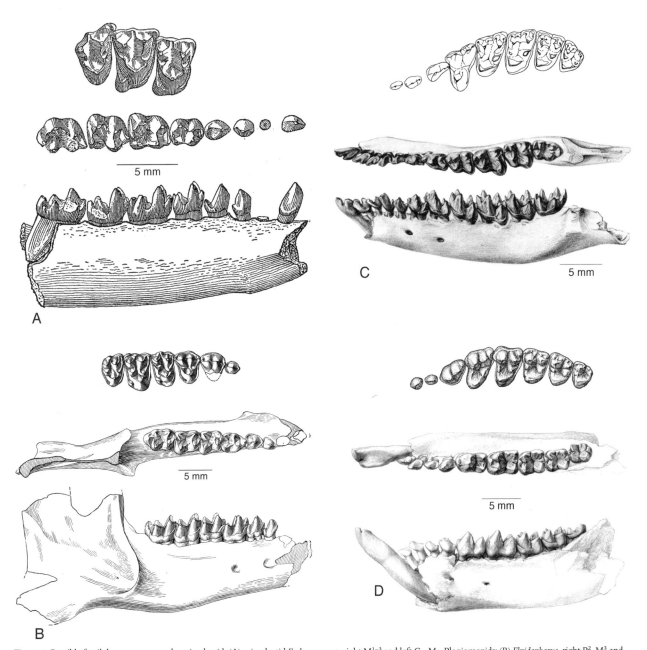

Fig. 10.8. Possible fossil dermopterans and a mixodectid: (A) mixodectid *Eudaemonema*, right M^{1-3} and left C_1–M_3. Plagiomenids: (B) *Elpidophorus*, right P^2–M^3 and P_2–M_3; (C) *Plagiomene*, left upper and lower dentitions; (D) *Worlandia*, left upper and lower dentitions. (A from Simpson, 1937a; B from Simpson, 1936; C from MacPhee et al., 1989, and Rose, 1982a; D from Bown and Rose, 1979, and Rose, 1982a.)

it is now known that *Heterohyus* also had a separate fibula (Kalthoff et al., 2004). Consequently, the Messel skeleton could belong to either *Heterohyus* or an archontan not previously recognized in that fauna, but without associated dentition, its identity remains in doubt. Identification as a plagiomenid, however, is especially problematic, because their postcrania have not been recognized, and plagiomenids are otherwise unknown from Europe.

Dermoptera and Primatomorpha

Although living dermopterans have traditionally been placed in their own order, recent studies suggesting a close tie with Paleogene Paromomyidae and Micromomyidae (mammals usually classified as Plesiadapiformes) have led some experts to substantially expand the concept of Dermoptera to include one or both of these families, or even all plesiadapiforms (Beard, 1993a,b; McKenna and Bell, 1997). This expanded Dermoptera has then either been included within the order Primates or, together with Primates, in the mirorder Primatomorpha. Such an arrangement remains controversial, however, because it overlooks significant contrasts in dental morphology and dental formula between paromomyids and galeopithecids, as well as the closer overall similarity between paromomyids and other plesiadapiforms. Moreover, recent morphological analyses find stronger evidence for Volitantia than for Primatomorpha (e.g., Silcox, 2001; Sargis, 2002). In the present account the Paromomyidae are considered in detail in the discussion of Plesiadapiformes below.

PRIMATES AND PLESIADAPIFORMES

Humans and our closest living relatives—apes, monkeys, tarsiers, lemurs, and lorises—as well as closely allied fossil taxa, constitute the order Primates (Table 10.2). The phylogenetic position of the extinct Early Tertiary Plesiadapiformes ("archaic primates") is less clear: they are usually classified either as primates or as the sister group of primates. When plesiadapiforms are included within the Primates, as they are in this book, the higher taxon Euprimates ("true primates") is often used to encompass all nonplesiadapiform primates.

Several different higher classifications of euprimates are currently in use (Fig. 10.9), which vary primarily in the position of tarsiers and their extinct relatives, the Omomyidae (together comprising the Tarsiiformes). In one scheme euprimates are divided into two evolutionary grades that are also considered clades, the lower primates or Prosimii (lemurs, lorises, and tarsiers), and the higher primates or Anthropoidea (monkeys, apes, and humans). But several lines of evidence, including genetics, placentation, and cranial anatomy, suggest that tarsiers are more closely related to anthropoids than to lemurs and lorises (e.g., R.D. Martin, 1993; Kay et al., 1997; Ross et al., 1998; Fleagle, 1999). Proponents of this view use Strepsirrhini for lemurs and lorises (tooth-combed prosimians), and Haplorhini to include anthropoids and tarsiiforms. A third arrangement recognizes three suborders: Prosimii, Tarsiiformes, and Anthropoidea (Shoshani, Groves, et al., 1996). The phylogenetic position of the Tertiary Adapoidea is another matter of continuing debate, some authors considering them to be archaic strepsirrhines (e.g., Szalay et al., 1987; Kay et al., 1997) and others arguing that they are more closely related to anthropoids than are tarsiers (Franzen, 1994; Simons and Rasmussen, 1996). The closest relative of Anthropoidea thus remains highly controversial.

Nonhuman primates occur today in South America, Africa, and Asia, primarily in tropical regions and usually in forests. Most are arboreal, but even terrestrial forms retain morphological evidence of their arboreal heritage. Some are specialized for leaping and vertical grasping, or for suspensory locomotion, whereas others are more generalized arboreal quadrupeds. Most living primates are herbivorous, various species preferring fruit, seeds, leaves, or grass. Many smaller species are partly or wholly insectivorous, and some of the smallest also feed on tree gum and sap.

Although the euprimate skeleton retains many primitive eutherian features (e.g., five digits on hands and feet, separate radius and ulna, unfused carpal bones), it also shows distinctive specializations for arboreal life. Cartmill (1972, 1992) postulated that some typical euprimate traits arose in association with visually oriented predation in an arboreal

Table 10.2. Classification of Primates

Order PRIMATES
  Suborder †PLESIADAPIFORMES
    †**Purgatoriidae**
    †**Microsyopidae**[1]
    †**Micromomyidae**
    †**Picromomyidae**
    †**Toliapinidae**[2]
    Superfamily †Paromomyoidea
      †**Palaechthonidae**
      †**Paromomyidae**
      †**Picrodontidae**
    Superfamily †Plesiadapoidea
      †*Chronolestes*
      †**Plesiadapidae**
      †**Carpolestidae**
      †**Saxonellidae**
  Suborder EUPRIMATES
    †*Altiatlasius,* †*Altanius*
    Infraorder STREPSIRRHINI
      †**Plesiopithecidae**[3]
    Parvorder LEMURIFORMES
      Superfamily †Adapoidea[4]
        †**Notharctidae**[5]
        †**Adapidae**
        †**Sivaladapidae**[5]
      Superfamily Lemuroidea
        Lemuridae
        Lepilemuridae
        Daubentoniidae
        Indriidae
        Cheirogaleidae
      Superfamily Loroidea (=Lorisoidea)
        Loridae (=Lorisidae)
        Galagidae
    Infraorder HAPLORHINI
    Parvorder TARSIIFORMES
      †**Omomyidae**
      Tarsiidae
    Parvorder ANTHROPOIDEA
      †*Afrotarsius*[6]
      †**Eosimiidae**
      †**Parapithecidae**
      †**Amphipithecidae**[7]
      †**Proteopithecidae**
    PLATYRRHINI[8]
      Superfamily Ceboidea
    CATARRHINI[8]
      Superfamily †Propliopithecoidea
        †**Oligopithecidae**
        †**Propliopithecidae**
        †**Pliopithecidae**
      Superfamily Cercopithecoidea
      Superfamily Hominoidea

Notes: Modified after multiple sources, mainly Fleagle, 1999, and Silcox et al., 2005. The dagger (†) denotes extinct taxa. Families and genera in boldface are known from the Paleocene or Eocene.

[1] Phylogenetic position uncertain; could be closely related to palaechthonids.
[2] Phylogenetic position uncertain; have been considered close to either microsyopids or euprimates.
[3] May be related to lorisoids.
[4] Phylogenetic position uncertain: may be outside Lemuriformes or Strepsirrhini; sometimes separated in the Adapiformes.
[5] Sometimes considered a subfamily of Adapidae.
[6] Phylogenetic position uncertain; could be a tarsiid.
[7] May be a subfamily of Notharctidae.
[8] Unranked taxon between parvorder and superfamily necessitated by other ranks used here.

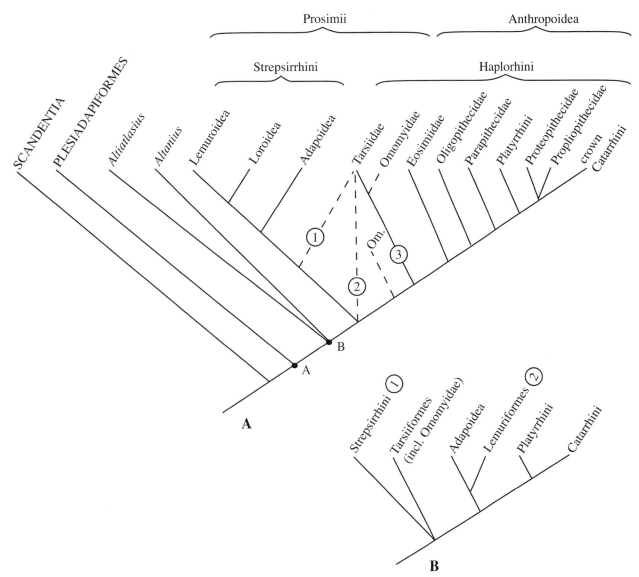

Fig. 10.9. Primate relationships. (A) Based mainly on Kay et al. (1997). Order Primates is variously recognized at node A or node B. Three possible positions of Tarsiidae are indicated, option 3 considered the most probable. Omomyidae (shown in two positions) is variously considered the sister taxon of Tarsiidae or of all Haplorhini. Positions of Oligopithecidae and Proteopithecidae differ from the classification in Table 10.2. (B) Two alternative views, placing Adapoidea or Adapoidea + Lemuriformes as the sister taxon of Anthropoidea and showing differing positions of extant strepsirrhines (=lemuriforms). Option 1 is based on Simons and Rasmussen (1996), and 2 is based on Bloch et al. (1997). According to option 1 Strepsirrhini is restricted to lemurs and lorises and does not include adapoids.

setting. Especially characteristic of euprimates are relatively large orbits that tend to face more anteriorly than laterally, promoting stereoscopic vision (see Fig. 10.10). The orbit always consists of a closed ring of bone, including a complete postorbital bar formed by the junction of processes of the frontal and zygomatic (jugal) bones. The postorbital bar provides a protective bony lateral margin for the eye socket and forms a convenient boundary between it and the temporal fossa. In more advanced primates the back of the orbit becomes closed by a bony postorbital septum (a unique feature of haplorhines known as postorbital closure), which physically separates the orbit from the temporal fossa. The euprimate braincase is relatively large, sometimes very large, compared to that in most other mammals, reflecting especially the increasing size of the cerebrum. Enlargement of the braincase often occurs at the expense of the facial region. The snout is usually relatively short and broad, but is longer in primitive euprimates such as lemurs. The auditory bulla, which forms the tympanic floor, is an outgrowth of the petrosal bone, and this petrosal bulla is considered diagnostic of euprimates.

The dental formula of the most primitive fossil euprimates is 2.1.4.3/2.1.4.3, but many species reduced this number, and no Recent primates have more than three premolars per quadrant. The molars are tribosphenic and usually relatively low-crowned. Depending on dietary preference, they may be secodont (insectivorous), bunodont (frugivorous), or selenodont (folivorous). Regardless of these distinctions, in the lower molars the paraconid is usually reduced or absent and the talonid basins are typically large and

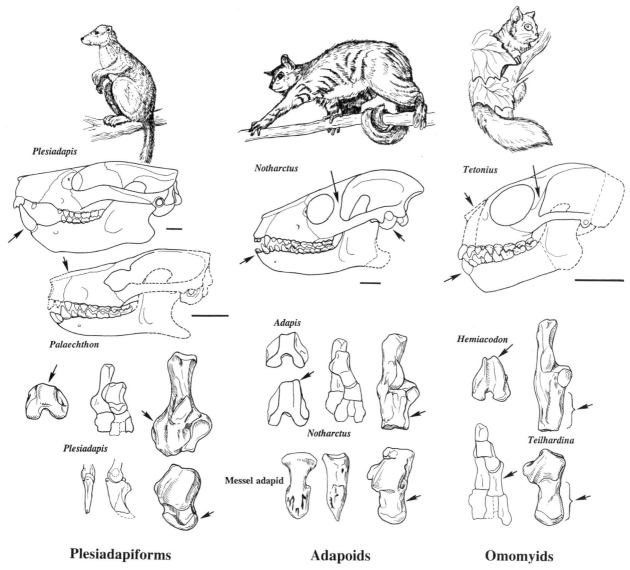

Fig. 10.10. Comparison of some anatomical features of plesiadapiforms and euprimates. Key features are indicated by arrows. Euprimates (adapoids and omomyids) differ from plesiadapiforms in having a larger braincase and a postorbital bar. Most plesiadapiforms have relatively long snouts and enlarged central incisors; adapiforms tend to have long snouts but small incisors, whereas omomyids have shorter snouts and either small or enlarged incisors. The postcranial skeleton of most euprimates is hind limb dominated and modified for arboreal running and leaping, with a high patellar trochlea on the distal femur and elongate tarsals (especially the calcaneus and astragalus). The digits of euprimates are terminated by nails, rather than claws as in plesiadapiforms. (From Rose, 1995.)

broad. The upper molars usually have a reduced stylar shelf and a hypocone that is sometimes large, making the molars quadrate. The incisors vary from small to moderate in size and pointed or spatulate (some tarsiiforms, adapoids, and anthropoids), to large and procumbent (some omomyids and the aye-aye), to long, slender and procumbent, forming a dental comb (modern strepsirrhines). The canines are almost always prominent, and in anthropoids and some fossil prosimians they are sexually dimorphic. The premolars of most euprimates are relatively simple, although P^4_4 became hypertrophied in some fossil forms and submolariform in others.

The limb skeleton of euprimates emphasizes mobility: the shoulder, elbow, hip, and ankle joints, in particular, have evolved to promote flexibility, an advantage in the arboreal environment. The head of the radius typically is round, allowing 180° of supination. The hallux and the pollex are usually opposable. Nails are always present on the hallux and almost always on the other digits (except in marmosets, which have secondarily evolved clawlike unguals on all other digits). The presence of nails can be inferred in fossils by the unique shape of the ungual phalanges. Broad ungual phalanges with nails and apical pads are thought to have arisen as an adaptation for increased stability when foraging on small branches (Cartmill, 1972; Hamrick, 1998).

Despite a rich fossil record, the origin of Primates is surprisingly nebulous. The oldest known plesiadapiforms come from the early Paleocene and the oldest euprimates from the earliest Eocene (or possibly the latest Paleocene), so it is generally assumed that the order originated in the Late Cretaceous or very early Paleocene. Recent estimates of the time of divergence of euprimates using models based on the fossil record vary from 55–63 million years ago (Gingerich and Uhen, 1994) to 72–90 million years ago (Tavaré et al., 2002), whereas molecular analyses generally suggest an origin at least 80–90 million years ago (e.g., Arnason et al., 2000; Douady and Douzery, 2003; Springer et al., 2003). Both the source group and place of origin of primates remain unknown. North America was unquestionably a major center of diversification (if not origin) of plesiadapiforms, though perhaps not the only one. Eastern Asia, India, and Africa have each been suggested as possible sites of origin of Euprimates.

Plesiadapiformes

Plesiadapiforms were a diverse group of small arboreal mammals that flourished during the first 20 million years of the Cenozoic (Paleocene–middle Eocene). They ranged in size from tiny *Picromomys* (10 g), which was smaller than many shrews, to the marmot-sized species of *Plesiadapis* (3–5 kg). Because they lack several diagnostic characters of euprimates, including a petrosal bulla, a well-developed internal carotid arterial system, a postorbital bar, an opposable hallux, and nail-bearing terminal phalanges (see Fig. 10.10), some authorities exclude them altogether from Primates (Cartmill, 1974; R.D. Martin, 1990, 1993). Nevertheless, plesiadapiforms closely resemble euprimates in molar structure and in certain arboreal postcranial adaptations, including a divergent hallux (Szalay et al., 1987; Van Valen, 1994b). On this basis they are widely regarded as "archaic primates" and have conventionally been included in Primates as a suborder or infraorder (e.g., Simons, 1972; Szalay and Delson, 1979; Conroy, 1990; Shoshani, Groves, et al., 1996). Accordingly, they are the implied sister taxon of Euprimates, a relationship upheld by a recent phylogenetic analysis (Silcox, 2001). At the same time it is generally acknowledged that almost all known plesiadapiforms are dentally too derived to have been directly ancestral to euprimates. Furthermore the molar resemblance is closest between certain derived plesiadapiforms and early euprimates that cannot be directly descended from them, a clear indication that these detailed similarities are homoplastic. Some authors (e.g., Rose, 1995; Fleagle, 1999) have considered Plesiadapiformes to be a separate order closely related to Primates. Beard (1993a), however, transferred plesiadapiforms from Primates to Dermoptera, citing postcranial resemblances, and grouped all of these in a new mirorder Primatomorpha.

In many respects plesiadapiforms are more primitive than euprimates, suggesting to some experts that the group may be a paraphyletic assemblage of the most primitive primates. Consequently, McKenna and Bell (1997) and Silcox et al. (2005) have abandoned the term but continue to include the various plesiadapiform families within Primates. The presence of complex enlarged upper incisors could be a uniquely derived trait of plesiadapiforms (Rose et al., 1993), but incisor morphology is still unknown for some families and may be fundamentally different in one supposed plesiadapiform family (Microsyopidae). Nevertheless, it has not been possible to link any known plesiadapiform directly to euprimates and, as noted above, nearly all are too specialized in some respect to be closely related. Consequently it may be premature to reject the possibility that plesiadapiforms are a monophyletic group.

Plesiadapiforms are known from all three northern continents but were especially common and diverse in western North America. There are more than 40 genera representing 11 families: Purgatoriidae, Plesiadapidae, Palaechthonidae, Paromomyidae, Carpolestidae, Picrodontidae, Saxonellidae, Micromomyidae, Picromomyidae, Microsyopidae, and Toliapinidae (Fig. 10.11). An additional family of possible plesiadapiforms, Azibiidae, was recently named from northern Africa. Most of these families are known primarily or solely from dentitions. Thus it is not surprising that their interrelationships and the question of whether some of the families even belong in Plesiadapiformes remain unsettled. Certain of the families can, however, be united with reasonable confidence. The first six families listed above were already established in North America during the early Paleocene, Purgatoriidae in the Puercan, and the other five by the Torrejonian. The fossil record of the remaining five families begins in the late Paleocene or early Eocene, although they must have existed before then. Plesiadapiforms occupied an ecological niche approximating that of Eocene rodents and euprimates. Their decline during the Eocene appears to correspond with the initial radiation of rodents, and may be attributable, at least in part, to competition with them (Van Valen and Sloan, 1966; Maas et al., 1988). There are even stronger resemblances to certain extant phalangeroid marsupials (Cartmill, 1974), suggesting analogous dietary preferences and locomotor behavior.

Most plesiadapiforms had a long snout with a small braincase and laterally directed orbits that are confluent posteriorly with the temporal fossa (i.e., there is no postorbital bar). Composition of the auditory bulla is difficult to establish with certainty in the few available skulls, but in the paromomyid *Ignacius* it seems to consist of the entotympanic bone, not the petrosal as in euprimates (Kay et al., 1990, 1992; Bloch and Silcox, 2001). Kay et al. (1992) believed that the endocranial arterial pattern also differs from that of euprimates, and resembles that of extant Dermoptera, in lacking any significant blood supply from the internal carotid artery. However, subsequent discoveries indicate that a small promontorial branch of the internal carotid was present, as in euprimates (Bloch and Silcox, 2001). Some other details of the ear region are also euprimate-like.

The generalized tribosphenic molars of most plesiadapiforms resemble those of early euprimates in having bunodont cusps, relatively low trigonids, broad talonid basins, and narrow stylar shelves (Fig. 10.12). The upper molars in many types have a crest joining the protocone to the hypocone or hypocone shelf, called a postprotocingulum or "nannopithex fold." The lower third molars in both groups are elongated. This derived dental complex suggests a shift

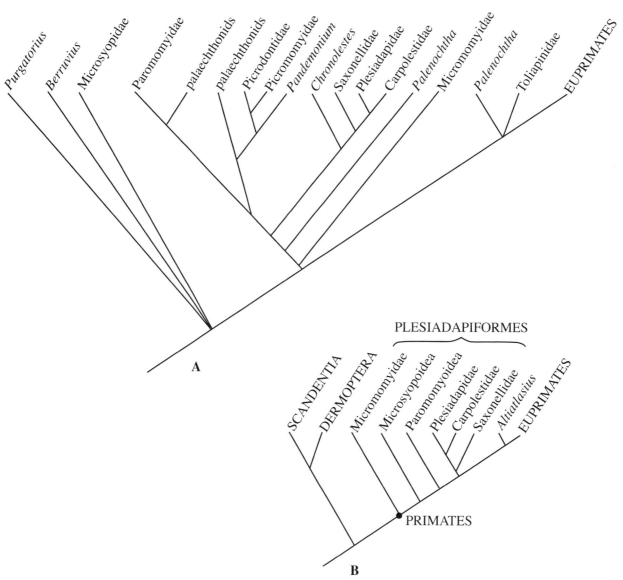

Fig. 10.11. Plesiadapiform relationships: (A) simplified after Silcox (2001); (B) modified after Bloch and Boyer (2003). Palaechthonids and the palaechthonid *Palenochtha* are paraphyletic basal taxa that appear in several places in cladogram A.

from insectivory to a more omnivorous or herbivorous diet. In many plesiadapiform lineages the molars remained conservative while the antemolar dentitions became specialized. Particularly characteristic of plesiadapiforms are the enlarged central incisors (I^1 and I_1). The procumbent lower incisor occludes with a more vertical upper incisor with multiple pronglike cusps and a basal cusp that served as a stop for the lower incisor (see Gingerich, 1976, and Fig. 10.14C).

Parallel dental specializations were common among plesiadapiforms. Reduction in number or size of the anterior teeth behind the enlarged incisor was a common trend in most lineages, sometimes resulting in a diastema between the incisor and the posterior premolars (Plesiadapidae and Paromomyidae). An enlarged, pointed last lower premolar (P_4) evolved independently in paromomyids and micromomyids, and a hypertrophied, multicusped, bladelike premolar arose separately in carpolestids (where it was P_4) and saxonellids (P_3). The diminutive body size and sharp molar cusps and crests of some plesiadapiforms suggest that they were primarily insectivorous, whereas larger species (plesiadapids and some microsyopids) have dental modifications indicative of omnivory or herbivory, including fruit and seed eating. In other lineages, flatter molars coupled with long, pointed, procumbent incisors suggest a dietary preference for tree exudates.

The postcranial skeleton of plesiadapiforms (Fig. 10.13) is known in just a few genera representing four families (Plesiadapidae, Carpolestidae, Paromomyidae, and Micromomyidae). It is most completely known in *Plesiadapis*, which had a moderately robust skeleton bearing the anatomical hallmarks of arboreal adaptation (strong crests and processes on limb elements; mobile elbow and ankle joints; and sharp, laterally compressed claws). In the absence of other specializations, *Plesiadapis* can be interpreted as a generalized climber that lacked the acrobatic abilities of many living euprimates. Carpolestids were also adapted for arboreality but had a more gracile skeleton specialized for terminal-branch foraging (Bloch and Boyer, 2002). Paromomyids and micromomyids also had more delicate limb bones than did *Plesiadapis*. Beard (1990, 1993a,b) listed certain osteological specializations similar to those in living dermopterans as evidence that these fossil taxa had a patagium, or gliding membrane. These traits and cranial features noted earlier were also considered synapomorphies indicating relationship to extant Dermoptera (Kay et al., 1992; Beard, 1993a). Further consideration of these issues is presented in the discussion of Paromomyidae below.

Purgatoriidae

Early Paleocene *Purgatorius* (Fig. 10.12) from western North America is the oldest and most primitive plesiadapiform and the sole member of the family Purgatoriidae. A supposed Upper Cretaceous record, founded on a single tooth, is the only basis for the claim that the primate fossil record extends back into the Cretaceous. However, this tooth is of questionable identity and comes from a partly reworked assemblage that was deposited in the early Paleocene, casting doubt on its Cretaceous age (Lofgren, 1995). *Purgatorius* is known only from its dentition, which differs from that of other plesiadapiforms in having sharper cusps and retaining the most primitive lower dental formula (3.1.4.3) known for Plesiadapiformes (the upper dental formula is unknown). Almost all later plesiadapiforms lost at least one incisor (I_3) and one premolar (P_1). Judged from alveoli, the first two lower incisors of *Purgatorius* were relatively large and procumbent, while the third was reduced; the canine remained primitively relatively large (Clemens, 2004). P_2–P_4 are two-rooted, and the lower and upper fourth premolars are submolariform, as in primitive palaechthonids. The upper molars have well-developed conules and strong buccal, anterior, and posterior cingula. *Purgatorius* or a closely related form probably lies near the ancestry of all other plesiadapiforms and is one of only two plesiadapiforms that are dentally primitive enough to be ancestral to Euprimates as well, because basal euprimates retained four premolars (Van Valen, 1994b; Rose, 1995; Fleagle, 1999).

Palaechthonidae

Palaechthonidae are primitive plesiadapiforms known mainly from the late early Paleocene (Torrejonian) of western North America, though a few genera survived into the early Tiffanian (Gunnell, 1989). They resemble *Purgatorius* in many details of the dentition, including the usual presence of semimolariform fourth premolars, but they are more advanced in having only two lower incisors and, usually, three lower premolars, including a single-rooted P_2. Semimolariform fourth premolars occur in the Torrejonian genera *Palaechthon* (Fig. 10.12) and *Plesiolestes* and were presumably primitive for the family. Variation among known genera suggests a trend toward simpler premolars. Other family characteristics include relatively bunodont molars, mesiodistal compression of the molar trigonids and reduction of the paraconid, addition of a mesoconid cusp on the cristid obliqua (see Fig. 2.2), and enlargement of the hypoconulid lobe of M_3. Some of these traits also occur in other plesiadapiform families. The Torrejonian palaechthonid *Anasazia* is the only plesiadapiform as primitive as *Purgatorius* in retaining four premolars (Van Valen, 1994b).

Palaechthonid genera were included in the Paromomyidae in much of the older literature; more recently they have sometimes been allocated to the Microsyopidae or to an expanded Purgatoriidae. In most respects palaechthonids are more primitive than other plesiadapiforms except *Purgatorius*, and they probably represent a paraphyletic assemblage. In a comprehensive phylogenetic analysis of plesiadapiforms, Silcox (2001) found that palaechthonids emerged as primitive sister taxa of many other clades of plesiadapiforms.

Plesiadapidae

Plesiadapidae were among the most common members of North American and European Paleocene mammal faunas. They are the best-known plesiadapiforms (e.g., Gingerich, 1976), being represented by nearly all of the skeleton. Hundreds of jaws and many skulls have been found. Their abundance and rapid rate of evolution make them especially useful in biostratigraphy. Plesiadapids were rather generalized plesiadapiforms, superficially resembling squirrels: the dentition was unspecialized except for enlarged, chisel-like incisors and a diastema in progressive species; and the skeleton was arboreally adapted, with generalized limb proportions and a long tail. Besides the arboreal features mentioned earlier, it may be noted that the elbow joint allowed substantial supination, the tarsus was flexible (but not elongated, as in Eocene euprimates), and the hallux was divergent. Plesiadapids had simple lower premolars and distinctive submolariform upper posterior premolars with a central conule between the buccal cusps and the protocone. The skull of *Plesiadapis* (Figs. 10.10, 10.13A) had widely flaring zygomae and a moderately long, narrow snout consisting largely of expanded premaxillae. The auditory bulla has been reported to be composed of the petrosal, but no individuals are known that are young enough to verify which bones are actually involved. The internal carotid artery was evidently greatly reduced, to judge from the variable arrangement of minute grooves on the promontorium. *Plesiadapis* had a much smaller brain than in any extant primate (EQ [encephalization quotient] = 0.25; Gingerich and Gunnell, 2005).

The oldest and most primitive well-known plesiadapid is Torrejonian *Pronothodectes* (Fig. 10.14B), with a dental formula of 2.1.3.3/2.1.3.3. Subsequent plesiadapids lost the second lower incisor and sometimes the canines and P_2 as well, resulting in a conspicuous diastema at the front of the jaw like that of rodents. Late Paleocene *Nannodectes* (Fig. 10.13C) and *Plesiadapis* were likely descendants of

Pronothodectes. Plesiadapis was common in both North America and Europe, and was probably ancestral to the deep-jawed seed-eater *Chiromyoides* (also present on both continents) and the marmot-sized herbivore *Platychoerops* (from the early Eocene of Europe). A possible plesiadapid, *Asioplesiadapis*, was recently reported from the early Eocene at Wutu, China (Fu et al., 2002). Although it resembles *Plesiadapis* in its incisor morphology and in having a lower dental formula of 1.0.3.3, its very small size and relatively primitive molars suggest that it represents a lineage distinct from those in Europe and North America.

Van Valen (1994b) interpreted the Puercan genus *Pandemonium*, based on several isolated teeth from Montana, as a link between *Purgatorius* and plesiadapids. Phylogenetic analysis of dental characters indicates that *Pandemonium* occupies a position closer to the base of all Plesiadapoidea, a superfamily encompassing Carpolestidae and Saxonellidae as well as Plesiadapidae (Silcox et al., 2001). All three families have a central conule on P^4 and share details of incisor and molar morphology.

Carpolestidae

Carpolestids were small, mouse-sized animals that lived during the Paleocene (Torrejonian-Clarkforkian) in western North America and the late Paleocene to early Eocene in China. They are more primitive than all other plesiadapiforms except *Purgatorius* in retaining three upper incisors and, in one species, three lowers as well (Fox, 1993; Bloch and Gingerich, 1998). Despite this plesiomorphous condition, they are derived in many other respects: no carpolestid is known to have had more than three premolars, and most have only one or two lower premolars. In most aspects of dental morphology carpolestids resemble plesiadapids more than any other plesiadapiforms, indicating that they are sister groups (Silcox et al., 2001). This conclusion contradicts the recent assignment of Carpolestidae to the Tarsiiformes by McKenna and Bell (1997).

Carpolestids are among the most distinctive Paleogene mammals because of their hypertrophied, bladelike P_4 with multiple aligned cusps (Fig. 10.15B). This tooth superficially resembles the bladelike premolars of *Saxonella* (Fig. 10.15A), multituberculates, and various marsupials. All these animals have an enlarged, procumbent incisor followed by several greatly reduced teeth preceding the bladelike tooth, a complex Simpson (1933) called "the plagiaulacoid dentition," in reference to the Mesozoic multituberculates in which it first evolved. The upper premolars (P^{3-4}) that opposed the blade in carpolestids were typically larger than the molars and highly specialized, with a buccal row of multiple cusps, a central crest with a conule often flanked by accessory cuspules or crests, and a lingual row with two or three cusps. According to Biknevicius (1986), these unique premolars held food in place while the sectorial P_4 first sliced it apart and then sheared medially across the premolars.

Increasing hypertrophy and complexity of P_4 and P^{3-4} can be seen in the North American genera *Elphidotarsius*,

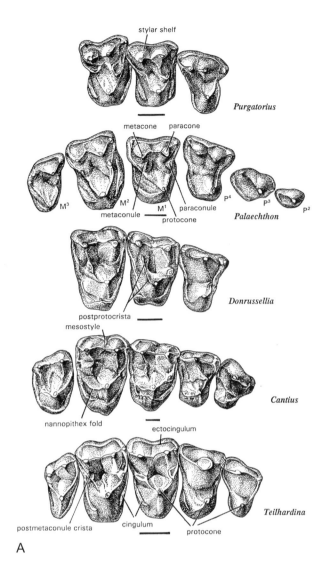

Carpodaptes, and *Carpolestes*, which are found more or less sequentially in Torrejonian, Tiffanian, and Clarkforkian deposits, respectively (Rose, 1975a). Late Tiffanian *Carpomegadon* had the largest P_4 but not the most complex, although it predates *Carpolestes* and probably originated independently from *Carpodaptes* (Bloch et al., 2001). The most primitive species of *Elphidotarsius* had the smallest and least specialized P_4 and P^{3-4}, the latter with only two buccal cusps and a central conule, as in *Pronothodectes*.

Two carpolestids have recently been described from Asia, the first to be found outside North America. Late Paleocene (Gashatan) *Subengius*, based on isolated teeth from Inner Mongolia, China, has the least specialized P_4 of any carpolestid (T. Smith et al., 2004). With just three apical cusps preceded by a tiny accessory cuspule, it is slightly more primitive than P_4 of the much older *Elphidotarsius*. *Carpocristes* from the early Eocene of Wutu, China, is similar to *Carpodaptes* but has even more elaborate upper premolars, with multiple crests flanking the conule (Beard and Wang, 1995). These two Asian carpolestids appear to represent different lineages that originated much earlier from North American

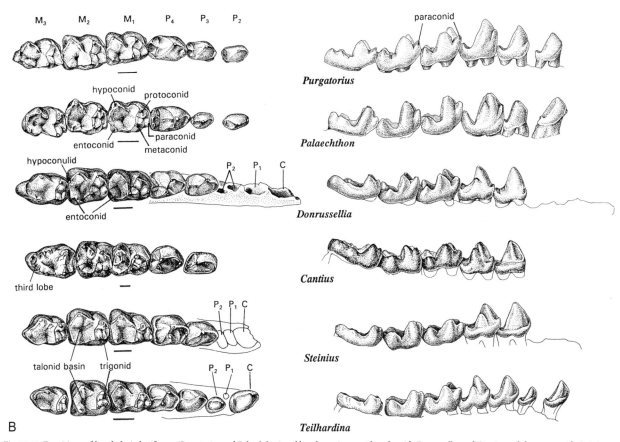

Fig. 10.12. Dentitions of basal plesiadapiforms (*Purgatorius* and *Palaechthon*) and basal euprimates (the adapoids *Donrussellia* and *Cantius* and the omomyids *Steinius* and *Teilhardina*): (A, *opposite*) right upper teeth; (B) left lower teeth. Scale bars = 1 mm. (From Rose, 1995.)

immigrants. *Parvocristes,* a supposed carpolestid from the early Eocene of Pakistan (Thewissen et al., 2001a), was founded on two fragmentary teeth that do not appear to pertain to the Carpolestidae.

The first known skeleton of a carpolestid (*Carpolestes;* Fig. 10.13B) was recently reported by Bloch and Boyer (2002). It reveals that carpolestids were arboreal mammals that lack specializations for leaping and were instead adapted for foraging in terminal branches (Plate 4.3). A surprising discovery was the presence of a nail-bearing, opposable hallux, making *Carpolestes* the only plesiadapiform known to have had these primatelike modifications. Differences in the form of the terminal phalanx from that of Eocene euprimates, however, suggest that the hallucal nail of carpolestids was acquired independently.

Chronolestes, from the same early Eocene site in China as *Carpocristes,* has been interpreted as the most primitive carpolestid and the sister taxon of all other carpolestids (Beard and Wang, 1995). It resembles carpolestids in having slightly enlarged fourth premolars and reduced teeth anterior to P_4 but lacks the characteristic polycuspidate bladelike P_4 of carpolestids. P^{3-4} and the upper central incisor are also simpler than in carpolestids or *Pronothodectes*. P^4, however, exhibits the typical plesiadapoid central conule. The lower dental formula of *Chronolestes* is 2.1.3.3, as in primitive plesiadapoids, but the anterior teeth are clearly derived relative to those of primitive carpolestids and plesiadapids. *Chrono-*

lestes is one of the latest-surviving plesiadapoids. Its unique suite of highly primitive and derived features suggests that it is a relict of a basal plesiadapoid lineage rather than a carpolestid (Silcox et al., 2001).

Saxonellidae

Saxonella (Fig. 10.15A), the sole member of the Saxonellidae, was a mouse-sized plesiadapiform distinguished by its bladelike P_3, which superficially resembles the P_4 of carpolestids. Its hypertrophied P^3 also resembles that of derived carpolestids, but peculiarities of structure indicate that the similarity was achieved independently (Fox, 1991). *Saxonella* is known from the late Paleocene of Europe and western Canada and is one of several plesiadapiforms that dispersed across the North Atlantic corridor during the Paleocene. Its multipronged upper central incisor resembles that of paromomyids a little more than it does those of plesiadapids or carpolestids (Rose et al., 1993). *Saxonella* has been variously linked with carpolestids, paromomyids, or plesiadapids. Current evidence favors the view that it is a specialized offshoot of the plesiadapoid stem.

Micromomyidae

The family Micromomyidae comprises three genera of very small plesiadapiforms (about 15–50 g) from the late

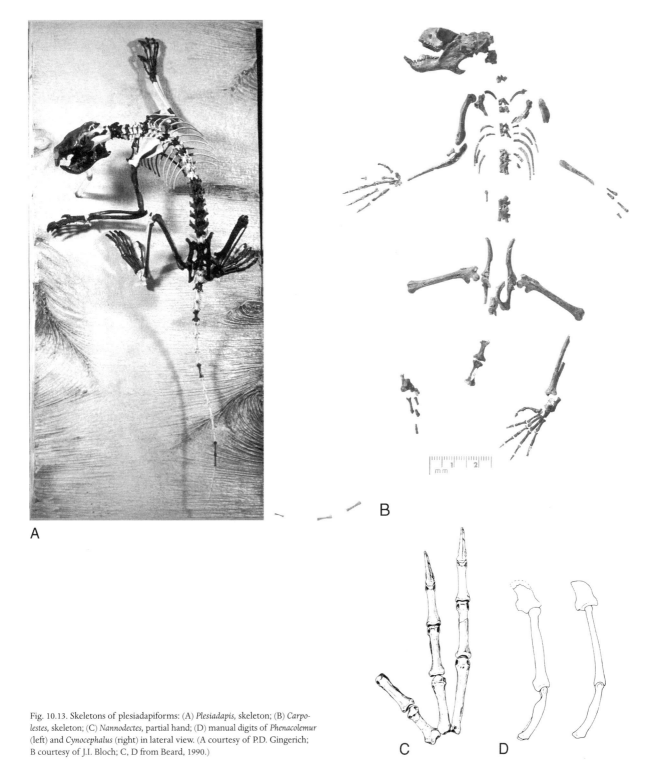

Fig. 10.13. Skeletons of plesiadapiforms: (A) *Plesiadapis*, skeleton; (B) *Carpolestes*, skeleton; (C) *Nannodectes*, partial hand; (D) manual digits of *Phenacolemur* (left) and *Cynocephalus* (right) in lateral view. (A courtesy of P.D. Gingerich; B courtesy of J.I. Bloch; C, D from Beard, 1990.)

Paleocene–early Eocene of western North America (Fox, 1984b; Beard and Houde, 1989; Gunnell, 1989; Rose et al., 1993). All are characterized by enlarged fourth premolars; P^4 is semimolariform and P_4 swollen, tall, and pointed. *Micromomys* and *Chalicomomys* are closely similar in having very primitive molars with relatively sharp cusps and no distinct hypocone, whereas in *Tinimomys* (Figs. 10.14A, 10.16) the cusps are more bunodont and a small hypocone is present. Their dental morphology and small size suggest that insects were a significant part of their diet. Micromomyids are dentally among the most primitive plesiadapiforms, resembling *Purgatorius* and primitive palaechthonids in molar structure more than other plesiadapiforms. However, all known species had reduced the lower dental formula to 1.1.3.3 or less, implying a long ghost lineage.

The auditory region, which often provides evidence of relationship, has proven controversial. Different specimens suggest either that the internal carotid artery had well-

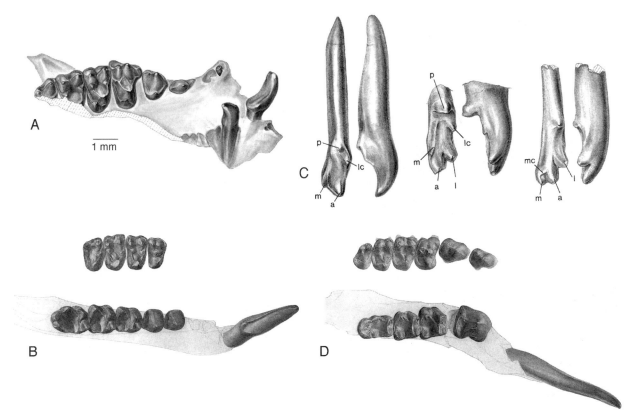

Fig. 10.14. Dentitions of plesiadapiforms (not to scale): (A) *Tinimomys*, palate with right P²–M³ and left I¹⁻². (B) *Pronothodectes*, right upper and left lower teeth; (C) right I¹ of (left to right) micromomyid, plesiadapid, and paromomyid (key: a, anterocone; l, laterocone; lc, lateroconule; m, mediocone; mc, mediocrista; p, posterocone); (D) *Phenacolemur*, right upper and left lower teeth. (A, C from Rose et al., 1993; B, D from Simpson, 1955.)

developed promontorial and stapedial branches in the tympanic cavity, as in microsyopids (Gunnell, 1989), or that the internal carotid was greatly reduced, as in plesiadapids and paromomyids (MacPhee et al., 1995). Isolated limb elements attributed to micromomyids are elongate and gracile, resembling those of paromomyids (Beard, 1993a,b).

The relationships of micromomyids are uncertain. They have variously been considered to be members of a broadly conceived Paromomyidae (Szalay and Delson, 1979), the sister taxon of a stricter Paromomyidae (Beard, 1993b), or closely related to Microsyopidae (Fox, 1984b; Gunnell, 1989). However, the combination of dental features displayed suggests that micromomyids could represent a separate plesiadapiform lineage stemming from *Purgatorius*.

Picromomyidae

Perhaps related to micromomyids are the early Eocene Picromomyidae, two genera of shrew-sized animals with broad, very low-crowned molars, and a peculiarly enlarged P_4 with a wide, flat talonid (Rose and Bown, 1996). The lower dental formula of *Picromomys* is probably 1.0.3.3. The first two lower molars in *Picromomys* (Fig. 10.16) have an accessory trigonid cusp unknown in any other plesiadapiform. Its jaw is foreshortened, and only two single-rooted teeth filled the space between P_4 and the large, horizontal incisor. With a body weight estimated at about 10 g, *Picromomys* is one of the smallest known primates, much smaller than any extant primate. Its diminutive size and odd dentition suggest that it fed on larvae, nectar, and gum, like the extant pygmy gliding possum *Acrobates*, to which it is dentally convergent.

Picrodontidae

The tiny Paleocene Picrodontidae are unique among plesiadapiforms in having highly modified molars rather than premolars. The upper and lower first molars of *Picrodus* (Fig. 10.16) are much larger than the others, and all have crenulated enamel and are broad, with low relief. The lower molars have small trigonids and greatly expanded talonid basins, accentuated on M_1. Picrodontids were originally thought to be bats, because the molars are superficially similar to those of fruit bats, but most experts now agree that they are plesiadapiforms. Nevertheless it seems probable that, like some bats, they were fruit or nectar feeders. Relationships of picrodontids to other plesiadapiforms are obscure, although Szalay (1968) argued that they are closely allied with Paromomyidae.

Paromomyidae

The Paromomyidae comprise about a half-dozen genera from the Paleocene–middle Eocene (Torrejonian-Uintan)

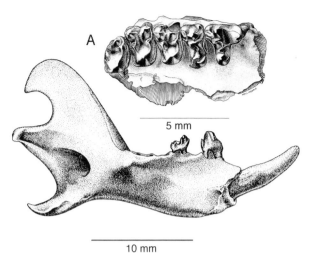

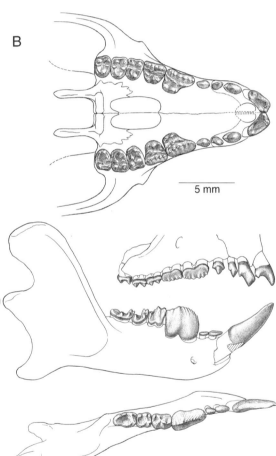

Fig. 10.15. Saxonellid and carpolestid dentitions: (A) *Saxonella*, right P³–M³ and left dentary with I$_1$, P$_3$, and M$_1$ (medial view); (B) *Carpolestes*, palate and right dentary. (A from Fox, 1991; B from Bloch and Gingerich, 1998.)

of North America and the early-middle Eocene (Ypresian-Lutetian) of Eurasia. They have relatively broader and flatter molars than most other plesiadapiforms and a distinctive P$_4$ with a tall, single-cusped trigonid and a small talonid basin. The plesiomorphic Torrejonian genus *Paromomys* retains a primitive lower dental formula (2.1.3.3), but all subsequent genera show significant reduction of the dental formula (to 2.0.2.3 in *Acidomomys*, 1.0.2.3 in *Elwynella* and some *Ignacius*, and 1.0.1.3 in other *Ignacius* as well as in all *Phenacolemur* and *Arcius*; Bown and Rose, 1976; Rose and Bown, 1982; Bloch et al., 2002). In addition, all genera except *Paromomys* have a conspicuous diastema between the incisor(s) and P$_3$ or P$_4$. P$_4$ of *Ignacius* (Torrejonian-Uintan; Fig. 10.16) is usually no larger (and often smaller) than the first molar, whereas P$_4$ of Tiffanian-Wasatchian *Phenacolemur* (Fig. 10.14D) is typically larger than the first molar. Paromomyid molars are quadrate, the lowers having wide talonid basins and the uppers having expanded hypocone shelves. The cusps are low and bunodont and usually subordinate to the low crests that join them. A close parallel to the paromomyid dentition is seen in some living petaurid marsupials, suggesting a similar diet rich in tree gum and sap. In addition, the pointed fourth premolar may have been used to open fruits or seeds. Based on incisor morphology, however, Godinot (1984) suggested that the European genus *Arcius* was insectivorous.

The skull is best known in early Eocene *Ignacius graybullianus*. It had a long, narrow snout; a low sagittal crest; widely flaring zygomae; and very inflated auditory bullae composed mainly of the entotympanic. The disposition of the internal carotid artery in the ear region has been a matter of considerable controversy. The artery was initially inferred to have been degenerate, as in plesiadapids and flying lemurs (Kay et al., 1992), but additional fossils revealed a shallow groove that may have housed a small promontorial branch (Bloch and Silcox, 2001). Using ultrahigh resolution X-ray computed tomography, Silcox (2003) discovered a bony tube in the position of this groove, which must have contained the promontorial artery and/or the internal carotid nerve. The anatomy is very similar to that of euprimates and scandentians, but unlike that of dermopterans.

Eocene species of *Ignacius* and *Phenacolemur* have slender limb bones and exhibit specializations of the tarsus and manus, including relatively long intermediate phalanges (Fig. 10.13D). Beard (1990, 1993a,b) interpreted these features to be indicative of gliding ability and synapomorphies shared with extant Dermoptera. But characters supporting such a relationship are rather few and have met with criticism. Extant dermopterans are highly autapomorphous, so it is not surprising that features supporting paromomyid-dermopteran affinity are not numerous. Nevertheless, the details of the basicranium, wrist, phalanges, and ankle enumerated by Beard provide a compelling prima facie case for relationship. As just noted, however, new evidence from the basicranium suggests that the intrabullar internal carotid arterial pattern differs from that of dermopterans. Furthermore, reappraisal of wrist anatomy suggests that the arrangement of carpal bones may be less similar to that of *Cynocephalus* than previously thought (Stafford and Thorington, 1998). It is also noteworthy that the phalanges of paromomyids resemble those of gliding rodents as well as those of the dermopteran *Cynocephalus* (Hamrick et al., 1999), raising the possibility of convergence. It is well known that

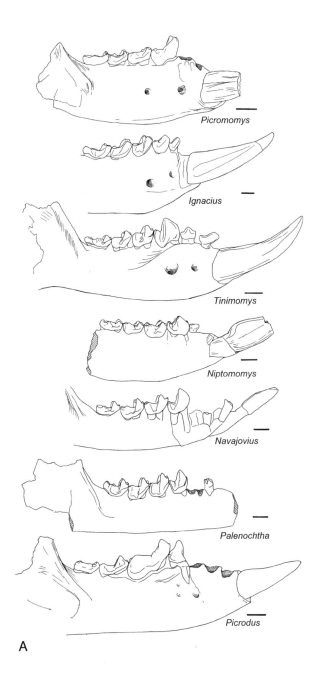
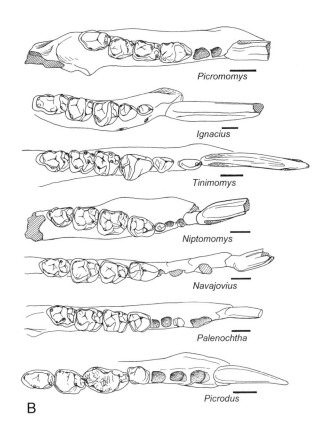

Fig. 10.16. Plesiadapiform right lower dentitions: (A) lateral view; (B) crown view. Scale bars = 1 mm. (From Rose and Bown, 1996.)

paromomyids are much more similar in cranial shape and dental anatomy to other plesiadapiforms than to *Cynocephalus*. If paromomyids are the sister group of living dermopterans, their last common ancestor, presumably in the early Paleocene, must have already acquired all the derived traits present in Eocene paromomyids, but corroborating fossil evidence is lacking.

The gliding hypothesis has also been challenged. Paromomyid limb elements are relatively shorter and more robust than those of extant gliders, perhaps rendering them incapable of gliding, at least in the way living gliders do (Runestad and Ruff, 1995). In addition, Hamrick et al. (1999) observed that paromomyid phalangeal structure and proportions imply vertical climbing and grasping (clinging to tree trunks), but the phalanges lack the proportions of "mitten-gliders" such as *Cynocephalus*. Consequently, if paromomyids had a patagium (gliding membrane), it apparently did not incorporate the fingers.

The weight of the evidence therefore indicates that paromomyids were clearly arboreally adapted, but it has not been conclusively demonstrated that they were gliders. A partial skeleton of Clarkforkian *Acidomomys*, which includes substantial parts of the forelimbs (Bloch and Boyer, 2001), may help to settle the debate when it has been studied.

Microsyopidae

The late Paleocene–middle Eocene Microsyopidae (sensu McKenna and Bell, 1997) are similar in some ways to other plesiadapiforms but different enough that they have variously been considered to be divergent plesiadapiforms, more closely related to euprimates, or not particularly close to either group. They resemble plesiadapiforms in having an enlarged, procumbent I_1 and in their basic molar structure, except that the uppers lack a nannopithex fold (a crest between the protocone and the hypocone). But they differ in retaining a primitive auditory region with well-developed internal carotid branches and an unossified bulla (presumably membranous rather than formed by the petrosal bone). Additionally, where the upper central incisor is known it is simple, rather than multicusped as in other plesiadapiforms. Among the latter, microsyopids have been considered dentally close to palaechthonids and are sometimes linked with them in a superfamily Microsyopoidea (Bown and Gingerich, 1973; Gunnell, 1989).

A synapomorphic feature that unites microsyopids is the peculiar shape of the enlarged lower incisor, usually described as lanceolate (i.e., laterally compressed and leaf- or spearpoint-shaped). Based on this feature and molar structure, Tiffanian *Navajovius* (Fig. 10.16) is the oldest and most primitive known microsyopid. The lower dental formula is ambiguous but has been interpreted as 1.1.3.3 (Gunnell, 1989). Most later microsyopids lost the canine. *Navajovius* was succeeded in the latest Paleocene and early-middle Eocene by *Arctodontomys* and *Microsyops*. The latter was relatively common in Wasatchian faunas. Successive species of *Microsyops* evolved progressively more molariform fourth premolars.

Middle Eocene *Megadelphus* (Bridgerian) and *Craseops* (Uintan), with estimated body weights of 4–6 kg (Gunnell, 1989), were among the largest plesiadapiforms. *Craseops* had molariform fourth premolars and dilambdodont upper molars, perhaps reflecting an increasingly folivorous diet. At the other end of the size spectrum, the latest Paleocene–middle Eocene uintasoricine microsyopids, such as *Niptomomys* and *Uintasorex*, were among the smallest plesiadapiforms, weighing only 20–60 g (Gunnell, 1989; Rose et al., 1993).

Toliapinidae

Hooker et al. (1999) proposed a new family of plesiadapiforms, Toliapinidae, for several very tiny mammals known mainly from isolated teeth from Europe. Some of them rival *Picromomys* as the smallest known plesiadapiforms. Toliapinids appear to be closely related to microsyopids, although Silcox (2001) considered them to be the sister taxon of euprimates (with plesiadapiforms being the next outgroup). Among taxa transferred to the family by Hooker et al. are *Berruvius* and *Avenius*, formerly included in Microsyopidae, and *Altiatlasius*, a north African dental taxon that has been interpreted by others to be the oldest euprimate (see below). Toliapinids are likely to remain problematic until they are known from more than fragmentary dentitions.

Azibiidae

Tabuce et al. (2004) recently employed the name Azibiidae (formerly a subfamily of Adapidae) for two genera, *Azibius* and the new genus *Dralestes*, from the Eocene of northern Africa, and transferred the family to Plesiadapiformes. Only isolated teeth (*Dralestes*) and one lower jaw fragment (*Azibius;* Sudre, 1979) are known, making the assessment of relationships difficult. They have some features in common with plesiadapiforms, including a nannopithex fold on the upper molars, but the overall morphology is sufficiently unusual that plesiadapiform affinity is open to question. If confirmed as members of this group (by more complete specimens) they would be the only known African plesiadapiforms except possibly *Altiatlasius* (whose plesiadapiform affinities are very doubtful).

Plesiadapiform Origins

The origin of Plesiadapiformes is obscure because most taxa—including the oldest and most primitive members—are known only from dentitions, and knowledge of potential Cretaceous precursors is so limited. It is usually stated that plesiadapiforms (and euprimates as well) evolved from insectivores, in the broad sense, but such a vague postulate is not very enlightening. It is true that dental morphology of *Purgatorius* and other primitive plesiadapiforms generally resembles that of erinaceomorph insectivorans, which differ from many other "insectivores" in having lower-crowned teeth with less acute cusps and broad talonid basins; however, these general attributes characterize several other primitive eutherians as well. Moreover, close relationship to erinaceomorphs is not supported by other anatomical evidence, such as the basicranium (Novacek et al., 1983; MacPhee et al., 1988), leaving the question of plesiadapiform origins currently unresolved.

Fossil Euprimates

The oldest fossil primates that are clearly allied with extant forms appear abruptly in basal Eocene deposits of North America, Europe, and Asia, without antecedent transitional forms that would indicate their ancestry. They were part of a wave of immigration, also involving the first appearance of artiodactyls and perissodactyls, which coincides with a brief, very warm interval at the beginning of the Eocene (the Initial Eocene Thermal Maximum). These earliest primates are already divisible into two prosimian clades, the lemurlike Adapoidea and the tarsier- or galagolike Omomyidae, which are believed by some researchers to exemplify the basic euprimate dichotomy into Strepsirrhini and Haplorhini (e.g., Kay et al., 2004). However, convincing demonstration of a direct link between any specific adapoid or omomyid and any extant primate family has proven elusive. Therefore, even though it remains quite possible that Adapoidea and Omomyidae are paraphyletic stem taxa of living primates, this relationship has yet to be established.

Indeed, molecular evidence suggests that strepsirrhines had already diverged from other primates by the early Eocene at the latest (and perhaps as early as the Late Cretaceous, 70–75 Ma), and that the split between lemurs and lorises predates known Eocene primates (Yoder et al., 1996; Yoder and Yang, 2004). This hypothesis gains some support from recently discovered fossils that extend the record of definitive strepsirrhines back to the Early Tertiary—more than twice the age of the oldest known tooth-combed prosimian fossils only a few years ago (see the section on Earliest Strepsirrhines below).

Adapoids and omomyids are demonstrably euprimates, based on characteristics of the skull, teeth, and postcranial skeleton (Fig. 10.10). These include the presence of a petrosal auditory bulla, a bony postorbital bar, a relatively large brain, an opposable hallux (implied by a prominent peroneal process on the hallucal metatarsal and a saddle-shaped joint between this bone and the entocuneiform), and digits with nails rather than claws (indicated by distinctive flat terminal phalanges). Primitive members of both groups have hands and feet specialized for arboreal grasping and climbing and elongate hind limbs adapted for occasional leaping, as in many extant prosimians. These features immediately separate them from plesiadapiforms and unite them with extant primates (e.g., Dagosto, 1988, 1993; Rose, 1995). Adapoids and omomyids differ in enough aspects of their anatomy to indicate that they had been evolving independently for some time when they first appeared in the basal Eocene, but their overall similarity and the very close dental resemblance among the most primitive members of both clades suggest that they did not diverge before the Paleocene (Rose and Bown, 1991). However, the fossil record so far has provided no direct evidence as to when, where, and how this divergence took place.

Even though the euprimate status of Adapoidea and Omomyidae is not in doubt, their broader relationships among Euprimates are controversial (see Fig. 10.9). Some authors assign them to their own infraorders (Adapiformes, Omomyiformes) to emphasize their differences from living primates and the lack of incontrovertible evidence of direct relationship. Others highlight shared (but not always derived) features with extant primates and reduce them to families within the Lemuriformes (or Lemuroidea) and Tarsiiformes (or Tarsioidea). Because adapoids and omomyids are the oldest euprimates, and for a long time were the only known Eocene euprimates, most paleoprimatologists have also sought the origin of anthropoids among either adapoids or omomyids. Substantial arguments can be and have been made to support either case, and the controversy continues.

About 75 genera and 180 species of adapoids and omomyids are currently known, with omomyids slightly more taxonomically diverse than adapoids (Gebo, 2002; Gunnell and Rose, 2002). These figures are several times greater than the number of prosimian taxa alive today. They thrived in Laurasia during the Eocene, where they often rank among the most common elements of mammalian faunas. A few species have been reported from northern Africa, where they were evidently much rarer. Most were gone by the end of the Eocene, but a few kinds existed during the Oligocene, and the Asian sivaladapid adapoids persisted into the late Miocene.

Other euprimate clades appeared later in the Eocene. True tarsiers (Tarsiidae) and anthropoids made their first appearance in the record in middle Eocene fissure-fillings in southern China (Beard et al., 1994), or possibly even earlier, if north African *Algeripithecus* and *Tabelia* are correctly attributed to Anthropoidea. In addition, rare forms that appear to be more closely related to crown strepsirrhines than any of those mentioned so far have been reported in the past decade from the middle and late Eocene of north Africa and the early Oligocene of southern Asia. All of these forms are further discussed below.

Several very early genera that are not clearly assignable to any of these euprimate groups have, in recent years, been claimed to be the oldest or most primitive euprimate. All are known only from teeth, which contributes to their uncertain status. The most important of them is *Altiatlasius* (Fig. 10.17), based on a series of isolated teeth from the late Paleocene of Morocco, which resemble those of both euprimates and plesiadapiforms. It was first considered to be an omomyid closely allied with anthropoids (Sigé et al., 1990) or a "protosimiiform" (i.e., a protoanthropoid; Godinot, 1994), but was subsequently assigned to the new plesiadapiform family Toliapinidae (Hooker et al., 1999). Silcox (2001) returned *Altiatlasius* to the Euprimates, and Beard (2004) considered it the oldest known anthropoid. If its euprimate status and Paleocene age are upheld, it could suggest an African origin for Euprimates; but the diversity of opinions about its affinities clearly indicates that more complete specimens are needed to determine its relationships with confidence. *Altanius* (Fig. 10.17), from the earliest Eocene of Mongolia, is another problematic form. It is known from nearly complete jaws, which reveal a very primitive dental formula (2.1.4.3) and cheek tooth resemblances to both omomyids and certain plesiadapiforms (Dashzeveg and McKenna, 1977; Rose and Krause, 1984; Gingerich et al., 1991; Rose et al., 1994). Because of its curious mix of derived and very primitive characters, it has been considered a primitive omomyid or an aberrant plesiadapiform, but it is probably best interpreted as an early offshoot of the euprimate stem. A skull or diagnostic associated postcrania would obviously clarify its relationships. *Decoredon* and *Petrolemur*, based on fragmentary dentitions from the Paleocene of China, have been proposed as a basal omomyid and an adapoid, respectively (Tong, 1979; Szalay and Li, 1986). The first is inadequately known to determine its true affinities, whereas the second is not clearly primate and might be an arctocyonid. Both could eventually prove to be important to euprimate origins, but like the other taxa discussed here, more complete fossils are needed to confirm their relationships.

Adapoidea

Among the best known of all fossil primates, adapoids are represented by thousands of jaws, dozens of skulls, and several nearly complete skeletons (Figs. 10.18–10.21). In many aspects of their anatomy, they bear a remarkable resemblance to extant strepsirrhines.

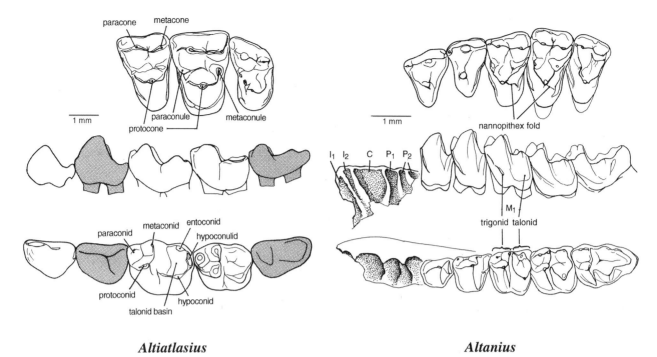

Altiatlasius   *Altanius*

Fig. 10.17. Left dentition of basal euprimates: late Paleocene *Altiatlasius* (M^{1-3} and P_3, M_{1-2}) and earliest Eocene *Altanius* (P^3–M^3 and P_3–M_3). (From Rose, 1995.)

The primitive dental formula of adapoids is 2.1.4.3/2.1.4.3, which is retained in most species. The spatulate, more or less vertical lower incisors, with I_2 often larger than I_1, are often cited as a resemblance to anthropoids and are quite unlike those of plesiadapiforms. Adapoid premolars are primitively simple except for P_4, which is often semimolariform. P_1 is single-rooted and P_{2-4} are two-rooted. In contrast to omomyids, the antemolar teeth are never mesiodistally compacted. Adapoid molars show considerable variation. They may have either low, rounded cusps adapted for frugivory (the primitive state; e.g., *Cantius* and *Protoadapis*) or sharp crests specialized for leaf eating (e.g., *Notharctus* and *Adapis*). Some small species have sharper cusps and may have been partly insectivorous (e.g., *Anchomomys*). The upper molars are variable with respect to such features as hypocones, lingual cingula, and mesostyles. Lower molars usually have small trigonids with reduced paraconids and large, broad talonid basins. Fusion of the mandibular symphysis is common but not universal among adapoids.

Where known, the skull is usually relatively long-snouted with small orbits (e.g., *Adapis*, *Europolemur*, *Notharctus*; Figs. 10.10, 10.18, 10.21), suggesting that they were diurnal (Gingerich and Martin, 1981). Middle Eocene *Pronycticebus* and *Godinotia*, from Europe, however, had short snouts and large orbits, suggesting nocturnality (Simons, 1962; Martin, 1990; Godinot, 1998; Franzen, 2000). A free ectotympanic ring supported the eardrum within the auditory bulla, as in living lemurs. The promontorial and stapedial branches of the internal carotid artery vary in size (even in the same species) and are usually enclosed within bony tubes for much of their length (MacPhee and Cartmill, 1986; Rose et al., 1999). Some notharctid adapoids were evidently sexually dimorphic, as indicated by a more pronounced sagittal crest and larger canines in the presumed males (Krishtalka et al., 1990; Alexander, 1994; Gingerich, 1995).

There are two distinct types of postcranial skeletons among adapoids, reflecting at least two different locomotor modes. In Notharctidae—North American *Notharctus* (Fig. 10.18) and *Smilodectes,* and European *Europolemur*—the hind limbs were markedly longer than the forelimbs (intermembral index 60–61) and the femur was longer than the tibia. The joints were very flexible, and the elbow was configured to allow extensive supination. The hands and feet were adapted for grasping, with an opposable hallux and digits tipped by nails rather than claws (more accurately described as transitional between claws and nails; Godinot, 1992b). In these respects, as well as size, they closely resembled extant arboreal lemurs (Gregory, 1920; Rose and Walker, 1985; Dagosto, 1993; Franzen and Frey, 1993; Plate 6). Limb proportions and some joint surfaces are similar to those in the agile, vertical clinging and leaping lemurs, but in other features they compare more closely with active arboreal lemurs that run along branches and leap less frequently. In contrast, the adapids *Adapis, Palaeolemur,* and *Leptadapis* lack leaping specializations and instead had forelimbs and hind limbs of more equal length. Their limb modifications have been interpreted to indicate either slow arboreal progression, like lorises, or quadrupedal running, like small platyrrhines (Dagosto, 1983; Godinot, 1991; Bacon and Godinot, 1998). Most adapoids probably weighed 0.5–7 kg, and thus many were larger than most plesiadapiforms and omomyids and comparable in size to living lemurs; however, some species of *Anchomomys* may have weighed less than 100 g (Godinot, 1998; Fleagle, 1999).

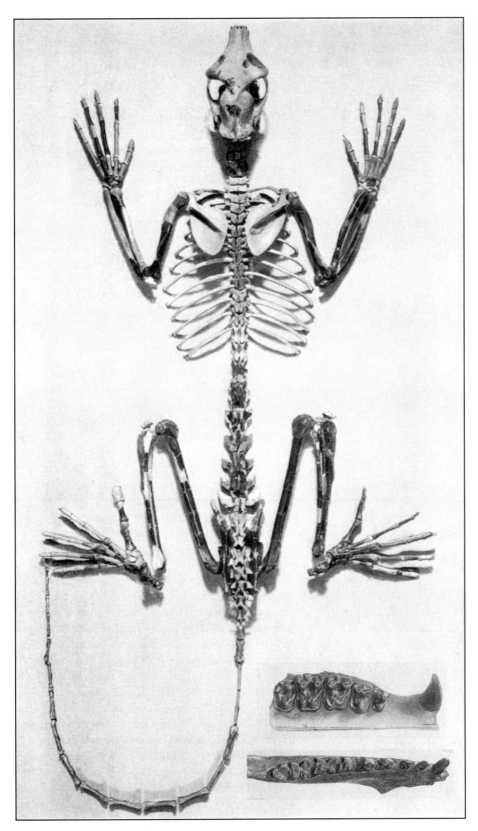

Fig. 10.18. Middle Eocene adapoid *Notharctus*. Right upper and left lower dentitions shown at bottom. (From Gregory, 1920.)

The striking similarity of the skeletons of adapoids to those of various living lemurs and lorises (Gregory, 1920; Koenigswald, 1979; Dagosto, 1993) has led some researchers to regard them as primitive strepsirrhines, or as the sister group of Strepsirrhini. Beyond this overall resemblance, adapoids and extant strepsirrhines do share some possibly uniquely derived features, including conformation of the tibiotalar joint and the presence of a grooming claw on the

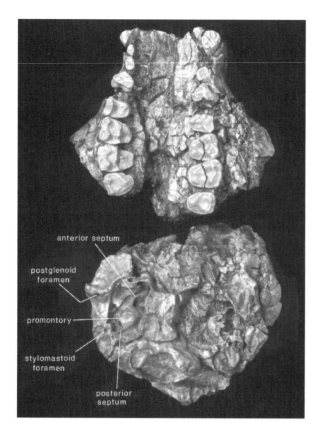

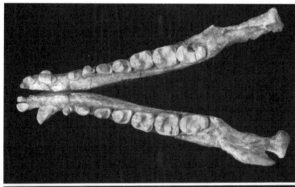

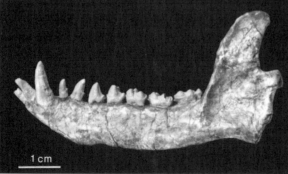

Fig. 10.19. Skull and mandible of early Eocene *Cantius*. (From Rose et al., 1999.)

son's) organ (Rosenberger et al., 1985). But experts disagree as to whether adapoids were true strepsirrhines, which are defined by the presence of soft-tissue characteristics of the nose, namely, a rhinarium—glandular, naked skin around the nostrils. Even if they were, it has been argued that strepsirrhinism may not be a derived condition. A rhinarium is present in many nonprimates and may have characterized omomyids as well, whereas the *absence* of a rhinarium in haplorhines appears to be a derived state (Beard, 1988; Martin, 1990).

Thus most characters shared by adapoids and strepsirrhines appear to be primitive for euprimates. Furthermore, no known adapoid had a lower tooth-comb, the principal characteristic that unites extant strepsirrhines (Schwartz and Tattersall, 1987; Rose, 1995; Rasmussen and Nekaris, 1998; Fleagle, 1999; Gebo, 2002). Instead the lower incisors have spatulate crowns and range in orientation from somewhat procumbent to more or less vertical, and the canines are almost always separate, prominent, and projecting (Rosenberger et al., 1985; Rose et al., 1999). Gingerich (1975) hypothesized that the close-packed, procumbent lower incisors and canines of late Eocene *Adapis parisiensis* represented an incipient stage in formation of a tooth-comb, but this conjecture remains to be convincingly demonstrated. Regardless of whether adapoids had the strepsirrhine condition or a tooth-comb, shared derived traits might still indicate that they are the stem group or the sister group of strepsirrhines.

There are three principal subdivisions of adapoids, now regarded as families, but long ranked as subfamilies of a single family Adapidae: the lemurlike Notharctidae, the more lorislike Adapidae, and the Sivaladapidae (often considered a subfamily of one of the others; Godinot, 1998; Gebo, 2002). Historically, most Old World adapoids were assigned to the subfamily Adapinae, whereas Notharctinae primarily encompassed New World forms. The two groups were easily distinguished by the origin of the hypocone—from the postcingulum in adapids (true hypocone), or budding off the protocone in notharctids (sometimes called a "pseudohypocone," e.g., Simons, 1972). However, with the realization that differences in postcranial anatomy as well as dentition indicate at least two distinct clades of European "adapids," one of which is more closely allied with notharctids, a more phylogenetic classification has emerged over the past 10–15 years (e.g., Franzen, 1987, 1994; Thalmann et al., 1989). About two-thirds of the 30–35 genera and 80–85 species of adapoids are currently included in Notharctidae. Although, or perhaps because, the systematics and evolution of adapoids have attracted so much interest, there is still considerable difference of opinion on the proper family attribution of several genera, mostly from the Old World (compare McKenna and Bell, 1997; Godinot, 1998; Fleagle, 1999; Gebo, 2002).

European *Donrussellia* and Euroamerican *Cantius* are the most primitive known adapoids (Figs. 10.12, 10.19). Both are usually included in Notharctidae, but *Donrussellia* is so plesiomorphous that it is better regarded as the sister group or stem taxon of all other adapoids (e.g., Godinot, 1992a).

second toe (Koenigswald, 1979; Dagosto, 1988; Gebo 2002). Both groups also have a median gap between the upper incisors, a hallmark of the strepsirrhine condition, which in living primates is associated with the vomeronasal (Jacob-

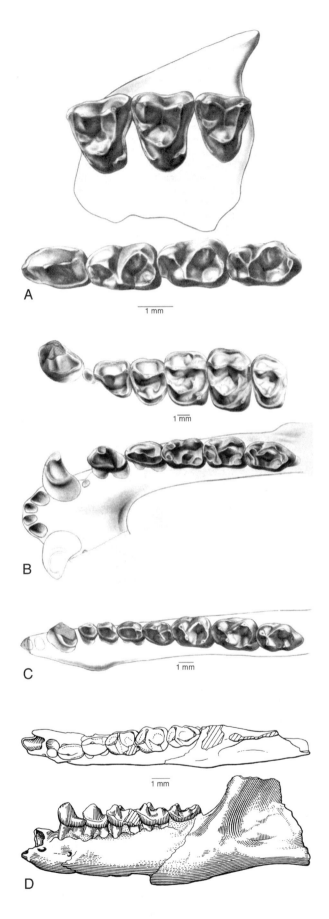

Donrussellia is known only from a few jaws and isolated teeth from the early Eocene of France (e.g., Godinot et al., 1987; Godinot, 1998). The lower premolars are unreduced, and the cheek teeth show little basal inflation. A small hypocone is variably present on M^{1-2} and when expressed, originates from the postcingulum with no connection to the protocone (an adapid characteristic). Though initially thought to be an omomyid, *Donrussellia* shares certain dental synapomorphies with adapoids, including the absence of the crest running posteriorly from the metaconule on the upper molars (postmetaconule crista; Godinot, 1992a).

Notharctidae. Members of this family were abundant and diverse in the Eocene of North America and Europe. The subfamily Notharctinae was prevalent in the early and middle Eocene of North America, whereas most European forms represent the subfamily Cercamoniinae (=Protoadapinae). Possible notharctids are also known from northern Africa, Arabia, and southern Asia. Notharctids are the most primitive adapoids, and they seem to be linked by few clear synapomorphies.

Notharctines are among the best known of all fossil primates, owing in large part to Gregory's (1920) classic study of the skeleton of middle Eocene *Notharctus* (Fig. 10.18). A succession of studies have examined its skeleton in detail, some concluding that *Notharctus* was similar to extant vertical clingers and leapers (e.g., *Propithecus*) and others inferring more generalized quadrupedal habits including occasional leaping. Besides the anatomical features listed in the preceding section, it may be noted that the anatomy of the wrist and hand of *Notharctus* (short metacarpals and long phalanges adapted for grasping) is most comparable to that in extant quadrupedal lemurs (Hamrick, 1996; Hamrick and Alexander, 1996). The fourth digit of the hand of *Notharctus* is the longest, as in living lemurs. In sum, it is probable that *Notharctus* and its relatives were active arboreal primates that approached extant lemurs in many respects but were not yet fully specialized vertical clingers and leapers.

The most primitive notharctine, early Eocene *Cantius*, is the only member of this subfamily known from both North America and Europe. The North American fossils were for a long time called *Pelycodus*, but that name is now restricted to a rare, closely related notharctine. The incisors in *Cantius* were anteriorly inclined (Rose et al., 1999; Fig. 10.19), not vertical, as often stated. Its lower cheek teeth are more derived than those of *Donrussellia* in being relatively shorter and wider, while the uppers are less transverse; all are slightly more swollen near the base. The lower molars are bunodont with three trigonid cusps; the paraconid (reduced or lost in many subsequent adapoids) remains distinct, though a little less so than in *Donrussellia*. As in notharctines generally, M_3

Fig. 10.20. Adapoid dentitions: (A) *Anchomomys*, left M^{1-3} and right P_4–M_3; (B) *Mahgarita*, left upper and right lower dentitions; (C) *Microadapis*, right C_1–M_3; (D) *Djebelemur*, left dentary with P_3–M_3. *Djebelemur* may be a primitive strepsirrhine rather than an adapoid. (A, C from Szalay, 1974; B from Wilson and Szalay, 1976; D from Hartenberger and Marandat, 1992.)

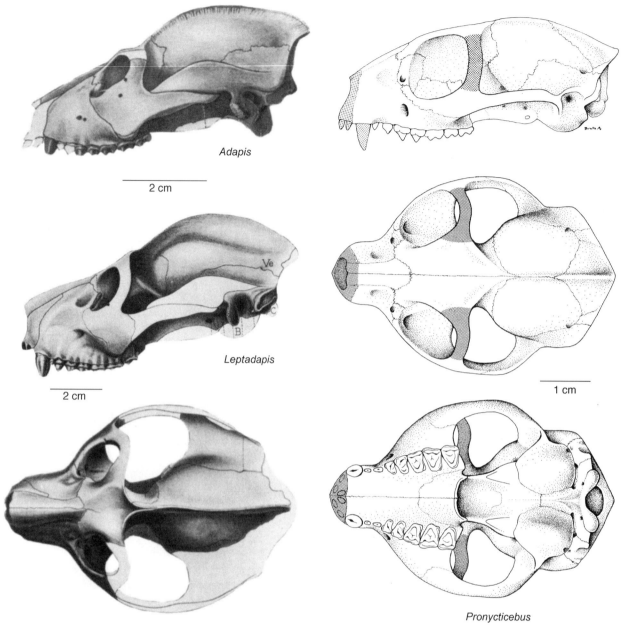

Fig. 10.21. Adapid skulls from western Europe: *Adapis* (lateral view), *Leptadapis* (=*Adapis magnus;* lateral and dorsal views), and *Pronycticebus* (lateral, dorsal, and ventral views). (*Adapis* and *Leptadapis* modified from Stehlin, 1912; *Pronycticebus* from Szalay, 1971.)

has an expanded hypoconulid lobe. In the upper molars the hypocone, which gets progressively larger in more derived species, is joined to the protocone by a nannopithex fold. This cusp is poorly differentiated in the most primitive *Cantius,* hence it is possible that this condition gave rise to forms with either a pseudohypocone or a true hypocone.

Species-level evolution of North American *Cantius* has been well studied and provides compelling evidence of gradual evolution (e.g., Gingerich and Simons, 1977; O'Leary, 1996). Trends in the evolution of *Cantius* include increasing size, progressive development of a mesostyle as well as a hypocone, and incipient selenodonty. These trends were further developed in middle Eocene *Notharctus* and *Smilodectes.* The mandibular symphysis is fused in *Notharctus* but not in other genera. The size and bunodont dentition of *Cantius* suggest that it was a frugivore, whereas the more selenodont molars of *Notharctus* and *Smilodectes* indicate that they included more leaves in their diet. *Cantius* probably lies in or near the ancestry of all other notharctines (Godinot, 1998).

European cercamoniines seem to be derived from either *Donrussellia* or *Cantius.* They existed throughout the Eocene. As noted above, *Donrussellia* is usually classified as a cercamoniine (e.g., Godinot, 1998; Fleagle, 1999; Gebo, 2002), but it probably represents an adapoid stem taxon. Several branches of cercamoniines have been identified. One is represented by early and middle Eocene *Protoadapis* (including *Cercamonius*) and *Europolemur.* These genera comprise moderately large species (1–3 kg; Fleagle, 1999) with dentitions similar to that of *Cantius*, except that the upper molars have a complete lingual cingulum and a cingular hypocone. There was a tendency to lose the first premolar and reduce P_2. Partial skeletons of *Europolemur* from Messel, Germany, have similar anatomy and proportions to *Notharctus* (e.g., Franzen and Frey, 1993).

Another branch of cercamoniines is represented by the European *Periconodon* and *Anchomomys*, which were much smaller (100–900 g). *Anchomomys* (Fig. 10.20A) had simple, sharp-cusped molars with either no hypocone or a small hypocone and was probably insectivorous. The slightly larger *Periconodon* had more rounded cusps and lingual cingula bearing a well-developed hypocone posteriorly and a cusp lingual to the protocone (called a pericone, for which the genus is named). *Omanodon* and *Shizarodon*, from the early Oligocene of Arabia, are small *Galago*-sized primates that may be related to *Anchomomys*.

Middle and late Eocene *Pronycticebus* (Fig. 10.21) was a mid-sized adapoid (about 1 kg) with relatively primitive, sharp-cusped cheek teeth, suggesting an insectivorous/frugivorous diet. The upper molars are triangular with a distinct cingular hypocone. Because of its large orbits and short snout, Godinot (1998) placed *Pronycticebus* in an unassigned subfamily separate from other adapoids; however, most other authors currently classify it as a cercamoniine. The skeleton of the closely allied *Godinotia* (formerly assigned to *Pronycticebus*), known from both Geiseltal and Messel in Germany, is generally similar to that of other notharctids but may have had somewhat longer forelimbs (Thalmann et al., 1989; Franzen, 2000).

Other noteworthy possible cercamoniines include *Djebelemur* from the early Eocene of Tunisia, *Aframonius* from the late Eocene of Egypt, *Caenopithecus* from the late Eocene of Europe, and *Mahgarita* from the late Eocene of Texas. Most authors agree that these genera are adapoids, but their precise relationships within that group are problematic. Nonetheless, they attest to the wide distribution of diverse adapoids during the Eocene. *Djebelemur* (Fig. 10.20D) is a tiny primate (about 100 g; Fleagle, 1999), based on a lower jaw that resembles cercamoniines in having a reduced paracristid and no paraconid on the lower molars. Alternatively, it may be a primitive strepsirrhine (Seiffert et al., 2005). *Aframonius* was much larger (1.6 kg; Fleagle, 1999) and, together with other possible cercamoniines from northern Africa, documents that multiple adapoid lineages were present in northern Africa by the late Eocene (Simons, 1997a). *Caenopithecus* is a large adapoid (3.5 kg; Fleagle, 1999), sometimes placed in the Adapidae (e.g., Godinot, 1998), whose quadrate upper molars have a cingular hypocone and a prominent mesostyle. The first premolar is lost and the last premolars are simple. *Mahgarita* (Fig. 10.20B) is an especially problematic form, which has been identified variously as cercamoniine, adapid, or closely related to Anthropoidea (Rasmussen, 1990). It has lost P^1_1 and has vestigial P^2_2, and the mandibular symphysis is fused. Although found in North America, its cingular hypocone clearly separates it from indigenous notharctines and allies it with European adapoids.

A final group that has recently been included in Notharctidae by some authors is the subfamily Amphipithecinae. It includes several controversial southeast Asian taxa known mainly from jaws (*Pondaungia, Amphipithecus*, and possibly *Siamopithecus* and *Myanmarpithecus*), which have more often been considered basal anthropoids (e.g., Jaeger et al., 1998; Ducrocq, 1999; Takai et al., 2001). They were relatively large, with estimated weights ranging from 4 to 9 kg (Ciochon et al., 2001). Their deep mandibles with short, stout lower premolars and broad, bunodont molars with low relief and, in some cases, very crenulated enamel (features suggesting hard-object feeding), have been cited as resemblances to primitive anthropoids (Jaeger et al., 1998; Fleagle, 1999; Beard, 2002). Unlike anthropoids, however, postorbital closure was absent or poorly developed in amphipithecids (Shigehara et al., 2002). In addition, some dental features, including the presence of a nannopithex fold and a "pseudohypocone," have been cited as synapomorphies with notharctines (Ciochon and Holroyd, 1994), although there is controversy over the homology of this cusp. Recently described skeletal remains of *Pondaungia*, if properly attributed, also seem to support notharctid affinities (Ciochon et al., 2001); however, a new talus from the Pondaung Formation of Myanmar that was tentatively attributed to *Amphipithecus* appears to have anthropoid traits (Marivaux et al., 2003). *Siamopithecus* is particularly enigmatic, because it has several probable autapomorphies that contrast with both notharctids and basal anthropoids, including a very long protocone slope of the upper molars and retention of only faint vestiges of cingula. It is unclear whether *Siamopithecus* even belongs to Amphipithecinae, whatever the affinities of that subfamily.

Adapidae. With the transfer of cercamoniines to the Notharctidae, the Adapidae as currently conceived is a relatively restricted (and more likely monophyletic) group of primarily late Eocene primates from Europe (*Adapis, Cryptadapis, Leptadapis, Microadapis*, and *Palaeolemur*; Gebo, 2002; Figs. 10.20C, 10.21). As noted in the previous section, some authors (e.g., Godinot, 1998) also include *Caenopithecus* and *Mahgarita* in Adapidae. Adapids typically retain the primitive euprimate dental formula of 2.1.4.3 both above and below. Adapids are characterized by submolariform fourth premolars, elongate lower molars with metastylids, and squared upper molars with lingual cingula and distinct hypocones. The molars closely resemble those of some living lemurids. Adapid cheek teeth are dominated by crests, indicating that they were folivores. Most adapids were moderately large Eocene prosimians, weighing 1–4 kg (Fleagle, 1999), about the size of living *Hapalemur* and *Lemur*. *Microadapis* was somewhat smaller (about 600 g), whereas *Leptadapis* may have reached weights of 8–9 kg (Gingerich and Martin, 1981) and was markedly more robust than living strepsirrhines.

Adapis was the first named fossil primate, described by Cuvier in 1821, although it was not recognized as a primate until much later. Adapids are particularly well known from the famous late Eocene Quercy Phosphorite deposits in France, which have produced many skulls and isolated skeletal remains. Long ago, Stehlin (1912) observed that there was considerable diversity in the Quercy adapid sample, but subsequent studies (e.g., Gingerich, 1981c) concluded that much of the variation in overall size, canine size, and prominence of sagittal and nuchal crests could be explained by sexual dimorphism. More recent analyses of the Quercy adapids again concluded that they exhibit considerable morphological and locomotor diversity, representing several species of the genera *Adapis, Leptadapis,* and *Palaeolemur* (Lanèque, 1992, 1993; Bacon and Godinot, 1998).

Postcrania from Quercy reflect a variety of quadrupedal arboreal locomotor patterns. Dagosto (1983) showed that *Adapis* and *Leptadapis* resemble lorises more than lemurs in most postcranial features, such as proximal and distal femoral anatomy and proportions of tarsal bones, suggesting that they were slow-climbing arboreal quadrupeds. Additional femora and tibiae from Quercy show variations suggesting that some species of *Adapis* were adapted for walking and running on branches, whereas others were perhaps better adapted for climbing (Bacon and Godinot, 1998). This inference is consistent with conclusions drawn from the wrist anatomy of *Adapis* (Hamrick, 1996). Both of these later studies concluded that adapids lack specializations typical of both vertical clinging and slow-climbing primates and are more accurately interpreted as generalized arboreal quadrupeds, similar to some small platyrrhines (Bacon and Godinot, 1998).

European adapids do not seem to have evolved from known earlier cercamoniines from Europe and were more likely immigrants from Africa or Asia in the middle Eocene (Godinot, 1998). *Adapoides* from the middle Eocene of China appears to be related to adapids, but it is more primitive in lacking metastylids on the lower molars and in having a small hypocone on the lingual cingulum (Beard et al., 1994). Its presence in deposits older than those yielding adapids in Europe is consistent with an Asian origin of the family. Adapids became extinct in Europe during the *Grande Coupure* (Fleagle, 1999).

Sivaladapidae. Very few adapoids or omomyids are known to have survived beyond the Eocene/Oligocene boundary, but the sivaladapids are an exception. Sivaladapids are mainly known from the late Miocene of southern and eastern Asia, where they are represented by three genera, the best known being *Sivaladapis* (Gingerich and Sahni, 1984). They are known only from dentitions, which are characterized by fully molariform fourth premolars, simple three-cusped upper molars with well-developed mesostyles and lingual cingula lacking either a hypocone or a pericone, and lower molars with the hypoconulid and entoconid closely appressed but separated by a distinct notch.

Although sivaladapids appear to be related to adapoids, their precise affinities within the group are unclear. Recent discoveries suggest that they may be closely allied with middle-late Eocene *Hoanghonius* and *Rencunius* through a stage like the late Eocene *Guangxilemur* (Qi and Beard, 1998; Marivaux et al., 2002). All three are poorly known forms from eastern China, based only on teeth or fragmentary dentitions. *Hoanghonius* has been variously called an omomyid or an adapid and has even figured in discussions of anthropoid origins. Adapid affinities seem most probable, but until better specimens are known its relationships will remain uncertain. *Hoanghonius* lacks a distinct paraconid on M_{2-3} and has a semimolariform P_4. M^1 has a complete lingual cingulum with pericone and hypocone cusps but no mesostyle. *Guangxilemur,* known from isolated teeth, seems to bridge the morphological gap between *Hoanghonius* and Miocene sivaladapids. It has a strong lingual cingulum with a pericone and a hypocone as well as a variably developed mesostyle. Late Eocene *Wailekia* from Krabi, Thailand, which was described as a basal anthropoid, is probably related to *Hoanghonius* (Qi and Beard, 1998).

Earliest Strepsirrhines

Several genera or species of Old World adapoids have been singled out as possible strepsirrhine ancestors. Nevertheless, in spite of the rich record of Eocene primates on the northern continents, no fossils that are clearly transitional between adapoids and crown strepsirrhines (true lemurs or lorises) are known. As mentioned earlier, molecular evidence in fact suggests that strepsirrhines diverged from other primates earlier than the oldest known euprimates (Yoder and Yang, 2004), yet until very recently, no demonstrable crown strepsirrhines were known from before the Miocene. A few recent discoveries, however, expand the temporal range of both lemuroid and lorisoid primates back to the Early Tertiary.

Lemuroid primates today are restricted to Madagascar and the Comoro Islands, and the only known fossils are from the Quaternary of that island (see the review by Godfrey and Jungers, 2002). Thus the discovery of what appears to be an early Oligocene lemuroid in Pakistan was quite unexpected. *Bugtilemur* (Fig. 10.22B) is based on isolated upper and lower teeth from the Bugti Hills of Balochistan (Marivaux et al., 2001). These teeth resemble those of extant cheirogaleid lemurs in details of molar structure, such as the presence of an oblique, mesiodistally compressed trigonid with a reduced paraconid and a long talonid basin with a very buccal cristid obliqua. The upper molars lack a hypocone and are mesiodistally elongate with a very narrow stylar shelf. The lower canine is relatively long and laterally compressed, suggesting that a tooth-comb may have been present; however, it was not yet fully developed, because the tooth is not as elongate or as procumbent as in extant lemuriforms.

Prior to the recent discovery of Eocene lorisoids, the oldest unequivocal lorisoids (lorises, pottos, and galagos) came from the early Miocene of eastern Africa (Phillips and Walker, 2002). Earlier reports of possible Eocene Lorisoidea (=Loroidea) from Europe and the Fayum of Egypt have not been confirmed; instead those specimens probably represent adapids or *Plesiopithecus* (see below), or perhaps an indeterminate primitive anthropoid. New dental material from late middle Eocene strata of the Fayum now demonstrates the existence of unequivocal lorisoids 20 million years earlier than any known previously. Three genera have been described, the galagids *Wadilemur* (initially described as an adapoid) and *Saharagalago,* and a possible lorisid, *Karanisia* (Seiffert et al., 2003, 2005; Fig. 10.22C). The upper molars resemble those of extant lorisoids in having a large trigon basin and a prominent posterolingual cingulum with a distinct hypocone. The lower molars have an oblique protocristid and a short paracristid with a buccally displaced paraconid. The lower canine of *Karanisia* is long and slender (more so than that of *Bugtilemur*), as in tooth-combed prosimians and, just as in living forms, it preserves microscopic grooves on its margins caused by hair-grooming.

Plesiopithecus (Fig. 10.22A), from the late Eocene quarry L-41 in the Fayum, is one of the most distinctive Eocene prosimians. It is represented by a skull and mandible comparable in size to those of a slow loris. The mandible contains an enlarged, procumbent anterior tooth that could be either an incisor or a canine (in which case there would be no incisors) and four simple teeth between the anterior tooth and the molars (either four premolars or a reduced canine and three premolars; Simons and Rasmussen, 1994). The lower molars compare closely with those of lorisoids, but they also resemble those of anthropoids in the reduction of the paraconid. The uppers lack a hypocone but have a prominent posterolingual cingulum, as in lorisoids (which usually have a hypocone as well). The third molars are reduced. The orbits are large, suggesting nocturnal habits. Based on cranial anatomy and molar morphology, *Plesiopithecus* has been interpreted as a strepsirrhine with similarities to lorisoids (Simons and Rasmussen, 1994). At the same time, it may differ from lorisoids in having four premolars, but the dental formula is ambiguous. *Plesiopithecus* differs from anthropoids in lacking postorbital closure and fusion of the mandibular symphysis. It seems to represent an early strepsirrhine branch without close ties to extant primates (Rasmussen and Nekaris, 1998).

Tarsiiformes

The tarsiiforms comprise two families, the Tarsiidae, which includes the extant genus *Tarsius,* and the much more diversified Early Tertiary Omomyidae. Omomyids are known only from the Eocene and Oligocene, whereas Tarsiidae are first known from the middle Eocene and survive today, with practically no fossil record from intervening strata.

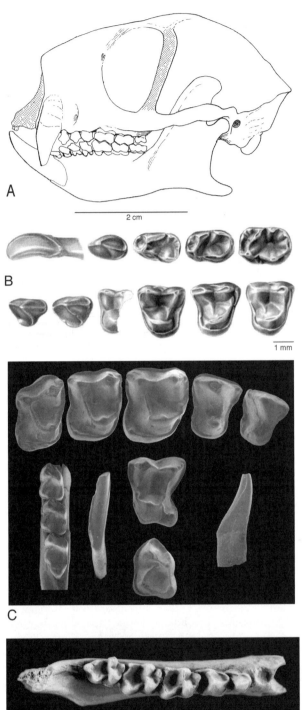

Fig. 10.22. Fossil strepsirrhines and tarsier: (A) *Plesiopithecus*, skull; (B) *Bugtilemur*, lower and upper teeth; (C) lorisoids *Karanisia* (right P^3–M^3, left M_{1-3}, and left I_1) and *Saharagalago* (single left upper and lower molars between views of I_1, not to scale); (D) Eocene tarsiid *Xanthorhysis*, right dentary with P_3–M_3. (A from Simons and Rasmussen, 1994; B from Marivaux et al., 2001; C from Seiffert et al., 2003; D from Beard, 1998b.)

Omomyidae. The Omomyidae consists of 35–40 genera and 85–90 species of Early Cenozoic primates. They flourished in the Eocene, particularly in North America, but they were also common in Europe and are known from Asia and possibly northern Africa as well. Most omomyids were an order of magnitude smaller than contemporary adapoids, weighing 50–500 g, but a few species reached weights of 1–2.5 kg (Gingerich, 1981a; Fleagle, 1999). Omomyids are commonly assigned to one of three subfamilies, the mainly North American Anaptomorphinae and Omomyinae, and the exclusively European Microchoerinae, which some primate experts regard as a distinct family.

The dentition of omomyids resembles that of living tarsiers, except for the anterior teeth. In omomyids the canines are typically relatively reduced and the anterior incisor is often enlarged, whereas in tarsiers this situation is reversed. The most primitive omomyids, Holarctic *Teilhardina* and North American *Steinius*, had the same lower dental formula as in basal adapoids (2.1.4.3), and the teeth show only subtle differences from those of the most primitive adapoid *Donrussellia*, corroborating a close common ancestry (Fig. 10.12). Omomyids have relatively simple, low-crowned molars, the uppers with three primary cusps and smaller, distinct conules. A nannopithex fold often runs posteriorly from the protocone, but the hypocone is primitively small or absent and when present arises from the postcingulum. The stylar shelf is usually narrow and relatively featureless, but a mesostyle is present in some species. The lower molars tend to be shorter and broader than in adapoids, with compressed trigonids and large talonid basins. The last molar is reduced in some lineages but not in others. The incisors are typically pointed and procumbent, and I_1 is usually somewhat, and often substantially, larger than I_2. The canines are reduced in all but the most primitive species, and even in those they are relatively smaller than in adapoids. The antemolar teeth of omomyids are typically mesiodistally compacted and reduced in number. Because of their similar morphology, however, it is often difficult to ascertain which teeth have been lost. Four premolars are retained only in some species of *Teilhardina* and *Steinius*; all other omomyids reduce the number to two or three in each quadrant. P_1 and P_2 (when present) are typically reduced and one-rooted. The premolars generally have simple crowns, although P_4 is tall and pointed in some species and lower and semimolariform in others; P^{3-4} are essentially bicuspid and P^4 is sometimes enlarged. Molar enamel varies from smooth in some species to wrinkled (crenulated) in others. Many of these dental features seem to have evolved repeatedly among omomyids, making it a challenge to untangle their precise interrelationships.

Although most omomyid species are represented only by dentitions, the skull is known in several genera (early Eocene *Teilhardina*, *Tetonius*, and *Shoshonius*; middle Eocene *Omomys* and *Nannopithex*; and late Eocene *Necrolemur* and *Rooneyia*; Fig. 10.23, Plate 4.4). Compared to adapoids, omomyids had shorter faces, narrower snouts, and relatively larger orbits (Fig. 10.10). The orbits (except in *Rooneyia*) are comparable in size to those of galagos and other extant nocturnal prosimians, although only *Shoshonius* approaches tarsiers in orbital diameter (Martin, 1990; Beard et al., 1991). *Rooneyia* had somewhat smaller orbits and was probably diurnal. The auditory bulla is large and inflated in some omomyids (*Tetonius* and *Shoshonius*), and the ectotympanic ring (which supports the tympanic membrane) is extended laterally as a bony external auditory tube, as in *Tarsius* but unlike the free intrabullar tympanic ring of adapoids. Both intrabullar branches of the internal carotid artery (promontorial and stapedial) are present and are enclosed within bony tubes (e.g., Simons and Russell, 1960; Szalay, 1976; Beard and MacPhee, 1994), a relatively unusual condition that is typical of most adapoids as well. The mandibular symphysis of omomyids is never fused, in contrast to adapoids.

Postcranial anatomy is less well known in omomyids than in adapoids, but fragmentary remains (mostly isolated bones) have been attributed to a diversity of genera, including the anaptomorphines *Absarokius*, *Arapahovius*, *Teilhardina*, and *Tetonius*; the omomyines *Hemiacodon*, *Omomys*, and *Shoshonius*; and most microchoerines (especially *Necrolemur*). Based on these fossils, omomyids are interpreted to have been active arboreal quadrupeds adept at leaping, comparable in size and locomotor behavior to extant *Cheirogaleus* (dwarf lemurs) and *Galagoides demidoff* (Demidoff's bushbaby; Dagosto, 1993; Thalmann, 1994; Dagosto et al., 1999; Anemone and Covert, 2000). This behavior is indicated by their slender, elongate hind limbs; semicylindrical femoral head (approaching the morphology in galagos and tarsiers); and deep distal femur with its elevated, narrow, and well-defined patellar groove bounded laterally by a high, rounded rim. In most of these features omomyids resemble notharctids. Certain differences indicate that omomyids were better or more frequent leapers than were notharctids: the tarsal elements are longer and the tibia and fibula were slender, elongate, and either fused or closely appressed (indicating a long fibrous rather than synovial joint) for the distal third of their length. However, a long ischium and relatively longer forelimbs compared to hind limbs suggest that they were not as specialized as comparably sized living vertical clingers and leapers (the forelimbs of omomyids are about two-thirds as long as the hind limbs, compared to only half as long in *Tarsius* and *Galago*, and the femur of omomyids is relatively shorter compared to the humerus; Dagosto et al., 1999). Microchoerines are more specialized than anaptomorphines and omomyines in having a distally fused tibiofibula, a relatively shorter humerus (59% of femoral length vs. 65% in *Shoshonius*), and more elongate tarsals (Dagosto, 1985, 1993).

The oldest known omomyid, *Teilhardina*, belongs to the subfamily Anaptomorphinae and appears abruptly at the base of the Eocene in North America, Europe, and Asia (Szalay, 1976; Bown and Rose, 1987; Rose et al., 1994; Ni et al., 2004; Smith et al., in press). It is the only early primate genus known from all three northern continents. *Teilhardina* is more primitive than any other anaptomorphine in several dental features, including the presence of four premolars and a relatively unreduced canine in the earliest species. Most authors recognize at least 14 anaptomorphine genera.

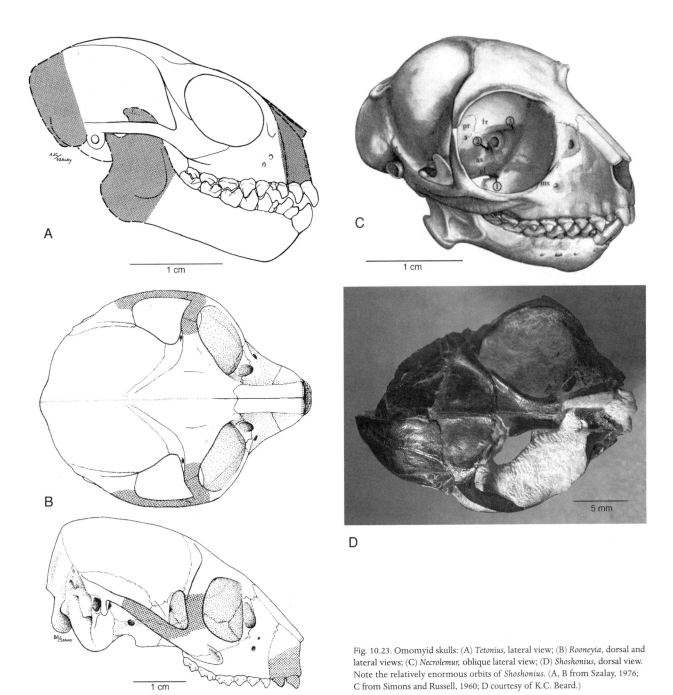

Fig. 10.23. Omomyid skulls: (A) *Tetonius*, lateral view; (B) *Rooneyia*, dorsal and lateral views; (C) *Necrolemur*, oblique lateral view; (D) *Shoshonius*, dorsal view. Note the relatively enormous orbits of *Shoshonius*. (A, B from Szalay, 1976; C from Simons and Russell, 1960; D courtesy of K.C. Beard.)

The cheek teeth tend to be basally inflated, with cusps set in from the margin. The anaptomorphines *Tetonius* (Fig. 10.24A), *Absarokius*, and *Anemorhysis* represent three distinct lineages that evolved from *Teilhardina* or a closely related form in western North America. Large samples of these anaptomorphines from the Bighorn Basin, Wyoming, show gradual transformation between nominal species in each lineage and document the independent evolution of several features characteristic of omomyid dentitions (Rose and Bown, 1984; Bown and Rose, 1987; Fig. 10.25). Recurring trends among anaptomorphines include the enlargement of I_1, enlargement or molarization of the fourth premolars, reduction in size and number of premolars and mesiodistal compression of the remaining anterior teeth, reduction of third molars, and wrinkling of enamel (Rose et al., 1994).

The subfamily Omomyinae is usually considered to include 15–20 genera, which generally differ from anaptomorphines in having a relatively smaller and less procumbent I_1, less reduction and compression of anterior teeth, relatively unreduced third molars, and little basal inflation of the molars, which results in more peripheral cusps (Szalay, 1976; Bown and Rose, 1987). The differences are subtle, however, and conflicting characters make assignment of several genera controversial. Early Eocene *Steinius*, the oldest and most primitive omomyine, appears in the fossil record at least a million years later than *Teilhardina* but is in some ways more

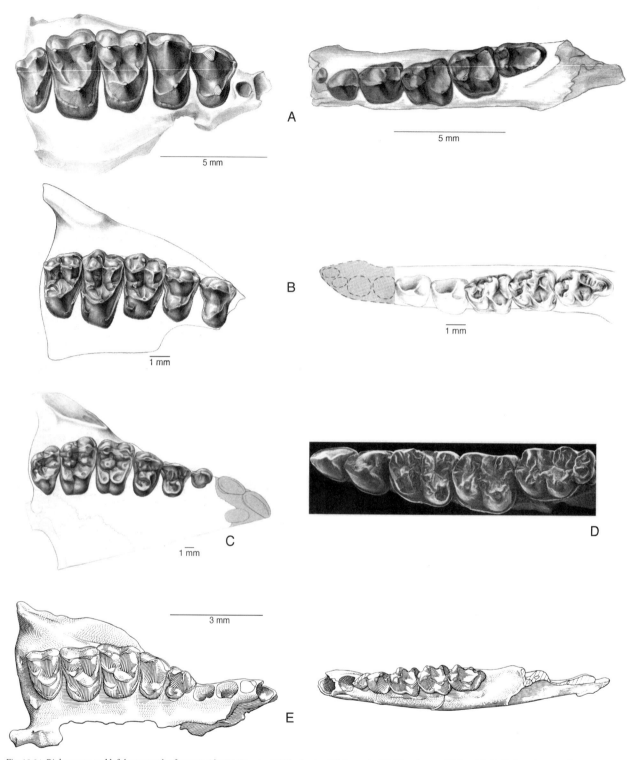

Fig. 10.24. Right upper and left lower teeth of omomyids: (A) *Tetonius;* (B) *Shoshonius;* (C) *Rooneyia;* (D) *Microchoerus;* (E) *Pseudoloris.* (A from Bown and Rose, 1987; B, C from Szalay, 1976; D from Hooker and Weidmann, 2000; E from Godinot, 1988.)

primitive (and in others more derived; Rose and Bown, 1991). It appears to be most closely related to *Omomys* and *Jemezius.* Omomyines were more diverse in size (ranging in weight from 100 g to 2.5 kg) and dental morphology than were anaptomorphines and largely supplanted the latter in middle and late Eocene faunas (Gunnell, 1995; Fleagle, 1999). Common trends in omomyines include development of molar mesostyles and metastylids (e.g., *Shoshonius;* Fig. 10.24B), crenulated enamel, and molarization of P_4. Although omomyids were predominantly North American, two omomyines with North American affinities are known from Asia (*Asiomomys* and a species of *Macrotarsius*).

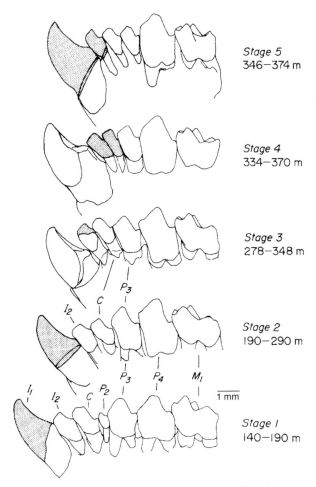

Fig. 10.25. Evolution of the lower dentition in the omomyid *Tetonius-Pseudotetonius* lineage. Stages represent typical morphs from specified stratigraphic intervals in the Bighorn Basin of Wyoming. Fossils document loss of P_2, gradual enlargement of I_1, and reduction of teeth between I_1 and P_4. (From Bown and Rose, 1987.)

Two of the latest occurring omomyids are also among the most problematic. Though usually classified as omomyines, their affinities are uncertain. Late Eocene *Rooneyia* (Figs. 10.23B, 10.24C) from Texas, known from a skull, has upper molars with rounded cusps and large hypocones. Aspects of its dental and cranial anatomy are unusual compared to other omomyids, and it is possible that *Rooneyia* is not an omomyid at all. Since its discovery it has been variously interpreted as a basal tarsiiform or a possible anthropoid, as well as an omomyid, and in a recent analysis it fell closest to adapoids (Kay et al., 2004). The second problematic taxon, late Oligocene *Ekgmowechashala* from South Dakota and Oregon, is unique in its low-crowned, bunodont teeth with accessory cusps and highly crenulated enamel. These peculiar features led McKenna (1990) to propose that *Ekgmowechashala* is a plagiomenid dermopteran rather than a primate. The relationships of these two aberrant forms are likely to remain controversial until more complete fossils are found.

The European Microchoerinae includes four or five genera of small middle and late Eocene omomyids with enlarged central incisors and a reduced dental formula (2.1.3.3/2.1.2.3). The primitive *Nannopithex* closely resembles North American anaptomorphines such as *Tetonius* and probably evolved from an anaptomorphine (e.g., Rose, 1995). Its upper molars lack a hypocone but exhibit the distinctive crest or "nannopithex fold" running distal to the protocone, which was named for this taxon. Later microchoerines (*Microchoerus* and *Necrolemur*; Fig. 10.24D) evolved bunodont molars with large cingular hypocones and crenulated enamel (in parallel with North American omomyids) and were probably frugivores (Fleagle, 1999). As noted above, microchoerines were more specialized for arboreal leaping than were other omomyids.

To assess the probable diets of omomyids, Strait (2001) used a combination of body mass and relative molar shearing crest length compared to these parameters in extant primates. Although several omomyids had been considered probable insectivores in earlier accounts, based on gross dental morphology, Strait's analysis placed nearly all anaptomorphines and microchoerines (with shorter total crest length) clearly in the range of extant frugivores. According to her analysis, most small omomyines (including *Omomys*, *Shoshonius*, and *Washakius*) and the tiny microchoerine *Pseudoloris* (<100 g; Fig. 10.24E) were probably faunivorous—that is, they ate insects and other small invertebrates.

Omomyids have been considered to be closely allied with or ancestral to living tarsiers, based on their relatively large orbits, tubular ectotympanic bone (external auditory meatus), and other cranial details, as well as their elongate tarsal bones and closely apposed or fused tibia and fibula. In all these characters omomyids contrast with adapoids and resemble tarsiers. They differ from tarsiers in other ways, including incisor and canine proportions, orbital structure, and endocranial arterial pattern, raising the possibility that the resemblances could be convergent. Nevertheless, one of the most recent phylogenetic analyses found that omomyids are a paraphyletic group and that crown haplorhines (Tarsiidae and Anthropoidea) are nested within Omomyidae (Kay et al., 2004).

Tarsiidae. Until fairly recently, the fossil record of tarsiiforms was virtually limited to omomyids. Since the mid-1980s, however, dental remains of several species that appear to be unequivocal tarsiids have been found. Two are known from the middle Eocene of China. Isolated teeth from southern China were described as a new species of the living genus *Tarsius*, demonstrating that tarsiers very similar to extant forms existed by the middle Eocene (Beard et al., 1994). *Tarsius eocaenus* is the smallest species of the genus and must have weighed much less than 100 g (extant tarsiers weigh about 80–150 g; Nowak, 1999). If the attribution to the living genus is correct, *Tarsius* is the most ancient extant primate genus and one of only a few living mammal genera known to extend as far back in time as the Eocene. A second genus of tarsiid, *Xanthorhysis* (Fig. 10.22D), was subsequently found in Shanxi Province (Beard, 1998b). It is within the size range of living tarsiers. *Xanthorhysis* differs from other tarsiids only in minor details, such as having slightly longer, narrower, and lower-crowned molars. It differs from omomyids in having a larger canine. *Xanthorhysis* is dentally not very different from the primitive anthropoid *Eosimias*, but unlike the latter it lacks an expanded anterolateral root of P_{3-4} and anteriorly shifted molar entoconids. In addition, the molar talonids of *Xanthorhysis* are wider than the molar trigonids, unlike *Eosimias*. The two Chinese Eocene tarsiers are evidence of the antiquity of Tarsiidae and suggest that the family may prove to be as ancient as Omomyidae and Adapoidea.

Afrotarsius, from the early Oligocene of the Fayum, Egypt, is based on a single fragmentary jaw with only three complete teeth, making its affinities difficult to interpret. It was described as a tarsiid (Simons and Bown, 1985) and would be the only known non-Asian tarsiid; but it was subsequently considered either the sister taxon of Anthropoidea (Fleagle and Kay, 1987; Kay and Williams, 1994) or a member of a new family of basal anthropoids (Ginsburg and Mein, 1987; Beard, 2002). In the most recent analysis, its position was ambiguous: it could be either a tarsiid or a stem haplorhine (Kay et al., 2004). Its molars resemble those of *Tarsius* and the probable basal anthropoid *Eosimias* in retaining a distinct paraconid and a hypoconulid well separated from the entoconid. Rasmussen et al. (1998) recently reported a fused tibia-fibula from the Fayum, which they attributed to *Afrotarsius*. If correctly allocated, it would provide strong evidence of tarsiid affinities for *Afrotarsius*, but its reference to *Afrotarsius* remains to be confirmed (Simons, 2003).

The omomyid genera *Necrolemur* and *Shoshonius* have sometimes been considered to be particularly close to tarsiid origins, or even referable to the Tarsiidae, in part because of their large orbits and specialized postcranial skeletons. As noted earlier, however, only in *Shoshonius* do the orbits approach the size of those in *Tarsius*. The tarsal modifications of both omomyid genera more closely resemble those of cheirogaleids, and their teeth are more similar to those of other omomyids. Consequently, the general consensus is that they are not true tarsiers.

Anthropoidea

The source of Anthropoidea (higher primates) has perennially been one of the most hotly debated issues in primate paleontology. Much progress has been made in the past decade or so, as a result of a wealth of exciting new fossils from northern Africa and eastern Asia. Nevertheless, specialists remain divided as to whether Anthropoidea are more closely related to adapoids, omomyids, or tarsiers (see recent discussions by Fleagle, 1999, and Dagosto, 2002). Anatomical features can be cited in support of each argument, but the polarities of some of these features are controversial, and a compelling case that would positively exclude one of these alternatives has yet to be made. Current evidence seems to favor a tarsiiform relationship. It is also possible that anthropoids represent another major clade tracing back to the origin of Primates, yet even so, anthropoids must be closer to one of the prosimian groups than the others. One conclusion that seems indisputable is that Anthropoidea evolved much earlier than was thought only a decade ago.

Earliest North African Anthropoids. Aside from the controversial late Paleocene *Altiatlasius* mentioned earlier, probably the oldest fossils that have been attributed to Anthropoidea are isolated molars from late early Eocene or early middle Eocene deposits of Algeria and Tunisia. Named *Algeripithecus* and *Tabelia* (see Fig. 10.27C), these molars are distinctive for their small size, indicating a body mass of less than 300 g, and their bunodonty (Godinot and Mahboubi, 1992, 1994), the latter feature suggesting to these authors that an adapoid origin of Anthropoidea is unlikely. These genera may be related to parapithecids, primitive anthropoids well known from the late Eocene–early Oligocene of Egypt. *Algeripithecus* and *Tabelia* are tantalizing in view of their antiquity, but it is difficult to make definitive arguments based on isolated teeth; hence their significance will remain moot until more complete evidence is found.

Early Asian Anthropoids. More complete but also controversial are several fossils from the middle and late Eocene of Asia. Particularly significant is *Eosimias* (Fig. 10.26A) from the middle Eocene of China, the oldest potential anthropoid known from good dental material. It is placed in its own family, Eosimiidae. Based on molar area, *Eosimias* was about the size of a small tarsier (90–180 g; Beard et al., 1996). It resembles primitive anthropoids in having small, vertical lower incisors with I_1 smaller than I_2, large canines, obliquely oriented P_{3-4}, molars with broad trigonids, and a moderately deep mandible with a deep (though unfused) symphysis and a rounded mandibular angle (Beard et al., 1994, 1996). *Eosimias* is more primitive than other anthropoids, however, in having well-developed molar paraconids and a distal hypoconulid (not twinned with the entoconid, as is typical in primitive anthropoids), features that give the molars a close resemblance to those of tarsiiforms. In addition, the incisor crowns are intermediate between the pointed crowns of some omomyids and the spatulate crowns of adapoids and most primitive anthropoids. Like *Algeripithecus*, the dental morphology of *Eosimias* seems to conflict with an adapoid ancestry of anthropoids (Beard et al., 1996). However, it has also led to speculation by some paleoprimatologists that *Eosimias* is not an anthropoid (e.g., Godinot, 1994; Simons, 1995b).

Tarsal elements attributed to *Eosimias* show a combination of features found in omomyids and early anthropoids and differ from those of adapoids (Gebo et al., 2000). Although many of these haplorhine traits could be primitive, at least a couple of them seem to be synapomorphies of anthropoids. The tarsal elements suggest that *Eosimias* evolved from an arboreal leaper but that, like anthropoids, it used horizontal postures more often than do prosimian leapers.

Both dental and tarsal evidence thus seems to link *Eosimias* with both Anthropoidea and Tarsiiformes. If *Eosimias* is a

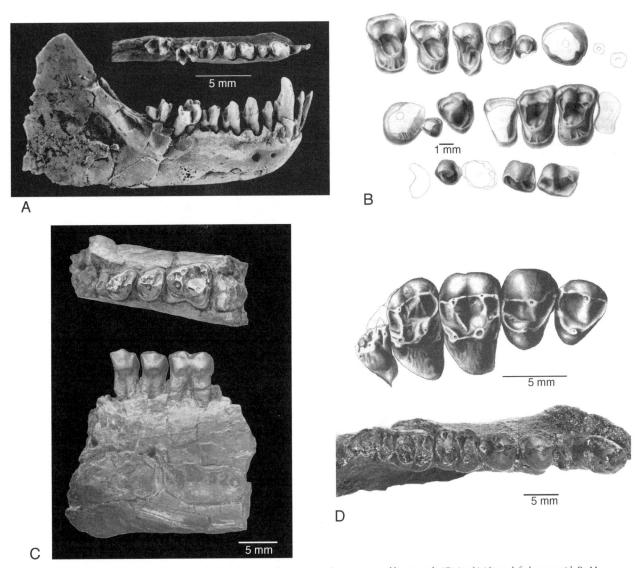

Fig. 10.26. Basal anthropoid dentitions from Asia: (A) *Eosimias*, right dentary; (B) *Bahinia*, upper and lower teeth; (C) *Amphipithecus*, left dentary with P_3–M_1; (D) *Siamopithecus*, right upper and lower teeth. (A courtesy of K.C. Beard; B from Jaeger et al., 1999; C courtesy of R.L. Ciochon; D courtesy of S. Ducrocq.)

basal anthropoid, its anatomy suggests that Anthropoidea is the sister group or descendant of Tarsiiformes. At the same time, these fossils indicate that *Eosimias* lacks some key anthropoid traits and is intermediate between omomyids (or tarsiiforms) and anthropoids in others. Additional fossils that would shed light on its affinities, such as well-preserved crania, are greatly needed.

Jaeger et al. (1999) described a second eosimiid genus, *Bahinia* (Fig. 10.26B), from the late middle Eocene of Myanmar (formerly Burma). It is based on associated upper and lower dentitions (dental formula 2.1.3.3, as in *Eosimias*) and represents a larger animal with an estimated weight of 400 g. Like *Eosimias*, *Bahinia* has simple tricuspid upper molars with a continuous lingual cingulum but no hypocone or distinct conules, slightly oblique P_3 and P_4, and large canines. M_1 retains a small paraconid. Although *Bahinia* is closely similar to *Eosimias*, it should be noted that several of these characters are plesiomorphic. A much smaller eosimiid, represented by a calcaneus comparable in size to that of *Eosimias*, has been reported from deposits of a similar age at Pondaung (Gebo et al., 2002). Recently Beard and Wang (2004) described additional eosimiid fossils from the late middle Eocene of China. They include a new genus, *Phenacopithecus*, and additional upper teeth that show resemblances to those of the most primitive known euprimates, suggesting a very ancient pedigree for Anthropoidea.

Other discoveries from the late middle Eocene of southeast Asia could represent another family of primitive anthropoids, Amphipithecidae, composed of *Amphipithecus*, *Pondaungia*, and *Siamopithecus* (Fig. 10.26C,D). Recent discoveries in Myanmar suggest that *Amphipithecus* and *Pondaungia* are sexual dimorphs of the same species (Jaeger et al., 2004). As noted earlier, there are two competing interpretations of these fossils. Amphipithecids have long been viewed as primitive anthropoids, based on their deep mandible and broad, flat molars, but they are quite distinct from eosimiids. If this hypothesis is correct, they may be related to the Oligocene anthropoid *Aegyptopithecus*, which would suggest that faunal exchange was ongoing between southeast Asia and northern Africa during the late Eocene (Ducrocq, 1999). Alternatively, as noted above, it has been argued on both dental and postcranial evidence that these genera are more closely related to notharctid adapoids. The matter seems insoluble based on existing fossils. Here, too, more complete specimens, preferably skulls or good skeletons, are needed to clarify relationships.

Fayum Anthropoids. The richest record of early anthropoids comes from the Fayum Depression, southwest of Cairo, Egypt, which has produced 11 genera and 18 species of primitive anthropoids from a 2.5-million-year sequence spanning the Eocene/Oligocene boundary (about 33.2–35.5 Ma; Kappelman et al., 1992; Simons, 1995c; Fleagle, 1999). These early anthropoids are assigned to at least four family-level groups, Parapithecidae, Propliopithecidae, Proteopithecidae, and Oligopithecidae, the latter sometimes considered a subfamily of propliopithecids. Despite this diversity, the Fayum anthropoids share a number of characters that secure their status as primitive anthropoids. The dental formula is either 2.1.3.3 or 2.1.2.3 both above and below. Where preserved, the incisors are spatulate and vertically implanted, I_1 is smaller than I_2, and I^1 is larger than I^2. The canines are sexually dimorphic. The lower premolars are slightly obliquely oriented and the anterior one tends to be larger, with a well-developed wear facet ("honing facet") that occludes with and sharpens the back of the upper canine. The molars are bunodont. The uppers usually have a well-developed lingual cingulum, which gives rise to a rounded hypocone, often quite large, and occasionally a pericone. The lower molar paraconids are typically weak or absent, the talonid basins tend to be broad, and the hypoconulid is sometimes twinned with the entoconid. The mandibular symphysis is usually fused, though not in the earliest form, *Catopithecus*.

Where skulls are known, they reveal the diagnostic anthropoid traits of postorbital closure (a bony partition separating the orbit from the temporal fossa) and fusion of the frontal bones (obliteration of the metopic suture; Simons, 1995a,b, 1997b; Simons and Rasmussen, 1996). The snout is of moderate length and the orbits are relatively small, indicating diurnal habits. The Fayum anthropoids resemble platyrrhines in having a bony tympanic ring fused to the lateral wall of the auditory bulla rather than the tubular ectotympanic characteristic of catarrhines and tarsiers. Inside the bulla, the promontorial artery tends to be large and the stapedial artery absent. The exact phylogenetic positions of the Fayum anthropoids continue to be debated, but there is widespread agreement that these fossils represent critical taxa from near the beginning of the anthropoid radiation, some belonging to the initial catarrhine diversification and others representing a stage before the platyrrhine-catarrhine split (i.e., outside crown anthropoids).

The oldest Fayum anthropoids published so far come from the late Eocene Quarry L-41 (the richest Eocene site in Africa), which has produced five different anthropoid genera. Best known is *Catopithecus* (Oligopithecidae; Fig. 10.27A,B), which is represented by numerous jaws, skulls, and a few fragmentary limb bones. *Catopithecus* was a tamarin-sized primate (400–800 g) with a catarrhine-like dental formula (2.1.2.3/2.1.2.3) in which only two premolars remained in each quadrant (Simons and Rasmussen, 1996). Its teeth resemble those of adapoids and extant squirrel monkeys (*Saimiri*), suggesting a mixed diet of fruit and insects. *Catopithecus* is more primitive than propliopithecids in having an unfused mandibular symphysis, a paraconid on M_1, and a relatively small brain with large olfactory bulbs. The degree of postorbital closure approximates that in platyrrhines. The slightly younger *Oligopithecus* is known only from fragmentary dentitions, which are similar in most respects to those of *Catopithecus*. Oligopithecids appear to occupy a position either near the base of catarrhines (e.g., Simons and Rasmussen, 1996; Rasmussen, 2002; Seiffert et al., 2004), or possibly near the base of anthropoids prior to the catarrhine-platyrrhine dichotomy (e.g., Ross et al., 1998).

Also from L-41 are two primitive genera constituting the family Proteopithecidae: *Proteopithecus* and *Serapia*. *Proteopithecus* is represented by skulls, lower jaws, and fragmentary postcrania, all of which resemble those of platyrrhines (Miller and Simons, 1997; Simons, 1997b; Simons and Seiffert, 1999). It was the size of a small marmoset. The dentition is primitive, with one more premolar than in *Catopithecus* (dental formula 2.1.3.3/2.1.3.3), transverse upper molars with a small hypocone, a small paraconid on M_1, and twinned hypoconulid-entoconid. As in *Catopithecus* (and anthropoids generally), the skull shows postorbital closure and complete fusion of the frontals. Hind limb elements, including a slender tibia slightly longer than the femur, indicate that *Proteopithecus* was an agile arboreal runner and leaper that was similar to a marmoset. Simons and Seiffert (1999) considered *Proteopithecus* to be the most probable sister taxon of Platyrrhini among all Fayum primates. *Serapia*, formerly considered a parapithecid, is very similar to *Proteopithecus* but slightly larger and probably belongs in this family (Simons et al., 2001). Kay et al. (2004) allocated both of these genera to the Parapithecidae.

Another L-41 anthropoid whose affinities are less clear is *Arsinoea*. Known only from a single lower jaw fragment, the teeth are generalized but distinctive in having lingual molar paraconids and relatively flat molar crowns, with crenulated enamel on M_2 (Simons, 1992; Simons et al., 2001).

Finally, L-41 has produced the oldest and most primitive known member of the Parapithecidae, *Abuqatrania*, based on two lower jaw fragments (Simons et al., 2001). Many authors have positioned parapithecids as basal Old World anthropoids (catarrhines), but Fleagle and Kay (1987) considered them to be the most primitive anthropoids, diverging from the anthropoid stem before the split between platyrrhines and catarrhines (see also Seiffert et al., 2004).

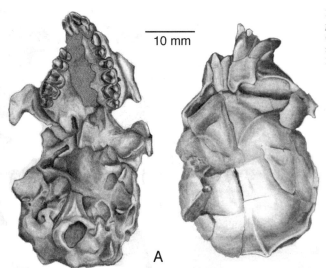
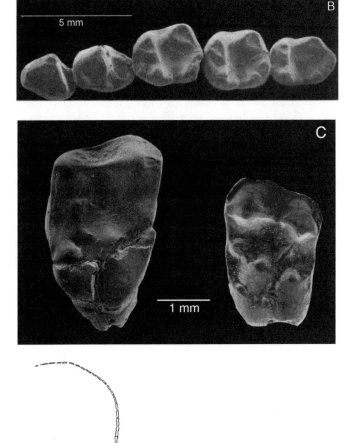
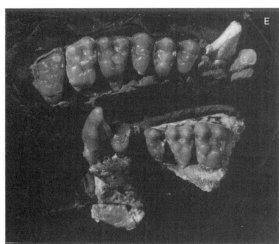
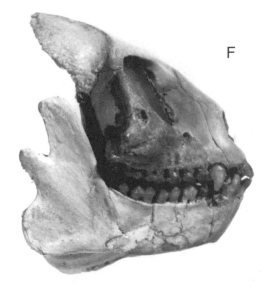

Fig. 10.27. Basal anthropoids from Africa: (A) *Catopithecus*, skull; (B) *Catopithecus*, right P_3–M_3; (C) *Tabelia* (left) and *Algeripithecus*, upper molars; (D) *Apidium*, skeletal restoration; (E) *Apidium*, upper dentition; (F) *Apidium*, restored skull. (A, B from Simons and Rasmussen, 1996; C courtesy of M. Godinot; D from Fleagle, 1999; E, F courtesy of E. L. Simons.)

If eosimiids are basal anthropoids, however, parapithecids would have to be considered as the most primitive African anthropoids. Also assigned to Parapithecidae are three other Fayum genera from higher strata, *Qatrania, Parapithecus,* and *Apidium* (Fig. 10.27D–F). *Apidium* is the most common Fayum primate and one of the best known. These genera ranged in size from about 300 g to 1.7 kg, roughly the size of *Saimiri*. Parapithecids are distinctive in much of their anatomy. They retain three premolars in both upper and lower jaws (2.1.3.3/2.1.3.3), the upper premolars bearing a central cusp between the paracone and protocone, and the lowers with a small, distal metaconid. The cheek teeth, especially the molars, are typically very bunodont with bulbous cusps and sometimes accessory cuspules, suggesting a frugivorous diet. The upper molars have large conules and hypocone, and the lowers may have a mesoconid on the cristid obliqua. The incisors are small and spatulate except in *Parapithecus,* which is unique in having lost the permanent incisors.

Much of the limb skeleton is known in early Oligocene *Apidium,* although from isolated elements. They show that *Apidium* had relatively short forelimbs (intermembral index 65), a longer tibia than femur (crural index 111), a deeper distal femur than any other anthropoid, and close appression of the tibia and fibula for the distal 40% of their length (Fleagle and Kay, 1987; Fleagle and Simons, 1995). These specializations indicate that *Apidium* was an active arboreal quadruped adapted for leaping from horizontal rather than vertical supports—that is, it was more monkeylike than prosimian-like. In anatomy and habits it was most similar to small platyrrhines like the squirrel monkey *Saimiri*.

The last family of Fayum anthropoids to appear in the record is the Propliopithecidae, the oldest unequivocal catarrhines, which are first known from several early Oligocene Fayum quarries. Two genera are usually recognized, *Propliopithecus* (including *Moeripithecus*) and *Aegyptopithecus* (Fig. 10.28). Both were much larger (4–8 kg; Fleagle, 1999) and anatomically more derived than the families that first appear at L-41. Propliopithecines have a catarrhine dental formula of 2.1.2.3/2.1.2.3. P_3 has a well-developed honing facet and P_4 is semimolariform with a large metaconid. The molars are bunodont and inflated, suggesting a primarily frugivorous diet. The upper molars have a large hypocone, and the hypoconulid of the lower molars is centrally situ-

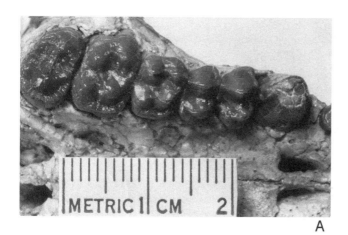
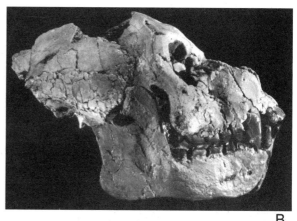
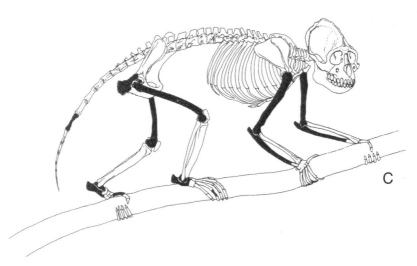

Fig. 10.28. Early Oligocene anthropoid *Aegyptopithecus:* (A) right upper dentition; (B) skull; (C) skeletal restoration; black elements are the only known postcrania. (A, B courtesy of E. L. Simons; C from Fleagle, 1999.)

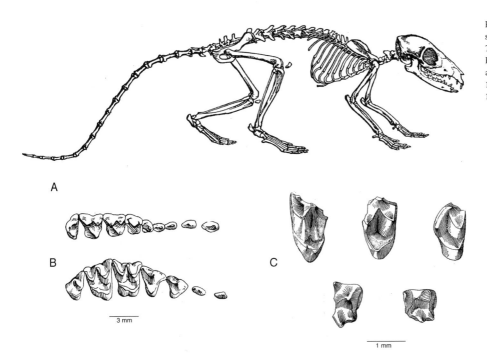

Fig. 10.29 Scandentia: (A) extant *Tupaia*, skeleton; (B) *Ptilocercus* (above) and *Tupaia*, right upper teeth; (C) middle Eocene *Eodendrogale*, fragmentary upper and lower molars. (A from Gregory, 1951, after de Blainville; B from Gregory, 1922; C from Tong, 1988.)

ated, not twinned with the entoconid, as in proteopithecids and oligopithecids. The mandibular symphysis is fused. The skull of *Aegyptopithecus* has full (catarrhine-like) postorbital closure, and the orbits are relatively small, indicating diurnal activity patterns. Males have a prominent sagittal crest, which enlarged with age. The brain was smaller and more primitive than in living catarrhines, however. Isolated limb bones of propliopithecids are platyrrhine-like and indicate that they were arboreal quadrupeds. *Propliopithecus* had a grasping foot, suggesting suspensory postures (Fleagle and Simons, 1982a; Fleagle, 1999). The larger *Aegyptopithecus* was similar in size and many aspects of limb anatomy to the howler monkey *Alouatta*, though more robust, and was probably a slow arboreal quadruped (Fleagle and Simons, 1982b; Simons, 1995c). Propliopithecines appear to lie near the base of crown catarrhines, closer to extant Old World monkeys and apes than are either proteopithecids or oligopithecids (Fleagle, 1999).

SCANDENTIA

The tree shrews, represented today by the southeast Asian family Tupaiidae (Fig. 10.29), were long considered primitive primates, based on studies by the eminent British anatomist Sir W.E. Le Gros Clark. Subsequent research, however, concluded that most resemblances to primates are either primitive or convergent (e.g., Luckett, 1980), and that tree shrews are not primates. The term Scandentia has been revived as an order for tupaiids. The skeleton of tree shrews is rather generalized and adapted for scansorial and arboreal locomotion. The dental formula is 2.1.3.3/3.1.3.3, in contrast to primitive primates, which have one more premolar. The upper incisors are caniniform, whereas the lower ones form a toothcomb superfically similar to that in strepsirrhine primates but excluding the canine. Tree shrews are omnivorous, preferring insects, and their molars have a primitive tribosphenic pattern.

Although the primitive anatomy of tupaiids suggests an ancient divergence from other archontans, almost nothing is known of the fossil record of tree shrews. Unequivocal tupaiids are not known until the Miocene. Only a single Paleogene species has been referred to Tupaiidae, *Eodendrogale* (Fig. 10.29C) from the middle Eocene of China, which is represented by just a few isolated teeth (Tong, 1988). The upper molars resemble those of extant *Dendrogale* in being dilambdodont and lacking a hypocone. Confirmation that *Eodendrogale* is a tupaiid, however, will require more conclusive evidence. Various other early Tertiary fossils, including mixodectids, microsyopids, and adapisoriculids, have at times been suggested to be either tupaiids or close relatives of tupaiids; but those hypotheses were usually based on plesiomorphic dental resemblance, and tupaiid affinities for any of them are very doubtful.

11

"Edentates"
Xenarthra and Pholidota

THE CONCEPT OF THE EDENTATA has varied considerably over the past two centuries. Cuvier originally used Edentata to encompass armadillos, anteaters, sloths, pangolins, and aardvarks—animals linked by a tendency to reduce or lose the teeth in association with a dietary preference for ants and termites. Most of these animals also possess a robust, fossorial skeleton. Subsequent authors added ground sloths, glyptodonts, taeniodonts, palaeanodonts, gondwanatheres, and various other taxa to the Edentata. Many of these were later removed from the concept of a natural Edentata. For example, aardvarks are assigned to the order Tubulidentata, taeniodonts to Taeniodonta, and gondwanatheres to an uncertain position unrelated to edentates. The interrelationships of the remaining groups—Xenarthra, Pholidota, and Palaeanodonta, as well as such extinct genera as *Eurotamandua* and *Ernanodon*—have been the subject of considerable debate (Table 11.1). Although the monophyly of each of the extant orders (Xenarthra and Pholidota) is not seriously in doubt (see reviews by Gaudin, 2003; Rose et al., 2005), the evidence that all these "remaining" taxa constitute a larger monophyletic group is very weak.

Throughout their history Xenarthra have been largely restricted to the Neotropics, although a diversity of xenarthrans immigrated into North America late in the Cenozoic. In contrast, Pholidota have been almost entirely restricted to the Old World, with present-day forms limited to the tropics of Asia and Africa. However, as we shall see, there are significant Early Tertiary exceptions, or possible exceptions, to both ranges. Palaeanodonts, an Early Tertiary group of small, fossorially adapted mammals with variously reduced dentitions, have been widely regarded to be closely related to Xenarthra or Pholidota, or both. Long known only from western North America, they are now known from Europe and Asia as well.

Table 11.1. Classification of "Edentates"

```
Order XENARTHRA
    Suborder CINGULATA
        Superfamily Dasypodoidea
            Dasypodidae
            †Peltephilidae
        †Superfamily Glyptodontoidea
            †Pampatheriidae
            †Palaeopeltidae
            †Glyptodontidae
    Suborder PILOSA
        Infraorder VERMILINGUA
            Myrmecophagidae¹
            Cyclopedidae
        Infraorder PHYLLOPHAGA
                †Pseudoglyptodon
            †Entelopidae
            Bradypodidae
            Parvorder †MYLODONTA
                †Scelidotheriidae
                †Mylodontidae
            Parvorder MEGATHERIA
                †Megatheriidae
                †Nothrotheriidae
                Megalonychidae
Mirorder CIMOLESTA
    Order PHOLIDOTA
        †Eomanidae
        †Patriomanidae
        Manidae
    ?Order PHOLIDOTA
        Suborder †PALAEANODONTA
            †Escavadodontidae
            †Epoicotheriidae
            †Metacheiromyidae
        Suborder †ERNANODONTA
            †Ernanodontidae
```

Notes: Modified after Rose et al., 2005. The dagger (†) denotes extinct taxa. Families and genera in boldface are known from the Paleocene or Eocene.

¹ Known from the middle Eocene if *Eurotamandua* belongs here (otherwise first known from the Miocene); alternatively, *Eurotamandua* may be an eomanid pholidotan.

In a series of papers, Novacek and colleagues (Novacek, 1986a; Novacek and Wyss, 1986; Novacek et al., 1988) proposed that Xenarthra, Pholidota, and possibly Palaeanodonta compose a monophyletic group Edentata, reviving an idea originated by Matthew (1918). Derived features shared by the three groups include the reduction or loss of teeth and a robust postcranial skeleton in which the limbs are short and the forelimbs have exaggerated crests and processes associated with powerful digging musculature. However, these resemblances could have arisen convergently in response to similar lifestyles. Novacek (1986a) and Novacek and Wyss (1986) added several skull features and one soft-tissue feature to the characters supporting a Xenarthra-Pholidota clade. Reevaluation of the anatomical evidence, however, led to the conclusion that most or all of these features are either ambiguous, inaccurate, of uncertain polarity, or homoplasious, thus casting doubt on a close relationship between Xenarthra and Pholidota (Rose and Emry, 1993). Still, the possibility of a relationship between xenarthrans and pholidotes, perhaps linked by palaeanodonts, cannot be entirely ruled out.

Both morphological and molecular data generally favor the interpretation that Xenarthra is a primitive eutherian clade that diverged very early from all other placentals, or Epitheria, whereas Pholidota and palaeanodonts belong to an unrelated epithere clade (Cimolesta) and are possibly allied with Carnivora (e.g., Emry, 1970; Honeycutt and Adkins, 1993; Rose and Emry, 1993; McKenna and Bell, 1997; Madsen et al., 2001). Some molecular evidence suggests that Xenarthra together with Afrotheria are the most primitive placentals (Madsen et al., 2001; Delsuc et al., 2002). The plesiomorphic nature of Xenarthra has been inferred from their physiology as well as several anatomical features, including the presence of ossified sternal ribs and a septomaxillary bone in the nose that is apparently homologous with that of monotremes and some Mesozoic mammals (McKenna, 1975a; Zeller et al., 1993). The primitiveness of these features, however, is open to question (Gaudin et al., 1996). No transitional fossil forms linking xenarthrans with any other mammalian group, with the possible exception of palaeanodonts, have ever been identified.

Palaeanodonts were widely accepted as primitive xenarthrans following early studies by Osborn (1904), Matthew (1918), and Simpson (1931a). Simpson interpreted them as an early clade that evolved from a basal xenarthran, in part because the oldest palaeanodonts then known were geologically too young (Eocene) to be directly ancestral to South American xenarthrans, which had already been found in the late Paleocene (Riochican) of Argentina. But palaeanodonts are now known that predate the oldest known xenarthrans, nullifying this objection to a possible palaeanodont ancestry for Xenarthra.

The notion that palaeanodonts were related to Pholidota, and not Xenarthra, emerged from studies of the oldest definitive pholidotans (Emry, 1970; Storch, 1978) and has been generally accepted, based on numerous derived traits shared by the two groups. In 1918, Matthew in fact proposed that palaeanodonts were related to both Xenarthra and Pholidota, a view concordant with Novacek's concept of Edentata. Characters of the auditory region, however, appear to be more consistent with Simpson's view of a close relationship between palaeanodonts and xenarthrans (Patterson et al., 1992). Recently adding to the debate are some new limb elements of the oldest known xenarthrans from Itaboraí, Brazil, which resemble those of palaeanodonts in certain derived features (Bergqvist et al., 2004).

Thus the relationships of palaeanodonts and other "edentates" are not fully resolved. Except for features of the ear region, however, the early pholidotans *Eomanis* and *Patriomanis* share many more features with palaeanodonts, especially Metacheiromyidae, than with xenarthrans, suggesting that palaeanodonts are the sister group (or even the direct ancestor) of Pholidota, as implied by Emry (Storch, 2003; Rose et al., 2005; Fig. 11.1).

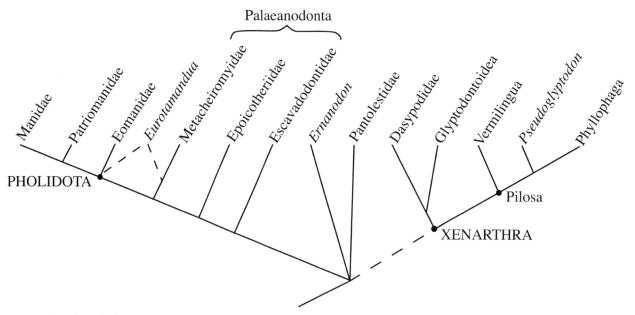

Fig. 11.1. Relationships of palaeanodonts, Pholidota, and Xenarthra. Dashed lines indicate the tenuous relationship of Xenarthra to other taxa and the uncertain affinities of *Eurotamandua* to Metacheiromyidae or Eomanidae. Pholidota, *sensu stricto*, is limited to living and fossil pangolins, but a broader concept of the order includes palaeanodonts. Leptictidae (not shown) may be related to the nonxenarthran clade on the left side of the diagram (pholidotes, palaeanodonts, pantolestids, and *Ernanodon*). (Modified after Gaudin, 2003, and Rose et al., 2005.)

XENARTHRA

Xenarthra consists of the armadillos and the extinct glyptodonts (suborder Cingulata), the living tree sloths and extinct ground sloths (Phyllophaga), and the anteaters (Vermilingua), the latter two groups often combined as sister taxa in the Pilosa. The unifying feature of Xenarthra is the presence of accessory joints (xenarthrous articulations) between adjacent vertebrae in the lumbar and sometimes in the lower thoracic region (Fig. 11.2), in addition to the zygapophyses found in all other mammals. This peculiar configuration occurs in all living and fossil xenarthrans except glyptodonts, in which the lumbar vertebrae are autapomorphously fused. It is least developed, but still evident, in sloths. No other mammals are xenarthrous, with the possible exception of the Jurassic noneutherian *Fruitafossor* (see Chapter 4). Hence it is assumed that the condition evolved only once (at least in Eutheria), in the last common ancestor of Xenarthra. Gaudin and Biewener (1992) suggested that xenarthrous vertebral joints originated as an adaptation for strengthening the vertebral column against extension and lateral bending associated with fossorial habits.

Another pervasive character of Xenarthra is the reduction or loss of teeth, as suggested by the name Edentata. Extant xenarthrans lack incisors and canine teeth, and the cheek teeth are simple homodont pegs that lack enamel. They are often reduced in number (as in sloths) or lost altogether (anteaters), but some armadillos have supernumerary teeth (as many as 20 in each quadrant in the giant armadillo *Priodontes*). In spite of the typical dental reduction, xenarthran diets are unusually varied. Only in vermilinguans, which have lost all teeth, is the diet restricted to ants and termites. Although some armadillos also prefer ants, most are omnivorous, and some types seem to have been more strictly herbivorous or carnivorous. Sloths are and were herbivorous (folivorous), whereas glyptodonts, which had hypsodont, lobate cheek teeth, are believed to have been grazers.

The skeleton of most xenarthrans is robust and specialized for digging. These characteristics are especially true of the forelimb, in which the manus is short and stout, with

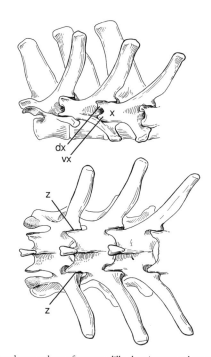

Fig. 11.2. Lumbar vertebrae of an armadillo showing normal zygapophyses (z) and xenarthrous joints (x, dx, vx). (From Rose and Emry, 1993.)

enlarged claws on two or three central digits. Also distinctive of xenarthrans is an expanded ischium with bony connections to the sacrum and often to the first few caudal vertebrae. The skeleton of some anteaters is secondarily modified for climbing, whereas living tree sloths have elongate and gracile limbs adapted for arboreal climbing and suspensory postures. Sloths are also unusual in being among the only mammals to have a variable number of cervical vertebrae (six to nine).

The oldest known xenarthrans come from the late Paleocene (Itaboraian-Riochican) of Brazil and Argentina. Though very fragmentary, the remains are diagnostic, consisting of dermal scutes (or osteoderms) and astragali already recognizable as those of armadillos (Scillato-Yané, 1976; Cifelli, 1983a). Glyptodonts, as well as mylodont and megatheriid sloths, also possessed dermal ossicles, and the presence of osteoderms in these oldest known xenarthrans strengthens this character as a diagnostic specialization of the order.

No xenarthrans have been reported from the rich early Paleocene site of Tiupampa, Bolivia. Their absence suggests the possibility that the stem xenarthran had not yet reached South America at that time, which would be consistent with a North American origin of Xenarthra and possible palaeanodont ties. Alternatively, it may simply mean that Xenarthra were not present or have not yet been found at Tiupampa but already existed elsewhere on the continent. The origin of Xenarthra remains one of the great mysteries of mammalian evolution.

Cingulata

Armadillos and glyptodonts are the only mammals protected by bony external armor, composed of a mosaic of dermal scutes overlain by keratinous epidermal scales. In armadillos, this carapace is flexible as a result of a variable number of mobile bands. Some of the scutes from the late Paleocene of Itaboraí, Brazil, have been assigned to the new dasypodid genus *Riostegotherium* (Oliveira and Bergqvist, 1998; Fig. 11.3A). There is enough variation in osteoderms and postcrania from Itaboraí and Riochican strata of Patagonia to indicate that a modest diversity of dasypodids already existed by the late Paleocene (Oliveira and Bergqvist, 1999). However, we still know virtually nothing about the origin of armadillos—when, where, or from what group they evolved. The only nonxenarthran fossils with possible ties are palaeanodonts, and the link remains weak at best.

It is in the Casamayoran (nominal early Eocene, but perhaps younger) of Argentina that the first well-preserved xenarthran remains are found, representing the dasypodid *Utaetus* (Simpson, 1948). It appears to have been very much like extant armadillos, particularly *Euphractus*, in already having xenarthrous vertebral articulations; a bony connection between the ischium and the sacrum (actually, "sacralized" caudals, sometimes called pseudosacrals); and ever-growing, peglike, cylindrical teeth with gabled occlusal wear (Fig. 11.3C). There were 10 lower teeth on each side, the first two much smaller and interpreted as incisors, the others prob-

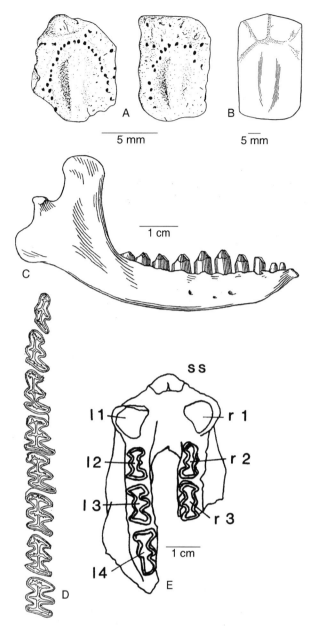

Fig. 11.3. Primitive cingulates (A–D) and sloth (E): (A) *Riostegotherium*, osteoderms; (B) *Machlydotherium*, osteoderm; (C) *Utaetus*, right dentary; (D) *Glyptodon*, lower teeth (crown view); (E) *Pseudoglyptodon*, mandible (from above). Key: l 1, left caniniform tooth; l 2–l 4, left cheek teeth; r 1, right caniniform tooth; r 2–r 3, right cheek teeth; ss, symphyseal spout. (A from Oliveira and Bergqvist, 1998; B, C from Simpson, 1948; D from Scott, 1903–1904; E from Engelmann, 1987.)

ably homologous with the canine and seven postcanines. Unlike later armadillos, however, *Utaetus* still had thin enamel variably present on the lingual and buccal surfaces of the teeth, and the cervical vertebrae were separate, not co-ossified. The skeleton was adapted for digging, as indicated by the large acromion process on the scapula and the prominent ulnar olecranon process. The caudal margin of the scapula is thickened, forming an incipient secondary spine.

Eocene (Casamayoran-Divisaderan) dasypodids were taxonomically relatively diverse, with as many as a dozen genera reported (Simpson, 1948; Vizcaíno, 1994; McKenna and

Bell, 1997). Most are represented only by scutes, however, so very little is known about them except that they were armored, like living armadillos. Not until the early Miocene are several kinds of armadillos known from reasonably complete skulls and skeletons. Later Tertiary armadillos achieved considerable diversity in size, skull shape, and dental adaptation.

Glyptodonts had an even more extensive and rigid bony carapace than armadillos. Many later Tertiary glyptodonts were gigantic and evolved bizarre skeletal modifications to support the armor. Glyptodontids were particularly diverse in the Miocene to Pleistocene, when about 60 genera existed.

The earliest definitive glyptodont (*Glyptatelus*) comes from the Mustersan (?middle Eocene) and, like the oldest dasypodids, is based on osteoderms (Simpson, 1948). Glyptodont osteoderms are readily distinguished from those of armadillos by their rosette-shaped pattern. A possible glyptodont astragalus was reported from the late Paleocene Itaboraian (Cifelli, 1983a), but Bergqvist and Oliveira (1995) believe that it lacks derived glyptodont features and is better identified as dasypodoid. Relatively complete glyptodonts are not known until the Miocene (Santacrucian; Scott, 1903–1904; Simpson, 1980). Santacrucian glyptodonts had eight cheek teeth in each jaw, all trilobate except the first two or three. Aside from their clearly derived hypsodont and enamelless crowns, these teeth are less reduced (more complex) than those of even the oldest known dasypodids and hence could be interpreted as more primitive. Therefore they might provide insight into the dental morphology of basal xenarthrans (Engelmann, 1987). It is also possible that this tooth morphology represents a derived condition.

Additional diversity of Eocene cingulates is indicated by two other kinds of bony plates that differ from those typical of dasypodids or glyptodonts. Relatively large and thick osteoderms from the Casamayoran and Mustersan have been assigned to *Machlydotherium* (Fig. 11.3B), whereas other Mustersan plates with almost no ornamentation are referred to *Palaeopeltis* (Simpson, 1948). Although named more than a century ago by Ameghino, both genera are still so poorly known that there is no consensus on their relationships. *Machlydotherium* has sometimes been considered a primitive pampatheriid, although those glyptodont relatives are otherwise unknown until the Miocene. *Palaeopeltis* would probably be regarded as a relative of armadillos based on its osteoderms, but the snout of Deseadan *Pseudorophodon*, now generally considered a synonym of *Palaeopeltis*, is dorsoventrally deep, like that of glyptodonts. It differs from all other xenarthrans in having a continuous arcade of cylindrical teeth, even in the front, where other xenarthrans have lost the teeth. Hoffstetter (1982) suggested that these peculiar genera might represent armored ground sloths.

Pilosa

Phyllophaga

The two living genera of sloths are the remnants of a once diverse and thriving radiation, with about 90 extinct genera known mainly from the Miocene through Pleistocene. Most extinct sloths—so-called "ground sloths," though not all were terrestrial—were large animals, and some Pleistocene megatheriid sloths were among the largest of all land mammals. The living two-toed sloth *Choloepus* is related to megalonychid sloths, whereas the three-toed sloth *Bradypus* is believed either to be related to megatheriids (Webb, 1985) or to be the sister group of all other sloths (Gaudin, 1995).

The oldest known probable fossil sloth is *Pseudoglyptodon*, from the Eocene/Oligocene boundary (Tinguirirican) of Chile and the late Oligocene (Deseadan) of Bolivia (Engelmann, 1987; Wyss et al., 1994). Like Miocene and later sloths, *Pseudoglyptodon* has a deep mandible with a strong, spoutlike symphysis. The dental formula is reduced to 5/4, with the anteriormost tooth caniniform. The generic name refers to the morphology of the cheek teeth, which are high-crowned and trilobed like those of glyptodonts, in contrast to the cylindrical, peglike teeth of later sloths (Fig. 11.3D,E). As in other xenarthrans the teeth lack enamel. The presence of relatively unreduced, lobate cheek teeth also in the early Miocene pilosan *Entelops* suggests that this morphology could represent the plesiomorphic dental condition for xenarthrans, or at least for pilosans (Engelmann, 1987). Otherwise, as distinctive as *Pseudoglyptodon* is, it offers little insight on the origin of Phyllophaga, and its relationships to later sloths are ambiguous.

The oldest forms that seem clearly related to later sloths are the Deseadan genera *Orophodon* and *Octodontotherium*. They have bilobed cheek teeth, like mylodontid sloths, and also share several derived cranial features with mylodontids (Patterson et al., 1992; Gaudin, 1995).

Older remains that seem to belong to pilosans, and possibly sloths, have recently been found in the middle Eocene La Meseta Formation of Seymour Island, Antarctica. These very fragmentary fossils include an ungual phalanx of a sloth, or possibly an anteater, and a hypselodont, enamelless caniniform tooth resembling that of a sloth (Carlini et al., 1990; Marenssi et al., 1994; Vizcaíno and Scillato-Yané, 1995). If their identity and age can be corroborated, these fossils would not only confirm the presence of xenarthrans in Antarctica but would also extend the range of sloths farther back in the Eocene.

Asiabradypus, based on a jaw fragment from the late Paleocene of Kazakhstan, was interpreted as a primitive sloth (Nessov, 1987). The fragment has large alveoli and two apparently simple teeth. If phyllophagan affinities were confirmed, it would be by far the oldest sloth and the only one known from the Old World. Regardless of its unanticipated stratigraphic and geographic occurrence, it is much smaller than the jaws of known sloths, and the simple teeth, upon reexamination, appear to be broken tooth roots. Consequently it is very unlikely that *Asiabradypus* represents a xenarthran.

Vermilingua and Eurotamandua

The anteaters are a small xenarthran clade that appears to have been geographically limited to South America and

Central America throughout its history—with one prominent possible exception. The oldest Neotropical anteaters appear in the early Miocene (Santacrucian *Protamandua*), already looking much like extant forms. Their origin and early history is unknown—or seemed to be until the discovery of an apparent anteater in middle Eocene deposits of Messel, Germany. *Eurotamandua*, based on a complete, articulated skeleton from Messel (Fig. 11.4, Plate 7.1), has been interpreted as the oldest known anteater and was even placed in the same family as the three living genera, Myrmecophagidae (Storch, 1981). It resembles the living anteater *Tamandua* in having a tubular, toothless skull, and robust limbs with large foreclaws. According to Storch, *Eurotamandua* possesses several other traits considered diagnostic of Xenarthra, including a large facial exposure of the lacrimal bone, an accessory auditory bulla, a secondary scapular spine, and ischiosacral synostosis. Some of these features are ambiguous, however, and contrary to initial reports, *Eurotamandua* does not possess xenarthrous vertebral joints, the key synapomorphy of the order (Szalay and Schrenk, 1998; Storch, 2003).

The presence of a true xenarthran in the middle Eocene of Europe—especially one apparently so similar to recent forms—is unexpected, to say the least. Not only does *Eurotamandua* pose an obvious paleobiogeographic dilemma as the only known Old World xenarthran (although some other vertebrates from Messel are also thought to have Neotropical affinities; see Peters and Storch, 1993), it also seems chronologically out of place, for Vermilingua are otherwise unknown before the Miocene.

Consequently, the xenarthran status of *Eurotamandua* is highly controversial. It might seem that a complete skeleton should be relatively easy to interpret, but like other Messel skeletons, that of *Eurotamandua* was compressed into oil shale (and is now embedded in epoxy resin for preservation), crushing some bones, distorting others, and obscuring many critical details. Most joint surfaces are only partly visible because the skeleton is articulated. Radiographic study has helped but has not solved all questions. In much of the skeleton *Eurotamandua* resembles palaeanodonts, as well as the primitive pholidotan *Eomanis*, also from Messel, suggesting a closer relationship to one or both of these taxa than to Xenarthra (Rose, 1988, 1999b; Shoshani et al., 1997). Similarities include numerous synapomorphic features of the limb skeleton. This resemblance is underscored by the recent reinterpretation of the skeleton of *Eomanis krebsi* as a juvenile specimen of *Eurotamandua* (Szalay and Schrenk, 1998). After restudy of the skeleton of *Eurotamandua*, however, Szalay and Schrenk (1998) rejected any special relationship between *Eurotamandua* and Xenarthra, Pholidota, or Palaeanodonta and proposed instead that *Eurotamandua* belongs to a new order, Afredentata, whose affinities are uncertain but might be with Xenarthra or Palaeanodonta. Thus, despite the existence of extraordinarily complete skeletons of *Eurotamandua* and similar forms from Messel, detailed studies have not led to a consensus as to its phylogenetic

Fig. 11.4. *Eurotamandua* from the middle Eocene of Messel, Germany. (From Storch and Richter, 1992.)

position. It seems that the debate may not be resolved unless additional skeletons of *Eurotamandua* are found.

PHOLIDOTA

Pholidota is one of the smallest and most unusual orders of mammals. Pholidota in the strict sense includes only the bizarre scaly anteaters, or pangolins, comprising but a handful of genera. A single family, Manidae, is recognized for the extant forms, and the seven or eight living species are variously included in from one to four genera, some of which are also known from late Tertiary or Quaternary fossils (McKenna and Bell, 1997; Wilson and Reeder, 2005). They live today in sub-Saharan Africa and southeast Asia. The three recognized Tertiary genera of pangolins (Eocene *Eomanis* and *Patriomanis,* and Oligo-Miocene *Necromanis*) were assigned to a new family, Patriomanidae, by Szalay and Schrenk (1998). Storch (2003) recently separated *Eomanis* in the new family Eomanidae. The extinct suborder Palaeanodonta appears to be related and is tentatively included in Pholidota.

Living pangolins include both terrestrial and arboreal forms. All are heavily built and myrmecophagous, which has resulted in a number of highly distinctive anatomical features. Among them are a rather smooth, tubular skull; edentulous jaws; a very reduced mandible lacking coronoid and angular processes; and an extraordinarily long, protrusile tongue that originates on the xiphisternum (Grassé, 1955b). The skeleton is fossorially adapted for tearing into ant and termite colonies, and the body has an outer covering of keratinous, imbricated scales for protection from insects and predators. The forelimbs are stout, with a long, prominent deltopectoral crest; large entepicondyle; long olecranon process; and large, curved, deeply fissured ungual phalanges. Although the manus is pentadactyl, the lateral digits are much reduced, rendering it functionally tridactyl. The scaphoid and lunar are fused, as in carnivorans. In the hind limb the femur lacks a third trochanter, because the superficial gluteal muscles insert very distally, just above the lateral condyle. Unlike almost all other mammals, the astragalus has a circular, concave head (i.e., the navicular articulation). Although this trait appears to be a synapomorphy of the living genera, it is not diagnostic of Pholidota, because the astragalar head of Tertiary genera is subspherical to cylindrical, as in most other primitive placentals (Rose and Emry, 1993).

Eomanidae and Patriomanidae

Although fossil pholidotans are very rare, the order was evidently widespread across the Holarctic continents in the Eocene. The oldest known pholidotan is middle Eocene *Eomanis,* known from several skeletons from the oil shales of Messel (Storch, 1978; Storch and Richter, 1992; Fig. 11.5). *Eomanis* was slightly smaller than Recent manids but similar in having a tubular skull with delicate, edentulous dentaries, and robust forelimbs with large claw-bearing ungual phalanges adapted for digging. Keratinous scales were found with one skeleton, providing compelling evidence of pholidotan affinities (Koenigswald et al., 1981). *Eomanis* differs from other pholidotans, however, in retaining a clavicle and having a prominent acromion process on the scapula, a more expanded supinator crest on the humerus, unfused scaphoid and lunar bones in the wrist, a femoral third trochanter proximal to midshaft (reflecting more proximal attachment of superficial gluteal muscles), and unfissured ungual phalanges. Most of these are plesiomorphic characters.

The skull and jaws of *Eomanis* indicate that it should have been an obligate myrmecophage, yet surprisingly, only one of the several Messel skeletons that preserve gut contents reveals any insect chitin (Storch and Richter, 1992). Instead the gut contained substantial amounts of leaf fragments and other plant matter, a most improbable diet for *Eomanis*. Storch and Richter (1992) postulated that it may have initially been herbivorous but, lacking teeth, required comminuted vegetable matter, which it obtained from leaf-cutter ants; only later did it become myrmecophagous. This hypothesis does not explain why it evolved definitive myrmecophagous characters prior to such a diet. It seems more likely that plant material was ingested coincidently with ants or termites, whose remains were, for unknown reasons, not preserved. For the present, the predominance of plant material in the gut contents of *Eomanis* remains a mystery.

Eomanis closely resembles primitive palaeanodonts in much of its skeletal anatomy (but is more specialized in being completely edentulous), suggesting a close common ancestry or even an ancestor-descendant relationship. The possibility of such a relationship was bolstered by Storch's (2003) recent description of a thick medial ridge, or "medial buttress," on the back of the dentary of *Eomanis,* a feature previously considered diagnostic of palaeanodonts. As discussed above, the supposed anteater *Eurotamandua,* also from Messel, is closer in much of its anatomy to *Eomanis* and palaeanodonts than to xenarthrans. *Eurotamandua* could be a pholidotan as well.

Patriomanis (Fig. 11.6), from the latest Eocene (Chadronian) of North America, was described from fragmentary remains (Emry, 1970), but additional material, including a virtually complete, articulated skeleton with the skull and mandible, is now known (Gaudin and Emry, 2002; Emry, 2004). It is the only known New World pangolin. Like *Eomanis* and living pangolins, *Patriomanis* is completely edentulous. The palate is characterized by a shallow, longitudinal groove, a hallmark of myrmecophagous mammals, presumably associated with a protrusile tongue. *Patriomanis* is more derived than *Eomanis* in having several features characteristic of extant manids, including a spoutlike mandibular symphysis and a pair of laterally projecting, canine-like prongs (not teeth, but bony processes) on the mandibular symphysis (incipient in *Eomanis*); a fused scapholunar; fissured ungual phalanges; and embracing (interlocking) zygapophyses on the lumbar vertebrae. Nevertheless *Patri-*

Fig. 11.5. Primitive pholidotan *Eomanis* from the middle Eocene of Messel, Germany. (Photograph courtesy of G. Storch; restorations from Storch, 1978.)

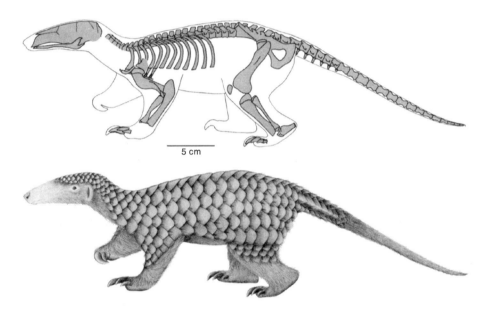

omanis is primitive compared to extant pangolins in a number of features, including retention of a third trochanter on the femur and a convex astragalar head (concave in manids).

An undescribed late Eocene pholidotan from Mongolia demonstrates that pangolins were also present in Asia early in their history (Rose and Emry, 1993; Gaudin and Emry, 2002). It is larger and more robust than *Patriomanis* and more primitive in having a more proximal femoral third trochanter. Like *Patriomanis*, the Mongolian pangolin shares several features with present-day pangolins, including a fused scapholunar, fissured unguals, and embracing lumbar zygapophyses.

It seems clear that by the end of the Eocene the basic pangolin anatomy was established. Pangolin skeletons changed relatively little after that, the most obvious modifications being appearance of a concave head (navicular facet) on the astragalus and further distal migration (and ultimate loss) of the third trochanter as the gluteal attachment moved to the distal femur.

Palaeanodonta

This intriguing Paleogene group includes some of the most derived fossorial mammals that ever existed. They were never abundant but were not uncommon and achieved modest diversity in North America; they are also known rarely from Europe and China. Nearly all palaeanodonts are placed in one of two families: Epoicotheriidae, which were dentally more conservative but evolved more fossorially specialized skeletons, or Metacheiromyidae, with greatly reduced dentitions but less modified skeletons (e.g., Simpson, 1931a; Rose et al., 1991). Relatively complete, articulated skeletons are known for several genera, suggesting that they were preserved in burrows. All are characterized by short, very robust limb skeletons and broad, low skulls that are wedge-shaped and have extensive lambdoidal crests for attachment of neck muscles. Where known, there is only a single, small lower incisor, and the edentulous mandibular symphysis is somewhat spoutlike, superficially resembling

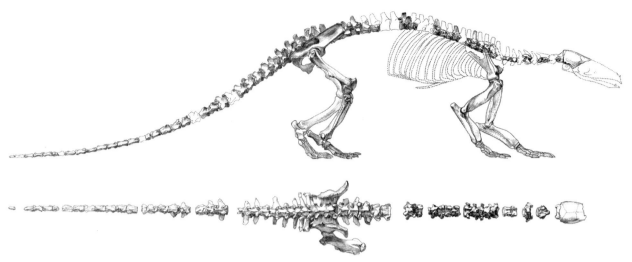

Fig. 11.6. *Patriomanis*, a late Eocene pholidotan from Wyoming. (Courtesy of R.J. Emry.)

that of sloths. Prominent canine teeth are always present, typically triangular in cross-section and developing distinctive honing wear facets. In most forms the postcanine teeth are reduced both in number and complexity. Palaeanodonts probably fed on insects and other small invertebrates, and some were probably myrmecophagous.

Escavadodon is the oldest and most primitive palaeanodont (Rose and Lucas, 2000; Fig. 11.7). Known from a single skeleton from the early Paleocene (Torrejonian) of the San Juan Basin, New Mexico, it was a small animal, about the size of the prairie dog *Cynomys*. Its postcranial skeleton shows fossorial adaptations similar to those of other palaeanodonts but less well developed. The dentition is primitive, with a large canine and seven postcanine teeth, including trituberculate molars and molariform posterior premolars reminiscent of those of Leptictidae. For these reasons, *Escavadodon* has been placed in a family of its own. Previous studies suggested that palaeanodonts may be closely related to or derived from pantolestan cimolestans (Szalay, 1977; Rose, 1978), based on similarities in tooth wear, tarsal anatomy, and particular fossorial specializations. But many aspects of the anatomy of *Escavadodon* seem to be intermediate between leptictids and palaeanodonts, suggesting that palaeanodonts may share a common ancestry with Leptictidae. The available evidence is most consistent with the interpretation that palaeanodonts and pantolestans are sister taxa, and that leptictids are more distantly related (Rose and Lucas, 2000).

In the late Paleocene (late Tiffanian) of Wyoming are found the oldest representatives of both Epoicotheriidae (*Amelotabes*) and Metacheiromyidae (*Propalaeanodon*), the former based on a lower jaw and the latter on a lower jaw and a humerus (Rose, 1978, 1979; Fig. 11.8). Both were primitive, like *Escavadodon*, in having a large canine followed by seven postcanines (presumably four premolars and three molars) in the lower jaw; but all the teeth are now separated by short diastemata, and the dentary is thickened posteromedially, derived traits of all other palaeanodonts. In *Amelotabes* all postcanines except P_1 are two-rooted. The preserved molars, though heavily worn, are still basically tribosphenic; however, the cusps are low and poorly defined, there are no cingula, and the cristid obliqua and postcristid are absent. The dentition is plesiomorphic relative to that of metacheiromyids, but later epoicotheres whose skeletons are known are postcranially more derived than metacheiromyids. *Propalaeanodon*, like subsequent metacheiromyids, had wider diastemata and smaller, peglike teeth, nearly all of which had single or bilobed roots. The humerus of *Propalaeanodon* (the only postcranial element known) possesses the distinctive elongate deltopectoral shelf and extensive supinator crest characteristic of palaeanodonts.

By the early Eocene (Wasatchian), epoicotheriids and metacheiromyids had further diverged dentally. Numerous skeletons are known from the Eocene, showing other differences between the two families. Metacheiromyids, including Clarkforkian-Wasatchian *Palaeanodon* (Fig. 11.8) and Bridgerian *Metacheiromys* (Fig. 11.7) and *Brachianodon*, were roughly the size of small to medium-sized armadillos (Simpson, 1931a; Gunnell and Gingerich, 1993). They had fewer postcanine teeth than in Paleocene palaeanodonts (five or fewer), and the teeth were reduced to simple pegs, so the distinction between premolars and molars is no longer obvious. The postcanines are concentrated just behind the canine, leaving the back of the ramus edentulous, although the length of this posterior diastema varies. *Metacheiromys* retained only a couple of vestigial postcanines (one in the maxilla, two in the dentary) and consequently was functionally edentulous and probably a committed myrmecophage. Metacheiromyids had a distinctly fossorial skeleton, including the following hallmarks: a large, bifid acromion process of the scapula; an elevated, shelflike deltopectoral crest extending about two-thirds the length of the humerus; a very broad distal humerus with projecting entepicondyle and broad supinator crest that may have a hooklike proximal extension; a long, medially inflected olecranon process; a distally expanded radius; short, stout metapodials and pha-

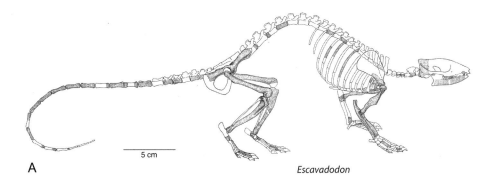

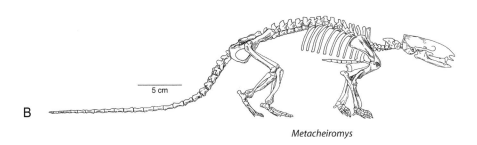

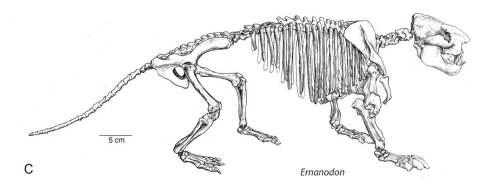

Fig. 11.7. Palaeanodonts and *Ernanodon*: (A) early Paleocene *Escavadodon*; (B) middle Eocene *Metacheiromys*; (C) late Paleocene *Ernanodon*. (A from Rose and Lucas, 2000; B after Simpson, 1931a; C from Ding, 1987.)

langes; and huge claw-bearing unguals. Metacheiromyids were long thought to be exclusively North American, but *Palaeanodon* has recently been recognized in the early Eocene of France (Gheerbrant, Rose, and Godinot, 2005). Among living mammals, the closest analogues of metacheiromyids are armadillos. The teeth are more reduced in metacheiromyids, however, which suggests that they were more strictly myrmecophagous.

Eocene epoicotheriids generally retained teeth to the back of the ramus and had more teeth than contemporaneous metacheiromyids. At the same time, epoicotheriid skeletons were more specialized for digging than those of metacheiromyids, which is manifested, for example, in relatively longer deltopectoral crests and olecranon processes and shorter manual metapodials and phalanges. *Tubulodon* (including *Alocodontulum* and *Pentapassalus;* see Fig. 11.10A) coexisted with *Palaeanodon* in the later Wasatchian and was about the same size but had relatively shorter limbs. It further differed in having six postcanine teeth, which were larger than those in *Palaeanodon* and retained traces of the original tribosphenic pattern. The enamel was very thin, however, and the cusps were often obliterated by heavy wear. *Auroratherium,* a primitive epoicothere that was dentally intermediate between *Amelotabes* and *Tubulodon,* was recently described from the early Eocene Wutu fauna of eastern China (Tong and Wang, 1997). It is the only palaeanodont known from Asia.

Later epoicotheriids reduced the postcanines to simple single-rooted pegs, usually five in number but still extending to the back of the ramus, in contrast to the configuration in metacheiromyids. An exception was the aberrant early Eocene *Dipassalus*, which, as its name suggests, had only two postcanines in the front half of the ramus (Rose et al., 1991). *Dipassalus,* Bridgerian *Tetrapassalus,* and Chadronian *Epoicotherium* were small palaeanodonts, about the size of the extant fairy armadillo *Chlamyphorus*. Chadronian (latest Eocene) *Xenocranium* was somewhat larger.

These late Eocene epoicotheriids evolved some of the most extreme fossorial skeletal adaptations known in mammals (Rose and Emry, 1983; Figs. 11.9, 11.10). In some of them the cervical vertebrae were fused, and the skull of *Xenocranium* had an upturned, spoon-shaped rostrum used

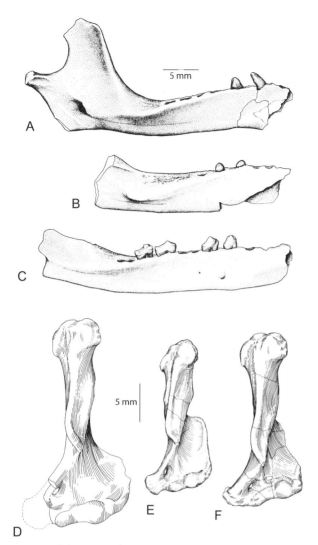

Fig. 11.8. Left dentaries (medial view, A–C) and left humeri (D–F) of palaeanodonts: (A) *Propalaeanodon;* (B) *Palaeanodon;* (C) *Amelotabes;* (D) *Palaeanodon;* (E) *Propalaeanodon;* (F) *Tubulodon.* Scale bars apply to all panels. (From Rose, 1979.)

for digging. The auditory region was specialized for low-frequency reception, presumably an adaptation for living underground. Forelimb crests and processes for muscle attachment were greatly exaggerated. In *Epoicotherium* and *Xenocranium* the deltopectoral crest extends almost to the distal end of the humerus, which is as wide across as the humerus is long. The entepicondyle projects far medially, and the supinator crest has an extraordinary hooklike process that extends proximally almost to the humeral head. The olecranon process is nearly as long as the entire ulnar shaft distal to the elbow joint. Additional modifications are present in the manus, including an enormous ungual phalanx on the central digit and a huge carpal sesamoid bone that increased the leverage of the digital flexor muscles. The tibia and fibula are fused at both ends, a condition found in many extant mammals that use the hind limbs for stability while digging. The anatomy of these small epoicotheriids strongly suggests that they were subterranean mammals that lived in burrows of their own construction and fed on grubs

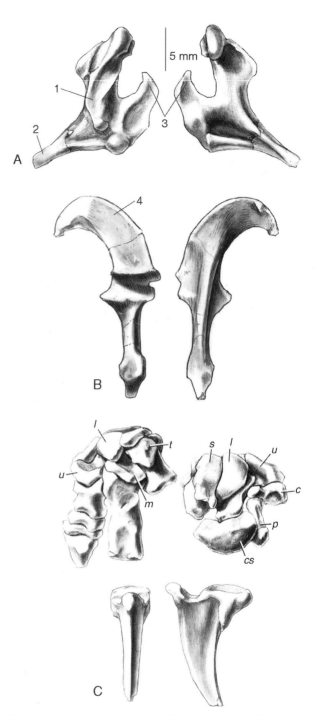

Fig. 11.9. Forelimb elements of specialized late Eocene palaeanodonts: (A) *Epoicotherium,* left humerus; (B) *Epoicotherium,* left ulna; (C) *Xenocranium,* partial right manus (dorsal and proximal views) and ungual phalanx of the middle digit (dorsal and lateral views). Key: 1, pectoral crest; 2, entepicondyle; 3, supinator crest; 4, olecranon; c, cuneiform; cs, carpal sesamoid; l, lunate; m, magnum; p, pisiform; s, scaphoid; t, trapezoid; u, unciform. (From Rose and Emry, 1983.)

and other subterranean invertebrates. Although they were molelike in lifestyle, their forelimb structure indicates that they dug by scratch-digging, unlike moles and similar to most other fossorial mammals.

An early Oligocene epoicotheriid very similar to *Epoicotherium* was recently described from Europe (Storch and

Fig. 11.10. Palaeanodonts: (A) *Tubulodon*, skull; (B) *Xenocranium*, restoration; (C) *Xenocranium*, skull. (A from Gazin, 1952; B, C from Rose and Emry, 1983.)

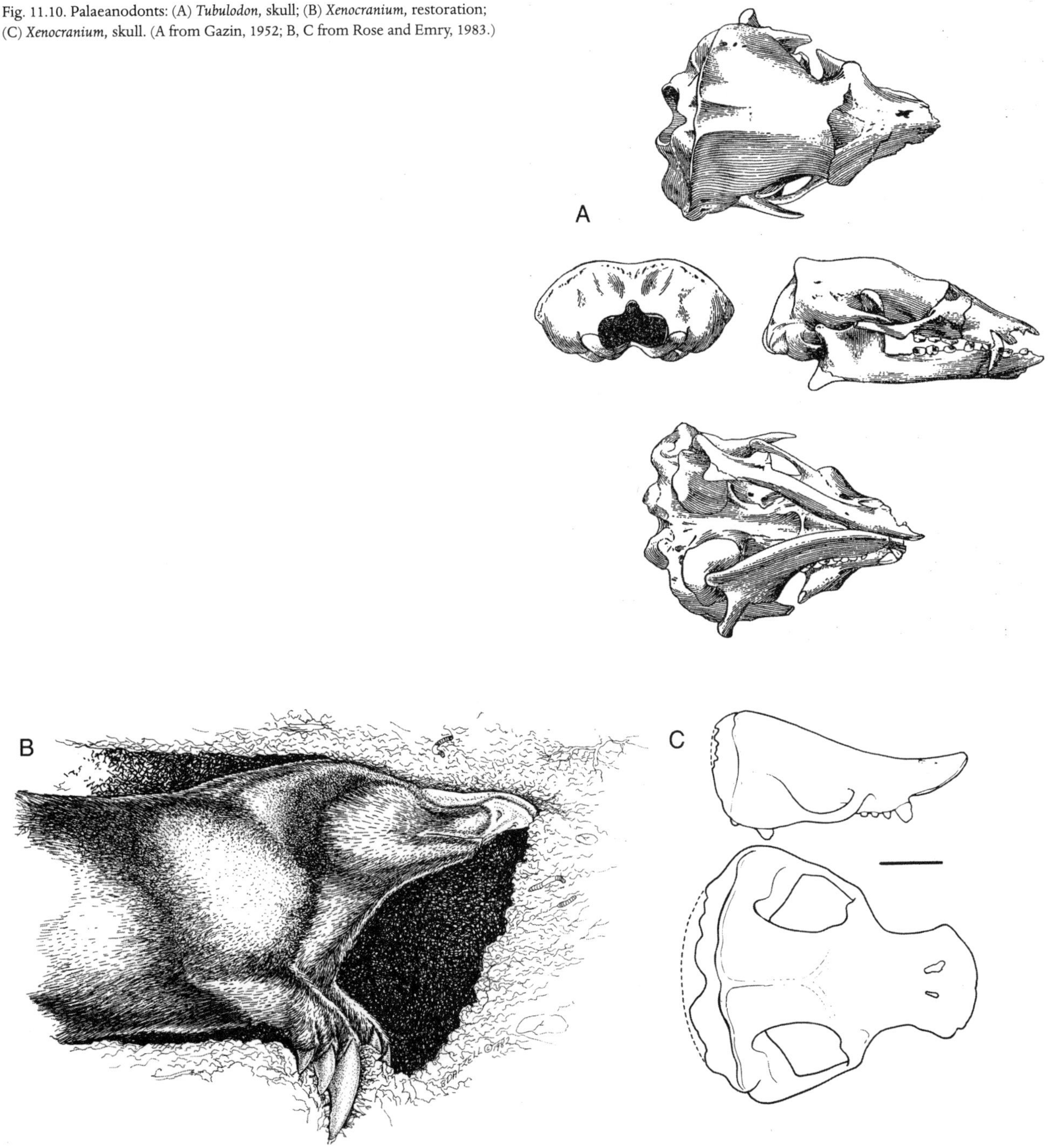

Rummel, 1999). *Molaetherium* is known from the dentary and the humerus, which are almost identical in size and form to those of *Epoicotherium*. The presence of such a subterranean form in Europe, well after the land connection with North America was interrupted, is a paleogeographic mystery. One hypothesis is that the North American and European forms evolved from a common Asian ancestor (Storch and Rummel, 1999), although no closely related form is known from Asia. The recent discoveries of *Palaeanodon* in France and a small palaeanodont (*Arcticanodon*) in late early Eocene deposits of Ellesmere Island (Rose et al., 2004) suggest that a North American source is equally or more likely.

Why palaeanodonts became extinct is also puzzling. Perhaps they were too narrowly adapted and were unable to adjust to environmental changes that accompanied the climatic shift near the Eocene/Oligocene boundary.

Ernanodonta

From the late Paleocene (Nongshanian-Gashatan) of Asia comes a bizarre mammal that was originally described as an edentate. *Ernanodon*, the sole member of its family and suborder, is known from a virtually complete skeleton from southern China (see Ding, 1987; Fig. 11.7) and partial skeletons from Mongolia (Kondrashov and Agadjanian, 2005). It is a little less than a meter in length (slightly larger than the largest known palaeanodont). It has a short, broad skull, a single lower incisor, large canines and single-rooted postcanines (except for a two-rooted M_2), and a spoutlike mandibular symphysis. The auditory bulla was apparently unossified. Ding (1987) reported the presence of ossified sternal ribs and incipiently developed xenarthrous processes, as in xenarthrans, but it has not been the demonstrated that *Ernanodon* had xenarthrous joints. The limb skeleton is very robust: the scapula has a very elevated spine and bifid acromion; the humerus has a broad, raised deltopectoral crest and very wide distal end with expanded supinator crest; the ulna has a prominent, inflected olecranon; and the metapodials and phalanges are short and broad. In most of these limb traits *Ernanodon* resembles palaeanodonts but is even more robust. The terminal phalanges of the manus are large and laterally compressed, but are more nearly the same size than in palaeanodonts, in which the third digit is usually larger than the others.

The phylogenetic position of *Ernanodon* is uncertain. Ding (1987) classified it in a new suborder of Edentata, which she used in the sense of Xenarthra only. McKenna and Bell (1997) placed it in a separate suborder of Cimolesta near the Pholidota, in which they included palaeanodonts. The unusual anatomy of *Ernanodon,* especially details of the forelimb, suggests that it is more closely related to palaeanodonts than to any other group.

Archaic Ungulates

THE COMPOSITION AND PHYLOGENETIC REALITY of a superordinal taxon Ungulata, for hoofed mammals, has been a topic of considerable debate over the past decade. The taxa that have conventionally been united in the Ungulata (Fig. 12.1A,B; Table 12.1) include the artiodactyls, perissodactyls, hyracoids, elephants (Proboscidea), and sirenians (the last three often grouped as Paenungulata), whales (Cetacea), and probably aardvarks (Tubulidentata), as well as the extinct desmostylians, arsinoitheres (Embrithopoda), South American ungulates (Meridiungulata), and the archaic ungulates (Condylarthra; e.g., Van Valen, 1978; Prothero et al., 1988; McKenna and Bell, 1997). Until quite recently, this conventional view of Ungulata was widely accepted. Other extinct groups have sometimes been added, including Tillodontia, Pantodonta, and Dinocerata, but their status as part of this clade is questionable (Lucas, 1993). Tillodonts and pantodonts were discussed earlier, with Cimolesta, whereas dinoceratans are included in this chapter.

Despite the general acceptance of a higher taxon Ungulata, it has been recognized that the monophyly of this group, as well as the branching sequence within it, are weakly based (e.g., Novacek, 1990). For example, synapomorphies that support Ungulata include only a small number of characters, some rather vague, such as a bunodont dentition with anteroposteriorly compressed trigonids; an astragalus with a short, robust head; an ectotympanic bulla (when a bulla is present); and a few other cranial features (Prothero et al., 1988). Most ungulates also share cursorial specializations of the postcranial skeleton, including the presence of hoofs, but these features were not yet present in some condylarths.

The traditional concept of Ungulata has been challenged recently by molecular evidence that unites African ungulates—elephants, sirenians, hyracoids, and aardvarks—

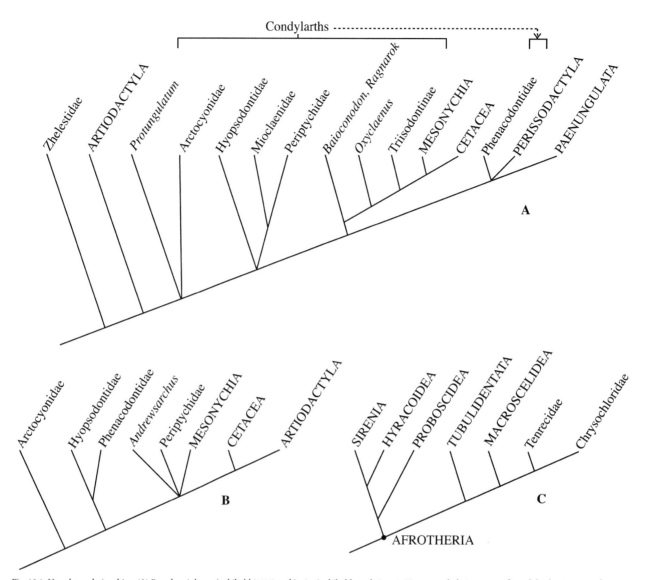

Fig. 12.1. Ungulate relationships. (A) Based mainly on Archibald (1998) and Janis, Archibald, et al. (1998). Note paraphyletic nature of condylarths. Paenungulata includes Hyracoidea, Sirenia, and Proboscidea. (B) Simplified after Thewissen et al. (2001b). (C) Afrothere relationships based on molecular data (mainly after Murphy et al., 2001; Springer et al., 2003). Meridiungulata and Tubulidentata were not included in the analyses summarized in A and B but would be nested within those cladograms. See Fig. 1.5 for broader relationships of Afrotheria based on molecular data.

in the clade Afrotheria together with elephant shrews (Macroscelidea) and the lipotyphlan families Chrysochloridae (golden moles) and Tenrecidae (tenrecs; see, for example, Stanhope et al., 1998; Madsen et al., 2001; Murphy et al., 2001; Van Dijk et al., 2001; Fig. 12.1C). Afrotheria excludes perissodactyls, artiodactyls, and whales, and of course does not consider extinct groups. This arrangement not only renders Ungulata polyphyletic but also breaks up the Lipotyphla.

Although molecular data supporting Afrotheria continue to accumulate, it is notable that no convincing morphological evidence corroborating this taxon has been found; only when molecular data are included in analyses is Afrotheria supported (e.g., Asher et al., 2003), probably, as mentioned earlier, because base pairs far outnumber morphological characters. In fact, the morphological evidence, both neontological and paleontological, still strongly supports the monophyly of several of the ungulate clades recognized by Prothero et al. (1988). The fossil record provides particularly compelling evidence for the monophyly of tethytheres (a subset of afrotheres composed of Proboscidea, Sirenia, and their extinct relatives) and still supports a close relationship between perissodactyls and Paenungulata (which consists of tethytheres + hyracoids; e.g., Hooker, 2005). Moreover, recently discovered fossil evidence of probable stem macroscelideans from the Paleocene and Eocene of western North America conflicts with a monophyletic Afrotheria and with the supposed African origin of the group (Zack et al., 2005; see the section on Hyopsodontidae below). Therefore, although it is important to remain open to new phylogenetic arrangements, the conventional classification of Ungulata is

Table 12.1. Synoptic classification of Ungulates, with emphasis on archaic ungulates

Grandorder UNGULATA
    Order †CONDYLARTHRA[1]
        †Arctocyonidae
        †Quettacyonidae
        †Periptychidae
        †Phenacodontidae
        †Hyopsodontidae
        †Mioclaenidae[2]
        †Didolodontidae[2]
    Order TUBULIDENTATA
    Order †DINOCERATA
        †Prodinoceratidae
        †Uintatheriidae
    Order †ARCTOSTYLOPIDA[3]
    Order ARTIODACTYLA
    Mirorder †MERIDIUNGULATA[4]
        Amilnedwardsia[5]
    Order †NOTOUNGULATA
      Suborder †NOTIOPROGONIA
        †Henricosborniidae
        †Notostylopidae
      Suborder †TOXODONTIA
        †Isotemnidae
        †Notohippidae
        †Leontiniidae
        †Toxodontidae
        †Homalodotheriidae
      Suborder TYPOTHERIA
        †Oldfieldthomasiidae
        †Interatheriidae
        †Archaeopithecidae
        †Mesotheriidae
        †Campanorcidae
      Suborder HEGETOTHERIA[6]
        †Archaeohyracidae
        †Hegetotheriidae
    Order †LITOPTERNA[2]
        †Protolipternidae
        †Notonychopidae[5]
        †Macraucheniidae
        †Adianthidae
        †Proterotheriidae
    Order †ASTRAPOTHERIA
        †Eoastrapostylopidae
        †Trigonostylopidae[7]
        †Astrapotheriidae
    Order †PYROTHERIA
        †Colombitheriidae
        †Pyrotheriidae
    Order †XENUNGULATA
        †Carodniidae
  Mirorder CETE
    Order †MESONYCHIA
    Order CETACEA
  Mirorder ALTUNGULATA
    Order PERISSODACTYLA
    Order PAENUNGULATA
      Suborder HYRACOIDEA
      Suborder TETHYTHERIA
        Infraorder †EMBRITHOPODA
        Infraorder SIRENIA
        Infraorder †DESMOSTYLIA
        Infraorder PROBOSCIDEA

Notes: Modified after McKenna and Bell (1997). See Tables 13.1, 14.1, and 14.2 for details of Altungulata, Cete, and Artiodactyla, respectively. The dagger (†) denotes extinct taxa. Families and genera in boldface are known from the Paleocene or Eocene.

[1] Condylarthra is almost certainly paraphyletic, as are several of its constituent families.
[2] Mioclaenids, didolodontids, and litopterns were united in a new order Panameriungulata by Muizon and Cifelli (2000).
[3] Possibly related to Notoungulata.
[4] Monophyly of Meridiungulata has not been demonstrated and is uncertain.
[5] Placed in the new order Notopterna by Soria (1989a,b).
[6] Probably nested within Typotheria.
[7] Placed in a separate order Trigonostylopoidea by Simpson (1967).

provisionally maintained here to underscore the morphological evidence and its current incongruence with molecular results.

OLDEST UNGULATE RELATIVES

The oldest fossils that have been thought to be related to ungulates are a series of genera called zhelestids (Fig. 12.2), based on jaws and teeth from the Late Cretaceous (about 85 Ma) of western Asia (Uzbekistan: *Aspanlestes, Eoungulatum, Kumsuperus, Parazhelestes, Sorlestes,* and *Zhelestes*), North America (*Alostera, Avitotherium,* and *Gallolestes*), and Europe (*Labes* and *Lainodon;* Archibald, 1996; Nessov et al., 1998). Zhelestids have relatively wide, basined, and low-crowned cheek teeth compared to those of most other Cretaceous mammals. These dental modifications are often thought to signal a shift away from a strictly insectivorous diet to a partly herbivorous one, and they characterize a diversity of

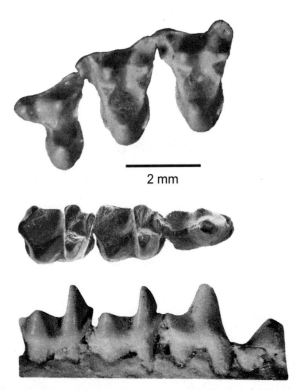

Fig. 12.2. Dentition of the zhelestid *Aspanlestes*. (From Nessov et al. 1998, courtesy of J. D. Archibald.)

archaic as well as more advanced ungulates that became abundant in the Early Tertiary. Archibald (1996) therefore proposed the higher taxon Ungulatomorpha to recognize the potential relationship between zhelestids and later ungulates. In view of the recent molecular evidence that Ungulata may be polyphyletic, Averianov and Archibald (2005) abandoned the name Ungulatomorpha. However, they retained Zhelestidae as a paraphyletic group of primitive Laurasiatheria, the molecular taxon that includes artiodactyls and perissodactyls but not paenungulates (see Fig. 1.5).

Nevertheless, these dental traits are also seen to some extent in some other Late Cretaceous eutherians, including Leptictida (*Gypsonictops*) and basal Lipotyphla (e.g., primitive erinaceomorphs, nyctitheriids), so the ungulate affinities of zhelestids are by no means certain. McKenna and Bell (1997) excluded zhelestids from relationship to ungulates, instead assigning most "zhelestid" genera to the Leptictida, and *Alostera* and *Avitotherium* to the Cimolesta. It seems plausible that some forms assigned here (e.g., *Aspanlestes*) could be primitive lipotyphlans. Zhelestids are more primitive than ungulates in having five premolars and a low mandibular condyle. Limb elements thought to belong to zhelestids, currently under study by F. S. Szalay and colleagues, may help to clarify their phylogenetic position.

Uncontested fossil Ungulata are first known from the earliest Puercan (basal Paleocene) of northeastern Montana and are allocated to the genera *Protungulatum*, *Oxyprimus*, *Baioconodon* (and the closely allied or synonymous *Ragnarok*), and *Mimatuta* (Van Valen, 1978; Archibald, 1982, 1998; Archibald and Lofgren, 1990; Luo, 1991). They are the oldest and most primitive members of the Condylarthra, an admittedly paraphyletic assemblage of basal ungulates. The first three are usually assigned to the Arctocyonidae (Van Valen, 1978; Luo, 1991, Muizon and Cifelli, 2000), which is the basal family of condylarths, whereas *Mimatuta* has been considered the basal member of the Periptychidae.

The age of these primitive ungulates is often said to be equivocal (either latest Cretaceous or earliest Paleocene), because of uncertainty about the age of the original sites yielding the fossils, the famous Bug Creek and Harbicht Hill localities in Montana. Although these sites were long believed to be of latest Cretaceous age, Lofgren (1995) has provided compelling evidence that they were deposited during the early Paleocene and contain a time-averaged assemblage of early Puercan and reworked Late Cretaceous fossils. In other localities not subject to the time averaging of these channel deposits, the archaic ungulate genera are present only above the K/T boundary. Consequently, they can be confidently dated as earliest Paleocene, and the first appearance of *Protungulatum* is now used to define the beginning of the Puercan North American Land-Mammal Age (Cifelli et al., 2004; Lofgren et al., 2004).

Like zhelestids, these early ungulates are known primarily from teeth (mostly molars), which are bunodont and low-crowned, the lowers with relatively well-developed talonid basins. The upper molars have a relatively narrow stylar shelf, strong conules, and prominent pre- and (especially) postcingula. *Protungulatum* (Fig. 12.3A) is widely regarded as

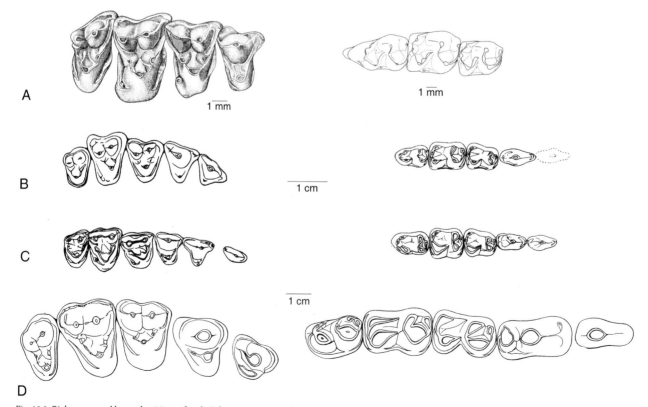

Fig. 12.3. Right upper and lower dentitions of early Paleocene arctocyonids: (A) *Protungulatum*; (B) *Oxyclaenus* (scale applies to B and C); (C) *Loxolophus*; (D) *Triisodon*. (A from Luo, 1991; B–D from Matthew, 1937.)

the morphotypic ungulate, but Luo (1991) concluded that *Oxyprimus* is actually slightly more primitive. Archibald (1998), however, classified *Oxyprimus* as a primitive hyopsodontid. These disagreements highlight the close similarity among these early Puercan genera and indicate that all lie near the base of the ungulate radiation.

Until recently it was generally accepted that these basal ungulates probably evolved from a cimolestid "insectivore," but the discovery of zhelestids in Asian strata 20 million years older than the Bug Creek ungulates provides an earlier and more likely source. (The origin of zhelestids, however, remains a mystery.) In fact, the ties between zhelestids and *Protungulatum* are conjectural at present and represent a hypothesis to be tested by further evidence.

Despite their placement near the beginning of the ungulate radiation, it must be remembered that *Protungulatum* and its allies are still very archaic compared with the earliest members of the extant ungulate orders, which are not known until the beginning of the Eocene. The teeth of these basal condylarths show only incipient stages of modification in the direction of later ungulates, and the few isolated skeletal elements that have been attributed to them (see Fig. 7.4A,B) lack any hint of the cursorial specializations that characterize most later ungulates.

CONDYLARTHRA: ARCHAIC UNGULATES

Primitive Early Tertiary mammals thought to be related to the extant ungulate orders have long been classified in an order Condylarthra, or grouped informally as condylarths. The principal families usually considered condylarths are Arctocyonidae, Mioclaenidae, Hyopsodontidae, Periptychidae, Phenacodontidae, and Didolodontidae. Mesonychia (Mesonychidae + Hapalodectidae), often included in Condylarthra, is discussed here with Cete. Cope's (1882) original concept of Condylarthra, which he initially considered to be a suborder of Perissodactyla, was limited mainly to Phenacodontidae, although subsequently (Cope, 1884) he expanded it to include Periptychidae and Meniscotheriidae, the latter now considered to be a subfamily of Phenacodontidae. Other families were added later by various authors, resulting in the paraphyletic assemblage now called condylarths. Some experts have removed various condylarth genera to their own higher taxa (e.g., Arctocyonia, Mesonychia, Phenacodonta) as apparent phylogenetic relationships to specific, more derived ungulate clades have emerged. Nonetheless, few of these proposed relationships have been conclusively demonstrated and few intermediates are known, hence it is both useful and convenient to retain the name "condylarths" for this basal assemblage until such ties are better established. (For a cogent justification of the term, see Muizon and Cifelli, 2000.) As many as 16 mammalian orders (6 extant and 10 extinct: Artiodactyla, Cetacea, Perissodactyla, Hyracoidea, Sirenia, Proboscidea, Desmostylia, Embrithopoda, Litopterna, Notoungulata, Astrapotheria, and less certainly Pyrotheria, Xenungulata, Dinocerata, Tillodontia, and Pantodonta) are thought to be derived from condylarths (e.g., Van Valen, 1978).

Condylarths represent a grade of ungulate evolution that is more advanced than Zhelestidae but less so than extant ungulate groups. They seem to have few, if any, derived characteristics that are not also found in one or more of the more advanced ungulate taxa. Most condylarths have rather generalized skulls with moderately long snouts and prominent sagittal and occipital crests. The dental formula is usually unreduced from the primitive condition (3.1.4.3/3.1.4.3), and the canines are generally prominent. The molars of most condylarths are broad, low-crowned, and bunodont, with narrow stylar shelves, low mesiodistally compressed trigonids, and rather broad talonid basins. The conules tend to be prominent, and a hypocone is variably developed. The incisors and premolars are usually simple, although the last premolar is often molarized. These dental features suggest an omnivorous diet that included an increasing proportion of vegetable matter. The skeleton is also generalized, with moderately robust limbs that are usually neither elongate nor much shortened, and pentadactyl feet, which bore hoofs in some forms but claws in others. Most were terrestrial, with some, such as phenacodontids, showing incipient cursorial features; others (many arctocyonids) were adapted for scansorial or arboreal lifestyles.

Condylarths were particularly diverse and abundant in North America during the Paleocene and Eocene (Archibald, 1998). In the Puercan and Torrejonian they account for 25–50% of both species and individuals in many local faunas (even more in the Hanna Basin of Wyoming; Rose, 1981; Williamson, 1996; Eberle and Lillegraven, 1998).

Because of their basal position among ungulates, there has been more than the usual amount of confusion regarding phylogenetic positions and relationships of many condylarth genera and families. The family assignment of many key genera is in dispute, not so much because they are poorly understood as because their primitive morphology makes different allocations defensible, depending on the characters emphasized. As Van Valen (1978) observed, early Paleocene representatives of different families are very similar and would likely be placed in the same family if they lacked presumed descendants.

Arctocyonidae

Long associated with Carnivora, Arctocyonidae (including Oxyclaeninae, which some authors elevate to family level) is now generally considered the stem family of condylarths or alternatively is placed in a separate order Procreodi (=Arctocyonia), which is the stem group of ungulates. Two dozen genera are known, mainly from the Paleocene of North America, as well as a few from Europe, but only a small number of species survived into the Eocene. Particularly characteristic are the genera *Loxolophus* (Figs. 12.3C, 12.4B), *Chriacus* (see Fig. 12.6), *Arctocyon* (Fig. 12.4A), and *Claenodon*. McKenna and Bell (1997) placed several poorly known Asian genera—*Zhujegale*, *Astigale*, *Khashanagale*, *Petrolemur*, and

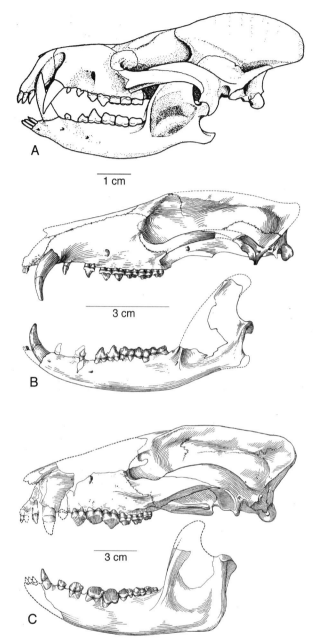

Fig. 12.4. Skulls of Paleocene arctocyonids and a periptychid: (A) *Arctocyon;* (B) *Loxolophus;* (C) *Periptychus.* (A from Russell, 1964; B, C from Matthew, 1937.)

Lantianius (formerly allocated to Anagalida, Primates, and Artiodactyla)—in the Arctocyonidae, but this reassignment has not yet been substantiated. Recently Thewissen et al. (2001a) assigned a poorly preserved jaw fragment from the Eocene of Pakistan to a new arctocyonid genus, *Karakia,* but here again better specimens are required to confirm assignment to this family. Otherwise, no definitive arctocyonids are known from Asia. Asia did, however, have an endemic group of arctocyonid-like condylarths (probably derived from Arctocyonidae), constituting the family Quettacyonidae, which occupied a similar dietary niche in Indo-Pakistan during the early Eocene (Gingerich et al., 1997, 1999).

Most arctocyonids were small to medium-sized mammals with moderately low-crowned, broad molars and simple, pointed premolars (Fig. 12.3). The molar cusps range from relatively acute in some forms (e.g., *Chriacus*) to bulbous and rounded or almost flat and indistinct in others (*Anacodon, Arctocyon,* and *Baioconodon*), suggesting omnivorous diets with a preference for meat in some taxa, fruit or other vegetable matter in others. The skulls of arctocyonids were moderately long, with prominent sagittal and occipital crests and a small braincase (Fig. 12.4A,B). The tympanic bulla was not ossified (Matthew, 1937). Some arctocyonids were generalized terrestrial animals, whereas others were specialized for climbing. The earliest arctocyonids, the aforementioned Puercan *Protungulatum* and its relatives, were small animals probably not much bigger than a tree squirrel. Isolated tarsal bones that Szalay and Decker (1974) ascribed to *Protungulatum* (see Fig. 7.4A,B) indicate that it was terrestrial.

By the end of the Paleocene arctocyonids had radiated into a diversity of forms. The largest arctocyonid was the dog- to bear-sized *Arctocyon* (probably =*Claenodon*), present in both Europe (Thanetian) and North America (Torrejonian-Tiffanian). The skull and substantial parts of the skeleton are known from both continents. Some species of *Arctocyon* and its close relative *Anacodon* had relatively flat teeth with markedly crenulated enamel. *Anacodon* is further distinguished by its enlarged upper canine protected by a bony flange at the front of the lower jaw, making it one of the oldest known saber-toothed mammals, although one whose molars suggest a predominantly herbivorous diet. *Arctocyon* and *Anacodon* were heavily built bearlike forms that were probably mainly terrestrial. However, their relatively flexible limb joints and sharp, curved terminal phalanges suggest that, like bears, they were also capable tree climbers (Matthew, 1937; Russell, 1964; Rose, 1990).

Chriacus and *Thryptacodon* were raccoon-sized Paleocene–early Eocene arctocyonids. They possessed an incisor tooth comb that was used to groom the fur, as indicated by minute grooves worn into the incisor enamel (Gingerich and Rose, 1979; Rose et al., 1981; Fig. 12.5). The tooth comb was strikingly similar, but not homologous, to that of extant strepsirrhine primates and evolved long before primate tooth combs. Both *Thryptacodon* and *Chriacus* were scansorial or arboreal animals (Rose, 1987, 1990; MacLeod and Rose, 1993). *Chriacus* is one of the best-known arctocyonids, being represented by almost the entire skeleton (Fig. 12.6, Plate 7.2). Its skeleton resembled those of living palm civets (Viverridae) and procyonids, such as the coatimundi *Nasua*. Highly mobile joints, including an elbow that allowed extensive supination and an ankle that permitted hind foot reversal, coupled with short, curved, laterally compressed claws and a long possibly prehensile tail, made *Chriacus* an adept climber similar to those living carnivorans (see Plate 6).

Chriacus has been considered close to the origin of Artiodactyla, based on dental similarity (Van Valen, 1971, 1978), but its arboreally adapted skeleton is hardly what would be expected in an artiodactyl predecessor. However, a small arctocyonid with *Chriacus*-like teeth from the Torrejonian of New Mexico differed from other known arctocyonids in

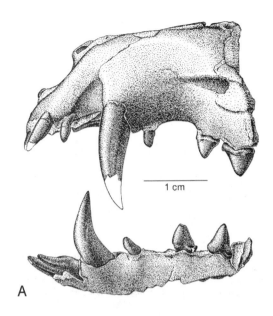

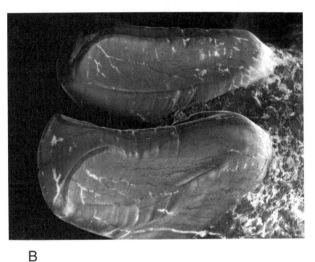

Fig. 12.5. Arctocyonid tooth combs: (A) anterior dentition of late Paleocene *Thryptacodon;* (B) scanning electron micrograph of incisors of an early Paleocene (Torrejonian) arctocyonid, showing grooves presumably generated by fur grooming. (A from Gingerich and Rose, 1979; B from Rose et al., 1981.)

having an incipiently cursorial hind limb (the only known part of its postcranial skeleton). It thus foreshadowed primitive artiodactyls in both hind limb and dental features (Rose, 1996a). Such a form could be a plausible sister taxon (or even ancestor) of Artiodactyla.

Arctocyonids are believed to lie close to the origin of all other families of condylarths as well as many other ungulates. Some arctocyonid genera are sometimes transferred to other families to emphasize these potential relationships. Particularly noteworthy in this regard are the Triisodontinae (sometimes elevated to family level), a group of mainly North American early Paleocene genera, including *Triisodon* and *Eoconodon,* whose molars are morphologically intermediate between those of other arctocyonids and Mesonychia (Fig. 12.3D). Mesonychians are characterized by laterally compressed cheek teeth with longitudinally aligned cusps.

Intermediate conditions of triisodontines include relatively narrow lower molars with a low, reduced paraconid and more closely approximated protoconid and metaconid, and reduced third molars. The upper molars are essentially tritubercular. Because of their mesonychid-like dental advances, triisodontines are sometimes included in Mesonychia (e.g., by McKenna and Bell, 1997; see Table 14.1); but unlike mesonychians, they retain small conules on the upper molars and a fully developed talonid basin on the lowers.

Triisodontines include some of the largest arctocyonid relatives. The remarkable *Andrewsarchus* (probably =*Paratriisodon*), from the middle Eocene of Mongolia and China, was probably a late-surviving triisodontine. It had a gigantic head approaching a meter in length, which far surpasses the skull size of any living carnivore (Osborn, 1924; Fig. 12.7). *Andrewsarchus* has also been interpreted as a mesonychid. The teeth are badly eroded in the holotype skull, making phylogenetic analysis difficult; however, the better-preserved teeth of *Paratriisodon* show that the latter, like triisodontines, retained conules on the upper molars and had basined lower molar talonids (Chow, 1959). Although apparently derived from arctocyonids by way of triisodontines, Mesonychia (Mesonychidae + Hapalodectidae) has been considered to be closely related to cetaceans, and is discussed in the present account under Cete and Cetacea in Chapter 14.

Paleocene *Desmatoclaenus,* with low, bunodont molars and a semimolariform P_4, has been considered a basal phenacodontid or transitional between Arctocyonidae and Phenacodontidae (West, 1976), but subsequent authors have assigned the genus to the Arctocyonidae (Van Valen, 1978; Cifelli, 1983b). As noted earlier, Archibald (1982, 1998) assigned *Oxyprimus* to the Hyopsodontidae, based on structure of its P_4 and lower molar talonids. He also considered *Oxyclaenus* and *Baioconodon* to be primitive Cete (see below) rather than arctocyonids. These varying opinions of the phylogenetic positions of primitive condylarths underscore both their overall dental similarity and the central role of early Paleocene condylarths, especially arctocyonids, in early ungulate evolution.

Mioclaenidae

Mioclaenids were primitive, small condylarths long considered to be a subfamily of Hyopsodontidae. Recently they have been linked with periptychids and didolodontids, because all three tend to possess relatively inflated posterior premolars (Archibald, 1998; Muizon and Cifelli, 2000). The molars of mioclaenids are bunodont and relatively low-crowned, with trigonids barely rising above the talonids. In most forms the cusps are bulbous, the protocone and protoconid are enlarged, the conules are well developed, and the stylar shelf is greatly reduced (Muizon and Cifelli, 2000). In contrast to most other condylarths, the hypocone is either rudimentary (North American genera) or absent (South American genera). On the lower molars the paraconid tends to be reduced and close to the metaconid, and the entoconid is often fused with the hypoconulid. There is

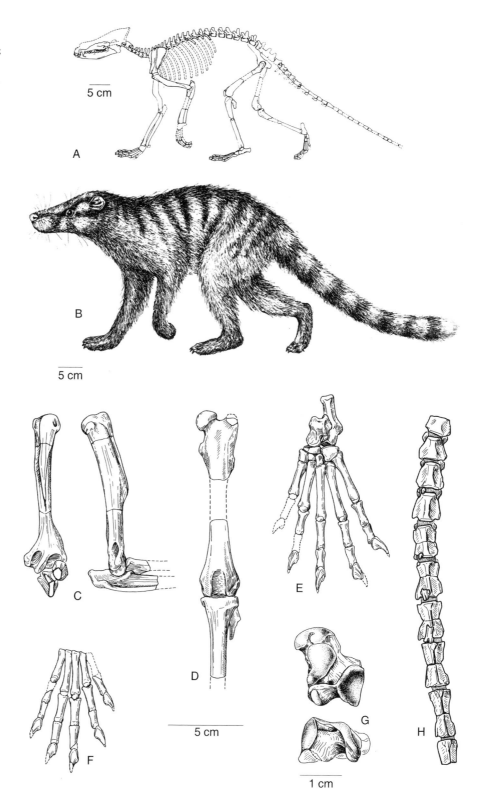

Fig. 12.6. Arctocyonid *Chriacus*: (A) skeleton; (B) restoration; (C) left humerus with proximal ulna and radius; (D) left femur and proximal tibia and fibula; (E) left pes; (F) right manus; (G) left astragalus; (H) proximal caudal vertebrae. (From Rose, 1987.)

typically a lingual notch separating the trigonid from the entoconid.

Mioclaenids are best known from several Paleocene faunas of western North America, beginning in the Puercan, and from early Paleocene assemblages from Tiupampa, Bolivia, and Punta Peligro, Argentina. There are eight North American genera, including *Litaletes* and *Promioclaenus* (Fig. 12.8A); five Tiupampan genera, of which *Tiuclaenus* and *Molinodus* are typical (see Fig. 12.14); and one Peligran genus (*Escribania*). Mioclaenids are among the most common mammals in some North American Torrejonian faunas (Rose, 1981; Williamson, 1996). *Abdounodus,* a bunodont form with

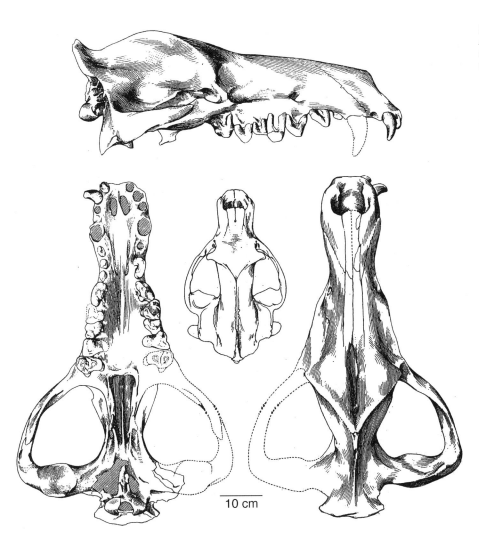

Fig. 12.7. Skull of middle Eocene *Andrewsarchus* compared to that of a grizzly bear (*Ursus arctos*, center). (From Osborn, 1924.)

inflated premolars from the early Eocene of Morocco, has been interpreted as a probable African mioclaenid (Gheerbrant et al., 2001).

Mioclaenids are believed to lie near the origin of some endemic South American ungulates. In a recent analysis focusing on Tiupampa mioclaenids, Muizon and Cifelli (2000) found that they share several dental features with didolodontids and litopterns, both strictly South American ungulate groups, the latter usually included in Meridiungulata. Based on these synapomorphies, they proposed a new order Panameriungulata for the three groups, concomitantly suggesting that Meridiungulata is not monophyletic.

Pleuraspidotherium and *Orthaspidotherium* from the late Paleocene of Europe are incipiently selenodont forms that were formerly allocated to the phenacodontid subfamily Meniscotheriinae (Russell, 1964) but more recently have been allied with mioclaenids (Van Valen, 1978). Van Valen proposed that they were derived from the Torrejonian mioclaenid *Protoselene*. The posterointernal cusp of the upper molars is interpreted as a displaced metaconule (rather than the hypocone), as in primitive artiodactyls. Skeletal remains attributed to *Pleuraspidotherium* (Thewissen, 1991) indicate a generalized plantigrade form that was probably somewhat scansorial. The few known limb fragments of *Protoselene* (Matthew, 1937) suggest a more terrestrial or even incipiently cursorial animal. No other significant postcrania of mioclaenids have been described. It seems that the relationships of the two European genera to other condylarths have not been conclusively settled.

Periptychidae

This primarily North American family includes about a dozen genera of small to medium-sized condylarths. A few tentatively referred specimens are reported from Asia and South America (McKenna and Bell, 1997). Except for one early Eocene European genus of dubious affinity (*Lessnessina*; see the discussion of Taeniodonta in Chapter 7), periptychids are limited to the Paleocene. They were particularly common in the San Juan Basin of New Mexico. The skeleton is relatively well known in the larger genera, *Periptychus* and *Ectoconus*, in which it is heavily built, with relatively short limbs and feet, the five digits terminated by narrow, hooflike unguals slightly fissured at the tip (see Matthew, 1937; Fig. 12.9). Remains of smaller types, such as the Puercan anisonchine *Mithrandir* (=*Gillisonchus*), are more gracile

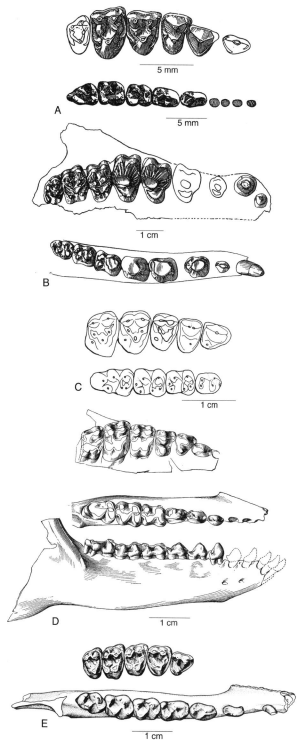

Fig. 12.8. Dentitions of condylarths (anterior to right): (A) mioclaenid *Promioclaenus* (=*Ellipsodon*), right P²–M³ and left P₃–M₃; (B) periptychid *Periptychus*; right upper and lower dentitions (C) didolodontid *Paulacoutoia*, right P³–M³ and left P₄–M₃; (D) hyopsodontid *Hyopsodus*, right upper and lower dentitions; (E) phenacodontid *Ectocion*, right P³–M³ and P₁–M₃. (A from Simpson, 1937a; B from Matthew, 1937; C from Cifelli, 1983b; D from Matthew, 1915b; E from Rose, 1981.)

and slightly more specialized. Skeletal attributes include a wide, ellipsoid proximal radius; a posteriorly directed lesser trochanter and well-defined, somewhat elevated patellar trochlea on the femur; a slightly elongate astragalar neck; and a distal peroneal tubercle on the calcaneus (Rigby, 1981). These skeletal features suggest that periptychids were generalized terrestrial animals, perhaps with limited digging ability.

The cheek teeth of periptychids are so distinctive that Periptychidae has been the most stable of the condylarth families, with little disagreement as to which genera are included. Like other condylarths, periptychids have bunodont teeth with bulbous cusps. In most forms the posterior premolars are enlarged and swollen. The molars also tend to be swollen at the bases, so that the cusp apices are set well in from the basal margin. These characteristics are especially true of the protocone, which is situated near the center of the upper molars and has an exceptionally long and high lingual slope. The hypocone, which arises from a strong posterolingual cingulum, also tends to be large or very large and is more lingual than the protocone, in some types (*Conacodon*) being directly lingual rather than posterolingual to it. Some genera also have an anterolingual cusp (the pericone). The lower molars have shortened talonids and, unlike those of other condylarths, often have a lingual cingulid (Archibald, 1998), whereas a buccal cingulid is only sometimes present.

Perhaps the most unusual feature is the presence of vertical ridges on the enamel of the cheek teeth, conspicuous in *Periptychus,* and faintly developed in some other genera (Figs. 12.4C, 12.8B). Although often characterized as crenulated or wrinkled, the condition in periptychids is very different from the wrinkled enamel of arctocyonids such as *Anacodon*. No recent mammals display this morphology, and its precise function remains a mystery. Together with the large, inflated premolars and typically heavy apical wear, these vertical ridges suggest a durophagous diet of some sort. Curiously, periptychids lack Hunter-Schreger bands (HSB) in the enamel (bands caused by decussating enamel prisms), which are present in arctocyonids, phenacodontids, and most larger eutherians and are thought to strengthen the enamel (Koenigswald et al., 1987; Stefen, 1999). Perhaps the unique ridges of periptychid cheek teeth provided an alternative means of strengthening the enamel in the absence of HSB.

The oldest and most primitive periptychids, earliest Paleocene *Mimatuta* and *Maiorana*, show only hints of the dental anatomy that characterizes later periptychids. Archibald (1998) adduced weak evidence—swollen premolars, barely evident in these earliest Paleocene genera—to support relationship between periptychids and mioclaenids, whereas Van Valen (1978) postulated that periptychids evolved directly from *Protungulatum*. This short-lived family became extinct before the end of the Paleocene and left no descendants.

Didolodontidae

This condylarth group is the only one that is endemic to South America. Didolodontids are known mainly from

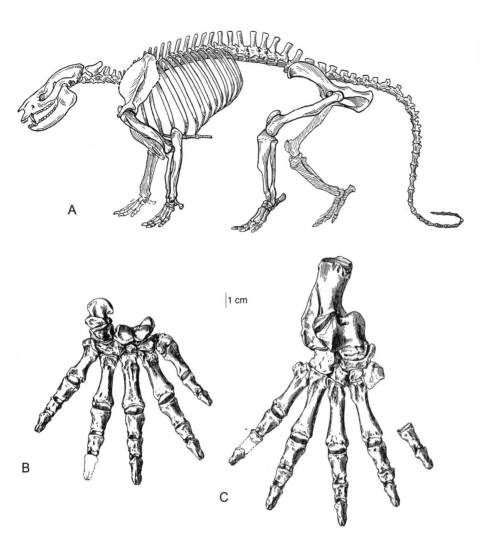

Fig. 12.9. Puercan periptychid *Ectoconus*: (A) skeleton; (B) right manus; (C) right pes. Scale pertains to B and C. (A from Gregory, 1951; B, C from Matthew, 1937.)

the late Paleocene of Brazil (Itaboraian *Lamegoia* and *Paulacoutoia*; Fig. 12.8C) and the early Eocene of Patagonia (Casamayoran *Didolodus*), but they persisted through the Oligocene (Deseadan *Salladolodus*). They are assigned to Condylarthra because of a remarkable dental similarity between didolodontids and some North American condylarths, especially phenacodontids and hyopsodontids (Simpson, 1948, 1980). Indeed, Marshall and Muizon (1988) suggested that didolodontids could only be distinguished from phenacodontids by their geographic occurrence. However, Cifelli (1993c) proposed that the resemblance is merely convergent and probably derived from more primitive condylarths. As mentioned earlier, the teeth of didolodontids are also very similar to those of mioclaenids (e.g., in being brachydont and very bunodont), but they differ in having a well-developed hypocone even on M^3 (Cifelli, 1983b).

Equally important is a close similarity to the teeth of primitive litopterns—so close, in fact, that one genus assigned to Didolodontidae based on its teeth (*Ernestokokenia*) was found to contain both didolodontid and litoptern species, based on ankle structure (Cifelli, 1983a). Some other genera long included in Didolodontidae, such as Itaboraian-Casamayoran *Asmithwoodwardia* (Simpson, 1948), are now generally considered to be primitive litopterns. Another interesting example is Miocene *Megadolodus*, long considered a didolodontid, based on its bunodont cheek teeth. Discovery of a nearly complete skeleton revealed that *Megadolodus* has diagnostic litoptern tarsal and vertebral specializations (Cifelli and Villarroel, 1997). Its teeth evidently became secondarily "generalized" for a diet of hard fruits, thereby converging on the dentition of didolodontids. As aforementioned, Muizon and Cifelli (2000) considered the dental resemblance among mioclaenids, didolodontids, and litopterns to reflect close relationship and erected a new order Panameriungulata to accommodate them. However, the further step of assigning Didolodontidae to the Litopterna, as proposed by some authors, now seems unjustified, because didolodontids lack the derived tarsal traits that characterize litopterns and have ankle specializations of their own (Cifelli, 1993c; see Fig. 12.23).

Two additional South American families may be noted here. The Sparnotheriodontidae (including Riochican-Casamayoran *Victorlemoinea* and its relatives) have usually been considered litopterns (Simpson, 1945, 1948; McKenna and Bell, 1997), but Cifelli (1983a, 1993c) classified them as didolodontoids, because isolated tarsal bones referred to

Victorlemoinea share derived traits with didolodontids rather than with litopterns. Nevertheless, *Victorlemoinea* has selenodont molars (see Fig. 12.25B), as in many litopterns, in contrast to the bunodont molars of didolodontids. If Cifelli is right, selenodonty must have evolved convergently in *Victorlemoinea*. Confident placement of sparnotheriodontids will depend on discovering definitively associated postcrania. Peligrotheriidae is a supposed condylarthran family including only *Peligrotherium,* which is based on a few jaw fragments from the early Paleocene of Argentina (Bonaparte et al., 1993). Additional teeth attributed to *Peligrotherium* have been interpreted to represent a late-surviving dryolestoid (Gelfo and Pascual, 2001; Rougier, Novacek, et al., 2003), but assignment to this genus is equivocal, because of the poor state of preservation of the original material. If the new specimens are properly referred to this genus, *Peligrotherium* is apparently not even eutherian.

Hyopsodontidae

Hyopsodontids comprise about 20 small (mostly weighing less than 1 kg) Paleocene and Eocene genera occurring mostly in North America and Europe and known primarily from jaws and teeth; there are also a few Asian and questionable North African records. Typical in the Paleocene of North America are *Litomylus* and *Haplaletes,* whereas *Haplomylus* was common in latest Paleocene–early Eocene strata. In Europe, *Louisina, Paschatherium,* and related forms were prevalent (Russell, 1964; Denys and Russell, 1981). None of them, however, approached the broad distribution (North America and Eurasia) or phenomenal abundance of *Hyopsodus,* the most common mammal in many early and middle Eocene local faunas in western North America (Gazin, 1968; Rose, 1981; Bown et al., 1994). *Hyopsodus* accounts for 25–30% of individuals in many early Eocene faunas. Tens of thousands of specimens (mostly teeth and jaw fragments) are known from the Bighorn Basin of Wyoming alone.

Like other condylarths, hyopsodontids have a primitive dental formula of 3.1.4.3/3.1.4.3, without significant diastemata. The following description is based on *Hyopsodus* (Gazin, 1968; Fig. 12.8D) but applies generally to most other hyopsodontids as well. The incisors are simple and pointed; the canines reduced; and the premolars generally simple except for P_4 (and sometimes P^4), which is submolariform. The molars are bunodont to incipiently selenodont, the latter condition better developed in the lower molars. The paraconid is typically small or absent on the lower molars, whereas the uppers have six cusps (both conules and hypocone well developed). The dentition indicates that hyopsodontids were herbivorous. The skull is known only in *Hyopsodus,* in which it is relatively elongate and generalized, with a superficial resemblance to that of the hedgehog *Erinaceus* (Gazin, 1968). In most regards it is primitive and shares no particular derived traits with more advanced ungulates.

Despite abundant dental remains, skeletal remains of hyopsodontids are surprisingly rare and are best known for Eocene *Hyopsodus* (Matthew, 1915b; Gazin, 1968)—which, it turns out, may not be typical for the family. Species of *Hyopsodus* ranged in size from smaller than a hedgehog to as large as the ground hog *Marmota. Hyopsodus* was proportioned like a weasel or the prairie dog *Cynomys,* having a long spine and short limbs with pentadactyl hands and feet (Fig. 12.10). Several features (e.g., supratrochlear foramen of humerus, restricted elbow, well-defined patellar groove) suggest terrestriality, but coupled with a nearly flat astragalar trochlea and short claws (not hoofs) they suggest that *Hyopsodus* may also have been capable of climbing or digging and apparently was not specialized for any particular mode of life. Rather, it was a generalist, as also indicated by its teeth and its abundance and ubiquity in the first half of the Eocene. Tarsal bones attributed to *Paschatherium* suggest that this European genus was more scansorially adapted than was *Hyopsodus* (Godinot et al., 1996).

Tricuspiodon, from the late Paleocene of the Paris Basin, is a peculiar form whose relationships remain problematic. It has variously been included in Hyopsodontidae (Van Valen, 1978; McKenna and Bell, 1997) or placed in its own family (D. E. Russell, 1980). Russell described new specimens and reevaluated its affinities, suggesting that it may be related to Phenacodontidae. The mandible is thick but shallow, with a fused symphysis. The lower incisors are procumbent and are followed by a large canine. Unlike in most other condylarths, the trigonid and talonid are columnar, moderately hypsodont, and lack cingula. The upper molars have peripheral bunodont cusps, small conules, a very small hypocone, and weak cingula. Thewissen (1991) assigned several isolated limb bones to *Tricuspiodon.* The humeri are short and robust, resembling those of the beaver *Castor.* The ulna is robust with a large olecranon, and the shape of the proximal radius would have limited supination. These bones suggest an animal with fossorial habits. If they are correctly attributed, *Tricuspiodon* is more likely to have evolved independently from an arctocyonid than to be closely related to phenacodontids, which were incipiently cursorial.

The late Paleocene–early Eocene apheliscines (*Phenacodaptes* and *Apheliscus*) are small mammals known mainly from teeth from western North America. They had triangular upper molars with small conules and hypocones, quadrate bunodont lower molars with greatly reduced paraconids, and enlarged, pointed fourth premolars. Apheliscine relationships have been challenging to decipher. Suggested relatives, judging from dental anatomy, include such diverse groups as artiodactyls (Jepsen, 1930) and pentacodontid pantolestans (Gazin, 1959), with most recent authorities favoring close ties with hyopsodontid or mioclaenid condylarths. Newly recognized postcrania support close relationship of *Apheliscus* to *Haplomylus* and to various European hyopsodontids (Louisininae). They also reveal numerous differences from *Hyopsodus* and suggest the possibility that Hyopsodontidae as currently recognized is polyphyletic (Zack et al., 2005).

Direct ties between hyopsodontids and noncondylarth mammals have proven difficult to establish, but several pos-

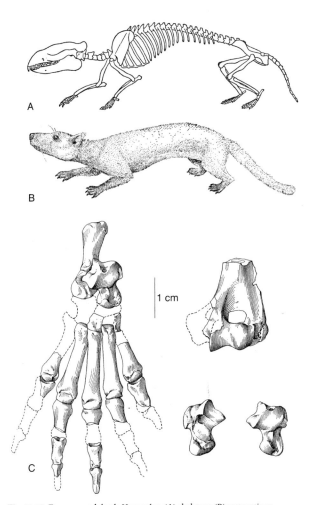

Fig. 12.10. Eocene condylarth *Hyopsodus:* (A) skeleton; (B) restoration; (C) right foot, distal left humerus, and right astragalus. (A, B from Gazin, 1968; C from Matthew, 1915b.)

Phenacodontidae

Although consisting of only a half-dozen or so genera, this family was an important constituent of Torrejonian-Bridgerian faunas of western North America. Phenacodontids were the dominant mid-sized mammals during the late Paleocene and early Eocene in North America, and they far outnumbered any other species in most Clarkforkian mammal assemblages (Rose, 1981). They were rare in Eurasia, however, only a single genus being recognized from each continent. Phenacodontids ranged in body mass from about 3 kg to more than 50 kg (Thewissen, 1990) and included some of the largest known condylarths.

Phenacodontids had swollen, bunodont to lophodont or selenodont cheek teeth with submolariform fourth premolars and quadrate molars, the last molar somewhat reduced (Figs. 12.8E, 13.2A). The molar cusps are either rounded (generally in larger species) or somewhat selenodont, a pattern accentuated by wear especially on lower molars. The upper molars have a well-developed hypocone and (except in *Tetraclaenodon*) a mesostyle, resulting in a W-shaped ectoloph. The lower molars have a reduced paraconid and a broad talonid basin and often have a metastylid cusp posterior to the metaconid. Lophodont or selenodont phenacodontids, such as *Ectocion* and *Meniscotherium,* were presumably more strictly herbivorous than were bunodont forms, with a diet rich in leafy vegetation.

The skull of phenacodontids was moderately elongate, with a modestly developed nuchal (occipital) crest and weak sagittal crest. *Phenacodus* had relatively large nasal openings and somewhat retracted nasal bones, suggestive of a short, tapirlike proboscis. The auditory bulla (known only in *Ectocion*) is formed from the ectotympanic. Phenacodontids were sexually dimorphic in canine size and possibly in extent of paranasal sinuses (Thewissen, 1990).

Skeletal remains are known for most phenacodontid genera, but relatively complete skeletons are known only for *Phenacodus* (Fig. 12.11). Limb anatomy indicates that phenacodontids were somewhat cursorial but not as specialized as perissodactyls or artiodactyls. The limbs were only slightly elongate. Furthermore, the feet were pentadactyl and digitigrade, with the lateral digits reduced in most genera and all digits tipped with broad, flat, hooflike unguals. The carpus has been described as "serial," meaning that the scaphoid (proximal row) articulates distally with the trapezium and trapezoid, whereas the lunate articulates solely with the magnum (Radinsky, 1966a). This configuration is in contrast to the "alternating" condition (typical of perissodactyls and artiodactyls), in which the scaphoid also articulates with the magnum, and the lunate also with the unciform (see Fig. 2.10). The distinction is not always clear, however, as intermediate conditions exist in phenacodontids (Radinsky, 1966a; Thewissen, 1990). Other cursorial adaptations include a high greater tuberosity and reduced deltopectoral crest on the humerus; a narrower distal humerus than in arctocyonids, with a supratrochlear foramen to maximize forearm

sibilities have been proposed. Various authors have noted dental resemblances to primitive litopterns, artiodactyls, and, most recently, macroscelideans (via *Haplomylus:* Simons et al., 1991; or via louisinines such as *Monshyus* and *Microhyus:* Tabuce, Coiffait, et al., 2001). The new postcrania of *Haplomylus* and *Apheliscus* show derived features of the limb skeleton indicative of cursorial or saltatorial locomotion. These include relatively gracile and elongate limb bones (especially the hind limb), the tibia longer than the femur, and extensive distal fusion of the tibia and fibula. The lesser trochanter projects posteromedially, and the distal femur is anteroposteriorly deep, with a narrow patellar groove. The calcaneus, astragalus, and cuboid are narrow and elongate; the astragalar trochlea is well grooved; and its medial side bears a distinctive depression (the cotylar fossa). Many of these specializations are shared specifically with macroscelideans (see Chapter 15 and Fig. 15.7), strengthening the likelihood of a close phylogenetic relationship between these groups (Zack et al., 2005). If this alliance is corroborated, it would challenge the monophyly and supposed African origin of the Afrotheria.

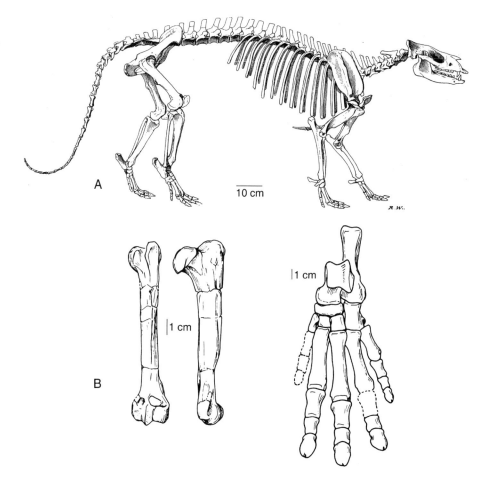

Fig. 12.11. The Eocene condylarth *Phenacodus*: (A) skeleton; (B) left humerus and left foot. (A from Osborn, 1898a; B from Rose, 1990.)

extension; a wide and uneven proximal radius that restricted supination (see Fig. 2.9); and a deeply grooved astragalar trochlea, which limited upper ankle movement to parasagittal flexion-extension. Except for loss of the centrale in the carpus, no limb elements were lost or co-ossified.

The oldest and most primitive phenacodontid is Torrejonian *Tetraclaenodon,* which is the least progressive form both dentally and postcranially. It resembles other phenacodontids and differs from arctocyonids in having hooflike ungual phalanges (Matthew, 1937). The slightly older and poorly known *Desmatoclaenus* is variously regarded as the stem phenacodontid (West, 1976) or as an arctocyonid near the base of Phenacodontidae (Van Valen, 1978; Thewissen, 1990).

Meniscotherium is the most selenodont phenacodontid, with crests dominating over cusps and upper molars exhibiting a strong W-shaped ectoloph and a well-developed mesostyle, characteristics that have led some researchers to classify it in a separate condylarth family (sometimes together with *Ectocion*). At first glance the selenodont pattern could be easily confused with that of some early artiodactyls or perissodactyls, but *Meniscotherium* differs from most of these in having six primary cusps on the upper molars, including both a metaconule and a hypocone. Furthermore, the postcranial skeleton is relatively well known in *Meniscotherium,* and any close relationship with the modern ungulate orders is belied by its phenacodontid-like limb structure (Gazin, 1965; Williamson and Lucas, 1992). Although the largest species of *Meniscotherium* is more robust than *Phenacodus,* the overall skeletal anatomy is quite similar. *Meniscotherium* differs from other phenacodontids in having a shallower astragalar trochlea and narrower terminal (ungual) phalanges, expanded at the tip, which resemble those of early euprimates. This unusual morphology has suggested to some researchers that the unguals may have borne nails rather than hoofs, but in this context the difference between a nail and a hoof is moot. Williamson and Lucas (1992) suggested that hyracoids are the best living analogue of *Meniscotherium.*

Phenacodontids have been considered to occupy a pivotal position in the evolution of higher ungulates. Citing evidence primarily from the dentition and limb skeleton, Radinsky (1966a) argued that Phenacodontidae is the most probable source of Perissodactyla. The resemblances between the two groups are indeed striking and extend through most of the skeleton. Although some of them may have evolved in parallel, a relationship seems difficult to deny. Subsequent more comprehensive morphological studies suggest that Phenacodontidae is the sister taxon to a wider ungulate group, Altungulata (see below), which includes tethytheres and hyracoids in addition to perissodactyls (Thewissen and Domning, 1992). This arrangement, however, is in conflict with the molecular-based Afrotheria. Moreover, the absence of a centrale in phenacodontids, a bone retained as a separate element in basal proboscideans and hyracoids, would seem to preclude known phenacodontids from a directly ancestral position to those groups. It has also been observed that perissodactyls have an alternating carpus, unlike the serial one of some phenacodontids. However, as noted above, the distinction between serial and alternating carpi is often ambiguous. Radinsky (1966a) noted that some smaller phenacodontids have an incipiently alternating carpus. The morphological evidence, then, currently weighs in favor of phenacodontids being a plausible—indeed, probable—stem group for Altungulata.

The new genus *Ocepeia*, based on lower teeth from the early Eocene of Morocco, appears to be closely related to phenacodontids (Gheerbrant et al., 2001). According to its describers, *Ocepeia* could be a stem taxon of phenacodonts + altungulates, a hypothesis that merits reassessment when more complete fossils are known.

Tingamarra

Before leaving the condylarths, it is necessary to mention one final genus, *Tingamarra*, based on a single lower molar from a formation of the same name in Australia (see Fig. 5.16D). Godthelp et al. (1992) referred it questionably to the condylarths. The tooth is generalized and bunodont, with a slightly wider trigonid than talonid, and it comes from a formation that has been dated at 54.6 million years, just after the Paleocene/Eocene boundary (Godthelp et al., 1992). If it had come from almost anywhere else in the world at that time, its occurrence would not be unusual. But *Tingamarra*, if truly a condylarth, would be the only nonvolant Early Tertiary terrestrial eutherian from the Australian continent. Woodburne and Case (1996) questioned the age of the formation, however, and suggested that *Tingamarra*, despite lacking a twinned hypoconulid and entoconid, is more likely to be a marsupial. More complete remains are needed to place *Tingamarra* confidently.

ARCTOSTYLOPIDA

Arctostylopids were small, rabbit-sized animals, which may have been analogous to present-day hyraxes (Cifelli and Schaff, 1998). This small clade of about 10 closely allied genera, all assigned to the family Arctostylopidae, is known mainly from jaw fragments from the late Paleocene and early Eocene of Asia (nine genera) and western North America (one genus). If not for their highly unusual dentitions, the arctostylopids might be considered too obscure to merit attention. The cheek teeth, however, bear an uncanny resemblance to those of notostylopid notoungulates, which are known only from South America. The resemblance is close enough that for most of the twentieth century arctostylopids were classified as notioprogonian notoungulates, but different enough that some recent authorities have challenged their allocation to Notoungulata.

The dentition of arctostylopids is unreduced (3.1.4.3 above and below) and forms an evenly graded series without diastemata (Fig. 12.12). As in notoungulates, the molars are lophodont, with a prominent straight ectoloph on the uppers and ectolophid (formed by the cristid obliqua) on the lowers. The upper molars are primitively triangular, with well-developed pre- and postprotocristae (similar to the crosslophs of notoungulates) and no conules; more progressive forms add a pseudohypocone (Cifelli et al., 1989). The lower molars are bicrescentic and often have a well-developed oblique loph (entolophid) running buccally from the entoconid toward the ectolophid. The ectoloph and ectolophid are better developed in more derived species, forming mesiodistally oriented vertical shearing crests similar to the carnassials of carnivorans (Cifelli and Schaff, 1998). The fourth premolars are submolariform and the anterior lower premolars have three mesiodistally aligned cusps.

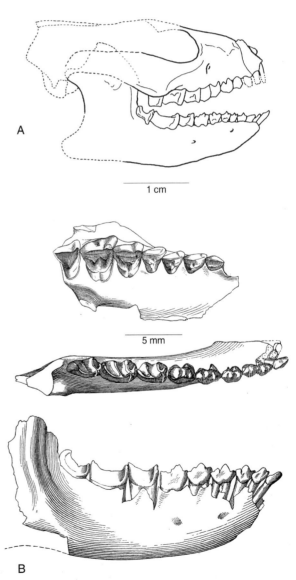

Fig. 12.12. Asian late Paleocene arctostylopids: (A) restored skull of *Gashatostylops*; (B) right upper and lower teeth of *Palaeostylops*. Compare Fig. 12.18. (A from Cifelli and Schaff, 1998; B from Matthew and Granger, 1925a.)

Arctostylopids first appear at the beginning of the Nongshanian (late Paleocene) in Asia. *Asiostylops* is the oldest and most primitive arctostylopid, as indicated by its simple triangular upper molars and the poorly developed entolophid on the lowers (Cifelli et al., 1989). Together with *Sinostylops* and *Bothriostylops* it forms the sister clade to North American *Arctostylops* and several related Asian genera, including *Palaeostylops* and *Gashatostylops*. The close similarity of the last three genera is indisputable evidence of late Paleocene faunal interchange between Asia and North America.

Arctostylops, a rare arctostylopid from the late Paleocene (late Tiffanian and Clarkforkian) of Wyoming, was the first named genus (Matthew, 1915c). The only other North American arctostylopid is an unnamed early Eocene species, based on a single tooth from southern Wyoming (Zack, 2004). A partial skeleton of *Arctostylops* has been known for several years (Bloch, 1999) but has yet to be described. Bloch indicated that the humerus suggests cursorial habits, whereas the tarsus shows certain probable notoungulate features. Cifelli et al. (1989), however, found that isolated tarsal bones they allocated to Asian *Gashatostylops* more closely approach those of the anagalidan *Pseudictops*. The resemblance to *Pseudictops* suggests the possibility that arctostylopids could be related to Anagalida; alternatively, those tarsals may actually belong to *Pseudictops* or a related anagalidan rather than to an arctostylopid (Kondrashov and Lucas, 2004). Arctostylopid postcrania obviously merit further study before definitive conclusions can be made.

The relationships of arctostylopids are equivocal and probably will remain so until the skeleton is better known. Long considered to be notoungulates, they were variously thought to be either ancestral to South American notoungulates or derived from them. But arctostylopids lack the distinctive crochet of notoungulate upper molars, and their pseudohypocone seems not to be homologous with the true hypocone of notoungulates (Cifelli et al., 1989). Consequently, the dental resemblances to notoungulates have been interpreted to be convergent, and the new order Arctostylopida was erected by Cifelli et al. to accommodate them. Van Valen (1988) suggested that both Arctostylopida and Notoungulata evolved from primitive astrapotheres, which implies that the two groups are not so far apart. It may be premature to rule out a relationship between arctostylopids and notoungulates before a thorough analysis that includes postcranial features.

MERIDIUNGULATA: ENDEMIC SOUTH AMERICAN UNGULATES

The native South American ungulates—Astrapotheria, Litopterna, Notoungulata, Pyrotheria, and Xenungulata—evolved essentially in isolation from Holarctic faunas during most of the Cenozoic, but all are now extinct. The inference that they may share a common origin, most likely from primitive didolodontid or mioclaenid condylarths, led to the concept of the mirorder Meridiungulata to encompass these five orders (McKenna, 1975a; Cifelli, 1993c; McKenna and Bell, 1997). The strongest support for this association, however, seems to be their geographic occurrence. Derived characters that would unite the mirorder as a monophyletic group have not been identified (Cifelli, 1993c). Dental or auditory similarities found in some members do not extend to all and in some cases are demonstrably convergent. At the same time, other relationships have been postulated, such as a xenungulate-pyrothere association with northern uintatheres (e.g., Gingerich, 1985; Schoch and Lucas, 1985; Lucas, 1993) or specific associations of particular meridiungulate groups to the exclusion of others (e.g., astrapotheres + notoungulates; Van Valen, 1988). These alternative hypotheses are generally no more convincing than meridiungulate monophyly and have not been widely adopted.

Recently Muizon and Cifelli (2000) adduced evidence that meridiungulates may not be monophyletic, and that instead South American mioclaenids (subfamily Kollpaniinae, initially described as marsupials), didolodontids, and litopterns constitute a monophyletic group for which they proposed the name Panameriungulata (Fig. 12.13). Nonetheless, they did not rule out the possibility that notoungulates and perhaps other South American ungulates may ultimately prove to be part of the same clade. In fact, there is no clearly preferable hypothesis to the origin of all South American ungulates from one or more North American "condylarths." For this reason they are grouped together in this chapter as meridiungulates, with the important caveat that their monophyly has yet to be demonstrated.

Perutherium, based on a few molar fragments from supposed Late Cretaceous deposits at Laguna Umayo, Peru, was once thought to be the oldest South American ungulate. However, the mammal-bearing levels at Laguna Umayo are now thought to be late Paleocene in age (Sigé et al., 2004). Furthermore, although the dental morphology of *Perutherium* seems consistent with ungulate affinities, it is so fragmentary that it has been variously considered a didolodontid, an arctocyonid, a periptychid, a basal notoungulate, or even a marsupial. The younger age and ambiguous identification of these teeth reduces their importance.

By the early Paleocene, however, unquestioned primitive ungulates are present in South America. Sediments at Tiupampa, Bolivia, have yielded a diversity of mioclaenids, such as *Tiuclaenus* (Fig. 12.14), as well as a possible notoungulate tooth, the oldest record of that order (Muizon and Cifelli, 2000). Mioclaenids, common in the Paleocene of North America, and the problematic *Peligrotherium* have also been reported from the Punta Peligro local fauna of Patagonia (early Paleocene, Peligran; Bonaparte et al., 1993). *Peligrotherium* was the basis for a new condylarthran family Peligrotheriidae, but as noted above, it may not even be eutherian.

Recent analyses of anatomical traits suggest that convergence has been much more common among South American ungulates than previously supposed. Proposals of relationship between North American Arctostylopidae and Notoungulata, and North American Dinocerata and Xenungulata, are plausibly based on homoplasy rather than synapomorphy (Cifelli, 1993c).

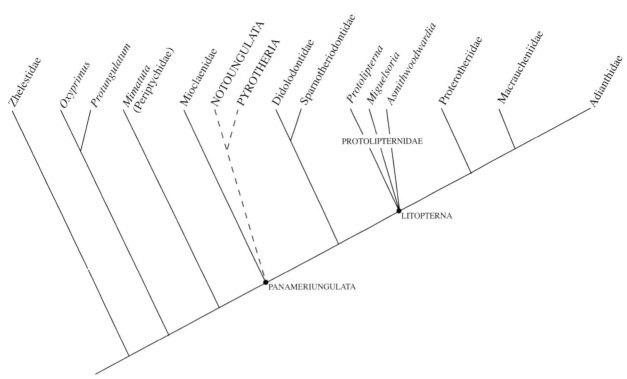

Fig. 12.13. Panameriungulata relationships (simplified after Cifelli, 1993c, and Muizon and Cifelli, 2000). Dashed lines indicate that allocation of Notoungulata and Pyrotheria to Panameriungulata is uncertain.

Notoungulata

With 14 families and more than 150 genera (McKenna and Bell, 1997), the notoungulates—meaning "southern hoofed mammals"—were by far the most diverse and successful meridiungulates (Fig. 12.15), filling many of the ecological niches occupied by artiodactyls and perissodactyls on the northern continents. Notoungulates were present throughout the Cenozoic, from the late Paleocene to the late Pleistocene, although diversity declined markedly from the Miocene onward. As noted above, a possible notoungulate tooth is known from the early Paleocene of Tiupampa, but this record remains to be corroborated by more complete specimens. Dental and tarsal evidence suggests that notoungulates evolved from mioclaenids during the Paleocene (Muizon et al., 1998) and experienced a rapid radiation. Nine notoungulate families already existed by the Casamayoran, and all but two were present by the end of the Eocene.

Early Tertiary notoungulates are generally characterized by a primitive dental formula (3.1.4.3/3.1.4.3), the teeth usually closely spaced without diastemata (Casamayoran notostylopids are an exception). The cheek teeth are typically lophodont, but with an arrangement of crests different from that in northern ungulates (Simpson, 1948; Paula Couto, 1979; Cifelli, 1993c). The upper molars have a strong, straight ectoloph; an oblique protoloph; and a more transverse metaloph, somewhat reminiscent of the pattern in ceratomorph perissodactyls. There is a tendency to add extra crests running toward the center of the upper molars; especially characteristic is one called the crochet, which extends anterobuccally from the metaloph (Fig. 12.16). Some types add an antecrochet from the protoloph or one or more

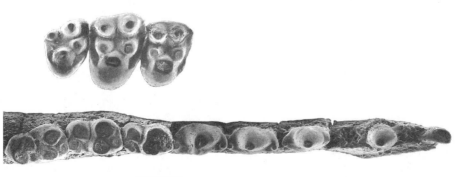

Fig.12.14. *Tiuclaenus*, an early Paleocene mioclaenid from Bolivia: right upper molars and left lower dentition. (From Muizon and Cifelli, 2000; courtesy of C. De Muizon.)

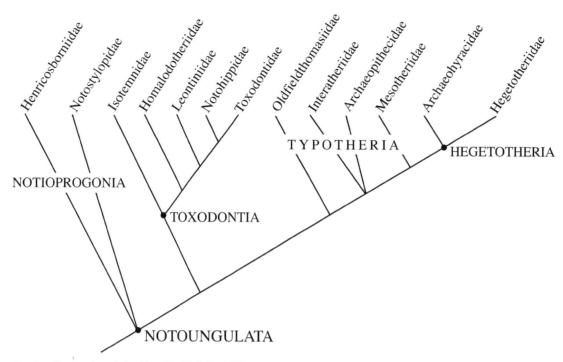

Fig. 12.15. Notoungulate relationships. (Simplified after Cifelli, 1993c.)

cristae from the ectoloph. The lower molars have two main crescentic crests, a metalophid and a hypolophid, together with a shorter transverse entolophid derived from the entoconid. The paraconid is typically absent and the paracristid short. The ectocingulum tends to be weak or absent on both upper and lower teeth. Besides increasing lophodonty, common trends in notoungulates include increasing hypsodonty, and even hypselodonty, as well as the evolution of rodent-like incisors.

The skull lacks a postorbital bar, and the posterior end of the zygomatic arch arises high on the back of the skull. The most diagnostic aspect of the skull is the complex ear region (Patterson, 1936; Simpson, 1948, 1980; Cifelli, 1993c). The tympanic cavity is connected to a large epitympanic sinus in the squamosal and often to a separate hypotympanic sinus inferior to the tympanic cavity. An inflated ectotympanic bulla surrounds the tympanic cavity and extends laterally as an external auditory tube with a characteristic ventral crest. Behind the bulla is a fossa for a projection of the hyoid bone (the stylohyoid). Although somewhat similar bullae have evolved in various rodents and marsupials, this particular complex of auditory features is unique to notoungulates.

The postcranial skeleton of Paleocene-Eocene notoungulates, with a few exceptions, is very poorly known. Casamayoran *Thomashuxleya* is one of the few notoungulates known from most of the skeleton (see Simpson, 1967; Fig. 12.17). It had generalized proportions and in this regard was not unlike the condylarth *Phenacodus*. Ankle bones representing the stem families Henricosborniidae and Oldfieldthomasiidae have been identified from the late Paleocene of Itaboraí, Brazil (Cifelli, 1983a; see Fig. 12.23B). They are similar to tarsals of other (younger) primitive notoungulates.

Though relatively generalized, the astragali are distinctive in having a moderately long, constricted neck; a roughly hemispherical head; a medial projection on the body; a prominent dorsal foramen and posterior sulcus; and partly confluent sustentacular and navicular facets. The astragalar trochlea is slightly grooved, with a high lateral rim, and the calcaneal peroneal tubercle is distally situated. Together these features suggest a generalized terrestrial habit. The feet of notoungulates were almost invariably mesaxonic (with the axis running through the third digit) and, in primitive forms, usually five-toed. In many lineages the lateral digits became smaller or were lost, leaving three functional digits. Some Miocene typotheres retained only two equal-sized toes (digits III and IV) on the hind feet and were therefore paraxonic like artiodactyls.

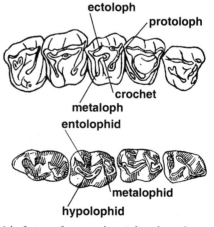

Fig. 12.16. Molar features of notoungulates, indicated on right upper and lower cheek teeth of *Henricosbornia*. (Modified from Simpson, 1948.)

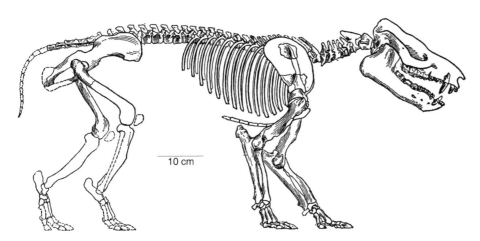

Fig. 12.17. *Thomashuxleya*, a primitive Casamayoran toxodont. (From Simpson, 1967.)

Notoungulates have traditionally been divided into four suborders, the paraphyletic Notioprogonia and the monophyletic Toxodontia, Typotheria, and Hegetotheria (Simpson, 1948, 1967; McKenna and Bell, 1997).

Notioprogonia

This suborder includes two families, Henricosborniidae and Notostylopidae, which are assigned here not because they uniquely share any derived features, but rather because they do not clearly belong to the other clades. Notioprogonia is therefore a convenient assemblage of primitive notoungulates and not a natural group. Henricosborniids (e.g., *Henricosbornia, Othnielmarshia*), which are known only from the late Paleocene and early Eocene (Itaboraian-Casamayoran), are dentally the most primitive notoungulates, most with generalized, low-crowned teeth (Fig. 12.16), and a dental formula of 3.1.4.3/3.1.4.3 (Simpson, 1948, 1980; Cifelli, 1993c). They are believed to lie close to the source of all other notoungulates. The skull is known only in the late Paleocene henricosborniid *Simpsonotus*. According to Pascual et al. (1978), it lacks an epitympanic sinus and a tympanic crest, suggesting that it may be the most primitive known notoungulate. If verified, this would imply that these auditory specializations—which have been considered synapomorphies of Notoungulata—evolved within the order. *Simpsonotus* has an oddly derived anterior dentition: I^3 is caniniform and much larger than I^{1-2} and the upper canine, whereas in the mandible I_3 is the largest incisor, I_1 is missing, and I_2 is tiny or absent. These modifications indicate that *Simpsonotus* represents a divergent lineage.

Notostylopids (e.g., Casamayoran *Notostylops, Boreastylops*) are somewhat more specialized than henricosborniids. The skull has enlarged but rooted anterior incisors (I^1 and I_2) and usually an anterior diastema created by the reduction or loss of the third incisor, canine, and first premolar (Simpson, 1948; Paula Couto, 1979; Fig. 12.18), giving the skull a superficial resemblance to that of *Plesiadapis*. The posterior premolars are molariform, but the molars remain quite primitive. Both *Notostylops* and *Henricosbornia* are known to show considerable intraspecific variability in dental formula and details of crown morphology (Simpson, 1948, 1980). Endocasts of *Notostylops* indicate that it had a relatively small brain (EQ = 0.36–0.46), with large olfactory bulbs and midbrain, and a relatively small neocortex (Radinsky, 1981).

For many years the northern family Arctostylopidae (known from the Paleocene–early Eocene of Asia and North America) was included within Notioprogonia, but

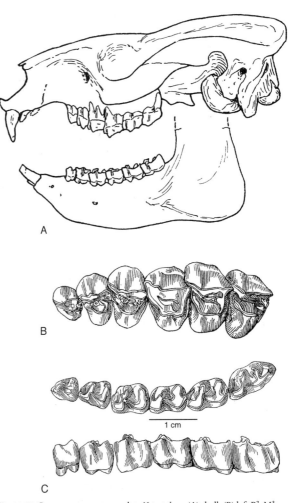

Fig. 12.18. Casamayoran notoungulate *Notostylops*: (A) skull; (B) left P^2–M^3; (C) left P_2–M_3 in crown and buccal views. (From Simpson, 1948.)

their resemblances to primitive notoungulates are now believed to be convergent. Accordingly, they were discussed in the previous section.

It is pertinent here to digress briefly to consider the unusual names of many notoungulates and other South American fossil mammals. The eminent late-nineteenth–early-twentieth-century Argentine paleontologist Florentino Ameghino had a penchant for naming genera after prominent scientists, usually combining first name and surname. In so doing, he created a veritable Who's Who of great naturalists of that era. Besides genera mentioned in this section on South American ungulates, they include the notoungulates *Carolodarwinia*, *Edvardocopeia*, *Edvardotrouessartia*, *Guilielmoscottia*, and *Maxschlosseria*, the litoptern *Ricardolydekkeria*, and various invalid names, such as *Ernestohaeckelia*, *Guilielmofloweria*, *Josepholeidya*, and *Ricardowenia*. Many subsequent researchers have continued the custom, with such names as *Bryanpattersonia*, *Colbertia*, *Miguelsoria*, *Santiagorothia*, and *Simpsonotus*. Other of Ameghino's generic names reflect his belief that many northern placental groups, including primates and horses, originated in South America. Despite their names, however, *Archaeopithecus*, *Notopithecus*, *Notohippus*, *Archaeohyrax*, and *Progaleopithecus* are all now known to be notoungulates. For more on Ameghino's contributions to vertebrate paleontology, see Simpson (1948, 1984).

Toxodontia

Toxodonts include the largest notoungulates, and most share a number of dental, auditory, and tarsal specializations. However, the five families are a rather eclectic lot. Three of them had evolved by the middle Eocene (Mustersan).

The oldest and most primitive toxodonts were the Isotemnidae, comprising 12 genera; they were also the most common Paleocene-Eocene toxodonts. Isotemnids are weakly linked with other toxodonts by only a few dental features (Cifelli, 1993c). Riochican-Casamayoran *Isotemnus* is the oldest known genus; several others, including *Thomashuxleya* (Fig. 12.17), *Pleurostylodon* (Fig. 12.19A), and *Pampatemnus*, are known from the Casamayoran. The family is united primarily by primitive features (Simpson, 1980), such as a complete dentition with unreduced canines and no diastemata in early forms, and is therefore probably paraphyletic or polyphyletic. However, the upper molars are derived relative to *Henricosbornia* in having one or two additional transverse crests (cristae) extending lingually from the ectoloph (Simpson, 1980; Cifelli, 1993c). The cheek tooth pattern in isotemnids is sufficiently primitive to be basal to all other notoungulates except notioprogonians (Cifelli, 1993c). Isotemnids generally were larger animals with larger canines (polarity uncertain) than other early notoungulates.

The postcranial skeleton of the sheep-sized isotemnid *Thomashuxleya* is the best-known early notoungulate skeleton. It is relatively robust, with generalized limbs showing little distal elongation (Simpson, 1967). The feet are penta-

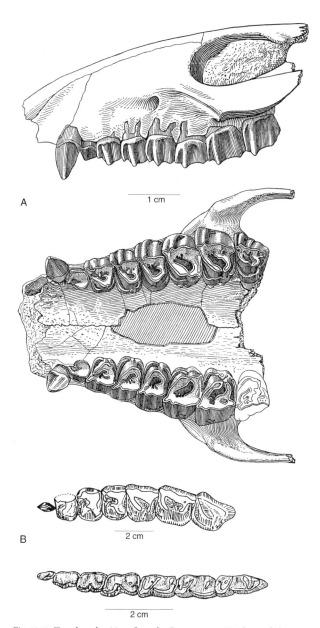

Fig. 12.19. Toxodont dentitions from the Casamayoran: (A) *Pleurostylodon*, snout in lateral and palatal views; (B) *Pampahippus*, left P^1–M^3 and P_1–M_3. (A from Simpson, 1967; B from Bond and Lopez, 1993.)

dactyl, with the central digit largest, and as in nearly all other notoungulates the digits were tipped with hoofs.

Another toxodont family present in the Casamayoran was the Notohippidae, the earliest of which were *Pampahippus* and *Plexotemnus*, both from Argentina. Notohippids are generally distinguished from isotemnids by having more hypsodont cheek teeth (Simpson, 1980). *Pampahippus* (Fig. 12.19B), however, is low-crowned, with a close-packed tooth series (no diastemata), and retains paraconids on the lower molars, features suggesting that it is a very primitive toxodont (Bond and Lopez, 1993). *Pampahippus* is known from the Lumbrera Formation at Pampa Grande, northwestern Argentina, which is generally regarded as early Eocene in age. A recent magnetostratigraphic study, however, suggested that it could be late Paleocene (Marshall et al., 1997). The Mustersan notohippid *Eomorphippus* had moderately hypsodont cheek teeth, whereas Deseadan *Rhynchippus* and *Eurygenium* had very high-crowned teeth (Simpson, 1967; Shockey, 1997). By the Santacrucian *Notohippus* had also acquired cementum on the crowns, converging on grazing equids (Simpson, 1980).

The only other toxodonts known from the early Tertiary are the Leontiniidae. Both this family and the Toxodontidae, which did not appear until the Deseadan, are nested within the paraphyletic Notohippidae (Shockey, 1997). Notohippids and leontiniids share several derived features of the tarsus and auditory region with toxodontids (Cifelli, 1993c). The oldest and most primitive leontiniid, *Martinmiguelia,* comes from the Mustersan (Bond and Lopez, 1995); leontoniids are otherwise unknown until the Deseadan and later. *Martinmiguelia* retained a relatively primitive dentition (upper dental formula 3.1.4.3), which was brachydont and without diastemata except for small spaces around the very small canines. Unfortunately, details of cheek tooth crown morphology are obliterated by heavy wear in the holotype specimen. The upper incisors are unusual in having labial cingula, and I^2 is larger than the other incisors. The best-known leontiniid is Deseadan *Scarrittia,* which had mesaxonic feet with reduced lateral digits.

Typotheria

Typotheres were small to medium-sized notoungulates whose primitive members also had complete, brachydont dentitions without diastemata (Fig. 12.20). The upper molars are more complex than those of notioprogonians in generally having two cristae, a longer crochet, and sometimes an antecrochet (Simpson, 1967). These crests typically result in a "face" pattern of **fossettes** on the upper molars (e.g., in Casamayoran *Notopithecus;* Fig. 12.21)—rings of enamel surrounding either an open depression or cementum in lightly worn teeth (Simpson 1967, 1980; Cifelli, 1993c). The trigonid and talonid of the lower molars are typically separated by a deep buccal groove or hypoflexid. Common trends among typotheres include increasing hypsodonty; enlargement of medial incisors; and development of a diastema through the reduction or loss of lateral incisors, canine, and anterior premolars.

McKenna and Bell (1997) recognized five families of typotheres (Oldfieldthomasiidae, Interatheriidae, Archaeopithecidae, Campanorcidae, and Mesotheriidae), all of which had diverged by the end of the Eocene; the first four were already present by the Casamayoran. The best-known early Tertiary typotheres belong to the families Oldfieldthomasiidae and Interatheriidae, for both of which skulls are known. Oldfieldthomasiidae (Fig. 12.20A,B) are considered the most primitive typotheres. At least eight genera are known, dating from the late Paleocene through middle Eocene (Itaboraian-Mustersan, and possibly Divisaderan). The closed dental series forms a more or less continual mor-

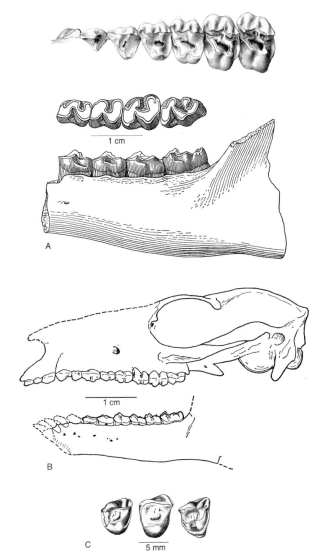

Fig. 12.20. Typothere dentitions: (A) *Oldfieldthomasia,* left upper and lower teeth; (B) *Oldfieldthomasia,* reconstructed skull; (C) *Itaboraitherium,* upper molars. (A, B from Simpson, 1967; C from Paula Couto, 1970b.)

phological gradation from the first incisor to the last molar (the canines were relatively small and incisiform). The brain of *Oldfieldthomasia* was primitive, like that of *Notostylops* (Radinsky, 1981).

Interatheres (e.g., *Notopithecus;* Fig. 12.21) were common and diverse small typotheres. They are known back at least to the Casamayoran and possibly the Riochican. Synapomorphies that unite interatheres include lower incisors and canine that are bifid lingually; I^2-P^1 that are narrow and bladelike; and deep buccal and lingual grooves separating the molar trigonid and talonid, which give the molars a bilobed appearance (Cifelli, 1993c). In addition, the maxilla forms part of the orbit to the exclusion of the jugal. Oligo-Miocene interatheres generally had hypsodont cheek teeth and include the only notoungulates with paraxonic feet and reduced lateral digits. Unfortunately, no postcrania of pre-Deseadan interatheres are known.

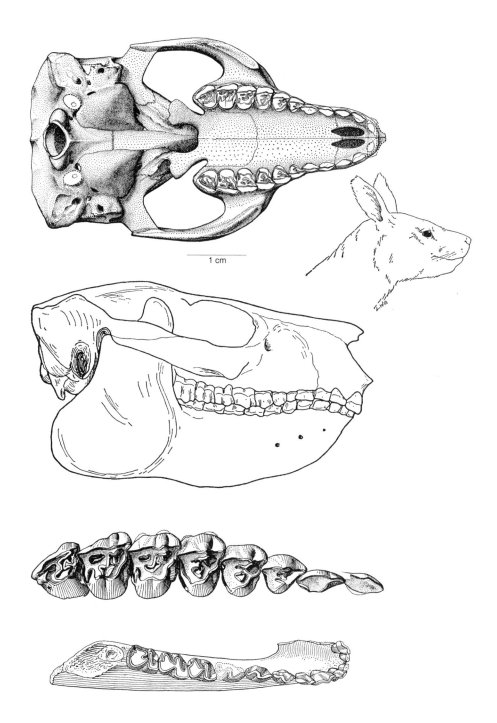

Fig. 12.21. *Notopithecus,* an Eocene typothere. (From Simpson, 1967.)

Casamayoran archaeopithecids (*Archaeopithecus* and *Acropithecus*), whose family name means "ancient ape," were small typotheres with moderately high-crowned cheek teeth, pointed incisors, and a deep mandible that deepens posteriorly. Ameghino considered them to be related to primates, despite this unusual morphology and only the most superficial resemblance to apes. Their relationships to other typotheres are still poorly understood.

One group of typotheres, the Mesotheriidae, became dentally quite specialized, some forms losing several anterior teeth while the remaining teeth became hypselodont (ever-growing). Though best known from Miocene and later beds, a few upper teeth, referred to the subfamily Trachytheriinae, have been reported from the early Tertiary. They demonstrate that at least as early as the nominal middle Eocene (Divisaderan) the teeth were already ever-growing and have a smooth buccal surface without ridges or cingula (Simpson et al., 1962).

Hegetotheria

This derived group of mostly small notoungulates appears to be phylogenetically nested within the Typotheria, possibly as the sister group of Mesotheriidae (Cifelli, 1993c). If this hypothesis is correct, the conventional Typotheria is paraphyletic. Hegetotheres are best known from the later

Tertiary. Only a few genera of hegetotheres, mostly based on teeth, have been reported from the Paleocene or Eocene. The oldest is *Eohyrax* (Archaeohyracidae), which is known from the Casamayoran and questionably from the Riochican. It had moderately hypsodont cheek teeth (Simpson, 1967). Archaeohyracids (Fig. 12.22A) are the more primitive sister group of the hegetotheriids. Hegetotheriids were hypselodont, with large, gliriform incisors and usually an anterior diastema due to loss of one or two incisors, the canine, and the first premolar. However, no diastema was yet present in Divisaderan *Ethegotherium* (Fig. 12.22B), the only pre-Oligocene hegetotheriid. It retained a primitive dental formula (3.1.4.3/3.1.4.3), with only slight gaps between the front teeth (Simpson et al., 1962). The cheek teeth and I^1 of *Ethegotherium* were apparently hypselodont and, as in other hegetotheriids, I^1 was enlarged and the molar crowns were greatly simplified. Later hegetotheres became rabbitlike, with much longer hind legs than forelegs (Simpson, 1980).

Litopterna

Litopterns were cursorial ungulates in which the limb skeleton became highly specialized, while the dentition, at least in early forms, remained relatively conservative and didolodont-like (Cifelli, 1993c). Although skeletons of early litopterns are unknown, cursorial adaptation is already evident in the tarsal bones of late Paleocene forms (Fig. 12.23C). As many as five families are known from the Paleocene and Eocene (Protolipternidae, Notonychopidae, Proterotheriidae [or Anisolambdidae], Macraucheniidae, and Adianthidae), but the phylogenetic position of several of the basal forms is in dispute. The skeleton is relatively well known only in the later proterotheres and macraucheniids, in which the clavicle was apparently absent and the hands and feet are mesaxonic and already reduced to three digits. Some Miocene proterotheres showed a remarkable convergence toward horses, one lineage even reducing the feet to a single functional digit (*Thoatherium*) long before this was achieved by equids. In addition, the wrist elements show a

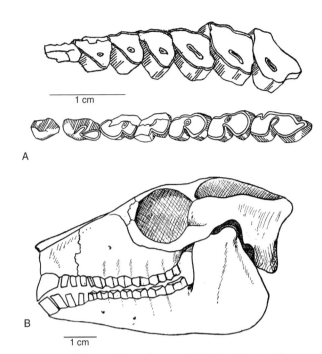

Fig. 12.22. Middle Eocene hegetotheres: (A) left P^1–M^3 and P_1–M_3 of the Mustersan archaeohyracid *Bryanpattersonia*; (B) skull of the Divisaderan hegetotheriid *Ethegotherium*. (A from Simpson, 1967; B from Simpson et al., 1962.)

"reversed alternating" arrangement, in which the cuneiform articulates with the magnum, preventing the usual contact between lunar and unciform (Cifelli, 1993c). Litopterns (apart from protolipternids) were evidently browsers, with relatively low-crowned dentition characterized by molarized P^4_4, selenodont lower molars, and bunolophodont upper molars with a W-shaped ectoloph (Cifelli, 1983b). The dental formula was primitively unreduced (3.1.4.3 above and below), but some later forms lost one or more of the anterior teeth.

The most primitive well-known litopterns—the mainly late Paleocene (Itaboraian) *Protolipterna*, *Miguelsoria*, and *Asmithwoodwardia* (Fig. 12.24)—have been assigned to the

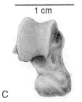

Fig. 12.23. Tarsal bones of South American ungulates: (A) *Paulacoutoia*, a didolodontid condylarth, left calcaneus and astragalus; (B) *Colbertia*, an oldfield-thomasiid notoungulate, left calcaneus and right astragalus; (C) *Miguelsoria*, a basal litoptern, right calcaneus and astragalus; (D) *Tetragonostylops*, a primitive astrapothere, left calcaneus and astragalus. (From Cifelli, 1983a.)

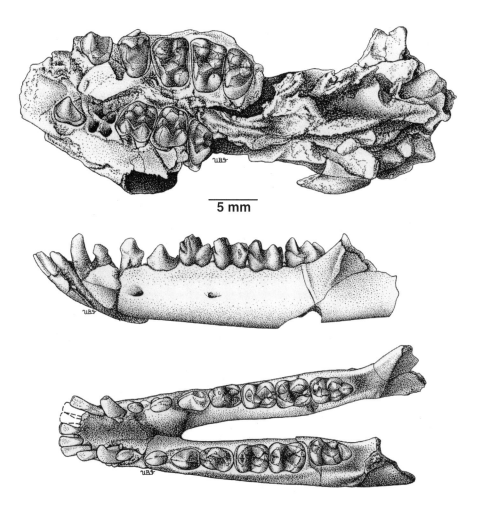

Fig. 12.24. Skull (palatal view) and mandible of *Asmithwoodwardia*, a primitive litoptern. (From Paula Couto, 1979.)

paraphyletic family Protolipternidae (Cifelli, 1983b, 1993c). The cheek teeth of these genera are brachydont and bunodont, like those of didolodontids and hyopsodontids. Earlier authors, in fact, routinely assigned them to one of those two condylarth families. They were subsequently recognized as litopterns from derived features of isolated ankle bones that were assigned to *Miguelsoria* and *Protolipterna* on account of their size and relative abundance (see Cifelli, 1983a,b; Fig. 12.23C). These tarsals are narrow and distally elongate, the astragalus with a narrow and well-defined trochlea and a semicylindrical head—synapomorphies of litopterns. Such features suggest cursorial or, considering their small size, possibly saltorial locomotion.

Possibly older than the protolipternids are the poorly known notonychopids, initially based on *Notonychops* from the late Paleocene Rio Loro Formation of Argentina (Soria, 1989b). Represented only by the dentition, *Notonychops* lacks the first premolar (dental formula 3?.1?.3.3/3.1.3.3) and has selenodont lower molars, upper premolars with prominent parastyles and metastyles, and upper molars with prominent parastyles and no hypocone on M^3. These features give the teeth a superficial resemblance to those of the tillodont *Esthonyx*. Soria (1989a,b) assigned Notonychopidae to a new order, Notopterna, in which he also included *Amilnedwardsia* and *Indalecia,* based on supposed dental differences from

litopterns. Notopterna were characterized as brachydont with litoptern-like lower molars, notoungulate-like uppers, and an unusual ear region, and were thought to have had a separate origin from other Meridiungulata. Subsequent authors have generally regarded them as litopterns (e.g., Bonaparte and Morales, 1997; McKenna and Bell, 1997), although McKenna and Bell left *Amilnedwardsia* unassigned in Meridiungulata, and Cifelli (1993c) considered *Indalecia* to be a didolodontoid. Bonaparte and Morales (1997) named a new genus, *Requisia,* based on a very fragmentary dentition from the early Paleocene of Punta Peligro, Argentina. They referred it to the Notonychopidae and regarded it as the oldest known litoptern.

Proterotheriidae were more derived than these primitive litopterns in molarizing the last two premolars and having clearly lophodont or selenodont cheek teeth (Fig. 12.25A). Three or four genera were already present by the late Paleocene (Itaboraian; *Anisolambda* and *Paranisolambda* are typical) and two more by the middle Eocene. Paleocene and Eocene proterotheres are usually grouped in the subfamily Anisolambdinae, a convention followed here. In a recent systematic revision of litopterns, however, Soria (2001) restricted the family Proterotheriidae to more derived taxa of Deseadan and later age and elevated the Paleocene and Eocene Anisolambdinae to family level. I^2 of anisolambdines

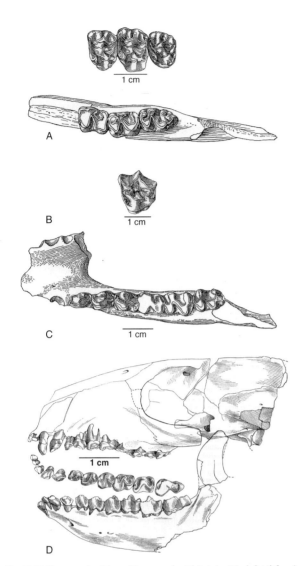

Fig. 12.25. Litoptern dentitions: (A) proterotheriid *Anisolambda*, left M^{1-3} and M_{1-3}; (B) sparnotheriodontid *Victorlemoinea*, left upper molar; (C) macraucheniid *Polymorphis*, partial mandible with left P_2-M_3; (D) adianthid *Adiantoides*, skull and mandible. (A–C from Simpson, 1948; D from Simpson et al., 1962.)

is tusklike, with a triangular cross-section (Cifelli, 1993c). Paula Couto (1979) described their primitive molars as intermediate between those of bunodont didolodontids and more selenodont forms, such as *Victorlemoinea*.

Macraucheniids were the latest surviving litopterns, persisting into the late Pleistocene. The best-known representative is probably Pleistocene *Macrauchenia*, a large, camel-like form with a moderately long neck, a high nasal opening, a short proboscis, and very hypsodont teeth. Whether macraucheniids were present as early as the late Paleocene depends on the true affinities of *Victorlemoinea* (Fig. 12.25B) and other Sparnotheriodontidae. As noted above (in the section on didolodontid condylarths), these taxa could be either didolodontoids (based on characteristics of referred tarsal elements) or litopterns (based on their selenodont molars). In the latter case, they have been interpreted as a subfamily of Macraucheniidae (e.g., McKenna and Bell, 1997). If they are excluded from Macraucheniidae, however, the first occurrence of the family is not until the Mustersan (*Polymorphis*; Fig. 12.25C). All these early litopterns had low-crowned, moderately selenodont cheek teeth.

The adianthids or "pygmy litopterns" (Cifelli and Soria, 1983) are the last of the Early Tertiary litoptern families to appear in the fossil record, being first known from the early Eocene (Casamayoran). They were small, gracile forms with relatively derived, selenodont lower cheek teeth and lophate upper molars with three main fossettes formed by enlarged conule cristae (Fig. 12.25D). Here again there is confusion concerning Eocene genera. *Indalecia* and *Adiantoides*, initially considered aberrant adianthids, were transferred from Litopterna to Didolodontoidea by Cifelli (1993c), who considered their litoptern-like selenodont cheek teeth to be convergent. (Soria, 1989a, had included *Indalecia* in his new order Notopterna.) Earlier, Cifelli (1983b) observed similarities between Casamayoran *Proectocion* (classified by Simpson, 1948, as a didolodontid) and adianthids and transferred it to the Adianthidae. Most authors regard all three of these Casamayoran genera to be the oldest adianthids. These disagreements emphasize the dental resemblances between didolodontids and primitive litopterns. Once again, more complete fossils, particularly dentitions associated with hind limb skeletons, would help to resolve the controversy. Except for these Eocene genera, adianthids are unknown until the Deseadan.

Astrapotheria

The 16 genera of astrapotheres (McKenna and Bell, 1997) constitute one of the most bizarre orders of mammals. They existed in South America from the late Paleocene (Itaboraian) into the middle Miocene (Friasian). Most astrapotheres were large, somewhat rhinoceros-like animals, but this description hardly does justice to their oddity. The name, which translates as "lightning beasts," is probably an allusion to their size (analogous to North American *Brontotherium*—"thunder beast"—named just a few years before *Astrapotherium*).

Astrapotheres are characterized by lophodont molars and tusklike canine teeth (Fig. 12.26), which became very large and ever-growing in the later, more derived forms. The upper molars lack an ectocingulum and are dominated by well-developed ectoloph and protoloph. Additional lophs formed in some derived taxa (e.g., a metaloph and a crochet-like loph from the ectoloph in Mustersan *Astraponotus*). The lower molars have two cross-lophs, including a high protocristid, and eventually became almost selenodont. As a result of these modifications the cheek teeth resemble those of notoungulates, but the similarity is thought to have arisen independently (Simpson, 1980; Cifelli, 1993c). The cheek teeth of astrapotheres are generally similar to those of rhinocerotoids and even share details of enamel microstructure with them, suggesting that their teeth functioned in a similar way (Rensberger and Pfretzschner, 1992).

The postcranial skeleton, best known in late Oligocene and Miocene astrapotheres, is moderately robust and more

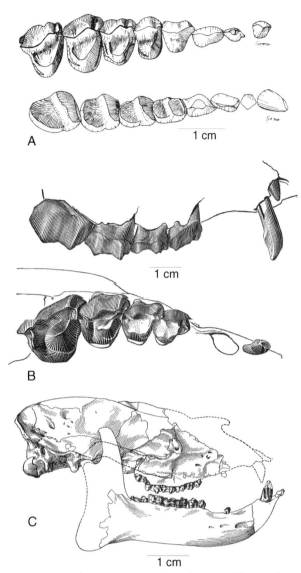

Fig. 12.26. Astrapotheres: (A) *Eoastrapostylops*, right upper and lower teeth; (B) *Scaglia*, right upper teeth; (C) *Trigonostylops*, skull. The suggestion of retracted nasals, shown here, is now known to be erroneous. (A from Soria and Powell, 1981; B, C from Simpson, 1967.)

prisingly small and weak, which suggested to Scott (1937) that these large astrapotheres were amphibious.

Three families of astrapotheres are currently recognized, late Paleocene Eoastrapostylopidae, Paleocene-Eocene Trigonostylopidae (considered a separate order by Simpson, 1967), and the Eocene-Miocene Astrapotheriidae. The oldest known astrapotheres are late Paleocene *Eoastrapostylops* (the only known eoastrapostylopid, from the Rio Loro Formation, Riochican of Argentina) and *Tetragonostylops* (the oldest trigonostylopid, Itaboraian of Brazil). *Eoastrapostylops* (Fig. 12.26A) is also considered to be the most primitive known astrapothere (Soria and Powell, 1981; Soria, 1987; Cifelli, 1993c). It has relatively low-crowned, lophoselenodont cheek teeth. The fourth premolars are molariform, and P^4–M^3 are triangular and lack hypocones. Trigonostylopids differ from other astrapotheres in details of the ear region, which prompted Simpson (1967) to assign them to a separate order; but shared similarities in the dentition and tarsus have led subsequent authors to include them in Astrapotheria (e.g., Soria, 1982, 1987; Soria and Bond, 1984; Cifelli, 1983a, 1993c). Eocene *Trigonostylops* (Fig. 12.26C) is one of only a few eutherian taxa that have been found in Early Tertiary strata of Antarctica (Hooker, 1992b).

The Eocene (Casamayoran) astrapotheriids *Scaglia* (Fig. 12.26B) and *Albertogaudrya* were sheep-sized to small tapir-sized animals, and were already among the larger South American mammals of their day; *Trigonostylops* was considerably smaller. The order is best known from the grotesque late Oligocene–early Miocene *Astrapotherium*, an elephantine beast about 3 m long that had lost the upper incisors (the lower incisors cropped against a horny plate in the upper jaw) and developed hippolike, ever-growing canine tusks. The anterior premolars were reduced or lost, creating a diastema between the tusks and the moderately hypsodont cheek teeth, which is especially conspicuous in the mandible. The nasal bones were short and retracted, indicating a moderately developed proboscis (Scott, 1928). The small Eocene astrapothere *Trigonostylops* had small, rooted tusks and diastemata, but it did not have retracted nasals (Soria and Bond, 1984) and therefore probably lacked a proboscis.

Pyrotheria

Pyrotheres are a small and rare group of medium-sized to large mammals known only from the Eocene and Oligocene of South America. The name, meaning "fire beasts," is presumably a reference to the provenance of the first known specimens, which came from volcanic ash–bearing sediments (Simpson, 1980). Most pyrotheres are known only from jaw fragments. The cheek teeth are quadrate and bunodont in the most primitive forms (the colombitheriids *Proticia* and *Colombitherium*; Fig. 12.27A,B) and distinctly bilophodont in derived members. The posterior premolars are fully molariform. A diastema separates the premolars from the large, procumbent, tusklike incisors, a pair in each upper jaw and a single incisor in each side of the mandible (Patterson, 1977;

or less graviportal in the largest forms, although the long bones (especially in the hind limb) are longer and more slender than might be expected (Scott, 1928; Simpson, 1967; Cifelli, 1993c). The neck is moderately long and massive compared to the rest of the vertebral column, to support the large head and proboscis (see below). The manus and pes are pentadactyl and contain short and stout podial and metapodial elements. Especially characteristic are the relatively flat astragalus, with a short neck and flat head that articulates with both navicular and cuboid (the alternating condition), and the calcaneus with its greatly enlarged peroneal tubercle (Fig. 12.23D; Cifelli, 1983b, 1993c). Despite these graviportal modifications, the hind limbs appear sur-

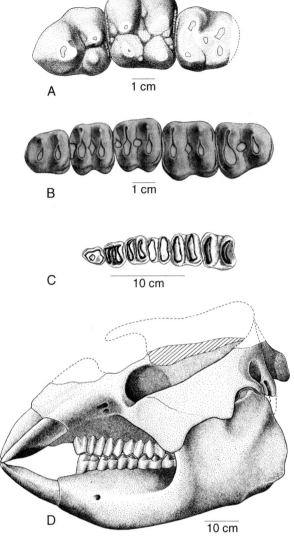

Fig. 12.27. Pyrotheres (anterior to left): (A) *Proticia*, right P_3–M_1; (B) *Colombitherium*, left P^3–M^3; (C) *Pyrotherium*, left P^2–M^3; (D) *Pyrotherium*, skull. (A, D from Patterson, 1977; B from Hoffstetter, 1970; C from MacFadden and Frailey, 1984.)

suggesting that it may actually have come from late Oligocene rocks and might be a sirenian rather than a pyrothere. All other Eocene pyrotheres show some degree of bilophodonty, though less well developed than in *Pyrotherium* (Fig. 12.27C).

Colombitherium, based on a tapir-sized maxillary dentition from the middle(?) Eocene of Colombia, has almost square upper cheek teeth with weakly developed cross-lophs, foreshadowing the strong lophs of later pyrotheres (Hoffstetter, 1970). Casamayoran *Carolozittelia* has usually been regarded as a pyrothere because of its bilophodont M^3, but other cheek teeth resemble those of the xenungulate *Carodnia*; hence its allocation to Pyrotheria is not certain (Cifelli, 1993c).

The best-known pyrothere is also the latest occurring— Deseadan *Pyrotherium* (Fig. 12.27C,D), which is known from the skull—nearly a meter long!—and part of the skeleton. The external nares were high on the rostrum, implying that a short proboscis was present, as in astrapotheres. The strongly bilophodont cheek teeth and tusklike incisors led some early workers, including Florentino Ameghino and Frederick Loomis, to suggest a relationship to proboscideans. The resemblance appears to be only superficial and is now regarded as convergent. Neverthless, Shockey and Anaya Daza (2004) recently observed that *Pyrotherium* shares several tarsal features with the African embrithopod *Arsinoitherium* (see Chapter 13), including an alternating tarsus; a broad, flat astragalus and a dorsoventrally flattened calcaneal tuber; a concave ectal (posterior calcaneal) facet on the calcaneus; continuous ectal and sustentacular facets; reduced calcaneocuboid contact; and no astragalar neck. The last four traits are unique to these two genera. In view of their geographic separation, it is easy to conclude that these are simply graviportal specializations that evolved convergently, but the remote possibility of a phylogenetic relationship between these two large, bilophodont forms is intriguing.

The bunodont dentitions of primitive pyrotheres suggested to McKenna (1980b) that they could be related to didolodontid condylarths. There is a substantial size difference, however, and much of the resemblance could be primitive (Cifelli, 1993c). Nonetheless, Cifelli postulated that the enigmatic Casamayoran genus *Florentinoameghinia* could be a structural intermediate between didolodontids and pyrotheres. The affinities of this genus have perplexed generations of paleontologists. The possibility that it represents a sirenian rather than a meridiungulate (McKenna, 1980b; Sereno, 1982) should be reexamined if new specimens come to light.

Patterson (1977) found derived features in the ear region of pyrotheres to be so similar to those of notoungulates that he regarded Pyrotheria as a suborder of Notoungulata. A subsequent study of basicranial anatomy upheld a close relationship between pyrotheres and notoungulates (McGehee and Gould, 1991). Although the dental anatomy seems to contradict this (Simpson, 1980), it remains the only reasonably well-founded phylogenetic hypothesis for pyrotheres (Cifelli, 1993c).

MacFadden and Frailey, 1984; Cifelli, 1993c); the canines are absent. The half-dozen genera included in the Pyrotheria, assigned to the families Pyrotheriidae and Colombitheriidae, are appropriately segregated in their own order, pending better understanding of their relationships.

Most pyrotheres are considered to be from the Eocene, although the precise age is often uncertain, and with the recent revisions in South American Land-Mammal Ages even these general estimates may be inaccurate. Eocene pyrotheres are known only from fragmentary jaws and teeth, making positive identification difficult. The most primitive and perhaps the oldest putative pyrothere (?early Eocene) is *Proticia*, known from a single mandibular fragment with P_3–M_1 from Venezuela (Patterson, 1977). Its teeth are very bunodont and show almost no indication of the bilophodonty that characterizes later species. Sánchez-Villagra et al. (2000) recently questioned the age and affinities of *Proticia*,

Xenungulata

This peculiar order, proposed for the enigmatic late Paleocene genus *Carodnia* (Fig. 12.28), is characterized by bilophodont M^{1-2} and M_{1-2}, similar to those of southern pyrotheres, and more complex lophate third molars, resembling those of northern uintatheres (Paula Couto, 1952c, 1978; McKenna, 1980b). The similarities to either group are quite restricted, however, leaving their relationships highly ambiguous. Van Valen (1988) suggested that they could be derived from arctocyonids, but intermediate stages are unknown. The bones of the manus and pes of *Carodnia* are short and robust and the digits terminated in broad, flat, unfissured hooflike unguals (Paula Couto, 1952c; Cifelli, 1993c). They do not compare closely to those of other meridiungulates, with the possible exception of *Pyrotherium*.

Subsequent discovery of a smaller and slightly more primitive xenungulate genus, *Etayoa,* in upper Paleocene strata of Colombia, strengthens the distinctiveness of Xenungulata. Unlike *Carodnia,* *Etayoa* lacks lophate molar talonids. Distinct lophodonty is also absent in basal pyrotheres, such as *Proticia,* suggesting that bilophodonty evolved separately in xenungulates and pyrotheres (Villarroel, 1987). *Etayoa* is also less like uintatheres in third molar morphology than is *Carodnia,* suggesting that this resemblance, too, arose independently. Villarroel noted possible dental similarities between *Etayoa* and primitive astrapotheres.

DINOCERATA

The uintatheres of North America and Asia, order Dinocerata, are without doubt among the most bizarre mammals of the Early Tertiary. Whether they properly belong in a discussion of ungulates is an unsettled question at present, but most authors agree that regarding them as aberrant ungulates is preferable to other alternatives. Uintatheres existed for a relatively short period, appearing in the late Paleocene (Tiffanian) and becoming extinct by the late Eocene. The most derived uintatheres attained enormous size and grotesque appearance, vaguely rhinolike but with three pairs of large bony protuberances on the head, which inspired the ordinal name, meaning "terrible horns" (Fig. 12.29). Although they spanned a considerable size range, from the medium-sized late Paleocene–early Eocene *Prodinoceras* and *Bathyopsis* to the rhino-sized middle Eocene *Eobasileus,* even the smaller uintatheres were ponderous animals for their time. Two families are usually recognized, Prodinoceratidae for the oldest and most primitive genus, *Prodinoceras* (with many synonyms, including *Probathyopsis* and *Mongolotherium;* Dashzeveg, 1982; Schoch and Lucas, 1985), and Uintatheriidae for five or so other genera (Lucas and Schoch, 1998b). Some authors assign all dinoceratans to a single family.

The teeth of uintatheres are highly distinctive and remarkably uniform across genera (Fig. 12.30). The primitive dental formula is 3.1.3–4.3 both above and below, and includes large saberlike upper canines, molarized posterior premolars, and moderately low-crowned lophodont molars

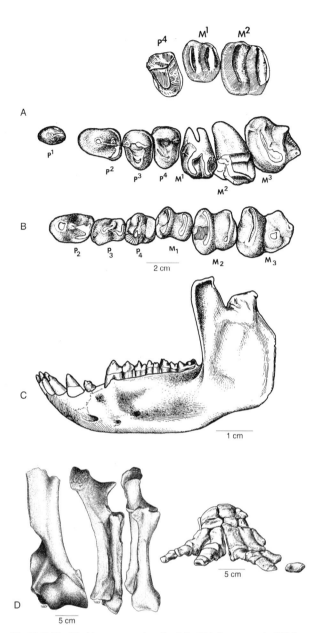

Fig. 12.28. The Riochican xenungulate *Carodnia:* (A) left upper teeth; (B) left lower teeth; (C) mandible; (D) partial right humerus, radius and ulna, and manus. (From Paula Couto, 1978.)

that increase in size posteriorly. The molariform upper teeth (P^3–M^3) are dominated by the protoloph (=paraloph) and metaloph, which run lingually from the paracone and metacone to converge at the protocone, forming a large V-shaped crest that points lingually. There are broad anterior and posterior cingula that may be continuous around the lingual border. The lower cheek teeth are dominated by the high metalophid (=protolophid) that runs between the protoconid and metaconid. A much lower crest (the cristid obliqua) extends from the hypoconid toward the prominent metastylid, which is a characteristic of uintathere lower teeth. The paraconid and paralophid are vestigial or absent. This dental pattern is highly conservative in uintatheres, with a few notable exceptions. The upper incisors were lost in Uintatheriidae, and the lower incisors, unicuspid in Pro-

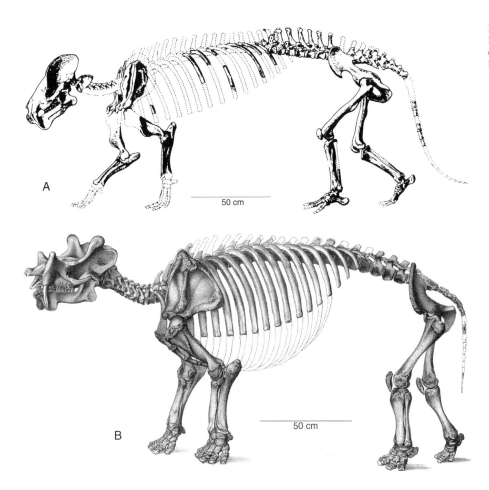

Fig. 12.29. Uintatheres: (A) late Paleocene–early Eocene *Prodinoceras*; (B) middle Eocene *Uintatherium*. (A from Flerov, 1967; B from Marsh, 1886.)

dinoceratidae, became multicuspate. Similarly, the lower canines became smaller and more incisiform as the upper canines got larger (except in Asian *Gobiatherium*, which lost the upper canines). In these modifications uintatheriids are somewhat convergent to ruminant artiodactyls (Flerov, 1967). Although no animals today have cheek teeth similar to those of uintatheres, the pronounced lophodonty is a hallmark of a herbivorous diet (despite the saberlike canines).

The mandibular symphysis was solidly fused in all uintatheres, and prominent inframandibular flanges developed anteriorly to protect the long upper canines. Both the canines and the flanges are evident even in the earliest uintatheres but are strongly sexually dimorphic in size, being much smaller in presumed females and juveniles (Flerov, 1967; Thewissen and Gingerich, 1987). *Gobiatherium*, which lacked upper canines, also lacked inframandibular flanges; instead its mandible was shallow and the symphysis rather spoutlike (Osborn and Granger, 1932). For these reasons it is sometimes placed in a separate subfamily.

The skull of *Prodinoceras* is primitive, exhibiting a very strong sagittal crest and lacking all but the faintest rudiments of horns (Flerov, 1967). It is not unlike the skull of arctocyonid condylarths but is larger, up to a half-meter long. The skulls of advanced uintatheres (e.g., *Uintatherium*, *Eobasileus*), which reached nearly a meter in length, are unique and instantly recognizable by the three pairs of prominent bony "horns" situated on the nasals, the maxillae, and the parietals. An additional pair of low frontal protuberances is present just above the front of the orbits. In these animals, the skull was relatively deep, with a broad depression in the skull roof between the temporal crests and anterior to the occipital crest. Incipient protuberances were present in the most primitive uintatheriid, *Bathyopsis*, which is both morphologically and stratigraphically intermediate between *Prodinoceras* and later uintatheres. Here again *Gobiatherium* differs from all other uintatheres. It had a very flat skull with peculiar arched nasals bearing small protuberances in presumed males, but no other trace of horns—another indication of its divergence from other uintatheres.

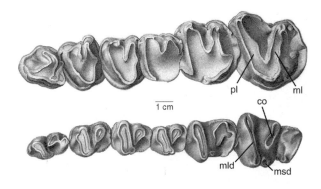

Fig. 12.30. Left P^2–M^3 and right P_2–M_3 of *Uintatherium*. Key: co, cristid obliqua; ml, metaloph; mld, metalophid; msd, metastylid; pl, protoloph. (Modified from Marsh, 1886.)

Flerov (1967) speculated that the high nasals and spoutlike symphysis of *Gobiatherium* may have been associated with semiaquatic habits.

The postcranial skeleton of all uintatheres was massive. The limbs were robust, the intermediate elements shorter than the proximal ones, and the metapodials typically short and stout. The manus and pes were pentadactyl and, where known, the digits were short and terminated in broad hoofs. *Gobiatherium* differed in having somewhat longer and more slender metapodials and phalanges. *Prodinoceras* is more primitive than later uintatheres in having a prominent third trochanter on the femur, a shallow-grooved astragalus with a short neck, a plantigrade foot that was somewhat bearlike, and a long, well-developed tail.

The huge middle Eocene uintatheriids were graviportal, with short, robust limb bones, the femur longer than the tibia. In *Uintatherium* and its close relatives the third trochanter was absent and the tail was short. The feet were elephantine in form but digitigrade (Marsh, 1886; Flerov, 1967). The tarsal elements were wide and proximodistally compressed; the astragalus was particularly modified, having a nearly flat trochlea and no neck. These terminal uintatheres had a barrel-shaped rib cage and short legs with dense (osteosclerotic) long bones, features suggesting that they were semiaquatic (Turnbull, 2002). They also had a disproportionately large pelvis, perhaps associated with a large hindgut. Turnbull believes this trait indicates that the large uintatheres were hindgut fermenters (see also Janis, 1989), like equids and sirenians, and that slow, thorough hindgut digestion compensated for their relatively small teeth. They were among the largest Eocene land mammals, achieving weights of at least 1,450 kg (Turnbull, 2002) and perhaps much more (Lucas and Schoch, 1998b, estimated that some uintatheres attained weights of 3,000–4,500 kg). Competition with large perissodactyls during the later Eocene may have contributed to their extinction.

There can be little doubt of the monophyly of Dinocerata, based on their numerous peculiar and unique features. The broader relationships of uintatheres, however, are controversial. They have conventionally been regarded as ungulates, sometimes together with Pantodonta and other taxa in the now abandoned order Amblypoda (e.g., Osborn, 1898b; Romer, 1966) or in a larger ungulate assemblage, Simpson's (1945) superorder Paenungulata (see below). Several subsequent authors adopted this assignment, along with the implication that uintatheres are ungulates (e.g., McKenna and Manning, 1977; Cifelli, 1983a; Prothero et al., 1988). Van Valen (1978, 1988) proposed that Dinocerata was derived from the basal ungulate family Arctocyonidae (Condylarthra), and McKenna and Bell (1997) maintained Dinocerata within the grandorder Ungulata, an assignment tentatively adopted here.

Lucas (1993; see also Tong and Lucas, 1982; Schoch and Lucas, 1985; Lucas and Schoch, 1998b), however, proposed that dinoceratans are most closely allied with the South American Pyrotheria (in which he included the Xenungulata), the two together forming a clade Uintatheriamorpha. He further argued that uintatheriamorphs are not ungulates at all but instead are related to pseudictopid anagalidans, specifically *Pseudictops* (which is thought to be related to lagomorphs and rodents). Pseudictopids resemble uintatheres in having molariform posterior premolars and V-shaped lophs on the upper teeth and mesiodistally compressed trigonids with broad protolophids (=metalophids) on the lower molars. The uintathere-pseudictopid hypothesis has not garnered much support, however, perhaps because of a substantial difference in size and the specialized rabbitlike skeleton of pseudictopids, which has little in common with the generalized, robust skeleton of basal uintatheres. These conflicting features suggest that the dental similarities are only superficial.

Various authors (e.g., Gregory, 1910; Wheeler, 1961) have noted similarities between uintatheres and pantodonts, both in the dentition and in the graviportal skeleton and tarsal anatomy, but few recent authorities have found these resemblances convincing, and they are now usually ascribed to convergence. However, the long-known similarity of the third molars of dinoceratans and *Carodnia*, a Paleocene (Itaboraian-Riochican) member of the exclusively South American Xenungulata, has led to continuing claims of a special relationship between these two groups (McKenna, 1980b; Gingerich, 1985; Schoch and Lucas, 1985; Van Valen, 1988; Lucas, 1993), but resemblances in the rest of the dentition are less compelling. As mentioned before, late Paleocene *Etayoa* is less similar to dinoceratans, suggesting that *Carodnia* is merely convergent to uintatheres.

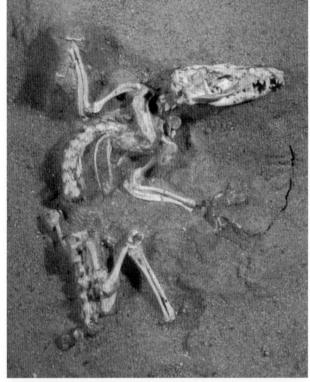

Plate 1.1. (*top*) *Ptilodus,* a Paleocene multituberculate. (From Jenkins and Krause, 1983; restoration by L. L. Sadler.)

Plate 1.2. (*bottom*) *Pucadelphys,* a Paleocene marsupial from Tiupampa, Bolivia. (From Marshall et al., 1995, courtesy of C. de Muizon.)

Plate 1.3. (*top*) *Ukhaatherium,* a primitive Late Cretaceous eutherian from Mongolia. (Courtesy of M. J. Novacek.)

Plate 1.4. (*bottom*) *Zalambdalestes,* a primitive Late Cretaceous eutherian from Mongolia. (Courtesy of M. J. Novacek.)

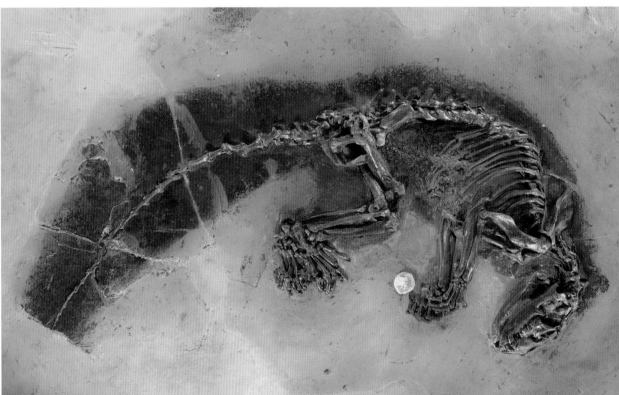

Plate 2.1. (*top*) *Palaeosinopa,* an early Eocene pantolestid from the Green River Formation, Wyoming. (From Rose and Koenigswald, 2005; photograph by G. Oleschinski.)

Plate 2.2. (*bottom*) *Kopidodon,* a middle Eocene paroxyclaenid from Messel, Germany. (Courtesy of the Forschungsinstitut Senckenberg, Frankfurt, Germany.)

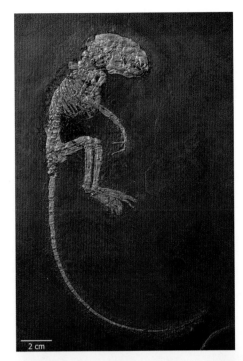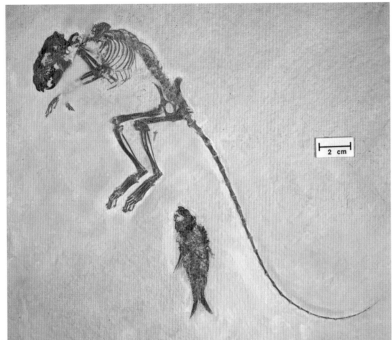

Plate 3.1. (*top left*) *Heterohyus,* a middle Eocene apatemyid from Messel, Germany. (Courtesy of W. von Koenigswald and the Hessisches Landesmuseum Darmstadt, Germany.)

Plate 3.2. (*top right*) *Apatemys* from the lower Eocene part of the Green River Formation, Wyoming. (From Koenigswald et al., 2005; photograph by J. Weinstock.)

Plate 3.3. (*bottom*) Restoration of the early Eocene pantodont *Coryphodon* (left) and the tapiromorph *Homogalax* in what is now the Bighorn Basin, Wyoming. (Courtesy of the artist, Utako Kikutani.)

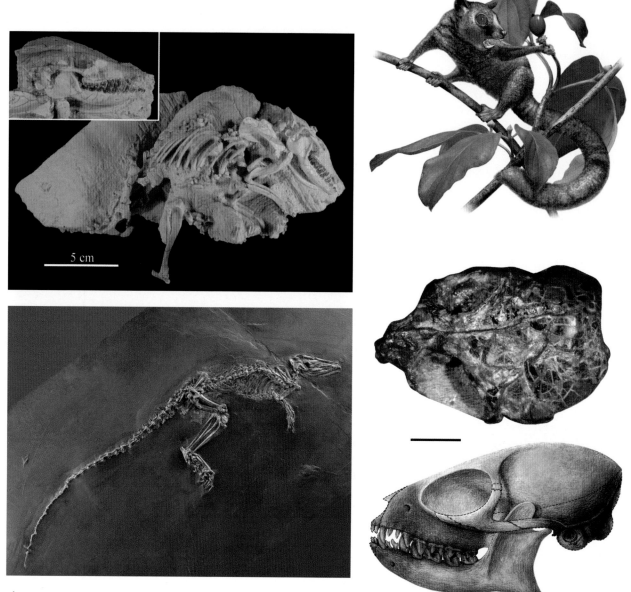

Plate 4.1. (*top*) *Leptictis,* an early Oligocene leptictid from Wyoming. (Photograph by T. Smith).

Plate 4.2. (*bottom*) *Leptictidium,* a middle Eocene leptictidan from Messel, Germany. (Courtesy of G. Storch and the Forschungsinstitut Senckenberg, Frankfurt, Germany.)

Plate 4.3. (*top*) *Carpolestes,* a late Paleocene plesiadapiform from Wyoming. (Courtesy of the artist, D. M. Boyer.)

Plate 4.4. (*bottom*) *Teilhardina,* skull and restoration of the basal euprimate *T. asiatica* from the early Eocene of China. Scale bar is 5 mm. (Courtesy of X.-J. Ni and C.-K. Li.)

Plate 5.1. (*opposite, top left*) *Icaronycteris,* a bat from the early Eocene of Wyoming. (From Simmons and Geisler, 1998.)

Plate 5.2. (*opposite, bottom left*) *Archaeonycteris,* an Eocene bat from Messel, Germany. (Courtesy of G. Storch and the Forschungsinstitut Senckenberg, Frankfurt, Germany.)

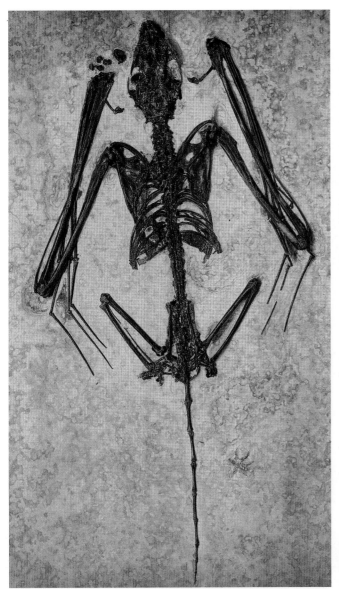

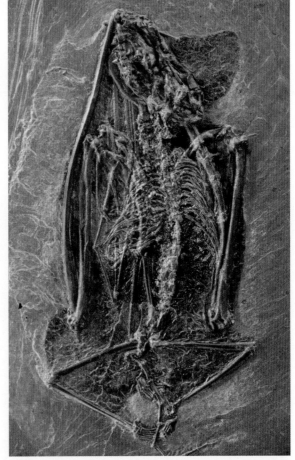

Plate 5.3. (*top right*) *Hassianycteris*, an Eocene bat from Messel, Germany. (Courtesy of G. Storch and the Forschungsinstitut Senckenberg, Frankfurt, Germany.)

Plate 5.4. (*bottom right*) *Tachypteron*, an Eocene bat from Messel, Germany. (Courtesy of G. Storch and the Forschungsinstitut Senckenberg, Frankfurt, Germany.)

Plate 6. Early Eocene mammals in what is now the Bighorn Basin, Wyoming. At right, a troop of the adapoid primate *Cantius* forages in a laurel tree, while another *Cantius* pair and a pair of the arctocyonid *Chriacus* are seen in a sycamore at left, above a pair of the basal artiodactyl *Diacodexis*. In the center a giant bird (*Diatryma*) attacks a herd of dawn horses *Hyracotherium*, while at lower left the mesonychid *Pachyaena* scavenges a carcass. (Courtesy of the artist, Utako Kikutani.)

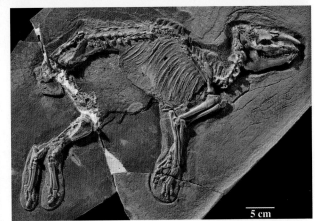

Plate 7.1. (*top*) *Eurotamandua,* a middle Eocene possible anteater or palaeanodont relative from Messel, Germany. (Courtesy of G. Storch and the Forschungsinstitut Senckenberg, Frankfurt, Germany.)

Plate 7.2. (*bottom*) *Chriacus,* an early Eocene arctocyonid from Wyoming. (From Rose, 1990, courtesy of the artist, Elaine Kasmer.)

Plate 7.3. (*top*) *Hyrachyus,* a middle Eocene rhinocerotoid from Messel, Germany. (Courtesy of the Hessisches Landesmuseum, Darmstadt, Germany.)

Plate 7.4. (*bottom*) *Saghatherium,* a primitive hyracoid from the early Oligocene of Libya. (From Thomas et al., 2004, courtesy of E. Gheerbrant.)

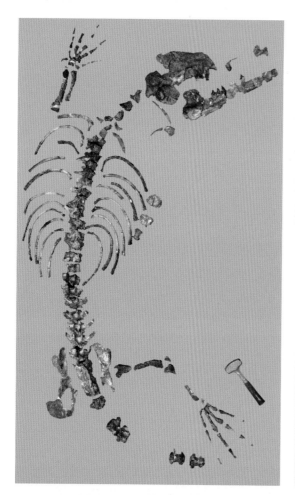
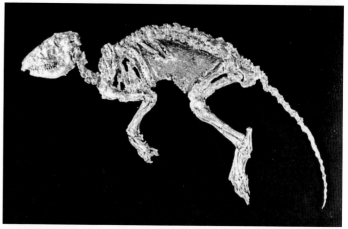

Plate 8.1. (*top left*) *Ambulocetus,* a middle Eocene archaeocete whale from Pakistan. (From http://darla.neoucom.edu/DEPTS/ANAT/Hans/AmbulocetusPhoto.jpg.)

Plate 8.2. (*bottom left*) *Rodhocetus,* a middle Eocene archaeocete whale from Pakistan. (From Gingerich, ul Haq et al., 2001, courtesy of the artist, John Klausmeyer.)

Plate 8.3. (*top right*) *Aumelasia,* a middle Eocene dichobunid artiodactyl from Messel, Germany. (Courtesy of J. L. Franzen.)

Plate 8.4. (*bottom right*) *Anthracobunodon,* a primitive middle Eocene artiodactyl from Geiseltal, Germany. (Restoration by Pawel Major, from Erfurt and Altner, 2003, courtesy of J. Erfurt.)

13

Altungulata
Perissodactyls, Hyraxes, and Tethytheres

BASED ON ANATOMICAL EVIDENCE, the "modern" ungulates appear to be divisible into two major clades, the artiodactyls and cetaceans (treated in Chapter 14), and the perissodactyls, hyracoids, and tethytheres (sirenians, proboscideans, and their extinct kin), which are the subject of this chapter. The latter clade was given the name Altungulata by Prothero and Schoch (1989). Altungulata is essentially the same as Pantomesaxonia, as used by Fischer (1986), Prothero et al. (1988), and some subsequent authors, but markedly different from the original concept of Pantomesaxonia; consequently, the latter name has not been widely accepted. Molecular evidence supports the artiodactyl-cetacean clade, but not Altungulata, unless perissodactyls are excluded. The taxonomic ranks of these clades vary in different accounts. The less inclusive taxa (i.e., Artiodactyla, Cetacea, Perissodactyla, Hyracoidea, Sirenia, and Proboscidea) have generally been recognized as orders, but in the cladistic classification of McKenna and Bell (1997) only Artiodactyla and Perissodactyla are considered orders, whereas the others are accorded the lesser ranks of suborder, infraorder, or parvorder (see Table 13.1).

Altungulata is united by numerous derived dental, osteological, and soft-tissue features (Prothero et al., 1988; Thewissen and Domning, 1992; Fischer and Tassy, 1993; Gheerbrant, Domning, and Tassy, 2005). They include bilophodonty, large third molars, molarized posterior premolars (only the deciduous premolars in sirenians), presence of an elongate thoracic region with at least 19 thoracic vertebrae, absence of a clavicle, and similar development of fetal membranes. There is controversy, however, over the homology and distribution of some of these supposed synapomorphies. Consequently, the reality of this clade is open to question and has been seriously challenged by molecular studies, which separate the perissodactyls (in

Table 13.1. Classification of Altungulata

Mirorder ALTUNGULATA
    †*Radinskya*, †*Olbitherium*
  Order PERISSODACTYLA
    Suborder HIPPOMORPHA
      Superfamily Equoidea
        Equidae
        †Palaeotheriidae
    Suborder TAPIROMORPHA
      †Isectolophidae
      Infraorder CERATOMORPHA
        Superfamily Tapiroidea
          †Helaletidae
          †Deperetellidae
          †Lophialetidae
          Tapiridae
        Superfamily Rhinocerotoidea
          †Hyrachyidae[1]
          †Hyracodontidae
          †Amynodontidae[2]
          Rhinocerotidae
      Infraorder †ANCYLOPODA
        †Eomoropidae
        †Chalicotheriidae
        †Lophiodontidae[3]
    Suborder †TITANOTHERIOMORPHA
      Superfamily †Brontotherioidea[4]
        †Brontotheriidae
        †Anchilophidae
  Order PAENUNGULATA (=URANOTHERIA)[5]
    Suborder HYRACOIDEA[6]
      †Pliohyracidae
      Procaviidae
    Suborder TETHYTHERIA
      Infraorder †EMBRITHOPODA
        †Phenacolophidae
        †Arsinoitheriidae
      Infraorder SIRENIA
        †Prorastomidae
        †Protosirenidae[7]
        Dugongidae
        Trichechidae
      Infraorder †DESMOSTYLIA
        †Desmostylidae
      Infraorder PROBOSCIDEA
        †Anthracobunidae[8]
        †Phosphatheriidae[9]
        †Numidotheriidae
        †Moeritheriidae
        †Barytheriidae
        †Deinotheriidae
        †Palaeomastodontidae
        †Phiomiidae
        †Hemimastodontidae
        †Mammutidae
        †Gomphotheriidae
        Elephantidae

Notes: Perissodactyla mainly after Hooker (2005); paenungulates modified after McKenna and Bell (1997). The dagger (†) denotes extinct taxa. Families and genera in boldface are known from the Paleocene or Eocene.

[1] Sometimes included in Hyracodontidae.
[2] Sometimes included in Rhinocerotidae.
[3] Sometimes included in Tapiroidea.
[4] May be closely allied with Equoidea; anchilophids may be palaeotheriid equoids.
[5] This order, together with Macroscelidea, Tubulidentata, and the lipotyphlan families Tenrecidae and Chrysochloridae, compose the Afrotheria of molecular systematics.
[6] Considered by some authors to be the sister taxon of Perissodactyla.
[7] Sometimes included in Dugongidae.
[8] Allocation to Proboscidea is controversial; may be stem tethytheres or belong to another ungulate clade.
[9] Sometimes included in Numidotheriidae.

Laurasiatheria) from the remaining altungulates (in Afrotheria; e.g., Murphy et al., 2001).

The monophyly of hyracoids + tethytheres, however, has been supported by most recent morphological and molecular research (e.g., Shoshani, 1986, 1993; Novacek and Wyss, 1986; Springer and Kirsch, 1993; Lavergne et al., 1996; Stanhope et al., 1998; Amrine and Springer, 1999; Murphy et al., 2001; Nikaido et al., 2003; Gheerbrant, Domning, and Tassy, 2005). This clade is usually called Paenungulata (Fig. 13.1). Simpson (1945), however, used Paenungulata in a much broader sense, to include Pantodonta, Dinocerata, and Pyrotheria as well—groups no longer believed to be related to hyracoids and tethytheres. Consequently, McKenna and Bell (1997) proposed the new name Uranotheria for the more restricted group of hyracoids and tethytheres. The name Paenungulata (in the restricted sense) is in much wider use and, therefore, is applied to this clade here.

Although molecular data are virtually unanimous in supporting Paenungulata, some morphological studies have concluded that perissodactyls and hyracoids are more closely allied to each other than either is to tethytheres (Fischer, 1986, 1989; Prothero et al., 1988; Prothero and Schoch, 1989; Fischer and Tassy, 1993). These studies consider several proposed paenungulate synapomorphies to be homoplasies and cite such features as mesaxonic foot symmetry and the presence of a eustachian sac as probable synapomorphies of a perissodactyl + hyracoid clade. They place the latter clade as the sister group of Tethytheria.

The monophyly of Tethytheria—consisting of extant Proboscidea and Sirenia, and extinct Desmostylia and Embrithopoda—is strongly supported by anatomical and most molecular data, but there is no consensus as to the precise interrelationships of the constituent groups (e.g., Domning et al., 1986; Tassy and Shoshani, 1988; Court, 1990, 1992c; Fischer and Tassy, 1993; Springer and Kirsch, 1993; Domning, 1994; Ray et al., 1994; Savage et al., 1994; Lavergne et al., 1996; Murphy et al., 2001). Some recent molecular research, however, suggests that hyracoids form a clade with either sirenians or proboscideans to the exclusion of the other, thus contradicting tethythere monophyly (e.g., Amrine and Springer, 1999; Madsen et al., 2001). This confusion may reflect roughly contemporaneous divergence of all these groups from the paenungulate stem. Here Tethytheria is considered to be monophyletic and to include Proboscidea, Sirenia, Desmostylia, and Embrithopoda, following Domning et al. (1986), Ray et al. (1994), and Gheerbrant, Domning, and Tassy (2005). Tethytheres are united by anteriorly shifted orbits, cheek teeth that are at least somewhat bilophodont, and a few other features. Auditory characters of the petrosal bone that have been considered important in linking Proboscidea and Sirenia, however, may

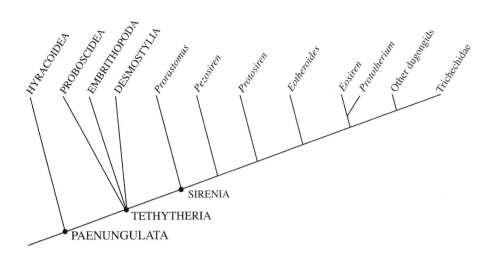

Fig. 13.1. Relationships of Paenungulata, emphasizing Sirenia. (Based mainly on Domning, 1994.)

be homoplasious (Court and Jaeger, 1991). The early Eocene Anthracobunidae, discussed below with Proboscidea, are possibly the sister taxon of all other tethytheres (Gingerich, Russell, and Wells, 1990).

Late Paleocene *Radinskya* (originally referred questionably to the Phenacolophidae) and the Paleocene-Eocene Phenacodontidae (see the section on Condylarthra in Chapter 12) have been considered to be pivotal taxa that lie close to the base of Altungulata (Fig. 13.2A,B). *Radinskya* is known only from a partial skull and upper dentition, limiting assessment of its exact relationships. The upper molars are quadrate with a rhomboid outline and a weak, π-shaped crown pattern formed by the incipient ectoloph, protoloph, and metaloph. This arrangement closely resembles the crown pattern of early perissodactyls, but the strong conules and some other characters suggest relationship to phenacolophids (McKenna et al., 1989; Hooker and Dashzeveg, 2003). *Radinskya* may be the sister taxon of all other Altungulata or may be closer to the origin of Perissodactyla than is any phenacodontid (McKenna et al., 1989).

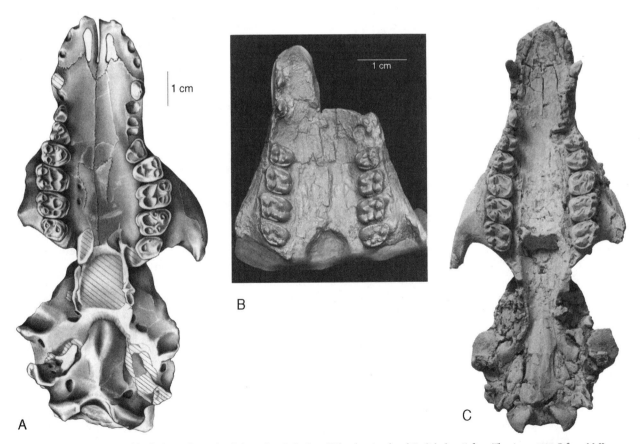

Fig. 13.2. Palatal views: (A) condylarth *Phenacodus*; (B) basal altungulate *Radinskya*; (C) basal perissodactyl *Cardiolophus*. (A from Thewissen, 1990; B from McKenna et al., 1989; C from Gingerich, 1991.)

Another taxon that appears to be relevant to the origin of Altungulata or Perissodactyla is the recently described *Olbitherium* from the early Eocene Wutu Formation of China (Tong et al., 2004). Its lower teeth resemble those of perissodactyls slightly more than phenacodontids, and the upper molars are quadrate with an incipient π-shaped pattern. Apart from these two genera, Phenacodontidae appears to be the closest outgroup to Altungulata. In view of the unsettled relationships of the major groups of Altungulata, however, it may be premature to contend that one of these is closer to the origin of Perissodactyla (Thewissen and Domning, 1992; Fischer and Tassy, 1993).

PERISSODACTYLA

The perissodactyls, or "odd-toed ungulates," are represented by today's horses, tapirs, and rhinoceroses. They are but a remnant of a much wider Tertiary radiation that also included the extinct chalicotheres and titanotheres. Perissodactyls include several of the largest land mammals ever known. Among living perissodactyls, tapirs are the most conservative and provide a reasonably good model of what the early perissodactyl postcranial skeleton was like, but even tapirs have modified the skull and dentition from the primitive condition (though less so than horses or rhinos).

Perissodactyls appeared abruptly at the beginning of the Eocene (Wasatchian, Ypresian, and Bumbanian land-mammal ages) across the northern continents, probably having evolved during the Paleocene from phenacodontid condylarths. Early perissodactyls were advanced compared to phenacodontids in having molars with a greater emphasis on transverse shearing and a more cursorially specialized skeleton (Radinsky, 1969). They soon became among the most common animals in Holarctic faunas and by the end of the early Eocene had differentiated into all the principal clades (horses, tapirs + rhinos, chalicotheres, and brontotheres). Several possible late Paleocene perissodactyls have been reported, but no reliable pre-Eocene records have been corroborated. It seems clear that, although they probably originated in the Paleocene, perissodactyls were not a significant faunal element before the Eocene.

Anatomical evidence indicates that the most likely source of Perissodactyla is a phenacodontid condylarth. Radinsky (1966a) considered Torrejonian *Tetraclaenodon* to be the only phenacodontid primitive enough to have given rise to perissodactyls, largely because it lacks a mesostyle, as does *Hyracotherium*, then thought to be the archetypal perissodactyl. Subsequently, Hooker (1989, 1994) proposed that isectolophid tapiroids or brontotheres could be more plesiomorphic. This hypothesis, in turn, would allow derivation from a more derived phenacodontid with a mesostyle, such as *Ectocion* or *Lophocion*. Although some additional evidence has emerged supporting a basal position for certain tapiromorphs (see the section on ceratomorphs below), the fossil record currently does not indicate an unequivocal choice for the most primitive perissodactyl.

Living perissodactyls are medium-sized and large terrestrial herbivores. The perissodactyl skull (Fig. 13.2C) has an elongate snout that holds a primitive number of teeth in most horses and tapirs (3.1.4.3/3.1.4.3; the canines and P^1 are lost in some equids) and a reduced number in rhinos (incisors, canines, and one premolar sometimes absent). A postcanine diastema develops in many lines. The premolars are often molarized, and the molariform teeth are lophodont or have complex enamel patterns. They are low-crowned in browsers (most rhinos and tapirs) and high-crowned in grazers (horses). In association with their herbivorous diets, perissodactyls have evolved an enlarged cecum (the proximal part of the large intestine), which serves as a fermentation chamber in which bacteria break down cellulose (Janis, 1976). This digestive specialization contrasts with the foregut fermentation that evolved in artiodactyls. The presence of cecal digestion in all extant perissodactyls suggests that it had already evolved when perissodactyls first appeared in the fossil record.

The skeleton is specialized for running, as is evident even in the most primitive perissodactyls (Figs. 13.3, 13.4). The limbs are often but not always elongated; the clavicle is absent; the greater tuberosity of the humerus and greater trochanter of the femur project high above the proximal articulations of these bones; the deltopectoral and supinator crests of the humerus are relatively reduced (except in some rhinos); the humeral epicondyles are reduced; the proximal radius is reoriented anterior to the ulna, articulating with the full breadth of the distal humerus; the ulna is strongly concave posteriorly; the carpal bones interlock in an alternating arrangement; the distal femoral articulation is transversely narrow and anteroposteriorly deep; and the feet are mesaxonic (their plane of symmetry runs through digit 3), with a reduced number of digits terminated by broad, flat hoofs. There are typically either three toes (in rhinos and the hind foot of tapirs) or one (in living horses). Tapirs retain a fourth toe, a reduced digit V, on the forefoot. The ankle of perissodactyls is highly distinctive. The calcaneus and astragalus articulate in a manner that restricts movement between them. The astragalar trochlea (which articulates with the tibia) is deeply grooved and obliquely oriented; the astragalar neck is short; and the head (navicular facet) is gently grooved—saddle-shaped—or almost flat, rather than convex as in most other placentals. The joint modifications mentioned promote flexion-extension and restrict lateral motion. As in artiodactyls, the stance is unguligrade. The ilium of perissodactyls is greatly expanded, and the femur differs from that of artiodactyls in having a prominent third trochanter.

Following Radinsky's studies of perissodactyl evolution in the 1960s, three suborders of perissodactyls were generally recognized: Hippomorpha for the horses and brontotheres, Ceratomorpha for the tapirs and rhinos, and Ancylopoda for the chalicotheres. A close relationship between tapiroids and rhinocerotoids is now well established; but exactly how brontotheres and chalicotheres, as well as several basal perissodactyls often labeled as "tapiroids," relate to

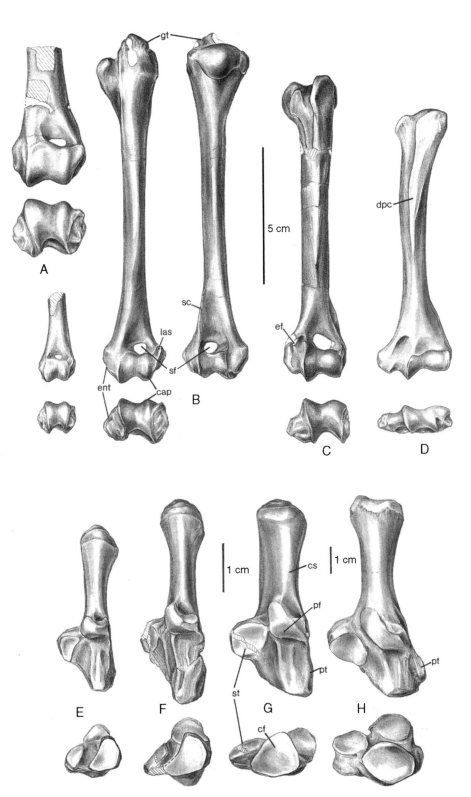

Fig. 13.3. Perissodactyl skeletal traits illustrated on left humeri (above) and left calcanei (below) of basal members, compared with condylarths: (A) distal humeri of *Heptodon* (above) and *Hyracotherium* (below); (B) *Homogalax*; (C) condylarth *Phenacodus*; (D) arctocyonid *Chriacus*; (E) *Hyracotherium*; (F) *Cardiolophus*; (G) *Homogalax*; (H) *Phenacodus*. Key: cap, capitulum; cf, cuboid facet; cs, calcaneal shaft; dpc, deltopectoral crest; ef, entepicondylar foramen; ent, entepicondyle; gt, greater tuberosity; las, lateral articular shelf; pf, proximal astragalar facet; pt, peroneal tubercle; sc, supinator crest; sf, supratrochlear foramen; st, sustentaculum tali. (From Rose, 1996b.)

other perissodactyls is controversial and not yet well resolved. Alternative hypotheses are that chalicotheres are more closely related to ceratomorphs than to other perissodactyls (Hooker, 1989; Prothero and Schoch, 1989; Colbert and Schoch, 1998; Froehlich, 1999), that chalicotheres + brontotheres make up a separate clade of ceratomorphs (McKenna and Bell, 1997, resurrecting a much older view), or that brontotheres belong to an independent perissodactyl clade separate from equoids (Hooker, 1989; Prothero and Schoch, 1989; Janis, Colbert, et al., 1998). The latest view places brontotheres as the lowest branch of Perissodactyla, with Ancylopoda as the sister taxon of a hippomorph-

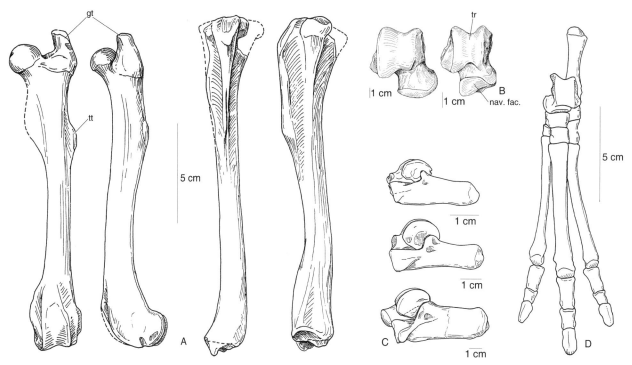

Fig. 13.4. Perissodactyl skeletal traits illustrated on hind limb bones of basal members: (A) left femur and tibia of *Homogalax* in anterior and lateral views; (B) right astragalus of *Hyracotherium* (right) compared to that of the condylarth *Phenacodus* (left); (C) lateral view of calcaneus and astragalus of (top to bottom) *Hyracotherium*, *Homogalax*, and *Phenacodus*; (D) left foot of *Hyracotherium*. Key: gt, greater trochanter; nav. fac., navicular facet; tr, trochlea; tt, third trochanter. (A, C, D from Rose, 1996b; B modified from O'Leary and Rose, 1995.)

ceratomorph clade (Hooker and Dashzeveg, 2004). These differences of opinion stem in part from the difficulty of distinguishing between synapomorphic and homoplastic resemblances in primitive perissodactyls, which is a consequence of the extent of parallelism that occurred early in their radiation (Radinsky, 1969). They also reflect the use of "different character set[s] and distinctly different polarization of those characters" (Froehlich, 1999: 151).

Holbrook's (1999, 2001) analyses of dental, cranial, and postcranial characters were unable to resolve the relationships among several groups of perissodactyls, finding instead a basal polytomy consisting of six branches: horses, brontotheres, chalicotheres, *Cardiolophus*, *Homogalax*, and a clade composed of tapirs, rhinos, and *Isectolophus* (Fig. 13.5). The three genera in this list constitute the family Isectolophidae (McKenna and Bell, 1997; Colbert and Schoch, 1998), which is usually regarded as the most primitive family of tapiroids or tapiromorphs. However, in their recent analysis, Hooker and Dashzeveg (2004) found isectolophids to be stem members of the Ancylopoda. Holbrook's analy-

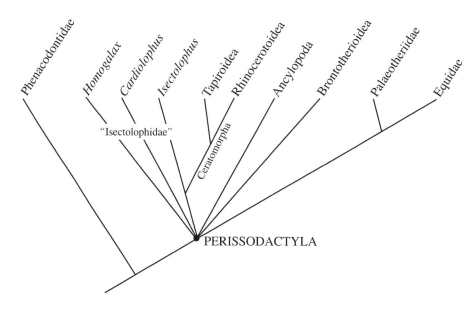

Fig. 13.5. Relationships of perissodactyls, modified mainly after Holbrook (1999).

sis underscores the primitive nature and relative homogeneity of the basal members of the major perissodactyl clades. Because of the unsettled higher taxonomy of perissodactyls, they are discussed here in the following sequence: equoids, brontotheres, ceratomorphs, and ancylopods.

Equoids

This group includes the equids (horses) and the closely related palaeotheres, which largely filled the early equid niche in Europe. They include some of the first vertebrate fossils to be formally named. The oldest equids (traditionally called *Hyracotherium* and closely allied forms) are known from the early Eocene of North America and Europe, first appearing during the brief, very warm episode that marks the onset of the Eocene (the Initial Eocene Thermal Maximum). Thereafter, equids were a strictly North American radiation until they dispersed throughout the Old World in the Miocene. They reached South America in the Pliocene. Equids survived in all of these regions at least into the late Pleistocene, but natural populations are restricted to Asia and Africa today. Palaeotheriidae are known only from the Eocene through early Oligocene of Europe. Unfortunately, the proper family affiliation of many of the early equoid genera is controversial (see below), making it difficult to characterize the families. The most important disagreements are noted in the following discussion.

Equidae

Hyracotherium (=*Eohippus*, the "dawn horse") has been widely considered to be the oldest and most primitive equid. It was first described from the London Clay by Sir Richard Owen in 1840, but it was later found to be far more abundant in western North America. (Despite the widespread use and recognition of the generic name *Hyracotherium*, for reasons discussed below it may be an incorrect name for dawn horses. There is no consensus, however, on what name[s] *should* be applied to them. Hence, for simplicity, *Hyracotherium* is used here to refer to these primitive equids.) *Hyracotherium* is one of the most familiar Eocene mammals. It was long considered to be the morphotypic perissodactyl, but recent discoveries suggest that some forms traditionally called tapiromorphs may be equally or more primitive. *Hyracotherium* shares derived conditions of the optic foramen and foramen ovale (the cranial openings for the optic and mandibular nerves) with later horses, justifying its assignment to the Equidae (MacFadden, 1992).

Hyracotherium (Figs. 13.6A, 13.7A, 13.8A, Plate 6) is the best-known basal perissodactyl, and its anatomy serves as a reasonable model of a primitive perissodactyl. At first glance, dawn horses may appear to have little in common with present-day equids. They were small—the oldest species about the size of a housecat—and their limbs, although long and specialized for the early Eocene, seem short by modern standards. The ulna and the fibula are complete elements, unlike their reduced counterparts in modern equids. There are four toes on the forefeet and three on the hind feet, unlike the monodactyl condition of present-day horses. The teeth are low-crowned and still show evidence of the primitive tribosphenic pattern. The molar cusps are joined by weak oblique transverse crests, the protoloph and metaloph on the uppers (interrupted by the paraconule and metaconule, respectively) and the protolophid and hypolophid on the lowers. The paraconid is weak or absent, and the upper molars are squared by the addition of a strong hypocone. The posterior premolars are somewhat molarized, but not fully molariform as in many later equids (e.g., they lack a hypocone). However, the limb skeleton already possesses most of the cursorial specializations that characterize extant perissodactyls, including the diagnostic ankle morphology and elongate metatarsals, as well as joint modifications in the elbow, wrist, and ankle to restrict limb movement to a parasagittal plane. These features indicate that *Hyracotherium* was a browser and was among the swiftest runners of its day. The abundance of *Hyracotherium* fossils at many sites suggests that it was gregarious. It appears to have been sexually dimorphic in skull and canine size and may have been polygynous, like modern equids (Gingerich, 1981d).

Recent phylogenetic analyses have shown that *Hyracotherium* is a paraphyletic genus, the type species of which (the rather rare European *H. leporinum*) may actually be more closely related to palaeotheres than to equids (Hooker, 1994; Froehlich, 1999, 2002). The latter conclusion is of more than esoteric interest for, if correct, it means that the genus *Hyracotherium* would no longer properly apply to primitive equids. Consequently, in the most recent phylogenetic analysis of early equids, Froehlich (2002) employed six different generic names for species that have conventionally been allocated to *Hyracotherium* (the resurrected names *Pliolophus, Eohippus,* and *Protorohippus,* and three new genera, *Sifrhippus, Minippus,* and *Arenahippus*). A seventh genus, *Xenicohippus,* characterized by robust premolars, was considered a basal brontothere by McKenna and Bell (1997), but was reaffirmed as an equid by Froehlich. Regardless of this proliferation of names, the anatomical differences among these early Eocene equids, and between them and some other basal perissodactyls, are so minor that even experts have difficulty distinguishing them. They clearly lie at or near the base of the equoid radiation and are also close to the perissodactyl stem. The subtle differences between these and other basal perissodactyls have nevertheless been used to ally particular species or genera with different perissodactyl families.

Equid evolution was essentially a simple generic succession of browsers during the Early Tertiary, with few offshoots, the primary advance being the progressive molarization of the premolars in the more or less successive genera *Orohippus, Epihippus, Mesohippus* (Fig. 13.8C), and *Miohippus* (MacFadden, 1992). Late Eocene *Haplohippus* may represent a minor side branch. It was only after *Miohippus* that the equid radiation branched into multiple, diverse lineages. Lophodonty of the cheek teeth also increased, but

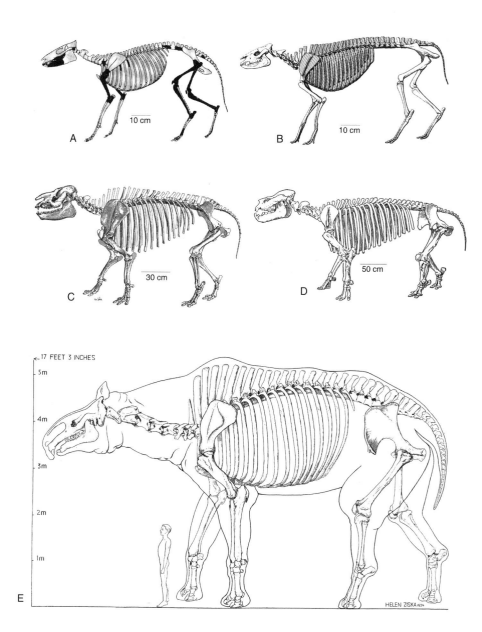

Fig. 13.6. Skeletons of early perissodactyls: (A) equid *Hyracotherium*; (B) tapiroid *Heptodon*; (C) brontothere *Palaeosyops*; (D) rhinocerotoid *Metamynodon*; (E) rhinocerotoid *Paraceratherium* (=*Baluchitherium*). (A from Gingerich, 1989; B from Radinsky, 1965a; C from Osborn, 1929; D from Gregory, 1951; E from Granger and Gregory, 1935.)

significant hypsodonty did not evolve until the Miocene, in *Merychippus* and allied genera. Accompanying the dental changes were increasing size, as well as derived modifications in the skull, jaw, and limb skeleton, including reduction of the ulna, tibia, and fourth toe (digit V) of the manus.

The presence of Early Tertiary equids in Asia remains equivocal. Supposed records (e.g., Dashzeveg, 1979) have been subsequently reidentified as brontotheres, chalicotheres, or isectolophids (Hooker, 1994; Hooker and Dashzeveg, 2004; but see Lucas and Kondrashov, 2004, for a contrary opinion).

Palaeotheriidae

Palaeotheres constitute the principal Eurasian (mainly European) radiation of equoids and are the sister taxon of Equidae (Hooker, 1994; Froehlich, 1999). The composition of the Palaeotheriidae is controversial, however, being variously applied to only some or virtually all European equoids, including, as aforementioned, the type species of *Hyracotherium*. According to the more restricted usage, several traditional palaeothere genera (particularly *Propachynolophus*, *Pachynolophus*, *Anchilophus*, and *Propalaeotherium*; Figs. 13.8B, 13.9) should instead either be considered equids or be placed in a separate family, Pachynolophidae, which may or may not belong to the Equoidea. Hooker (1994), for instance, excluded Pachynolophidae from Equoidea and considered them closer to isectolophids. The dentition of forms such as *Propalaeotherium* is indisputably very similar to that of basal equids (compare Fig. 13.8A and 13.8B). Nevertheless, these taxa may represent the initial divergence of Palaeotheriidae, and for simplicity this more inclusive concept of palaeotheres (Froehlich, 1999) is adopted here.

Palaeotheres differ from equids in subtle dental features and in having a less prominent greater trochanter on the femur and a less expanded ilium. These differences suggested to Franzen (1989) that they are too primitive to have descended from conventional *Hyracotherium*. Alternatively these traits just might be derived in palaeotheres. Trends in palaeothere evolution include increasing lophodonty, progressive molarization of premolars, development of a mesostyle and a W-shaped ectoloph leading to dilambdodont molars, a deepening nasal incision (indicating a tapirlike proboscis), reduction of the nasal process of the premaxilla, elongation of the cervical vertebrae, and shortening of the hind limbs (resulting in longer metacarpals than metatarsals). The dental features typical of later palaeotheres are incipiently developed in the early to middle Eocene genera *Propachynolophus*, *Pachynolophus*, and *Propalaeotherium* (Savage et al., 1965; Remy, 2001). Some of the most derived palaeotheres (e.g., middle Eocene to early Oligocene *Palaeotherium* and *Plagiolophus*; see Fig. 13.13D) were as large as modern horses and evolved relatively hypsodont, dilambdodont cheek teeth and, in some genera, molarized premolars, in parallel with equids (Fig. 13.8D; e.g., Remy, 1985). The upper molars are squared, as in other equoids, and M^3 is longer than wide. Except for the dentition, however, they were more tapirlike than horselike. *Anchilophus* was a middle and late Eocene form that had relatively brachydont cheek teeth and fully molariform posterior premolars. *Hallensia*, best known from the middle Eocene site at Messel, Germany, is variously considered to be a palaeothere or a basal perissodactyl. Its molars are bunodont and similar to those of phenacodontid condylarths (which led to its initial allocation to that family), but its skeleton possesses the hallmark perissodactyl tarsal morphology (Franzen, 1990).

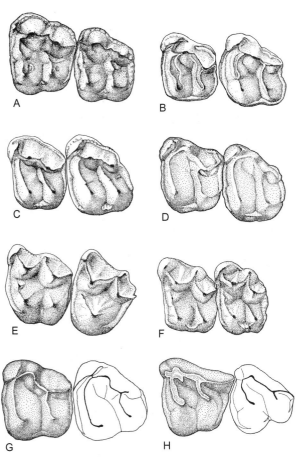

Fig. 13.7. Left M^{2-3} of early perissodactyls: (A) equid *Hyracotherium*; (B) chalicothere or lophiodontid *Paleomoropus*; (C) tapiromorph *Homogalax*; (D) tapiroid *Heptodon*; (E) brontothere *Eotitanops*; (F) brontothere *Lambdotherium*; (G) rhinocerotoid *Hyrachyus*; (H) rhinocerotoid *Hyracodon*. Not to scale. (From Radinsky, 1969.)

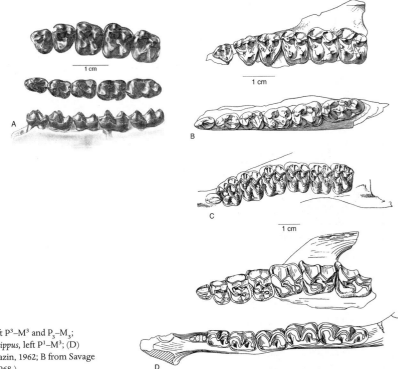

Fig. 13.8. Dentition of equoids: (A) *Hyracotherium*, left P^3–M^3 and P_3–M_3; (B) *Propalaeotherium*, left P^2–M^3 and P_2–M_3; (C) *Mesohippus*, left P^1–M^3; (D) *Palaeotherium*, left P^1–M^3 and right P_3–M_3. (A from Gazin, 1962; B from Savage et al., 1965; C from Osborn, 1918; D from Franzen, 1968.)

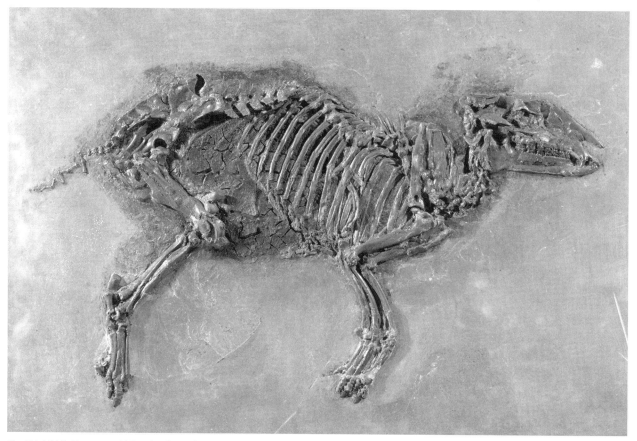

Fig. 13.9. Middle Eocene equoid *Propalaeotherium* from Messel, Germany. Size about 30 cm high at the shoulder. (Courtesy of the Forschungsinstitut Senckenberg, Frankfurt.)

Brontotheriidae

Brontotheres (family Brontotheriidae), also known as titanotheres, include some of the largest Tertiary land mammals. Most of them lived during the Eocene in North America and Asia, and a few types reached southeastern Europe. Brontotheres became extinct in North America at the end of the Eocene and slightly later in Eurasia. There was a clear trend toward size increase in brontothere evolution, with the latest forms achieving elephantine proportions.

Brontotheres had relatively low-crowned cheek teeth and small, nonmolariform to submolariform premolars (Figs. 13.7E,F, 13.10). The lower molars are essentially selenodont and the third molar retains a prominent hypoconulid. The upper molars differ from those of other perissodactyls in their bunoselenodont pattern, consisting of a strong, W-shaped ectoloph, as in chalicotheres and some equoids, but no cross-lophs; instead the protocone and hypocone are isolated and bunodont. Upper molar conules are primitively present but lost in later forms. This dental morphology constrained brontotheres to a browsing diet. Advanced brontotheres reduced their incisors and probably compensated for this reduction by having a prehensile upper lip.

Other trends in evolution of the family included shortening the face and lengthening the postorbital part of the skull, accompanied by the evolution of large, bony fronto-

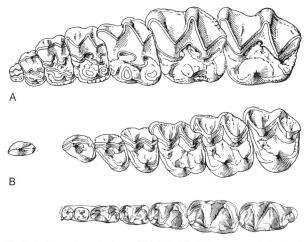

Fig. 13.10. Titanothere dentitions: (A) left P^1–M^3 of *Megacerops*; (B) left P^1–M^3 and right P_3–M_3 of *Eotitanops*. Not to scale. (From Osborn, 1929.)

nasal horns and an elevated occiput in members of the subfamilies Brontotheriinae and Embolotheriinae (Osborn, 1929; Janovskaja, 1980; Mader, 1998; Fig. 13.11). The horns, which were blunt and evidently hide-covered rather than sheath-covered like those of bovids, show considerable variation. Although this variation was once thought to reflect great species diversity, it is now believed to be related primarily to intraspecific sexual dimorphism (Lucas and Schoch,

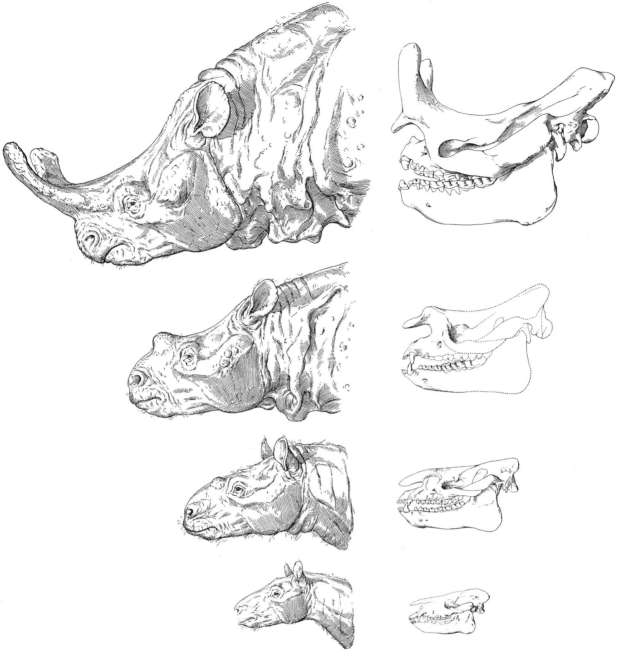

Fig. 13.11. Heads and skulls of a succession of brontotheres, according to Osborn (1929). From bottom to top are *Eotitanops*, *Manteoceras*, *Protitanotherium*, and *Megacerops*. Drawn to scale; skull of *Megacerops* is 85 cm long.

1989b; Mihlbachler et al., 2004). The number of valid species or genera is unresolved, but certainly there were far fewer than the 37 species recognized by Osborn (1929) in his classic monograph on the group. The horns were probably used for display, species recognition, and intraspecific combat (Osborn, 1929; Mader, 1998). Stanley (1974) reasoned that derived brontotheres probably used the horns for head-on ramming, much like sheep. In the gigantic Chadronian genera *Brontops* and *Megacerops* (=*Brontotherium*) the horns evolved to enormous size and were sometimes distinctly bifurcated, perhaps to allow interlocking with those of an opponent, comparable to the combat of cervids and some bovids. Other brontotheres may have used their laterally directed horns to inflict injury by swinging the head sideways into the body of an opponent.

The skeleton of brontotheres was comparatively robust, and became massive and graviportal in the largest Chadronian species (Fig. 13.12). The feet retained the primitive perissodactyl condition of a four-toed manus and three-toed pes, with short, stout elements in the graviportal forms. The latter types, which were analogous to modern rhinos, developed long spinous processes on the anterior thoracic vertebrae in association with muscles and ligaments to hold up the head.

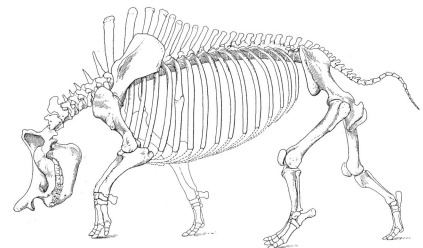

Fig. 13.12. Late Eocene brontothere *Brontops* (about 2 m tall at the withers and 4 m long). Unlike the posture shown here, the head was probably held almost horizontal during head-on ramming (Stanley, 1974). (From Osborn, 1929.)

Late Wasatchian *Lambdotherium* (sometimes placed in its own family) and *Eotitanops* from western North America have generally been considered to be the earliest brontotheres (Fig. 13.7E,F). Both have a prominent W-shaped ectoloph and very weak cross-lophs on the upper molars, presumably remnants from a more lophodont ancestor. The status of both these genera, however, is in question. McKenna and Bell (1997) regarded *Eotitanops* to be a junior synonym of *Palaeosyops,* but most other authors maintain their separation. Some researchers have suggested that *Lambdotherium* may be more closely related to palaeotheres than to brontotheres (Mader, 1998; Lucas and Holbrook, 2004). However, the evidence presented so far is not very compelling, perhaps because *Lambdotherium* is very primitive in many features. *Palaeosyops* was a somewhat larger brontothere, particularly characteristic of the Bridgerian (Fig. 13.6C). These early brontotheres had the basic brontothere molar pattern but retained primitive skull proportions and lacked horns.

Brontotheres diversified extensively during the Uintan-Chadronian in North America and the equivalent time period in Asia. It was during this interval that they evolved horns and attained huge size. The brontotheres' demise at the end of the Eocene in North America and slightly later in Eurasia was perhaps related to their inability to adapt to changing climate and vegetation. In particular, their rather conservative, unspecialized dentition may have been poorly suited to these shifts.

In a phylogenetic analysis of basal perissodactyls, Froehlich (1999) once again found support for a hippomorph clade containing brontotheres. Both Froehlich (1999) and Hooker and Dashzeveg (2004) have suggested that brontotheres represent the most primitive perissodactyls. If this hypothesis is borne out, bunolophodont molars with upper molar mesostyles could be plesiomorphic for the order.

Ceratomorpha

Tapiroids and rhinocerotoids are united under this heading. The two groups are viewed as either sister taxa, or ancestor-descendant taxa, with rhinocerotoids probably descended from a tapiroid. The broader term Tapiromorpha

(=Moropomorpha) is sometimes applied to the ceratomorphs together with isectolophids and some other incipiently lophodont basal perissodactyls that were formerly considered tapiroids. The exact relationships of these forms—*Cardiolophus, Homogalax* (see Plate 3.3), *Cymbalophus,* and *Orientolophus*—including whether they are tapiromorphs, equoids, ancylopods, or undifferentiated basal perissodactyls, is still being debated by experts. What can be said with some confidence is that, like *Hyracotherium,* they lie near the base of the perissodactyl radiation. In addition, they display a slightly lophodont molar condition, as would be expected in basal ceratomorphs. Significantly, one or more of these forms is known from early Eocene deposits of each of the three northern continents.

Tapiromorphs differ from equoids in having more lophodont cheek teeth, but the differences among the earliest taxa are quite subtle. The cross-lophs on the upper molars are higher and more continuous, with little or no evidence of conules (Fig. 13.7C,D), and the protolophid and hypolophid of the lower molars are stronger and slightly reoriented. These traits apply even to the four basal genera listed above, suggesting that some degree of lophodonty may have been primitive for perissodactyls—a possibility supported by the incipient lophodont condition in *Radinskya* and *Olbitherium.* Limb elements of *Homogalax* and *Cardiolophus* are more robust and slightly less cursorially specialized than those of *Hyracotherium,* suggesting that those genera could lie closer to the base of Perissodactyla than does *Hyracotherium* (Rose, 1996b; Figs. 13.3, 13.4).

Tapiroidea

The most primitive definitive tapiroid, early Eocene *Heptodon* (Helaletidae; Fig. 13.6B) from western North America and Asia, already shows most of the anatomical features characteristic of extant tapirs, except for the retracted nasals and accompanying features associated with a proboscis in the living forms (Fig. 13.13). However, there is a general trend toward reduction of the nasals and a higher nasal incision during tapiroid evolution. Compared to *Homogalax,* early Eocene *Heptodon* has much stronger, uninterrupted, oblique cross-lophs on the molars, as in living tapirs, but unlike the latter the posterior premolars are not fully molariform. The upper molar crests in *Heptodon* (Figs. 13.7D, 13.14A) and many other ceratomorphs describe a π-shaped pattern. This configuration was the beginning of the bilophodont pattern, further developed in middle to late Eocene *Helaletes* (Fig. 13.14B) and middle Eocene–early Oligocene *Colodon,* that characterizes the cheek teeth of later tapirs and suggests a largely folivorous diet. The nasal incisure is noticeably more retracted in those genera than in *Heptodon,* and the limbs in all three are slightly more derived (cursorially adapted) in the direction of modern tapirs than are those of *Homogalax.* Nevertheless, the differences between these Eocene tapiroids and extant forms are not great, and the changes that took place between the Eocene and the present are relatively minor (Radinsky, 1965a).

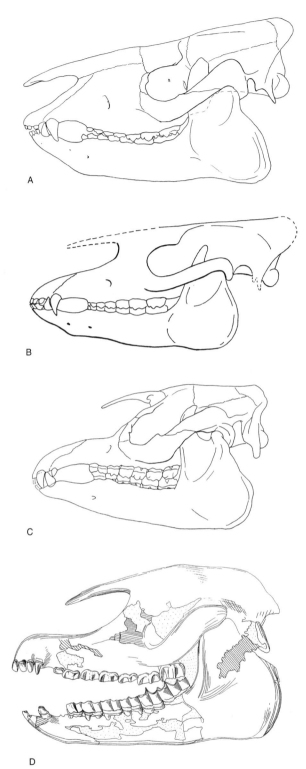

Fig. 13.13. Tapiroid and palaeothere skulls: (A) Eocene tapiroid *Heptodon;* (B) Eocene tapiroid *Lophialetes;* (C) extant tapiroid *Tapirus;* (D) Eocene–early Oligocene palaeothere *Palaeotherium.* Not to scale. (A–C from Radinsky, 1965a; D from Franzen, 1968.)

Protapirus, first known from the late Uintan of western North America, is the oldest true tapir (Tapiridae; Colbert and Schoch, 1998). Like helaletids, it had a deep nasal incisure, indicating the presence of a short proboscis, as in living tapirs (Radinsky, 1963).

Fig. 13.14. Right dentitions of ceratomorphs: (A) tapiroid *Heptodon*, P^2–M^3 and P$_3$–M$_3$; (B) tapiroid *Helaletes*, P^1–M^3 and P$_2$–M$_3$; (C) rhinocerotoid *Hyracodon*, P^1–M^3 and P$_2$–M$_3$. (A, B from Radinsky, 1963; C from Radinsky, 1967a.)

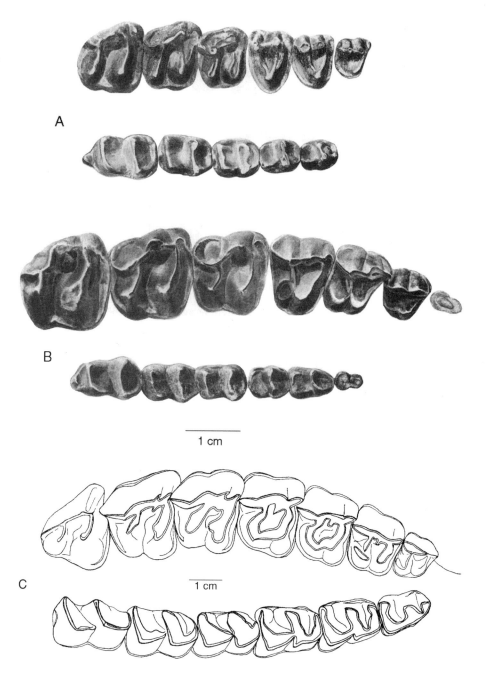

Two other tapiroid, or perhaps stem ceratomorph, families are known from Asia, Deperetellidae and Lophialetidae, which are mainly of middle to late Eocene age. These animals had lophodont molars (strongly bilophodont in deperetellids) with a reduced and lingually shifted metacone (Radinsky, 1965b; Reshetov, 1979; Dashzeveg and Hooker, 1997). The premolars are submolariform to molariform in deperetellids but less molarized in lophialetids. The nasal incisure of *Lophialetes* was almost as deep as in modern tapirs, indicating a well-developed proboscis (Reshetov, 1979). The skeleton was cursorially adapted, as in other tapiroids.

Rhinocerotoidea

Rhinocerotoids first appeared in the late early Eocene of North American and Europe, in the form of *Hyrachyus* (Prothero et al., 1989). The teeth and skeleton of *Hyrachyus* (Plate 7.3) resemble those of tapiroids such as *Heptodon* so closely that Radinsky (1967b) assigned this genus to the tapiroid family Helaletidae. Although *Hyrachyus* is now generally considered to be a basal rhinocerotoid (Emry, 1989; Schoch, 1989; Holbrook, 1999), its clear resemblance to both groups indicates either that rhinocerotoids evolved from a tapiroid (Radinsky, 1966b, 1969) or that tapiroids and rhinocerotoids share a common ancestor. The molars of *Hyrachyus* (Fig. 13.7G) have slightly higher cross-lophs than those of *Heptodon,* and, as in other rhinocerotoids, M_3 lacks a hypoconulid, a cusp that forms a large third lobe in many perissodactyls. But the M_3 hypoconulid was independently lost in some tapiroids as well (Radinsky, 1966b), weakening its utility as a diagnostic trait of rhinocerotoids. The upper molars show the beginning of the π-shaped pattern that became strongly developed in later rhinocerotoids. Rensberger and Koenigswald (1980) found that some time after the early Eocene, the enamel of fossil rhinos and most other perissodactyls except equoids underwent a reorientation of prism layers (compared to that in primitive equoids) from horizontal to vertical, which made the enamel more resistant to wear.

Although the general trend in rhinocerotoids was toward increasing size, not all were large. The smallest known ceratomorph was middle Eocene *Fouchia,* a relative of *Hyrachyus* (Emry, 1989). Its cheek teeth were as small as those in the housecat-sized early Eocene species *Hyracotherium sandrae* and *Pachynolophus hookeri.*

From the middle Eocene through the Oligocene, rhinocerotoids were common, and they diversified into at least three families in North America and Eurasia—Amynodontidae, Hyracodontidae, and Rhinocerotidae—which are most easily distinguished by the structure of M^3 (Prothero et al., 1989). The nasal horns (composed of keratin, not bone) for which the group is named were not present in most Early Tertiary representatives, first appearing in some Oligocene rhinocerotids. Prothero (2005) recently provided a comprehensive review of North American rhinos.

Amynodonts were medium-sized to large animals characterized by elongate upper molars with a strong ectoloph and a prominent, extended metastyle (both associated with an emphasis on vertical shearing), and large, sexually dimorphic canines (Wall, 1980, 1989, 1998). The molars were generally much larger than the premolars, and there was a trend toward hypsodonty. M^3 is quadrangular, not tapered in back as it is in hyracodontids. The skulls of amynodonts are characterized by a depression in front of the orbits (preorbital fossa), which may have housed a nasal diverticulum. *Metamynodon* (Fig. 13.6D) was a hypsodont form that had elevated orbits and relatively short, robust limbs, suggesting hippolike, semiaquatic adaptations. Its snout morphology suggests the presence of a prehensile upper lip. Other amynodonts evolved a short, tapirlike proboscis, as inferred from a shorter rostrum and higher skull, prominent muscle scars on the snout, reduced nasal bones, and a receding nasal incision. One such form was *Cadurcodon,* a subcursorial amynodont from the early Oligocene of Mongolia, whose deep, high nasal incisure indicates that it had a well-developed proboscis.

The 20 or so genera of hyracodonts (Hyracodontidae) were cursorial browsers with elongate limbs. Their upper molars had a shorter protoloph and metaloph than in *Hyrachyus* and a short, lingually inflected metastyle, which was further reduced or lost in some types (Figs. 13.7H, 13.14C). M^3 is usually triangular. Hyracodontids show considerable variation in incisor morphology and number, from the primitive retention of three small, spatulate incisors to the presence of only a single large, procumbent, tusklike incisor in each jaw (I^1 and I_1; Radinsky, 1966b). Although some hyracodonts were no larger than sheep, the hornless indricotheres (sometimes placed in their own family) hold the record as the largest land mammals ever known (Fig. 13.6E). *Paraceratherium* (probably including *Baluchitherium* and *Indricotherium*), known mainly from the late Eocene–Oligocene of Eurasia, stood 5–6 m at the shoulder, was over 7 m long, and, even by conservative estimates, weighed more than 11,000 kg (Fortelius and Kappelman, 1993)—twice as much as a large elephant!

The extant family Rhinocerotidae first appeared in the middle Eocene, probably descended from a hyracodont. Rhinocerotids are characterized by a pair of enlarged incisors, I^1 and I_2, the upper one chisel-like and the lower one procumbent and tusklike; the other incisors and canines were reduced or lost in more progressive taxa. Middle Eocene (Uintan) *Uintaceras,* from Utah and Wyoming, is a primitive rhinocerotid that has been considered to be either the earliest rhinocerotid (Prothero, 2005) or the sister taxon of rhinocerotids (Holbrook and Lucas, 1997). The upper incisors are labiolingually compressed, a synapomorphy of rhinocerotids, but the incisors are not preserved in place, so it is uncertain whether *Uintaceras* possessed the characteristic rhinocerotid "chisel/tusk" arrangement. The postcranial skeleton is robust, relatively noncursorial, and primitively retains a four-toed manus. One of the earliest known rhinocerotids, *Teletaceras,* had small I^1/I_2 tusks, a reduced premaxilla, and a reduced metacone, diagnostic traits of the family; but it retained an unreduced anterior dentition, including all three incisors above and below, and is also more primitive than other rhinocerotids in several other dental features (Hanson, 1989). *Teletaceras* is known from the middle and late Eocene (Duchesnean-Chadronian) of North America and the late Eocene of Asia. *Trigonias,* a common Chadronian rhinocerotid, also was primitive in retaining the full complement of anterior teeth except for I_3 and C_1, as well as a functional fifth digit on the manus (Prothero, 1998c). Although Prothero (1998c) suggested that rhinocerotids immigrated to North America from Asia in the Duchesnean (late middle Eocene), these fossils raise the possibility that dispersal could have been earlier and in the other direction. In the late Eocene (Chadronian) and Oligocene, rhinocerotids diversified on the northern continents and by the Miocene had supplanted the other rhinocerotoid families and spread to Africa.

Ancylopoda

The name Ancylopoda is used to unite the widespread Eocene-Pleistocene chalicotheres (Eomoropidae and Chalicotheriidae) and the European Eocene family Lophiodontidae, which is sometimes assigned to Tapiroidea. Ancylopods are a bizarre group of medium-sized and large perissodactyls that are unusual for having claws, or at least clawlike unguals, rather than hoofs (hence the name, which means "hook foot"). Claw-bearing ungual phalanges are known in chalicotheriids and in one of the oldest eomoropids, middle Eocene *Grangeria* (Lucas and Schoch, 1989a; Coombs, 1998), and are therefore assumed to have characterized all chalicotheres. In those animals the unguals are proximally deep, narrow, pointed, and deeply fissured at the tip, but almost flat on the plantar surface, hinting at their origin from a more hooflike condition (Fig. 13.15B). The ungual phalanges are unknown in lophiodontids; therefore their alliance with chalicotheres, which is based on dental similarity alone, is tenuous. Broader relationships of chalicotheres are controversial. They are variously considered to be allied with either ceratomorphs or brontotheres, or to compose a separate clade of perissodactyls. Hooker and Dashzeveg (2004) considered ancylopods to be closely related to isectolophids.

The upper molar structure of chalicotheres (Fig. 13.15A) is distinctive and distinguished from that of other perissodactyls by the presence of a continuous metaloph with no metaconule and a protoloph interrupted by a paraconule, both crests meeting a distinctive W-shaped ectoloph (Coombs, 1998). Some researchers cite the complete metaloph as evidence that ancylopods are related to tapiromorphs (which, however, lack a W-shaped ectoloph); others point to the W-shaped ectoloph as indicative of a relationship with brontotheres (which lack cross-lophs) and possibly equoids. Both features may well have arisen more than once in perissodactyls, however, complicating phylogenetic interpretations. The cheek teeth of chalicotheres are typically low crowned, and the premolars nonmolariform, suggesting a general browsing habit, but some species had more

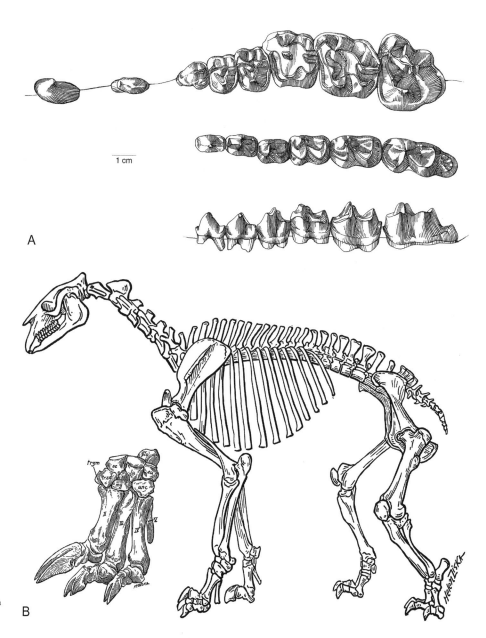

Fig. 13.15. Chalicotheres: (A) left upper and lower dentition of middle Eocene eomoropid *Litolophus*; (B) skeleton of the Miocene chalicotheriid *Moropus*; inset shows enlarged left manus. (A from Colbert, 1934; B from Gregory, 1951.)

elongate or hypsodont molars, indicating more specialized feeding.

Eomoropidae is a paraphyletic group of the most primitive chalicotheres, which is known from the Eocene of Asia and North America. Eomoropids are smaller than chalicotheriids and more primitive in retaining P^1 and a hypoconulid on M_3, and in having less simplified premolars, lower-crowned cheek teeth, less emphasis on the ectoloph, and unspecialized metapodials (Lucas and Schoch, 1989a). Early Eocene *Paleomoropus* (Fig. 13.7B) is the oldest North American chalicothere (Radinsky, 1964, 1969) and has generally been considered to be the oldest ancylopod. It is known only from three upper molars, which have a prominent parastyle and strong lophs, the protoloph with a distinct paraconule and the metaloph high and continuous without a metaconule. Although these features are synapomorphies of chalicotheres, *Paleomoropus* lacks a W-shaped ectoloph. As a result, its chalicothere status has been questioned, some authors including it instead in the related family Lophiodontidae. It should be noted that some other primitive ancylopods also lack a mesostyle (*Litolophus, Lophiaspis*).

Slightly older is *Protomoropus*, a new genus based on a few jaw fragments formerly assigned to *Hyracotherium* and *Homogalax*, from the Bumbanian of Naran Bulak, Mongolia (Hooker and Dashzeveg, 2004). *Protomoropus* is moderately lophodont and differs from other basal perissodactyls only by very subtle features of the molars. Indeed, Lucas and Kondrashov (2004) retained these specimens in *Homogalax* and *Hyracotherium*. This ambiguity is further indication of the very primitive morphology and close similarity of these basal perissodactyls. In this case, it also results from the fragmentary nature of the evidence.

Danjiangia, based on a skull and mandible from the late early Eocene of China, was described as a primitive chalicothere (Wang, 1995). Its upper molars are generally similar to those of *Paleomoropus* but differ in having a well-developed W-shaped ectoloph and lacking a high metaloph. Judging from these differences, and the presence of strong dilambdodonty, Hooker and Dashzeveg (2004) concluded that *Danjiangia* is more likely a primitive brontothere.

Based on their interpretation of *Protomoropus* as the most primitive chalicothere, Hooker and Dashzeveg (2004) inferred an Asian origin of chalicotheres. By the middle Eocene, undisputed chalicotheres are known from both Asia and North America. *Eomoropus* and *Grangeria* were small animals (present on both continents) with well-developed cross-lophs and a W-shaped ectoloph. Although the metapodials were less specialized than those of chalicotheriids, *Grangeria*, at least, had specialized phalanges, including claw-bearing unguals (Lucas and Schoch, 1989a). The closely allied *Litolophus* is dentally similar to *Grangeria* but has a straight ectoloph without a mesostyle (Radinsky, 1964).

Chalicotheriidae comprises the larger, more specialized chalicotheres of the Oligocene-Pleistocene. The oldest genus, *Schizotherium*, first appeared in the late Eocene of Asia. Dentally, the hypoconulid was lost from M_3 and the W-shaped ectoloph dominates the upper molars (Coombs, 1998). In Miocene *Moropus* (Fig. 13.15) the forelimbs were longer than the hind limbs, and the metacarpals longer than the metatarsals. These trends culminated in Miocene *Chalicotherium*, which was gorilla-like in body proportions, with much longer forelimbs than hind, and a much shorter tibia than femur (Zapfe, 1979). Coombs (1983) argued that chalicotheriids frequently adopted bipedal postures when browsing so the clawed forelimbs could be used to hook branches.

PAENUNGULATA

Hyracoidea

The Hyracoidea are represented today by the rabbit-sized dassies or hyraxes of Africa (referred to as conies in the Old Testament). They are considered to be related to tethytheres or perissodactyls or both (as discussed in the introduction to Altungulata, above), but their exact phylogenetic position is a matter of debate. Although only a few small species (1–5 kg) exist in Africa today, the order was much more diversified in the Early Tertiary, and some forms dispersed to Europe and Asia in the Miocene. Pickford et al. (1997) recognized five families of hyracoids (four of them extinct) placed in two suborders, based on dental, cranial, and astragalar characters. Because so few species have associated teeth and tarsal elements, however, this classification has not been widely accepted. Other authors place most fossil hyraxes in the Pliohyracidae and group extant hyraxes with a few Neogene genera in the Procaviidae (Rasmussen, 1989; McKenna and Bell, 1997; Gheerbrant, Domning, and Tassy, 2005). This simpler arrangement is followed here. As with so many groups, the precise composition of these families is controversial.

Extant hyraxes are superficially rodent- or rabbitlike in form, and they present a curious combination of features, some of which are typical of running mammals and others of climbers (Fig. 13.16). According to Kingdon (1974), they can run fast only for short distances. In fact, some inhabit rocky terrain and others climb trees; all are adept jumpers and climbers. Apparent cursorial specializations therefore seem to reflect phylogenetic heritage more than current adaptation.

Hyraxes have very short tails but a long presacral vertebral column, as is characteristic of tethytheres (Fischer and Tassy, 1993). As in many cursorial mammals, there is no clavicle. In marked contrast to typical runners, however, the feet are plantigrade, with specialized volar pads that enhance traction, and the knees and hips are habitually flexed. The ulna and radius are of similar size and may be firmly attached; the structure of the proximal radioulnar joint prohibits rotation. The femur retains a third trochanter, as in nearly all perissodactyls. The fibula is well developed, and the tibia and fibula are fused proximally and either fused or bound by a tight syndesmosis distally (Barnett and Napier, 1953). Like the earliest perissodactyls, modern hyraxes are mesaxonic, with four functional toes on the front feet (and

Fig. 13.16. Hyracoids: (A) extant hyrax *Procavia*; (B) restoration of the Oligocene hyracoid *Saghatherium* (shaded elements are poorly preserved or missing; see Plate 7.4). (A from Grassé, 1955a; B from Thomas et al., 2004.)

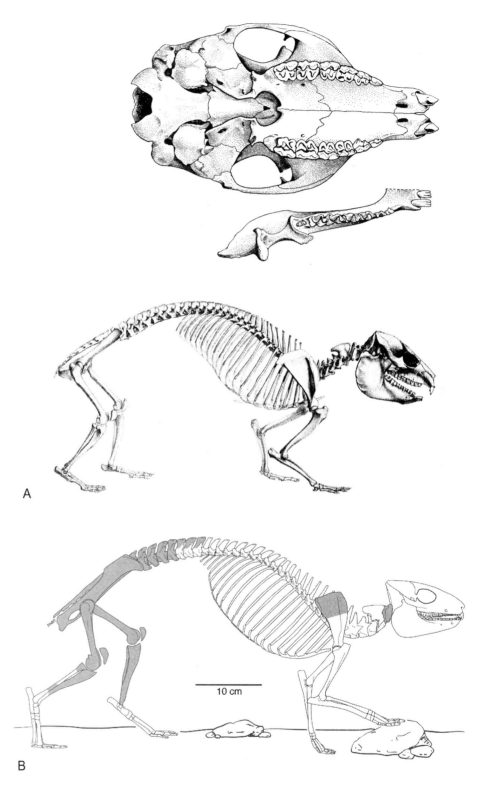

a vestigial pollex) and three on the hind; all have hooflike nails except for the inner (second) digit of the foot, which bears a long claw for grooming. The terminal phalanx bearing this claw is deeply fissured.

Hyraxes are capable of mid-carpal rotation (supination), an adaptation unique among mammals, which, together with the specialized volar pads, enables them to climb with agility despite the lack of claws. Both living and fossil hyracoids have a serial (taxeopode) arrangement of carpal and tarsal elements, meaning that the distal carpal elements are aligned with a single proximal element rather than interlocking with two. Proboscideans are similar to hyracoids in this regard, but they differ from perissodactyls, which have an alternating arrangement (Rasmussen et al., 1990). Indeed, taxeopody has sometimes been considered to be a synapomorphy of paenungulates (e.g., Novacek and Wyss, 1986).

In contrast to other ungulates (except the basal proboscidean *Numidotherium*), however, hyracoids retain a separate centrale in the carpus. The centrale has generally been lost in more advanced mammals, including phenacodontids, which are considered to lie near the ancestry of Altungulata.

The skull of present-day hyraxes is also superficially rabbitlike, with a lower jaw that is very deep and expanded posteriorly. There is a complete or nearly complete postorbital bar. The dental formula is 1.0.4.3/2.0.4.3. A wide diastema separates the incisors and cheek teeth where anterior teeth have been lost. The upper incisors are widely separated at the midline and grow continuously throughout life. They are triangular in cross-section, with thin enamel on the posterior surface that is worn away by occlusion with the lower incisors, an adaptation for maintaining a sharp edge. The lower incisors, which are used for grooming, are inclined and tricuspid when unworn. The cheek teeth of hyracoids are quite variable, in some forms hypsodont and either lophodont or selenodont (resembling miniature ceratomorph perissodactyls), in others low crowned and bunodont. Hyraxes are herbivorous.

Pliohyracids had an essentially primitive placental dental formula of 3.1.4.3/3.1.4.3; earlier reports that some fossil forms from Egypt may have had five premolars now appear to be incorrect (Thewissen and Simons, 2001). The most primitive hyracoids had bunodont or weakly lophodont molars and simple premolars. Molarization of the premolars and increasing lophodonty are common trends that evolved repeatedly in hyracoids (Court and Mahboubi, 1993).

Particularly distinctive of Pliohyracidae is a hollow chamber in the mandible that opens through a large, round hole on the inside of the jaw, usually below the last molar. The chamber is believed to be sexually dimorphic (like the incisors; Meyer, 1978), probably occurring only in males of most species (DeBlieux et al., 2001). It may have served as a resonating chamber for sound production. Living hyraxes have an expanded eustachian tube, which apparently serves to amplify vocalizations (Kingdon, 1974; Fischer, 1989), suggesting a parallel specialization.

The most ancient hyracoids are represented only by dentitions. *Seggeurius* (Fig. 13.17C), the oldest and most primitive known hyracoid, is a small, bunodont form with simple premolars from probable lower Eocene beds of the southern Atlas area of Algeria. This genus is probably also present in the early Ypresian of Ouled Abdoun, Morocco, which has yielded the oldest known proboscideans (Gheerbrant et al., 2003). *Seggeurius* differs from all other hyracoids in having virtually no paracristid on the lower molars. Although this condition would normally be considered derived, its occurrence in such an ancient form suggests that it could be primitive for Hyracoidea (Court and Mahboubi, 1993). This possibility is strengthened by the observation that the paracristid progressively enlarged in successive samples of *Thyrohyrax*. The upper molars of *Seggeurius* are already essentially quadritubercular, with a large hypocone and a more or less W-shaped ectoloph, as in extant hyracoids. Of similar age is a small species of *Titanohyrax*, reported from the

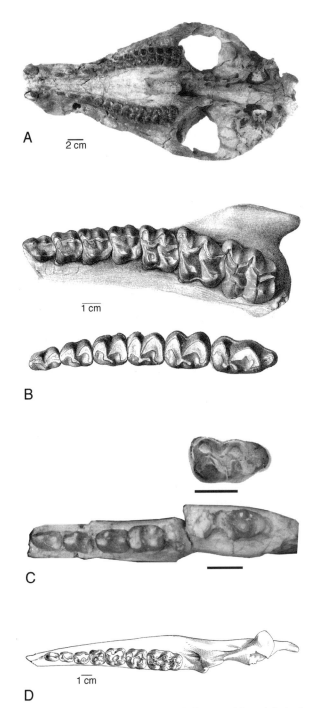

Fig. 13.17. Fossil hyracoids: (A) *Megalohyrax*, skull; (B) *Megalohyrax*, left P^1–M^3 and right P_2–M_3; (C) *Seggeurius*, right M_3 and left P_2–M_1 and M_3 (scale bars = 5 mm); (D) *Bunohyrax*, right P_1–M_3. (A from Thewissen and Simons, 2001; B from Andrews, 1906; C from Gheerbrant, Domning, and Tassy, 2005; D from Matsumoto, 1926.)

?early Eocene of Chambi, Tunisia (Court and Hartenberger, 1992). Both of these early Eocene hyracoids have a mesoconid cusp on the lower molars and a long preprotocrista that joins the parastyle on the upper molars, suggesting that these features are primitive for Hyracoidea. In other respects, these two contenders for oldest known hyracoid exhibit rather different adaptations, suggesting that the origin of the group must be considerably older still.

By the middle Eocene, hyracoids had already diversified considerably. The assemblage from Gour Lazib, Algeria, includes four taxa that show an astonishing range of size and adaptation for so early in hyracoid evolution, from the small bunodont *Microhyrax* (3–4 kg) to the giant selenodont *Titanohyrax* (675 kg; Schwartz et al., 1995). *Microhyrax* resembles *Seggeurius* in having simple premolars, quadritubercular upper molars, and lower molars lacking a paraconid, but its molars are presumably more derived in being bunolophodont and having a short paracristid. *Microhyrax* may be regarded as the sister taxon of all other pliohyracids (Tabuce, Mahboubi, and Sudre, 2001), or may be excluded from Pliohyracidae and, together with *Seggeurius*, left as an unassigned plesiomorphic hyracoid (Gheerbrant, Domning, and Tassy, 2005).

Pliohyracidae reached the zenith of their diversity and abundance during the late Eocene–early Oligocene in the Fayum Depression of Egypt, which has produced a rich record of early hyracoids (Rasmussen and Simons, 1988, 1991). At least nine genera have been recognized, seven of which are already present in the lowest levels of the Jebel Qatrani Formation (at the upper Eocene quarry known as L-41). Some of them, like *Bunohyrax* (Fig. 13.17D) and *Geniohyus*, were bunodont forms whose teeth have been mistaken for those of pigs or anthracotheres. The most common Fayum hyracoid is *Saghatherium*, a form about twice the size of living hyraxes, or roughly 9 kg (Schwartz et al., 1995). Its teeth are bunoselenodont: having low, rounded cusps joined by crescent-shaped shearing crests, particularly evident on the buccal side of the crowns. About the same size was *Thyrohyrax*, whose more strongly lophodont teeth suggest a browsing habit. Tapir-sized *Megalohyrax* (Fig. 13.17A,B) is known through the entire Fayum sequence. It had more generalized bunoselenodont teeth and, like most pliohyracids, submolariform premolars. *Titanohyrax* is also known throughout the Fayum sequence. It was a specialized selenodont form with molarized premolars and well-developed ectolophs on the upper molars, features recalling those of various perissodactyls and suggesting a more folivorous diet. *Titanohyrax ultimus*, known only from teeth, was the largest known hyracoid, estimated to have weighed 1,000 kg (Schwartz et al., 1995).

Among the most unusual early hyraxes from L-41 was the recently described *Antilohyrax*, which was similar in size to the living springbok *Antidorcas* (Rasmussen and Simons, 2000; De Blieux and Simons, 2002). Its upper central incisors were enlarged as small tusks, as in other hyracoids, and occluded with a low, sickle-shaped I_2. The lower central incisors (I_1) had pectinate (comblike) crowns with eight to ten tines, superficially resembling the incisors of the living colugo (*Cynocephalus*). The premolars are molariform, and all the cheek teeth are selenodont and increase in size posteriorly. The dentition suggests a folivorous browsing habit, but the function of the bizarre lower incisors is unknown. As in extant hyraxes, the mandible deepens posteriorly, but it is longer and shallower than in living forms. Hind limb elements attributed to *Antilohyrax* include a tibia and very slender fibula fused for almost their entire length, as well as an astragalus with a short neck and a saddle-shaped navicular facet somewhat like that in early perissodactyls (Rasmussen and Simons, 2000).

Until very recently, early fossil hyracoids were known only from teeth, skulls, and a few isolated postcranial bones from the Fayum of Egypt. Thomas et al. (2004) reported the first nearly complete skeletons of *Saghatherium* from the early Oligocene of Libya (Fig. 13.16, Plate 7.4). They are very similar to present-day hyraxes but somewhat larger. The carpus and tarsus are serial (a characteristic of paenungulates) and constructed much as in extant hyraxes, implying that *Saghatherium* was already capable of mid-carpal and mid-tarsal supination. In addition, there is a deep cotylar fossa for the tibial malleolus on the medial side of the astragalus, a feature seen also in Fayum hyracoids. The cotylar fossa is often considered a characteristic of hyracoids but is more likely a paenungulate synapomorphy. The limb skeleton also has many traits associated with cursorial locomotion and digitigrade stance and is similar in many respects to that of primitive perissodactyls. *Saghatherium* is more primitive than extant hyraxes in retaining separate radius/ulna and tibia/fibula, and fissured, clawlike terminal phalanges on all digits of the manus.

Hyracoids were the dominant Paleogene terrestrial ungulates in northern Africa, which was separated from Eurasia by the intervening Tethys Sea. Their early success is probably attributable to the absence of artiodactyls before the late Eocene and perissodactyls before the Miocene. Hyracoid diversity declined later in the Tertiary, following the immigration of these northern ungulates into Africa (Schwartz et al., 1995).

Tethytheria

Proboscidea

The elephants and their extinct relatives, including mammoths, mastodons, gomphotheres, deinotheres, barytheres, and others, make up the Proboscidea. Elephants are the largest extant land mammals, reaching heights of 3–4 m and weights of 2,500–6,000 kg. The name Proboscidea, of course, derives from the long, flexible, trunk, or proboscis, of derived forms, but many early proboscideans lacked a trunk. Evolution of the trunk has been explained as one solution to the problem of feeding and drinking in animals of increasing body size. As body mass and skull size increased over time, it became biomechanically advantageous to shorten the neck and skull, as well as to straighten the limbs (see below), thereby further elevating the head. In such circumstances, a trunk provides an efficient and safe way to procure food and water (Savage and Long, 1986).

The two living species of elephants (Elephantidae), with restricted distributions in Africa and southeast Asia, are all that remains of a once much wider radiation. Ten extinct families are recognized by McKenna and Bell (1997). The order apparently originated in Africa in the Paleocene or

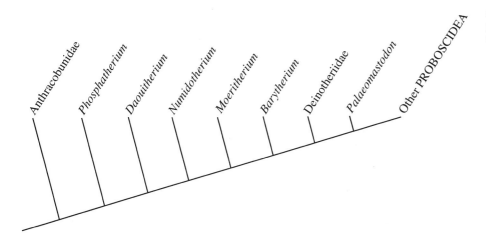

Fig. 13.18. Relationships of primitive proboscideans. (Mainly after Gheerbrant et al., 2002.)

possibly earlier, although potential primitive early members (anthracobunids) are also known from the Eocene of southeast Asia. In the Miocene, diverse proboscideans dispersed throughout the northern continents and eventually reached South America as well.

The skeleton of present-day elephants is massive and may account for 15% of the body weight (Jones, 1984). The huge skull is very short and high and is extensively pneumatized (filled with air sinuses) to lessen the weight. It is especially high in back, providing broad attachment for the strong neck muscles needed to support the trunk and tusks. The external ears are very large, serving both for display and as thermoregulatory organs. The orbit and temporal openings adjoin, with no intervening postorbital bar. The bony nasal opening, from which the trunk extends, is high on the face, between the small eyes. Below the nasal opening are the premaxillae, containing the tusks, which are greatly enlarged, ever-growing second incisors composed of dentine (ivory); thin enamel is present only at the tip and is rapidly worn off. The other incisors and the canines are normally missing.

During their lifetime elephants retain six cheek teeth in each quadrant, but typically only one is fully in place and functional at a given time. The premolars and then the molars erupt sequentially. That the six cheek teeth consist of three unreplaced deciduous premolars and three molars (the permanent premolars are lost), not six molars, as sometimes stated, is supported by studies of both fossil and extant proboscideans (Maglio, 1973; Roth and Shoshani, 1988). As the tooth in place becomes heavily worn, it is pushed out and replaced from behind by the next tooth in succession. Each tooth is a highly modified structure (far removed from the generalized tribosphenic condition), consisting of transverse plates, or laminae, of enamel surrounding dentine, with cementum between the laminae. The more posterior teeth are successively larger and more hypsodont.

The limbs of elephants are graviportal. The ilium and scapula are broad, providing extensive area for the attachment of limb muscles. In contrast to most mammals, the acetabulum faces downward rather than laterally. The femur and humerus are robust and much longer than the distal limb segments. For mechanical efficiency, the knee and elbow joints are fully extended when supporting the body, so that the legs are straight and columnar. The tibia and fibula are separate, as are the radius and ulna, and the configuration of the radioulnar joint is such that the radius is fixed in pronation and cannot rotate. The feet are digitigrade and pentadactyl, with spreading digits that are short and robust, supported by a volar pad. The toes end in hoof- or naillike structures. Like other tethytheres, elephants have a relatively long thoracic region.

The earliest proboscideans come from the Eocene of northern Africa and southeast Asia. They are assigned to as many as seven families: the Asian Anthracobunidae and the exclusively African Phosphatheriidae, Numidotheriidae, Moeritheriidae, Barytheriidae, Palaeomastodontidae, and Phiomiidae (McKenna and Bell, 1997; Gheerbrant, Sudre, et al., 2005; Fig. 13.18, see Table 13.1). The first three families are decidedly more plesiomorphous than the others in several key characters, whereas the last two are similar to each other and more like advanced proboscideans. Phosphatheriidae and Numidotheriidae are the oldest and most primitive unequivocal proboscideans. Anthracobunidae are dentally even more primitive, but their inclusion in Proboscidea is debatable. Late Eocene *Moeritherium* was long considered to be the archetypal proboscidean but is now generally regarded as an early offshoot. Most archaic proboscideans were around the size of a tapir (about 200 kg) or larger.

Anthracobunidae includes eight genera known primarily from dentitions from the early and middle Eocene of Asia. Many recent authors have considered them to be the most primitive Proboscidea, closely allied with moeritheriids (West, 1984; Gingerich, Russell, and Wells, 1990; Kumar, 1991; Fischer and Tassy, 1993; Ray et al., 1994; Ginsburg et al., 1999), but their allocation to this order is not at all secure (Tassy, 1988; Court, 1995; Gheerbrant, Domning, and Tassy, 2005). In the past 25 years or so, various anthracobunids have been assigned to the Artiodactyla, Perissodactyla, or Sirenia. Shoshani, West, et al. (1996) included *Anthracobune* (but not other anthracobunids) in the Proboscidea, based on the presence of a large coracoid process on the scapula and a medial tubercle on the astragalus, which they considered

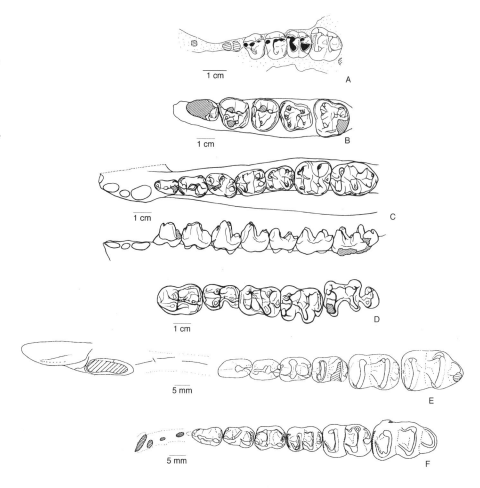

Fig. 13.19. Left dentitions of primitive proboscideans: (A) *Phosphatherium*; (B, C) anthracobunid *Pilgrimella*; (D) anthracobunid *Jozaria*; (E) *Numidotherium*; (F) *Daouitherium*. A–B are upper teeth; C–F are lowers. Proboscidean affinities of anthracobunids are controversial; they may be basal tethytheres or may pertain to another ungulate clade. (A from Gheerbrant et al., 1996; B–D from Wells and Gingerich, 1983; E, F from Gheerbrant et al., 2002.)

to be proboscidean synapomorphies. The same characters were cited by Tassy (1996), however, as evidence that anthracobunids are the sister group of Proboscidea. The embrithopod *Arsinoitherium* also has an enlarged coracoid, but whether it is homologous with that of proboscideans is controversial (Shoshani, West, et al., 1996).

Anthracobunids are more primitive than other proboscideans in retaining four premolars in both upper and lower jaws as well as the lower canine (dental formula 3.1.4.3/3.1.4.3) and in lacking enlarged, tusklike incisors and any significant diastema in the lower series (upper anterior teeth are unknown; West, 1984). These features support the interpretation that Anthracobunidae occupy a more basal position among tethytheres, outside the Proboscidea (e.g., Wells and Gingerich, 1983; Tassy, 1996; Ginsburg et al., 1999). Molar morphology in anthracobunids (Fig. 13.19B–D) ranges from bunolophodont in the most primitive member, *Pilgrimella*, to more strongly bilophodont in *Jozaria*. The teeth are less lophodont than those of Eocene numidotheres (see below), the oldest uncontested proboscideans, and are not unlike those in various primitive artiodactyls and perissodactyls. The posterior premolars are submolariform. Most anthracobunids had a size range roughly comparable to that of modern suids (100–275 kg; Gheerbrant, Domning, and Tassy, 2005), although the most primitive representatives were much smaller. The dentition and occurrence suggest that anthracobunids were amphibious animals that frequented marshes along the northern Tethyan shoreline (Wells and Gingerich, 1983; Kumar, 1991).

The oldest proboscidean for which we know much of the skeleton is *Numidotherium*, which was based on a skull found in lower to middle Eocene deposits of El Kohol, Algeria, in the 1980s (Mahboubi et al. 1984, 1986; see Fig. 13.21A). A second species comes from upper Eocene strata of Libya (Court, 1995). That *Numidotherium* is unequivocally a proboscidean is demonstrated by several derived features, including the pneumatic structure of the skull, the anterior position of the orbits (a tethythere trait), the enlargement of the second incisors as tusks, the absence of the upper and lower first premolars, and the diastema between the anterior teeth and cheek teeth (Fig. 13.19E). In addition, the skull is deep dorsoventrally and has elevated external nares, which suggest the presence of a tapirlike proboscis (Shoshani, West, et al., 1996). *Numidotherium* is more primitive than other proboscideans (except anthracobunids, *Phosphatherium*, and *Moeritherium*) in retaining a full complement of upper incisors ($I^1–I^3$) as well as the upper canine (dental formula 3.1.3.3/2.0.3.3). The lower cheek teeth are rather low crowned and distinctly bilophodont, like those of *Barytherium*, whereas those of *Moeritherium*, *Palaeomastodon*, and *Phiomia* have stronger cusps and weaker crests and are more bunolophodont (Tobien, 1978). The ascending ramus of the mandible in *Numidotherium* is situated lateral to the toothrow and rises vertically a little anterior to the back of M_3, or may even tilt slightly anteriorly.

Numidotherium had a more primitive inner ear structure than *Moeritherium* and later proboscideans, being adapted for high-frequency sound rather than the more specialized, low-frequency adaptation of extant proboscideans (Court, 1992a). The postcranial skeleton also shows plesiomorphic features. The humerus retains an entepicondylar foramen, which was lost in later proboscideans, including *Moeritherium*. The carpus was arranged serially and, to judge from facets on adjoining elements, plesiomorphously retained a separate centrale (as in hyracoids). *Numidotherium* seems to have had plantigrade feet and a sprawling stance, unlike the parasagittal, columnar limbs of modern elephants; however, it is uncertain if this stance represents retention of primitive eutherian posture or is an autapomorphy (Court, 1994). Thus most characters suggest that *Numidotherium* is sufficiently primitive to be the sister taxon or ancestor of all later proboscideans (Domning et al., 1986). However, *Numidotherium* is more derived than the younger *Moeritherium* in having the femur longer than the tibia, as in later graviportal proboscideans (Tassy, 1996).

Older and more primitive than *Numidotherium* is *Phosphatherium* from Ouled Abdoun, Morocco, which is assigned either to the family Numidotheriidae or to its own family, Phosphatheriidae (Gheerbrant et al., 1996, 1998; Gheerbrant, Sudre, et al., 2005). Though initially reported to be of late Paleocene age, it is now believed to date from the early Eocene (earliest Ypresian; Gheerbrant et al., 2003). *Phosphatherium* was originally based on only two very fragmentary upper dentitions, but several upper and lower jaws and partial skulls were recently described (Gheerbrant, Sudre, et al., 2005; Figs. 13.19A, 13.20). The new specimens reveal many primitive features, including a relatively long and narrow snout with a prominent postorbital process on the frontal. The nasal opening is at the front of the snout, not retracted as in most proboscideans; hence *Phosphatherium* lacked a proboscis. The dental formula is the most primitive for undoubted Proboscidea (3.1.4.3/2.1.3.3). The canines and P^1 are retained, and the lower central incisor (probably I_1, but possibly I_2) is slightly enlarged. Diastemata are lacking except between P^1 and P^2. The mandibular symphysis is unfused. Tethythere affinities are indicated by the anterior position of the orbit; bilophodont upper teeth with a weak centrocrista, reminiscent of the dilambdodont condition of primitive Altungulata; and the vertically oriented coronoid process of the mandible. Superimposed on this pattern are a few derived proboscidean dental traits, including the absence of molar conules and the presence of a crest (distocrista) on the upper molars that joins the hypocone to the postentoconule. Based on these features and its much smaller size (body mass estimated to be 10–15 kg), *Phosphatherium* is considered to be the most primitive known proboscidean (Gheerbrant et al., 1996; Gheerbrant, Sudre, et al., 2005).

Also from Ouled Abdoun, Morocco, comes a larger (about 80–170 kg) and dentally more advanced proboscidean, *Daoui-*

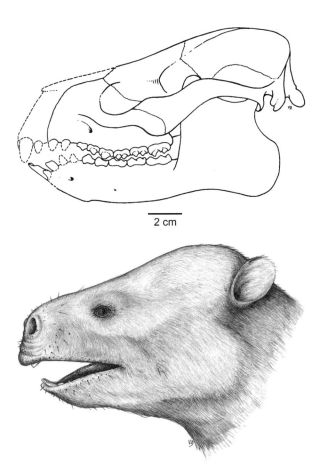

Fig. 13.20. Skull and restoration of *Phosphatherium*, an early Eocene proboscidean. (From Gheerbrant, Sudre, et al., 2005.)

therium (Fig. 13.19F), which appears to be structurally intermediate between *Phosphatherium* and *Numidotherium* (Gheerbrant et al., 2002). *Daouitherium* is primitive in having four lower teeth anterior to P_2, without significant diastemata between them. Although their homologies are uncertain, at least the first two seem to be incisors, and the anterior one may be somewhat enlarged. The posterior premolars are molariform and they and the molars are distinctly bilophodont. The cheek teeth increase in size posteriorly. As in *Numidotherium*, the anterior border of the ascending ramus of the mandible is vertical. The presence of lophodont cheek teeth in the three oldest undoubted proboscideans indicates that this morphology is the plesiomorphic dental condition for Proboscidea and suggests that anthracobunids are still more primitive (Gheerbrant, Domning, and Tassy, 2005).

Moeritherium (Figs. 13.21B, 13.22A) is known from relatively complete skeletal remains from the late Eocene–early Oligocene of the Fayum in Egypt and has also been found in other sites across northern Africa. It was initially described as a proboscidean (Andrews, 1906) and was widely accepted as such through the first half of the twentieth century. Subsequently *Moeritherium* was excluded from the Proboscidea by several authors, usually on the basis of primitive traits. Plesiomorphic features include the relatively low skull with a long sagittal crest, long snout, and anteriorly placed nasal opening, suggesting that it lacked a proboscis (Shoshani,

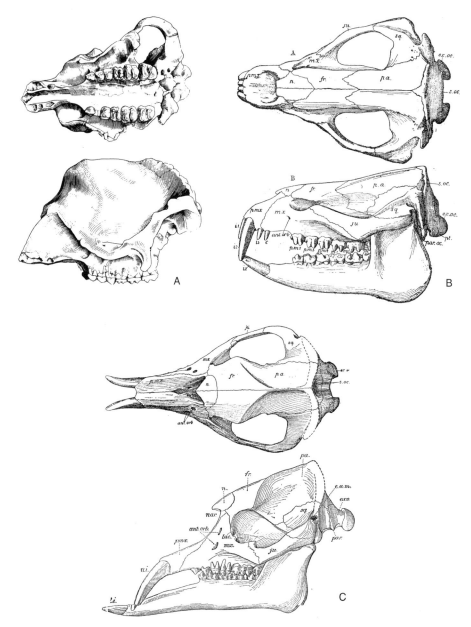

Fig. 13.21. Skulls of primitive proboscideans: (A) *Numidotherium*; (B) *Moeritherium*; (C) *Palaeomastodon*. Not to scale. Key: ant. orb., antorbital (infraorbital) foramen; e.a.m., external auditory meatus; ex. oc. or exo., exoccipital; fr., frontal; ju, jugal; lac., lacrimal; mx, maxilla; n., nasal; nar., external nares; pa., parietal; par. or par. oc., paroccipital process; pmx, premaxilla; pt., post-tympanic process of squamosal; s. oc., supraoccipital; sq., squamosal. (A from Mahboubi et al., 1984; B, C from Andrews, 1906.)

West, et al., 1996). *Moeritherium* now once again seems securely placed in the order, based on numerous derived traits shared with proboscideans, such as enlarged second incisors; incipient cranial pneumatization; a large, hooked coracoid process on the scapula; and a medial tubercle on the astragalus (e.g., Tassy, 1981; Domning et al., 1986; Shoshani, West, et al., 1996). However, its mosaic of features—including bunodont to bunolophodont cheek teeth, primitive retention of three upper incisors and an upper canine (dental formula 3.1.3.3/2.0.3.3 as in *Numidotherium*), lack of extensive cranial pneumatization, absence of a trunk, and presence of an elongate thoracolumbar region—make its position among primitive Proboscidea controversial. It has variously been considered to be the most primitive known proboscidean (e.g., Tassy, 1996) or a uniquely derived early offshoot (Court, 1995). *Moeritherium* was semiaquatic and about the size of a pig or a tapir.

Also from late Eocene and early Oligocene beds of the Fayum come the closely related genera *Palaeomastodon* (Figs. 13.21C, 13.22B) and *Phiomia* (Andrews, 1906). These taxa resemble later proboscideans in having the external nasal opening shifted back in front of the orbits, a high occiput, and a single incisor above and below enlarged as tusks (I^2_2). The upper tusks are downturned, and the lowers are horizontal and extend in front of the upper tusks. There is a long diastema between the tusk and the cheek teeth in the mandible; the upper diastema is much shorter. The molars are either bilophodont or trilophodont, and are sometimes described as zygodont (with cusps arranged in transverse pairs; Tobien, 1978). There is a broad, high nuchal crest, presumably associated with the longer rostrum and tusks. The postcranial skeleton is robust and generally similar to that in later proboscideans. Some species of *Palaeomastodon* approached the size of the present-day Asian elephant. *Palaeo-*

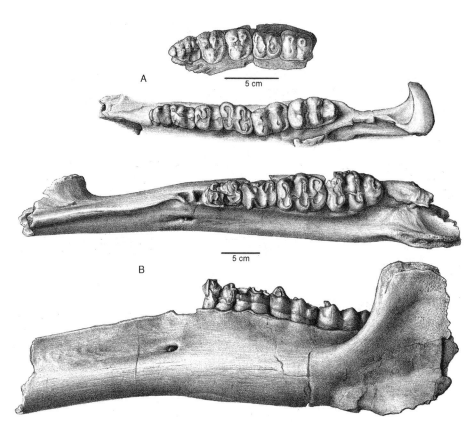

Fig. 13.22. Left dentitions of late Eocene–early Oligocene proboscideans: (A) *Moeritherium*; (B) *Palaeomastodon*. (From Andrews, 1906.)

mastodon and *Phiomia* are considered more closely related to modern elephants than is any other group of archaic proboscideans (e.g., Tassy, 1988, 1996; Court, 1995).

Barytherium (Barytheriidae) represents an unusual lineage from the late Eocene–early Oligocene of northern Africa. It was placed in its own order by H.F. Osborn and other early workers, but is now usually included in the Proboscidea, based on such features as the high, pneumatized skull with elevated nasal opening, enlarged I^2, hooklike scapular coracoid process, and medial tubercle of the astragalus (e.g., Shoshani, West, et al., 1996; Tassy, 1996). However, its exact phylogenetic position is uncertain (Domning et al., 1986; Court, 1995). The dental formula is 2.0.3.3/2.0.3.3. Its upper teeth are wider and the molars more strongly bilophodont than in contemporary proboscideans. The lower jaw and symphysis are massive and deep and contain a large, procumbent lower tusk (I_1, not I_2, as in other proboscideans), which is separated from the cheek teeth by a long diastema (Shoshani, West, et al., 1996). The few known postcranial elements are particularly robust, even compared to other proboscideans. Loss of the I^3 and the upper canine, and a few other features, suggest that *Barytherium* could be more closely related to later proboscideans than is *Moeritherium* (Domning et al., 1986; Shoshani, West, et al., 1996; Tassy, 1996). Barytheres were large mammals, ranging from the size of tapirs (about 200 kg) to that of small elephants (about 3,000–3,600 kg; Shoshani, West, et al., 1996).

Proboscidea and the closely related Desmostylia may stem from a form similar to the late Paleocene Chinese phenacolophid *Minchenella*. Phenacolophids are often classified in the Embrithopoda (Wells and Gingerich, 1983; Ray et al., 1994) but they might actually be stem tethytheres (Gheerbrant, Domning, and Tassy, 2005). Relationships remain moot, however, as only part of the dentition of *Minchenella* is known, and only lower jaws have been described. As noted earlier, anthracobunids could also be stem tethytheres. The potential relationship between phenacolophids and anthracobunids has not been thoroughly explored, but known fossils may not be adequate to enable a useful assessment. Better understanding of phenacolophids and anthracobunids is likely to shed light on the origin of proboscideans and other tethytheres.

Embrithopoda

The unusual extinct infraorder Embrithopoda (meaning "heavy-footed") comprises about six genera in two families, Phenacolophidae and Arsinoitheriidae, known principally from the late Paleocene to late Eocene of Asia, eastern Europe, and northern Africa. Most of the genera are known primarily from jaws and teeth. Embrithopoda are best known from the terminal member, the elephantine *Arsinoitherium*, from the lower Jebel Qatrani Formation, latest Eocene of the Fayum Depression, Egypt. Many complete skulls have been found, and the skeleton is substantially known, although only from isolated elements. *Arsinoitherium* was a bizarre, graviportal animal, with pillarlike limbs analogous to those of elephants (Fig. 13.23). The skull was huge, reaching a length of 80 cm. A pair of immense, hollow, bony horn cores, almost as long as the skull itself, projects from

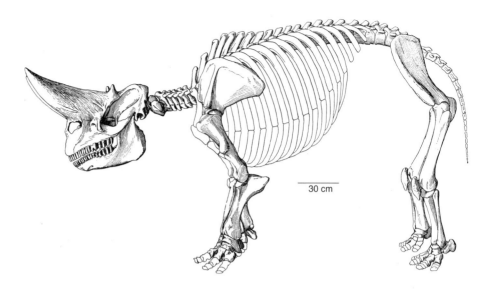

Fig. 13.23. Late Eocene embrithopod *Arsinoitherium*. (From Andrews, 1906.)

the nasal bones in front of the orbits. Vascular grooves on the surface indicate that the horns were covered with skin. A second, much smaller pair, also hollow, extended from the frontals above the orbits. Both upper and lower dentitions are unreduced (3.1.4.3) and without diastemata, in contrast to most other ungulates. The teeth of *Arsinoitherium* are high crowned, and the molars attained a uniquely specialized bilophodont condition in which the lingual cusps of upper molars were reduced and cusps usually oriented buccally were displaced lingually (Court, 1992b; Fig. 13.24). This structure may have evolved from a dilambdodont condition, remnants of which can be seen in the upper molars of the middle Eocene embrithopod *Palaeoamasia* from Turkey (Sen and Heintz, 1979; Fig. 13.25A).

Articulated skeletons of *Arsinoitherium* have not been found, but owing to Andrews's (1906) reconstruction of the skeleton from isolated bones (Fig. 13.23), and Court's (1993) thorough analysis of the skeleton, the postcranial anatomy of *Arsinoitherium* is relatively well understood. The feet were pentadactyl, spreading, and hoofed. *Arsinoitherium* had reduced limb girdles (especially the pelvis) with rather weak surrounding musculature. Although the muscles controlling fore and aft motion of the front legs were well developed, humeral features associated with adducting the forelimb were weak. As in other graviportal mammals, the humerus and femur were longer than the more distal segments. The knee-joint was extraordinarily low, implying a short stride length. The sacral vertebrae were unfused, indicating weak support for the rear legs, and the structure of the cervical vertebrae suggests that the neck typically was held up so that the head was well above the shoulders. All these features suggest that, contrary to being forest or open-land denizens analogous to rhinos or uintatheres, as often portrayed, *Arsinoitherium* was more likely a semiaquatic swamp dweller that occasionally emerged for slow, if ungainly, foraging on land (Court, 1993). Court's extensive analyses of cranial and postcranial skeletal anatomy of *Arsinoitherium* linked embrithopods with Proboscidea and Sirenia and sug-

gested that Embrithopoda and Proboscidea are sister taxa (Court, 1990, 1992c, 1993; see also Fischer and Tassy, 1993). Cranial anatomy also supports the monophyly of Pantomesaxonia (i.e., Altungulata as used here; Court, 1992c).

Arsinoitheres were long known only from the Fayum, but over the past two decades dentitions of older and more primitive arsinoitheriids have been found at several sites of probable middle and late Eocene age in eastern Europe and southwestern Asia (Turkey). They are assigned to three genera, the oldest and most primitive of which is middle Eocene *Palaeoamasia*. Its teeth are more primitive and lower crowned than those of *Arsinoitherium*, but still strongly bilophodont, and its skull lacked horns (Sen and Heintz, 1979). The occurrence of these fossils in lignites that formed in swamps and lakes is consistent with the interpretation of arsinoitheres as semiaquatic animals. The closely allied genus *Crivadiatherium* is known from slightly younger Eocene strata of Romania (Radulesco and Sudre, 1985). *Hypsamasia* is a poorly known arsinoitheriid based on fragmentary teeth from Turkey (Maas et al., 1998).

The late Paleocene Asian Phenacolophidae, a rather obscure group that includes the genera *Phenacolophus* (Fig. 13.25B) and *Minchenella*, has been regarded as the earliest and most primitive branch of Embrithopoda (McKenna and Manning, 1977; McKenna and Bell, 1997). Alternatively, phenacolophids generally, and *Minchenella* in particular, may occupy a much more pivotal position as basal tethytheres close to the origin of Proboscidea (Ray et al., 1994; Gheerbrant, Domning, and Tassy, 2005). Upper and lower cheek teeth are relatively low crowned and lophodont, each with two slightly oblique transverse crests, but less bilophodont than in arsinoitheriids. The upper molars are quadrate, with distinct hypocone, conules, and mesostyle (Matthew and Granger, 1925a). The posterior premolars are increasingly molariform.

Radinskya (see the discussion at the beginning of this chapter), from the late Paleocene of southern China, was initially referred questionably to Phenacolophidae (McKenna et al., 1989), which it closely resembles except for the ab-

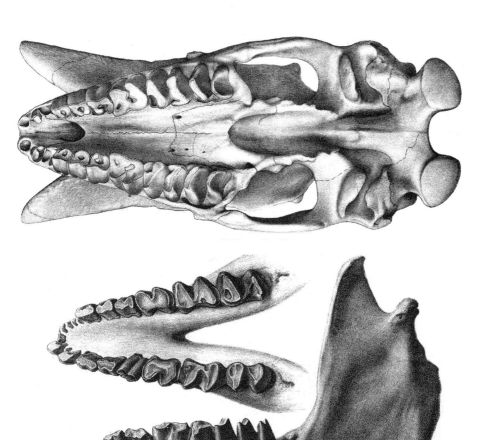

Fig. 13.24. Skull and mandible of *Arsinoitherium*. (From Andrews, 1906.)

sence of upper molar mesostyles. It was also thought to be possibly related to perissodactyls. *Radinskya* was subsequently considered a basal altungulate (McKenna and Bell, 1997). Hooker and Dashzeveg (2003) argued that *Radinskya* shares derived dental traits with embrithopods, and that its resemblances to primitive perissodactyls are convergent. More recently Holbrook (2005) identified new cranial traits that again suggest that *Radinskya* is pertinent to perissodactyl origins. The unstable phylogenetic position of *Radinskya* underscores the generalized altungulate dental morphology of the sole known specimen and suggests that, whatever its precise relationships, *Radinskya* is an important form for understanding the early diversification of altungulates.

Although phenacolophids have been considered to lie near the origins of Proboscidea, Perissodactyla, or all altungulates, they remain very poorly known, with most specimens being restricted to the dentition. They seem likely to occupy a pivotal position in the evolution of altungulates or tethytheres, but until they are better known, their relationships will remain controversial.

Sirenia

Sirenians, or sea cows, comprising the manatees and dugongs, are large, aquatic herbivorous mammals whose

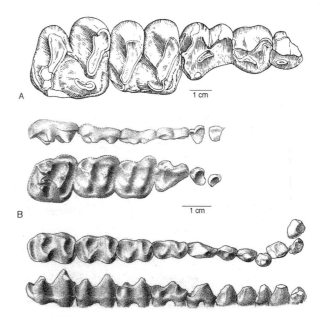

Fig. 13.25. Right dentitions of embrithopods (anterior to the right): (A) upper teeth of middle Eocene *Palaeoamasia*; (B) upper and lower teeth of late Paleocene *Phenacolophus*. (A from Sen and Heintz, 1979; B from McKenna and Manning, 1977.)

closest living relatives are the elephants (Thewissen and Domning, 1992; Lavergne et al., 1996). Only two genera survive today, although sirenians were much more diverse in the past. The oldest fossil sirenians come from lower or middle Eocene marine deposits of the western Atlantic, but the group probably diverged from other tethytheres in the Tethyan region during the Paleocene. By the middle Eocene, they had dispersed to become essentially worldwide in tropical marine waters. The present-day distribution of sirenians is more restricted: manatees (*Trichechus*) live along the Atlantic and Gulf coasts of southeastern North America and northern South America and in tropical rivers of Africa and South America, whereas dugongs (*Dugong*) are limited to coastal Indo-Pacific marine habitats.

Four families are generally recognized, Prorastomidae, Protosirenidae, Dugongidae, and Trichechidae, the first three of which were already distinct by the middle Eocene. The fourth, Trichechidae (manatees), probably did not originate before the late Eocene and are actually unknown until the late Oligocene; they appear to have evolved from dugongids in the broad sense. Dugongidae is the largest family of sirenians and is demonstrably paraphyletic, as trichechids are nested within the family (Domning, 1994). A recent study by Sagne (2001), however, derives trichechids directly from protosirenids.

Extant sirenians bear superficial resemblance to whales in having almost no hair except for facial bristles; a short neck; paddlelike front limbs; a vestigial pelvis and no hind limbs; and a horizontal tail fin used to propel them (slowly) through shallow, near-shore waters. Their eyes are small, and there are no external ears. They are large and may attain weights in excess of 1000 kg. The clavicle is absent, and the forelimb joints remain relatively mobile, except that the radius-ulna and some of the carpal bones are usually fused. Like many ungulates, and other tethytheres in particular, sirenians have a relatively long thoracic region, with 14–21 pairs of ribs. Particularly characteristic of sirenians are the swollen (pachyostotic) and very dense (osteosclerotic) ribs that function for ballast, for unlike cetaceans, the lungs do not collapse when diving (Domning and de Buffrénil, 1991). Pachyostosis and osteosclerosis are rare conditions among mammals but were already prominent in the oldest known sirenians.

In all but the most primitive sirenians the rostrum of the skull and the mandibular symphysis are conspicuously downturned (typically at an angle of 20–70° to the occlusal plane of the cheek teeth), to facilitate feeding on bottom vegetation, such as seagrasses and algae. Interspecific variations in rostral deflection are probably related to feeding preferences and the degree of commitment to bottom feeding, the most specialized bottom feeders having the most sharply downturned snouts (Domning, 1978a). The infraorbital foramen, mandibular canal, and mental foramina are very large, to convey large sensory nerves (branches of cranial nerve V) and vessels to the mobile snout and lips. The mandibular foramen, on the medial side of the jaw at the posterior end of the mandibular canal, is so large that (except in prorastomids) it exposes the back of the dental capsule (the part of the mandible where the teeth are developing). The ascending ramus is typically vertical. Sirenian skulls are characterized by robust zygomae; a large and retracted nasal opening; reduced nasal bones; prominent pterygoid processes; and a distinctive, dense skullcap composed of the strongly fused parietal and supraoccipital bones. (The remaining parts of the occipital are fused together but usually not synostosed to the skullcap.) Despite a small external auditory opening, the middle-ear bones are very large—indeed, the largest of any mammal. The petrosal bone (periotic) is also large and, in contrast to other mammals except cetaceans, is not attached to the basicranium (except in *Prorastomus*), instead being housed in a socket in the squamosal (Domning, 1994, 2001b).

The molars of most sirenians are bunodont and somewhat bilophodont, as in other primitive tethytheres. Living dugongs have a reduced number of teeth, usually including in adults only a single upper tusk (I^1, not I^2 as it is in proboscideans) and two or three molars in both the upper and lower jaws; however, three deciduous premolars precede the molars in the jaw, and vestigial lower incisors are occasionally present under the corneous plate that covers the symphysis (Petit, 1955). Dugong molars have very thin enamel, which typically wears away quickly, as does the bunodont crown pattern. Manatees, however, have evolved a very unusual situation in which they produce a virtually endless supply of molar teeth. Incisors, canines, and premolars are missing in adults, but there are typically seven or eight molariform cheek teeth in each jaw quadrant. Worn teeth are shed at the front of the toothrow, and new teeth (forming continuously in the dental capsule behind the toothrow) are accommodated at the back by mesial drift of the toothrow (Domning, 1982; Domning and Hayek, 1984). Domning argued that the possession of numerous, small teeth that are replaced throughout life is an alternative to hypsodonty, which evolved in manatees as an adaptation to exploit abrasive aquatic grasses.

Much has been made of the seemingly primitive dental formula of Eocene sirenians (3.1.5.3/3.1.5.3), because of the presence of a fifth premolar (lost in almost all other eutherians). Although the presence of five premolars in Eocene forms could reflect the primitive eutherian condition (McKenna, 1975a), this interpretation conflicts with the preponderance of other evidence, which nests Sirenia within the Paenungulata (and well within the Ungulata). It would also require the independent loss of P5 in all other ungulates (or retention of dP5 as M1 and loss of M3). Because the most primitive ungulates had four premolars, the presence of a fifth premolar is more likely an autapomorphy of Sirenia (Domning et al., 1982, 1986; Domning, 1994). Savage et al. (1994) postulated that the similarity of the deciduous P_4 of diverse ungulates suggests that the presence of five premolars and three molars in Sirenia evolved by reversal of the loss of P5 and M3, which McKenna (1975a) inferred to have occurred in early eutherians. In any case, all later sirenians again reduced the number of premolars. Except for a tusklike pair of upper incisors (I^1), the incisors and canines are also absent or reduced in post-Eocene sirenians. In modern sea cows they are replaced by a horny pad used to crop vegetation.

Sirenia appeared almost full-blown by the middle Eocene and became both anatomically and taxonomically diverse very early in their history. Their early evolution is thought to have been associated with the distribution of sea grasses and the structural evolution of the Tethys Sea (Domning et al., 1982; Domning 2001a).

Prorastomids, from the early or middle Eocene (about 50 Ma) of Jamaica, are the oldest and most primitive undoubted sirenians. For more than a century the holotype skull of *Prorastomus* (Fig. 13.26A), first described by Owen in 1855, was the only known specimen of this basal genus; but a few other fragments, including ribs, have recently been reported. Although ribs of many mammals are relatively uninformative, sirenian ribs, as already noted, are highly distinctive. These fossils indicate that *Prorastomus* had swollen, dense ribs; a dense skull roof; a large, high nasal opening; parallel anterior tooth rows; and a long, narrow mandibular symphysis—all sirenian synapomorphies—but the large rostrum and symphysis show almost no hint of the downward deflection typical of later sirenians (Savage et al., 1994). Based on these traits, *Prorastomus* is considered the sister taxon and potential common ancestor of all other Sirenia (Domning, 1994; Savage et al., 1994). Although comparatively small for a sirenian, *Prorastomus* was a medium-sized Eocene mammal, with a skull about the size of that of a sheep. In analogy with the more derived middle Eocene *Protosiren* (see below), *Prorastomus* was probably amphibious.

Its dense ribs, as well as its occurrence in marine sediments, suggest at least partly aquatic habits; but the likelihood that more derived middle Eocene sirenians could still move on land suggests that *Prorastomus* may well have been partly terrestrial, too (Savage et al., 1994).

A second genus of prorastomid, *Pezosiren* (Fig. 13.27), was recently reported from early middle Eocene beds of Jamaica (Domning, 2001c). This pig-sized form is more derived than *Prorastomus* in having a slightly downturned mandibular symphysis but more primitive in retaining a sagittal crest. Other primitive features of *Pezosiren* include a small mandibular foramen and the absence of the large, projecting pterygoid processes characteristic of later sirenians. Like other early tethytheres, it had a long thoracic region, consisting of 20 thoracic vertebrae. Most remarkable, however, are skeletal remains that indicate that this early sirenian was a quadruped whose limbs and spinal column were still fully able to support its weight on land. Like other sirenians, however, *Pezosiren* had a pachyosteosclerotic skeleton (especially the ribs) and a high nasal opening, and it lacked paranasal air sinuses—all indicating that it was already primarily aquatic. Supposed sirenians older than prorastomids have been reported from lower Eocene strata of Argentina, Hungary, India, and Pakistan, but the age and/or sirenian status of each of these records is in question.

By the middle Eocene a diversity of sirenians is known from marine deposits from the western Atlantic to India. Important recent discoveries have come from Egypt and Pakistan (e.g., Domning and Gingerich, 1994; Gingerich, Arif, et al., 1995). *Protosiren* (Protosirenidae) represents a structural grade intermediate between prorastomids and later

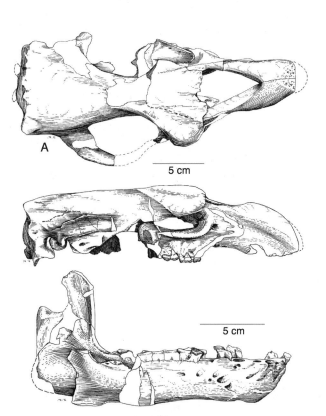

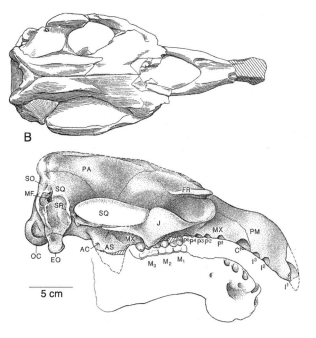

Fig. 13.26. Skulls of primitive sirenians: (A) *Prorastomus;* (B) *Protosiren.* Key: AC, alisphenoid canal; AS, alisphenoid; EO, exoccipital; FR, frontal; J, jugal; MF, mastoid foramen; MX, maxilla; OC, occipital condyle; PA, parietal; PM, premaxilla; SO, supraoccipital; SQ, squamosal; SR, sigmoid ridge. (A from Savage et al., 1994; B from Gingerich, Domning, et al., 1994.)

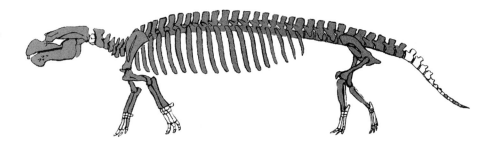

Fig. 13.27. Restoration of the middle Eocene prorastomid sirenian *Pezosiren* from Jamaica. (From Domning, 2001c.)

sirenians, although it is contemporary with the oldest dugongids (Domning, 1994). Both deciduous and permanent teeth were present at the P5 locus. Protosirenids had a broad Tethyan distribution, ranging from eastern North America to Egypt, Europe, and the Indian subcontinent. The snout and mandibular symphysis in protosirenids are moderately downturned, and the molars are low crowned and bilophodont, indicating that *Protosiren* was a bottom feeder that probably preferred sea grasses (Fig. 13.26B). Confirmation of such a diet resulted from recent stable isotope analyses of its teeth, which also indicated that *Protosiren* preferred marine habitats (MacFadden et al., 2004). *Protosiren* had dense ribs but was not fully pachyostotic (Gingerich, Arif, et al. 1995). At least one species also had a large pelvis with a well-developed acetabulum and a large obturator foramen, indicative of functional hind limbs (Savage et al., 1994; Gingerich, Arif, et al., 1995). These features suggest that *Protosiren* could still move on land and was not yet fully aquatic; however, it had only one sacral vertebra and a relatively loose sacroiliac articulation, which suggest that any terrestrial locomotion was less than agile. The holotype skull of *P. fraasi* has an unusual depression in front of the large nasal opening, perhaps for muscle attachment for a small proboscis, as in primitive proboscideans (Gingerich, Domning, et al., 1994).

Middle Eocene *Eotheroides* of North Africa is generally considered the oldest and most primitive known dugongid. It is a suitable structural, if not actual, ancestor for all later dugongids and in many ways is intermediate between *Protosiren* and other dugongids. Primitive dugongids (*Eotheroides*, *Eosiren*, and *Prototherium*) were advanced beyond *Protosiren* in having a more downturned rostrum (but less so than in some more derived dugongids) and no external hind limbs (Savage, 1976; Domning 1978b), although the pelvis was still well developed (Andrews, 1906). A pair of small tusks was present in the premaxilla. Both *Eotheroides* and the closely allied upper Eocene *Prototherium* from southern Europe still retained the primitive sirenian dental formula (3.1.5.3), but in *Eotheroides*, at least, the tooth in the fifth premolar locus appears to be an unreplaced dP5 (Domning, 1978b). More advanced dugongids lost the canines and reduced the number of incisors and cheek teeth (Savage, 1976). The molars of *Eotheroides* are strongly bilophodont. Following these Eocene forms, the fossil record of sirenians is poor until the late Oligocene (except for early Oligocene *Halitherium* of Europe), when additional dugongids and the oldest known trichechid are recorded.

Anatomical evidence, including incipiently bilophodont teeth, rostral displacement of the orbits, and other cranial characters, supports a close relationship between Sirenia and Proboscidea (Savage et al., 1994). Also closely allied are the Desmostylia, which first appear in the upper Oligocene. All three of these orders have been allocated to the Tethytheria (named for their presumed site of origin along the margins of the Tethys Seaway), which is also considered monophyletic (McKenna, 1975a). As detailed above, Embrithopoda may also be part of this group. Interrelationships among these four orders are unresolved. Prorastomids are markedly more primitive than other known sirenians in lacking several synapomorphies of other members of the order. They also retain primitive conditions of the auditory region and atlas vertebra (Savage et al., 1994) typical of condylarths. Present evidence favors derivation of Sirenia from a lophodont terrestrial herbivore, such as an anthracobunid.

14

Cete and Artiodactyla

ARTIODACTYLS AND CETACEANS CONSTITUTE the second great clade of extant ungulates. The broader relationships of artiodactyls to other placentals are not well documented either by fossils and morphology or by molecular data; but both lines of evidence have recently converged to support a close link between Cetacea and Artiodactyla.

CETE AND CETACEA

The whales and dolphins (Cetacea) are the most fully aquatic mammals and include the largest mammals that ever evolved—up to 30 m in length and weighing in excess of 100,000 kg. Some 80 species, representing 14 families, are alive today (Fordyce and Muizon, 2001). They range throughout the marine realm from the tropics to the poles, in shallow coastal regions to depths of more than 1,000 m; a few forms inhabit freshwaters in Asia and South America (large rivers such as the Ganges and the Amazon).

In the first half of the twentieth century it was thought that modern whales (odontocetes and mysticetes) were diphyletic, because intermediates linking them to a common ancestor seemed to be lacking. However, evidence accumulated over the past 30 years confirms that modern Cetacea are monophyletic (Van Valen, 1968; Fordyce and Barnes, 1994; Fordyce, 2002). The broader relationships of Cetacea among mammals have been a more vexing question. Both morphological and molecular evidence indicates that cetaceans are highly modified ungulates, but their precise phylogenetic position among ungulates is a matter of lively debate. Although some studies have suggested that, among extant mammals, whales are most closely

Fig. 14.1. Relationships of Cete and Cetacea: (A) paleontological view, modified after Uhen (1998); (B) molecular view, based on Madsen et al. (2001) and Springer et al. (2003); (C) paleontological view in light of new fossil evidence, based mostly on Thewissen et al. (2001b) and Gingerich (2005). Positions of Mesonychia and Triisodontinae are in dispute.

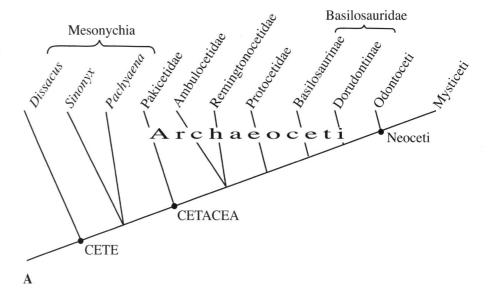

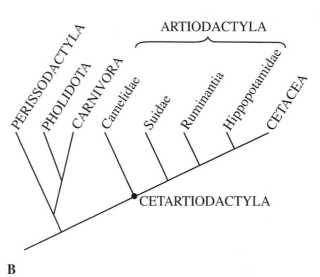

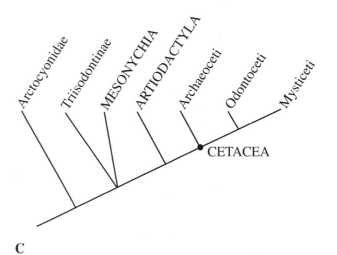

allied with perissodactyls and tethytheres, most current research, both morphological and molecular, strongly indicates that their closest living relatives are artiodactyls (Fig. 14.1).

Consideration of extinct groups, however, has suggested another, possibly closer, relative. Several morphological analyses have pointed to the extinct Mesonychia as the sister taxon of Cetacea, with Artiodactyla as the closest extant group (e.g., Van Valen, 1966, 1978; O'Leary, 1998; Luo and Gingerich, 1999; O'Leary and Geisler, 1999; O'Leary and Uhen, 1999; Geisler, 2001; Gatesy and O'Leary, 2001; Fig. 14.1A, Table 14.1). Derived characters linking the most primitive fossil whales and mesonychians include various basicranial features and unusual cheek teeth, which are simplified and tritubercular above and narrow with modified cusps aligned mesiodistally below. These specialized teeth contrast with those of primitive artiodactyls, which retain primitive tribosphenic molars, as in basal condylarths. Based on this evidence, some classifications therefore unite Mesonychia and Cetacea in an ordinal level taxon called Cete (Thewissen, 1994; McKenna and Bell, 1997). Recent fossil discoveries have revealed that mesonychians and primitive whales also share a suite of postcranial features, including paraxonic symmetry of the hind feet, hooflike terminal phalanges, and various other limb specializations usually associated with cursoriality. To complicate matters, however, primitive whales also share these postcranial features, as well as certain tarsal traits, with early artiodactyls (see below).

In fact, molecular data increasingly indicate that whales are not just related to, but are nested deep within Artiodactyla, and are the sister group either of hippopotami (e.g., Springer and Kirsch 1993; Graur and Higgins, 1994; Gatesy et al., 1996, 1999; Gatesy, 1997; Milinkovitch et al., 1998; Ursing and Arnason, 1998; Nikaido et al., 1999; O'Leary, 1999; Madsen et al., 2001; Murphy et al., 2001; Fig. 14.1B) or of hippopotami + ruminants, to the exclusion of other artiodactyls (Shimamura et al., 1997). This conflicts not only with the fossil evidence for a mesonychian-cetacean clade but also with anatomical, especially osteological, data indicating that hippopotami are suiform artiodactyls whose probable sister group is either the Tertiary family Anthracotheriidae (e.g., Colbert, 1935; Coombs and Coombs, 1977a; Gentry and Hooker, 1988; Luckett and Hong, 1998; Boisserie et al., 2005) or the doliochoerine peccaries (Tayassuidae; Pickford, 1983, 1989). Digestive tract anatomy, which shows important differences among ungulates, also places hippopotami firmly within Artiodactyla and less closely related to Cetacea (Langer, 2001).

Recently discovered fossil whales contribute significant new evidence to this debate—critical anatomical features that appear to confirm a closer relationship between Cetacea and Artiodactyla than between Cetacea and Mesonychia (Gingerich, ul Haq, et al., 2001; Thewissen et al., 2001b). They also show that either scenario of relationship would require extraordinary convergence or reversal in either dental or tarsal anatomy. The new fossils of Eocene archaeocete whales from Pakistan (the pakicetids *Pakicetus* and *Ichthyolestes* and the protocetids *Rodhocetus* and *Artiocetus*) possess

Table 14.1. Classification of Cete

Mirorder CETE
    Order †MESONYCHIA
        †Triisodontidae[1]
        †Mesonychidae
        †Hapalodectidae
    Order CETACEA[2]
        Suborder †ARCHAEOCETI
            †Pakicetidae[3]
            †Ambulocetidae[3]
            †Protocetidae
            †Remingtonocetidae
            †Basilosauridae
        Suborder ODONTOCETI[4,5]
        Suborder MYSTICETI[5]
            †Llanocetidae[6]
            †Aetiocetidae
            †Mammalodontidae
            †Kekenodontidae
            †Cetotheriidae
            Balaenopteridae
            Balaenidae

Notes: Modified after McKenna and Bell (1997), and Fordyce and Muizon (2001). The dagger (†) denotes extinct taxa. Families in boldface are known from the Paleocene or Eocene.

[1] Here considered a subfamily of Arctocyonidae (see Chapter 12).
[2] Increasing evidence suggests Cetacea is nested in Artiodactyla and could be the sister group of Hippopotamidae.
[3] Sometimes considered a subfamily of Protocetidae.
[4] Includes multiple families (not listed here), mostly Miocene and later; see McKenna and Bell (1997) and Fordyce and Muizon (2001). The oldest record is an unidentified odontocete from approximately the Eocene-Oligocene boundary (Barnes and Goedert, 2000).
[5] Odontoceti and Mysticeti are united in the Neoceti (=Autoceta) by Fordyce (2002).
[6] Approximately the Eocene-Oligocene boundary.

relatively well-developed limbs with paraxonic hind feet and other cursorial features typical of both artiodactyls and mesonychians. In addition, their ankle bones reveal striking and unexpected resemblances to those of artiodactyls—in particular, a double-trochlea astragalus and comparable modifications of the calcaneus (see Fig. 14.11). These fossils imply that whales are either highly specialized artiodactyls (as suggested by molecular evidence) or the sister taxon of artiodactyls. If this relationship is valid, then mesonychid-cetacean dental resemblances must be convergent.

Significantly, *Rodhocetus* appears to be more primitive than even the oldest known artiodactyl *Diacodexis* in having a shallower astragalar trochlea and retaining a vestigial astragalar foramen, a stronger clavicle, and a larger third trochanter on the femur. Consequently, if cetaceans are artiodactyls, they must have diverged from all other artiodactyls at an earlier and more primitive stage than any artiodactyl currently known (Rose, 2001b). Such an early divergence makes a close relationship between Cetacea and hippopotami even more problematic, as the fossil record of hippos extends back only about 15 million years before the present.

The two scenarios of cetacean relationship have different implications for the timing of the origin of Cetacea. A common ancestry of mesonychians and whales would indicate that Cetacea must have separated from Mesonychia by the earliest Paleocene (the age of the oldest known mesonychians) or perhaps even in the latest Cretaceous, and millions of years of cetacean evolution would be missing (O'Leary and Uhen, 1999). If Cetacea evolved from a mesonychian (rendering Mesonychia paraphyletic), the "first whale" might barely predate the early Eocene—which is currently the age of the oldest known fossil whale. Alternatively, common ancestry with Artiodactyla, which first appear at the beginning of the Eocene, implies that Cetacea probably originated in the Paleocene; whereas origin from an unknown artiodactyl would not necessarily require a long cetacean ghost lineage. Using a likelihood model that does not depend on ancestry to estimate the antiquity of Cetacea, Gingerich (2005) placed the origin of the order close to the Paleocene/Eocene boundary.

Until quite recently, phylogenetic analyses did not demonstrate conclusively whether mesonychians or artiodactyls are more closely allied to Cetacea (Gatesy and O'Leary, 2001). New phylogenetic analyses incorporating more taxa and more morphological characters than were used in earlier studies—including the new tarsal evidence from Eocene cetaceans—conclude that Cetacea are indeed nested within Artiodactyla as the sister taxon of hippopotami (Geisler and Uhen, 2003, 2005). If correct, this hypothesis could indicate that hippos have a long ghost lineage (back at least to the early Eocene), suggesting that many of their detailed resemblances to suoids and other artiodactyls evolved independently. The precise position of such a cetacean-hippo clade within Artiodactyla is not yet clear. One recent morphological analysis found that hippos are nested within the Anthracotheriidae, and that this clade is the sister taxon of Cetacea (Boisserie et al., 2005), the former reinforcing a long-held paleontological hypothesis and the latter supporting the clade Cetartiodactyla first suggested by molecular systematists. Alternatively, some of Geisler and Uhen's (2005) analyses indicated that the Asian Eocene family Raoellidae is more closely related to Cetacea than is Hippopotamidae. Clearly, we are only at the early stages of understanding the interrelationship between Cetacea and Artiodactyla. Phylogenetic assessments are likely to remain volatile as new evidence accumulates.

Mesonychia (=Acreodi)

Mesonychians are an extinct group of Holarctic, apparently carnivorous, terrestrial ungulates, which were widely accepted as the probable ancestor or sister group of Cetacea for most of the past 40 years. As related in the previous section, recently discovered fossils have led many experts to reevaluate this hypothesis. Even if mesonychians prove not to be the sister group of Cetacea, their many anatomical resemblances to early cetaceans and artiodactyls offer compelling evidence that they could still be closely related to a cetartiodactyl clade. If Mesonychia is the sister group of Cetartiodactyla, dental resemblances to primitive Cetacea probably evolved in parallel, but shared cursorial postcranial specializations could be synapomorphic.

Mesonychians are first known from the Torrejonian (late early Paleocene) and persisted through the Eocene in North America and Asia. They range in size from *Hapalodectes*, smaller than a housecat, to the gigantic *Pachyaena* and *Harpagolestes* (Fig. 14.2), some of which were bear-sized and had skulls a half-meter long. Despite their broad range of sizes, they are dentally quite uniform, all mesonychians sharing a similar, highly specialized lower dentition and simplified upper dentition. *Ankalagon,* from the Torrejonian of New Mexico, was one of the largest early Paleocene mammals and probably the biggest carnivore of its day, with a head 30 cm long. Substantial variation in jaw depth and canine size suggests that *Ankalagon* was sexually dimorphic (O'Leary et al., 2000). Even the largest mesonychians are dwarfed by *Andrewsarchus* (see Fig. 12.7) from the middle Eocene of Mongolia, a truly monstrous form whose head approached a meter in length! Although often considered a mesonychian in older accounts, *Andrewsarchus* is now usually assigned to the closely allied arctocyonid subfamily Triisodontinae (perhaps the source of mesonychians).

All mesonychians had large canines, simple, tritubercular upper molars (probably secondarily simplified), and very laterally compressed lower cheek teeth whose main cusps are longitudinally aligned. The paraconid is small and low, and the metaconid is either reduced and twinned with the protoconid or lost altogether. The unbasined talonid is low and bladelike, consisting mainly of the hypoconid. The third molars are typically reduced in size or lost in some forms. In contrast to other placental carnivores, the cusps of mesonychians are typically blunt rather than acute. Despite their bladelike appearance, the lower cheek teeth lack well-developed shearing facets like those seen on the carnassials of carnivorans and creodonts. Instead the molar cusps often exhibit heavy apical wear, more characteristic of mammals with a hard or abrasive diet than one of flesh.

The postcranial skeleton is poorly known in Paleocene representatives but well known in many Eocene types (Fig. 14.3). Early Eocene mesonychians possess a curious mixture of cursorial features found variously in later carnivorans, artiodactyls, and perissodactyls (O'Leary and Rose 1995; Rose and O'Leary 1995; Fig. 14.4), as well as primitive cetaceans. Limb proportions range from tapirlike to wolflike. The humerus, though robust, has a high greater tuberosity; reduced muscular crests; and a deep, sometimes perforate olecranon fossa. The radius was oriented anterior rather than lateral to the ulna, allowing little rotation. The hands and feet were paraxonic, with a much reduced first digit arranged similar to the dew claw of dogs. The digits were terminated not by claws but by fissured, hoof-bearing ungual phalanges. Both the wrist and the ankle have specializations to restrict rotational movements that converge on those in modern ungulates. In particular, the astragalus articulates with both the cuboid and the navicular, the latter facet being shallowly concave mediolaterally, foreshadowing the distal trochlea found in artiodactyls and primitive cetaceans (see Fig. 14.11). Middle Eocene *Synoplotherium* and especially *Mesonyx* were more gracile and carried these trends

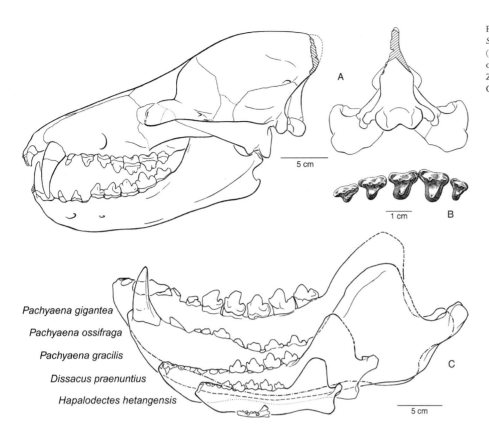

Fig. 14.2. Mesonychians: (A) skull of *Sinonyx*, lateral and posterior views; (B) Left P^3–M^3 of *Dissacus*; (C) size range of lower jaws of mesonychians. (A from Zhou et al., 1995; B from Matthew, 1937; C from O'Leary and Rose, 1995.)

further, for example in having relatively longer and more mediolaterally compact hind feet. Even the oldest known mesonychians (late early Paleocene), however, were to some degree cursorial.

The brain of *Mesonyx* was only about half the size of that expected in a present-day terrestrial mammal of similar body mass, but its small size was about average for Eocene carnivores (Radinsky, 1977). *Pachyaena*'s brain was relatively less than half as big (Gingerich, 1998), a comparatively small brain even in the Eocene!

Mesonychians are thought to have descended from Early Paleocene (Puercan-Torrejonian) triisodontine arctocyonids (see Chapter 12 and Fig. 12.3D), because the latter show incipient dental trends that became more fully developed in mesonychians. For this reason some authors (e.g., McKenna and Bell, 1997) classify them as a family of Mesonychia (see Table 14.1). They had relatively narrow lower molars and triangular upper molars, the third molars reduced, and all with rounded, blunt cusps. The initial stages of longitudinal alignment of cusps can be seen, portending the arrangement of cusps on the teeth of Torrejonian *Dissacus* and *Ankalagon*, the oldest mesonychians. Triisodontines were more primitive in retaining basined molar talonids with distinct buccal and lingual cusps. Unfortunately, we know virtually nothing about the postcranial skeleton of triisodontines. The only non-North American triisodontine is middle Eocene *Andrewsarchus* from Asia (see Chapter 12).

As discussed above, morphological evidence until recently suggested that Mesonychia was the sister group of Cetacea. Among Mesonychia, *Hapalodectes* (Hapalodectidae) was singled out as a potential cetacean ancestor because of dental similarity to the most primitive cetaceans (laterally compressed teeth with longitudinally arranged cusps, and mesial notches, or "reentrant grooves," on lower molars) and the presence of vascularized embrasure pits situated lingually between the upper molars (Szalay 1969b), a peculiar feature present also in primitive cetaceans. These remain intriguing resemblances—even though *Hapalodectes* is an order of magnitude smaller than early archaeocetes, and new fossil evidence (detailed below) supports a cetacean-artiodactyl relationship.

Cetacea

The living whales, porpoises, and dolphins are the most aquatically specialized mammals. They are classified in two suborders, Mysticeti (baleen whales) and the much more diverse Odontoceti (toothed whales), which together compose the Neoceti (=Autoceta, or crown-group Cetacea). A third suborder, Archaeoceti, encompasses the archaic toothed whales of the Eocene. Archaeoceti, which gave rise to both modern suborders, is a paraphyletic assemblage of primitive whales that lack many of the derived characteristics of the modern suborders. Examples of the three suborders are shown in Figure 14.5.

Based on analysis of mitochondrial DNA it has been suggested that odontocetes are not monophyletic, because sperm whales are more closely related to mysticetes than to other toothed whales (e.g., Milinkovitch et al., 1993). Reassessment of both morphological and molecular evidence,

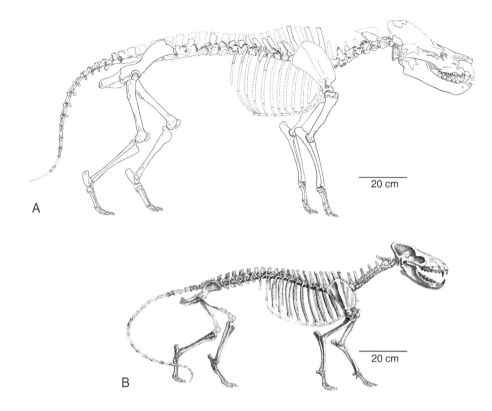

Fig. 14.3. Skeletons of mesonychians: (A) *Pachyaena*; (B) *Mesonyx*. (A from Zhou et al., 1992; B from Scott, 1888.)

however, firmly upholds the monophyly of each of the modern suborders and the allocation of sperm whales to the Odontoceti (Fordyce and Barnes, 1994; Heyning, 1997; Luckett and Hong, 1998; Fordyce and Muizon, 2001; Nikaido et al., 2001). Cranial traits, including shape of the maxilla, a mesorostral groove between the premaxillae, and amastoidy, support monophyly of the Neoceti (Fordyce, 2002).

Living cetaceans are among the most distinctive and instantly recognizable mammals, not only because of their size, but also because of the extent to which their anatomy has been modified in connection with their wholly aquatic lifestyle. Modern cetaceans have a streamlined, fusiform, and often elongate body with no hair except for a few facial vibrissae; they are insulated by a thick layer of blubber rather than by fur. The neck is very short and inflexible, and the cervical vertebrae are sometimes fused. In contrast to terrestrial mammals, the postcervical vertebrae are poorly differentiated by region, and there is no sacrum. The tail consists of distinctive horizontal fins, or flukes, which are supported by connective tissue but lack any specialized bony skeleton except for the rectangular caudal vertebrae. Whales swim by undulating the tail dorsoventrally, using the flukes as propulsive structures. Some whales can attain speeds of 30 km/hr (Rohr et al., 2002). Substantially higher speeds have been reported (Johannessen and Harder, 1960) but have not been confirmed. The limbs are reduced in all living whales. The clavicle is lost, and the other forelimb elements are shortened and modified into paddles for steering. This is accomplished partly by immobilizing the elbow joint, as well as by adding extra phalanges to the digits. Only vestigial pelvic and hind limb bones are present. They lack attachment to the vertebral column and are not evident externally.

Cetaceans have also undergone an extraordinary remodeling of the skull (Fig. 14.6) compared to most mammals, associated particularly with posterior shifting of the external nares, which open high on the head. The nares are joined as a single blowhole in odontocetes, but remain double in mysticetes. Changes in feeding and jaw mechanics, degeneration of olfactory ability, and development of acute hearing (and echolocation in odontocetes) have also influenced cranial restructuring. The eyes are small, the orbits continuous posteriorly with the temporal fossa (i.e., there is no postorbital bar), and there is no external ear.

Rostral elements (premaxillae and maxillae) are typically elongate and projected, or "telescoped," backward over other skull bones to approach or meet the forwardly shifted occipital bone. Although telescoping is characteristic of all living cetaceans, it is achieved differently in odontocetes and mysticetes, suggesting independent origin in the two groups. In odontocetes, for example, the maxillae extend back and laterally over the orbits, forming the supraorbital processes; along with other bones around the blowhole they are usually asymmetrical. In mysticetes, however, the maxilla extends under the orbit to form an infraorbital process, and the skull is symmetrical.

The two modern suborders also differ fundamentally in their feeding apparatus. Odontocetes have homodont, peglike teeth—often considerably more than the standard eutherian number (more than 60 per quadrant in some dolphins)—adapted for feeding on fish and squid. In contrast, mysticetes have evolved horny baleen plates ("whale-

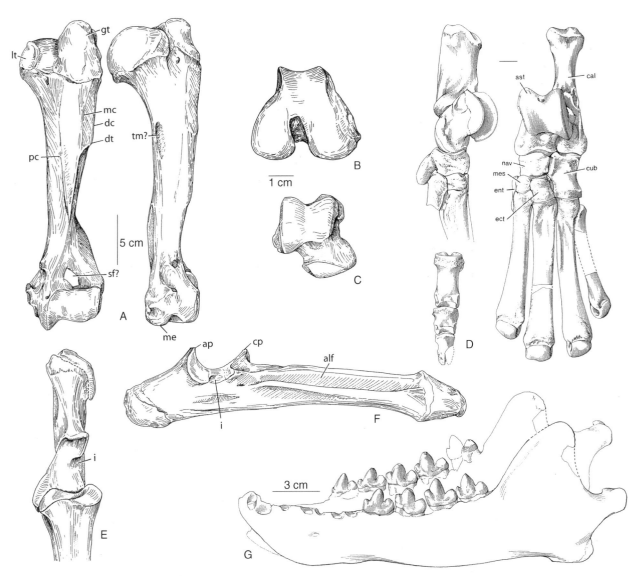

Fig. 14.4. Bones of early Eocene *Pachyaena*: (A) left humerus (anterior and medial views); (B) distal articulation of left femur; (C) right astragalus; (D) left foot (scale bar = 1 cm); (E) proximal left ulna and radius (anterior); (F) right ulna; (G) mandible. Key: alf, anterolateral fossa; ap, anconeal process; ast, astragalus; cal, calcaneus; cp, coronoid process; cub, cuboid; dc, deltoid crest; dt, deltoid tuberosity; ect, ectocuneiform; ent, entocuneiform; gt, greater tuberosity; i, lateral incisure in semilunar notch; lt, lesser tuberosity; mc, medial crest; me, medial epicondyle; mes, mesocuneiform; nav, navicular; pc, pectoral crest; sf, supratrochlear foramen; tm, teres major tuberosity. (D, G modified from Matthew, 1915a; A–C, E, F from O'Leary and Rose, 1995.)

bone") suspended from the maxillae, which are used for filter-feeding on zooplankton.

Living whales have very specialized ear regions that, in odontocetes at least, allow sophisticated echolocation when submerged. In mysticetes, the ectotympanic bullae are swollen and fused to the petrosal (periotic) bones, but in contrast to other mammals, they are only loosely attached to the skull. Particularly characteristic of Cetacea (and considered a synapomorphy of all living and fossil whales) is the thick and dense (pachyosteosclerotic) medial portion of the bulla, known as the involucrum. The malleus (the middle-ear ossicle in contact with the eardrum in terrestrial mammals) is fused to the sigmoid process of the ectotympanic, a derivative of the anterior part of the ectotympanic ring (which supports the eardrum in terrestrial mammals). The eardrum itself has been modified into a ligament that no longer functions significantly in hearing (Luo, 1998), and the external auditory canal is very small or closed. The dense petrotympanic complex is contained within a fatty capsule and is further insulated from the rest of the skull by a specialized system of air sinuses filled with an oil-mucus foam (Fraser and Purves, 1960). The density of the petrotympanic compared with the surrounding foam-filled sinuses, the separation of the ear bones from the basicranium and from each other, and (in odontocetes) the asymmetry of the skull bones combine to provide cetaceans with exceptional hearing under water. In addition to these modifications, sound waves do not enter the ear through the external auditory canal, as in terrestrial mammals. Instead, sound is conducted to the ear region through a fat pad housed in the enlarged mandibular foramen at the back of the lower jaw, which is in contact with the ectotympanic (Thewissen and

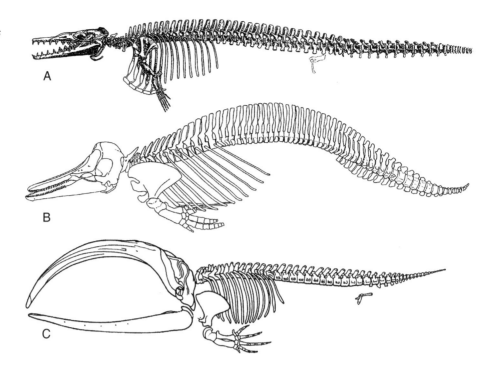

Fig. 14.5. Whale skeletons: (A) archaeocete, late Eocene *Zygorhiza*; (B) odontocete, extant *Lagenorhynchus*; (C) mysticete, extant *Balaena*. (From Fordyce and Muizon, 2001.)

Hussain, 1993; Fordyce and Barnes, 1994). Curiously, the mandibular foramen in some living mysticetes (balaenids and balaenopterids) is relatively small (implying a much smaller fat pad), and how sound is transmitted in these animals is unknown.

All living whales communicate by emitting a wide range of sounds, but only the odontocetes have evolved the ability to echolocate, enabling them to track prey and avoid obstacles with great precision. To achieve this they generate high-frequency clicks in their complex nasal passages. The clicks are focused and projected through a specialized fatty organ called the "melon," situated in front of the blowhole; it is the melon that gives odontocetes their characteristic domed forehead (e.g., Norris, 1968; Mead, 1975). Cranial asymmetry in odontocetes—the primitive condition for living odontocetes—may have resulted from selection for producing sound unilaterally from only one nasal sac (Heyning, 1989). Judging from a depressed area at the back of the premaxillae, most fossil odontocetes apparently had a melon and could echolocate, but their symmetrical skulls suggest a less specialized ability than in living species.

Archaeoceti

The oldest and most primitive whales belong to the Archaeoceti, a paraphyletic group known only from the Eocene. Five families of archaeocetes are recognized here (modified after Thewissen, 1998; Thewissen and Williams, 2002). The most primitive archaeocetes come from early and middle Eocene sediments deposited along coastal regions of the ancient Tethys Sea (India, Pakistan, and Egypt), as well as Nigeria and southeastern North America. They are members of the families Pakicetidae, Ambulocetidae, and Protocetidae (Fig. 14.7). Some authorities consider the first two of these to be subfamilies of protocetids (Fordyce and Barnes, 1994; McKenna and Bell, 1997). Two other families of archaeocetes are known from slightly younger beds in the same region: Remingtonocetidae from the middle Eocene and Basilosauridae (including Dorudontinae) from middle and upper Eocene strata. Dorudontinae is sometimes accorded family rank.

The first archaeocetes were recognized more than a century ago, but a recent surge in field work has resulted in new discoveries that greatly broaden our understanding of the origin, diversity, and early evolution of cetaceans. Half of the more than two dozen recognized genera have been described since 1990. Most are monotypic, however, suggesting that archaeocete generic diversity may be inflated. The essentially tropical distribution of early archaeocetes suggests that whales originated in warm Tethyan waters (Fordyce and Barnes, 1994), and the Indo-Pakistan region of the eastern Tethys was evidently a major center of origin and early radiation of Cetacea (Gingerich et al., 1998).

Archaeocetes were good-sized Eocene mammals, with estimated body weights generally more than 100 kg. Except for basilosaurines, however, they were small for whales, with skulls usually much less than a meter in length and total lengths of only a few meters. The snout is typically elongate, but there is no telescoping of the skull bones (see Fig. 14.6A). Diastemata separate the anterior teeth. The mandibular symphysis is long and was usually tightly joined by ligaments, or in some cases fused. Most archaeocetes retained a primitive eutherian dental formula of 3.1.4.3 in both upper and lower jaws, although basilosaurids were more advanced in having only two upper molars. The cheek teeth generally have rugose enamel and mesiodistally aligned cusps, as in

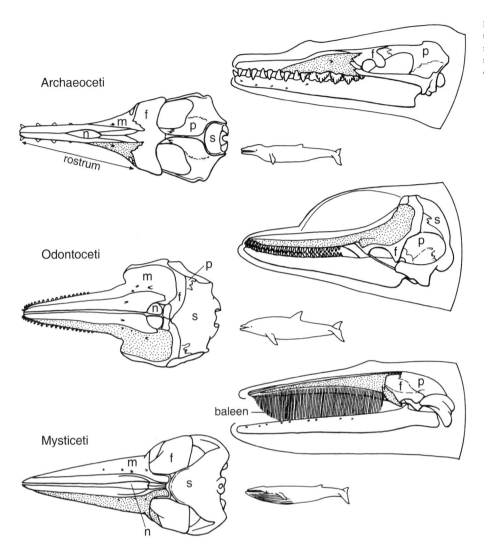

Fig. 14.6. Whale skulls, showing telescoping in odontocetes and mysticetes. Key: f, frontal; m, maxilla; n, position of nares; p, parietal; s, supraoccipital. (From Fordyce, 1982.)

mesonychians. Although living cetaceans have relatively large brains, the brains of archaeocetes are estimated to have been comparable in size to that of the mesonychian *Mesonyx*, roughly half the size predicted for an extant terrestrial mammal of similar body size (Gingerich, 1998).

It has long been realized that whales must have evolved from terrestrial quadrupeds, but transitional forms were not known until recently. In the past decade, however, spectacular new fossils have documented several intermediate stages in this transition. A number of archaeocetes are now known that retained well-developed hind limbs capable of bearing weight on land, together with auditory anatomy in between that of terrestrial mammals and fully aquatic whales. Other archaeocetes, however, show progressively more reduced limbs, making them obligate aquatic animals. Sedimentologic evidence and isotopic data from the bones themselves further document the shift from terrestrial and freshwater to marine shelf habitats. While these remarkable fossils are clarifying our understanding of how whales evolved, their mosaic of terrestrial and aquatic anatomical features is at the same time blurring the distinction between cetaceans and their terrestrial ancestors (Berta, 1994).

The earliest stages in the transition from terrestrial mammals to aquatic whales are represented by the Pakicetidae, which are known from the early Eocene (Ypresian equivalent) of northern Pakistan and India (Gingerich, Russell, and Shah, 1983; Thewissen and Hussain, 1998; Thewissen et al., 2001b). These are the oldest and most primitive archaeocetes. The cheek teeth of pakicetids are reminiscent of those of mesonychians but differ in having more medial molar paraconids and larger and more pointed posterior premolars, thought to have been associated with eating fish. *Himalayacetus* from northern India is the oldest known cetacean, dating from about 53.5 million years ago (Bajpai and Gingerich, 1998). It is known only from a fragmentary dentition, which resembles those of both pakicetids and ambulocetids. A slightly younger pakicetid, *Nalacetus*, also known only from teeth, is considered to be dentally the most primitive cetacean. *Nalacetus* had already acquired the long molar shearing facets characteristic of archaeocetes (but absent from mesonychid molars), suggesting that it was an aquatic predator (O'Leary and Uhen, 1999).

Pakicetus (Fig. 14.8A), from fluvial deposits in northern Pakistan and India dated at about 49 million years ago, is the

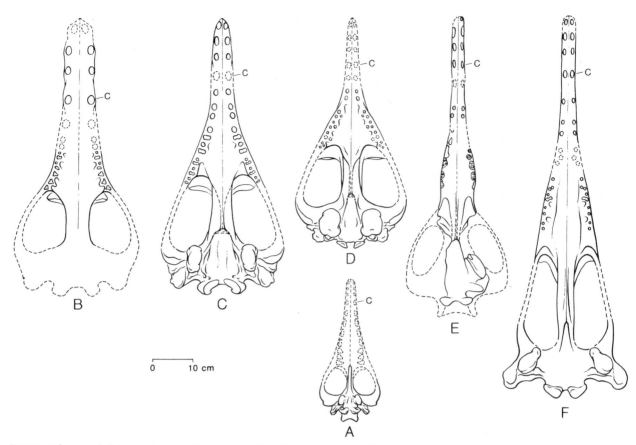

Fig. 14.7. Archaeocete skulls in ventral view: (A) *Pakicetus* (Pakicetidae); (B) *Takracetus* (Protocetidae); (C) *Rodhocetus* (Protocetidae); (D) *Gaviacetus* (Protocetidae); (E) *Remingtonocetus* (Remingtonocetidae); (F) *Dalanistes* (Remingtonocetidae). Key: c, position of canine. (From Gingerich et al., 1998.)

oldest relatively well-known whale (Gingerich, Russell, and Shah, 1983; Thewissen and Hussain, 1998; Thewissen et al., 2001b). Its lower jaw is elongate, indicating a long rostrum. The simple premolars are tall and pointed with a single cusp and are separated by short diastemata; they are distinctly larger than the molars. These traits contrast with the situation in mesonychians. As in mesonychians, however, the upper molars are tricuspid, with the paracone much larger than the metacone; the lower molars are narrow, with the trigonid higher and wider than the talonid; and (as in hapalodectids) the front of the lower molars is notched to receive the talonid of the more mesial tooth. Enamel microstructure is similar to that of primitive ungulates, including mesonychids, phenacodontids, and early artiodactyls (Maas and Thewissen, 1995).

Recently discovered skeletal remains of two pakicetid genera—*Pakicetus* and *Ichthyolestes*—possess cursorial features, including rather slender and elongate limbs, a long tail, and ankle specializations known otherwise only in artiodactyls. As already mentioned, these features suggest a closer relationship to Artiodactyla than to any other group. Although they may also seem to imply that pakicetids were primarily terrestrial (Thewissen et al., 2001b), not all evidence is consistent with that interpretation. For example, some limb elements of *Pakicetus* are very dense (osteosclerotic), perhaps serving as ballast in a subaqueous environment (Thewissen and Williams, 2002).

In addition, the ear region of *Pakicetus* is intermediate between that of terrestrial mammals and later whales. *Pakicetus* resembles mesonychians and other terrestrial mammals in having a functional external auditory canal, a complete tympanic ring to support the eardrum, septa within the auditory bulla, and a small mandibular foramen, features suggesting that it was better adapted for receiving airborne sound than for underwater hearing (Bajpai and Gingerich, 1998; Luo, 1998; Luo and Gingerich, 1999). In contrast to later cetaceans, the bulla is not fused to the petrosal (as it is in mysticetes), the petrosal remains attached to the basicranium, and there are no peribullar air sinuses. At the same time, however, *Pakicetus* has a robust incus intermediate in form between those of terrestrial mammals and whales and an involucrum (thickened medial rim of the bulla), a diagnostic cetacean feature associated with aquatic hearing (Gingerich, Russell, and Shah, 1983; Thewissen and Hussain, 1993). This mosaic of auditory features suggests that *Pakicetus* was amphibious.

The oldest known marine whale is *Ambulocetus* (Ambulocetidae; Fig. 14.9A, Plate 8.1) from slightly younger (earliest Lutetian) shallow marine beds of Pakistan. The skeleton is striking in preserving well-developed front and hind legs

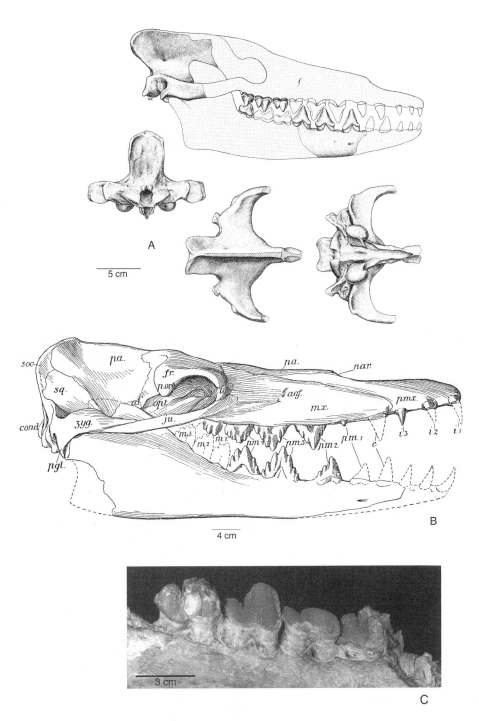

Fig. 14.8. Archaeocetes: (A) *Pakicetus*, reconstructed skull based on posterior part and jaw fragments; (B) *Dorudon* (=*Prozeuglodon*); (C) *Georgiacetus*, lower dentition. Key: al, alisphenoid; ao.f., antorbital (infraorbital) foramen; cond, occipital condyle; fr., frontal; ju., jugal; la, lacrimal; mx., maxilla; na., nasal; nar., external nares; opt, optic foramen; pa., parietal; pgl., postglenoid process; pmx., premaxilla; p.orb, postorbital process of frontal; soc, supraoccipital; sq., squamosal; zyg., zygomatic process of squamosal. (A from Gingerich, Russell, and Shah, 1983; B from Andrews, 1906; C from Hulbert et al., 1998.)

adapted for locomotion both on land and in the water (Thewissen et al., 1994; Thewissen, Madar, and Hussain, 1996). The limb bones were rather short compared to those of land mammals, but better developed than in any other cetaceans except pakicetids. The digits were long and spreading, and the hands and feet were flipperlike and much longer than the middle segments of the limbs. The hind feet were paraxonic (with the plane of symmetry passing between the longer third and fourth digits), as in both mesonychians and artiodactyls. A fused ecto-mesocuneiform was present, as in early artiodactyls but not mesonychians (Madar et al., 2002).

The elbow, wrist, and phalangeal joints were mobile, but the radius could not be rotated (a resemblance to terrestrial runners). The knee and ankle joints, like the elbow, could only flex and extend in a sagittal plane, and the digits ended in hoof-bearing ungual phalanges—specializations also seen in cursorial mammals, including mesonychids and artiodactyls. The tail was apparently rather long, more like that of mesonychids than whales, suggesting that tail flukes had not yet developed. Thewissen et al. (1994) postulated that *Ambulocetus* swam by undulating the vertebral column dorsoventrally, as in modern cetaceans, while propelling the body

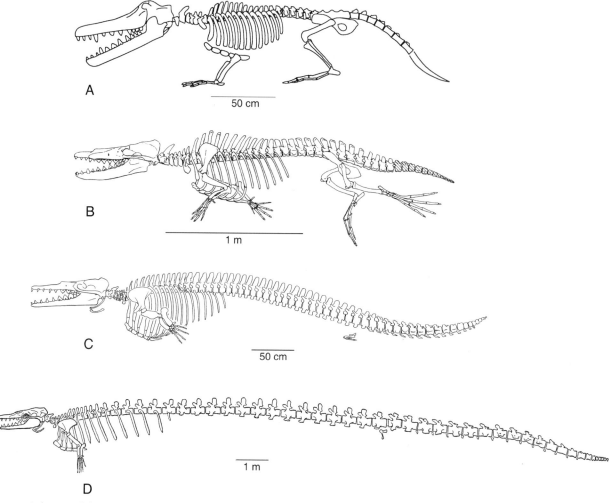

Fig. 14.9. Archaeocete skeletons: (A) *Ambulocetus* (Madar et al., 2002, now believe that the thoracolumbar length was probably longer than shown in this reconstruction); (B) *Rodhocetus;* (C) *Dorudon;* (D) *Basilosaurus.* (A from Thewissen, Madar, and Hussain, 1996; B from Gingerich, ul Haq, et al., 2001; C from Gingerich and Uhen, 1996; D from Gregory, 1951.)

with the hind limbs and maneuvering with the forelimbs. On land, however, their locomotion was probably awkward and somewhat like that of seals.

The skull of *Ambulocetus* had a long rostrum with laterally facing orbits elevated above the level of the snout and braincase, unlike other cetaceans. In extant vertebrates, elevated orbits are sometimes an adaptation for stalking terrestrial prey while remaining submerged (Thewissen, Madar, and Hussain, 1996). Buoyancy of the skull was enhanced by exceptionally large paranasal sinuses. The teeth of *Ambulocetus* are similar to those of *Pakicetus*. As in *Pakicetus*, the ectotympanic bone is dense, with a thickened involucrum, but the petrotympanic complex differs in being somewhat freed from attachment to the skull, as in later whales. The mandibular foramen is larger than in *Pakicetus* but not as large as in protocetids. These modifications suggest that *Ambulocetus* was better adapted for underwater hearing than was *Pakicetus*. *Ambulocetus* has been interpreted as an amphibious ambush predator analogous to crocodiles (Thewissen, Madar, and Hussain, 1996).

By the middle Eocene, archaeocetes had lost both the external auditory canal and the tympanic ring, suggesting that a tympanic ligament had replaced the eardrum (Luo and Gingerich, 1999). This change would have greatly limited sound reception on land and marks a further step in the transition to aquatic life.

The paraphyletic middle Eocene Protocetidae is the most diverse and widespread family of archaeocetes, with about nine genera that ranged from southeastern North America to Nigeria, Egypt, and Indo-Pakistan. At least one genus (*Eocetus*) is known from both sides of the Atlantic (Uhen, 1999). Much of the skeleton of the type genus, *Protocetus,* has been known for nearly a century and long served as a standard for primitive cetaceans. Protocetids had short necks and laterally facing orbits covered by a supraorbital shield. Variations in tooth shape and root number, as well as in the length and shape of the rostrum, suggest that protocetid genera differed in feeding habits, whereas variations in sacral and hind limb anatomy reflect different degrees of aquatic adaptation. Protocetid genera vary in the number and extent of

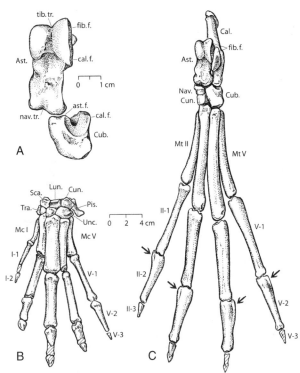

Fig. 14.10. Left foot skeleton of early whales: (A) *Artiocetus*, astragalus and cuboid; (B) *Rodhocetus*, manus; (C) *Rodhocetus*, pes. Arrows point to flanges for attachment of digital abductors, which spread the digits, allowing the wide, webbed feet to be used as paddles (see Fig. 14.9B). Key: Ast., astragalus; ast.f., astragalar facet; Cal., calcaneus; cal.f., calcaneal facet; Cub., cuboid; Cun., cuneiform; fib.f., fibular facet; Lun., lunate; Mc, metacarpal; Mt, metatarsal; Nav., navicular, nav.tr., navicular trochlea; Pis., pisiform; Sca., scaphoid; tib.tr., tibial trochlea; Tra., trapezium; Unc., unciform. (From Gingerich, ul Haq, et al., 2001.)

earlier, the recently discovered ankle bones of *Rodhocetus* and its close relative *Artiocetus* (Figs. 14.10, 14.11)—the first archaeocete tarsals directly associated with skulls—closely resemble those of artiodactyls (Gingerich, ul Haq, et al., 2001). *Georgiacetus* (Fig. 14.8C), whose cheek teeth are particularly similar to those of mesonychians, had a specialized vertebral column, indicating that it was the most fully aquatic protocetid.

Remingtonocetids (named for the eminent Smithsonian cetologist Remington Kellogg; Fig. 14.7E,F) were archaeocetes with very long, narrow snouts, and cheek teeth situated well in front of the small orbits (Fordyce and Barnes, 1994). They are found in middle Eocene shallow shelf marine sediments of Indo-Pakistan. The snout shape, together with auditory specializations and a long sinuous body, suggest aquatic habits (Thewissen and Williams, 2002). However, the external nasal opening was still at the front of the snout, and the neck was long, as in land mammals. In addition, remingtonocetids have a solid sacrum made up of four fused vertebrae, a strong sacroiliac joint, and short but still well-developed hind limbs, suggesting significant terrestrial capability. These seemingly contradictory features indicate that remingtonocetids were another group of transitional amphibious archaeocetes (Gingerich, Arif, and Clyde, 1995).

Geochemical evidence from the sediments that contain the oldest cetaceans and oxygen isotope analysis of their tooth enamel (which is believed to reflect the isotopic

fusion of the sacral vertebrae, differences that relate to the flexibility of the lumbo-sacro-caudal region and swimming efficiency (Buchholtz, 1998). The radius of protocetids is much shorter than the humerus, but the tibia is longer than the femur, as in cursorial mammals. The hind limbs of most protocetids, though more reduced than in *Ambulocetus,* were still strong enough to support their weight on land (Gingerich et al., 1998). But others, such as *Protocetus* and *Georgiacetus,* lack an articulation between the sacrum and pelvis, indicating that their hind limbs could not support the body weight. These genera were probably restricted to aquatic environments. In this regard they are more derived than remingtonocetids.

Rodhocetus, from Pakistan, and *Georgiacetus,* from southeastern United States, are now more completely known than other protocetids (Gingerich, Raza, et al., 1994; Hulbert et al., 1998; Gingerich, ul Haq, et al., 2001). *Rodhocetus* (Figs. 14.7C, 14.9B, 14.10B,C, Plate 8.2) had a short neck, unfused sacrum, and strong tail, as in aquatic cetaceans, but also retained an iliosacral articulation, a well-developed femur, and high neural spines at the withers, as in terrestrial ungulates. This combination of features suggests that *Rodhocetus* was still capable of terrestrial locomotion but was more aquatic than the other archaeocetes discussed above. As mentioned

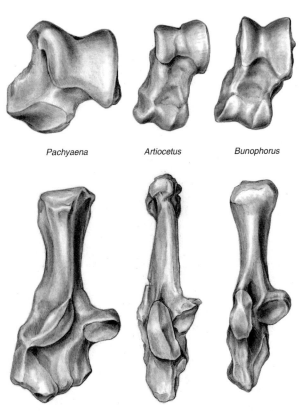

Fig. 14.11. Tarsals of the basal cetacean *Artiocetus* compared to those of the mesonychid *Pachyaena* and the basal artiodactyl *Bunophorus*. (From Rose, 2001b.)

composition of water ingested during life) provide a rare glimpse into the physiology of these ancient whales (Thewissen, Roe, et al., 1996). As noted above, *Pakicetus* has been found only in freshwater deposits, whereas *Ambulocetus* occurs in near-shore marine beds. The oxygen isotope composition of the enamel in both forms, however, is like that of extant freshwater cetaceans, indicating that they both ingested freshwater. Perhaps *Ambulocetus* was not yet capable of drinking seawater, as modern cetaceans do. A more intriguing possibility is that *Ambulocetus* spent the early part of its life (when the teeth formed) in freshwater. *Indocetus*, a protocetid from middle Eocene beds only a little younger than *Ambulocetus*, has a marine isotopic signature indicating tolerance of seawater ingestion. With the bond to freshwater no longer necessary, cetaceans were free to disperse across broad oceanic expanses—a possible explanation for the wide distribution of middle Eocene archaeocetes.

The most derived archaeocetes belong to the family Basilosauridae, which comprises two subfamilies, Basilosaurinae and Dorudontinae. Basilosaurids (specifically, the more generalized dorudontines) are considered to be the sister group and probable direct ancestor of Odontoceti + Mysticeti (e.g., Barnes and Mitchell, 1978; Fordyce, 2002; Uhen, 2004). Basilosaurids were the first fully aquatic cetaceans. They were widely distributed during middle and late Eocene time, being known from such distant locations as the southeastern United States, Senegal, Egypt, Indo-Pakistan, and even New Zealand (the dorudontine *Zygorhiza* is one of very few archaeocetes reported from the southern hemisphere; Köhler and Fordyce, 1997). Basilosaurids are characterized by highly derived, transversely compressed triangular cheek teeth with multiple accessory cusps. The upper molars lack protocones, and M^3 is absent. They are believed to have been fish eaters. Unlike other archaeocetes but like extant whales, they have expanded basicranial air sinuses (Fordyce and Muizon, 2001). Both basilosaurines and dorudontines had a very large mandibular foramen, as in extant odontocetes, suggesting that they had efficient underwater hearing.

Basilosaurinae, exemplified by *Basilosaurus* (=*Zeuglodon*; Fig. 14.9D) from Egypt and the Gulf Coast of North America, had a short neck but very long lumbar and caudal vertebrae and an unfused sacrum, giving it a long, almost serpentine body that reached lengths in excess of 15 m. Only small "floating" vestigial hind limbs remained, including a reduced pelvis with poorly defined acetabulum, a small femur, short fused tibia-fibula, fused tarsus, and reduced digits (Gingerich, Smith, and Simons, 1990).

The more diverse Dorudontinae had shorter vertebral centra but were otherwise generally similar to basilosaurids, with an extended thoracolumbar region (17 thoracic vertebrae and 20 lumbars in *Dorudon*; Uhen, 2004), flipperlike forelimbs, and greatly reduced hind limbs that cannot have been of any use in terrestrial locomotion. Anatomy of the caudal vertebrae indicates the presence of a tail fluke. Some dorudontines reached the size of *Basilosaurus*, but most, such as *Zygorhiza* and *Dorudon* (Figs. 14.5A, 14.8B, 14.9C,

Fig. 14.12. Restoration of the Eocene basilosaurid *Dorudon*. (From Uhen, 2004.)

14.12), were much smaller (about 5 m long). As in basilosaurines, the Egyptian *Ancalecetus* had an immobile elbow and carpus, which would seem to have limited its ability to maneuver (Gingerich and Uhen, 1996).

Mysticeti and Odontoceti (=Neoceti)

Archaeocetes were the dominant whales of the Eocene. The latest Eocene dorudontine basilosaurids *Zygorhiza* and *Saghacetus* have been suggested to be sister taxa of Neoceti (Uhen, 1998). Several fossils are now known that represent the earliest Neoceti, or possibly transitional stages between archaeocetes and neocetes.

The oldest known representative of one of the modern suborders is *Llanocetus*, based on a skull and jaw from approximately the Eocene/Oligocene boundary of Seymour Island, Antarctica (Fordyce, 1989; Mitchell, 1989). *Llanocetus* is placed in its own family, referred to the Mysticeti. Unlike the more advanced baleen-bearing mysticetes, *Llanocetus* still had teeth. It is recognized as a primitive mysticete based on cranial features, including vascular grooves on the palate, which suggest that it also had baleen (R.E. Fordyce, pers. comm.). Another transitional stage, at least structurally, is represented by the late Oligocene Aetiocetidae (*Aetiocetus* and *Chonecetus*). Aetiocetids were toothed whales that display a combination of archaeocete and mysticete features but, judging from the absence of vascular grooves on the palate, probably did not yet have baleen (Fordyce and Muizon, 2001). The late Oligocene "Cetotheriidae," an assemblage of extinct baleen whales that cannot be definitively assigned to Balaenidae or Balaenopteridae, are the earliest known toothless, baleen-bearing mysticetes.

Several undoubted odontocetes have been described from upper Oligocene sediments. Until recently these were the oldest known odontocetes, but their diversification into several families by that time implied a much earlier initial divergence from archaeocetes (Fordyce and Barnes, 1994). This implication appears to be borne out by the recent report of a primitive odontocete from near the Eocene/Oligocene boundary of the Pacific Northwest (Barnes and Goedert, 2000). It has anteriorly placed nares and a heterodont dentition. These earliest known Neoceti conform closely to likelihood estimates of the time of origin of both modern clades (Gingerich, 2005).

The evolution of baleen in mysticetes and echolocation in odontocetes during the Oligocene were pivotal specializations that led to the extinction of archaeocetes and their replacement by modern whales. Both molecular and fossil evidence suggest that the initial radiation of modern toothed and baleen whales occurred during the early Oligocene, 28–33 million years ago (Nikaido et al., 2001). Extant families, such as sperm whales (Odontoceti, Physeteridae) and right whales (Mysticeti, Balaenidae), appeared during the Oligocene. Changing oceanic currents near the end of the Eocene may have promoted the early radiations of the modern suborders.

ARTIODACTYLA

Artiodactyls, often referred to as even-toed or cloven-hoofed ungulates, today comprise the pigs, peccaries, and hippopotami (Suiformes); camels (Tylopoda); and chevrotains, giraffes, deer, cattle, sheep, goats, and antelope (Ruminantia). Results of two recent phylogenetic analyses of artiodactyls based on morphology are shown in Fig. 14.13. As mentioned earlier, recent evidence indicates that Cetacea are closely related to, if not members of, the Artiodactyla (see Fig. 14.1B). Pending resolution of their precise phylogenetic position, they have been discussed separately in the preceding section. Because there are no known fossil links between Cetacea and artiodactyls apart from the general resemblance of their tarsal elements, their independent treatment here does not affect the discussion of artiodactyl evolution.

Artiodactyls are the most diverse and abundant larger land mammals in existence today. Their success can be attributed largely to their cursorially specialized skeletons and herbivorously adapted dentitions. Even the less specialized forms, which are evidently secondarily generalized, show evidence of a more cursorial ancestry. Although perissodactyls were more common in the Eocene, artiodactyls have been the dominant hoofed mammals since the Miocene and far outnumber perissodactyls today.

Running adaptations abound throughout the artiodactyl skeleton (Fig. 14.14), even in the oldest known members of the order. Paramount among them is the unique double-pulley (trochleated) astragalus, a hallmark of the order that is present in all members and unknown outside of Artiodactyla, except in the earliest Cetacea. The astragalus is deeply grooved at both ends, the proximal end forming a tight joint with the tibia and the distal end articulating similarly with the navicular and cuboid, which are fused in ruminants. This arrangement, in which the astragalus glides in a sagittal plane on the sustentaculum, facilitates flexion and extension of the tarsal joints while prohibiting lateral movements. Fusion of other tarsal bones (ectomesocuneiform, cubonavicular) enhances this function. The limbs of artiodactyls are modified to lengthen stride and limit motion to a sagittal plane. As in other extant ungulates (except the aardvark), the clavicle is absent in all living forms. Limb elements are typically slender and elongate, especially distally (except in hippos and most pigs). In most extant artiodactyls the ulna and fibula are reduced distally and fused to the radius and tibia, respectively (again, suiforms are an exception). The femur lacks a third trochanter, the usual site of insertion of the superficial gluteal muscle; instead the muscle merges with the biceps femoris (forming the gluteobiceps) and inserts farther distally, on the proximal tibia and into deep fascia and ligaments near the knee (Getty, 1975). Artiodactyls differ from perissodactyls in having paraxonic symmetry in the feet, the axis passing between digits III and IV. Digits II and V are somewhat or greatly reduced in primitive and generalized forms and absent in advanced forms. The first digit is lost in all living forms (although a vestige is present in some primitive fossil types), hence there is always an even number of toes, which are terminated by hoofs. In camels and most ruminants the third and fourth metapodials fuse into a cannonbone.

Although the artiodactyl astragalus could plausibly be derived from various condylarth astragali (Schaeffer, 1947), no transitional stages from a more generalized eutherian condition are known (with the possible exception of mesonychians). Hence the origin of this morphology—and of Artiodactyla itself—remain among the great conundrums of mammalian evolution.

The antorbital portion of the skull is typically elongate in artiodactyls, and the orbits are usually surrounded by a complete or nearly complete bony rim, which sometimes protrudes above the otherwise relatively flat dorsal profile of the skull. Many ruminants possess bony outgrowths of the skull, such as the antlers of deer (exposed bone), the skin-covered ossicones of giraffes, or the horns of bovids, which are covered by a keratinous sheath. At the back of the skull, ruminants retain the primitive eutherian condition in which the mastoid process of the petrosal is exposed between the squamosal and the exoccipital. In most suiforms, however, the squamosal grows backward to meet the exoccipital, obscuring the mastoid process, a condition known as amastoidy. This distinction was formerly thought to divide the artiodactyls into two groups (suiforms vs. selenodont forms); but with the addition of Early Tertiary fossils, the situation now appears much more complex, and this taxonomic scheme has been generally abandoned. Prothero et al.

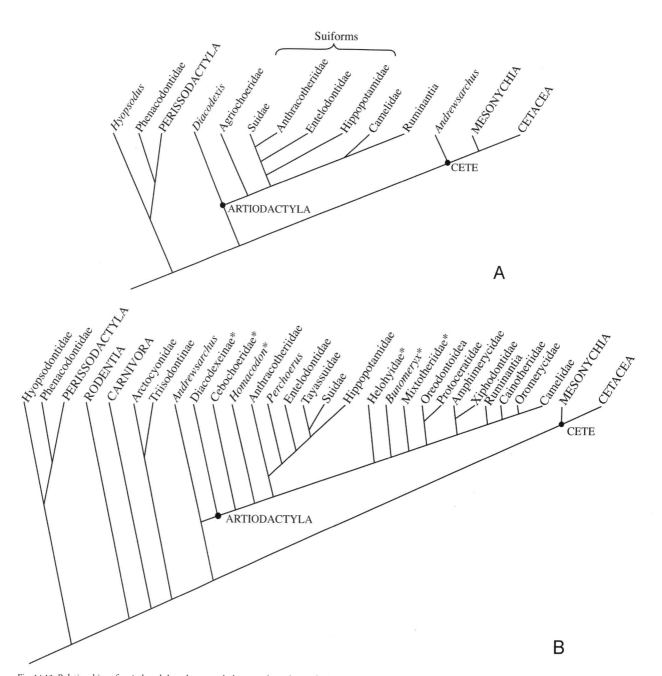

Fig. 14.13. Relationships of artiodactyls based on morphology: (A) hypothesis of relationships among artiodactyls, and position of Artiodactyla among eutherians, simplified after O'Leary and Geisler (1999); (B) more detailed assessment, simplified after Geisler (2001). Asterisks in B denote taxa grouped as "dichobunoids" by McKenna and Bell (1997). More recent studies suggest Mesonychia lies outside Artiodactyla + Cete.

(1988) listed several other details of cranial anatomy that distinguish artiodactyls from other ungulates.

Ruminants and camels have selenodont and often hypsodont cheek teeth, adapted for grazing or browsing, whereas suiforms have lower-crowned, bunodont cheek teeth suited for an omnivorous diet. Although some artiodactyls have a true hypocone on the upper molars, most lack a hypocone and instead have an enlarged metaconule that is shifted posterolingually, approximating the position of a hypocone. The permanent premolars (except the last one in selenodont forms) are not molarized, as they tend to be in perissodactyls.

However, primitive artiodactyls have a submolariform deciduous P_4 that is long and trilobate, an unusual morphology that is considered an important synapomorphy of Artiodactyla (Luckett and Hong, 1998). Three incisors are present in each quadrant of the dentition of primitive artiodactyls, but camels and ruminants lack some or all of the upper incisors. Well-developed canines are primitively present. In suiforms and the more primitive ruminants the canines may be large and projecting and are often larger in males, but in most ruminants the upper canine is lost and the lower canine tends to be incisiform. Selenodont artio-

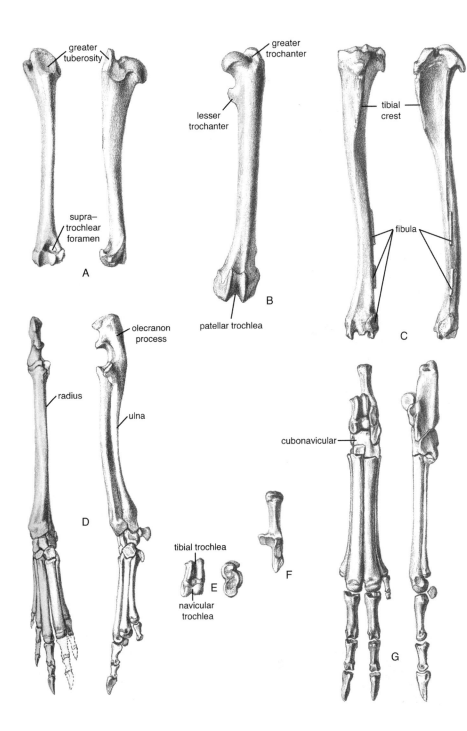

Fig. 14.14. Characteristic features of the limb bones of artiodactyls, illustrated on left elements of the middle Eocene–Oligocene basal ruminant *Hypertragulus*: (A) humerus—note high greater tuberosity, supratrochlear foramen, and greatly reduced deltopectoral and supinator crests; (B) femur—high greater trochanter, deep patellar trochlea, posteromedially directed lesser trochanter, third trochanter absent; (C) tibia—short and elevated tibial crest, deep distal articulation for astragalus; fibula is reduced and closely appressed; (D) forearm—radius anterior and closely appressed to ulna, ulna concave posteriorly; (E) astragalus—deep tibial trochlea proximally, (cubo)navicular trochlea distally; (F) calcaneus—narrow and elongate, with fibular facet laterally and gliding sustentacular facet; (G) hind foot—paraxonic symmetry with reduced lateral metapodials and digits; narrow, hoof-bearing ungual phalanges; cubonavicular, fused in ruminants only. (Modified from Scott, 1940.)

dactyls have modified stomachs for foregut fermentation, the most specialized condition being the four-chambered stomach of ruminants.

Despite the obvious specializations of most living artiodactyls, it is evident that Artiodactyla is an ancient ungulate clade, and its precise branching sequence from "archaic ungulates" is unresolved. Prothero et al. (1988) considered Artiodactyla to be the sister taxon of all other ungulates (including condylarths), which would suggest a very ancient pedigree, perhaps in line with molecular results that propose a Cretaceous origin. Other researchers, however, would draw Artiodactyla from some condylarth group, but which one is uncertain. One long-held view, based on dental similarity, is that artiodactyls are closely related to or derived from hyopsodontid condylarths or their close relatives, mioclaenids (e.g., Simpson, 1937a). There is slightly stronger evidence that artiodactyls originated from small oxyclaenine arctocyonids similar to *Chriacus* or *Tricentes,* which resemble the oldest known artiodactyls in morphology of the cheek teeth and hind limb long bones (Van Valen, 1971; Rose, 1996a; Fig. 14.15). If this inference is correct, Artiodactyla would have branched subsequent to Arctocyonidae, probably during the Paleocene and would be nested within that paraphyletic family—posing no small taxonomic dilemma. In summary, available evidence does not point conclusively to the source of Artiodactyla.

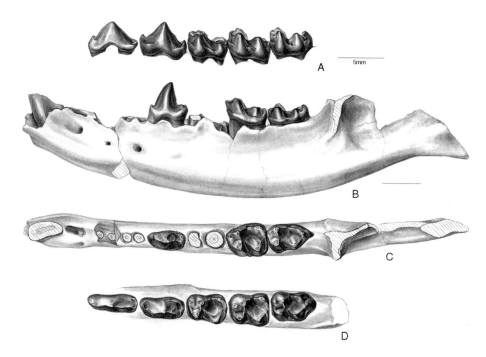

Fig. 14.15. Lower left dentitions of (A, D) early Eocene artiodactyl *Diacodexis* compared to (B, C) that of a small Paleocene arctocyonid. Scale bars = 5 mm. (From Rose, 1996a.)

Primitive Bunodont Artiodactyls

The most primitive known artiodactyls are Eocene bunodont and bunoselenodont forms often grouped in the family Dichobunidae or the superfamily Dichobunoidea (McKenna and Bell, 1997; Stucky, 1998). This grouping is essentially equivalent to the gradistic suborder Palaeodonta, as used by Romer (1966). Although sometimes included in the Suiformes (Simpson, 1945; McKenna and Bell, 1997), Dichobunoidea is clearly a paraphyletic assemblage of genera, which probably includes the ancestors or sister taxa of most, if not all, other artiodactyls. When relationships between particular dichobunoids and later taxa are better established, it may be desirable to break up the group to reflect these ties, but no consensus on such a rearrangement currently exists.

Consequently, McKenna and Bell's classification of Dichobunoidea is adopted here, with minor modifications, although the superfamily is not assigned to a suborder (see Table 14.2). In their scheme, the Dichobunoidea includes four families: a broadly conceived Dichobunidae (including diacodexeines, homacodonts, dichobunines, and leptochoerines), Helohyidae, Cebochoeridae, and Mixtotheriidae. Here the latter family is excluded from Dichobunoidea and instead is discussed in the section on primitive selenodont artiodactyls below. Two other archaic families whose relationships are even less secure—Raoellidae and Choeropotamidae—are discussed following dichobunoids. The current literature on primitive artiodactyls reveals considerable disagreement about the interrelationships of these taxa as well as various other primitive artiodactyls. We are faced with this dilemma because there was considerable parallelism among these basal lineages, and the polarity of morphological characters is far from certain. This ambiguity does not mean that all dichobunoids are poorly known. On the contrary, several genera are known from relatively complete skeletons.

Dichobunidae

Diacodexis (Figs. 14.15A,D, 14.16), which is usually assigned either to the Dichobunidae or to the Diacodexeidae, is the oldest known artiodactyl and is usually considered to be the most primitive. It appeared abruptly at the beginning of the Eocene (in basal Wasatchian/Ypresian sediments) in Euramerica and roughly equivalent strata in Asia, with little indication of its geographic or phylogenetic source. *Diacodexis* occupies a position near the base of the artiodactyl radiation (Gazin, 1955; Krishtalka and Stucky, 1985) and is therefore paraphyletic, but its precise phylogenetic position relative to subsequent artiodactyl lineages is controversial. For example, some species of *Diacodexis* appear to be closely related to other clades of dichobunids (Krishtalka and Stucky, 1985; Stucky, 1998), whereas another species of the genus has been interpreted to be the sister taxon of the more advanced neoselenodont artiodactyls (Gentry and Hooker, 1988). Regardless of these differences, *Diacodexis* provides an appropriate model of artiodactyl anatomy near the base of the order.

The dentition of *Diacodexis* and most other dichobunids is generalized. A primitive complement of teeth is present (3.1.4.3/3.1.4.3), with short diastemata separating the anterior premolars from adjacent teeth. The incisors are small and unspecialized, including the upper incisors, which are almost always replaced by a horny pad in ruminants. In some forms, such as *Diacodexis* and *Homacodon,* the canines are of moderate size and projecting, as in other primitive placentals, but in others, such as *Messelobunodon,* the lower

Table 14.2. Classification of Artiodactyla

```
Order ARTIODACTYLA
    Superfamily †Dichobunoidea
        †**Dichobunidae**
        †**Cebochoeridae**
        †**Mixtotheriidae**[1]
        †**Helohyidae**
    Suborder SUIFORMES
        †**Raoellidae**[2]
        †**Choeropotamidae**[2]
        Superfamily †Anthracotherioidea
            †**Haplobunodontidae**[3]
            †**Anthracotheriidae**
        Superfamily Suoidea
            **Suidae**
            **Tayassuidae**
            †Sanitheriidae
            Hippopotamidae[4]
        Superfamily †Anoplotherioidea
            †**Dacrytheriidae**
            †**Anoplotheriidae**[1]
            †**Cainotheriidae**[1]
        Superfamily †Oreodontoidea[1]
            †**Agriochoeridae**
            †**Oreodontidae** [=Merycoidodontidae]
        Superfamily †Entelodontoidea
            †**Entelodontidae**[2]
    Suborder TYLOPODA
        †**Xiphodontidae**
        Superfamily Cameloidea
            **Camelidae**
            †**Oromerycidae**
        Superfamily †Protoceratoidea
            †**Protoceratidae**
    Suborder RUMINANTIA
        Infraorder Tragulina
            †**Amphimerycidae**[1]
            †**Hypertragulidae**
            Tragulidae
            †**Leptomerycidae**
            †Bachitheriidae
            †**Lophiomerycidae**
        Infraorder Pecora
            †**Gelocidae**
            Superfamily Cervoidea
                Moschidae
                Antilocapridae
                †Palaeomerycidae
                †Hoplitomerycidae
                Cervidae
            Superfamily Giraffoidea
                †Climacoceratidae
                Giraffidae
            Superfamily Bovoidea
                Bovidae
```

Notes: Modified after McKenna and Bell (1997), and Webb and Taylor (1980). The dagger (†) denotes extinct taxa. Families in boldface are known from the Paleocene or Eocene.

[1] Sometimes assigned to Tylopoda.
[2] Allocation to Suiformes is uncertain and is not followed in this chapter.
[3] Included in Choeropotamidae by Hooker and Weidmann (2000).
[4] Cetacea might be the sister-taxon of Hippopotamidae.

canine is reduced and incisiform, and P_1 is caniniform. The other premolars are simple, elongate, and pointed, and P_3 is longer than P_4. The molars of dichobunids are brachydont and bunodont, although they vary somewhat, *Diacodexis* having more acute cusps and Euramerican *Bunophorus* blunt, rounded cusps. The lower molars of *Diacodexis* have three distinct trigonid cusps and a broad, basined talonid, and the uppers are tritubercular with distinct conules and a strong postcingulum but no hypocone. M_3 has an extended hypoconulid lobe. Common dental trends among dichobunids include the reduction or loss of the lower molar paraconid and development of a hypocone or an enlarged, distally shifted metaconule in place of a hypocone on the upper molars. Dichobunids were thus equipped to consume a generalized diet of fruits, seeds, leaves, and shoots.

Some dichobunid lines superimposed incipient dental specializations on this primitive diacodexeine plan. Leptochoerines (e.g., *Leptochoerus, Stibarus*), first known from the middle Eocene (Uintan), have robust posterior premolars, often larger than the first molar, and molars that decrease in size posteriorly. In contrast, the mainly North American homacodontines have more delicate premolars and incipiently selenodont molars with subcrescentic cusps. The early Eocene homacodontine *Hexacodus* is considered the earliest selenodont (or better, "preselenodont") genus (Stucky, 1998). Middle Eocene homacodontines were more clearly selenodont and some, such as *Pentacemylus,* had all but lost the hypocone, replacing it with a large posterolingual metaconule (Gazin, 1955). The primarily European dichobunines are very similar to homacodontines; however, Eurasian *Dichobune* (see Fig. 14.19B) had both a hypocone and a metaconule. The dental differences seen among dichobunids are thought to represent the initial stages of various more advanced artiodactyl clades. For example, Stucky (1998) considered homacodontines to be the probable source group for both tylopods and ruminants.

The skulls of dichobunids were also rather primitive. In most forms the skull is long and low, with a weak sagittal crest and somewhat stronger, though narrow, nuchal crest. The basicranium of *Diacodexis* and *Homacodon* resembles that of condylarths such as *Hyopsodus* more than that of advanced artiodactyls (Coombs and Coombs, 1982). As in eutherians generally (as well as in camels and ruminants), the mastoid process of the petrosal is exposed at the back of the skull between the squamosal and exoccipital bones. This morphology contrasts with suiforms, in which the squamosal projects posteriorly to meet the exoccipital, obscuring the mastoid (amastoidy). The homacodontine *Bunomeryx* has a deep mastoid fossa on the petrosal, a derived condition suggesting relationship with tylopods (Norris, 1999). This feature is also found in anoplotherioid and xiphodont artiodactyls, which some authorities consider to be tylopods (see below).

In sharp contrast to the primitive skull and dentition, the postcranial skeleton of dichobunids was highly specialized for cursorial-saltatorial locomotion. The skeleton is best known in *Diacodexis* (Fig. 14.16), middle Eocene

Fig. 14.16. *Diacodexis,* the oldest known artiodactyl. Above, skeleton and restoration (*Diacodexis* was probably digitigrade rather than unguligrade as shown here). Below, selected bones: (A) left distal humerus and proximal radius; (B) distal end of right femur; (C) right femur; (D) right tibia, ends of fibula, and proximal tarsus (lateral and anterior views); (E) right tibia of extant *Tragulus* for comparison; (F) left metatarsals (Mt). (Modified from Rose, 1982b.)

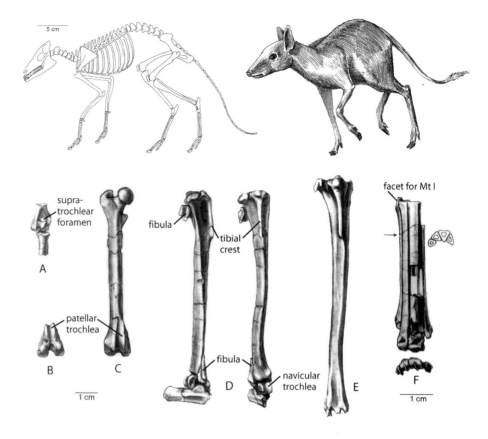

Messelobunodon (Fig. 14.17), and *Aumelasia* (Plate 8.3), the last two from Messel, Germany (Franzen, 1981, 1983, 1988; Rose 1982b, 1985; Thewissen and Hussain, 1990). Other dichobunids appear to have been very similar. *Diacodexis* was only about the size of a rabbit—smaller than any living artiodactyl, and a little smaller than the Messel genera. The clavicle was vestigial in *Diacodexis* and lost in all other taxa, as in many cursorial mammals. The limbs of these dichobunids were slender and elongate, although the forelimbs were conspicuously shorter than the hind limbs. The more distal segments were disproportionately lengthened compared to those of more primitive mammals: the tibia was noticeably longer than the femur, and the central metatarsals were at least two-thirds as long as the femur. Crests and processes for muscle attachment were reduced and the joints were modified so as to restrict motion to parasagittal flexion-extension. The astragalus already exhibits the characteristic artiodactyl double trochlea, although the distal trochlea is set at a slight angle to the proximal one (a plesiomorphic condition) rather than being directly in line, as in advanced artiodactyls. The cuboid and navicular, which are fused in ruminants and *Leptochoerus,* remained separate, but the ecto- and mesocuneiforms are variably fused. The fibula was a thin splint sometimes fused distally with the tibia, foreshadowing the vestigial malleolar bone of more advanced artiodactyls. *Diacodexis* and *Messelobunodon* had long, slender metapodials, the lateral ones shorter and of smaller caliber than the central ones and tucked behind them in an arcuate arrangement. The manus of *Diacodexis* was pentadactyl and mesaxonic (Thewissen and Hussain, 1990), whereas the hind foot was paraxonic with four functional digits and a vestigial hallux. The toes bore small, narrow hoofs; however, the orientation of the phalangeal articular surfaces suggests that *Diacodexis* was digitigrade, not yet unguligrade. Nevertheless, dichobunids were surely among the most fleet-footed mammals during much of the Eocene, exceeding even early perissodactyls in their cursorial specializations.

Helohyidae

Helohyids were somewhat piglike dichobunoids that ranged in size from that of dichobunids to considerably larger. They lived in Asia and North America during the middle Eocene. Helohyids had prominent canines and molars with bunodont cusps, inflated crowns, and sometimes wrinkled enamel. The upper molars were squared off by enlargement and displacement of the metaconule, but a small hypocone was usually also present; the paraconule was reduced and there was no mesostyle (e.g., *Helohyus;* see Fig. 14.19A). The lower molars increase in size posteriorly, and the paraconid is small or absent. In such forms as *Gobiohyus,* diastemata separated the anterior premolars from one another and from adjacent teeth (Coombs and Coombs, 1977b). The snout is variable in length, long in *Lophiohyus* (sometimes considered a synonym of *Helohyus*) but shorter in *Achaenodon.* Both genera have a more prominent sagittal crest than in dichobunids. *Achaenodon* reached the size of a

Fig. 14.17. Middle Eocene dichobunid artiodactyl *Messelobunodon* from Messel, Germany. About 60 cm long. (Courtesy of the Forschungsinstitut Senckenberg, Frankfurt.)

pig, with a head 35 cm long, and was the largest middle Eocene artiodactyl in North America. Compared to that of a pig, the skull of *Achaenodon* had a shorter snout, a more prominent sagittal crest, and a larger temporal fossa, somewhat resembling later hemicyonine bears in these respects. *Achaenodon* had short forelimbs and longer hind limbs. Both fore- and hind feet were four-toed, with unfused metapodials.

The composition and relationships of Helohyidae are in dispute. This chapter follows McKenna and Bell (1997) and Stucky (1998) in including achaenodonts and allying Helohyidae with dichobunids. Coombs and Coombs (1977b) excluded achaenodonts and placed Helohyidae in the Anthracotherioidea, a view that could be construed as supporting a dichobunid ancestry of anthracotherioids. However, they also included several genera now placed in Raoellidae (see below). Stucky explicitly rejected any close relationship between helohyids and anthracotherioids.

Cebochoeridae

Cebochoerids were small to medium-sized bunodont artiodactyls that lived during the Eocene and early Oligocene in Europe (Sudre, 1978b). Although similar in many respects to dichobunids and probably descended from that group, they are apparently amastoid and have shorter snouts, a mandible that deepens posteriorly, and somewhat piglike teeth. The molars are bunodont and basally inflated. The lowers are narrow, with the paraconid vestigial or absent and the hypoconulid posteriorly displaced. Like some dichobunids, later cebochoerids had an incisiform lower canine and caniniform P_1. The upper molars are quadrate, about as long as they are wide and with four primary cusps, including a large posterolingual metaconule in place of a hypocone (see Fig. 14.23A). The paraconule is small and mesially shifted, and cingula are weak or absent. The lingual cusps of the uppers and buccal cusps of the lowers are weakly selenodont. One of the best-known cebochoerids is *Gervachoerus* (Fig. 14.18A), considered a synonym of *Cebochoerus* by some authorities. It is known from partial skeletons from Geiseltal, Germany, which generally resemble those of dichobunids but have the femur and tibia of approximately the same length and an exceptionally long tail (Erfurt and Haubold, 1989). It was a little less than a meter long, including the tail, which accounts for half the length. In their phylogenetic analysis of artiodactyls and cetaceans, Geisler and Uhen (2005) found *Cebochoerus* to be closely related to a cetacean-hippopotamid clade.

Raoellidae

Raoellidae comprises a small group of early and middle Eocene genera endemic to Indo-Pakistan. Most members are poorly known and based primarily on teeth. Both the upper and lower molars have four main cusps—uppers have a hypocone, and lowers lack a paraconid—and are weakly bilophodont or bunolophodont (Kumar and Sahni, 1985; Thewissen et al., 1987; Thewissen et al., 2001a). They have broad crowns with relatively low relief and sometimes intricately wrinkled enamel (e.g., *Khirtharia*, Fig. 14.19C). The dietary implications of this unusual molar structure have not been explored. Relationships of raoellids remain obscure, but they probably evolved from a dichobunid. Geisler and Uhen (2005) concluded that raoellids could be more closely related to Cetacea than are hippopotami.

Choeropotamidae

Choeropotamus, a poorly known form from the Eocene of Europe, was long considered the only member of the family Choeropotamidae. Its molars are bunodont to

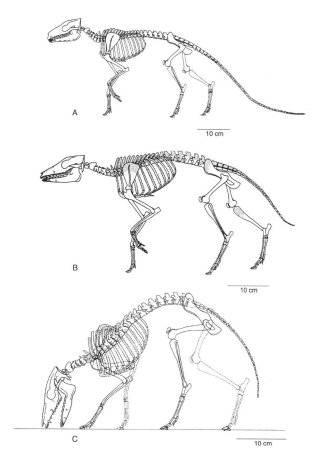

Fig. 14.18. Primitive artiodactyl skeletons: (A) cebochoerid *Gervachoerus*; (B) choeropotamid *Haplobunodon*; (C) choeropotamid *Anthracobunodon*. (A, B from Erfurt and Haubold, 1989; C from Erfurt, 2000.)

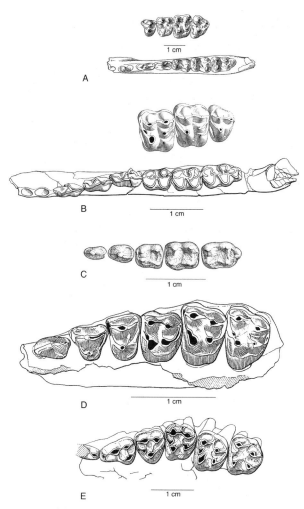

Fig. 14.19. Left dentitions of primitive bunodont artiodactyls: (A) helohyid *Helohyus*, P^4–M^3 and P_3–M_3; (B) dichobunid *Dichobune*, M^{1-3} and P_2–M_3; (C) raoellid *Khirtharia*, P_3–M_3; (D) basal choeropotamid *Cuisitherium*, P^2–M^3; (E) choeropotamid *Anthracobunodon*, P^2–M^3. (A from Sinclair, 1914; B from Sudre, 1988; C from Thewissen et al., 1987; D from Sudre and Lecompte, 2000; E from Sudre, 1978b.)

bunoselenodont, resembling those of homacodontines. In the lower dentition, the canine is incisiform, the first premolar is caniniform, and there is a long diastema between P_1 and P_2 (Hooker and Weidmann, 2000). Upper molars of *Choeropotamus* are squared, with four main cusps. The posterolingual one, apparently the metaconule (the hypocone being absent), is bulbous and as big as the protocone. There is also a smaller paraconule and a well-developed mesostyle (Sudre, 1978b).

According to Hooker and Weidmann (2000; see also Hooker and Thomas, 2001), these dental traits also characterize the genera usually assigned to the mainly European Eocene family Haplobunodontidae (including *Amphirhagatherium*, *Anthracobunodon*, *Haplobunodon*, and *Masillabune*; Fig. 14.19E), hence these authors subsumed the latter family into the Choeropotamidae (see also Sudre and Lecomte, 2000). These features are also present in various other early artiodactyls, particularly certain dichobunids and cebochoerids, highlighting the difficulty of separating shared derived traits from those that evolved independently. It may also be noted that several "haplobunodontid" genera, including the oldest genus, early Eocene (Ypresian) *Cuisitherium* (Fig. 14.19D), lack upper molar mesostyles (Sudre and Lecomte, 2000).

If "haplobunodontids" are indeed choeropotamids, knowledge of the family increases very substantially, as nearly complete, articulated skeletons are known for several genera found in the middle Eocene deposits at Geiseltal and Messel in Germany (*Anthracobunodon*, *Haplobunodon*, and *Masillabune*; Tobien, 1980; Erfurt and Haubold, 1989; Erfurt, 2000; Fig. 14.18B,C, Plate 8.4). The postcranial skeleton of these genera is similar to that of dichobunids. As in the latter, the limbs are slender and relatively long (particularly the middle and distal segments), and the forelimbs are shorter than the hind limbs. Most crests and processes are reduced (the femoral third trochanter is absent), and joints are modified for running, as in dichobunids. The fibula of choeropotamids is separate from the tibia and relatively strong. Both manus and pes are decidedly paraxonic, with the third and fourth metapodials longer than the lateral ones. Nevertheless, a reduced pollex is present, and it is also possible that a vestige of the first metatarsal was retained in *Anthracobunodon* (Erfurt, 2000). As in *Diacodexis*, narrow hoofs are present, and the caudally facing articular surface suggests a digitigrade (rather than unguligrade) stance. The tail was slightly more than a quarter of the total length, a little shorter than in *Diacodexis* and much shorter than in *Cebochoerus* but longer than in extant artiodactyls. The bunoselenodont molars of choeropotamids imply a mixed diet of fruits and leaves, which is confirmed to some extent by leafy stomach contents preserved in a skeleton of *Masillabune* (Tobien, 1980).

Although the affinities of *Choeropotamus* have long been uncertain, "haplobunodontids" have usually been considered to be closely related to anthracotheriids, the two families comprising the Anthracotherioidea. The expanded Choeropotamidae (including haplobunodontids) contrast with Anthracotheriidae in having the mastoid process of the petrosal exposed at the back of the skull, an incisiform canine, and in lacking distinct size increase of upper molars from M^1 to M^3. On this basis, Hooker and Thomas (2001) questioned their relationship to anthracotheriids. The status of Anthracotherioidea therefore should be reevaluated.

Relationships of Dichobunoids to Later Artiodactyls

By the late Eocene more than 20 artiodactyl families (excluding whales) had differentiated, representing all major clades of artiodactyls; most of them probably trace their origin to dichobunoids. Extensive parallel acquisition of progressive characters has made it difficult to untangle their interrelationships. As a consequence, artiodactyl classifications vary widely, but there is general agreement that an early divergence occurred between most bunodont taxa (suiforms) and selenodont forms (tylopods and ruminants). This early divergence also tends to separate lineages that were postcranially more generalized from the more specialized cursorial lineages, the latter generally associated with selenodont forms. Where to place the paraphyletic dichobunids (as well as several other early families) in such a scheme is controversial (Fig. 14.13 depicts one recent view). Gentry and Hooker (1988) considered them to be primitive sister taxa of selenodont artiodactyls, whereas some authors include them within Suiformes, presumably based on primitive characters. Still others consider various dichobunids to be sister taxa of all later artiodactyls, including suiforms (Stucky, 1998; Geisler, 2001). Despite these differences, there is general agreement that dichobunids occupy a position near the base of the artiodactyl radiation. Dichobunoids are similar to selenodont artiodactyls and most other mammals in having an exposed mastoid at the back of the skull (the primitive eutherian condition), unlike suiforms, which have the derived amastoid condition.

Suiformes (=Suina)

Pigs, peccaries, hippos, and related bunodont fossil groups are usually grouped as suiforms. They are characterized by relatively low-crowned, bunodont cheek teeth; an amastoid skull; and relatively short limbs that primitively retain four separate toes. Nonetheless, they can run swiftly. Extinct groups usually included in the suiforms are anthracotherioids and, less confidently, entelodonts.

Peccaries and pigs are first known from the late Eocene or perhaps slightly earlier. Peccaries are primarily a New World radiation, whereas pigs have always been restricted to the Old World. Nonetheless, they are quite similar in many respects, suggesting that they are sister taxa. Features supporting this view include the structure of the external auditory meatus, an orbital ethmoid exposure, and passage of cervical nerve branches through the neural arches of vertebrae C3–C6 (Wright, 1998). The origin of pigs and peccaries must be from among the Eocene bunodont artiodactyls, but no more precise ancestry has been identified.

Hippos do not appear in the fossil record until the Miocene (*Kenyapotamus*), apparently associated with the decline of anthracotheres, but they may well have diverged from other artiodactyls considerably earlier. Their ancestry is uncertain. Most authors have accepted Colbert's (1935) assessment that hippos evolved from anthracotheres similar to *Merycopotamus* (based largely on similar jaw shape), but it has also been suggested that they could have evolved from Old World peccaries (Doliochoerinae; Pickford, 1983, 1989). In a recent phylogenetic analysis, Boisserie et al. (2005) rejected close relationship to peccaries and concluded that the late Miocene anthracotheres *Merycopotamus* and *Libycosaurus* (which may be congeneric) are indeed the closest known relatives of hippos. Like hippos, they have elevated orbits and a wide mandibular symphysis and have lost the pollex. But these anthracothere genera occur too late to be ancestral to hippos. In fact, the fossil record provides little direct evidence of how hippopotami originated.

Eocene Cebochoeridae and Choeropotamidae (see above), although dentally piglike, were probably derived from dichobunids. Their various dental resemblances to suiforms are now generally considered to be convergent (Ducrocq, 1994; Ducrocq et al., 1998).

Anthracotheriidae

Anthracotheres were relatively large, suidlike animals that radiated in Eurasia, Africa, and North America beginning in the middle and late Eocene. They survived until the early Miocene in North America, and later in other areas. The diversity of middle and late Eocene genera in Asia suggests that the group originated there, perhaps from helohyid dichobunoids (Coombs and Coombs, 1977a). Alternatively, they may have evolved from Choeropotamidae ("haplobunodontids").

The skull of anthracotheres is amastoid, with an incomplete postorbital bar and a prominent sagittal crest. The snout is typically long and narrow, accentuated in many forms by anterior diastemata (e.g., Chadronian-Orellan *Bothriodon*). In the early anthracothere *Heptacodon* (Duchesnean-

Whitneyan; Fig. 14.20A), however, the snout is much shorter and lacks diastemata. The mandibular symphysis is elongate, and the dentary has an expanded angle. Most anthracotheres retain a primitive dental formula, and many have large canines. Anthracothere molars are brachydont and either selenodont or bunoselenodont (with crescentic or selenodont buccal cusps and bunodont lingual cusps). They increase in size posteriorly. The upper molars are square, usually with five cusps, including a well-developed paraconule and large posterolingual metaconule, and a W-shaped ectoloph—features that also characterize many other primitive selenodont artiodactyls. A distinctive feature is the reduction or absence of the postprotocrista and development of a transverse valley that divides the front and back halves of each molar (Coombs and Coombs, 1977a). The lower molars lack a paraconid, and M_3 typically has an elongate hypoconulid lobe, characteristics of several other primitive artiodactyls. The premolars are relatively simple, and P_4 has strong crests between the protoconid and the paraconid and metaconid (Kron and Manning, 1998). Heavy wear on the first molars before eruption of the third molars suggests a highly abrasive diet.

Anthracotheres had a robust postcranial skeleton, with relatively short limbs comparable to those of pigs, hippos, and oreodonts (Fig. 14.20B,C). The forelimbs and hind limbs were of roughly equal length. The radius and ulna were closely appressed but not co-ossified, and the fibula remained separate and strong. There were five digits on the forefoot and four on the hind foot, all with relatively short, separate metapodials. The arrangement is typically paraxonic, but in early Oligocene *Elomeryx* the third digit of the manus is longer, resulting in mesaxonic symmetry of the manus (Scott, 1940). Later anthracotheres became large and piglike or even hippolike, and there has been a general tendency to consider them to be closely related to hippopotami (as discussed above) or the sister taxon of suoids (pigs, peccaries, and hippos). The occurrence of many North American anthracotheres in channel deposits has suggested a somewhat semiaquatic habit. The mainly European Eocene family Haplobunodontidae has often been allied with anthracotheriids in the Anthracotherioidea but is now included in Choeropotamidae (see above).

Entelodontidae

Entelodonts (Fig. 14.21) are a Holarctic group of piglike artiodactyls usually allocated to the Suiformes because of their dental morphology as well as a general resemblance of the skull and skeleton to those of pigs, peccaries and, to a lesser extent, hippos. However, many of the features they share with these and other primitive artiodactyls, including an unreduced dental formula (3.1.4.3 above and below) and low-crowned, bunodont molars, are plesiomorphic. At the same time, entelodonts have numerous unusual derived features not seen in other artiodactyls. Most entelodonts were a good deal larger than other contemporary artiodactyls, ranging from the size of a peccary to as big as a bison. They

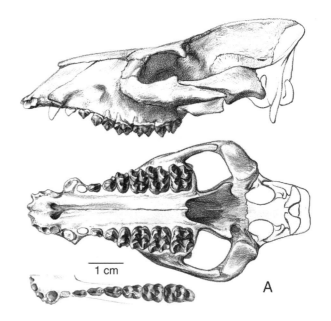

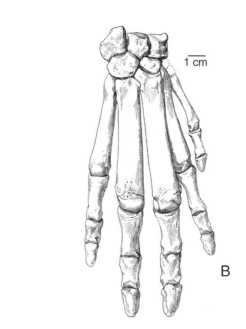

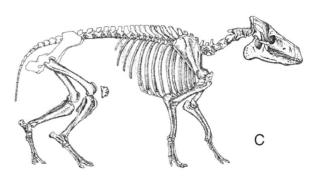

Fig. 14.20. Anthracotheres: (A) *Heptacodon,* skull and left lower teeth; (B) *Bothriodon,* right manus; (C) *Elomeryx.* (A, B from Scott, 1940; C from Scott, 1894.)

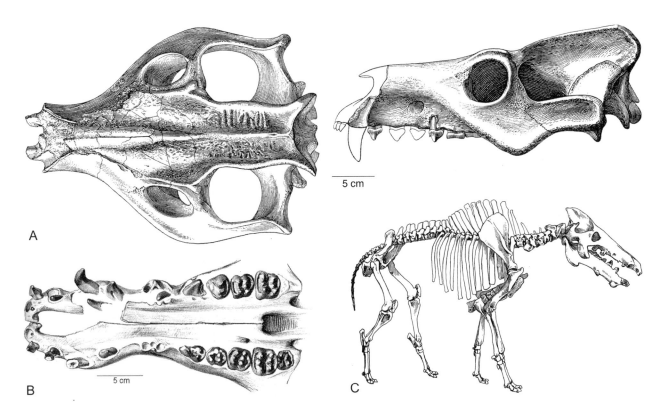

Fig. 14.21. Entelodonts: (A) *Brachyhyops*, skull; (B) *Archaeotherium*, palate (C) *Archaeotherium*, skeleton. About 1.5 m high at the withers. (A from Colbert, 1938; B, C from Scott, 1940.)

existed from the middle Eocene to early Miocene in Asia and North America and during the Oligocene of Europe, being particularly successful in the early Oligocene.

Entelodonts are distinctive for their relatively enormous skulls—up to a half-meter long in late Eocene to early Miocene *Archaeotherium* and reaching nearly a meter in early Miocene *Dinohyus* (Joeckel, 1990). Particularly characteristic of entelodont skulls are the widely flaring zygomatic arches with prominent suborbital flanges that project ventrolaterally. The mandible is characterized by its long, strong symphysis and expanded angle, and especially by the mental and mandibular tubercles protruding from the ventral border of the jaw below the canine and P_{3-4} in most forms. These flanges and tubercles, as well as variations in jaw depth and thickness and canine size, were probably sexually dimorphic and/or age-related and may have been associated with display (Scott, 1940; Effinger, 1998). Other characteristics of entelodont skulls include a long and narrow rostrum and a surprisingly small braincase. The orbits are completely encircled by bone, and in some types the frontal and parietal bones are rugosely sculptured (Colbert, 1938; L.S. Russell, 1980). Although the coronoid process of the mandible is reduced, there are very large temporal fossae and a prominent sagittal crest, suggesting well-developed temporalis muscles. The mandibular condyles are low and subcylindrical.

The incisors of entelodonts are rather long and pointed; the canines prominent; and the premolars simple, laterally compressed, and pointed. The canines and premolars typically show heavy apical wear reminiscent of that found in modern carnivorans, particularly those that use these teeth to crush bone (Joeckel, 1990), and it seems probable that entelodonts also included bones in their diet, at least occasionally. Entelodont molars are relatively small and quadrate, with four main cusps, including a hypocone above. The upper molar conules and lower molar paraconid and hypoconulid are smaller than the other cusps. Several features of the jaw, temporomandibular joint, and skull indicate that entelodonts were capable of a wide gape, perhaps associated with omnivory and scavenging (Joeckel, 1990).

The limbs of entelodonts are robust but moderately long and show typical artiodactyl cursorial modifications, except that the tibia is shorter than the femur. Entelodonts are primitive in retaining a separate fibula but specialized in having co-ossified radius and ulna and didactyl forefeet and hind feet; the lateral metapodials are reduced to mere nubbins (Scott, 1940). The anterior thoracic spines are exceptionally long, for attachment of muscles and ligaments to support the head, producing a shoulder hump similar to that in extant bison.

The oldest and most primitive known entelodonts are the middle–late Eocene genera *Eoentelodon* from China and *Brachyhyops* (Fig. 14.21A) from North America. Some researchers exclude the latter genus from Entelodontidae, because it has a short face, relatively poorly developed jugal flanges, and lacks tubercles on the mandible; but some

Archaeotherium also lack tubercles and have small flanges (Wilson, 1971b). *Brachyhyops* is probably best considered a primitive, short-snouted entelodont.

As noted above, the relationships of entelodonts are poorly understood, most resemblances to other groups being based on primitive characters. Some features, however, are uniquely derived (cranial and mandibular flanges and tubercles), whereas others, such as a fused radioulna and two-toed feet, are comparatively derived for bunodont artiodactyls. The origin of the family is equally uncertain. Although often sought among helohyid dichobunoids (*Helohyus* or *Achaenodon*), the latter make poor candidates for entelodont ancestry, as they tend to lose the hypocone and also differ in other regards (Coombs and Coombs, 1977b). For the present we must admit that their alliance with suiforms is rather weak; entelodonts could be derived independently from dichobunid stock or might represent a separate clade of archaic artiodactyls.

Tayassuidae

This family is often grouped together with Suidae and Hippopotamidae in the superfamily Suoidea. Tayassuidae, which include the extant New World peccaries, are first known from the late Eocene–early Oligocene of western North America (*Perchoerus, Thinohyus*; Fig. 14.22A,B), Thailand (*Egatochoerus*), and south China (Scott, 1940; Ducrocq, 1994; Wright, 1998; Liu, 2001). By the Oligocene, they were present in Europe (*Doliochoerus* at Quercy, France), and they later reached Africa and South America as well. The earliest Old World forms are older than the North American records, suggesting an Asian origin of the family, but they

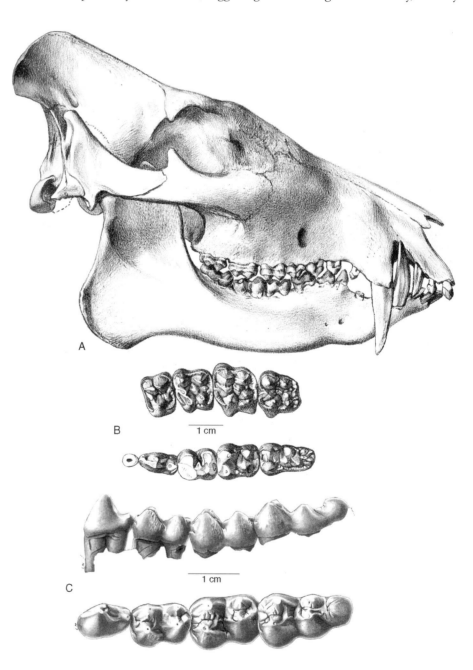

Fig. 14.22. Primitive suoids: (A) tayassuid *Perchoerus*, skull; (B) *Perchoerus*, left P^4–M^3 and P_4–M_3; (C) suid *Siamochoerus*, left P_4–M_3. (A, B from Scott, 1940; C from Ducrocq et al., 1998.)

are very fragmentary, and their tayassuid status is controversial (Wright, 1998). Indeed, the Old World forms differ from suids mainly in primitive features, and some experts do not believe they are true tayassuids.

Peccaries are generally much smaller (less than 40 kg) than suids and later became more cursorially adapted than suids, with functionally two-toed feet (lateral digits vestigial) and a partly fused cannonbone, while suids remained four-toed. In some species the radius and ulna are fused. The teeth are comparatively generalized (for artiodactyls), brachydont and typically bunodont. The molars are quadrate, with four main cusps and, in contrast to suids, relatively smooth enamel with few accessory cuspules. Some extinct forms, such as Late Cenozoic *Platygonus,* had bilophodont molars. The prominent upper canines point downward (not upward like those of suids) and hone against the back of the lower canines; in many Tertiary forms they are sexually dimorphic (Wright, 1998). Peccaries are primarily herbivorous and today occupy a broad range of habitats, from forest to desert. *Perchoerus* and *Thinohyus* were generally similar to later peccaries, differing in relatively minor ways, such as having distinct cranial sutures (usually fused or indiscernible in later peccaries), hollow auditory bullae, simpler premolars, wider upper molars, and a narrower braincase than extant forms.

Suidae

The true pigs, an Old World group, differ from peccaries in being larger (up to 275 kg) and having molars with complex, wrinkled enamel and accessory cusps. The third molars are elongate, and the large, ever-growing upper canines form tusks that typically curve dorsally. They retain a primitive placental dental formula (3.1.4.3/3.1.4.3). Most are forest dwellers.

Suids are first known in southeast Asia from fragmentary dental remains. *Eocenchoerus,* from the late middle or late Eocene (Duchesnean equivalent) of southern China, has been identified as the oldest known suid, based on its M^3, which is longer than wide and has four rounded cusps and a small talon or posterobuccal lobe (Liu, 2001). *Siamochoerus* (Fig. 14.22C), known from dental remains from the upper Eocene of Thailand and south China, is identified as a suid (subfamily Palaeochoerinae) rather than a peccary because of its buccally inflated molars and elongate M_3 with a complex talonid (Ducrocq et al., 1998). At this stage, however, the differences from peccaries are comparatively minor. Possible suids have been reported from the late Eocene of Europe (McKenna and Bell, 1997), but their status is dubious. The oldest definitive suid in Europe is *Palaeochoerus,* from the Oligocene of Quercy (Ginsburg, 1974). Palaeochoerines are more primitive than other suids and may be the basal stock for tayassuids as well as later suids (Liu, 2001). The presence of both palaeochoerines and tayassuids in southeast Asia in the late Eocene increases the probability that they shared a common ancestor in Asia earlier in the Eocene.

Primitive Selenodont Artiodactyls

Several families of selenodont or bunoselenodont artiodactyls that first appeared in the Eocene, and whose precise affinities are controversial, are considered here. Some authorities refer all of them to the Suiformes or other primitive suborders (Sudre, 1978b; McKenna and Bell, 1997; see Table 14.2), while others adopt a broad concept of Tylopoda that includes these families (Gentry and Hooker, 1988; Hooker and Weidmann, 2000). The distinction presumably reflects whether selenodonty and various postcranial specializations are viewed as synapomorphies shared with "higher" artiodactyls or as convergences acquired independently in multiple lineages. Wilson (1974) cautioned that selenodonty by itself is no guarantee of close relationship to tylopods, however, because this condition seems to have evolved many times in artiodactyls.

Mixtotheriidae

Mixtotherium (Fig. 14.23B), from the Eocene of Europe, is now usually assigned to its own family, Mixtotheriidae, whose broader affinities remain uncertain. It has been assigned to Cebochoeridae (Simpson, 1945; Sudre, 1972), allied with Dichobunoidea (Sudre, 1978b), or grouped with oreodonts or cainotheres in the Tylopoda (Gentry and Hooker, 1988; Hooker and Weidmann, 2000). Mixtotheres share certain features with cebochoerids (amastoidy; a short snout; posteriorly deepening mandible; and a large, posterolingual metaconule) and cainotheres (posteriorly deepening mandible and a tendency toward selenodonty and molarization of the fourth premolars; Hooker and Weidmann, 2000). However, their bunoselenodont molars are phenotypically closer to those of primitive anoplotheriids. As in the latter, the upper molars are characterized by a W-shaped ectoloph with a well-developed mesostyle, and there are five cusps including a small paraconule and a large, posterolingual metaconule that takes the place of a hypocone. The postcranial skeleton is unknown.

Anoplotherioidea

This superfamily has been used to unite the European selenodont families Anoplotheriidae, Dacrytheriidae (sometimes included within Anoplotheriidae), and Cainotheriidae (Sudre, 1977, 1978b; McKenna and Bell, 1997). Some authors exclude cainotheres but include Xiphodontidae (Gentry and Hooker, 1988; Hooker and Weidmann, 2000), but the latter are more widely accepted as primitive tylopods. Like those of mixtotheres, the upper molars in all these families have a dilambdodont ectoloph with a prominent mesostyle.

Anoplotheriidae. Anoplotheriids, known from middle Eocene–Oligocene deposits, had brachydont, bunoselenodont molars. The upper molars are triangular in *Ephelcomenus* and *Robiacina* (Fig. 14.24A) and more quadrate in *Diplobune* and *Anoplotherium,* with selenodont paracone,

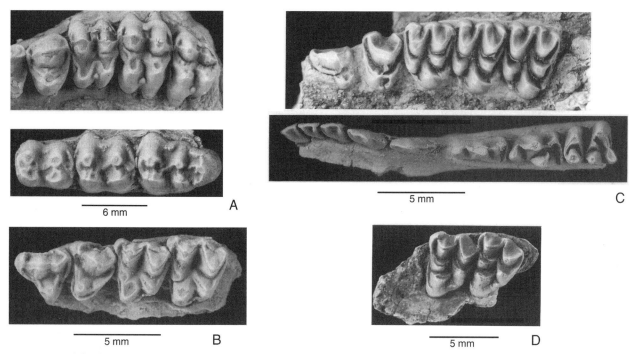

Fig. 14.23. Upper left and lower right dentitions of primitive artiodactyls: (A) cebochoerids *Acotherulum* (P^4–M^3, above) and *Cebochoerus* (M_{1-3}, below); (B) *Mixtotherium*, P^3–M^2; (C) basal cainothere *Paroxacron*, P^3–M^3 and I_2–M_1; (D) amphimerycid ruminant(?) *Pseudamphimeryx*, M^{2-3}. (From Hooker and Weidmann, 2000.)

metacone, and conules (the metaconule much larger than the paraconule); a bunodont protocone; and a buccally projecting mesostyle. The dentary tends to deepen posteriorly, the angle is expanded, and the coronoid process reduced—common trends in artiodactyls. In the lower dentition, the incisors are procumbent, the canine small and incisiform, and there are no appreciable diastemata. The paraconid of the lower molars is reduced or merges with the metaconid. The mastoid process of the petrosal is exposed at the back of the skull.

Where known, the postcranial skeleton of anoplotheres is relatively robust (humerus with well-developed tuberosities and epicondyles, radius strong and separate, astragalus short and broad with distinct cuboid and navicular facets, and tail sometimes long). The front and hind legs are of similar length, and the radius and tibia are noticeably shorter than the humerus and femur, respectively. Based on similar modification of the astragalocalcaneal joint, Heissig (1993) suggested that anoplotheres and oreodonts are sister taxa. Although anoplotheres such as *Anoplotherium* and *Diplobune* do have somewhat clawlike hoofs, as in some agriochoerid oreodonts, the forefeet and hind feet of anoplotheres were specialized, three-toed structures with subequal digits III and IV and a shorter, divergent digit II (Sudre, 1983), unlike any oreodont (or any other artiodactyl, for that matter). It was long thought that this peculiar foot structure supported a web and was therefore an adaptation for semiaquatic habits, but Sudre (1983) thought that it was more likely used for terrestrial locomotion. He postulated that the second digit may even have been opposable in one species, perhaps enabling *Diplobune* to forage in trees.

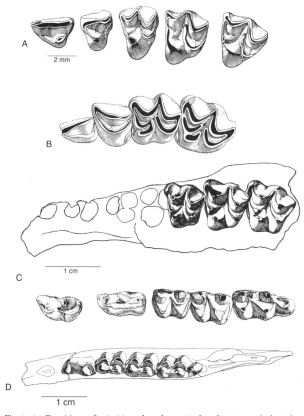

Fig. 14.24. Dentitions of primitive selenodont artiodactyls: (A) anoplotheriid *Robiacina*, left P^3–M^3; (B) tylopod *Xiphodon*, left P^3–M^2; (C) dacrytheriid *Dacrytherium*, left M^{1-3} and P_3–M_3; (D) ruminant(?) *Pseudamphimeryx*, right P_4–M_3. (A, B, D from Sudre, 1978b; C from Sudre, 1978a.)

Dacrytheriidae. The closely allied Dacrytheriidae were similar to anoplotheres but generally smaller. The lower premolars are typically narrow and elongate. *Dacrytherium* (Fig. 14.24C) is distinctive in having a large antorbital fossa in the maxilla, which may have housed a gland, but of what sort is uncertain (Delmont, 1941). The mastoid is exposed.

Although usually considered a dacrytheriid, *Tapirulus* lacks an antorbital fossa and has quadrate, bilophodont molars with four principal cusps, reminiscent of those of tapiroid perissodactyls, as the generic name suggests. The upper molar ectoloph, unlike that of other anoplotherioids, is straight or only weakly dilambdodont (Sudre, 1978b). Its allocation to this family (and perhaps even to Anoplotherioidea) is questionable, and Hooker and Thomas (2001) recently transferred it to the Choeropotamidae. Also of unsettled affinity is early Eocene *Cuisitherium,* initially allocated to Dacrytheriidae but now regarded as the oldest known choeropotamid (Sudre and Lecomte, 2000; Hooker and Thomas, 2001). These taxonomic shifts are further indication of the homoplasy of many dental traits in early artiodactyls.

Cainotheriidae. Cainotheres were small, gracile artiodactyls that lived from the late Eocene through the middle Miocene in Europe. The postcranial skeleton, like that of *Diacodexis,* is strikingly rabbitlike, with much longer hind limbs than forelimbs (see Fig. 14.30), and the skull even has large bullae and a fenestrated snout, as in rabbits. The selenodont molars and distinctive tarsal and foot morphology, however, affirm their artiodactyl affinities. The close postcranial resemblance to *Diacodexis,* along with such primitive features as retention of a complete fibula and the first metacarpal (Hürzeler, 1936), suggest that cainotheres derive from a very primitive artiodactyl. Sudre (1977) considered cainotheriids and anoplotheriids to be sister taxa derived from a form near middle Eocene *Robiacina*. Hooker and Weidmann (2000), however, found mixtotheres to be the more probable sister group of cainotheres.

The oldest cainotheres, *Oxacron* and *Paroxacron* (Fig. 14.23C), appear in the late Eocene. Even at this stage the upper molars show the diagnostic pattern of five crescentic cusps, with the conules large and equal in size and the protocone (unlike that of other early artiodactyls) shifted posteriorly almost to the position of a hypocone. The upper molar ectoloph is strongly dilambdodont; the premolars are simple; the canines are incisiform; and the lower incisors are small, equal in size, and more or less spatulate (Hooker and Weidmann, 2000).

Oreodontoidea

Oreodonts (Oreodontoidea, =Merycoidodontoidea) were a highly successful group of mid-Tertiary selenodont artiodactyls endemic to North America. They first occur in the middle Eocene and persisted through the Miocene. They are the most abundant mammalian fossils found in many Oligocene and early Miocene deposits of the Great Plains, but are rare after the early Miocene. Oreodonts (Fig. 14.25) are characterized by primitive postcranial skeletal features and relatively derived dentitions, a combination that has led to diverse views of their relationship to other artiodactyls. Despite their typically selenodont molars, primitive skeletal features—including separate cuboid and navicular bones in the ankle—leave no doubt that they are not ruminants; but

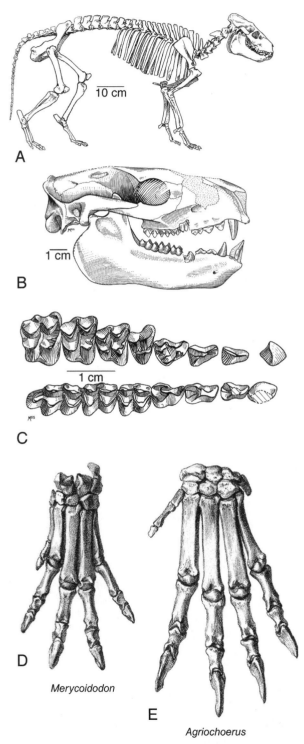

Fig. 14.25. Oreodonts: (A) *Merycoidodon;* (B) *Protoreodon;* (C) *Protoreodon,* right upper and lower dentition; (D, E) left manus of two oreodonts. (A, D, E from Scott, 1940; B, C from Wilson, 1971a.)

how they relate to nonruminants is not at all clear. McKenna and Bell (1997) allocated oreodonts to the Suiformes. Others consider them to be tylopods (Gentry and Hooker, 1988; Lander, 1998), but the oldest oreodonts predate the first appearance of other tylopods (which were presumably immigrants) and lack obvious synapomorphies with other tylopods, aside from selenodonty. Oreodonts are generally thought to have evolved from homacodontine dichobunids (Golz, 1976; Lander, 1998). Two families are recognized, the primitive Agriochoeridae and the more derived and diverse Oreodontidae (=Merycoidodontidae).

The most primitive oreodonts (both agriochoerids and oreodontids) had complete dentitions, including brachydont, bunoselenodont molars. The upper molars have a W-shaped ectoloph and prominent styles, a small paraconule (lost in more advanced oreodonts), a large lingual metaconule, and no hypocone. The upper canine is prominent, whereas the lower canine is incisiform and P_1 caniniform, characteristics perhaps inherited from a dichobunid ancestor. The lower incisors and canine are procumbent. Diastemata are short, if present at all. The skull has a prominent sagittal crest and an incomplete postorbital bar. The postcranial skeleton is robust, with a short neck, a long tail, and short limbs that retain relatively primitive structure. The ulna and the fibula are separate, strong elements. The distal trochlea of the astragalus is angled relative to the proximal trochlea, as in dichobunids. Early agriochoerids had five-toed front and hind feet, but the hallux was lost in all other oreodonts, and the pollex was lost in most oreodontids. A similar suite of features also characterizes anthracotheres, which are not considered to be closely related, but oreodonts have various derived dental and cranial features not found in anthracotheres.

Agriochoerids differ from oreodontids primarily in primitive traits, suggesting that the family is paraphyletic. A potentially derived character that distinguishes most agriochoerids is the presence of a molariform P^4, but even this trait varies, and its polarity is ambiguous. More derived members of both families are easier to distinguish. *Agriochoerus* (Fig. 14.25D) was a long-lived genus that had deep, curved, laterally compressed claws rather than hoofs (an autapomorphy). This unusual characteristic has prompted speculation that it was either a digger or a tree climber (Scott, 1940). Its habits remain elusive: Coombs (1983) found few features supporting fossorial habits and reluctantly opted for the second alternative, but noted that limited limb flexibility would have made *Agriochoerus* ill-adapted for climbing. The earliest and most primitive agriochoerid, *Protoreodon* (Fig. 14.25B), retained small hoofs and is considered the probable sister taxon or ancestral stock for other agriochoerids as well as oreodontids (Wilson, 1971a; Golz, 1976; Lander, 1998; Theodor, 1999).

Most oreodontids had more hypsodont and selenodont cheek teeth than did agriochoerids and are more derived in having four-toed front and hind feet, a complete postorbital bar, and a preorbital (lacrimal) fossa, perhaps for housing a facial gland, as in cervids. The tympanic bulla, and sometimes the external auditory tube as well, became large and inflated in several lineages. Most oreodontids are thought to have been forest-dwelling browsers, perhaps living in herds.

One of the most peculiar oreodontids was middle Eocene (Duchesnean)–early Miocene *Leptauchenia*, a short-faced form with the orbits and external auditory tube elevated and the mandibular angle expanded. Although *Leptauchenia* has sometimes been compared with hippos, the shape of its mandible is more like that of hyracoids. In contrast to most oreodontids, the lacrimal fossae are very small, but there are large, paired nasofacial vacuities extending from above to well in front of the orbits. *Leptauchenia* was long interpreted to have been a semiaquatic animal (Scott, 1940), but analysis of limb proportions indicates that it was probably an open-habitat rock climber similar to the extant hyrax *Procavia* or the rock cavy *Kerodon* (Wilhelm, 1993; Lander, 1998).

Parallel trends in different lineages, together with apparently frequent character reversals, have complicated efforts to unravel oreodont phylogeny. These phenomena have resulted in considerable homoplasy and have led to dramatically discordant views of which genera are valid and how they relate to one another. For example, McKenna and Bell (1997), using a modified version of Schultz and Falkenbach's (1968) classification, recognized more than 50 oreodont genera. There is a general consensus, however, that oreodonts have been taxonomically oversplit and that there are far fewer valid genera (CoBabe, 1996; Stevens and Stevens, 1996; Lander, 1998).

Tylopoda

The suborder Tylopoda is a group of selenodont artiodactyls including modern camels and their extinct relatives, which is more or less intermediate between suiforms and ruminants in many features. They share with ruminants several derived conditions, including reduction or loss of the upper incisors; loss of the upper molar paraconule; fusion of the ecto- and mesocuneiforms in the ankle (but not the cuboid and navicular); and a multichambered, ruminating stomach (Webb and Taylor, 1980). Although this last trait cannot be directly assessed in the fossil record, comparative anatomy of fossil and modern forms suggests that some specialization of the foregut must have been present in extinct primitive tylopods (and oreodonts, too, if they are related to tylopods), providing limited capability of foregut fermentation (Langer, 1974; Janis, 1976).

Despite these similarities to ruminants, there seem to be few uniquely derived features that characterize tylopods as a distinct group; hence they could be a paraphyletic assemblage of nonruminant selenodont artiodactyls (Janis, Effinger, et al., 1998). This paucity of synapomorphies helps to explain the mercurial content of the group. As mentioned earlier, several of the families discussed in the previous section (i.e., anoplotheres, cainotheres, mixtotheres, and oreodonts), as well as one family included below in ruminants (Amphimerycidae), are sometimes considered to be tylopods; and another family long placed with ruminants (Protoceratidae) is now generally assigned to Tylopoda.

Geisler's recent phylogenetic analysis (2001; see Fig. 14.13B), however, found this broad concept of Tylopoda to be a paraphyletic assemblage. His preferred cladogram indicated that protoceratids and oreodonts are sister taxa and xiphodonts and amphimerycids are sister taxa, but that both of these clades lie outside Ruminantia and crown Tylopoda. Except for xiphodontids, the groups considered here to be tylopods appear to have been confined to North America during their Early Cenozoic history.

Xiphodontidae

Xiphodonts were a European group of small, camel-like artiodactyls known from the middle to late Eocene. They possibly survived into the early Oligocene. They had narrow, very elongate lower premolars, and their molars were fully selenodont but still brachydont. The uppers had four or five cusps (when a small paraconule was present) and a strong W-shaped ectoloph (Fig. 14.24B). The skull is mastoid, and the auditory bullae are large. *Xiphodon* (the only genus in which the skeleton is known) had slender, elongate limbs and was functionally didactyl, with only vestiges of lateral digits (Sudre, 1978b). Xiphodonts are usually allied with tylopods, but their relationships are not secure. They are the only non-North American Paleogene family allocated to Tylopoda, and it remains possible that they arose independently.

Oromerycidae

Oromerycids were primitive tylopods that lived during the middle and late Eocene (Uintan-Chadronian) in North America. They were about the size of small to mid-sized deer. A half-dozen genera are recognized, Duchesnean-Chadronian *Eotylopus* (Fig. 14.26A) being the best known (Scott, 1940). As in camels, the auditory bulla is inflated and filled with cancellous bone (Prothero, 1998a). The snout was narrow, and there was a full complement of teeth with no diastemata. Compared to other tylopods, oromerycids had lower-crowned and less selenodont cheek teeth and lacked horns or cranial tuberosities (Prothero, 1998a). The radius and ulna are fused in *Eotylopus,* as in camelids, but not in the earliest forms (Uintan *Protylopus;* Golz, 1976), and the metapodials are similar to those in protoceratids (Scott, 1940). Otherwise the limbs are relatively unspecialized; consequently, there is little that separates them from primitive camels or protoceratids. Like those animals, they were probably forest browsers. Oromerycids differ from other tylopods in lacking the derived position of the vertebrarterial canal (i.e., passing through the vertebral arch); instead the canal passes through the transverse processes, as in placentals generally.

Protoceratidae

Protoceratids (Fig. 14.26B,C; see also Fig. 14.29A) were camel-like artiodactyls, whose more derived members were

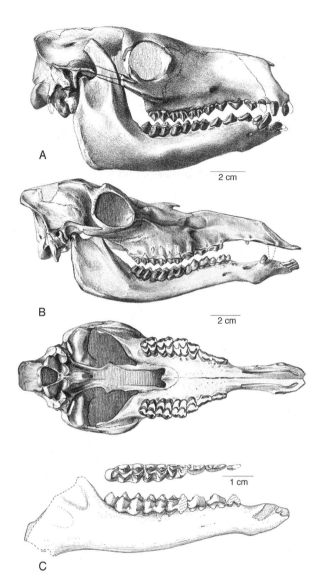

Fig. 14.26. Primitive tylopods: (A) middle–late Eocene oromerycid *Eotylopus;* (B) early Oligocene *Protoceras;* (C) middle Eocene protoceratid *Leptoreodon.* (A, B from Scott, 1940; C from Wilson, 1974.)

adorned with several pairs of tuberosities or horns, sometimes forked, on the maxillae, orbits, and parietals (Frick, 1937). Although long included in Ruminantia, protoceratids are now usually grouped with camels, because they are more primitive than ruminants in several features, including retention of a functionally four-toed forefoot and separate cuboid and navicular bones in the ankle (Patton and Taylor, 1973). Prothero (1996b) considered protoceratids to be the sister group of camelids, but this inference contrasts markedly with a recent phylogenetic analysis by Geisler (2001; see Fig. 14.13B). The forefoot had four toes with the lateral ones somewhat reduced, whereas the hind foot was essentially two-toed (the metatarsals remaining separate) with only remnants of the lateral metatarsals (Prothero, 1998b). Possible synapomorphic features linking them with camels include elongated limbs (though not as long as in camels or ruminants) and passage of the vertebral arteries through the base of the vertebral arch of the neck vertebrae, rather than more anteriorly through their transverse processes, as is the usual condition in placentals (Webb and Taylor, 1980).

A dorsally situated vertebrarterial canal also occurs in xiphodonts and is lacking in oromerycids, which are usually considered to be more closely related to camelids (Gentry and Hooker, 1988), suggesting that it may not be a reliable indication of relationship. Other protoceratid resemblances to camels (e.g., unfused metacarpals) are likely to be plesiomorphic. Moreover, some details of the basicranium of protoceratids are more like those of ruminants than those of camels, suggesting that protoceratids could, after all, be closely related to ruminants (Joeckel and Stavas, 1996; Norris, 2000). Like ruminants, protoceratids lost their upper incisors. According to Scott (1940), the auditory bullae, though inflated, do not contain cancellous bone, in contrast to camels. These conflicting features leave the phylogenetic position of protoceratids in doubt.

A half-dozen primitive protoceratid genera have been reported from the middle and late Eocene (Uintan-Chadronian), the most primitive being *Leptotragulus* and *Leptoreodon* (Fig. 14.26C). The group persisted into the Pliocene. Unlike the later members, however, Eocene protoceratids were relatively small and lacked horns and resembled other primitive selenodont forms, such as oromerycids, oreodonts, and basal ruminants (Prothero, 1998b).

Camelidae

Camels are tylopods with upper molars that are selenodont and transversely compressed (Prothero, 1996b). The cheek teeth of Recent camels are relatively high crowned, but those of many extinct forms are much less so. The outer crest of the upper molars (the ectoloph) is straight, and the paraconule is lost, leaving four main cusps. The snout is usually long and narrow, with diastemata around the first premolar. Camels are structurally advanced relative to oromerycids and protoceratids in having a longer neck and markedly longer and more slender limbs, characteristics already evident in the basal camelids *Poebrotherium* (Fig. 14.27; see also Fig. 14.29B) and *Paratylopus* (Scott, 1940). The radius and ulna are fused, and the fibula is reduced to small proximal and distal remnants, the former co-ossified with the tibia and the latter forming a separate malleolar bone. The central metapodials are long and partly fused in later camels, although apparently not always in the most primitive forms (Scott, 1940), and they always diverge somewhat distally (in contrast to the ruminant condition). The lateral metapodials are reduced to tiny nonfunctional vestiges. The metatarsals and phalanges of *Poebrotherium* are long and slender, resembling those of ruminants more than those of extant camels. It was probably unguligrade; the digitigrade stance of present-day camels is a derived condition (Janis et al., 2002).

The early history of camels is restricted to North America until the late Miocene, though by the Late Cenozoic they became nearly cosmopolitan. Only a few basal genera were present, but rare, in middle and late Eocene faunas (Honey

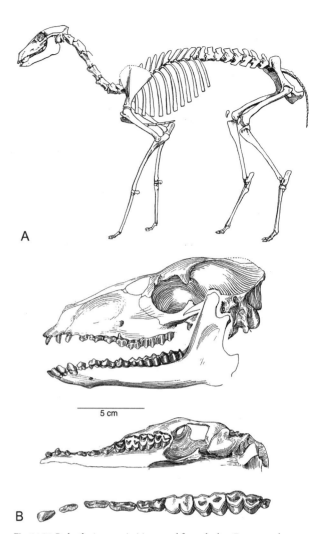

Fig. 14.27. *Poebrotherium*, a primitive camel from the late Eocene–early Oligocene: (A) skeleton; (B) skull and left upper and lower dentition. (A from Scott, 1940; B from Prothero, 1996b.)

et al., 1998). The oldest known camelid is Uintan *Poebrodon* (known only from teeth), but *Poebrotherium* and *Paratylopus*, first known from the Chadronian, are known from skulls and skeletons.

Ruminantia

The most derived artiodactyls, ruminants are cursorial animals with selenodont, sometimes hypsodont cheek teeth, no upper incisors, and an incisiform lower canine. They are further characterized by a fused cubonavicular and a three- or four-chambered stomach (Webb and Taylor, 1980; Janis, 1987; Scott and Janis, 1993; Webb, 1998). Ruminants can be subdivided into the primitive (and paraphyletic) Tragulina, including the still extant tragulids and their extinct hornless relatives (amphimerycids, hypertragulids, leptomerycids, bachitheriids, and lophiomerycids), and the more advanced Pecora, which includes the horned ruminants (Webb and Taylor, 1980; Janis, 1987). Some authors (e.g., Geraads et al., 1987) use a more restricted concept of Tragulina. The interrelationships of basal ruminants, including which is the most primitive and which the closest to pecorans, remain unsettled, in part because several of the critical taxa are known only from teeth. Nevertheless, the diversity of basal ruminants in existence by the end of the Eocene indicates that the ruminant radiation was already well under way at that time.

Amphimerycidae

This small family, questionably allocated to Ruminantia, consists of only two genera (*Amphimeryx* and *Pseudamphimeryx*; Figs. 14.23D, 14.24D) known from the early Eocene–early Oligocene of Europe. They have roughly triangular selenodont upper molars with large conules and a prominent mesostyle (producing a W-shaped ectoloph), not unlike many other primitive selenodont forms. The precise relationship of amphimerycids to other primitive ruminants is problematic. They have been variously considered to be very primitive ruminants, the sister taxon of ruminants, or tylopods (Webb and Taylor, 1980). The presence of a fused cubonavicular in *Amphimeryx* would seem to affirm its placement in Ruminantia (Sudre, 1978b; Gentry and Hooker, 1988), if the attribution of the bone to this genus had not been questioned (Webb and Taylor, 1980; Sudre, 1984). *Amphimeryx* had long, slender limbs, with particularly elongate metapodials (Sudre, 1978b).

Hypertragulidae

First known from the late Uintan (middle Eocene) of North America (*Simimeryx*; Webb, 1998), hypertragulids were relatively rare, rabbit-sized artiodactyls with hypsodont, selenodont cheek teeth. They became extinct in the Miocene. Hypertragulids have an antorbital fossa on the narrow rostrum, as in *Leptomeryx* (Scott, 1940). The postorbital bar is primitively incomplete, but was complete in late Eocene–early Oligocene *Hypisodus*. The long, slender limbs of hypertragulids, which reached an extreme in tiny *Hypisodus* (even smaller than a rabbit), were adapted for cursorial-saltatorial locomotion. The forelimbs were markedly shorter than the hind limbs. Hypertragulids had a fused cubonavicular—the hallmark of Ruminantia—as well as co-ossified radius and ulna, but otherwise *Hypertragulus* (Fig. 14.28, 14.29D) closely resembled *Diacodexis* in most of the postcranial skeleton. It remained primitive in having a pentadactyl forefoot, a four-toed hind foot, and a "bent" astragalus, in which the proximal and distal trochleae are not aligned. Moreover, hypertragulids are primitive in retaining P^1, an incomplete postorbital bar (*Hypertragulus*), a lateral exposure of the mastoid, a trapezium in the carpus, and a complete fibula. For these reasons, hypertragulids are widely considered to be the most primitive unequivocal ruminants (Webb and Taylor, 1980; Scott and Janis, 1993; Webb, 1998).

Leptomerycidae

Although they appear in the fossil record a little earlier than hypertragulids and much earlier than Tragulidae (which

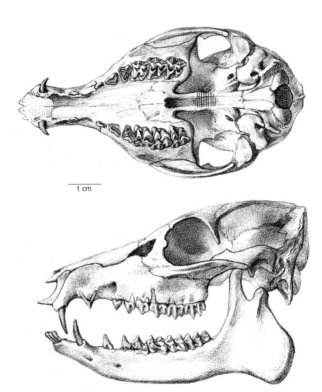

Fig. 14.28. Skull of the primitive ruminant *Hypertragulus*. (From Scott, 1940.)

do not appear until the Miocene), leptomerycids are in some ways more derived than both of those groups (Webb and Taylor, 1980). In leptomerycids the first metacarpal is lost, together with the trapezium in the carpus, and the magnum and trapezoid are fused. Leptomerycid limbs are distally more elongate than those of hypertragulids, the fibula is reduced to a small proximal remnant fused to the tibia and a distal malleolar bone (as in camels and all more advanced ruminants including tragulids), and the astragalar trochleae are aligned.

The oldest and best-known leptomerycids are Asian *Archaeomeryx* and North American *Leptomeryx* (Fig. 14.30), first known from the middle Eocene. *Leptomeryx* was abundant during the Oligocene, when it was an important forest browser (Webb, 1998). In *Leptomeryx* metatarsals III and IV, together with small remnants of the lateral metatarsals, are precociously fused into a cannonbone (Scott, 1940; Fig. 14.29C). Metatarsals III and IV of *Archaeomeryx*, however, remain separate, a factor that led Gentry and Hooker (1988) to exclude it from Leptomerycidae and to regard *Archaeomeryx* as the sister taxon of all other ruminants. Other experts consider leptomerycids, including *Archaeomeryx*, to be the closest relative of pecorans (Webb and Taylor, 1980; Guo et al., 2000). Geraads et al. (1987) used a more restricted concept of Tragulina that included only Leptomerycidae (excluding *Archaeomeryx*) and Tragulidae, evidently considering the metatarsal cannonbone to be a synapomorphy with pecorans. They regarded other taxa often assigned to Tragulina to be primitive stem ruminants (plesions). Janis (1987), however, interpreted the cannonbone of *Leptomeryx* to be homoplasious with that of pecorans.

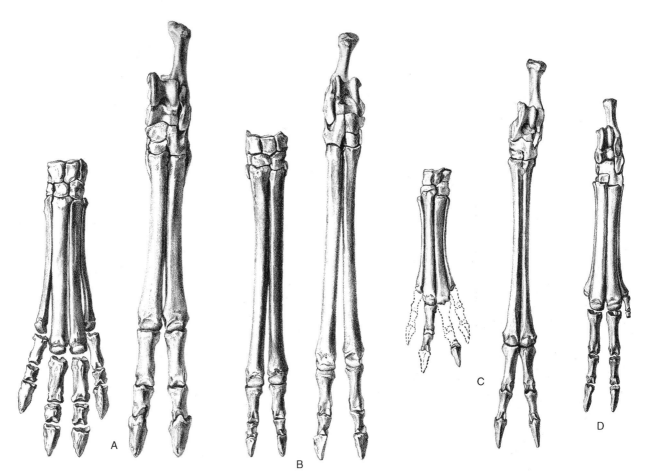

Fig. 14.29. Fore- and hind feet of primitive tylopods and ruminants (not to scale): (A) primitive tylopod *Protoceras;* (B) primitive tylopod *Poebrotherium;* (C) ruminant *Leptomeryx;* (D) ruminant *Hypertragulus.* (From Scott, 1940.)

Other Primitive Ruminants

Lophiomerycids are primitive ruminants long believed to be closely related to the basal pecoran family Gelocidae. First known from the late middle Eocene of China (*Zhailimeryx;* Guo et al., 2000), lophiomerycids reached Europe in the Oligocene and survived through the Oligocene there and in Asia. They were dentally among the most primitive ruminants (with open molar trigonids lacking a premetacristid), lacked a postorbital bar, and retained four-toed feet with unfused metapodials III and IV (Brunet and Sudre, 1987; Geraads et al., 1987; Guo et al., 2000). A cubonavicular is present, but the astragalar trochleae are primitively offset, as in hypertragulids. Janis (1987) considered lophiomerycids to be the sister group of Pecora, whereas Geraads et al. (1987) regarded *Lophiomeryx* as a basal ruminant more primitive than leptomerycids and tragulids. The many plesiomorphic traits of lophiomerycids suggest that the second alternative is more likely.

Oligocene *Bachitherium* (Bachitheriidae) from Europe is a primitive ruminant that is skeletally more advanced than lophiomerycids in having relatively longer distal segments. Metatarsals III and IV are fused for most of their length (Geraads et al., 1987). Its phylogenetic position is controversial. Janis (1987) considered *Bachitherium* to be dentally more primitive than pecorans and included it in the Tragulina, but Geraads et al. (1987) classified it as a primitive pecoran.

Pecora

Derived ruminants, including the horned ruminants, are sometimes united in the higher taxon Pecora. The term can be confusing, because it was previously equated with what is now regarded as Ruminantia (e.g., Scott, 1940; Romer, 1966). More recently Pecora has been used for the Gelocidae, the Moschidae, and the Eupecora, which includes ruminants with horns, antlers, or ossicones. Pecorans are the dominant artiodactyls today, but they did not appear until the late Eocene. They are characterized by the loss of the stapedial artery, a split paraconid on lower premolars, relatively long forelimbs, and an astragalus with parallel trochleae (Webb and Taylor, 1980).

Usually considered to be the oldest and most primitive pecorans are the Gelocidae, a paraphyletic assemblage placed here on the basis of derived astragalar and dental morphology and other features (Webb and Taylor, 1980; Janis, 1987). Gelocids are first known from late Eocene–Oligocene beds of Eurasia; later members reached North America and Africa in the Miocene. The composition of Gelocidae and its taxonomic position (either tragulines or basal pecorans) are unstable. Some genera (including *Indomeryx* and *Prodremotherium*) have been transferred to a new family, Prodremotheriidae, which includes the oldest known pecorans (Guo et al., 1999). Early Oligocene *Dremotherium* from Europe, assigned to the Moschidae (musk deer), is the oldest known member of an extant family of ruminants.

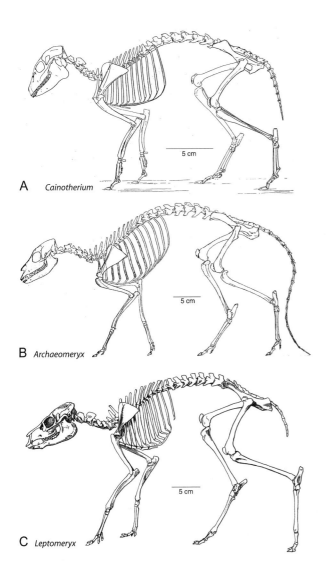

Fig. 14.30. Primitive artiodactyls: (A) European cainothere *Cainotherium*; (B) Asian primitive ruminant *Archaeomeryx*; (C) North American primitive ruminant *Leptomeryx*. (A from Hürzeler, 1936; B from Colbert, 1941; C from Scott, 1940.)

15

Anagalida
Rodents, Lagomorphs, and Their Relatives

McKENNA AND BELL (1997) used the name Anagalida as a grandorder to unite rodents, lagomorphs, and their closest relatives (which are collectively called Glires by most authorities, but not by McKenna and Bell) with Cretaceous Zalambdalestidae, early Paleogene Anagalidae and Pseudictopidae, and Macroscelidea (elephant shrews; Table 15.1). Possible relationships of these groups are depicted in Figure 15.1. Except for macroscelideans, which are exclusively African today, anagalidans are either restricted to Asia or are known very early in their history from Asia. In recent years evidence for close relationship between rodents and lagomorphs has increased, but the basis for uniting the other taxa listed above with one another, or with rodents and lagomorphs, remains relatively weak. For example, macroscelideans have otherwise been associated with ungulates, particularly in molecular studies, which place them together with paenungulates in the Afrotheria. As discussed in Chapter 12, fossil evidence suggests that certain hyopsodontid condylarths are closely related to macroscelideans. In a recent study focusing on the skull of *Zalambdalestes*, however, Wible et al. (2004) found support for a clade containing all the groups listed above except *Zalambdalestes*, which in most of their analyses separated as a primitive branch of Eutheria. Nevertheless, two of their analyses (like several earlier studies) placed zalambdalestids near Glires. Consequently, despite the uncertainty about their monophyly, it is convenient to discuss these groups together in this chapter.

Lucas (1993 and elsewhere) advocated a close relationship between Pseudictopidae and Uintatheriamorpha (used by Schoch and Lucas, 1985, to encompass Dinocerata, Pyrotheria, and Xenungulata), based on several dental resemblances. This hypothesis is contradicted by nondental anatomy (Van Valen, 1988), however, and has

Table 15.1. Classification of Anagalida

Superorder ANAGALIDA
    †Zalambdalestidae[1]
    †Anagalidae
    †Pseudictopidae
  Order MACROSCELIDEA
    Macroscelididae
  Grandorder GLIRES
    Mirorder DUPLICIDENTATA
      Order †MIMOTONIDA
        †Mimotonidae
      Order LAGOMORPHA
        Ochotonidae
        Leporidae
    Mirorder SIMPLICIDENTATA
      †Sinomylus
      Order †MIXODONTIA
        †Eurymylidae
      Order RODENTIA
        †Alagomyidae
        †Laredomyidae
        Suborder SCIUROMORPHA
          †Paramyidae[2]
          †Ischyromyidae
          †Sciuravidae[3,4]
          †Cylindrodontidae[3,4]
          †Theridomyidae
          Sciuridae
          Gliridae[3,5]
          Superfamily Aplodontoidea
            Aplodontidae
            †Allomyidae[6]
            †Mylagaulidae
          Superfamily Castoroidea
            Castoridae
            †Eutypomyidae
        Suborder MYOMORPHA
          †Protoptychidae[3]
          Infraorder MYODONTA
            †Armintomyidae
            †Simimyidae
            Superfamily Dipodoidea
              Zapodidae
              Dipodidae
            Superfamily Muroidea
              Cricetidae
              Muridae
          Infraorder GEOMORPHA
            †Eomyidae
            Superfamily Geomyoidea
              †Florentiamyidae
              †Heliscomyidae
              Geomyidae
              Heteromyidae[7]
        Suborder ANOMALUROMORPHA
          Pedetidae
          Superfamily Anomaluroidea
            †Zegdoumyidae[8]
            Anomaluridae
        Suborder Uncertain
          Superfamily Ctenodactyloidea[3]
            †Cocomyidae[9]
            †Chapattimyidae
            †Yuomyidae[9]
            †Tamquammyidae[9]
            †Gobiomyidae
            Ctenodactylidae
        Suborder HYSTRICOGNATHA[10]
          †Phiomyidae
          †Myophiomyidae
          †Diamantomyidae
          †Thryonomyidae
          Infraorder CAVIOMORPHA
            Agoutidae[11]

Notes: Modified after McKenna and Bell (1997). The dagger (†) denotes extinct taxa. Families and genera in boldface are known from the Paleocene or Eocene.

[1] Phylogenetic position uncertain; could be basal eutherians.
[2] Often considered a subfamily of Ischyromyidae, as they are in this chapter.
[3] Phylogenetic position uncertain.
[4] These families were united in the new suborder Sciuravida by McKenna and Bell (1997).
[5] Placed in a separate infraorder within Myomorpha by McKenna and Bell (1997).
[6] Often included in Aplodontidae.
[7] Often included in Geomyidae.
[8] Could be related to Gliridae.
[9] Included in Chapattimyidae by McKenna and Bell (1997).
[10] Only families with Eocene or early Oligocene records are listed; see McKenna and Bell (1997) for complete list.
[11] Eocene/Oligocene boundary (Tinguirirican).

not been generally accepted. It is likely that the dental similarities are convergent. The latter groups are here discussed with Ungulates.

PRIMITIVE ASIAN ANAGALIDANS AND POSSIBLE ANAGALIDANS

Zalambdalestidae

Zalambdalestids are a family of early eutherians from the Late Cretaceous of Mongolia and Uzbekistan that might be related to rodents, lagomorphs, elephant shrews, and several extinct groups (Novacek et al., 1988; McKenna and Bell, 1997; Archibald et al., 2001). The skull and skeleton of zalambdalestids (Fig. 15.2) show a rather striking general resemblance to those of extant elephant shrews, which is particularly remarkable in view of their antiquity. This resemblance was regarded as convergent by Kielan-Jaworowska (1978). A recent analysis by Archibald et al. (2001) found that zalambdalestids share numerous dental characters with Glires, but subsequent phylogenetic analyses (Meng et al., 2003; Wible et al., 2004; Asher et al., 2005) have placed *Zalambdalestes* with other primitive Cretaceous eutherians rather than with Glires.

As observed earlier (see Chapter 4), zalambdalestids have a long, narrow snout, with a long diastema between the incisors and canine, except in *Kulbeckia*, the oldest and most primitive member. The dental formula of *Zalambdalestes* (Fig. 15.2B) and *Barunlestes* is 3.1.3–4.3/3.1.3–4.3; *Kulbeckia* may have one or two more upper incisors and retained four upper premolars (Archibald et al., 2001; Archibald and Averianov, 2003). Despite their primitive dental formula, the dentition of zalambdalestids shows many advances over typical Cretaceous eutherians. I^2 is enlarged, and I^1 and I^3 are small. In the lower jaw the anterior incisors are enlarged and procumbent, and the enamel is restricted to the anterior (buccal) surface and apparently lacked Hunter-Schreger

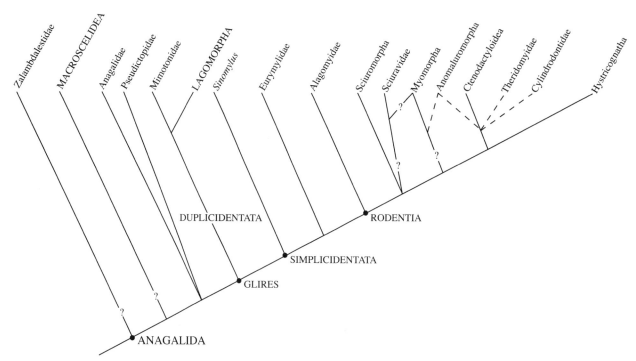

Fig. 15.1. One of many possible schemes of relationships among Glires and potentially related groups, based mainly on Flynn et al. (1986), Korth (1994), and McKenna and Bell (1997). Question marks and dashed lines indicate particularly uncertain positions. Some recent evidence suggests that theridomyids and cylindrodontids are more closely related to sciuromorphs (specifically ischyromyids) than to ctenodactyloids.

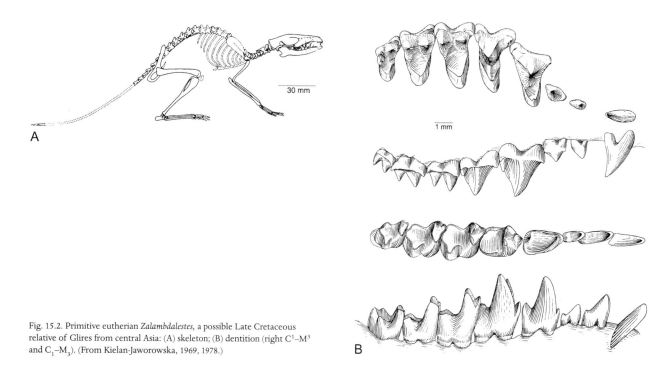

Fig. 15.2. Primitive eutherian *Zalambdalestes*, a possible Late Cretaceous relative of Glires from central Asia: (A) skeleton; (B) dentition (right C^1–M^3 and C_1–M_3). (From Kielan-Jaworowska, 1969, 1978.)

bands (HSB). P_4 is submolariform, the lower molars have reduced paraconids and mesiodistally compressed trigonids, and the molar talonids are broad and basined. The upper molars are transverse and essentially tricusped, with tall protocones, reduced stylar shelves, and no pre- or postcingula. M^3 is reduced. As pointed out by Archibald et al. (2001), several of these features resemble conditions in Glires.

The skeleton of zalambdalestids displays many specializations for running and jumping, including joints specialized for flexion-extension and much longer hind limbs than forelimbs (intermembral index about 63; Kielan-Jaworowska, 1978). The tibia and fibula are fused high up on the shaft, the astragalus is deeply grooved, and the metatarsals are very elongate, with a reduced hallux. In all these

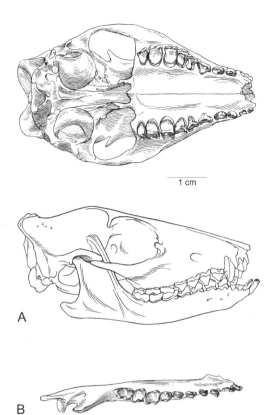
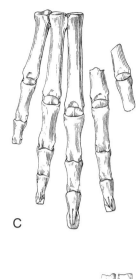
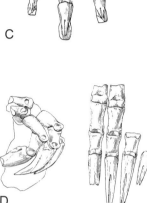

Fig. 15.3. Early Oligocene anagalid *Anagale:* (A) skull; (B) right dentary; (C) hind foot; (D) forefoot. (From Simpson, 1931b.)

features zalambdalestids resemble rabbits or elephant shrews, and their locomotion was probably similar. Overall, the skeleton of zalambdalestids is extraordinarily advanced compared to that of other Late Cretaceous mammals.

Anagalidae

Anagalids are an enigmatic Early Tertiary family endemic to Asia. Simpson (1931b) established this family for the early Oligocene genus *Anagale* (Fig. 15.3), which was then believed to be related to tree shrews (Scandentia) and thus pertinent to the origin of primates. Since then, a dozen or so genera have been added to the family (Hu, 1993; McKenna and Bell, 1997), nearly all from the Paleocene of China. Most genera are known only from jaws. The type genus, *Anagale,* is still the most completely known, being represented by a complete skull, mandible, and partial skeleton (the only substantial postcranial remains known for an anagalid; Simpson, 1931b). It seems to have been a rather generalized terrestrial quadruped, capable of digging with the forelimbs, as inferred from its prominent ulnar olecranon process; short and stout manual phalanges; and long, curved, deeply fissured ungual phalanges. Anagalids are now considered to be closely related to Glires and not particularly related to Scandentia or Primates (McKenna, 1963). This conclusion is partly based on the auditory bulla of *Anagale,* which is composed of the ectotympanic (as in Glires) rather than the entotympanic (as in Scandentia) or the petrosal (as in euprimates; Novacek et al., 1988).

Anagalids are characterized by a relatively primitive dental formula, 3?.1.4.3/3.1.4.3, without significant diastemata, although the first premolar is often set off by short gaps. The incisors in *Anagale* and early Paleocene *Eosigale* (Fig. 15.4) are small and unspecialized, the lowers somewhat inclined and the uppers more nearly vertical. Except for their slight procumbency, however, they show little specialization in the direction of Glires. Unlike Glires, but as in pseudictopids and macroscelideans, the canines are moderately to well developed (a plesiomorphic feature). The last premolars are submolariform, the others simple. The lower molars have tall, mesiodistally compressed trigonids, which quickly wore down almost to the level of the talonids. The upper molars are less transverse than those of pseudictopids and lagomorphs, but as in the former the main cusps of the upper molars are joined by a V-shaped crest, and there are prominent pre- and postcingula. The lingual side of the upper molars (the protocone) is moderately hypsodont, as is the buccal side of the lowers, a tendency often developed to a greater extent in other anagalidans. This unilateral hypsodonty is particularly evident in the Paleocene genera *Hsiuannania* and *Qipania,* in which the enamel extends buccally down onto the roots of the lower molars (Hu, 1993). In these genera the lingual cusps of the lower molars are markedly higher than the buccal cusps. The cheek teeth of anagalids are commonly heavily worn, with wide exposures of dentine on occlusal surfaces, which suggests that they had thin enamel and/or harsh, gritty diets.

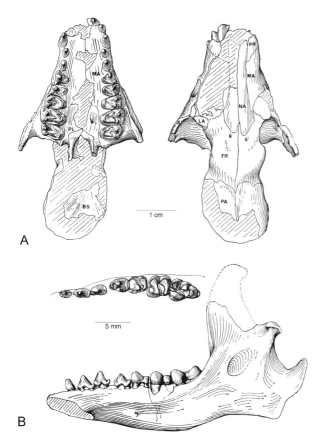

Fig. 15.4. Paleocene anagalid *Eosigale*: (A) snout; (B) right P_1–M_3 and medial view of dentary. Key: BS, basisphenoid; FR, frontal; JU, jugal; LA, lacrimal; MA, maxilla; NA, nasal; PA, parietal; PL, palatine; PR, premaxilla. (From Hu, 1993.)

Pseudictopidae

This endemic Asian family, known only from the Paleocene, is generally thought to be related to Anagalidae and thus possibly to Glires. The several included genera are known primarily from incomplete dentitions; the best-known form is late Paleocene *Pseudictops* (Fig. 15.5) itself, from the Gobi Desert of Mongolia. Van Valen (1964) suggested a special relationship between *Pseudictops* and lagomorphs. The lower molars of *Pseudictops* have elevated trigonids, and the upper molars are transversely wide, with a tall, sectorial protocone at the vertex of a V-shaped crest. They are unilaterally hypsodont (the upper molars lingually hypsodont, the lowers buccally hypsodont), a resemblance to anagalids (McKenna, 1963; Van Valen, 1964). This condition applies equally to the last two premolars, which are submolariform. The cheek teeth do bear a resemblance to those of lagomorphs in being moderately high crowned and having crowns dominated by transverse lophs and basins, but they are rooted, not ever-growing. Furthermore, *Pseudictops* has a primitive dental formula (3.1.4.3 both above and below), including relatively unreduced canines and only a short diastema between the first and second premolars. Sulimski (1968) illustrated the unusual incisor and canine crowns of *Pseudictops*, which are serrated on their posterior margins, quite unlike the incisors of Glires.

Skeletal remains of *Pseudictops* and other pseudictopids indicate that they were terrestrial animals adapted for running or perhaps hopping. Such habits are reflected by a relatively straight ulna with a tightly curved semilunar notch; a femur with a high greater trochanter; and limbs that are somewhat distally elongate, with the fibula separate but closely appressed to the tibia (Sulimski, 1968; Ding and Tong, 1979). The astragalar trochlea is moderately grooved, and the calcaneus and metatarsals are relatively long. There is a well-developed calcaneofibular articulation, as in lagomorphs (Szalay, 1985). The terminal phalanges are only slightly curved and distally somewhat expanded and fissured at the tip; but they are of somewhat different shape and much less deeply fissured than in *Anagale*. There is a superficial resemblance to the hind limb of some lagomorphs (i.e., leporids), but few specific features point to a close relationship. For now, the phylogenetic position of pseudictopids remains cryptic.

The only non-Asian form that has been linked with pseudictopids is *Mingotherium*, a late Paleocene taxon based on a single upper molar from South Carolina (Schoch, 1985). Although it bears a superficial resemblance to *Pseudictops* because the pre- and postprotocristae form a V-shaped crest, the paracone and metacone are bunodont and separate, unlike the tall, sectorial paracone and metacone and high, sharp centrocrista of *Pseudictops*. In addition, the *Mingotherium* molar is about three times larger than those of *Pseudictops*. The affinities of *Mingotherium* are obscure, but they probably do not lie with pseudictopids.

MACROSCELIDEA

This ordinal-level taxon comprises the endemic African elephant shrews, family Macroscelididae. Of the dozen or so genera, some are known as far back as the Eocene, and four or five are extant. They are mouse- to rat-sized animals (25–500 g), with large ears and eyes. Elephant shrews are named for their long, mobile snouts, but until recently, no one seriously contended that they could be related to elephants. Molecular studies have changed that assumption.

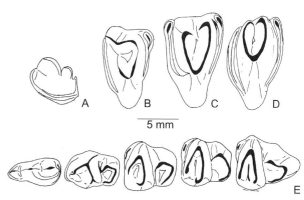

Fig. 15.5. Teeth of Paleocene *Pseudictops*: (A) lower incisor; (B) upper premolar; (C, D) upper molars; (E) right P_3–M_3. (From Lucas, 1993.)

The skull of elephant shrews has a long, pointed rostrum. The orbital rim is incomplete posteriorly. Two or three muscles for moving the snout originate just anterior to the orbit (Whidden, 2002). The auditory bulla is highly distinctive, being a compound structure composed of multiple elements, principally entotympanic, ectotympanic, and alisphenoid (MacPhee et al., 1989). Elephant shrews show a few reductions from the primitive placental dental formula (their dental formula is 1–3.1.4.2/3.1.4.2–3; notably, almost all extant species lack third molars), and the teeth are quite modified from the primitive tribosphenic condition. The lower incisors are small, equal in size, and in some species have bilobed crowns. The premolars are mesiodistally elongate, and the molars are bunodont to weakly selenodont and relatively low crowned to moderately hypsodont. The last premolars (P^4_4) are molariform and larger than the molars. P^4 and the anterior upper molars are quadrate (often longer than wide) and quadrituberculate, with a large hypocone and no cingula. Recent species eat mainly insects and other invertebrates, but some Neogene species were probably herbivorous.

Elephant shrews have gracile, elongate hind limbs, which enable them to run and bound at considerable speed (Rathbun, 1984). The hind limbs are substantially longer than the forelimbs (intermembral index, 62–75; Evans, 1942), and the tibia is much longer than the femur (crural index, or [tibia length/femur length] × 100, is 123–145; Evans, 1942). The femur is distally narrow, with an elevated patellar groove, and the tibia and fibula are fused for more than half their length. The foot (including the tarsal elements) is narrow and elongate; the first digits of both the manus and the pes are reduced or lost. The forelimbs are much shorter and less cursorially adapted and are often used for digging.

The unusually specialized teeth and skeletons of living elephant shrews have provided little insight to their broader relationships. Several authors have suggested that elephant shrews are related to Glires (McKenna, 1975a; Novacek and Wyss, 1986; Shoshani and McKenna, 1998; Meng and Wyss, 2001), but few of the supposed synapomorphies shared with Glires are known in the Eocene-Oligocene members. Except for a general resemblance between the postcranial skeletons of elephant shrews and rabbits, which is almost surely convergent, morphology does not provide very compelling support for a link between Macroscelidea and Glires.

Prior to their current rather tenuous attribution to the Anagalida, macroscelidids were considered peripheral members of a broadly construed Insectivora, or of an ordinal-level taxon Menotyphla (also containing tree shrews), which is now considered to be an unnatural group. Recent molecular studies point to very different relationships; namely, that macroscelidids are part of the endemic African order Afrotheria, which also includes tethytheres, hyracoids, aardvarks, and the former lipotyphlan families Chrysochloridae and Tenrecidae (Stanhope et al., 1998; Madsen et al., 2001; Murphy et al., 2001). However, apart from the fact that some Miocene macroscelidid cheek teeth (Myohyracinae) became

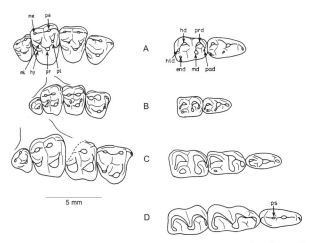

Fig. 15.6. Right P^4–M^3 (left column) and left P_4–M_1 or P_3–M_1 (right column) of macroscelideans and the condylarth *Haplomylus*: (A) *Haplomylus*; (B) *Chambius*; (C) *Herodotius*; (D) *Rhynchocyon*. Key: end, entoconid; hd, hypoconid; hld, hypoconulid; hy, hypocone; md, metaconid; me, metacone; ml, metaconule; pa, paracone; pad, paraconid; pl, paraconule; pr, protocone; prd, protoconid; ps, protostylid. (From Butler, 1995.)

hypsodont and converged rather closely on those of hyraxes, there is little anatomical evidence that supports a special relationship to Afrotheria.

Nor has the fossil record of elephant shrews been very helpful in elucidating the interrelationships of Macroscelidea —until quite recently. Knowledge of the early radiation of elephant shrews has been restricted to a few very fragmentary fossils from the late Eocene and early Oligocene of northern Africa. The oldest suspected macroscelidean, *Chambius* (Fig. 15.6B), was based on jaw fragments from the early Eocene of Tunisia. It was identified as an early macroscelidid because of its bunodont upper molars, reduced third molars, and submolariform fourth premolars (Hartenberger, 1986; Butler, 1995). Although these characters are suggestive of macroscelidean affinity, they are not necessarily diagnostic. In other ways the teeth of *Chambius* are primitive (e.g., retaining hypoconulids on lower molars, conules and cingula on uppers), and they closely resemble teeth of certain hyopsodontid condylarths. This resemblance suggests either that *Chambius* is actually a condylarth rather than a macroscelidid or that there is a close relationship between Hyopsodontidae and Macroscelidea.

The oldest definitive macroscelideans come from upper Eocene strata of northern Africa. *Herodotius* (Fig. 15.6C), based on late Eocene jaws from the Fayum of Egypt (Simons et al., 1991), was recognized as a macroscelidid on the basis of its long, shallow dentary; submolariform fourth premolars; quadrate and quadrituberculate upper molars (without conules but with buccal cingula); and reduced third molars. Its low-crowned lower molars are weakly selenodont. The macroscelidid subfamily Herodotiinae (considered by some researchers to include *Chambius*) is more primitive than subsequent macroscelidids in having less molarized fourth premolars and retaining third molars. A third herodotine, *Nementchatherium*, was recently described, based on isolated teeth from the later Eocene of Algeria (Tabuce, Coiffait, et al., 2001).

By the early Oligocene, more derived macroscelidids had evolved. *Metoldobotes,* also from the Fayum, had lost M_3, and P_4 was distinctly larger than M_1 but still submolariform (Patterson, 1965). Ankle bones found in the same Fayum quarries and probably referable to *Metoldobotes* show diagnostic macroscelidean anatomy. Several Miocene-Recent genera of elephant shrews are known from eastern and southern Africa, but their precise relationships to Early Tertiary forms is uncertain. A recently discovered skull of *Metoldobotes,* as yet undescribed, may help to clarify its relationships to other macroscelideans.

Several recent studies offer new fossil evidence bearing on macroscelidean relationships. Support for the macroscelidean affinities of *Chambius* has come from studies of the dental anatomy of hyopsodontid condylarths, such as the European louisinine *Microhyus* (Tabuce, Coiffait, et al., 2001) and North American *Haplomylus* (Simons et al., 1991; Butler, 1995; Fig. 15.6A), which are postulated to lie close to the source of Macroscelidea. Potentially more significant are limb skeletons of the diminutive North American late Paleocene–early Eocene hyopsodontids *Haplomylus* and *Apheliscus* (Fig. 15.7), which share specialized cursorial-saltatorial features with macroscelideans (see Chapter 12). Particularly notable are the slender and elongate distally fused tibia-fibula, and the cotylar fossa (for the tibial malleolus) on the astragalus. Additional modifications of the distal humerus, distal femur, and tarsus enhanced speed and stride length and helped to restrict motion to a parasagittal plane. Close resemblances in these details to the anatomy of elephant shrews constitutes further evidence that Macroscelidea evolved from Holarctic hyopsodontids (Zack et al., 2005). This hypothesis conflicts with the molecular view that Afrotheria arose in Africa.

GLIRES

The name Glires, originally proposed by Linnaeus in 1758, was long used to unite Lagomorpha, Rodentia, and their close relatives (e.g., Gregory, 1910; Simpson, 1945). The concept of Glires fell into disfavor in the middle of the twen-

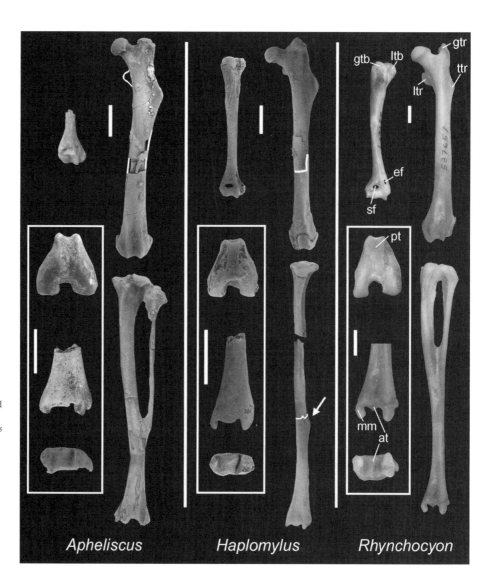

Fig. 15.7. Right humerus, left femur, and left tibia-fibula of the early Eocene hyopsodontids *Apheliscus* and *Haplomylus* compared with those of the extant elephant shrew *Rhynchocyon*. Arrow indicates level of fibular fusion. Scale bars = 5 mm. Key: at, anterior tubercle; ef, entepicondylar foramen; gtb, greater tuberosity; gtr, greater trochanter; ltb, lesser tuberosity; ltr, lesser trochanter; mm, medial malleolus; pt, patellar trochlea; sf, supratrochlear foramen; ttr, third trochanter. (From Zack et al., 2005.)

tieth century, but new fossils from the Paleocene of Asia, as well as studies of dental development and fetal membranes, have revived the notion that rodents and lagomorphs are closely related (e.g., Luckett and Hartenberger, 1985). Monophyly of Glires is strongly supported in a recent comprehensive phylogenetic analysis based on more than 200 morphological characters (Meng et al., 2003). Most recent molecular studies also support a monophyletic Glires (e.g., Madsen et al., 2001; Murphy et al., 2001; Huchon et al., 2002; but see Misawa and Janke, 2003, for a contrary view).

Although living rodents and lagomorphs are easy to distinguish, the distinction becomes increasingly blurred in older and more primitive fossil Glires. Moreover, these fossils, rather than clarifying the precise origin of each order, cloud the issue, because the early glires clades possess combinations of features that make their exact relationships controversial (see Meng and Wyss, 1994). The name Glires is used here to refer to rodents and lagomorphs, and their closest extinct relatives, which are assigned to the families Eurymylidae and Mimotonidae, respectively.

The most obvious derived character shared by all Glires is the presence of a pair of enlarged, laterally compressed, ever-growing incisors, the lower one extending back in the dentary to below M_3. The enamel is restricted to the anterior (labial) surface, as a mechanism to maintain a chisel-like cutting edge. Chisel-like, ever-growing incisors with limited enamel are not unique to Glires, however. They occur in such diverse groups as multituberculates, tillodonts, toxodontid notoungulates, hyracoids, primates (*Daubentonia*), and argyrolagid marsupials, and in some of these they resemble rodent incisors quite closely. Unlike in those groups, however, the enlarged incisors in Glires have been shown to be deciduous I^2_2, with I^1_1 and permanent I^2_2 having been lost (Luckett, 1985). Glires are also characterized by the presence of a conspicuous diastema behind the incisors, created by the loss of the canines and anterior premolars (P^1 and P_{1-2}), and elongate incisive foramina in the palate. Unilateral hypsodonty of the upper cheek teeth (higher lingually) is also generally characteristic, except in some basal rodents.

The oldest and most primitive Glires are assigned to two families, Eurymylidae (order Mixodontia) and Mimotonidae (order Mimotonida), from the Paleocene and Eocene of Asia. Some authors (e.g., Dashzeveg and Russell, 1988; Averianov, 1994) consider these two families to be closely related and assign both to the order Mixodontia (which was originally named for Eurymylidae alone; Sych, 1971), but such a grouping is almost surely a paraphyletic grade, based on shared primitive features (Meng, Wyss, et al., 1994). The original, more restrictive concept of Mixodontia is adopted here. Using shared derived features, existing evidence generally supports grouping the Eurymylidae with rodents in the Simplicidentata, and the Mimotonidae with lagomorphs in the Duplicidentata.

Enamel microstructure has played a significant role in the assessment of relationships among Glires, especially rodents (see discussion of tooth enamel in the section on rodents). The microstructure of the enamel of the ever-growing incisors has been extensively studied, and it is now well established that rodent incisors are almost always characterized by two layers of enamel, whereas lagomorphs generally have a single layer. The inner layer of rodent enamel, as well as the single layer of lagomorph enamel, consists of decussating prisms that form Hunter-Schreger Bands (HSB), a pattern believed to be an adaptation for strengthening enamel and preventing cracks (Fortelius, 1985; Koenigswald and Clemens, 1992). Recent studies of the incisor enamel of basal Glires (eurymylids and mimotonids) show that there is variation in the microstructure, some members of each family having single-layered enamel and others having double-layered enamel (e.g., Martin, 1999c). Nevertheless, in nearly all cases the single layer, or the inner of two layers, consists of thick HSB, whereas the outer layer is radial enamel. The double-layered pattern, so typical of rodents, thus seems to have arisen very early among Glires.

Duplicidentata

The mirorder Duplicidentata unites the lagomorphs and the Early Tertiary mimotonids, both of which are characterized by the possession of two pairs of upper incisors, an enlarged gliriform anterior pair and a second, smaller pair behind those. Paleocene *Mimotona* (Fig. 15.8) is the oldest mimotonid. The recent phylogenetic analysis by Meng et al. (2003) strongly supports the monophyly of this mirorder.

Lagomorpha

Lagomorpha is a relatively homogeneous group of small mammalian herbivores, comprising about 80 living species in two families, Leporidae (rabbits and hares) and Ochotonidae (pikas). They are widely distributed, occurring naturally on all continents except Australia and Antarctica. Lagomorphs, particularly leporids, are characterized by peculiar lacelike fenestrations in several skull bones, especially the rostral part of the maxillae and the basicranium. They also have large incisive fossae. The auditory bullae, composed of the ectotympanic, are relatively large and inflated and surround a bony external auditory tube. The orbits are large in rabbits but smaller in pikas.

The dental formula of extant duplicidentates is 2.0.3.2–3/1.0.2.3. As suggested by the group name, there are two upper incisors on each side, whose morphology and arrangement distinguish them instantly from all other mammals. The large anterior incisor (dI^2) has a distinct longitudinal groove on its anterior surface, causing the occlusal edge to be notched. In contrast to simplicidentates, there is a second, small, peglike upper incisor (I^3) positioned immediately behind the enlarged incisor, and P^2 is retained. In extant forms all the teeth are hypselodont (i.e., hypsodont and ever-growing) and adapted for feeding on grass and other vegetation. The cheek teeth have simple crown patterns, consisting of transverse crests and basins. Characteristic of the upper molars of lagomorphs is a lingual groove, or hypostria, between the hypocone and the pericone (an anterolingual

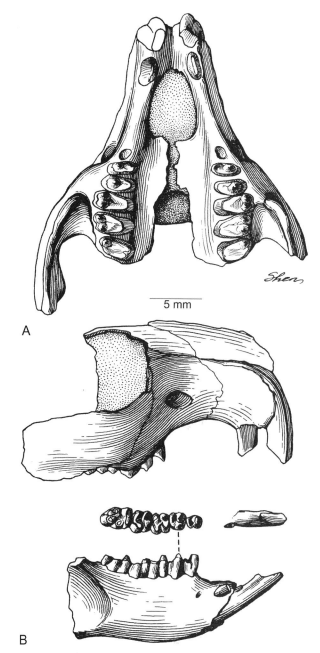

Fig. 15.8. Paleocene *Mimotona*: (A) snout; (B) right lower dentition. (From Li and Ting, 1993.)

cusp), which often extends buccally into the tooth. Lagomorphs have well-developed masseter and pterygoid muscles (and correspondingly, a large angular region of the dentary) and a reduced temporalis (and coronoid process), and the temporomandibular joint is high above the toothrow—all characteristics of herbivores. Their chewing cycle includes a significant transverse component, in contrast to the propalinal chewing characteristic of rodents. In many respects, lagomorphs—especially leporids—are functional equivalents of miniature ungulates, and they may well have been prevented from achieving larger body sizes by competition with small artiodactyls.

Rabbits are terrestrial and adapted for running and hopping. The forelimbs are shorter than the hind limbs, and the individual elements are slender and elongate. The distal femur is deep and narrow. The fibula is gracile and distally fused to the tibia for more than half its length. The elbow and ankle joints are mediolaterally compressed and constructed to limit motion to flexion and extension in a parasagittal plane. The forefoot has five digits and the hind foot four, but both are functionally four-toed. The tarsus and metatarsals are elongate. All lagomorphs have a short tail. Pikas are generally less than half the length of rabbits and an order of magnitude smaller in body mass (Nowak, 1999). Although also cursorially adapted, they are much less specialized for running. Their limbs are shorter overall, but the forelimbs are relatively longer than in leporids. They retain a strong clavicle, which is vestigial in rabbits. The tarsals and metatarsals are not elongate, in contrast to leporids. Many lagomorphs are also good burrowers.

The earliest fossil lagomorphs come from the Eocene. *Valerilagus*, based on a few teeth from the ?early or middle Eocene of Kyrgyzstan (Shevyreva, 1995; Averianov and Godinot, 1998), has been considered to be the oldest known lagomorph, but better specimens are needed to confirm its identity. Undoubted lagomorphs appear slightly later in the middle Eocene of Asia (*Gobiolagus, Lushilagus, Shamolagus,* and *Strenulagus;* Tong, 1997) and North America (*Mytonolagus* and *Procaprolagus*). These animals are usually included in the Leporidae, but some might represent plesiomorphic lineages that predate the leporid-ochotonid dichotomy (Dawson, 1967). Most of them are more primitive than later lagomorphs in details of crown morphology and in having rooted cheek teeth and less reduced third molars; nevertheless, the general dental resemblance to present-day lagomorphs is striking. The diversity and antiquity of early lagomorphs and their sister group, Mimotonidae, in Asia strongly suggests an Asian origin of Lagomorpha.

By the late Eocene and early Oligocene, well preserved remains of several leporid taxa are known, especially from Chadronian and Orellan deposits of western North America (Wood, 1940; Dawson, 1958; Gawne, 1978). *Chadrolagus* and *Palaeolagus* (Fig. 15.9) had hypselodont cheek teeth and fenestrated rostra as in present-day leporids. Skeletal remains of *Palaeolagus* and *Megalagus* are more similar to those of extant leporids than to those of ochotonids, although individual elements are generally more robust than in leporids. The forelimb, in particular, was less cursorially specialized. They must have been subcursorial animals capable of hopping.

Desmatolagus, first known from the late Eocene of Asia (Meng and Hu, 2004), is usually considered to be the oldest ochotonid (e.g., McKenna and Bell, 1997). Later species from North America and Europe have been referred to the genus, but they may instead be leporids. Like later ochotonids, *Desmatolagus* has relatively low-crowned cheek teeth (higher crowned in some species), a nonmolariform P^3, and a variably present premolar foramen on the palate lingual to P^4. But it lacks certain dental features, such as a well-developed hypostria, found in later lagomorphs. For this reason, McKenna (1982) considered *Desmatolagus* to be the sister taxon of ochotonids + leporids. Ochotonids became

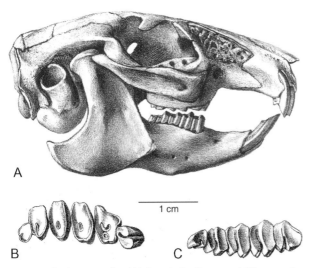

Fig. 15.9. *Palaeolagus*, an early rabbit from the late Eocene and Oligocene of North America (anterior to the right): (A) skull; (B) right upper dentition; (C) right lower dentition. (From Wood, 1940).

diverse during the Miocene but declined thereafter, and only the genus *Ochotona* survives today.

Mimotonidae

For many years the relationships of lagomorphs were obscure, and their origins were sought in a diverse array of primitive mammals, including periptychid condylarths, zalambdodont insectivores, anagalids, and even the pantodont *Coryphodon* (Wood, 1957; Russell, 1958; Van Valen, 1964). None of these suggestions was very convincing, however. More recently, Early Tertiary mimotonids of Asia have been found to show much more detailed similarity to lagomorphs. Like lagomorphs, mimotonids have a notched upper front incisor (dI^2) and a large incisive fossa near the front of the palate, both believed to be derived traits (Li and Ting, 1993). The reduced dental formula of mimotonids (2.0.3.3/2.0.2–3.3) is plesiomorphic for Glires in retaining two incisors in both upper and lower jaws. But like lagomorphs, mimotonids have a second pair of smaller upper incisors (I^3) *behind* the enlarged pair (dI^2), rather than lateral to it as in most other mammals (Fig. 15.8). The cheek teeth of mimotonids are unilaterally hypsodont and rooted, and the crowns are more primitive than those of lagomorphs, more closely resembling basal rodents in this regard. Mimotonids are also more primitive than lagomorphs in lacking the complex P$_3$ that is diagnostic of Lagomorpha.

Mimotonid incisor enamel may be either single-layered or double-layered, with an inner portion composed of HSB and an outer layer of radial enamel (Martin, 1999c, 2004). Leporids have single-layered enamel consisting of HSB, a pattern which is now thought to have been derived from the double-layered condition by the loss of the external layer of radial enamel.

The postcranial skeleton of mimotonids was adapted for running or jumping, as in leporids. It is best known in *Gomphos*, recently reported from the Paleocene/Eocene boundary (probably earliest Eocene) of Mongolia (Meng et al., 2004; Asher et al., 2005). It resembles leporids in having slender, elongate hind limbs and similarly specialized, relatively long and narrow ankle bones (calcaneus and astragalus). Unlike lagomorphs, however, the fibula is separate from the tibia, and there is no calcaneal canal, a feature found in fossil lagomorphs and in extant ochotonids (Bleefeld and Bock, 2002). Similar ankle bones are known in other mimotonids as well (Averianov, 1991). *Mimolagus*, formerly considered to be a basal lagomorph of probable Oligocene age, based on its ankle structure (Bleefeld and McKenna, 1985; Dashzeveg and Russell, 1988), is now excluded from Lagomorpha and regarded as a mimotonid because it lacks calcaneofibular contact (Szalay, 1985; Averianov, 1994). *Mimolagus* further differs from both *Mimotona* and lagomorphs in lacking the characteristic notch in the large upper incisor (dI^2).

Simplicidentata

The sister group of duplicidentates, simplicidentates are more derived in retaining only a single enlarged, evergrowing incisor in each quadrant (dI$^2_2$). Simplicidentata includes the orders Mixodontia (consisting of only the late Paleocene–middle Eocene family Eurymylidae) and Rodentia. Recent phylogenetic analysis weakly supports monophyly of Simplicidentata (Meng et al., 2003).

Sinomylus (Fig. 15.10), based on a snout from the late Paleocene of China, may be the most primitive known simplicidentate and the sister taxon of Mixodontia + Rodentia (McKenna and Meng, 2001). The molars are eurymylid-like and, as in many other anagalidans, lingually hypsodont. *Sinomylus* is very similar to the eurymylid *Heomys*, but its molars differ in having a smaller hypocone and smaller conules, which results in a somewhat more lophodont condition. Like other simplicidentates, *Sinomylus* retains only a single incisor (dI^2) on each side, with enamel restricted to the anterior surface. There is a long diastema between the incisor and the cheek teeth. *Sinomylus* is more primitive than other simplicidentates, including eurymylids, however, in having single-layered radial incisor enamel without HSB (Martin, 1999c) and in retaining small P^2 and ?P^1.

Eurymylidae

Eurymylids were long considered lagomorph relatives, based on superficial cranial and dental resemblances, but more recently they were shown to share derived features of the tibia and ankle with primitive rodents such as *Paramys* (Li and Ting, 1993). They are now generally believed to be the sister group of rodents (e.g., Li et al, 1987; Meng and Wyss, 2001; Meng et al., 2003). An important synapomorphy shared with rodents is the presence of a distinct posterior process on the distal tibia, which articulates with the back of the astragalar trochlea. Moreover, they are more specialized than duplicidentates and resemble rodents in having lost both I$^3_3$ and P^2 (dental formula 1.0.2.3/1.0.2.3) (Averianov,

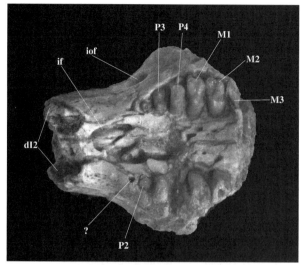

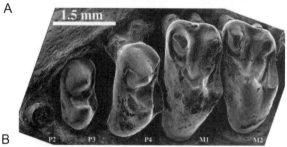

Fig. 15.10. *Sinomylus*, a basal simplicidentate from the late Paleocene of China: (A) skull; (B) upper teeth. Key: if, incisive foramen; iof, infraorbital foramen. (From McKenna and Meng, 2001; courtesy of M.C. McKenna.)

1994; Meng and Wyss, 2001; Meng et al., 2003). Eurymylid molars are possibly apomorphic in being slightly more lophodont and having larger hypocones than the most primitive rodents (alagomyids). Early Eocene *Rhombomylus* (Fig. 15.11) is the best-known eurymylid, being represented by many skulls of different ontogenetic stages, as well as much of the skeleton (Meng et al., 2003). Postcranial anatomy indicates that *Rhombomylus* and its close relative *Matutinia* (Ting et al., 2002) were generalized, primarily terrestrial creatures similar to the earliest known rodents.

Eurymylids, including *Rhombomylus, Matutinia, Eurymylus,* and *Heomys*, may have either single- or double-layered enamel, of which the outer layer may be very thin (Martin, 1999c, 2004; Meng and Wyss, 1994; Meng et al., 2003). The inner layer usually consists of thick HSB that are two to four prisms wide, a condition called "pauciserial," whereas the outer layer is radial enamel. This double-layered enamel is very similar to that of basal rodents. According to Martin (2004), the enamel of *Eurymylus* is more primitive than that of the others in consisting of a single layer of radial enamel, as in *Sinomylus*. The condition in *Heomys* is uncertain (Meng and Wyss, 1994). Some cranial features are intermediate between those of primitive eutherians and those of early rodents. *Heomys* was evidently close to the base of Rodentia, but opinions vary on whether to include it in Rodentia or Eurymylidae. Like eurymylids, but unlike rodents, it retains P_3 and an unreduced P^3.

Early Eocene *Decipomys* from Mongolia appears to be related to eurymylids but is more derived in having a lower dental formula of 1.0.1.3, as in rodents (Dashzeveg et al., 1998). However, it differs from both eurymylids and rodents in having double-layered enamel that lacks HSB. The outer layer is radial enamel, whereas the inner layer is tangential enamel, an apomorphic condition. *Decipomys* provides further evidence of the diversity of early Glires.

Rodentia

Rodents are the most successful and diversified living mammals, making up more than one-third of all mammalian species. They are nearly worldwide in distribution. All rodents are characterized by a reduced dental formula, which is primitively 1.0.2.3/1.0.1.3; many forms lost the premolars as well. They are perhaps most readily recognized by the distinctive single pair of enlarged, ever-growing incisors in the upper and lower jaws, with restricted and often pigmented enamel; but as noted in the preceding sections, this trait is not limited to rodents. A conspicuous diastema separates the incisors from the cheek teeth, which in most lines show at least some development of lophodonty. Rodents have long, shallow mandibular fossae on the base of the skull that allow the dentary to move considerably forward and backward. It moves forward during molar occlusion (mastication) and incisal occlusion (gnawing), and retracts so the molars can come together to begin the power stroke of chewing. In primitive rodents mastication still had a significant transverse component, as in many other placentals; in more derived rodents the tooth crowns had transverse lophs that quickly wore flat, and lower jaw movement was propalinal (anterior) during the power stroke (Butler, 1985).

The cheek teeth of primitive rodents exhibit a basic tribosphenic pattern, whereas derived forms may diverge considerably from that pattern, making cusp homology very difficult to determine. Even some early rodents develop crests and cusps atypical in other mammals, which has led to a dental terminology different from that in other mammals. For instance, an anteroconid may form de novo to occupy the position of the lost paraconid, a mesoconid often develops on the ectolophid (the cristid obliqua of other mammals), and cross-lophs may enter the talonid from the entoconid (a hypolophid) and the mesoconid (a mesolophid).

Most lines of evidence—morphological, paleontological, and molecular—support the monophyly of Rodentia (Luckett and Hartenberger, 1993; Meng and Wyss, 2001; Meng et al., 2003). Molecular data a few years ago suggested that caviomorphs—specifically, guinea pigs—are not true rodents, which would require that their similarities to other rodents be convergent (Graur et al., 1991; D'Erchia et al., 1996). But reexamination of the molecular evidence has reaffirmed the monophyly of rodents (e.g., Cao et al., 1994; Frye and Hedges, 1995; Murphy et al., 2001; Huchon et al., 2002).

Molecular studies that have considered divergence times have generally suggested that rodents diverged from other mammals well back in the Cretaceous, more than 100 mil-

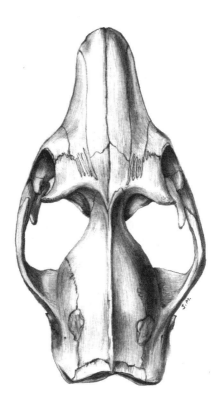
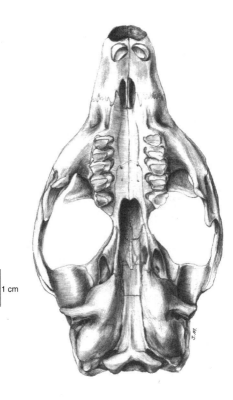

Fig. 15.11. Skull of *Rhombomylus*, an early Eocene eurymylid. (From Meng et al., 2003.)

lion years ago (Frye and Hedges, 1995; Kumar and Hedges, 1998), in sharp conflict with the fossil record. Both the oldest known rodents and more primitive Glires date from the Paleocene, more than 40 million years after the supposed diversification of rodent suborders, according to those molecular accounts. All currently available morphological and fossil evidence therefore points to the origin of rodents after the Cretaceous/Tertiary boundary (Meng at al., 2003). A significant recent study of three nuclear genes concluded that rodents and lagomorphs diverged near the Cretaceous/Tertiary boundary and that rodents initially radiated near the Paleocene/Eocene boundary (Huchon et al., 2002), thus bringing the molecular and paleontological interpretations into close agreement.

Based primarily on craniodental morphology, alagomyids, ischyromyids, and Eocene ctenodactyloids are widely considered to be the most primitive known rodents (Wood, 1962; Dawson et al., 1984; Li et al., 1989; Dashzeveg, 1990; Meng, Wyss, et al., 1994; B.-Y. Wang, 1994; Dawson and Beard, 1996; McKenna and Bell, 1997), although which group is the most primitive is still debatable. Using a crown-group definition of rodents, Meng and Wyss (2001; see also Wyss and Meng, 1996) recently excluded alagomyids from Rodentia and also suggested that ischyromyids and cocomyid ctenodactyloids may not be members of the crown clade, instead employing the broader taxon Rodentiaformes (equivalent to what has widely been called Rodentia) to encompass the outgroups + crown-group Rodentia. However, there is a broad consensus that ischyromyids are part of the clade (Sciuromorpha) that includes the extant families Sciuridae, Aplodontidae, and Castoridae (e.g., Simpson, 1945; Hartenberger, 1985; Korth, 1994; McKenna and Bell, 1997; Hartenberger, 1998), thus securing their allocation to crown-group Rodentia. Moreover, ischyromyids, ctenodactyloids, and alagomyids share derived features with rodents that are not found in eurymylids (Dawson and Beard, 1996). Consequently they are regarded here to be the oldest and most primitive known rodents.

Rodents are an extremely complicated clade because of their extraordinary taxonomic diversity, far exceeding that of any other order, which has resulted in more than the usual amount of within-group homoplasy during their evolution. Deciding which anatomical features are phylogenetically significant (synapomorphous) and which evolved convergently or in parallel has been an arduous task. This complexity has made the unraveling of relationships more challenging than in most other orders and has resulted in a plethora of rodent classifications and little overall consensus. Partly responsible for the confusion is the likelihood that multiple higher-level clades diverged from a common ancestor very soon after rodents originated (Hartenberger, 1998).

Among the features that have played important roles in rodent classification are the arrangement of jaw muscles and resulting differences in cranial and mandibular morphology, as well as enamel microstructure. Although it is widely acknowledged that the different morphologies are likely to have originated multiple times convergently, most classifications still make use of these distinctions. For example, Carleton (1984), Wilson and Reeder (1993), and Vaughan et al. (2000) all divide rodents into two suborders, Sciurognathi and Hystricognathi. McKenna and Bell (1997) used five suborders, differing essentially by subdividing the sciurognaths into four suborders. The most recent edition of Wilson and Reeder's (2005) authoritative list employs a somewhat different arrangement of rodents in five suborders. Hartenberger (1998) proposed a new arrangement of six suborders that contrasts in some significant ways with previous classifications—for example, by breaking up the Myomorpha. Recently Marivaux et al. (2004) presented a phylogenetic analysis based on dental anatomy of Early Tertiary rodents, from which they concluded that there are two basic clades of rodents: a ctenodactyloid + hystricognath clade and a clade of ischyromyoids and their relatives, which includes forms grouped in this chapter as sciuromorphs as well as myomorphs and anomaluroids.

Zygomasseteric Anatomy in Rodents. The most prevalent trend in rodents has been modification of the masseter muscle to strengthen propalinal jaw movement (Fig. 15.12). The masseter of rodents has three parts, superficial, lateral, and medial (deep), which insert on the angle and masseteric fossa of the mandible (Wood, 1965). The attachment of the masseter muscle to the skull varies and is still considered phylogenetically significant, although multiple origins of derived states are probable.

Four basic arrangements are recognized. In the most primitive rodents, such as ischyromyids, the superficial masseter arises from the front of the zygomatic arch, the lateral masseter from the ventral border of the arch, and the medial masseter from the medial side of the arch, a condition referred to as *protrogomorphous*. This configuration is generally accepted as the plesiomorphic condition for rodents. In squirrels, beavers, and some other rodents, part of the lateral masseter originates on the front of the zygomatic arch, often extending onto the snout dorsal to the infraorbital foramen, and passes in a groove under the zygomatic to the angle of the mandible. This arrangement is called "sciuromorphous." In *hystricomorphous* rodents, such as porcupines and agoutis, the medial masseter is expanded and passes through the greatly enlarged infraorbital foramen to attach to the snout. *Myomorphous* rodents, such as mice and rats, combine aspects of the other two derived states, with an extended lateral masseter anterior to the zygomatic arch and part of the medial masseter passing through the infraorbital foramen (which is less enlarged than in hystricomorphs).

Two principal mandibular morphologies have arisen in rodents (Fig. 15.13). In the more primitive condition, *sciurognathy*, which occurs in squirrels, ischyromyids, and many other rodents, the angle of the mandible arises in the same vertical plane as the lower incisor. *Hystricognathy* refers to mandibles in which the angle arises distinctly lateral to the vertical plane of the incisor. This mandibular anatomy is typical of porcupines, mole-rats, and South American rodents.

Tooth Enamel of Rodents. Several characteristics of enamel microstructure have been widely used in assessing interrelationships of rodents and other Glires. However, there is still considerable difference of opinion concerning the significance of particular enamel characteristics in early Glires (e.g., polarity of double- vs. single-layered enamel, presence/absence of HSB, enamel types such as pauciserial). Like other aspects of anatomy, enamel microstructure among Glires was not immune to homoplasy (Meng and Wyss, 1994).

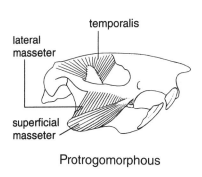

Protrogomorphous

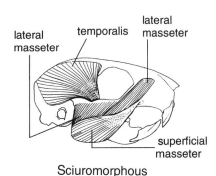

Sciuromorphous

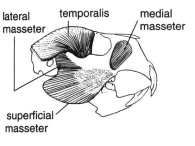

Hystricomorphous

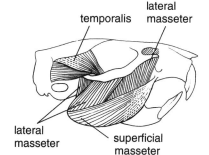

Myomorphous

Fig. 15.12. Zygomasseteric arrangements in rodents. (Modified from Vaughan et al., 2000.)

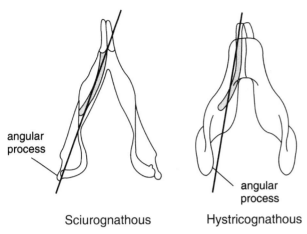

Fig. 15.13. Mandibular morphology in rodents. (Modified from Vaughan et al., 2000.)

Enamel is composed of prisms, each consisting of bundles of crystallites of hydroxyapatite. The prisms are deposited (by specialized cells called ameloblasts) in various arrangements that are characteristic of particular taxa. Rodent incisors, for example, have double-layered enamel. The inner layer consists of decussating prisms arranged in layers (HSB); however, in the outer layer the prisms abruptly change orientation, and all are parallel and run toward the occlusal surface of the incisor. This arrangement of prisms, which lacks HSB, is called radial enamel. In thin section, the change in orientation gives the appearance of two distinct layers, although individual prisms are actually continuous from the enamel/dentine junction to the occlusal surface. The width of the HSB (number of prisms per band) has been used to distinguish three types of enamel in the inner layer: *uniserial* (1 prism wide), *pauciserial* (usually 2–4 prisms wide), and *multiserial* (3–10 prisms wide; see, e.g., Wahlert, 1968, 1989; Koenigswald, 1995; T. Martin, 1993, 1997a; Fig. 15.14). There are additional differences among these types of enamel, for example concerning the interprismatic matrix (Martin, 1994).

A double-layered arrangement of enamel prisms is typical of most rodents, whereas other Glires, including many early ones, have a single layer (Flynn, 1994; Martin, 1999c). This double-layered pattern has often been considered to be an important synapomorphy of Rodentia, and could by itself support inclusion of both *Tribosphenomys* and *Heomys* (Fig. 15.15) in Rodentia (Flynn, 1994), but the composition of the layers is also important. Rodent incisor enamel nearly always has HSB in the inner layer and radial enamel in the outer layer (Martin, 1999c). Meng and Wyss (1994) observed double-layered enamel in *Tribosphenomys*, but were unable to confirm the presence of HSB.

As a further complication, it is now known that some eurymylids and mimotonids (basal Glires) have double-layered, rodentlike enamel with an inner layer of HSB and an outer one of radial enamel (Martin, 1999c, 2004). This finding suggests that the double-layered, rodentlike structure could be primitive for most Glires rather than a synapomorphy of Rodentia. Based on distribution among early Glires, pauciserial enamel appears to be the morphotypic condition for Rodentia (Sahni, 1985; Koenigswald and Clemens, 1992; T. Martin, 1993, 1997a; Meng, Wyss, et al., 1994). Koenigswald (1995) reasoned that the presence of single-layered enamel, as in rabbits (Leporidae), is not necessarily primitive, for the single layer consists entirely of derived enamel with HSB. This observation appears to be corroborated by Martin's (2004) recent report of double-layered enamel in an early Eocene lagomorph, which suggests that the single-layered condition arose through loss of the outer layer. The incisor enamel of pikas (Ochotonidae) has two or even three layers.

Oldest Known Rodents. The oldest known rodents come from the late Paleocene (Clarkforkian and Gashatan) of North America and Asia and belong to the families Alagomyidae (*Alagomys* and *Tribosphenomys*) and Ischyromyidae (*Acritoparamys* and *Paramys*; Wood, 1962; Ivy, 1990; Meng, Wyss, et al., 1994; Dawson and Beard, 1996). Alagomyids are known from both continents in the late Paleocene, whereas ischyromyids were then present only in North America but dispersed to Asia and Europe at the beginning of the Eocene. Like living rodents, members of both families have a single, enlarged, ever-growing incisor in each quadrant (assumed to be dI^2 and dI_2), and are derived compared to eurymylids and *Sinomylus* in having lost P_3 and greatly reduced P^3 (dental formula 1.0.2.3/1.0.1.3), increasing the wide diastema between the incisors and cheek teeth. Dawson and Beard (1996) observed that many of the oldest members of these two families are found in coals and carbonaceous sediments, suggesting that they may have preferred swampy habitats.

Alagomyids (*Alagomys* and *Tribosphenomys*; Fig. 15.15B,C) are generally considered to be the most primitive known rodents. They are known chiefly from teeth and jaws. Their lower cheek teeth are rhomboidal in occlusal view and generally resemble those of squirrels. P^4 and P_4 are molariform (although they could be deciduous, at least in *Tribosphenomys*), and the trigonids of P_4–M_3 are anteroposteriorly short, with large metaconids and generally no paraconid, another feature typical of rodent teeth. *Tribosphenomys* is more primitive than other rodents in retaining a small paraconid on the first lower molar, a distinct hypoconulid lobe on M_3, and larger buccal shelves on the upper molars. Alagomyids are very small and are considered more primitive than other rodents in several additional dental traits, such as lacking an anterior cingulum and a hypocone, and having stronger cusps (including well-developed conules) and weaker crests on the upper cheek teeth (Meng and Wyss, 2001). Compared to alagomyids, however, both *Sinomylus* and eurymylids are more lophodont and have larger hypocones, which are resemblances to paramyids. These features suggest that the more cuspate upper cheek teeth and absence of the hypocone in alagomyids could be apomorphic traits.

Enamel microstructure of alagomyids has also been considered primitive but is open to various interpretations. The incisor enamel of *Tribosphenomys* has two layers but is

Fig. 15.14. Enamel types in rodents based on width of Hunter-Schreger bands: (A) pauciserial (2–4 prisms wide): the Eocene ischyromyid *Ailuravus;* (B) uniserial (one prism wide): the Eocene muroid *Pappocricetodon;* (C) multiserial (3–10 prisms wide): the extant capybara *Hydrochaeris*. All sections are longitudinal, with the enamel-dentine junction to the left and the outer enamel surface to the right. (Scanning electron micrographs courtesy of D. Kalthoff.)

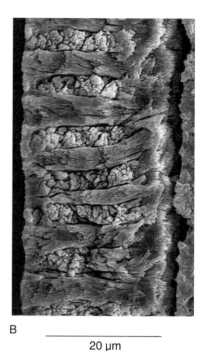

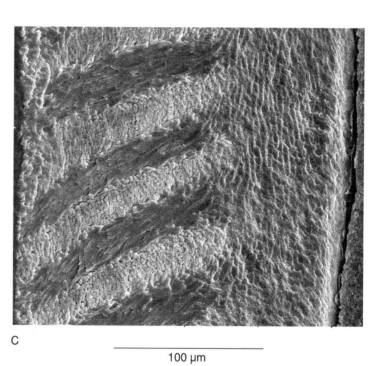

entirely radial in structure, lacking HSB. *Alagomys* has pauciserial HSB, but the Asian species of the genus has only single-layered enamel, whereas the American species has the typical rodent pattern of two layers, the outer one composed of radial enamel (Meng and Wyss, 1994; Dawson and Beard, 1996). Although radial enamel is generally regarded as the primitive state, the absence of HSB in *Tribosphenomys* could be an autapomorphy, especially considering the presence of HSB in basal Glires, *Alagomys,* and paramyids.

Fragmentary postcrania attributed to *Tribosphenomys* are very primitive. The astragalar trochlea and matching distal tibia are shallowly grooved (unlike paramyids). The tibia appears to lack a posterior process (a diagnostic rodent trait), although it may simply be damaged (Meng and Wyss, 2001).

Suborder Sciuromorpha. This group includes the living squirrels (Sciuridae), mountain beavers (Aplodontidae), true beavers (Castoridae), and their fossil relatives. They are in many ways the most conservative and primitive rodents except for alagomyids and include the most primitive extant members of the order. Relationships among sciuromorphs are shown in Figure 15.16.

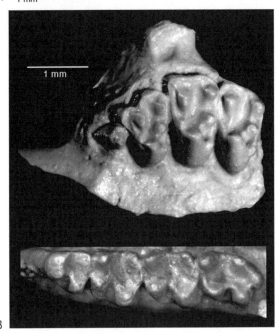

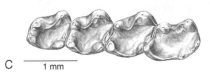

Fig. 15.15. (A) *Heomys*, a eurymylid close to the origin of rodents: left P³–M³; (B) basal rodent *Tribosphenomys*: left P³–M² and P$_4$–M$_3$; (C) basal rodent *Alagomys*: right P$_4$–M$_3$. (A from Dashzeveg and Russell, 1988; B courtesy of J. Meng; C from Tong and Dawson, 1995.)

Ischyromyidae. Ischyromyids are among the most primitive known rodents. They are squirrel-like in much of their dental and postcranial anatomy (Figs. 15.17A, 15.18A) and were long considered to be the stem group for all later rodents (Matthew, 1910a; Wood, 1962). Based on more recent discoveries, however, it is more likely that they are the paraphyletic stem group of sciuromorphs. Most of the roughly 30 genera have often been allocated to the family Paramyidae (e.g., Wood, 1962; Wahlert, 1974), but many current workers include this family within the Ischyromyidae, a convention adopted here for simplicity. Ischyromyids were the prevalent rodents of the late Paleocene and Eocene of North America and were also common in the Eocene of Europe and Asia. Clarkforkian species of *Acritoparamys*, *Paramys*, and *Microparamys* from Wyoming and Montana are among the oldest known rodents. Ischyromyids account for one-half to three-fourths of all rodent species in the early and middle Eocene of North America (Korth, 1994).

Ischyromyids are characterized by a relatively robust, protrogomorphous skull with a small infraorbital foramen and a median sagittal crest (see Fig. 15.21A). The auditory bullae were evidently unossified except in *Reithroparamys*, *Ischyromys*, and a few others. The mandible is sciurognathous. Most ischyromyids retain the primitive rodent dental formula (1.0.2.3/1.0.1.3) in which P³ is small and peglike, but some European forms lost P³. The inner layer of the incisor enamel is pauciserial in all forms except *Ischyromys*, in which it is uniserial. Pauciserial enamel is also found in eurymylids and is believed to be the primitive condition for rodents (T. Martin, 1993). The cheek teeth were low crowned and relatively generalized, for processing nuts, seeds, and fruit. The last premolars are molariform, and the upper molars are nearly quadrate and usually have well-developed anterior and posterior cingula, conules, stylar cusps, and hypocone (Fig. 15.19A). The lower cheek teeth are rectangular to rhomboidal in outline, and the talonid basins are wider than the trigonids and have peripheral cusps. The cusps are typically joined by weak crests (or lophs), and there is usually a distinct cristid obliqua (called the ectolophid in rodents). Some forms have accessory lophs or crenulated enamel. Common trends in the family include increasing size, simplification of the cheek teeth (or, less often, increasing complexity), and progressive lophodonty (Wood, 1962; Korth, 1994).

Excellent skeletal material of ischyromyids is known from the middle Eocene of the Bridger Formation of Wyoming and from Messel, Germany. Early Eocene remains are more fragmentary. The limb skeleton of early and middle Eocene North American ischyromyids, such as *Paramys* (Figs. 15.17A, 15.20), is very similar to that of extant squirrels, in several forelimb features being closer to arboreal species and in most hind limb traits more like terrestrial ones. This combination of features suggests that *Paramys* was well adapted for both tree climbing and ground locomotion (Wood, 1962; Rose and Chinnery, 2004).

The subfamily Ailuravinae includes *Ailuravus* (Fig. 15.18B) from the early and middle Eocene of Europe, a large, nearly marmot-sized rodent known from several skeletons from Messel. These remarkable fossils preserve outlines of the fur, showing a long, bushy tail, as well as gut contents, consisting mainly of leaves (Koenigswald et al., 1992b). The hind limbs were longer than the forelimbs, as in other ischyromyids, and the claws were sharp and similar to those of tree squirrels, indicating an arboreal lifestyle. The cheek teeth of *Ailuravus* had highly crenulated enamel (Wood, 1976). The oldest and most primitive ailuravine is early Eocene (Ypresian) *Euromys* (Escarguel, 1999). Ailuravines are also present in the Eocene of North America (*Mytonomys*, *Eohaplomys*). Based on certain primitive dental features, Hartenberger (1995) proposed transferring the Ailuravinae to the basal family Alagomyidae.

The early and middle Eocene ischyromyid *Reithroparamys* (Reithroparamyinae; Wood, 1962; Korth, 1994) has been

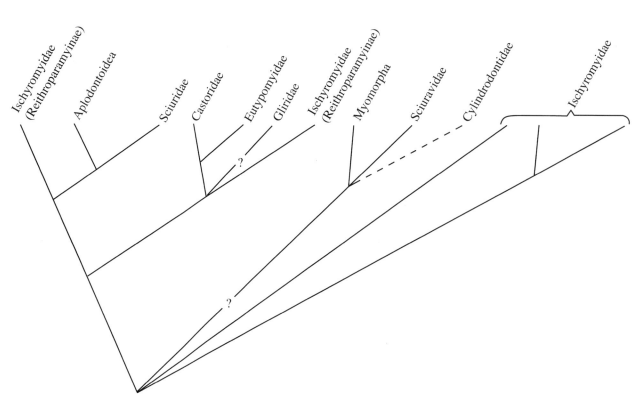

Fig. 15.16. Possible relationships among sciuromorphs, based mainly on Korth (1994) and McKenna and Bell (1997). Ischyromyidae are paraphyletic and appear in several parts of the cladogram. Dashed lines and question marks indicate uncertainties.

singled out as a taxon of potential significance to several groups of rodents. Besides ischyromyids, it has been associated with hystricognaths (Wood, 1975, 1985) and sciurids (as a separate family Reithroparamyidae; McKenna and Bell, 1997). Wood (e.g., 1975, 1985) included *Reithroparamys* and latest Paleocene–early Eocene *Franimys* in a new infraorder Franimorpha, based on their possession of "incipient hystricognathy," which he considered to be homologous with that of caviomorphs. Subsequent authors have found this proposal unconvincing, however. More recent studies emphasize shared derived features with both sciurids and aplodontids (Emry and Korth, 1989; Meng, 1990). *Reithroparamys* differs from most other ischyromyids in having a more gracile skeleton, ossified and septate bullae, and paired temporal crests rather than a median sagittal crest (Korth, 1994). Based on its relatively short forelimbs (intermembral index 57), long tibia (crural index greater than 100), and elongate metatarsals, Wood (1962) inferred that *Reithroparamys* was subricochetal to ricochetal. It was probably at least saltatorial, but living ricochetal mammals typically have more extreme limb proportions, with an intermembral index below 50 (see Chapter 2). Korth (1994) grouped the diminutive *Microparamys* and *Acritoparamys* in the same subfamily with *Reithroparamys* and considered them to be the most primitive ischyromyids. He considered reithroparamyines to be ancestral to Sciuridae, Aplodontoidea, and Castoroidea, as well as Gliridae.

Sciuravidae. This primitive family is known primarily from a half-dozen genera from the early and middle Eocene of North America. The family has been considered to be closely related to ischyromyids and was formerly included in the Protrogomorpha, a group now recognized to be linked mainly by plesiomorphic traits. Like ischyromyids, sciuravids have protrogomorphous skulls and a dental formula of 1.0.2.3/1.0.1.3, with brachydont, cuspate cheek teeth; a reduced P^3; and pauciserial incisor enamel, but these are all primitive resemblances (Korth, 1994). Sciuravids differ from ischyromyids in having incipiently lophodont molars, which became almost bilophodont in derived species. The uppers have a large hypocone, lophlike anterior and posterior cingula, reduced conules, and a metaloph that joins the hypocone rather than the protocone. The lower molars have a prominent mesoconid that tends to form a transverse mesolophid. The fourth premolars are submolariform and smaller than the first molars. It is not clear whether some of these differences—which are rather subtle—represent derived or more primitive states in sciuravids. The oldest sciuravid is *Knightomys*, first known from the early Wasatchian. The skull of early–middle Eocene *Sciuravus* is more gracile than that of ischyromyids and is similar in form to that of the basal ctenodactyloid *Cocomys* (Dawson, 1961). The auditory bullae are ossified but loosely attached to the skull. Common trends in the family, as in other early rodents, are toward increasing size, lophodonty, and hypsodonty.

Sciuravids are thought to have evolved from a primitive ischyromyid or possibly from a ctenodactyloid. They are considered to be the stem group from which myomorphs evolved (Korth, 1994; Wang and Dawson, 1994).

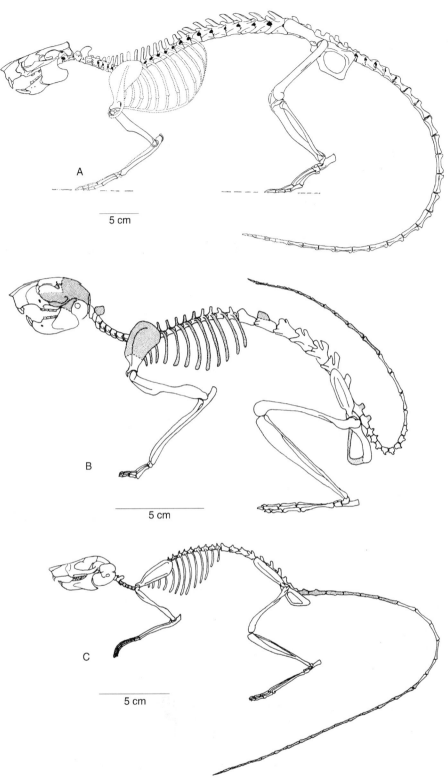

Fig. 15.17. Primitive rodent skeletons: (A) Eocene *Paramys;* (B) late Eocene *Douglassciurus,* a basal sciurid; (C) middle Eocene *Protoptychus.* (A from Wood, 1962; B, C from Korth, 1994.)

Cylindrodontidae. Cylindrodontids are a small family of primitive rodents best known from the Eocene of western North America. They have low, broad protrogomorphous skulls (known only in Chadronian members, such as *Ardynomys* and *Cylindrodon;* Fig. 15.21D), with a short, deep snout and dentary and large ossified bullae (Wood, 1937, 1974a). The incisor enamel is pauciserial in a few primitive forms (Wasatchian *Dawsonomys* and Bridgerian *Mysops*) and uniserial in all later cylindrodonts, including middle Eocene *Proardynomys* from Mongolia (Wahlert, 1968; Dashzeveg and Meng, 1998b). The dental formula was primitively 1.0.2.3/1.0.1.3, although P^3 is absent in some species of late Eocene *Cylindrodon* (Korth, 1994). The fourth premolar is submolariform, as in sciuravids and some ctenodactyloids. The

Fig. 15.18. Eocene rodents: (A) *Paramys*, an ischyromyid from North America and Eurasia; (B) *Ailuravus*, an ischyromyid from Messel, Germany; head–tail length = 1 m. (C) *Eogliravus*, a glirid from Messel; head–tail length = 12 cm. (A illustration by J. Matternes; B courtesy of Forschungsinstitut Senckenberg, Frankfurt; C courtesy of G. Storch.)

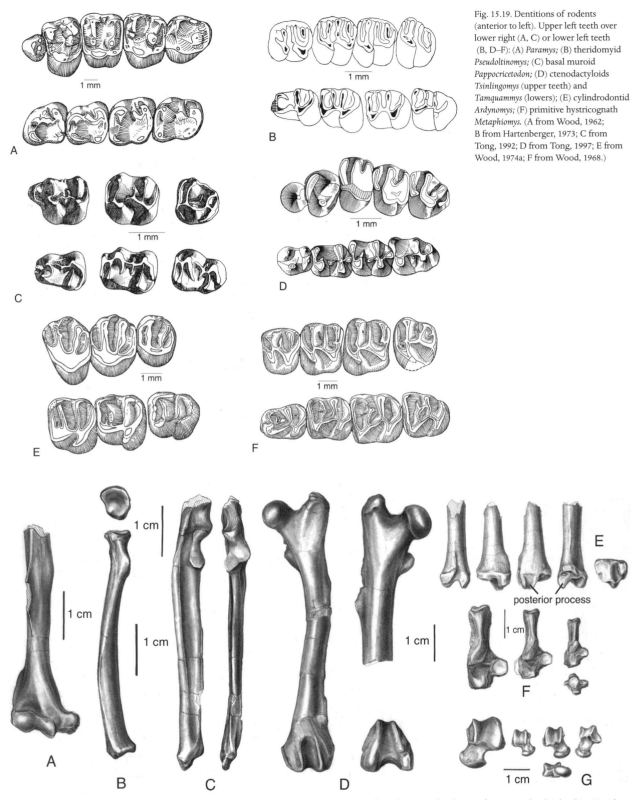

Fig. 15.19. Dentitions of rodents (anterior to left). Upper left teeth over lower right (A, C) or lower left teeth (B, D–F): (A) *Paramys*; (B) theridomyid *Pseudoltinomys*; (C) basal muroid *Pappocricetodon*; (D) ctenodactyloids *Tsinlingomys* (upper teeth) and *Tamquammys* (lowers); (E) cylindrodontid *Ardynomys*; (F) primitive hystricognath *Metaphiomys*. (A from Wood, 1962; B from Hartenberger, 1973; C from Tong, 1992; D from Tong, 1997; E from Wood, 1974a; F from Wood, 1968.)

Fig. 15.20. Limb elements of early Eocene ischyromyid rodents: (A) partial right humerus; (B) right radius; (C) right ulna; (D) femora; (E) distal right tibia; (F) right calcanei; (G) astragali. All represent *Paramys* except left images in F and G (cf. *Notoparamys*). (Modified from Rose and Chinnery, 2004.)

incisors are robust, and the cheek teeth are lophodont, with three or four cross-lophs, and are primitively brachydont (Fig. 15.19E). In more derived species they are hypsodont and ever-growing. The upper teeth approach a cylindrical shape in derived members, which gives the group its name. The conules and hypocone of the upper teeth and the mesoconid on the lowers are reduced or absent; a transverse mesolophid never develops, in contrast to sciuravids. Wood

(1980, 1984) considered the dentary of *Cylindrodon* to be incipiently hystricognathous, but based on more complete fossils it appears to be sciurognathous (Emry and Korth, 1996a).

Relationships of cylindrodonts are very unstable. Cranial foramina are similar to those of ischyromyids (Wahlert, 1974), and the family has often been placed in the Ischyromyoidea. The most primitive cylindrodonts (Wasatchian *Dawsonomys* and Bridgerian *Mysops*) were initially considered to be sciuravids but are also similar to ctenodactyloids, hence McKenna and Bell (1997) grouped these three family-level taxa together in a new order, Sciuravida. Dashzeveg and Meng (1998b), however, returned to the view that cylindrodonts are more likely related to sciuromorphs than to ctenodactyloids.

Aplodontoidea. This group consists of the aplodontids, including the extant mountain beavers of the Pacific Northwest (inaptly named, for they are neither mountain dwellers nor closely allied with true beavers, Castoridae) and the later Tertiary mylagaulids. The latter were specialized diggers that are first known from the late Oligocene. Mountain beavers are often considered to be the most primitive living rodents, in part because they are the only ones with a protrogomorphous skull; however, the cheek teeth (ever-growing and with complex occlusal patterns) and fossorial skeleton are derived compared to those of many sciurids. Besides being protrogomorphous, all aplodontoids have uniserial incisor enamel (Wahlert, 1968; Korth, 1994).

Uintan *Spurimus* and late Eocene–early Oligocene (Chadronian-Orellan) *Prosciurus* and *Pelycomys* are usually regarded as the oldest and most primitive aplodontoids. They are all North American members of the Prosciurinae, a mainly late Eocene–Oligocene group that is variously assigned to Aplodontidae or Allomyidae (itself often considered a subfamily of Aplodontidae). Prosciurinae is generally considered the sister taxon of other aplodontoids (e.g., Rensberger, 1975; Korth, 1994). They are more primitive than later aplodontoids in having narrower skulls, shallower jaws, low-crowned and simpler cuspate cheek teeth, and in lacking septate bullae (although the bullae are inflated).

Early prosciurines were dentally not very different from ischyromyids and were included in Ischyromyidae or Paramyidae by earlier authors. Indeed, Heissig (2003) recently excluded *Spurimus* from Aplodontidae after cladistic analysis of dental characters showed that it groups with ischyromyids rather than with aplodontids. The molars of prosciurines are bunolophodont to bunoselenodont, the uppers with a transverse central valley, a mesostyle, and a variable hypocone, and the lowers with three cross-lophs (Wood, 1980; Vianey-Liaud, 1985). The fourth premolars are molariform but no larger than the molars, in contrast to more advanced aplodontoids in which P^4 is the largest tooth. Later aplodontoids also evolved more complex crown patterns with enamel lakes (called fossettes and fossettids) and a W-shaped ectoloph on the upper molars. Prosciurines apparently evolved from ischyromyids, and several shared features suggest derivation from early–middle Eocene *Reithroparamys* (Emry and Korth, 1989; Korth, 1994). Aplodontoids reached Europe, presumably from North America via Asia, by the *Grande Coupure* (Vianey-Liaud, 1985); but they are unknown in Asia until later in the Oligocene.

Theridomyidae. While ischyromyids flourished in North America, the endemic theridomyids predominated in Europe during the Eocene (Luckett and Hartenberger, 1985). More than two dozen genera are known from the middle Eocene through the Oligocene, representing about six subfamilies (Hartenberger, 1973; McKenna and Bell, 1997), but the composition of these subfamilies is unstable. Half of the genera are already present in the middle Eocene, although their origin is not firmly established. Hartenberger (1973, 1990; see also Vianey-Liaud, 1985) proposed that theridomyids were immigrants derived from Asian ctenodactyloids. If this hypothesis is correct, theridomyids should not be classified as sciuromorphs. Indeed, adopting this interpretation, Hartenberger (1998) allocated them to a separate suborder together with anomalurids and zegdoumyids (see the section on Anomaluromorpha below). Subsequently, however, Escarguel (1999) concluded that theridomyids evolved in situ from European ischyromyids, such as early Eocene *Hartenbergeromys*. Theridomyids differ from other sciuromorphs in having a hystricomorphous zygomasseteric arrangement. Wood (1985) suggested that the hystricomorphous condition evolved in parallel to that in some other rodents; hence it is not necessarily an indication of relationship to ctenodactyloids (most of which are also hystricomorphous).

The cheek teeth of theridomyids were bunodont and essentially four-cusped in primitive forms (e.g., *Paradelomys, Protadelomys, Suevosciurus*) but became more lophate, with a pattern of four or five transverse lophs and crenulated enamel in more derived forms (*Elfomys, Pseudoltinomys, Theridomys*; Hartenberger, 1969; 1973; Vianey-Liaud, 1976, 1979b; Fig. 15.19B). The crests include a well-developed mesoloph on the upper molars and prominent anterolophid and mesolophid joined by an ectolophid (part of the modified cristid obliqua) on the lowers. Later forms became hypsodont. The dental formula is 1.0.1.3/1.0.1.3. The family is one of a few among rodents in which a transition of enamel structure is documented: the earliest forms have pauciserial enamel, whereas advanced forms have either uniserial enamel or pseudo-multiserial enamel (Wahlert, 1968; Martin, 1999a).

Early and middle Eocene *Masillamys*, best known from several skeletons from Messel, Germany, is a relative of *Hartenbergeromys* that may lie near the origin of theridomyids (e.g., Escarguel, 1999). A small squirrel-sized rodent, *Masillamys* was more primitive than theridomyids in having a sciuromorphous skull. It had short legs, as in diggers (quite different from its supposed reithroparamyine relatives), but lacks evidence of fossorial limb modifications (Koenigswald et al., 1992b). Unlike diggers it had a relatively long tail, though not bushy like that of *Ailuravus*. Its habits are unclear, but it may have been a generalized arboreal climber.

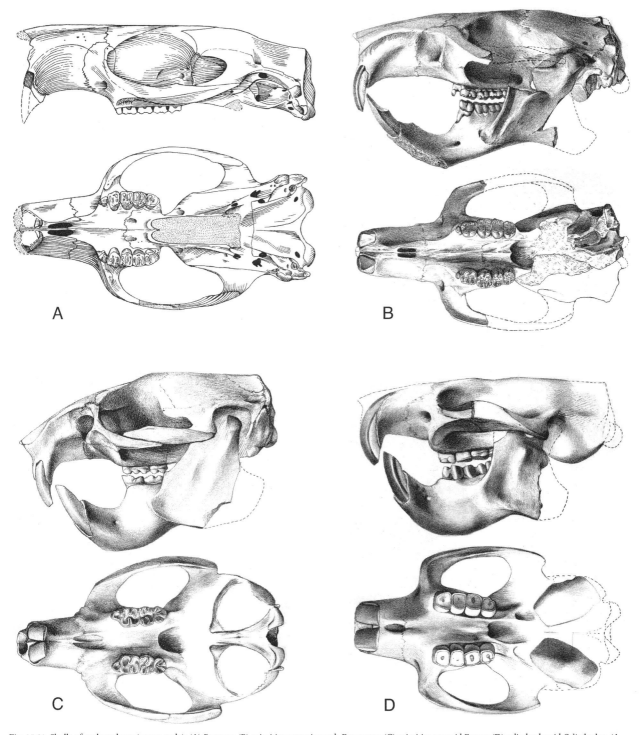

Fig. 15.21. Skulls of early rodents (not to scale): (A) *Paramys;* (B) primitive castorimorph *Eutypomys;* (C) primitive muroid *Eumys;* (D) cylindrodontid *Cylindrodon.* (A from Wood, 1962; B–D from Wood, 1937.)

Sciuridae. Squirrels are among the most successful rodents today, with about 50 genera and more than 250 species and a nearly cosmopolitan distribution. They are varied in their habits as well, including terrestrial, fossorial, arboreal, and gliding forms. Squirrels are first known from the late Eocene (Chadronian) of western North America, where two subfamilies are represented by the sciurine *Douglassciurus* (=*Douglassia,* =*Protosciurus jeffersoni*) and the cedromurine *Oligospermophilus.*

Douglassciurus (Fig. 15.17B) is known from a nearly complete skeleton, which compares closely with present-day tree squirrels, suggesting that the most primitive squirrels were arboreal (Emry and Thorington, 1982). The skull of *Douglassciurus* is primitive in being protrogomorphous, whereas almost all other sciurids are sciuromorphous. The snout is short and the auditory bullae are large and contain transbullar septae. The dentition is relatively primitive and ischyromyid-like. The cheek teeth are rooted, low crowned, and lophodont; they differ from those of later sciurids (which became progressively simpler through time) in having a larger hypocone and entoconid and a double metaconule. Nonetheless, *Douglassciurus* shares numerous apparently derived traits with sciurids, including uniserial enamel; inflated, septate bullae composed of the periotic and tympanic bones; enclosure of the stapedial artery in a bony tube; and similar basicranial foramina (Emry and Thorington, 1982; Emry and Korth, 1996b). Like primitive aplodontoids, *Douglassciurus* shares features with *Reithroparamys* that suggest that the latter genus lies near the origin of Sciuridae.

Not all researchers accept *Douglassciurus* as a primitive sciurid, however. Heissig (2003) cited the separate entoconid and entolophid, together with the protrogomorphous skull, as grounds for excluding *Douglassciurus* from Sciuridae. Vianey-Liaud (1985) argued that some traits shared with sciurids are also shared with aplodontids, whereas other similar features may be primitive. Consequently she suggested that *Douglassciurus* could be an aplodontoid rather than a sciurid. These authors consider Oligocene *Palaeosciurus* of Europe, which appears there at the *Grand Coupure*, to be the oldest sciurid. However, if *Douglassciurus* indeed has sciurid synapomorphies, primitive retentions are insufficient evidence to exclude it from the family.

Oligospermophilus is known mainly from dentitions, which are smaller than those of *Douglassciurus;* but associated forelimb bones indicate that *Oligospermophilus* was somewhat more robust, like extant ground squirrels (Korth, 1987). Cedromurines also appear to differ from other sciurids in having zygomasseteric anatomy approximating a myomorphous condition (Korth and Emry, 1991).

In the early Oligocene the oldest representatives of two sciurine clades appeared in Europe: *Heteroxerus* and *Palaeosciurus*. The extant members of these clades are ground squirrels. *Palaeosciurus* is the oldest and most primitive member of the marmot clade. Its limb proportions and robustness suggest that it was one of the earliest ground squirrels (Vianey-Liaud, 1974).

Castoroidea. This clade of sciuromorphs includes the Castoridae (beavers) and their plesiomorphic sister taxon, the mainly early and middle Eocene Eutypomyidae, which persisted into the Miocene (McKenna and Bell, 1997). All members of the clade have sciuromorphous skulls, sciurognathous mandibles, and uniserial enamel (Wahlert, 1968; Korth, 1994).

Eutypomyidae were a primarily North American Eocene group, only one genus of which reached Europe and Asia in the Miocene. They are characterized by relatively low-crowned molars with two main lophs, prominent anterior and posterior cingula, and variably developed small crests in the basins (Korth, 1994). Trends in the family were toward increasing size, hypsodonty, and crown complexity of the cheek teeth. The oldest and most primitive eutypomyid is early Eocene *Mattimys,* a very small rodent, based only on lower teeth (Korth, 1984). Its lower molars have well-developed mesoconids and small enamel swellings covering much of the shallow talonid basins. *Eutypomys* (Fig. 15.21B), known from the middle Eocene (Duchesnean) through the Oligocene, has a long snout and shares derived details of cranial foramina with castorids (Wahlert, 1977). Its cheek teeth are higher crowned and much more complex than in early eutypomyids. The postcranial skeleton of early Oligocene *Eutypomys* is relatively robust and ischyromyid-like (Wood, 1937). Dental similarities to *Microparamys* suggest that eutypomyids descended from ischyromyids (Dawson, 1966; Korth, 1984).

The true beavers (Castoridae) are not known until the latest Eocene (Chadronian) of North America, with the appearance of *Agnotocastor*. These earliest and most primitive beavers are already characterized by highly modified cheek teeth, consisting of an irregular enamel outline with buccal and lingual invaginations surrounding several small enamel lakes (fossettes and fossettids; Emry, 1972; Korth, 1994). The fourth premolars are the largest cheek teeth. As in later castorids, *Agnotocastor* has a prominent backward-projecting flange (digastric process) at the base of the mandibular symphysis (Korth, 2001). *Agnotocastor* is primitive, however, in retaining the stapedial artery (as in *Eutypomys*), as well as P^3, which are lost in all other castorids (Wahlert, 1977; Korth, 1994). By the early Oligocene castorids were present on all three northern continents (e.g., *Steneofiber* in Europe and *Propalaeocastor* in Asia), but they did not become diverse until the Arikareean (Korth, 1994). Castorid skeletons are robust. Some, like Oligo-Miocene *Palaeocastor* (a close relative of *Propalaeocastor*) were highly fossorial. Castorids paralleled eutypomyids in dental trends toward hypsodonty and increasing size; even the earliest forms are moderately high crowned (i.e., mesodont).

Gliridae (=Myoxidae). The glirids (dormice) are an Old World radiation of very small rodents that first appeared near the end of the early Eocene. They are best known from Europe, although most early forms are represented only or mainly by teeth. The molars are squared, brachydont, and lophate, with multiple transverse lophs and often crenulated enamel (Vianey-Liaud, 1994). In the upper molars the two main lophs form a V that joins the three trigon cusps, which are still evident in primitive taxa. The lower molars of Paleogene glirids have four main lophs and additional accessory crests, but the ectolophid, which is present on lower molars of their presumed ischyromyid ancestors, has been lost (Hartenberger, 1971, 1994).

The oldest known glirid, *Eogliravus* (Fig. 15.18C), is known from a skeleton from the middle Eocene of Messel. It was tiny (15 g), with sharp, curved claws, and already shows a relatively elongate tibia, presumably associated with arboreal leaping, and a bushy tail that may have served in balance or as a parachute (Escarguel et al., 2001).

At least four lineages of glirids, which differed in dental morphology or infraorbital condition, were already present in Europe by the late Eocene. Middle Eocene–Oligocene *Gliravus* was distinctly more lophodont than was *Eogliravus* and is the probable sister group of later glirids. In addition to being more primitive in masseteric anatomy than other late Eocene–Oligocene glirids, *Gliravus* retained P^3, which was usually lost in its contemporary *Glamys* (Vianey-Liaud, 1994; Freudenthal, 2004).

The relationships of glirids are controversial. It seems most likely, as documented by Hartenberger (1971), that Eocene glirids were derived from ischyromyids near *Microparamys*. *Eogliravus* has a larger hypocone but only slightly more lophate teeth than *Microparamys* and can be viewed as transitional to later glirids (Escarguel, 1999). Meng (1990) noted middle-ear similarities between extant glirids and *Reithroparamys*. These findings suggest that glirids are related to sciuromorphs (see Fig. 15.16), a relationship generally supported by molecular data as well. Nonetheless Wahlert (1978; Wahlert et al., 1993; a view also adopted by McKenna and Bell, 1997) has advocated myomorph ties (Fig. 15.22), based on the myomorphous skull of living glirids. However, the earliest glirids either have protrogomorphous skulls (*Gliravus, Bransatoglis*) or show a sciuromorphous or possibly intermediate "pseudomyomorphous" condition (*Glamys*), which suggests that the myomorphous condition characteristic of later glirids evolved independently from that in myomorphs (Vianey-Liaud, 1994). Myomorphs are thought to have passed through a hystricomorphous stage.

Laredomyidae. *Laredomys* is a puzzling genus known only from isolated teeth from the Uintan of Texas (Wilson and Westgate, 1991). Its low-crowned but strongly lophodont cheek teeth are somewhat reminiscent of those of glirids, phiomyids, and early caviomorphs, but differ enough in detail to suggest that any resemblance is convergent. *Laredomys* cannot at present be placed confidently in any higher taxon of rodents.

Suborder Myomorpha. The radiation of myomorph rodents—the highly successful and diversified mice, rats, pocket gophers, and their kin—had clearly begun by the start of the middle Eocene, as representatives of several distinct clades existed in both North America and Asia by that time. These early myomorphs include several lineages of jumping mice (dipodoids), true mice (muroids), and primitive eomyid gophers (Geomorpha), as well as the enigmatic protoptychids. By the late Eocene there was even greater diversity.

Although the composition of Myomorpha is still debated, there is general agreement that dipodoids and muroids comprise a monophyletic group (Myodonta), which probably shared a common ancestor with Geomorpha (eomyids and geomyoids). These two clades make up most of the Myomorpha. Myomorphs are widely considered to be closely related to Sciuravidae (e.g., Korth, 1994; Wang and Dawson, 1994). Korth (1994) further postulated that geomorphs and myodonts could have had separate origins from different sciuravids, which, if corroborated, would render Myomorpha polyphyletic. Possible relationships of myomorphs are shown in Figure 15.22. Most myomorphs are characterized by uniserial incisor enamel, a ring of uniserial enamel around the base of the molars (Koenigswald, 2004), reduction in size and number of premolars, molars that tend to be longer than wide, and myomorphous zygomasseteric anatomy. All of these features are in transitional stages in the most primitive representatives.

As mentioned above, Gliridae (=Myoxidae) are sometimes associated with myomorphs because of their myomorphous skulls, but this condition is believed to have arisen

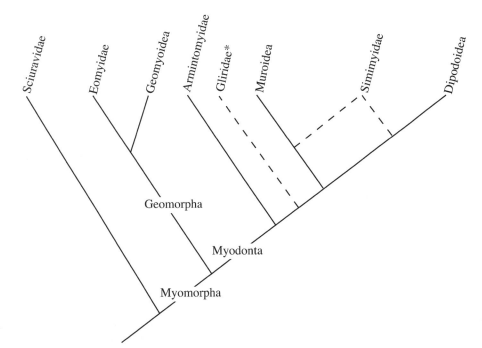

Fig. 15.22. Relationships among myomorphs, based mainly on Korth (1994) and McKenna and Bell (1997). The asterisk indicates an alternative position of Gliridae (here considered part of Sciuromorpha), as suggested by Wahlert (1978) and shown with a dashed line. (Compare with Fig. 15.16.)

directly from a protrogomorph, rather than through a hystricomorphous stage as is generally believed for myomorphs (Vianey-Liaud, 1985). Glirids also resemble myomorphs in enamel structure, but this trait, too, is thought to have arisen independently. Other evidence places glirids with sciuromorphs, and they are included under that heading in this chapter.

The oldest and most primitive known myomorph is *Armintomys*, from the early middle Eocene of Wyoming. It is also the oldest known hystricomorphous rodent and the oldest rodent that has transitional pauciserial-uniserial incisor enamel (Dawson et al., 1990). It retains a primitive upper dental formula (1.0.2.3; lower teeth are unknown), but with P^4 smaller than the molars and P^3 a vestigial peg, and it has sciuravid-like molars that are derived in being longer than wide. Dawson et al. (1990) considered *Armintomys* to be questionably the oldest dipodoid, but it is more primitive than other members of that clade in lacking a separate osseous canal for the infraorbital nerve and vessels, which is now considered to be a diagnostic feature of dipodoids (Wang and Dawson, 1994).

Dipodoidea. By the late early to middle Eocene (Bridgerian) definitive dipodoids can be recognized. Tiny *Elymys*, from the Bridgerian of Nevada, has a more derived dentition than *Armintomys* and is variously assigned to Zapodidae or Dipodidae (Emry and Korth, 1989; McKenna and Bell, 1997). This chapter follows current authors who use Zapodidae for early dipodoids and restrict Dipodidae to Miocene and later Old World forms. *Elymys* has a zapodid-like dental formula (1.0.1.3/1.0.0.3), with P^4 reduced to a peg, as is typical of dipodoids. The molars are very low crowned, lophate with large hypocones, and longer than wide. They differ from primitive muroids, however, in lacking a mesoloph and anterocone on M^1 and an anteroconid on M_1. Several other primitive zapodids, including *Aksyiromys, Primisminthus,* and *Banyuesminthus*, are known from isolated teeth from the middle and late Eocene of Asia (Tong, 1997; Emry et al., 1998). In these genera P^4 was retained, and the molars are square to elongate and have four conical cusps with variably developed oblique to transverse crests between lingual and buccal cusps. *Simiacritomys*, from the Duchesnean of California, has molars more like those of later dipodids, such as *Plesiosminthus*, with multiple transverse lophs (Korth, 1994).

Simimyidae, including middle Eocene *Simimys* and late Eocene *Nonomys*, were aberrant early myomorphs with an advanced, muroidlike dental formula (1.0.0.3/1.0.0.3) and a hystricomorphous skull structure precisely like that of dipodoids—that is, with a separate small foramen for the infraorbital nerve below the large opening that transmitted the medial masseter (Lillegraven and Wilson, 1975; Emry, 1981; Korth, 1994). Walsh (1997) reported that some populations of *Simimys* retain a vestigial P^4 or dP^4. Consequently, the phylogenetic position of simimyids is uncertain, and they have been variously allied with both superfamilies. Emry and Korth (1989) and Wang and Dawson (1994) consider the infraorbital canal to be more significant and, on that basis, regard *Simimys* as a dipodoid. The molars are generally longer than wide, low crowned, and cuspate with transverse crests (poorly developed in *Nonomys*); the third molars are reduced. Emry (1981) suggested that simimyids may represent an early myodont radiation close to the ancestry of both Dipodoidea and Muroidea.

Muroidea. Mice and rats are the most diverse and successful mammals, accounting for more than half of the living species of rodents and a quarter of all recent mammal species. Classification of muroids is highly variable; McKenna and Bell (1997) encompassed all muroids except Simimyidae in the family Muridae, whereas many other authors use multiple families. Muroids differ from dipodoids in lacking any premolars.

When multiple families are recognized, the oldest muroid rodents are generally assigned to the family Cricetidae, the stem group of Muroidea. Cricetids are small, plesiomorphic muroids with cuspate molars, primitively low crowned, in which the cusps are joined by transverse lophs that are connected by a longitudinal endoloph (upper molars) or ectolophid (lowers; Korth, 1994). M^1 is larger than M^2, which is larger than M^3. The first molars tend to be lengthened by expansion of an anterior cusp (an anteroconid on M_1, and an anterocone, which is equivalent to the parastyle, on M^1). The incisor enamel is uniserial, and the skull is either hystricomorphous or myomorphous.

The oldest known cricetids are based on isolated teeth and a number of jaws recently described from the middle Eocene of China and Kazakhstan (Tong, 1992, 1997; Wang and Dawson, 1994; Dawson and Tong, 1998; Emry et al., 1998). The best known is *Pappocricetodon* (Figs. 15.14B, 15.19C), but *Palasiomys* and *Raricricetodon* are slightly older and more primitive. These basal cricetids have brachydont molars that are longer than wide, with multiple transverse lophs (including mesolophs and mesolophids) and an anterior lobe and anterocone on M^1. Maxillary fragments of *Pappocricetodon* recently discovered in China reveal that it had a hystricomorphous zygomasseteric arrangement (Wang and Dawson, 1994). The most primitive species retain P^4, have equal-sized M^1 and M^2, and show little or no development of the anterior lobe of M^1, making them similar to primitive zapodids as well (Tong, 1997).

Eucricetodon, the oldest European cricetid, appears just after the *Grande Coupure* (early Oligocene), presumably an immigrant from Asia. It has a large M^1 with a pronounced anterior lobe. The skull was hystricomorphous in primitive species, and one lineage documents the transition to a myomorphous condition (Vianey-Liaud, 1979b). Together with the new evidence from *Pappocricetodon*, this finding supports the hypothesis that myomorphy is a modification of the hystricomorphous condition, which arose more than once in myomorphs (Vianey-Liaud, 1985).

The oldest cricetid in North America and the best known of all early cricetids is Chadronian-Orellan *Eumys* (Fig. 15.21C), represented by abundant remains found in the White River Formation of the Western Interior (Wood,

1937). The skull has a relatively short, deep rostrum and was myomorphous. The molars are brachydont with transverse crests. The upper molars decrease in size posteriorly and M^1 has a prominent anterocone. The lower molars vary in relative size among species, but M_1 always has a well-developed anteroconid. In these features, *Eumys* is clearly derived compared to the Asian middle Eocene cricetids.

Geomorpha. This myomorph clade includes the Geomyoidea (pocket gophers and their Tertiary relatives) and their plesiomorphic sister group, Eomyidae (Korth, 1994). These groups are united by a general dental similarity, uniserial enamel, and derived states of several cranial foramina (Wahlert, 1968, 1978). Unlike other myomorphs, however, they have a sciuromorphous zygomasseteric condition. Geomorphans have been primarily North American throughout their history, and they are generally thought to have evolved from sciuravids (Korth, 1994).

Eomyids first appeared in the middle Eocene (Uintan) of western North America (*Protadjidaumo* and *Metanoiamys*) and peaked in diversity during the late Eocene (Chadronian), when more than a dozen genera existed (Korth, 1994; Chiment and Korth, 1996; Walsh, 1997). The oldest Asian eomyids are late Eocene (Emry et al., 1997), whereas eomyids are not known in Europe until after the *Grande Coupure*. According to Emry et al., *Zaisaneomys*, a supposed early Eocene eomyid from Kazakhstan, is more likely younger (middle Eocene) and probably referable to either Zapodidae or Cricetidae. Early eomyids were small rodents with molars similar to those of sciuravids, but more lophate and with longitudinal crests (an endoloph on the uppers, an ectolophid on the lowers) uniting the cross-lophs, reminiscent of cricetids. They are more primitive than cricetids and other muroids, however, in retaining P^4_4, and in a few primitive forms P^3 (or dP^3) as well (*Yoderimys, Symplokeomys*, and *Metanoiamys*). Unworn molars are pentalophodont, but with wear the occlusal pattern approximates a trilophodont omega pattern (Korth, 1994). The molars of primitive forms were brachydont, whereas more advanced forms (e.g., late Eocene–early Oligocene *Paradjidaumo*) tended to have higher-crowned and more lophodont molars. Eomyids have a unique uniserial incisor enamel in which the HSB are oriented longitudinally rather than transversely as in other myomorphs and form two layers within the typical inner layer of enamel (Wahlert and Koenigswald, 1985).

The only definitive pre-Oligocene geomyoid is the diminutive *Heliscomys*, first known from Duchesnean (late middle Eocene) deposits of Saskatchewan and California. The dental formula of geomyoids is 1.0.1.3/1.0.1.3. Whereas the molars of other geomyoids are essentially bilophodont and trend toward hypsodonty, *Heliscomys* molars are very low crowned and simple, consisting of four bunodont cusps and a few small stylar cusps. Heliscomyids are considered an early branch of geomyoids and the sister taxon of other geomyoids (consisting of florentiamyids, heteromyids, and geomyids), which first appear in the early Oligocene (Korth et al., 1991). The snout of late Eocene *Heliscomys* was narrow and moderately long, with elongate incisive foramina.

Protoptychidae. The middle Eocene protoptychids—*Protoptychus* (Uintan) and *Presbymys* (Duchesnean)—are problematic rodents sometimes considered myomorphs because of their dipodid-like skeleton. *Protoptychus* (Fig. 15.17C) has a long, narrow snout; greatly inflated bullae; and an enlarged infraorbital foramen sometimes described as hystricomorphous. According to Turnbull (1991), however, it is closer to the myomorphous condition. Except for the infraorbital foramen, its cranial foramina are similar to those of ischyromyids (primitive), and its dentition (lophodont with cusps still evident) resembles that of ctenodactyloids. The enamel is pauciserial and resembles that of ischyromyids (Wahlert, 1973; Koenigswald, 2004). It has a distinctive high and narrow coronoid process and a very high mandibular condyle. The skeleton has elongated hind limbs, much longer than the forelimbs, a distally fused tibia-fibula, and a very long tail. The limb proportions are comparable to those in extant ricochetal myomorphs, such as jerboas and kangaroo rats (Turnbull, 1991). The affinities of *Protoptychus* remain uncertain. Relationships with dipodoids, ctenodactyloids, caviomorphs, ischyromyids, and sciuravids have been suggested.

Suborder Anomaluromorpha. McKenna and Bell (1997) grouped the saltatorial springhares (Pedetidae) and the gliding scaly-tailed squirrels (Anomaluridae) in this small, primarily African assemblage of rodents. The relationships of both families are poorly understood and highly controversial. Pedetids are unknown until the Miocene, but anomalurids and a possibly related family are present in the Eocene of North Africa, providing tantalizing glimpses of their potential affinities.

Isolated teeth from the late early Eocene of Algeria and Tunisia, representing several genera of the extinct family Zegdoumyidae, are the oldest African rodents (Vianey-Liaud et al., 1994; Vianey-Liaud and Jaeger, 1996). They have low-crowned, lophodont molars, the uppers with a large hypocone and transverse lophs but no longitudinal crest (ectoloph), and the lowers with trigonid and talonid of about the same elevation; a mesoconid often extended into a mesolophid; and no ectolophid. The fourth premolars are molariform and the incisor enamel is transitional pauciserial-uniserial (T. Martin, 1993). Vianey-Liaud and Jaeger (1996) proposed that the zegdoumyid *Glibemys* could be ancestral to anomalurids, and for this reason classified zegdoumyids as the oldest anomaluroids. They further postulated that zegdoumyids, sciuravids, and glirids could share a common ancestor among ischyromyids. Dawson et al. (2003) believe the dental evidence supports relationship of zegdoumyids to glirids, but that a relationship to anomalurids is less secure. Assignment of zegdoumyids to Anomaluromorpha is, therefore, tenuous.

The oldest known anomalurid is *Pondaungimys*, from the latest middle Eocene of Myanmar (Dawson et al., 2003). Its

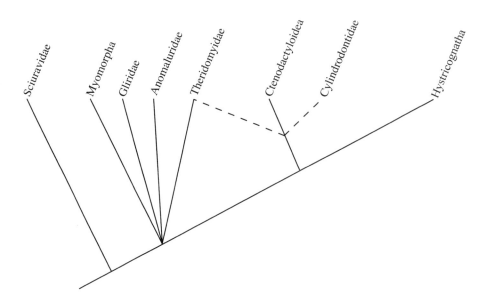

Fig. 15.23. Relationships among Sciuravida, based mainly on Flynn et al. (1986). Note that this arrangement differs from several others, particularly in position of Gliridae, Theridomyidae, and Cylindrodontidae. It does, however, reflect the consensus view that Ctenodactyloidea is the sister group of Hystricognatha.

molar teeth are pentalophodont and show complex enamel folds, and the lowers retain a complete ectolophid, missing in zegdoumyids. *Nementchamys,* also assigned to Anomaluridae, is based on isolated teeth from slightly younger late Eocene sediments of Algeria (Jaeger et al., 1985). Thus anomalurids were fairly widely distributed in the Old World soon after their first appearance. *Nementchamys* has molarized fourth premolars and molars with crenulated enamel and a pentalophodont pattern, including a complex mesoloph and mesolophid, a strong anterolophid, and a complete ectolophid. The dental anatomy of *Nementchamys* has suggested potential relationship to theridomyids or zegdoumyids, but these ties are weak, and the origin of anomalurids remains obscure.

Suborder Sciuravida. McKenna and Bell (1997) proposed this new suborder to encompass a mainly Paleogene radiation, consisting of the Sciuravidae, Cylindrodontidae, and the families here grouped as ctenodactyloids (Fig. 15.23). Unfortunately, there has been no phylogenetic analysis or assessment of anatomical characters to justify this arrangement. The first two families were long classified as protrogomorph rodents closely allied with ischyromyids (e.g., Wood, 1965) and are discussed here with sciuromorphs. There is mounting evidence that ctenodactyloids are related to Hystricognatha, hence they are considered in the next section.

Superfamily Ctenodactyloidea. Ctenodactyloids were the dominant rodents in Asia during the Paleogene and are among the most primitive of all rodents. Two-thirds of the more than 40 genera have been described in the past 20 years, many based only on jaws and teeth, and their interrelationships are still very uncertain. As many as six families —Cocomyidae, Chapattimyidae, Yuomyidae, Tamquammyidae, Gobiomyidae, and Ctenodactylidae—are recognized by some authors (e.g., B.-Y. Wang, 1994, 2001b); but this proliferation of taxa perhaps inflates the actual dental differences among them. Many of these families appear to be paraphyletic (e.g., Averianov, 1996; Dashzeveg and Meng, 1998a). McKenna and Bell (1997) included the first four families in the Chapattimyidae, and although they did not recognize Ctenodactyloidea, they united Chapattimyidae and Ctenodactylidae (together with sciuravids and cylindrodontids), in their new order Sciuravida. Ctenodactylidae includes the living gundis, which are restricted to Africa today, but the family originated in Asia during the Eocene and was especially diversified there during the Oligo-Miocene (Wang, 1997). The other families are known primarily from the Eocene of Asia.

Paleogene ctenodactyloids were very small, mouse-sized rodents. They were generally hystricomorphous (except early Eocene *Cocomys*) and sciurognathous, with a primitive dental formula of 1.0.2.3/1.0.1.3 and a reduced P^3. The incisor enamel is pauciserial in *Cocomys, Chapattimys,* and a few other genera, and multiserial in all others (T. Martin, 1993; Wang, 2001a). The cheek teeth of ctenodactyloids are distinctive (Fig. 15.19D) and have been well illustrated and described by many authors (e.g., Hussain et al., 1978; Shevyreva, 1989; Tong, 1997; Wang, 1997, 2001a,b; Dashzeveg and Meng, 1998a). They are variably lophodont and increase in size from front to back. The upper molars are usually wider than long, with well-developed conules (especially the metaconule) and hypocone, and a metaloph that joins the protocone and excludes the hypocone. The lower molars tend to be elongate, with a mesoconid (sometimes forming an ectolophid) and variably developed cross-lophs, including a hypolophid that joins the hypoconid and entoconid and isolates the hypoconulid on the postcingulid.

The condition of the fourth premolars is considered especially important for assessing ctenodactyloid interrelationships; nonetheless, the morphology of these teeth is not easy to evaluate. They are characterized as nonmolariform in some ctenodactyloids (cocomyids, gobiomyids, tamquammyids, and ctenodactylids) and submolariform to molariform in others. Although the nonmolariform condition has been considered primitive, because it occurs in the eurymylid *Heomys* (Dawson et al., 1984; Li et al., 1989), molariform fourth premolars characterize the oldest known rodents, late Paleocene alagomyids and ischyromyids (Dawson and Beard, 1996). Hence the polarity of this character is ambiguous (Dashzeveg and Meng, 1998a). Moreover the distinction between molariform and nonmolariform P4s can be subtle (nonmolariform premolars lack a metacone or a hypoconid and have a narrower talonid than trigonid), and intermediate conditions are known.

Ctenodactyloids were already diverse at their earliest appearance. As many as 13 ctenodactyloid genera are already known from the early Eocene Bumbanian Land-Mammal Age (Dawson, 2003). *Cocomys* (Fig. 15.24), known from the skull and mandible, is more primitive than other ctenodactyloids in being protrogomorphous (but with a relatively larger infraorbital foramen than in ischyromyids) and in lacking a hypolophid (Dawson et al., 1984; Li et al., 1989). The rostrum is short and robust and the auditory bullae are ossified, without septae, and are loosely attached to the skull. Its cuspate molars with relatively weak lophs and nonmolariform fourth premolars have been considered primitive for Rodentia.

Ctenodactyloids are generally thought to be closely related to the origin of Hystricognatha, including caviomorphs and Tertiary phiomorphs. All but the most primitive ctenodactyloids share with hystricognaths the derived traits of hystricomorphy and multiserial enamel (Hussain et al., 1978;

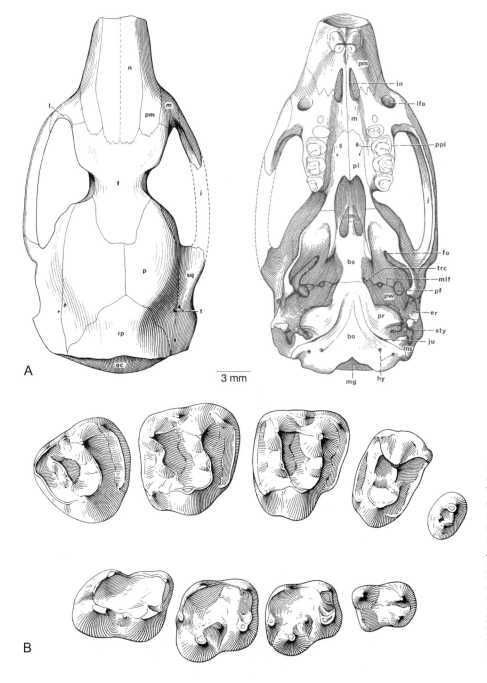

Fig. 15.24. Basal ctenodactyloid *Cocomys*: (A) skull; (B) right dentition. Key: bo, basioccipital; bs, basisphenoid; er, epitympanic recess; f, frontal; fo, foramen ovale; hy, hypoglossal foramen; in, incisive foramen; iof, infraorbital foramen; ip, interparietal; j, jugal; ju, jugular foramen; l, lacrimal; m, maxilla; mg, foramen magnum; mlf, middle lacerate foramen; ms, mastoid; n, nasal; oc, occipital; p, parietal; pf, pyriform fenestra; pl, palatine; pm, premaxilla; ppl, posterior palatine foramen; pr, promontorium; pw, epitympanic wing of petrosal; sq, squamosal; sty, stylomastoid foramen; t, temporal foramen; trc, transverse canal. (From Li et al., 1989.)

Flynn et al., 1986; T. Martin, 1993). Ctenodactyloids and phiomorphs also share a number of dental characters, as underscored by the attribution of the late Eocene Algerian *Protophiomys* to either Phiomyidae or Chapattimyidae (Jaeger et al., 1985; Flynn et al., 1986).

Suborder Hystricognatha (=Hystricomorpha). This suborder includes the extant Old World porcupines (Hystricidae), mole rats (Bathyergidae), rock rat (Petromuridae), and cane rats (Thryonomyidae), the New World porcupines (Erethizontidae), and the South American caviomorphs, most of which are unknown in the fossil record until the late Oligocene or after. A few basal forms, however, are known from near the Eocene/Oligocene boundary. All of these rodents share the derived traits of hystricomorphy, hystricognathy, and multiserial enamel.

In the late Eocene and early Oligocene of northern Africa (Fayum, Egypt) are found a diversity of primitive hystricognaths that Wood (1968) called thryonomyoids and Lavocat (1973) assigned to the Phiomorpha (which in his view also included Bathyergidae and Hystricidae). This paraphyletic group includes several families with late Eocene or early Oligocene representatives known chiefly from dentitions, including Phiomyidae (*Phiomys*), Myophiomyidae (*Phiocricetomys*), Diamantomyidae (*Metaphiomys*; Fig. 15.19F), and the earliest Thryonomyidae (*Gaudeamus* and *Paraphiomys*). These early genera share an overall dental similarity, and all were included in the Phiomyidae by Wood (1968). Most have quadrate molars with well-developed crests, three to five on the lower molars and three to six on the uppers. In most cases the deciduous fourth premolars were retained throughout life, as in extant *Petromus* and *Thryonomys*. Where known the skull is hystricomorphous, the jaw hystricognathous, and the incisor enamel multiserial. *Metaphiomys* and *Gaudeamus* are more strongly lophodont than is *Phiomys*. The molars of *Gaudeamus* have three well-developed oblique lophs, and its dP_4 is elongate and five-crested. Unlike other phiomorphs, *Phiocricetomys* has lost P_4, and its molars are bunodont with weak lophs and decrease in size from M_1 to M_3. *Protophiomys* from the late Eocene of Algeria is more primitive than Fayum phiomorphs in having weaker crests and transitional pauciserial-multiserial enamel (Jaeger et al., 1985). These northern African phiomorphs are the oldest known members of the Hystricognatha. Phiomorphs are now generally considered to be closely related to or derived from Eocene ctenodactyloids from Asia.

The oldest known definitive caviomorph is an unnamed agoutid known from a lower jaw from the Eocene/Oligocene boundary (Tinguirirican) of Chile (Wyss et al., 1993). It is characterized by moderately high-crowned cheek teeth with relatively flat occlusal surfaces bearing three enamel lakes or fossettids, a pattern usually associated with pentalophodont upper molars (as in phiomorphs).

The recently reported Santa Rosa local fauna from supposed middle or late Eocene strata of Peru contains an unexpected diversity of small rodents that appear to lie near the base of the caviomorph radiation (Campbell and Frailey, 2004; Frailey and Campbell, 2004). If the age is confirmed, they would be the oldest known South American rodents. No other rodent fossils are known from South America until the late Oligocene (Deseadan), when as many as 24 species in at least seven families are present (Patterson and Wood, 1982).

The origin of caviomorphs and the relationship between phiomorphs and other hystricognaths has been one of the major controversies in rodent evolution. Wood (e.g., 1974b, 1975, 1985) and Patterson and Wood (1982) forcefully argued in favor of a North American or Central American origin of caviomorphs from "franimorphs," a group of primitive, mostly ischyromyid rodents (no longer considered valid) that Wood interpreted as incipiently hystricognathous. Lavocat (e.g., 1969, 1973, 1974) and Hoffstetter (e.g., 1972, 1975), however, advocated an African origin from phiomorphs.

There is no dispute that phiomorphs and caviomorphs possess numerous derived traits in common, including multiserial enamel, hystricomorphy, and hystricognathy. But Patterson and Wood (1982) argued that caviomorphs and phiomorphs (=thryonomyoids) evolved separately from northern hemisphere "franimorphs," whereas Lavocat and Hoffstetter would derive caviomorphs directly from phiomorphs. Additional evidence that phiomorphs and caviomorphs share derived auditory structures (Lavocat and Parent, 1985), and that a specialized type of multiserial incisor enamel occurs in both (Martin, 1994), adds to mounting evidence in favor of a phiomorph ancestry of caviomorphs. A close relationship between phiomorphs and caviomorphs is also supported by recent molecular studies (Mouchaty et al., 2001). Similarities between phiomorphs and ctenodactyloids in turn suggest that caviomorphs may trace their origin ultimately to Asian ctenodactyloids.

16

Reflections and Speculations on the Beginning of the Age of Mammals

AS DETAILED IN THE PRECEDING CHAPTERS, the Early Cenozoic (Paleocene-Eocene) witnessed an extraordinary increase in mammalian diversity compared to the Late Cretaceous. The diversification began soon after the mass extinctions at the end of the Cretaceous, which included the disappearance of nonavian dinosaurs. Within a few million years of the Cretaceous/Tertiary (K/T) boundary 44 new mammalian families had appeared, another 41 new families are recognized by the end of the Paleocene, and 61 additional families in the early Eocene (see Fig. 1.1). During the Paleocene at least 20 placental orders appear in the fossil record for the first time, with seven more in the early or middle Eocene, by which time almost all the modern orders were established. Of the extant orders, only Tubulidentata, Scandentia, and Dermoptera were not certainly present by the middle Eocene; and their absence is most likely owing to the extremely sparse fossil record of these groups. It is also notable that, despite a rapidly improving fossil record, only three Cenozoic eutherian orders have recognized Cretaceous representatives: Leptictida, Lipotyphla, and Taeniodonta.

Consequently, although many molecular analyses suggest that extant mammal clades, including some placental orders, diverged deep in the Cretaceous, the fossil record has produced little evidence to support such early origins. However, the abrupt appearance of more than 20 orders—without evidence of how or where many of them evolved—and the still-limited exploration of the Cretaceous record strongly suggest that at least *some* of these orders may have originated in the Cretaceous in areas not yet sampled. Aside from these unknown stem groups and a handful of Cretaceous eutherians and marsupials that might be related to the Cenozoic radiations, the only Mesozoic mammals with direct Cenozoic descendants were multituberculates (still prevalent in the Paleocene), monotremes, dryolestoids, and gondwanatheres (these last three very rare in the Cenozoic). Nearly all other Mesozoic mammals belong to archaic groups not closely related to modern mammals.

Notwithstanding substantial recent advances in our knowledge of Mesozoic mammals (compare Kielan-Jaworowska et al., 2004, with Lillegraven et al., 1979), the record of Cenozoic mammals is undeniably richer and more diverse than that of the Mesozoic. Before recapitulating the highlights of the Early Cenozoic record, it is instructive to examine its relative completeness and reliability. It is important to realize that, although our sampling of Paleocene and Eocene faunas is geographically denser and more widespread than that of the Cretaceous, we may still be getting an incomplete, and possibly misleading, impression of Early Cenozoic faunas. A brief review of the strengths and principal deficiencies in our knowledge will make the point and suggest where future field efforts could be particularly enlightening.

EARLY CENOZOIC MAMMAL RECORD
Southern Hemisphere

Inadequate knowledge of Early Cenozoic faunas in the Southern Hemisphere poses one of the largest impediments to a synthesis of Early Cenozoic mammalian evolution. In South America, the record is rich in selected areas (principally Patagonia and a few isolated sites, such as Laguna Umayo, Peru; Tiupampa, Bolivia; Pampa Grande, Argentina; and Itaboraí, Brazil), but for all except Patagonia, the faunas are limited to narrow intervals of time. In large parts of the continent (e.g., the northern half) there is virtually no record at all. We have little understanding of how the early Paleocene Tiupampa fauna, which has hints of affinity with North American Paleocene faunas (with its marsupials, mioclaenids, pantodont, and cimolestid; Marshall et al., 1995), relates to the later Paleocene Itaboraian and Riochican faunas. In addition to an array of marsupials, the latter faunas include the oldest known xenarthrans, litopterns, notoungulates, and xenungulates, endemic clades no hint of which has been found at Tiupampa (or at Punta Peligro, Argentina, for that matter, but the fauna from Punta Peligro is very poorly known). Their origins remain obscure. They may well have originated in South America, probably from North American immigrants, although there is no direct evidence one way or the other. In fact, the striking endemicity of known Early Cenozoic faunas indicates that there was no contact with any other region (outside South America) after the early Paleocene until the arrival of primates and rodents from Africa around the Eocene/Oligocene boundary, or perhaps a little earlier. In any case, the known Cretaceous fauna of South America contains no taxa that appear to be related to Cenozoic mammals, except for gondwanatheres.

Nominal Eocene faunas in South America were essentially a continuation of the late Paleocene fauna, with particular proliferation of marsupials, litopterns, and notoungulates. During the late Paleocene and Eocene there were many lophodont forms, implying specialization for folivory.

Certainly one of the greatest obstacles to our understanding of South American Early Cenozoic faunas is the lack of a well-resolved chronology. Recent studies noted throughout the volume have greatly shaken our confidence in the traditional correlations of the South American Land-Mammal Ages (SALMAs), with only their sequence being relatively secure (and even here there are potential problems). The long-established Eocene SALMAs are now known to be younger than once thought, but how much younger is in dispute. This uncertainty in turn, casts doubt on the precise ages and durations of the Paleocene SALMAs.

For all its shortcomings, the South American Early Cenozoic record is far superior to that from Africa, Antarctica, or Australia. A single area, the Ouarzazate Basin of Morocco, provides a small window on the late Paleocene fauna of northwestern Africa (including palaeoryctids, pantolestans, primitive adapisoriculid lipotyphlans, and what may be the only known Paleocene euprimate, *Altiatlasius*), which, not surprisingly, suggests affinities with Europe. No other Paleocene mammal site is known in Africa. Sites of Eocene age, rarely well dated, are found across the Sahara in northern Africa from Morocco to Egypt. They provide glimpses, and in a few cases rich records, of a remarkable, largely endemic fauna, including proboscideans, hyracoids, arsinoitheres, and macroscelidids, as well as the somewhat more widely distributed sirenians, hyaenodontids, anthracotheres, archaeocetes, and both lorisoid and anthropoid primates. Faunas of the Fayum Depression in Egypt, which span from the middle Eocene into the Oligocene, are particularly noteworthy. The first five taxa just listed belong to the molecular clade Afrotheria, and of those all except macroscelidids make up the paenungulates, some of which had circum-Tethyan distribution.

South of the Sahara, scattered sites in Senegal, Nigeria, Somalia, and Tanzania have produced a small assortment of middle and late Eocene archaeocete whales, sirenians, primitive proboscideans (*Moeritherium*), a possible condylarth, and a bat, the last two known only from single specimens (Savage, 1969; Sudre, 1979; Savage and Russell, 1983; Gunnell et al., 2003). In addition, phiomorph rodents and postcrania of anthropoid primates and a macroscelidean were recently reported from the late Eocene or early Oligocene of Tanzania (Stevens et al., 2005). But so far these fossils offer little insight on the Eocene fauna not afforded by the much richer northern African fauna. No other Paleocene or Eocene mammals are known from sub-Saharan Africa. Our virtual ignorance of Early Cenozoic mammalian history in the southern half of Africa, including Madagascar, has been a serious handicap to understanding events during the beginning of the Age of Mammals. Thus it is of considerable significance that a late Eocene sirenian has now been reported from Madagascar, the first Paleogene mammal known from the island (Samonds et al., 2005), perhaps portending the discovery of Early Cenozoic Malagasy land mammals in the near future.

Until the past 15 years or so, nothing was known of Early Cenozoic mammals from Antarctica or Australia. Search efforts in the middle Eocene La Meseta Formation of the Antarctic Peninsula (Seymour Island) have yielded a few very fragmentary specimens of marsupials and placentals of Patagonian aspect, which reinforce the notion of trans-Antarctic dispersal of South American marsupials to Australia. The fossils include polydolopid and microbiotheriid marsupials and scraps tentatively identified as astrapotheres, litopterns, and xenarthrans (Woodburne and Case, 1996).

In Australia, the Tingamarra Local Fauna comprises a small number of jaws and teeth from apparent earliest Eocene sediments near Murgon, southeastern Queensland. These fossils, which include several marsupials, a bat, and a possible placental (which might, in fact, be a marsupial), constitute the only Cenozoic record of mammals from Australia prior to the late Oligocene (Archer et al., 1999). However, Woodburne and Case (1996) postulated that even these fossils might not predate the Oligocene, in which case there would be no Paleocene or Eocene mammals known from Australia.

North America

It is apparent from the foregoing that, apart from the endemic faunas of a few locations in South America and northern Africa, our understanding of the beginning of the Age of Mammals is based almost entirely on Holarctic faunas. Paleocene mammals are best known from North America, where a chronologic sequence of faunas spanning the entire epoch has long been well documented in the Western Interior, from Alberta and Saskatchewan to New Mexico and Texas. These faunas are dominated by condylarths, plesiadapiforms, and multituberculates, with lesser representation by carnivorans, pantodonts, taeniodonts, leptictids, cimolestids, mesonychians, and various other groups. Except for the Tiffanian Goler Formation in California (McKenna and Lofgren, 2003) and isolated Paleocene occurrences in Louisiana, South Carolina, and Maryland (Simpson, 1932; Schoch, 1985; Rose, 2000), however, knowledge of Paleocene mammals is confined to the Western Interior (Savage and Russell, 1983; Archibald et al., 1987; Lofgren et al., 2004). This rich record has tended to dominate our impression of mammalian evolution in the Paleocene. The occurrence of the enigmatic *Mingotherium* in South Carolina, however, is a reminder that the Western Interior does not hold all the answers.

Faunal exchange between North America and Asia during the late Paleocene is indicated by the similarity on both continents of uintatheres, arctostylopids, rodents, and tillodonts, as well as the pantodont *Coryphodon*, most or all of which probably dispersed from Asia (Rose, 1981; Beard, 1998a). The latter inference is drawn from the early diversity of these taxa in Asia as well as the absence of plausible ancestors in known North American faunas. Other closely allied taxa found in both North America and Europe (e.g., neoplagiaulacid multituberculates, plesiadapids, arctocyonids) indicate at least a filter connection with Europe. At least one mammal, the mesonychid *Dissacus*, clearly disseminated widely across all three continents during the Paleocene.

Various authors have entertained the possibility of Paleocene (or possibly earlier) mammalian dispersal from south to north across the seaway that separated North and South America (e.g., McKenna, 1980b; Gingerich, 1985). These hypotheses have been based on the similarity of didelphoid marsupials and the possibility of relationships between xenungulates and uintatheres, xenarthrans and palaeanodonts, and notoungulates and arctostylopids, respectively. Also suggestive of this dispersal route was the discovery of an early Paleocene pantodont at Tiupampa, perhaps the oldest known member of the order. Older marsupials are now known from North America, however, and more primitive (if not older) pantodonts from Asia, so it is more likely that these groups dispersed from North America to South America. The proposed relationships of the other taxa remain unsubstantiated. Consequently the evidence seems weak, at best, for any northward dispersal from South America during the Early Cenozoic.

A major immigration event involving the influx of the first euprimates, perissodactyls, artiodactyls, hyaenodontids, and probably the condylarth *Hyopsodus*, marked the beginning of the Eocene in North America and Europe, and Asia, too (except for hyaenodontids, which were present there in the latest Paleocene). It coincided with the Initial Eocene Thermal Maximum, which turned high-latitude land bridges into favorable corridors. In which direction the dispersal took place is a matter of dispute, but both northern Pacific and northern Atlantic bridges were probably involved. Beard (1998a) contends that dispersal was from Asia to North America, but Godinot and Lapparent de Broin (2003) provided arguments supporting dispersal from western Asia to Europe and from there to North America. Both hypotheses may be true. There is still no good evidence of where any of these clades originated.

Compared to the Paleocene, Eocene mammal faunas are more widely distributed in North America but still largely based on the Rocky Mountain region, especially Wyoming. There the Eocene witnessed a proliferation of perissodactyls, artiodactyls, euprimates, creodonts, carnivorans, and rodents. Phenacodontid condylarths declined, while *Hyopsodus* flourished. Notable, less abundant groups include bats, plagiomenids (possible dermopteran predecessors), tillodonts, uintatheres, palaeanodonts, and, near the end of the Eocene, the only New World pangolin. Outside the Western Interior, the principal localities for Eocene continental mammal faunas are in Baja California and Guanajuato, central Mexico; a few isolated spots on the U.S. west coast (southern California and Oregon); Ellesmere and Axel Heiberg Islands in Arctic Canada; and several sites on the Atlantic and Gulf coastal plain of the southeastern United States (in Texas, Arkansas, Mississippi, Alabama, Georgia, and Virginia; Robinson et al., 2004). Late Eocene marine mammals have been found from the Carolinas south to Florida and west to Texas (Savage and Russell, 1983). Sirenians and one land mammal, the rhinocerotoid *Hyrachyus,* are known from the early middle Eocene of the island of Jamaica in the Caribbean (Domning et al., 1997; Domning, 2001c). Most of the mammals from these disjunct peripheral areas are closely related to those of the Western Interior, except for the marine mammals of the U.S. east coast and Jamaica, which are of Tethyan aspect. Nevertheless, very little is known of Eocene mammal faunas in much of the eastern half and the northern half of the continent.

Faunal interchange with Europe continued through the early Eocene and with Asia throughout the Eocene. Direct exchange with Europe ceased at the end of the early Eocene when North Atlantic rifting interrupted the land connection.

Europe

In Europe the Paleocene record is much more limited. Sparse early Paleocene (Danian) mammals are known from only two localities: Hainin in Belgium and Fontllonga-3 in the Tremp Basin of northeastern Spain (Savage and Russell, 1983; Peláez-Campomanes et al., 2000). They include endemic multituberculates and eutherians of Euro-American aspect, such as condylarths, a marsupial or adapisoriculid, and a plesiadapiform (e.g., Vianey-Liaud, 1979a; Sudre and Russell, 1982; Sigé and Marandat, 1997). Fontllonga-3 dates from just after the K/T boundary, whereas Hainin seems to correlate with middle to late Torrejonian, leaving a large part of the early Paleocene mammal record of Europe completely unknown. Late Paleocene (Thanetian) faunas are known principally from the Paris Basin (late Thanetian) and Walbeck, northwestern Germany, which may be somewhat older than the Paris Basin sites (e.g., Weigelt, 1960; Russell, 1964). Several Thanetian genera have affinity with North American taxa and dispersed one way or the other via the North Atlantic connection, which acted as a filter at that time. Although these two areas have yielded a relatively diverse Thanetian fauna especially rich in small mammals (e.g., plesiadapiforms, lipotyphlans, and condylarths), the Paleocene of Europe so far lacks most of the larger mammals found in North America or Asia, such as pantodonts, taeniodonts, uintatheres, and large condylarths (except arctocyonids). It should be emphasized that our view of the European Paleocene is focused on western Europe, with the possible exception of a few mammal teeth from the Calcaires de Rona in Romania that could be of Thanetian age (Gheerbrant et al., 1999). For at least part of the Paleocene, Europe may have been fragmented into several emergent land masses; and we know almost nothing of Paleocene mammals of eastern Europe, Scandinavia, or the Tethyan coast.

The Eocene of Europe is much better known. Numerous local faunas throughout western Europe are the basis for 14 Eocene mammal reference levels (MP 7–20; compared to only six for the Paleocene, only two of which have produced mammals). As in North America, euprimates, artiodactyls, perissodactyls, and hyaenodontids first appeared at the onset of the Eocene (MP 7), but, again, the direction of dispersal is uncertain. Hooker and Dashzeveg (2003) argued that some taxa, such as perissodactyls and *Hyopsodus,* immigrated to Europe from Asia at this time. Dispersing from North America to Europe at the same time were rodents, tillodonts, palaeanodonts, and *Coryphodon,* which had been present in the Western Interior throughout the Clarkforkian. Major early Eocene (Ypresian) localities are still concentrated in western Europe: the Paris and London basins, Belgium (Dormaal), Portugal (Silveirinha), northern Spain, and the southern coast of France (e.g., Palette, Rians; Savage and Russell, 1983; Estravís Fernández, 1992). Early Eocene northwestern European faunas share greater similarity with North American faunas (more than 50% of the genera common to both regions) than they do with contemporary southern European faunas (see Chapter 1), underscoring that the North Atlantic connection was a corridor at that time and suggesting a physical barrier dividing Europe. Middle and late Eocene mammals are distributed a little more widely, being found in England, France, Germany, Spain, Switzerland, and a few scattered sites in Austria, Bulgaria, Hungary, and Romania. Of particular note are the Paris Basin localities (multiple levels); Bouxwiller, France (Lutetian); Messel and Geiseltal, Germany (Lutetian); Robiac, France (Bartonian); Egerkingen, Switzerland (Bartonian); and Headon Hill, England (Priabonian).

The famous Phosphorites of Quercy, southern France, which have produced thousands of mammal fossils over more than a century, pose a singular problem, because the fossils come from karstic (fissure) fillings that range in age mostly from late middle Eocene (Bartonian) through middle Oligocene (Remy et al., 1987; Legendre et al., 1997). The provenance of many of the earlier collections is uncertain, but careful collecting in the past few decades has yielded important, well-dated collections. Because it spans the Eocene/Oligocene boundary, the sequence of Quercy levels provides a detailed record of the earliest Oligocene faunal turnover known as the *Grande Coupure*. During this interval many European Eocene taxa were greatly diminished or became extinct (e.g., primates; palaeotheres; dichobunid, xiphodont, and amphimerycid artiodactyls), and new taxa, including feliform and musteloid carnivorans, sciurid, castorid, cricetid, and eomyid rodents, various ruminant and anthracotheriid artiodactyls, and rhinocerotoid perissodactyls, appeared as immigrants, probably from Asia (Russell and Tobien, 1986; Hooker et al., 2004).

Asia

The record of fossil mammals in Asia has grown at a rapid pace in recent years, but our knowledge of Paleocene faunas remains very limited. Compounding this limitation is the tenuous nature of the biochronology, particularly for older faunas. As in South America, intercontinental correlation of faunas that are largely endemic is challenging and, in the near absence of absolute dates, often based on "stage of evolution." Much of the following summary is based on Russell and Zhai (1987), supplemented by Tong et al. (1995) and Ting (1998).

Probable early Paleocene (Shanghuan) faunas are best represented from two sites in eastern China: the Nanxiong Basin, Guangdong Province, and the Qianshan Basin, Anhui Province. The fauna includes many endemics—pantodonts (bemalambdids and pantolambdodontids), anagalids, pseudictopids, and primitive glirans—as well as mesonychids and a viverravid that could be related to North American forms.

Late Paleocene sites are somewhat more widely distributed. Nongshanian sites are best known in several parts of eastern China, whereas the younger Gashatan interval is best known from Nei Mongol (northeastern China) and several sites in southern Mongolia, of which Gashato and Naran Bulak deserve special mention. The late Paleocene fauna is more diverse than that of the early Paleocene. In the Nongshanian there are endemics (the same groups as in the early Paleocene), the bizarre *Ernanodon,* phenacolophid tethytheres, *Radinskya,* and the first Cenozoic multituberculates in Asia. The first Asian rodent (*Tribosphenomys*) appeared in the Gashatan, contemporary with the oldest rodents in North America. Both Nongshanian and Gashatan faunas contain taxa closely allied with North American forms (uintatheres, arctostylopids, and mesonychids), but the correspondence seems closer in the Gashatan. Not only are *Oxyaena* and *Coryphodon* now present on both continents, but some Gashatan uintatheres, arctostylopids, and mesonychids are so similar to those from North America that they could also be congeneric. These occurrences strongly suggest broad exchange of faunas between Asia and North America near the Tiffanian/Clarkforkian boundary. Consequently, the presence of the hyaenodontid *Prolimnocyon* and an unidentified perissodactyl in the Gashatan Bayan Ulan fauna (Meng et al., 1998) was unexpected. The provenance of the perissodactyl, and even its identity as a perissodactyl, are in doubt, but the *Prolimnocyon* still constitutes the only pre-Eocene record of Hyaenodontidae.

Our image of Paleocene mammal faunas from Asia is thus drawn from assemblages from eastern and northern China and southern Mongolia. All of northern, western, and southeast Asia south of China, as well as the Indian subcontinent (which may not yet have joined Asia), are a void in our knowledge. Furthermore the relative uniformity of Shanghuan and Nongshanian faunas suggests that only limited intervals of the early and late Paleocene have been sampled.

Eocene mammal faunas are more widespread and diverse but still known mainly from the southeastern quadrant of Asia. In eastern Asia, early Eocene (Bumbanian) mammals are best known from the Bumban Member of the Naran Bulak Formation at Tsagan Khushu, Mongolia. They also come from several sites in China, most notably Wutu in Shandong Province, the upper part of the Lingcha Formation in Hunan, and the Turpan Basin in Xinjiang. These faunas include some taxa in common with those of the late Paleocene (eurymylids, arctostylopids, dinoceratans, pantodonts, and rodents), but they differ in containing perissodactyls and ctenodactyloid rodents, as well as *Hyopsodus,* a euprimate (*Altanius*), and rare artiodactyls (Dashzeveg et al., 1998; Tong and Wang, 1998; Kondrashov et al., 2004), most of which are usually indications that the Paleocene/Eocene boundary has been crossed. The Wutu fauna also includes several Paleocene relicts (the plesiadapoids *Carpocristes, Chronolestes,* and *Asioplesiadapis* and the neoplagiaulacid multituberculate *Mesodmops*), which led Beard and Dawson (1999) to suggest that it correlates with the Clarkforkian rather than the Wasatchian. Ting (1998), however, considered Wutu to be early Eocene and slightly younger than the Tsagan Khushu and Lingcha faunas. The Lingcha Formation was recently shown to date from the Carbon Isotope Excursion at the beginning of the Eocene (Bowen et al., 2002). Most of the constituents of these Bumbanian faunas were terrestrial mammals. The scarcity of arboreal forms, at least in the Mongolian faunas, suggests the lack of dense forests (contrary to the inferences for North America and Europe) during the early Eocene in this part of Asia (Meng and McKenna, 1998).

Other Asian early Eocene faunas are present at several sites in India and Pakistan. These sites have produced the oldest known cetaceans, the oldest Asian bats, and primitive anthracobunid tethytheres, as well as endemic quettacyonids, tillodonts, perissodactyls, artiodactyls, and adapoid primates (e.g., Gingerich, Arif, et al., 2001). The collision of India with Asia seems to have occurred near the Paleocene/Eocene boundary, when artiodactyls, perissodactyls, euprimates, and hyaenodontids first appeared, raising the intriguing possibility that mammals evolving on the drifting plate might have dispersed from India into Asia around the time of the collision (e.g., Krause and Maas, 1990). Much of the available faunal information, however, comes from near the suture zone and probably postdates collision. No Paleocene mammals are yet known from the region. Therefore, we can do little more than speculate until we have a better knowledge of the timing of the collision, the distribution of potential intervening island arcs, and the Late Cretaceous through early Eocene faunas of India. Although a certain degree of endemism is apparent in known early Eocene faunas, it is premature to evaluate the role of India with regard to the origin or dispersal of Paleocene and Eocene mammals.

Middle Eocene mammal faunas have been found at many sites in China, India, Japan, Kazakhstan, Kyrgyzstan, Mongolia, Myanmar, Pakistan, Turkey, and Uzbekistan. By the late Eocene, faunas are also known from Indonesia, Korea, and Thailand. These middle and late Eocene faunas are particularly rich in perissodactyls (especially brontotheres, rhinocerotoids, and endemic deperetellids and lophialetids), artiodactyls (especially raoellids, helohyids, and anthracotheres), and ctenodactyloid rodents. Other notable constituents of some of these faunas are the oldest lagomorphs and cricetid rodents, primitive zapodid rodents, archaeocete whales, tarsiid primates, and possible anthropoids (eosimiids).

Pantodonts made their last appearance during this interval in Asia. At the end of the Eocene, at least in central Asia, perissodactyl-dominated faunas adapted to warm, humid environments were replaced by faunas dominated by rodents and lagomorphs, which were adapted to cooler and drier conditions. This faunal turnover, which correlates with the *Grande Coupure* in Europe, has been dubbed the Mongolian Remodeling (Meng and McKenna, 1998).

Multiple episodes of faunal exchange took place between Asia and North America during the Paleocene and Eocene, probably dictated by periods of warming. Although it has often been assumed that the Turgai Strait and Obik Sea formed a complete barrier to dispersal between Asia and Europe during most of that time, recent studies suggest otherwise; as mentioned above, it has been argued that some Asian mammals crossed that "barrier" to reach Europe around the Paleocene/Eocene boundary, perhaps via ephemeral islands or land bridges. There also seems to have been some kind of limited biogeographic connection with northern Africa and the Arabian Peninsula. This link is most evident in the later Eocene, based on closely related rodents, anthracotheres, tethytheres, and anthropoid primates in the two regions (e.g., Ducrocq, 1997; Beard, 1998a), but the connection could be much older.

SYNOPSIS OF PALEOCENE AND EOCENE MAMMALS

Although mammals began to diversify almost immediately after the K/T boundary, several hundred thousand years passed before the radiation was really under way, a time lag perhaps related to floral evolution. In western North America there is evidence that the flora immediately after the K/T boundary was impoverished, and that angiosperms did not recover from the devastation caused by the bolide impact for several hundred thousand years (Wolfe and Upchurch, 1986; Wing, 1998a). This is roughly the same amount of time between the K/T boundary and the first significant increase in diversity of mammalian herbivores in the Western Interior, suggesting that coevolution of the flora and fauna was already important in the earliest Paleocene.

The mammalian faunas of the Paleocene and Eocene evolved in global greenhouse conditions, and most were therefore adapted to tropical and subtropical environments. Although these faunas were populated with many extinct, archaic clades, almost all the major modern clades first appeared during this interval. These Early Cenozoic faunas shaped the faunas that emerged during and after the global cooling near the end of the Eocene. Lineages that could adapt to the changing climatic and floral conditions survived to shape extant faunas.

The salient aspects of the evolution of the mammalian higher taxa of the Paleocene-Eocene are summarized here, based on the preceding chapters (where references are given). Detailed discussions of the current status of knowledge concerning the origins and relationships of extant placental orders can be found in Rose and Archibald (2005).

Nontribosphenic Mammals: Multituberculata and Dryolestoidea

Although primarily a Mesozoic radiation (during which they coexisted with dinosaurs for 100 million years), multituberculates survived for the first 25 million years of the Cenozoic, until the late Eocene. Their duration of about 125 million years exceeds that of any clade of Eutheria or Metatheria (in fact being equivalent to the known range of each of those infraclasses), and together with their broad distribution and relative abundance, ranks Multituberculata as one of the most successful of all mammalian clades. They were present on all three Holarctic continents during the Paleocene and were a major component of North American faunas, accounting for up to 20% of mammalian species and 25% of individuals in many assemblages. The principal Paleocene multituberculates were ptilodontoids, at least some of which were arboreal, and taeniolabidoids, which included fossorial, terrestrial, and arboreal forms. Only a few genera persisted into the Eocene. They were locally common in some early Eocene assemblages, but for the most part multituberculates declined markedly after the Paleocene and became extinct by the end of the Eocene.

Except for multituberculates, the nontribosphenic Mesozoic clades were long thought to have disappeared by the end of the Cretaceous. The reinterpretation of the enigmatic early Paleocene *Peligrotherium* (initially described as a condylarth) from Argentina as a relict eupantothere invalidates this assumption. *Peligrotherium* appears to be the only known Cenozoic representative of the Dryolestoidea. While this may seem surprising, it is much less so in view of the probable close relationship of *Peligrotherium* to the Late Cretaceous dryolestoids *Reigitherium* and *Mesungulatum* from the Los Alamitos Formation of Patagonia.

Monotremata

Monotremes have never been diverse or particularly successful, but they have endured as long as any other mammalian order—at least 125 million years. The only Early Cenozoic monotreme, *Monotrematum* from the early Paleocene of Argentina, is also the only one from outside the Australian region, but it dates from long after monotremes had diverged from other mammals. It probably indicates a once much wider Gondwanan distribution of monotremes, and therefore could hint at unimagined diversity in the order. The fossil record of monotremes is still so poorly known that almost any new fossil is likely to provide important new insights.

Metatheria

Although metatherians are known from the Lower Cretaceous of both North America and Asia, the somewhat older occurrence of *Sinodelphys* in China and the occurrence of two other primitive metatherian orders in Asia (Deltatheroida and Asiadelphia) suggest an Asian origin of the infraclass. Based on the presence of *Holoclemensia* in the Albian of Texas and the diversity of Late Cretaceous didelphimorphs in western North America, however, it is reasonable to conclude that this region was the geographic center of origin of Marsupialia. Only one or two marsupial genera are believed to have crossed the K/T boundary, giving rise to the small radiation of Early Cenozoic didelphimorphs in North America. Despite the existence of northern latitude land bridges joining North America to Asia and Europe, few marsupials reached the Old World, and they, too, were didelphimorphs. Their role in Holarctic faunas was minor. Were it not for the chance dispersal of one or two marsupials to South America in the Late Cretaceous or early Paleocene, the group would merit little attention during the beginning of the Age of Mammals.

Marsupial immigrants to South America blossomed into a highly successful Early Cenozoic radiation of marsupials that occupied a variety of ecological roles similar to those filled by placentals on the northern continents and Africa. Thus Sparassodonta were the dominant carnivorous mammals rather than carnivorans and creodonts, polydolopids converged on multituberculates and carpolestid plesiadapiforms, and argyrolagoids occupied parts of the rodent/lagomorph niche. Only later in the Cenozoic did some of these northern placental groups reach South America. The restricted distribution of Early Cenozoic localities means that we have only discovered glimpses of this marsupial radiation. Nevertheless, at least five families of marsupials already existed in the early Paleocene fauna at Tiupampa, Bolivia.

The real success story, however, was the dispersal of marsupials into and across Antarctica and thence to Australia during the Paleocene and Eocene. In addition to marsupials, several kinds of placentals crossed the possibly ephemeral land connection between the southern tip of South America and the Antarctic Peninsula, perhaps during the Initial Eocene Thermal Maximum (Reguero et al., 2002), but only marsupials crossed the seaway to Australia. Current evidence suggests that microbiotheres are the sister group of, or even nested within, Australian marsupials, and that the ancestral form(s) reached Australia in the Paleocene, perhaps via a filter rather than a sweepstakes route (Woodburne and Case, 1996).

Cimolesta

This eclectic assemblage of archaic eutherians includes some of the most bizarre Early Tertiary mammals (didelphodonts, apatotheres, taeniodonts, tillodonts, pantodonts, pantolestans, and pholidotans). All except Pholidota were restricted to the Paleogene, and most were extinct by the end of the Eocene. Broader relationships remain uncertain for most of them, but it has been proposed that the most plesiomorphic members, cimolestids, are a plausible stem group for the others grouped here, as well as for creodonts and carnivorans. As all of them, except creodonts, are known at least as far back as the early Paleocene (mostly in North America), they must have originated in the Late Cretaceous or evolved rapidly in the early Paleocene. Indeed, recent evidence suggests that Taeniodonta had already diverged in the Late Cretaceous. Whether the taxonomic arrangement followed here (based on McKenna and Bell, 1997) has any validity is a matter for future investigation. Other arrangements are possible and can be expected to emerge from ongoing and future studies.

The groups included under this heading attained a wide range of unusual specializations. Apatotheres were small, arboreal mammals that convergently evolved an insect-foraging strategy like that in the living aye-aye (*Daubentonia*). They were present in both North America and Europe but were never common. Taeniodonts, tillodonts, and pantodonts were medium-sized to large terrestrial herbivores. The earliest forms were small, however, indicating an origin in each case from a *Cimolestes*-sized animal. Taeniodonts and tillodonts seem to have dug with robust forelimbs and gnawed with enlarged (nonhomologous) front teeth, whereas most pantodonts were more generalized plant eaters. Taeniodonts were apparently confined to North America and must have originated there, whereas the other two groups ranged across the northern continents via both the Bering and North Atlantic corridors, but evidently reached Europe last.

The early occurrence and diversity of tillodonts in Asia suggests an Asian origin of the order, although Paleocene cimolestids are unknown there. Pantodonts are known from the early Paleocene in Asia and North America, and possibly just as early in South America, making their place of origin equivocal. If they are the sister group of tillodonts, as seems probable, then an Asian origin is likely. Pantodonts included some of the biggest Paleocene and early Eocene terrestrial mammals. *Coryphodon* was the most common and widely distributed large mammal of that interval. Pantodonts are generally hoofed and have sometimes been associated with ungulates, whereas tillodonts were at one time thought to have evolved from arctocyonids. The possibility that the tillodont-pantodont clade actually belongs with ungulates may be worth reconsideration.

Pantolestans also ranged across the northern continents and northern Africa but were relatively rare. Most became moderate-sized animals, specialized for digging, climbing, or semiaquatic life. Special resemblances to palaeanodonts and lepticids suggest that these three groups could be more closely related to one another than any of them is to cimolestans. Palaeanodonts and pholidotans are discussed later in this chapter.

Creodonta and Carnivora

Creodonts (mainly hyaenodontids) were the predominant Old World carnivorous mammals in the Early Cenozoic. Hyaenodontids became widespread through the northern continents, including northern Africa, during the Eocene. Oxyaenids were more common and diverse in North America. Although a few oxyaenid genera were present in Europe during the Eocene, most of them were probably immigrants from North America. *Oxyaena* was present in both North America and Europe and may have existed in Asia as well (Tong and Wang, 1998). Otherwise, middle Eocene *Sarkastodon*, also probably derived from North America, was the only Asian oxyaenid.

The monophyly of creodonts is not strongly based, but they are generally thought to have evolved from a *Cimolestes*-like form, which implies ghost lineages during at least the first half of the Paleocene for Oxyaenidae and nearly the entire Paleocene for Hyaenodontidae. A North American origin, although possible, is by no means certain. The oldest record of oxyaenids is in the Tiffanian of North America, their precise source unknown. Hyaenodontids first appear in the northern continents (Europe and North America, at least) as part of a wave of immigrants associated with the Initial Eocene Thermal Maximum. A probable latest Paleocene record of *Prolimnocyon* from Inner Mongolia implies an earlier occurrence (and perhaps origin) of hyaenodontids in Asia (Beard, 1998a), but the virtual absence of a Paleocene record of the family leaves its origin equivocal. Whether creodonts share an ancestry with Carnivora exclusive to other mammals is also moot.

Miacoids, or stem carnivorans, probably arose in North America, also from *Cimolestes*-like forms, in the early Paleocene or possibly before. Whether the P^4/M_1 carnassial complex diagnostic of Carnivora evolved only once or multiple times is now in question, raising the possibility of a diphyletic Carnivora. (The taxonomic dilemma could be avoided by restricting Carnivora to the crown-group, or the crown-group + its sister taxon, but this solution contributes nothing to resolving the real question of how many times the carnassial complex arose.) Miacoids are known on all three northern continents and were the only carnivorans through most of the Paleocene and Eocene. They are the presumed ancestral group for crown carnivorans, but only for one family, Canidae, have transitional stages been confidently identified. Whether or not miacids and viverravids represent stem caniforms and feliforms, respectively, the oldest unequivocal members of these two major extant carnivoran clades are unknown until the late middle Eocene (Duchesnean) or later.

During the Early Cenozoic, feliforms were primarily an Old World clade. The oldest feliforms occur in the late Eocene of Asia (the viverrid *Stenoplesictis*); viverrids and felids first appear in Europe (Quercy) just after the *Grande Coupure*, apparently having dispersed from Asia. Controversy over the taxonomic assignments of these basal feliforms suggests that they lie near the base of the clade. Felids did not reach North America until the Miocene, and viverrids never did. Nimravids (whose relationships are still controversial), however, were widespread across the northern continents (including North Africa in the Miocene), probably appearing first in the late Eocene of North America.

Among caniforms, the canids were strictly North American from their first appearance in the Duchesnean until the Miocene. Curiously, they apparently did not disperse to Asia until the Neogene. They appear to have evolved directly from miacids. Arctoids are also first known from the North American Duchesnean. By the end of the Eocene, amphicyonids, ursids, and musteloids were present in North America and had dispersed through Beringia to the Old World. Except for late Eocene amphicyonids in Europe and the ursid *Cephalogale* in Asia, however, they were not well represented in Eurasia until the early Oligocene (Quercy).

Carnivorans were not present on the southern continents during the Paleocene-Eocene, but they eventually reached Africa, Madagascar, and South America during the Neogene.

Insectivora

Leptictids and their Cretaceous relative *Gypsonictops* have been considered to be among the most primitive eutherians, based on their dental and cranial anatomy. The skeleton is unknown in *Gypsonictops*, but in leptictids and pseudorhyncocyonids it is specialized for running and hopping, reminiscent of that in Cretaceous zalambdalestids and extant elephant shrews. Although leptictids have been interpreted as the sister taxon of Lipotyphla, the postcranial skeleton suggests possible relationships to macroscelidids, Glires, pantolestids, and palaeanodonts. Further studies are clearly warranted.

In recent years the composition and higher relationships of Lipotyphla, as inferred from morphology and the fossil record, have been challenged by molecular evidence. Because the anatomical evidence for a lipotyphlan clade is weak, it may not be so difficult to abandon Lipotyphla in favor of the more likely monophyletic Eulipotyphla (moles, shrews, and hedgehogs). However, transfer of tenrecs and golden moles to the Afrotheria finds little support from morphology. Moreover the proposition that moles are the most primitive eulipotyphlans, whereas soricids and erinaceids are more derived sister taxa, appears to be contradicted by both the fossil record and the anatomy of extant members.

Lipotyphla is the only extant order for which potential Cretaceous representatives have been identified, but the relationships of these dental taxa (supposed soricomorphs *Otlestes, Batodon,* and *Paranyctoides*) with either Soricomorpha or Lipotyphla is far from secure. More complete fossils are desperately needed. Many Early Tertiary forms have also been assigned to the order, but in several cases (e.g., adapisoriculids, nyctitheres, apternodonts, various erinaceomorphs) there is still substantial disagreement concerning their relationships, including whether some of them belong to this clade at all. This confusion arises largely because most taxa are based solely on dentitions. Aside from the putative Cretaceous soricomorphs, the oldest eulipotyphlans are from the Torrejonian (the erinaceomorph *Adunator* and the soricomorph *Leptacodon*). Several families of supposed erinaceomorphs are known from the Paleocene and Eocene. The oldest presumed Erinaceidae (dental taxa) are from the Tiffanian, whereas the oldest representatives of an extant subfamily are middle Eocene galericines from Asia. A half-dozen families of soricomorphs are known from the Early Tertiary, but only one of them, Soricidae, survives today. Again, most extinct forms are known only from dentitions, making it difficult to be certain of their soricomorph ties. Definitive shrews (heterosoricines) are first known from the Uintan of North America and the early Oligocene of Europe. Slightly older Asian forms seem to bridge the gap between nyctitheres and shrews. Talpoids, which are more derived postcranially than other eulipotyphlans, first appear in the record in the late Eocene and are not common until after the *Grande Coupure*. These observations are difficult to reconcile with molecular studies that place talpids at the base of Eulipotyphla.

Archonta

If either Archonta or Euarchonta is monophyletic, the possibility that Late Cretaceous *Deccanolestes* from India represents a basal member raises major questions about the time and place of origin of this clade. If *Deccanolestes* is an archontan, Archonta (or its stem group) must have existed in Africa in the Cretaceous, and early evolution would have taken place on northward-drifting India (Krause and Maas, 1990). Evidence of Paleocene archontans in India would strengthen this argument, but no Paleocene mammals are yet known from the subcontinent. If plesiadapiforms are archontans, as strongly suggested by their anatomy, their presence on Holarctic continents in the Paleocene is contrary to the Indian origin hypothesis. It would be consistent, however, with an earlier African origin and dispersal from there in the Late Cretaceous or early Paleocene.

When Chiroptera first appear in the fossil record—as scattered occurrences in the early Eocene across Holarctica and possibly Australia—they are incontrovertibly bats. No transitional forms have been found, and the phylogenetic, geographic, and chronologic origins of the order remain unknown. Six families are already recognized in the early Eocene. This diversity, together with the obvious chiropteran anatomical specializations already present in the early Eocene, implies an extended period of their history (at least well down into the Paleocene) for which we have no recognized record. Although bats generally tend to be rare as fossils, the localities of Messel, Germany, and the Green River Formation of Wyoming have provided exceptional insight into the anatomy, flight behavior, and diet of the earliest bats.

The fossil record has shed little light on the origin or relationships of Scandentia and Dermoptera. The entire fossil record of Scandentia consists of a few teeth from the middle Eocene of China and a few other specimens from the late Tertiary of southern Asia, leaving little indication of their source or time of origin. The single known fossil galeopithecid, based on a jaw from Thailand, suggests that dermopterans have long been confined to southeast Asia. If either plagiomenids or (less likely) paromomyids are stem dermopterans, however, a North American or perhaps European origin is possible. Curiously, both of these families are constituents of the early to middle Eocene fauna of Ellesmere Island, within the Eocene Arctic Circle. Paromomyids are found in North America, Europe, and Asia (Tong and Wang, 1998), but plagiomenids apparently never left North America.

The history of primates is relatively well documented, but not their origin. The Paleocene record consists solely of plesiadapiforms (with the possible exception of North African late Paleocene *Altiatlasius*). Plesiadapiforms, as well as euprimates, are usually traced back to early Paleocene *Purgatorius*, although no transitional forms leading to euprimates have been identified, and the source of *Purgatorius* itself is completely unknown. Plesiadapiforms were diverse and abundant in North America, and probably Europe and Asia as well, but their record on those continents is neither as rich nor as diversified. Dispersal across high-latitude land bridges between North America and either Asia (e.g., carpolestids) or Europe (e.g., plesiadapids, saxonellids, paromomyids) is indicated.

Aside from the aforementioned *Altiatlasius*, euprimates appeared abruptly at the beginning of the Eocene and spread across the northern continents during the brief Initial Eocene Thermal Maximum. The oldest known euprimates, adapoids and omomyids, were the most abundant and widespread primates during the Eocene, but increasing evidence suggests that strepsirrhines, tarsiids, and anthropoids may also go back to (or nearly to) the beginning of Euprimates.

Xenarthra and Pholidota

Both morphological and molecular data support a monophyletic Xenarthra that lies near the base of Eutheria, either alone or as the sister taxon of Afrotheria (molecular data only). It may not be surprising, therefore, that its origin is unknown. All definitive Early Tertiary xenarthrans are from South America, and it appears that the order has always been confined to the New World (and Antarctica). The oldest xenarthran remains (armadillos) come from the late Paleocene (Itaboraian-Riochican). Even in the most primitive xenarthrans, the teeth are so modified and reduced that no hint of their ancestry is evident. The only group ever proposed as a plausible source, North American palaeanodonts, remains but a weak possibility.

Dasypodids were relatively diverse throughout the nominal Eocene. Other cingulates (glyptodonts and a possible pampathere), however, are not known until the late Eocene (Mustersan). Other xenarthrans appear even later in the record (a sloth with glyptodont-like teeth in the Tinguirirican, and modern-aspect anteaters in the Miocene). The few fragments of xenarthrans from the late Eocene of Antarctica are apparently of South American affinity. The fossil record of xenarthrans is obviously very incomplete.

The only serious challenge to the New World distribution of Xenarthra is *Eurotamandua* from Messel. Its unexpected occurrence in Europe, prior to South American anteaters, is attributed to a trans-Tethyan dispersal from Africa, which in turn depends on a broad South American–African distribution of xenarthrans before the separation of those continents in the Late Cretaceous (Storch, 1993b). No evidence of such a distribution, or of Xenarthra prior to the late Paleocene, has been found. Furthermore, despite its superficial resemblance to extant *Tamandua, Eurotamandua* shares numerous detailed features with palaeanodonts and *Eomanis,* which indicate a closer relationship to those taxa than to Xenarthra.

Palaeanodonts are known from the Torrejonian through the end of the Eocene in North America and from scattered occurrences in Europe and Asia. They appear to be related to pantolestids, which is the principal reason to ally them with Cimolesta (as proposed by McKenna and Bell, 1997). However, available evidence suggests that both palaeanodonts and pantolestids may have evolved from leptictids rather than from cimolestids. Detailed similarities throughout the skeleton indicate that palaeanodonts are the sister group, if not the direct ancestor, of Pholidota *sensu stricto.* Nevertheless, a few threads of evidence still suggest possible ties with Xenarthra; but this putative link is the only basis at present to support an Edentata uniting Xenarthra and Pholidota. Primitive pholidotans, though very rare, have been found in the middle or late Eocene of all three Holarctic continents.

Ungulata

One of the paramount questions here remains whether Ungulata is a monophyletic group. Current anatomical evidence supplies no definitive answer. The few traits unifying the group's members—a suite of rather vague dental attributes associated with herbivory and postcranial features related to cursorial locomotion—are clearly subject to homoplasy. However, morphological evidence does support several ungulate clades, including Altungulata, Tethytheria (more strongly than Paenungulata), Perissodactyla, and Artiodactyla (or Cetartiodactyla, depending on whether Cetacea are interpreted as the sister group of Artiodactyla or as nested within it). Molecular evidence generally rejects a monophyletic Ungulata, but supports subgroups of ungulates, including Perissodactyla, Paenungulata (more strongly than Tethytheria), and Cetartiodactyla. Perhaps the lack of clear support for Ungulata based on either morphological or molecular data is sufficient cause to abandon the term as a formal taxonomic name.

Another significant unsolved problem is the source of ungulates. Cretaceous zhelestids are a plausible—but not unequivocal—stem group of ungulates, and no intermediate stages are known that tie them into early Paleocene *Protungulatum* and its close relatives (the oldest mammals widely accepted as ungulates). In fact, before the discovery of zhelestids, *Protungulatum* and kin were considered to be derivatives of cimolestids.

Archaic Ungulates

Condylarths encompass seven families of primitive ungulates that are generally thought to include the ancestors or sister taxa of all other ungulates. They were immensely successful mammals during the Paleocene and Eocene. Most primitive are arctocyonids (a basically North American and European group), including *Protungulatum* and its close relatives, which are probably basal to most or all other condylarth families and to mesonychians and artiodactyls as well. Several forms that are potentially intermediate between arctocyonids and other groups have been identified, but most are known primarily or only from teeth. Better knowledge of other aspects of their anatomy is desirable before transferring these taxa to other clades. Mioclaenids, a mainly early Paleocene group known from both North and South America, appear to share derived dental features with South American didolodontids and litopterns and may be their stem group; they are the only Neotropical condylarths besides didolodontids. Didolodonts are currently thought to have evolved from mioclaenids and to be the sister group of litopterns. Periptychids are an odd group definitively present only in the Paleocene of North America. They may be related to mioclaenids but seem to have been a dead end. Hyopsodontidae were among the most common, diverse, and widespread of all condylarths. Although it is still possible that they are related to the origin of artiodactyls, the only broader relationship for which there is good evidence is the proposed link between apheliscine and louisinine hyopsodontids and Macroscelidea. An important implication of this potential relationship is that either Afrotheria originated outside Africa or Macroscelidea are not afrotheres, both of which conflict with molecular conclusions. Phenacodontids appear to play a central role in the origin of perissodactyls and paenungulates (a large subset of afrotheres). Their anatomy shows many of the specializations that were further developed in those advanced ungulate clades. Again, however, these morphological data conflict with molecular evidence, which sets perissodactyls far apart from paenungulates. Resolving these discrepancies will be one of the major challenges of paleomammalogy in the near future.

The origin of South American ungulates, which McKenna (1975a; McKenna and Bell, 1997) termed Meridiungulata, has always been a conundrum. Although there is little question of the monophyly of each of the individual clades (Litopterna, Notoungulata, Astrapotheria, Pyrotheria, and Xenungulata), the evidence to unite them all in a superorder is weak and seems to be based largely on their restriction to South America. They are usually held to have evolved from North American condylarths, but aside from the mioclaenid alliance with litopterns, no close relationships have been demonstrated. Meridiungulates radiated in South America to fill many of the niches occupied by perissodactyls, artiodactyls, and paenungulates on the northern continents. All are first known from the nominal late Paleocene, except pyrotheres (Eocene). A number of interesting proposed relationships of various meridiungulates (e.g., to Proboscidea, Sirenia, Dinocerata) may be worthy of reevaluation when the fossil record improves.

Arctostylopids have an unusual dentition resembling that of primitive notoungulates more than that of any other mammal. The predominance of arctostylopids in the late

Paleocene of Asia, however, strongly suggests that they originated there, rather than in South America (or North America); limited evidence suggests potential affinity with pseudictopids. But until cranial and postcranial anatomy is better known, it is premature to exclude the possibility of a notoungulate relationship.

Dinocerata are included in the Ungulata *faute de mieux*. They are known from the late Paleocene through middle Eocene of North America and Asia and probably originated in Asia, as no transitional forms have been found in North America. An arctocyonid ancestry has been proposed, as well as potential relationships with xenungulates, pantodonts, or pseudictopids (the last two not even ungulates), but none of these hypotheses is very compelling. Overall similarity suggests that pantodonts may be their closest relatives, but the resemblances could simply be convergent.

Altungulata

The fossil record points strongly toward phenacodontid condylarths as the sister group or source of altungulates (perissodactyls and paenungulates). Fossils that could represent transitional stages between phenacodontids and altungulates (e.g., *Radinskya, Olbitherium*) have been found in Asia but are known only from dentitions. The Asian Paleocene Phenacolophidae, though now assigned to Embrithopoda, may also (or instead) play an important role as stem altungulates or stem tethytheres; at present they are too poorly known to be sure of their precise phylogenetic position. More complete fossils of all these potential intermediates are desperately needed. Nevertheless, these fossils lead to the tentative conclusion that altungulates arose in Asia.

Although there is no dispute about the monophyly of Perissodactyla, there is disagreement over which perissodactyl group is most primitive (tapiroids, brontotheres, or equoids). The oldest perissodactyls appear suddenly as part of the wave of immigrants that define the base of the Eocene across the northern continents. Their geographic source is unknown. All five major groups of perissodactyls had diverged by the end of the early Eocene, but the earliest and most primitive members of most clades are anatomically very similar and thus difficult to distinguish. Perissodactyls were common on the three northern continents, but most groups were prevalent in only part of this range (e.g., equids in North America and Europe, palaeotheres in Eurasia, tapiroids and brontotheres in North America and Asia). Rhinocerotoids flourished throughout this range in the Early Cenozoic and include the largest known land mammals.

Hyracoids, the sister group of tethytheres, are a mainly African group whose Early Cenozoic record is entirely African. At that time they were the principal ungulates of northern Africa, filling the niches that were occupied elsewhere by artiodactyls and perissodactyls (which did not reach Africa until later). By the middle Eocene hyracoids had radiated into a much wider range of body size and diet than at any other time in their history.

The Asian Anthracobunidae are either the most primitive proboscideans or (more likely) very primitive tethytheres. Little is known about them besides dentitions, often incomplete, leaving their phylogenetic position ambiguous. Except for anthracobunids, all early proboscideans are African (suggesting they arose there), and the clade was already distinct by the early Eocene. Embrithopods were distinct even earlier (by the late Paleocene), if phenacolophids belong to this clade. Excluding phenacolophids, embrithopods are known from only a few middle and late Eocene genera from eastern Europe, Turkey, and northern Africa. Sirenians had the broadest Early Cenozoic distribution among tethytheres, being known from the Caribbean (Jamaica) through northern Africa to Indo-Pakistan during the Eocene. They, too, were distinct by the early Eocene.

Janis (1989) postulated that more than half of Early Tertiary ungulates were hindgut fermenters. Most of them were perissodactyls and paenungulates. Pantodonts and uintatheres may also have been hindgut fermenters, but their status as ungulates is questionable. The proportion of hindgut fermenters declined in higher latitudes to about one-quarter of ungulates with the dramatic cooling near the Eocene/Oligocene boundary.

Cete and Artiodactyla

Based on recent fossil evidence and molecular data, it now appears undeniable that Cetacea and Artiodactyla are sister taxa. Molecular data, however, increasingly indicate that Cetacea is the sister taxon of Hippopotamidae to the exclusion of other artiodactyls, whereas very little morphological evidence supports that conclusion. The combination of artiodactyl-like tarsal features (at a primitive stage) with highly specialized skulls and dentition is more consistent with Cetacea being the sister taxon of all Artiodactyla. However, nothing is known of the fossil record of hippopotami until the Miocene, and their proposed links with anthracotheres, although plausible, are far from compelling. Nonetheless, a recent analysis by Boisserie et al. (2005) nested hippos within anthracotheres and found this clade to be the sister group of Cetacea. Hence the possibility that a Cetacea + Hippopotamidae (or anthracotherioid) clade diverged from all others in the Paleocene remains viable and should be tested when pertinent new fossils are unearthed. Another recent study (Geisler and Uhen, 2005) indicated that raoellids, endemic Asian Eocene artiodactyls that occur together with the oldest fossil whales in India and Pakistan, could be the sister taxon of Cetacea. As is often the case, until raoellids are better known, their potential relationships will remain controversial.

Where does this leave Mesonychia? Until recently, it was the presumed sister group of Cetacea, and the numerous derived anatomical resemblances between mesonychians and basal archaeocetes seem too great to ignore. It is very improbable that artiodactyls, which primitively have generalized tribosphenic molars, evolved from an ancestor with the derived dentition of mesonychians. Therefore, if whales are nested within Artiodactyla or are the sister group of Artiodactyla, the specialized dentitions of mesonychians and archaeocetes must have evolved independently. Mesonychia might still be the sister group of Cetartiodactyla, as suggested by common features of the postcranial skeleton. A better knowledge of triisodontine arctocyonids, which could represent an intermediate stage between basal arctocyonids and mesonychians, is critical.

Current evidence indicates that Cetacea was a separate clade by the end of the early Eocene and possibly early in the early Eocene, if the date for *Himalayacetus* is reliable. The remarkable diversity of archaeocetes in Indo-Pakistan suggests that Cetacea originated along the margin of the eastern Tethys. Exceptional skeletons discovered in recent years document the transformation from quadrupedal terrestrial habits to a fully committed aquatic lifestyle. Marine Eocene archaeocetes became virtually cosmopolitan. Archaeocetes predominated until the end of the Eocene, when both clades of modern whales (mysticetes and odontocetes) emerged, probably in association with global cooling and changes in oceanic circulation.

Artiodactyls first appear in the record at the beginning of the Eocene (at least 2 Ma before Cetacea), together with perissodactyls and euprimates. Their source is unknown. They are generally thought to have evolved from condylarths, possibly hyopsodontids or mioclaenids, but an arctocyonid ancestry seems slightly more probable, based on current evidence. The earliest forms, dichobunoids, were small, bunodont, gracile animals with long limbs, resembling living tragulids or leporid lagomorphs. Dichobunoids are usually considered to be the stem group for virtually all other artiodactyls, but some authorities consider dichobunoids to be the sister group of selenodont artiodactyls (but not suiforms). By the end of the Eocene all major artiodactyl clades were recognizable, although ruminants had just begun to diversify, and no extant ruminant family had yet emerged.

Suiformes is used to unite a variable assortment of conservative, bunodont artiodactyls (see Table 14.1), but we lack a clear understanding of their relationships, and it seems improbable that it is a monophyletic group. At the center of the group are pigs and peccaries, first known from the late Eocene. They are assumed to have evolved from primitive bunodont artiodactyls, although the precise pedigree is unknown. Aside from suids, the most diverse suiforms are anthracotheres, first known from the middle Eocene and widespread across the Holarctic continents and northern Africa from the late Eocene to the Miocene. Based on early diversity, they may have originated in Asia, but there is no consensus on which bunodont artiodactyl group was the likely source. Anthracotheres are generally considered to include the antecedents of hippos, which have taken on special significance as the proposed sister group of Cetacea. Unfortunately the fossil record of hippos provides little direct evidence of their origin. Such ambiguities increase the likelihood that Hippopotamidae could represent a very ancient clade.

A host of selenodont artiodactyls of uncertain relationships characterizes the Eocene and Oligocene of North America (oreodonts) and Europe (mixtotheres, anoplotherioids, cainotheres, and xiphodonts). The difficulty of assigning these taxa underscores the extent of homoplasy widespread in early artiodactyls. Advanced selenodont artiodactyls—tylopods and ruminants—first appear in the middle Eocene. Tylopods were a North American radiation that did not disperse to the Old World until the Miocene. The earliest definitive ruminants occur in both North America and Asia, and their place of origin cannot yet be determined with confidence. Possible ruminants (amphimerycids) occur in the early Eocene of Europe, but ruminants were not common in Europe until after the *Grande Coupure*.

Anagalida

The superorder Anagalida has been used to unite Glirans (Simplicidentata and Duplicidentata) with the otherwise orphaned extinct families Anagalidae, Pseudictopidae, and possibly Zalambdalestidae, as well as the order Macroscelidea. This superorder is doubtfully a monophyletic group, although subsets of it probably are. All except Macroscelidea share an Asian origin, probably in the Paleocene (apart from Zalambdalestidae), which increases the probability of their monophyly. In the most recent analyses the Cretaceous Zalambdalestidae usually cluster with primitive Asian eutherians, but several unusual derived features in common with Glires make it difficult to completely reject the possibility of a relationship with the latter.

Anagalidae share certain characters with glirans, such as an ectotympanic bulla and unilateral hypsodonty, but they lack gliriform incisors and show little tendency in that direction. Except for Oligocene *Anagale*, they are frustratingly poorly known. Pseudictopids share some derived dental features with anagalids and lagomorphs and have a superficially lagomorph-like skeleton, but the incisors are not gliriform. The general similarities shared by these families and glirans are tantalizing, but the contrasts are troubling. A compelling case for a relationship will require stronger evidence. Once again it seems that only more complete fossils can hope to resolve the ambiguity.

Definitive members of the Macroscelidea are all African, and the oldest come from the Eocene. As noted above, apheliscine and louisinine hyopsodontids from North America and Europe, respectively, could be stem macroscelideans. If corroborated, this relationship would argue against inclusion of Macroscelidea in either Anagalida or Afrotheria.

Fossil evidence accumulated over the past 20 years or so, as well as molecular analyses, increasingly support the monophyly of Glires, consisting of two major clades, Simplicidentata (rodents and eurymylids) and Duplicidentata (lagomorphs and mimotonids). In each case the primitive, Early Tertiary sister taxon has some of the derived traits characteristic of the extant order, reflecting an intermediate stage in their evolution. Nearly all of the critical fossils are from eastern Asia (China and Mongolia), thus an Asian origin of both major clades is highly probable. The oldest glirans (*Mimotona* and the eurymylid *Heomys*) are first known from the early Paleocene of Qianshan, China, suggesting that the superordinal clades Glires, Duplicidentata, and Simplicidentata might trace their roots back to the earliest Paleocene or perhaps into the latest Cretaceous (in accord with one recent molecular study: Huchon et al., 2002). But there is no fossil evidence to support an origin deep in the Cretaceous, as asserted by some molecular systematists. Furthermore, the fossil evidence strongly suggests that Rodentia originated during the Paleocene and Lagomorpha in the Paleocene or early Eocene.

Duplicidentata and Lagomorpha are well-supported clades. The oldest lagomorphs come from the middle Eocene of North America and Asia (where they are more diverse and perhaps slightly older). The weight of the evidence indicates an Asian origin and immigration to North America. Simplicidentata (less well supported than Rodentia or the other major gliran clades) are first reported from late Paleocene sediments of Asia and North America; again the evidence points to an Asian origin and rapid dispersal to North America. Rodents did not reach Europe (at least the areas sampled) until the beginning of the Eocene. They thrived and diversified on the northern continents during the Eocene, although there was marked endemism (e.g., theridomyids and glirids in Europe, ctenodactyloids in Asia). Their ability to adapt well to a variety of environments and diets, while often remaining generalists, no doubt contributed to their extraordinary success. An unusual amount of homoplasy (and perhaps also intraspecific variability) in Eocene rodents has made untangling their relationships a particularly arduous task and has led to a lack of consensus among rodent experts.

The oldest rodents are late Paleocene alagomyids and ischyromyids. The squirrel-like ischyromyids were diverse and common across Holarctica into the early Oligocene. They are the probable stem group for other sciuromorph rodents, including castoroids (which first appear in the early Eocene of North America), glirids (early Eocene of Eurasia), aplodontoids (middle Eocene of North America), sciurids (late Eocene of North America), cylindrodonts (early Eocene of North America), and probably sciuravids and theridomyids. Alternatively the last two families might have descended from Asian ctenodactyloids.

Sciuravidae is generally considered the stem group of Myomorpha, the most successful rodent clade. The oldest known myomorph, early middle Eocene *Armintomys* from western North America, has numerous intermediate features that bridge the gap from sciuravids, suggesting that the group could have originated in North America. However, a variety of dipodoids and muroids had appeared in North America and Asia only slightly later in the middle Eocene, followed by the oldest geomorphans (Uintan of North America).

The mainly African anomaluromorphs are of interest here because they potentially include the oldest known African rodents, the late early Eocene zegdoumyids. Another interpretation, however, places zegdoumyids near glirids. The earliest anomalurids are known from the late middle Eocene of southeast Asia and the late Eocene of northern Africa. Pedetoids (the only other anomaluromorphs) are not known until the Miocene.

Ctenodactyloids were the predominant Paleogene rodents in Asia. They were already diverse at their initial appearance (Bumbanian), indicating a probable Paleocene origin; however, their derivation is unknown. Peculiarities of the dentition, as well as the widespread presence of hystricomorphy and multiserial enamel, have led most experts to conclude that they are the stem group of the extensive hystricognath rodent radiation, which includes the northern African Early Tertiary phiomyids and associated forms, hystricids, erethizontids, bathyergids, and New World caviomorphs. Recent discoveries suggest that caviomorphs arrived in South America from Africa, presumably by rafting, before the end of the Eocene.

The profound cooling and concomitant floral change at the end of Eocene precipitated or hastened the demise of some primitive rodents, but other rodent clades proliferated thereafter (Dawson 2003).

A FINAL NOTE

Contrary to frequent claims of its inadequacy (and despite the weaknesses enumerated above), the fossil record has, in many areas, preserved a remarkable portrait of the history of mammals. If we pause to consider what we knew (or thought we knew) only a generation ago, the gaps that have been filled in since then are staggering. Many of the greatest advances in our understanding of mammalian evolution and relationships have come from the discovery of new fossil evidence. These findings include the first skeletons of stem eutherians and metatherians from the Lower Cretaceous; the first Cretaceous taeniodont; possible Late Cretaceous stem ungulates and glirans; the enigmatic gondwanatheres; a Paleocene monotreme and an early Paleocene pantodont in South America; Eocene faunas from the Arctic and Antarctic that corroborate hypothesized Early Tertiary dispersal routes; middle Eocene anthropoids, tarsiers, and strepsirrhines; early Eocene proboscideans; quadrupedal Eocene sirenians; potential stem macroscelideans in North America and Europe; fossils that unite Glires; and transitional walking whales, whose anatomy demonstrates a close relationship to artiodactyls.

Such discoveries indicate that we have only scratched the surface of what the fossil record has to offer. We can anticipate that equivalent and even more astounding discoveries lie ahead. The current controversies concerning mammalian relationships and divergence times raised by molecular systematics make a better knowledge of the fossil record during the beginning of the Age of Mammals more relevant and important than ever before.

LITERATURE CITED

Adkins, R. M., and R. L. Honeycutt. 1991. Molecular phylogeny of the superorder Archonta. Proceedings of the National Academy of Sciences, USA 88:10317–10321.

Adkins, R. M., A. H. Walton, and R. L. Honeycutt. 2003. Higher-level systematics of rodents and divergence time estimates based on two congruent nuclear genes. Molecular Phylogenetics and Evolution 26:409–420.

Aimi, M., and H. Inagaki. 1988. Grooved lower incisors in flying lemurs. Journal of Mammalogy 69:138–140.

Akersten, W. A., H. Lowenstam, A. Walker, W. Traub, and A. Biknevicius. 2002. How and why do some shrews have red teeth? Journal of Vertebrate Paleontology 22:31A.

Alexander, J. P. 1994. Sexual dimorphism in notharctid primates. Folia Primatologica 63:59–62.

Allard, M. W., B. E. McNiff, and M. M. Miyamoto. 1996. Support for interordinal eutherian relationships with an emphasis on primates and their archontan relatives. Molecular Phylogenetics and Evolution 5:78–88.

Alroy, J. 1999. The fossil record of North American mammals: Evidence for a Paleocene evolutionary radiation. Systematic Biology 48:107–118.

Amrine, H., and M. S. Springer. 1999. Maximum-likelihood analysis of the tethythere hypothesis based on a multigene data set and a comparison of different models of sequence evolution. Journal of Mammalian Evolution 6:161–176.

Amrine-Madsen, H., K.-P. Koepfli, R. K. Wayne, and M. S. Springer. 2003. A new phylogenetic marker, apolipoprotein B, provides compelling evidence for eutherian relationships. Molecular Phylogenetics and Evolution 28:225–240.

Andrews, C. W. 1906. A Descriptive Catalogue of the Tertiary Vertebrata of the Fayum, Egypt. London, British Museum (Natural History), 324 p.

Anemone, R. L., and H. H. Covert. 2000. New skeletal remains of *Omomys* (Primates, Omomyidae): Functional morphology of the hindlimb and locomotor behavior of a

middle Eocene primate. Journal of Human Evolution 38: 607–633.
Aplin, K. P., and M. Archer. 1987. Recent advances in marsupial systematics with a new syncretic classification; pp. xv–lxxii *in* M. Archer (ed.), Possums and Opossums: Studies in Evolution. Surrey Beatty & Sons and the Royal Zoological Society of New South Wales, Sydney.
Archer, M., R. Arena, M. Bassarova, K. Black, J. Brammall, B. Cooke, P. Creaser, K. Crosby, A. Gillespie, H. Godthelp, M. Gott, S. J. Hand, B. Kear, A. Krikmann, B. Mackness, J. Muirhead, A. Musser, T. Myers, N. Pledge, Y.-Q. Wang, and S. Wroe. 1999. The evolutionary history and diversity of Australian mammals. Australian Mammalogy 21:1–45.
Archer, M., T. F. Flannery, A. Ritchie, and R. E. Molnar. 1985. First Mesozoic mammal from Australia—An early Cretaceous monotreme. Nature 318:363–366.
Archer, M., H. Godthelp, and S. J. Hand. 1993. Early Eocene marsupial from Australia. Kaupia-Darmstädter Beiträge zur Naturgeschichte 3:193–200.
Archer, M., P. Murray, S. Hand, and H. Godthelp. 1993. Reconsideration of monotreme relationships based on the skull and dentition of the Miocene *Obdurodon dicksoni*; pp. 75–94 *in* F. S. Szalay, M. J. Novacek, and M. C. McKenna (eds.), Mammal Phylogeny: Mesozoic Differentiation, Multituberculates, Monotremes, Early Therians, and Marsupials. Springer-Verlag, New York.
Archibald, J. D. 1982. A study of Mammalia and geology across the Cretaceous-Tertiary boundary in Garfield County, Montana. University of California Publications in Geological Sciences 122:1–286.
———. 1983. Structure of the K-T mammal radiation in North America: Speculations on turnover rates and trophic structure. Acta Palaeontologica Polonica 28:7–17.
———. 1996. Fossil evidence for a Late Cretaceous origin of "hoofed" mammals. Science 272:1150–1153.
———. 1998. Archaic ungulates ("Condylarthra"); pp. 292–331 *in* C. M. Janis, K. M. Scott, and L. L. Jacobs (eds.), Evolution of Tertiary Mammals of North America. Cambridge University Press, Cambridge, UK.
———. 2003. Timing and biogeography of the eutherian radiation: Fossils and molecules compared. Molecular Phylogeny and Evolution 28:350–359.
Archibald, J. D., and A. O. Averianov. 2001. *Paranyctoides* and allies from the Late Cretaceous of North America and Asia. Acta Palaeontologica Polonica 46:533–551.
———. 2003. The Late Cretaceous placental mammal *Kulbeckia*. Journal of Vertebrate Paleontology 23:404–419.
Archibald, J. D., and D. H. Deutschman. 2001. Quantitative analysis of the timing of the origin and diversification of extant placental orders. Journal of Mammalian Evolution 8:107–124.
Archibald, J. D., and D. L. Lofgren. 1990. Mammalian zonation near the Cretaceous-Tertiary boundary; pp. 31–50 *in* T. M. Bown, and K. D. Rose (eds.), Dawn of the Age of Mammals in the Northern Part of the Rocky Mountain Interior, North America. Geological Society of America Special Paper 243, Boulder.
Archibald, J. D., A. O. Averianov, and E. G. Ekdale. 2001. Late Cretaceous relatives of rabbits, rodents, and other extant eutherian mammals. Nature 414:62–65.
Archibald, J. D., W. A. Clemens, P. D. Gingerich, D. W. Krause, E. H. Lindsay, and K. D. Rose. 1987. First North American Land Mammal Ages of the Cenozoic Era; pp. 24–76 *in* M. O. Woodburne (ed.), Cenozoic Mammals of North America. University of California Press, Berkeley.
Argot, C. 2001. Functional-adaptive anatomy of the forelimb in the Didelphidae, and the paleobiology of the Paleocene marsupials *Mayulestes ferox* and *Pucadelphys andinus*. Journal of Morphology 247:51–79.
Arnason, U., A. Gullberg, A. Schweizer Burguete, and A. Janke. 2000. Molecular estimates of primate divergences and new hypotheses for primate dispersal and the origin of modern humans. Hereditas 133:217–228.
Asher, R. J., M. C. McKenna, R. J. Emry, A. R. Tabrum, and D. G. Kron. 2002. Morphology and relationships of *Apternodus* and other extinct, zalambdodont, placental mammals. Bulletin of the American Museum of Natural History 273:1–117.
Asher, R. J., J. Meng, J. R. Wible, M. C. McKenna, G. W. Rougier, D. Dashzeveg, and M. J. Novacek. 2005. Stem Lagomorpha and the antiquity of Glires. Science 307:1091–1094.
Asher, R. J., M. J. Novacek, and J. H. Geisler. 2003. Relationships of endemic African mammals and their fossil relatives based on morphological and molecular evidence. Journal of Mammalian Evolution 10:131–194.
Ashlock, P. D. 1971. Monophyly and associated terms. Systematic Zoology 20:63–69.
Aubry, M. P., W. A. Berggren, J. A. Van Couvering, J. Ali, H. Brinkhuis, B. Cramer, D. V. Kent, C. C. Swisher III, C. Dupuis, P. D. Gingerich, C. Heilmann-Clausen, C. King, D. J. Ward, R. W. O'B. Knox, K. Ouda, L. D. Stott, and M. Thiry. 2003. Chronostratigraphic terminology at the Paleocene/Eocene boundary; pp. 551–566 *in* S. L. Wing, P. D. Gingerich, B. Schmitz, and E. Thomas (eds.), Causes and Consequences of Globally Warm Climates in the Early Paleogene. Geological Society of America Special Paper 369, Boulder.
Averianov, A. O. 1991. Tarsals of Glires (Mammalia) from the early Eocene of Kirgizia. Géobios 24:215–220.
———. 1994. Early Eocene mimotonids of Kyrgyzstan and the problem of Mixodontia. Acta Palaeontologica Polonica 39:393–411.
———. 1996. Early Eocene Rodentia of Kyrgyzstan. Bulletin Muséum National d'Histoire Naturelle Paris, sér. 4, 18C:629–662.
Averianov, A. O., and J. D. Archibald. 2005. Mammals from the mid-Cretaceous Khodzhakul Formation, Kyzylkum Desert, Uzbekistan. Cretaceous Research 26:593–608.
Averianov, A. O., and M. Godinot. 1998. A report on the Eocene Andarak mammal fauna of Kyrgyzstan; pp. 210–219 *in* K. C. Beard and M. R. Dawson (eds.), Dawn of the Age of Mammals in Asia. Bulletin of Carnegie Museum of Natural History 34.
Averianov, A. O., and P. P. Skutschas. 2001. A new genus of eutherian mammal from the Early Cretaceous of Transbaikalia, Russia. Acta Palaeaontologica Polonica 46:431–436.
Averianov, A. O., J. D. Archibald, and T. Martin. 2003. Placental nature of the alleged marsupial from the Cretaceous of Madagascar. Acta Palaeontologica Polonica 48:149–151.
Ayala, F. J. 1997. Vagaries of the molecular clock. Proceedings of the National Academy of Sciences, USA 94:7776–7783.
Babot, M. J., J. E. Powell, and C. de Muizon. 2002. *Callistoe vincei*, a new Proborhyaenidae (Borhyaenoidea, Metatheria, Mammalia) from the early Eocene of Argentina. Géobios 35:615–629.

Bacon, A.-M., and M. Godinot. 1998. Analyse morphofonctionnelle des fémurs et des tibias des *"Adapis"* du Quercy: Mise en évidence de cinq types morphologiques. Folia Primatologica 69:1–21.

Bajpai, S., and P. D. Gingerich. 1998. A new Eocene archaeocete (Mammalia, Cetacea) from India and the time of origin of whales. Proceedings of the National Academy of Sciences, USA 95:15464–15468.

Barghusen, H. R., and J. A. Hopson. 1979. The endoskeleton: The comparative anatomy of the skull and the visceral skeleton; pp. 265–326 *in* M. H. Wake (ed.), Hyman's Comparative Vertebrate Anatomy. University of Chicago Press, Chicago.

Barnes, L. G., and J. L. Goedert. 2000. The world's oldest known odontocete (Mammalia, Cetacea) (abstract). Journal of Vertebrate Paleontology 20:28A.

Barnes, L. G., and E. Mitchell. 1978. Cetacea; pp. 582–602 *in* V. J. Maglio and H. B. S. Cooke (eds.), Evolution of African Mammals. Harvard University Press, Cambridge, Mass.

Barnett, C. H., and J. R. Napier. 1953. The rotatory mobility of the fibula in eutherian mammals. Journal of Anatomy (London) 87:11–21.

Barnosky, A. D. 1981. A skeleton of *Mesoscalops* (Mammalia, Insectivora) from the Miocene Deep River Formation, Montana, and a review of the proscalopid moles: Evolutionary, functional, and stratigraphic relationships. Journal of Vertebrate Paleontology 1:285–339.

Baskin, J. A. 1998. Mustelidae; pp. 152–173 *in* C. M. Janis, K. M. Scott, and L. L. Jacobs (eds.), Evolution of Tertiary Mammals of North America. Cambridge University Press, Cambridge, UK.

Baskin, J. A., and R. H. Tedford. 1996. Small arctoid and feliform carnivorans; pp. 486–497 *in* D. R. Prothero and R. J. Emry (eds.), The Terrestrial Eocene-Oligocene Transition in North America. Cambridge University Press, Cambridge, UK.

Baudry, M. 1992. Les tillodontes (Mammalia) de l'Éocène inférieur de France. Bulletin du Muséum National d'Histoire Naturelle, Paris, sér. 4, 14C:205–243.

Beard, K. C. 1988. The phylogenetic significance of strepsirhinism in Paleogene primates. International Journal of Primatology 9:83–96.

———. 1990. Gliding behaviour and palaeoecology of the alleged primate family Paromomyidae (Mammalia, Dermoptera). Nature 345:340–341.

———. 1993a. Phylogenetic systematics of the Primatomorpha, with special reference to Dermoptera; pp. 129–150 *in* F. S. Szalay, M. J. Novacek, and M. C. McKenna (eds.), Mammal Phylogeny: Placentals. Springer-Verlag, New York.

———. 1993b. Origin and evolution of gliding in Early Cenozoic Dermoptera (Mammalia, Primatomorpha); pp. 63–90 *in* R. D. E. MacPhee (ed.), Primates and Their Relatives in Phylogenetic Perspective. Plenum Press, New York.

———. 1998a. East of Eden: Asia as an important center of taxonomic origination in mammalian evolution; pp. 5–39 *in* K. C. Beard and M. R. Dawson (eds.), Dawn of the Age of Mammals in Asia. Bulletin of Carnegie Museum of Natural History 34.

———. 1998b. A new genus of Tarsiidae (Mammalia: Primates) from the middle Eocene of Shanxi Province, China, with notes on the historical biogeography of tarsiers; pp. 260–277 *in* K. C. Beard and M. R. Dawson (eds.), Dawn of the Age of Mammals in Asia. Bulletin of Carnegie Museum of Natural History 34.

———. 2002. Basal anthropoids; pp. 133–150 *in* W. C. Hartwig (ed.), The Primate Fossil Record. Cambridge University Press, Cambridge, UK.

———. 2004. The Hunt for the Dawn Monkey. Unearthing the Origins of Monkeys, Apes, and Humans. University of California Press, Berkeley.

Beard, K. C. and M. R. Dawson. 1999. Intercontinental dispersal of Holarctic land mammals near the Paleocene/Eocene boundary: Paleogeographic, paleoclimatic, and biostratigraphic implications. Bulletin de la Société Géologique de France 170:697–706.

Beard, K. C., and P. Houde. 1989. An unusual assemblage of diminutive Plesiadapiformes (Mammalia, ?Primates) from the early Eocene of the Clarks Fork Basin, Wyoming. Journal of Vertebrate Paleontology 9:388–399.

Beard, K. C., and R. D. E. MacPhee. 1994. Cranial anatomy of *Shoshonius* and the antiquity of Anthropoidea; pp. 55–97 *in* J. G. Fleagle and R. F. Kay (eds.), Anthropoid Origins. Plenum Press, New York.

Beard, K. C., and J. Wang. 1995. The first Asian plesiadapoids (Mammalia: Primatomorpha). Annals of Carnegie Museum 64:1–33.

———. 2004. The eosimiid primates (Anthropoidea) of the Heti Formation, Yuanqu Basin, Shanxi and Henan Provinces, People's Republic of China. Journal of Human Evolution 46:401–432.

Beard, K. C., L. Krishtalka, and R. K. Stucky. 1991. First skulls of the early Eocene primate *Shoshonius cooperi* and the anthropoid-tarsier dichotomy. Nature 349:64–67.

Beard, K. C., T. Qi, M. R. Dawson, B. Wang, and C.-K. Li. 1994. A diverse new primate fauna from middle Eocene fissure-fillings in southeastern China. Nature 368:604–609.

Beard, K. C., B. Sigé, and L. Krishtalka. 1992. A primitive vespertilionid bat from the early Eocene of central Wyoming. Comptes Rendus de l' Académie des Sciences Paris, sér. II, 314:735–741.

Beard, K. C., Y. Tong, M. R. Dawson, J. Wang, and X. Huang. 1996. Earliest complete dentition of an anthropoid primate from the late middle Eocene of Shanxi Province, China. Science 272:82–85.

Beck, R. A., A. Sinha, D. W. Burbank, W. J. Sercombe, and A. M. Khan. 1998. Climatic, oceanographic, and isotopic consequences of the Paleocene India-Asia collision; pp. 103–117 *in* M.-P. Aubry, S. G. Lucas, and W. A. Berggren (eds.), Late Paleocene–Early Eocene Climatic and Biotic Events in the Marine and Terrestrial Records. Columbia University Press, New York.

Benton, M. J. 1999. Early origins of modern birds and mammals: Molecules vs. morphology. BioEssays 21:1043–1051.

Berggren, W. A., and D. R. Prothero. 1992. Eocene-Oligocene climatic and biotic evolution: An overview; pp. 1–28 *in* D. R. Prothero and W. A. Berggren (eds.), Eocene-Oligocene Climatic and Biotic Evolution. Princeton University Press, Princeton.

Berggren, W. A., D. V. Kent, M.-P. Aubry, and J. Hardenbol, 1995a. Geochronology, Time Scales and Global Stratigraphic Correlation. SEPM (Society for Sedimentary Geology), Tulsa, Okla.

Berggren, W. A., D. V. Kent, J. D. Obradovich, and C. C. Swisher III. 1992. Toward a revised Paleogene geochronology; pp. 29–45 *in* D. R. Prothero and W. A. Berggren (eds.),

Eocene-Oligocene Climatic and Biotic Evolution. Princeton University Press, Princeton.

Berggren, W. A., D. V. Kent, C. C. Swisher III, and M.-P. Aubry. 1995b. A revised Cenozoic geochronology and chronostratigraphy; pp. 129–212 in W. A. Berggren, D. V. Kent, M.-P. Aubry, and J. Hardenbol (eds.), Geochronology, Time Scales and Global Stratigraphic Correlation. SEPM (Society for Sedimentary Geology), Tulsa, Okla.

Bergqvist, L. P., and E. V. Oliveira. 1995. Comments on the xenarthran astragali from the Itaboraí Basin (middle Paleocene) of Rio de Janeiro, Brazil. 12° Jornadas Argentinas de Paleontología de Vertebrados (Tucamán):14.

Bergqvist, L. P., E. A. L. Abrantes, and L. S. Avilla. 2004. The Xenarthra (Mammalia) of São José de Itaboraí Basin (upper Paleocene, Itaboraian), Rio de Janeiro, Brazil. Geodiversitas 26:323–337.

Berta, A. 1994. What is a whale? Science 263:180–181.

Biknevicius, A. R. 1986. Dental function and diet in the Carpolestidae (Primates, Plesiadapiformes). American Journal of Physical Anthropology 71:157–171.

Bleefeld, A. R., and W. J. Bock. 2002. Unique anatomy of lagomorph calcaneus. Acta Palaeontologica Polonica 47:181–183.

Bleefeld, A. R., and M. C. McKenna. 1985. Skeletal integrity of *Mimolagus rodens* (Lagomorpha, Mammalia). American Museum Novitates 2806:1–5.

Bloch, J. I. 1999. Partial skeleton of *Arctostylops* from the Paleocene of Wyoming: Arctostylopid-notoungulate relationship revisited. Journal of Vertebrate Paleontology 19(supplement to no. 3):32A.

Bloch, J. I., and D. M. Boyer. 2001. Taphonomy of small mammals in freshwater limestones from the Paleocene of the Clarks Fork Basin. University of Michigan Papers on Paleontology 33:185–198.

———. 2002. Grasping primate origins. Science 298:1606–1610.

———. 2003. Response to comment on "Grasping Primate Origins." Science 300:741c.

Bloch, J. I., and P. D. Gingerich. 1998. *Carpolestes simpsoni*, new species (Mammalia, Proprimates) from the late Paleocene of the Clarks Fork Basin, Wyoming. Contributions from the Museum of Paleontology, the University of Michigan 30:131–162.

Bloch, J. I., and M. T. Silcox. 2001. New basicrania of Paleocene-Eocene *Ignacius*: Re-evaluation of the plesiadapiform-dermopteran link. American Journal of Physical Anthropology 116:184–198.

Bloch, J. I., D. M. Boyer, P. D. Gingerich, and G. F. Gunnell. 2002. New primitive paromomyid from the Clarkforkian of Wyoming and dental eruption in Plesiadapiformes. Journal of Vertebrate Paleontology 22:366–379.

Bloch, J. I., D. C. Fisher, P. D. Gingerich, G. F. Gunnell, E. L. Simons, and M. D. Uhen. 1997. Cladistic analysis and anthropoid origins. Science 278:2134–2135.

Bloch, J. I., D. C. Fisher, K. D. Rose, and P. D. Gingerich. 2001. Stratocladistic analysis of Paleocene Carpolestidae (Mammalia, Plesiadapiformes) with description of a new late Tiffanian genus. Journal of Vertebrate Paleontology 21:119–131.

Bloch, J. I., K. D. Rose, and P. D. Gingerich. 1998. New species of *Batodonoides* (Lipotyphla, Geolabididae) from the early Eocene of Wyoming: Smallest known mammal? Journal of Mammalogy 79:804–827.

Bloch, J. I., R. Secord, and P. D. Gingerich. 2004. Systematics and phylogeny of late Paleocene and early Eocene Palaeoryctinae (Mammalia, Insectivora) from the Clarks Fork and Bighorn basins, Wyoming. Contributions from the Museum of Paleontology, the University of Michigan 31:119–154.

Boisserie, J.-R., F. Lihoreau, and M. Brunet. 2005. The position of Hippopotamidae within Cetartiodactyla. Proceedings of the National Academy of Sciences, USA 102:1537–1541.

Bonaparte, J. F. 1990. New Late Cretaceous mammals from the Los Alamitos Formation, northern Patagonia. National Geographic Research 6:63–93.

———. 1994. Approach to the significance of the Late Cretaceous mammals of South America. Berliner Geowissenschaftliche Abhandlungen E 13:31–44.

Bonaparte, J. F., and M. C. Barberena. 2001. On two advanced carnivorous cynodonts from the Late Triassic of southern Brazil. Bulletin of the Museum of Comparative Zoology 156:59–80.

Bonaparte, J. F., and J. Morales. 1997. Un primitivo Notonychopidae (Litopterna) del Paleoceno Inferior de Punta Peligro, Chubut, Argentina. Estudios Geologicos 53:263–274.

Bonaparte, J. F., A. G. Martinelli, C. L. Schultz, and R. Rubert. 2003. The sister group of mammals: Small cynodonts from the Late Triassic of southern Brazil. Revista Brasileira de Paleontologia 5:5–27.

Bonaparte, J. F., L. M. Van Valen, and A. Kramartz. 1993. La fauna local de Punta Peligro, Paleoceno inferior, de la Provincia del Chubut, Patagonia, Argentina. Evolutionary Monographs 14:1–61.

Bond, M., and G. Lopez. 1993. El primer Notohippidae (Mammalia, Notoungulata) de la Formación Lumbrera (Grupo Salta) del Noroeste Argentino. Consideraciones sobre la sistemática de la familia Notohippidae. Ameghiniana 30:59–68.

———. 1995. Los mamíferos de la formación Casa Grande (Eoceno) de la Provincia de Jujuy, Argentina. Ameghiniana 32:301–309.

Bonis, L. de. 1997. Précisions sur l'âge géologique et les relations phylétiques de *Mustelictis olivieri* nov. sp. (Carnivora, Mustelidae), carnassier de l'Oligocène inférieur (MP 22) des Phosphorites du Quercy (France). Géobios Mémoire Spécial 20:55–60.

Bonis, L. de, S. Peigné, and M. Hugueney. 1999. Carnivores féloides de l'Oligocène supérieur de Coderet-Bransat (Allier, France). Bulletin de la Société Géologique de France 170:939–949.

Bowen, G. J., W. C. Clyde, P. L. Koch, S. Ting, J. Alroy, T. Tsubamoto, Y.-Q. Wang, and Y. Wang. 2002. Mammalian dispersal at the Paleocene/Eocene boundary. Science 295:2062–2065.

Bown, T. M., and P. D. Gingerich. 1973. The Paleocene primate *Plesiolestes* and the origin of Microsyopidae. Folia Primatologica 19:1–8.

Bown, T. M., and M. J. Kraus. 1979. Origin of the tribosphenic molar and metatherian and eutherian dental formulae; pp. 172–181 in J. A. Lillegraven, Z. Kielan-Jaworowska, and W. A. Clemens (eds.), Mesozoic Mammals. The First Two-Thirds of Mammalian History. University of California Press, Berkeley.

Bown, T. M., and K. D. Rose. 1976. New early Tertiary Primates and a reappraisal of some Plesiadapiformes. Folia Primatologica 26:109–138.

———. 1979. *Mimoperadectes*, a new marsupial, and *Worlandia*, a new dermopteran, from the lower part of the Willwood

Formation (early Eocene), Bighorn Basin, Wyoming. Contributions from the Museum of Paleontology, the University of Michigan 25:89–104.

———. 1987. Patterns of dental evolution in early Eocene anaptomorphine primates (Omomyidae) from the Bighorn Basin, Wyoming. Paleontological Society Memoir 23:1–162.

Bown, T. M., and D. M. Schankler. 1982. A review of the Proteutheria and Insectivora of the Willwood Formation (lower Eocene), Bighorn Basin, Wyoming. U. S. Geological Survey Bulletin 1523:1–79.

Bown, T. M., and E. L. Simons. 1987. New Oligocene Ptolemaiidae (Mammalia: ?Pantolesta) from the Jebel Qatrani Formation, Fayum Depression, Egypt. Journal of Vertebrate Paleontology 7:311–324.

Bown, T. M., K. D. Rose, E. L. Simons, and S. L. Wing. 1994. Distribution and stratigraphic correlation of Upper Paleocene and Lower Eocene fossil mammal and plant localities of the Fort Union, Willwood, and Tatman formations, southern Bighorn Basin, Wyoming. U. S. Geological Survey Professional Paper 1540:1–103.

Boyer, D., and J. I. Bloch. 2003. Comparative anatomy of the pentacodontid *Aphronorus orieli* (Mammalia: Pantolesta) from the Paleocene of the western Crazy Mountains Basin, Montana. Journal of Vertebrate Paleontology 23(supplement to no. 3):36A.

Bralower, T. J., D. J. Thomas, J. C. Zachos, M. M. Hirschmann, U. Röhl, H. Sigurdsson, E. Thomas, and D. L. Whitney. 1997. High-resolution records of the late Paleocene thermal maximum and circum-Caribbean volcanism: Is there a causal link? Geology 25:963–967.

Bromham, L., M. J. Phillips, and D. Penny. 1999. Growing up with dinosaurs: Molecular dates and the mammalian radiation. Trends in Ecology and Evolution 14:113–118.

Brown, R. W. 1962. Paleocene flora of the Rocky Mountains and Great Plains. U. S. Geological Survey Professional Paper 375:1–119.

Brunet, M., and J. Sudre. 1987. Evolution et systematique du genre *Lophiomeryx* Pomel 1853 (Mammalia, Artiodactyla). Münchner Geowissenschaftliche Abhandlungen, Reihe A 10:225–242.

Bryant, H. N. 1991. Phylogenetic relationships and systematics of the Nimravidae (Carnivora). Journal of Mammalogy 72:56–78.

———. 1996. Nimravidae; pp. 453–475 in D. R. Prothero and R. J. Emry (eds.), The Terrestrial Eocene-Oligocene Transition in North America. Cambridge University Press, Cambridge, UK.

Buchholtz, E. A. 1998. Implications of vertebral morphology for locomotor evolution in early Cetacea; pp. 325–351 in J. G. M. Thewissen (ed.), The Emergence of Whales. Evolutionary Patterns in the Origin of Cetacea. Plenum Press, New York.

Bull, J. J., M. R. Badgett, H. A. Wichman, J. P. Huelsenbeck, D. M. Hillis, A. Gulati, C. Ho, and I. J. Molineux. 1997. Exceptional convergent evolution in a virus. Genetics 147:1497–1507.

Burk, A., M. Westerman, D. J. Kao, J. R. Kavanagh, and M. S. Springer. 1999. An analysis of marsupial interordinal relationships based on 12S rRNA, tRNA Valine, 16S rRNA, and Cytochrome *b* sequences. Journal of Mammalian Evolution 6:317–334.

Butler, P. M. 1956. The skull of *Ictops* and the classification of the Insectivora. Proceedings of the Zoological Society of London 126:453–481.

———. 1972. The problem of insectivore classification; pp. 253–265 in K. A. Joysey and T. S. Kemp (eds.), Studies in Vertebrate Evolution. Oliver and Boyd, Edinburgh.

———. 1977. Evolutionary radiation of the cheek teeth of Cretaceous placentals. Acta Palaeontologica Polonica 22:241–271.

———. 1978. A new interpretation of the mammalian teeth of tribosphenic pattern from the Albian of Texas. Breviora 446:1–27.

———. 1985. Homologies of molar cusps and crests, and their bearing on assessments of rodent phylogeny; pp. 381–401 in W. P. Luckett, and J.-L. Hartenberger (eds.), Evolutionary Relationships among Rodents. A Multidisciplinary Analysis. Plenum Press, New York.

———. 1988. Phylogeny of the insectivores; pp. 117–141 in M. J. Benton (ed.), The Phylogeny and Classification of the Tetrapods, Volume 2: Mammals. Clarendon Press, Oxford.

———. 1990. Early trends in the evolution of tribosphenic molars. Biological Reviews of the Cambridge Philosophical Society 65:529–552.

———. 1995. Fossil Macroscelidea. Mammal Review 25:3–14.

———. 1996. Dilambdodont molars: A functional interpretation of their evolution. Palaeovertebrata, Volume jubilaire D. E. Russell 25:205–213.

———. 1997. An alternative hypothesis on the origin of docodont molar teeth. Journal of Vertebrate Paleontology 17:435–439.

———. 2000. Review of the early allotherian mammals. Acta Palaeontologica Polonica 45:317–342.

Butler, P. M., and W. A. Clemens. 2001. Dental morphology of the Jurassic holotherian mammal *Amphitherium,* with a discussion of the evolution of mammalian post-canine dental formulae. Palaeontology 44:1–20.

Butler, P. M., and J. J. Hooker. 2005. New teeth of allotherian mammals from the English Bathonian, including the earliest multituberculates. Acta Palaeontologica Polonica 50:185–207.

Butler, P. M., and G. T. MacIntyre. 1994. Review of the British Haramiyidae (?Mammalia, Allotheria), their molar occlusion and relationships. Philosophical Transactions of the Royal Society of London B 345:433–458.

Campbell, K., and C. Frailey. 2004. Reviewing the Paleogene Santa Rosa local fauna of Amazonian Peru. Journal of Vertebrate Paleontology 24(supplement to no. 3):43A.

Cao, Y. J. A., T. Yano, and M. Hasegawa. 1994. Phylogenetic place of guinea pigs: No support of the rodent-polyphyly hypothesis from maximum-likelihood analyses of multiple protein sequences. Molecular Biology and Evolution 11:593–604.

Carleton, M. D. 1984. Introduction to rodents; pp. 255–265 in S. Anderson and J. K. Jones, Jr. (eds.), Orders and Families of Recent Mammals of the World. John Wiley and Sons, New York.

Carlini, A. A., R. Pascual, M. A. Reguero, G. J. Scillato-Yané, E. P. Tonni, and S. F. Vizcaíno. 1990. The first Paleogene land placental mammal from Antarctica: Its paleoclimatic and paleogeographical bearings; p. 325 in Abstracts IV International Congress of Systematic and Evolutionary Biology, University of Maryland.

Carroll, R. L. 1988. Vertebrate Paleontology and Evolution. W. H. Freeman and Co., New York.

Cartmill, M. 1972. Arboreal adaptations and the origin of the order Primates; pp. 97–122 in R. Tuttle (ed.), Functional and Evolutionary Biology of Primates. Aldine-Atherton, Chicago.

———. 1974. Rethinking primate origins. Science 184:436–443.

———. 1992. New views on primate origins. Evolutionary Anthropology 1:105–111.

Case, J. A., F. J. Goin, and M. O. Woodburne. 2004. "South American" marsupials from the Late Cretaceous of North America and the origin of marsupial cohorts. Journal of Mammalian Evolution 11:223–255.

Cassiliano, M. L., and W. A. Clemens. 1979. Symmetrodonta; pp. 150–161 in J. A. Lillegraven, Z. Kielan-Jaworowska, and W. A. Clemens (eds.), Mesozoic Mammals. The First Two-Thirds of Mammalian History. University of California Press, Berkeley.

Cavigelli, J.-P. 1997. A preliminary description of a *Leptictis* skeleton from the White River Formation of eastern Wyoming. Tate Geological Museum Guidebook 2:101–118.

Chiment, J. J., and W. W. Korth. 1996. A new genus of eomyid rodent (Mammalia) from the Eocene (Uintan-Duchesnean) of southern California. Journal of Vertebrate Paleontology 16:116–124.

Chow, M. 1959. A new arctocyonid from the upper Eocene of Lushih, Honan. Vertebrata PalAsiatica 3:133–138.

Chow, M., and T. Qi. 1978. Paleocene mammalian fossils from Nomogen Formation of Inner Mongolia. Vertebrata PalAsiatica 16:77–85.

Chow, M., and T. H. V. Rich. 1982. *Shuotherium dongi*, n. gen. and sp., a therian with pseudo-tribosphenic molars from the Jurassic of Sichuan, China. Australian Mammalogy 5:127–142.

Chow, M., and B. Wang. 1979. Relationship between the pantodonts and tillodonts and classification of the order Pantodonta. Vertebrata PalAsiatica 17:37–48.

Chow, M., J.-W. Wang, and J. Meng. 1996. A new species of *Chungchienia* (Tillodontia, Mammalia) from the Eocene of Lushi, China. American Museum Novitates 3171:1–10.

Christian, A. 1999. Zur Biomechanik der Fortbewegung von *Leptictidium* (Mammalia Proteutheria). Courier Forschungsinstitut Senckenberg 216:1–18.

Cifelli, R. L. 1983a. Eutherian tarsals from the late Paleocene of Brazil. American Museum Novitates 2761:1–31.

———. 1983b. The origin and affinities of the South American Condylarthra and Early Tertiary Litopterna (Mammalia). American Museum Novitates 2772:1–49.

———. 1985. Biostratigraphy of the Casamayoran, early Eocene, of Patagonia. American Museum Novitates 2820:1–26.

———. 1990a. Cretaceous mammals of southern Utah. I. Marsupials from the Kaiparowits Formation (Judithian). Journal of Vertebrate Paleontology 10:295–319.

———. 1990b. Cretaceous mammals of southern Utah. II. Marsupials and marsupial-like mammals from the Wahweap Formation (early Campanian). Journal of Vertebrate Paleontology 10:320–331.

———. 1993a. Early Cretaceous mammal from North America and the evolution of marsupial dental characters. Proceedings of the National Academy of Sciences, USA 90:9413–9416.

———. 1993b. Theria of metatherian-eutherian grade and the origin of marsupials; pp. 205–215 in F. S. Szalay, M. J. Novacek, and M. C. McKenna (eds.), Mammal Phylogeny: Mesozoic Differentiation, Multituberculates, Monotremes, Early Therians, and Marsupials. Springer-Verlag, New York.

———. 1993c. The phylogeny of the native South American ungulates; pp. 195–216 in F. S. Szalay, M. J. Novacek, and M. C. McKenna (eds.), Mammal Phylogeny: Placentals. Springer-Verlag, New York.

———. 1999. Tribosphenic mammal from the North American Early Cretaceous. Nature 401:363–366.

———. 2000. Cretaceous mammals of Asia and North America. Paleontological Society of Korea Special Publication 4:49–84.

———. 2001. Early mammalian radiations. Journal of Paleontology 75:1214–1226.

Cifelli, R. L., and J. G. Eaton. 1987. Marsupial from the earliest Late Cretaceous of western US. Nature 325:520–522.

Cifelli, R. L., and Z. Johanson. 1994. New marsupial from the Upper Cretaceous of Utah. Journal of Vertebrate Paleontology 14:292–295.

Cifelli, R. L., and S. K. Madsen. 1999. Spalacotheriid symmetrodonts (Mammalia) from the medial Cretaceous (upper Albian or lower Cenomanian) Mussentuchit local fauna, Cedar Mountain Formation, Utah, USA. Geodiversitas 21:167–214.

Cifelli, R. L., and C. de Muizon. 1997. Dentition and jaw of *Kokopellia juddi*, a primitive marsupial or near-marsupial from the medial Cretaceous of Utah. Journal of Mammalian Evolution 4:241–258.

Cifelli, R. L., and C. R. Schaff. 1998. Arctostylopida; pp. 332–336 in C. M. Janis, K. M. Scott, and L. L. Jacobs (eds.), Evolution of Tertiary Mammals of North America. Cambridge University Press, Cambridge, UK.

Cifelli, R. L., and M. F. Soria. 1983. Systematics of the Adianthidae (Litopterna, Mammalia). American Museum Novitates 2771:1–25.

Cifelli, R. L., and C. Villarroel. 1997. Paleobiology and affinities of *Megadolodus*; pp. 265–288 in R. F. Kay, R. H. Madden, R. L. Cifelli, and J. J. Flynn (eds.), Vertebrate Paleontology in the Neotropics: The Miocene Fauna of La Venta. Smithsonian Institution Press, Washington, D. C.

Cifelli, R. L., N. J. Czaplewski, and K. D. Rose. 1995. Additions to knowledge of Paleocene mammals from the North Horn Formation, central Utah. Great Basin Naturalist 55:304–314.

Cifelli, R. L., J. J. Eberle, D. L. Lofgren, J. A. Lillegraven, and W. A. Clemens. 2004. Mammalian biochronology of the latest Cretaceous; pp. 21–42 in M. O. Woodburne (ed.), Late Cretaceous and Cenozoic Mammals of North America. Columbia University Press, New York.

Cifelli, R. L., C. R. Schaff, and M. C. McKenna. 1989. The relationships of the Arctostylopidae (Mammalia): New data and interpretations. Bulletin of the Museum of Comparative Zoology 152:1–44.

Ciochon, R. L., and P. A. Holroyd. 1994. The Asian origin of Anthropoidea revisited; pp. 143–162 in J. G. Fleagle and R. F. Kay (eds.), Anthropoid Origins. Plenum Press, New York.

Ciochon, R. L., P. D. Gingerich, G. F. Gunnell, and E. L. Simons. 2001. Primate postcrania from the late middle Eocene of Myanmar. Proceedings of the National Academy of Sciences, USA 98:7672–7677.

Cirot, E., and L. de Bonis. 1992. Revision du genre *Amphicynodon*, carnivore de l'Oligocène. Palaeontographica Abteilung A 220:103–130.

Clark, J., and T. E. Guensburg. 1972. Arctoid genetic characters as related to the genus *Parictis*. Fieldiana Geology 26:1–76.

Clemens, W. A., Jr. 1966. Fossil mammals of the type Lance Formation, Wyoming. Part II. Marsupialia. University of California Publications in Geological Sciences 62:1–122.

———. 1970. Mesozoic Mammalian Evolution. Annual Review of Ecology and Systematics 1:357–390.

———. 1971. Mammalian evolution in the Cretaceous; pp. 165–180 *in* D. M. Kermack and K. A. Kermack (eds.), Early Mammals. Zoological Journal of the Linnean Society 50, supplement 1.

———. 1973. Fossil Mammals of the type Lance Formation Wyoming. Part III. Eutheria and summary. University of California Publications in Geological Sciences 94:1–102.

———. 1979. Marsupialia; pp. 192–220 *in* J. A. Lillegraven, Z. Kielan-Jaworowska, and W. A. Clemens (eds.), Mesozoic Mammals. The First Two-Thirds of Mammalian History. University of California Press, Berkeley.

———. 1980. *Gallolestes pachymandibularis* (Theria, incertae sedis; Mammalia) from Late Cretaceous deposits in Baja California del Norte, Mexico. Paleobios 33:1–10.

———. 1986. On Triassic and Jurassic mammals; pp. 237–246 *in* K. Padian (ed.), The Beginning of the Age of Dinosaurs. Faunal Change across the Triassic-Jurassic Boundary. Cambridge University Press, Cambridge, UK.

———. 2002. Evolution of the mammalian fauna across the Cretaceous-Tertiary boundary in northeastern Montana and other areas of the Western Interior; pp. 217–245 *in* J. H. Hartman, K. R. Johnson, and D. J. Nichols (eds.), The Hell Creek Formation and the Cretaceous-Tertiary Boundary in the Northern Great Plains: An Integrated Continental Record of the End of the Cretaceous. Geological Society of America Special Paper 361, Boulder.

———. 2004. *Purgatorius* (Plesiadapiformes, Primates?, Mammalia), a Paleocene immigrant into northeastern Montana: Statigraphic occurrences and incisor proportions. Bulletin of Carnegie Museum of Natural History 36:3–13.

Clemens, W. A., Jr., and Z. Kielan-Jaworowska. 1979. Multituberculata; pp. 99–149 *in* J. A. Lillegraven, Z. Kielan-Jaworowska, and W. A. Clemens (eds.), Mesozoic Mammals. The First Two-Thirds of Mammalian History. University of California Press, Berkeley.

Clemens, W. A., Jr., and W. von Koenigswald. 1993. A new skeleton of *Kopidodon macrognathus* from the middle Eocene of Messel and the relationships of paroxyclaenids and pantolestids based on postcranial evidence. Kaupia—Darmstädter Beiträge zur Naturgeschichte 3:57–73.

Clemens, W. A., Jr., and J. R. E. Mills. 1971. Review of *Peramus tenuirostris* Owen (Eupantotheria, Mammalia). Bulletin of the British Museum of Natural History (Geology) 20:87–113.

CoBabe, E. A. 1996. Leptaucheniinae; pp. 574–580 *in* D. R. Prothero, and R. J. Emry (eds.), The Terrestrial Eocene-Oligocene Transition in North America. Cambridge University Press, Cambridge, UK.

Colbert, E. H. 1934. Chalicotheres from Mongolia and China in the American Museum. Bulletin of the American Museum of Natural History 67:353–387.

———. 1935. Distributional and phylogenetic studies on Indian fossil mammals. IV. The phylogeny of the Indian Suidae and the origin of the Hippopotamidae. American Museum Novitates 799:1–24.

———. 1938. *Brachyhyops*, a new bunodont artiodactyl from Beaver Divide, Wyoming. Annals of Carnegie Museum 27:87–108.

———. 1941. The osteology and relationships of *Archaeomeryx*, an ancestral ruminant. American Museum Novitates 1135:1–24.

Colbert, M. W., and R. M. Schoch. 1998. Tapiroidea and other moropomorphs; pp. 569–582 *in* C. M. Janis, K. M. Scott, and L. L. Jacobs (eds.), Evolution of Tertiary Mammals of North America. Cambridge University Press, Cambridge, UK.

Collinson, M. E., and J. J. Hooker. 1987. Vegetational and mammalian faunal changes in the early Tertiary of southern England; pp. 259–303 *in* E. M. Friis, W. G. Chaloner, and P. R. Crane (eds.), The Origins of Angiosperms and Their Biological Consequences. Cambridge University Press, Cambridge, UK.

Conroy, G. C. 1990. Primate Evolution. W. W. Norton and Co., New York.

Coombs, M. C. 1983. Large mammalian clawed herbivores: A comparative study. Transactions of the American Philosophical Society 73:1–96.

———. 1998. Chalicotherioidea; pp. 560–568 *in* C. M. Janis, K. M. Scott, and L. L. Jacobs (eds.), Evolution of Tertiary Mammals of North America. Cambridge University Press, Cambridge, UK.

Coombs, M. C., and W. P. Coombs, Jr. 1977b. Dentition of *Gobiohyus* and a reevaluation of the Helohyidae (Artiodactyla). Journal of Mammalogy 58:291–308.

———. 1982. Anatomy of the ear region of four Eocene artiodactyls: *Gobiohyus,* ?*Helohyus, Diacodexis,* and *Homacodon*. Journal of Vertebrate Paleontology 2:219–236.

Coombs, W. P., Jr., and M. C. Coombs. 1977a. The origin of anthracotheres. Neues Jahrbuch für Geologie und Paläontologie, Monatshefte 10:584–599.

Cooper, A., and R. Fortey. 1998. Evolutionary explosions and the phylogenetic fuse. Trends in Ecology and Evolution 13:151–156.

Cooper, C. F. 1928. On the ear region of certain of the Chrysochloridae. Philosophical Transactions of the Royal Society of London B 216:265–283.

Cope, E. D. 1882. Contributions to the history of the Vertebrata of the lower Eocene of Wyoming and New Mexico, made during 1881. Proceedings of the American Philosophical Society 20:139–197.

———. 1884. The Vertebrata of the Tertiary formations of the West. Report of the U. S. Geological Survey of the Territories 3:1–1009.

Corbet, G. B., and J. E. Hill. 1991. A World List of Mammalian Species. Oxford University Press, Oxford, UK.

Court, N. 1990. The periotic of *Arsinoitherium* and its phylogenetic implications. Journal of Vertebrate Paleontology 10:170–182.

———. 1992a. Cochlea anatomy of *Numidotherium koholense*: Auditory acuity in the oldest known proboscidean. Lethaia 25:211–215.

———. 1992b. A unique form of dental bilophodonty and a functional interpretation of peculiarities in the masticatory system of *Arsinoitherium* (Mammalia, Embrithopoda). Historical Biology 6:91–111.

———. 1992c. The skull of *Arsinoitherium* (Mammalia, Embrithopoda) and the higher order interrelationships of ungulates. Palaeovertebrata 22:1–43.

———. 1993. Morphology and functional anatomy of the postcranial skeleton in *Arsinoitherium* (Mammalia, Embrithopoda). Palaeontographica Abteilung A 226:125–169.

———. 1994. Limb posture and gait in *Numidotherium koholense,* a primitive proboscidean from the Eocene of Algeria. Zoological Journal of the Linnean Society 111:297–338.

———. 1995. A new species of *Numidotherium* (Mammalia: Proboscidea) from the Eocene of Libya and the early phylogeny of the Proboscidea. Journal of Vertebrate Paleontology 15:650–671.

Court, N., and J.-L. Hartenberger. 1992. A new species of the hyracoid mammal *Titanohyrax* from the Eocene of Tunisia. Palaeontology 35:309–317.

Court, N., and J.-J. Jaeger. 1991. Anatomy of the periotic bone in the Eocene proboscidean *Numidotherium koholense*: An example of parallel evolution in the inner ear of tethytheres. Comptes Rendus de l' Académie des Sciences Paris, sér. II, 312:559–565.

Court, N., and M. Mahboubi. 1993. Reassessment of lower Eocene *Seggeurius amourensis*: Aspects of primitive dental morphology in the mammalian order Hyracoidea. Journal of Paleontology 67:889–893.

Coxall, H. K., P. A. Wilson, H. Pälike, C. H. Lear, and J. Backman. 2005. Rapid stepwise onset of Antarctic glaciation and deeper calcite compensation in the Pacific Ocean. Nature 433:53–57.

Crochet, J.-Y. 1974. Les insectivores des Phosphorites du Quercy. Palaeovertebrata 6:109–159.

———. 1984. *Garatherium mahboubii* nov. gen., nov. sp., marsupial de l'Éocène inférieur d'El Kohol (Sud-Oranais, Algérie). Annales de Paléontologie 70:275–294.

Crochet, J.-Y., and B. Sigé. 1996. Un marsupial ancien (transition Crétacé-Tertiaire) à denture évoluée en Amérique du Sud (Chulpas, formation Umayo, Pérou). Neues Jahrbuch für Geologie und Paläontologie, Monatshefte H 10:622–634.

Crompton, A. W. 1963. On the lower jaw of *Diarthrognathus* and the origin of the mammalian lower jaw. Proceedings of the Zoological Society of London 140:697–753.

———. 1971. The origin of the tribosphenic molar; pp. 65–87 in D. M. Kermack and K. A. Kermack (eds.), Early Mammals. Zoological Journal of the Linnean Society 50, supplement 1.

———. 1974. The dentitions and relationships of the southern African Triassic mammals, *Erythrotherium parringtoni* and *Megazostrodon rudnerae*. Bulletin of the British Museum of Natural History (Geology) 24:397–437.

Crompton, A. W., and F. A. Jenkins, Jr. 1967. American Jurassic symmetrodonts and Rhaetic "pantotheres." Science 155:1006–1008.

———. 1973. Mammals from reptiles: A review of mammalian origins. Annual Review of Earth and Planetary Sciences 1:131–155.

———. 1979. Origin of mammals; pp. 59–73 in J. A. Lillegraven, Z. Kielan-Jaworowska, and W. A. Clemens (eds.), Mesozoic Mammals. The First Two-Thirds of Mammalian History. University of California Press, Berkeley.

Crompton, A. W., and Z. Kielan-Jaworowska. 1978. Molar structure and occlusion in Cretaceous therian mammals; pp. 249–287 in P. M. Butler and K. A. Joysey (eds.), Development, Function and Evolution of Teeth. Academic Press, London.

Crompton, A. W., and Z.-X. Luo. 1993. Relationships of the Liassic mammals *Sinoconodon*, *Morganucodon oehleri*, and *Dinnetherium*; pp. 30–44 in F. S. Szalay, M. J. Novacek, and M. C. McKenna (eds.), Mammal Phylogeny: Mesozoic Differentiation, Multituberculates, Monotremes, Early Therians, and Marsupials. Springer-Verlag, New York.

Crompton, A. W., and P. Parker. 1978. Evolution of the mammalian masticatory apparatus. American Scientist 66:192–201.

Crompton, A. W., and A. Sun. 1985. Cranial structure and relationships of the Liassic mammal *Sinoconodon*. Zoological Journal of the Linnean Society 85:99–119.

Dagosto, M. 1983. Postcranium of *Adapis parisiensis* and *Leptadapis magnus* (Adapiformes, Primates). Folia Primatologica 41:49–101.

———. 1985. The distal tibia of Primates with special reference to the Omomyidae. International Journal of Primatology 6:45–75.

———. 1988. Implications of postcranial evidence for the origin of euprimates. Journal of Human Evolution 17:35–56.

———. 1993. Postcranial anatomy and locomotor behavior in Eocene primates; pp. 199–219 in D. L. Gebo (ed.), Postcranial Adaptation in Nonhuman Primates. Northern Illinois University Press, DeKalb.

———. 2002. The origin and diversification of anthropoid primates. Introduction; pp. 125–132 in W. C. Hartwig (ed.), The Primate Fossil Record. Cambridge University Press, Cambridge, UK.

Dagosto, M., D. L. Gebo, and K. C. Beard. 1999. Revision of the Wind River faunas, early Eocene of central Wyoming. Part 14. Postcranium of *Shoshonius cooperi* (Mammalia: Primates). Annals of Carnegie Museum 68:175–211.

Dashzeveg, D. 1979. On an archaic representative of the Equoidea (Mammalia, Perissodactyla) from the Eocene of central Asia [in Russian]. Joint Soviet-Mongolian Paleontological Expedition Transactions 8:10–22.

———. 1982. A revision of the Prodinoceratinae of central Asia and North America. Paleontological Journal 1982:91–99.

———. 1990. New trends in adaptive radiation of Early Tertiary rodents (Rodentia, Mammalia). Acta Zoologica Cracoviensia 33:37–44.

———. 1994. Two previously unknown eupantotheres (Mammalia, Eupantotheria). American Museum Novitates 3107:1–11.

———. 1996. Some carnivorous mammals from the Paleogene of the eastern Gobi Desert, Mongolia, and the application of Oligocene carnivores to stratigraphic correlation. American Museum Novitates 3179:1–14.

Dashzeveg, D., and J. J. Hooker. 1997. New ceratomorph perissodactyls (Mammalia) from the middle and late Eocene of Mongolia: Their implications for phylogeny and dating. Zoological Journal of the Linnean Society 120:105–138.

Dashzeveg, D., and Z. Kielan-Jaworowska. 1984. The lower jaw of an aegialodontid mammal from the Early Cretaceous of Mongolia. Zoological Journal of the Linnean Society 82:217–227.

Dashzeveg, D., and M. C. McKenna. 1977. Tarsioid primate from the Early Tertiary of the Mongolian People's Republic. Acta Palaeontologica Polonica 22:119–137.

Dashzeveg, D., and J. Meng. 1998a. New Eocene ctenodactyloid rodents from the eastern Gobi Desert of Mongolia and a phylogenetic analysis of ctenodactyloids based on dental features. American Museum Novitates 3246:1–20.

———. 1998b. A new Eocene cylindrodont rodent (Mammalia, Rodentia) from the eastern Gobi of Mongolia. American Museum Novitates 3253:1–18.

Dashzeveg, D., and D. E. Russell. 1988. Palaeocene and Eocene Mixodontia (Mammalia, Glires) of Mongolia and China. Palaeontology 31:129–164.

Dashzeveg, D., J.-L. Hartenberger, T. Martin, and S. Legendre. 1998. A peculiar minute Glires (Mammalia) from the early

Eocene of Mongolia; pp. 194–209 *in* K. C. Beard and M. R. Dawson (eds.), Dawn of the Age of Mammals in Asia. Bulletin of Carnegie Museum of Natural History 34.

Dawson, M. R. 1958. Later Tertiary Leporidae of North America. University of Kansas Paleontological Contributions, Vertebrata 22:1–75.

———. 1961. The skull of *Sciuravus nitidus*, a middle Eocene rodent. Postilla 53:1–13.

———. 1966. Additional late Eocene rodents (Mammalia) from the Uinta Basin, Utah. Annals of Carnegie Museum 38:97–114.

———. 1967. Lagomorph history and the stratigraphic record; pp. 287–316, *in* Essays in Paleontology and Stratigraphy, Raymond C. Moore Commemorative Volume. University of Kansas Department of Geology Special Publication 2, Lawrence.

———. 2003. Paleogene rodents of Eurasia; pp. 97–126 *in* J. W. F. Reumer and W. Wessels (eds.), Distribution and Migration of Tertiary Mammals in Eurasia. A Volume in Honour of Hans de Bruijn. DEINSEA 10 (Annual of the Natural History Museum of Rotterdam).

Dawson, M. R., and K. C. Beard. 1996. New late Paleocene rodents (Mammalia) from Big Multi Quarry, Washakie Basin, Wyoming. Palaeovertebrata 25:302–321.

Dawson, M. R., and L. Krishtalka. 1984. Fossil history of the families of Recent mammals; pp. 11–57 *in* S. Anderson and J. K. Jones, Jr. (eds.), Orders and Families of Recent Mammals of the World. John Wiley and Sons, New York.

Dawson, M. R., and Y. Tong. 1998. New material of *Pappocricetodon schaubi*, an Eocene rodent (Mammalia: Cricetidae) from the Yuanqu Basin, Shanxi Province, China; pp. 278–285 *in* K. C. Beard and M. R. Dawson (eds.), Dawn of the Age of Mammals in Asia. Bulletin of Carnegie Museum of Natural History 34.

Dawson, M. R., L. Krishtalka, and R. K. Stucky. 1990. Revision of the Wind River faunas, early Eocene of central Wyoming. Part 9. The oldest known hystricomorphous rodent (Mammalia: Rodentia). Annals of Carnegie Museum 59:135–147.

Dawson, M. R., C.-K. Li, and T. Qi. 1984. Eocene ctenodactyloid rodents (Mammalia) of eastern and central Asia. Carnegie Museum of Natural History Special Publication 9:138–150.

Dawson, M. R., M. C. McKenna, K. C. Beard, and J. H. Hutchison. 1993. An early Eocene plagiomenid mammal from Ellesmere and Axel Heiberg Islands, Canada. Kaupia—Darmstädter Beiträge zur Naturgeschichte 3:179–192.

Dawson, M. R., R. K. Stucky, L. Krishtalka, and C. C. Black. 1986. *Machaeroides simpsoni*, new species, oldest known sabertooth creodont (Mammalia), of the Lost Cabin Eocene; pp. 177–182 *in* K. M. Flanagan and J. A. Lillegraven (eds.), Vertebrates, Phylogeny, and Philosophy. Contributions to Geology, University of Wyoming, Special Paper 3, Laramie.

Dawson, M. R., T. Tsubamoto, M. Takai, N. Egi, S. T. Tun, and C. Sein. 2003. Rodents of the family Anomaluridae (Mammalia) from southeast Asia (middle Eocene, Pondaung Formation, Myanmar). Annals of Carnegie Museum 72:203–213.

Dawson, M. R., R. M. West, W. Langston, Jr., and J. H. Hutchison. 1976. Paleogene terrestrial vertebrates: Northernmost occurrence, Ellesmere Island, Canada. Science 192:781–782.

De Blieux, D. D., and E. L. Simons. 2002. Cranial and dental anatomy of *Antilohyrax pectidens*, a late Eocene hyracoid (Mammalia) from the Fayum, Egypt. Journal of Vertebrate Paleontology 22:122–136.

De Blieux, D. D., E. L. Simons, M. R. Baumrind, P. S. Chatrath, and G. E. Meyer. 2001. The internal mandibular chamber in fossil hyracoids (Mammalia): Patterns of sexual dimorphism in a novel mammalian feature. Journal of Vertebrate Paleontology 21(supplement to no. 3):44A.

Delmont, N. 1941. Un mammifère artiodactyle de l'Eocène: Le *Dacrytherium*. Annales de Paléontologie 29:29–50.

Delsuc, F., M. Scally, O. Madsen, M. J. Stanhope, W. W. de Jong, F. M. Catzeflis, M. S. Springer, and E. J. P. Douzery. 2002. Molecular phylogeny of living xenarthrans and the impact of character and taxon sampling on the placental tree rooting. Molecular Biology and Evolution 19:1656–1671.

Denison, R. H. 1938. The broad-skulled Pseudocreodi. Annals of the New York Academy of Sciences 37:163–256.

Denys, C., and D. E. Russell. 1981. Etude de la variabilité dentaire d'une population de *Paschatherium* (condylarthre hyopsodontidé), provenant de la localité Sparnacienne de Dormaal (Belgique). Bulletin d'Information des Géologues du Bassin de Paris 18:37–45.

De Queiroz, K., and J. Gauthier. 1992. Phylogenetic taxonomy. Annual Review of Ecology and Systematics 23:449–480.

D'Erchia, A. M., C. Gissi, G. Pesole, C. Saccone, and U. Arnason. 1996. The guinea-pig is not a rodent. Nature 381:597–600.

Dickens, G. R., J. R. O'Neil, D. K. Rea, and R. M. Owen. 1995. Dissociation of oceanic methane hydrate as a cause of the Carbon Isotope Excursion at the end of the Paleocene. Paleoceanography 10:965–971.

Ding, S.-Y. 1987. A Paleocene edentate from Nanxiong Basin, Guangdong. Palaeontologia Sinica 173:1–118.

Ding, S.-Y., and Y.-S. Tong. 1979. Some Paleocene anagalids from Nanxiong, Guangdong. Vertebrata PalAsiatica 17:137–145.

Ding, S.-Y., J. A. Schiebout, and M. Zhou. 1987. A skull of *Pantolambdodon* (Mammalia, Pantodonta) from Ningxia, North China. Journal of Vertebrate Paleontology 7:155–161.

Domning, D. P. 1978a. Sirenian evolution in the north Pacific Ocean. University of California Publications in Geological Sciences 118:1–176.

———. 1978b. Sirenia; pp. 573–581 *in* V. J. Maglio and H. B. S. Cooke (eds.), Evolution of African Mammals. Harvard University Press, Cambridge, Mass.

———. 1982. Evolution of manatees: A speculative history. Journal of Paleontology 56:599–619.

———. 1994. A phylogenetic analysis of the Sirenia. Proceedings of the San Diego Society of Natural History 29:177–189.

———. 2001a. Sirenians, seagrasses, and Cenozoic ecological change in the Caribbean. Palaeogeography, Palaeoclimatology, Palaeoecology 166:27–50.

———. 2001b. Evolution of the Sirenia and Desmostylia; pp. 151–168 *in* J.-M. Mazin and V. de Buffrénil (eds.), Secondary Adaptation of Tetrapods to Life in Water. Verlag Dr. Friedrich Pfeil, Munich.

———. 2001c. The earliest known fully quadrupedal sirenian. Nature 413:625–627.

Domning, D. P., and V. de Buffrénil. 1991. Hydrostasis in the Sirenia: Quantitative data and functional interpretations. Marine Mammal Science 7:331–368.

Domning, D. P., and P. D. Gingerich. 1994. *Protosiren smithae*, new species (Mammalia, Sirenia), from the late middle Eocene of Wadi Hitan, Egypt. Contributions from the Museum of Paleontology, University of Michigan 29:69–87.

Domning, D. P., and L.-A. C. Hayek. 1984. Horizontal tooth replacement in the Amazonian manatee (*Trichechus inunguis*). Mammalia 48:105–127.

Domning, D. P., R. J. Emry, R. W. Portell, S. K. Donovan, and K. S. Schindler. 1997. Oldest West Indian land mammal: Rhinocerotoid ungulate from the Eocene of Jamaica. Journal of Vertebrate Paleontology 17:638–641.

Domning, D. P., G. S. Morgan, and C. E. Ray. 1982. North American Eocene sea cows (Mammalia: Sirenia). Smithsonian Contributions to Paleobiology 52:1–69.

Domning, D. P., C. E. Ray, and M. C. McKenna. 1986. Two new Oligocene desmostylians and a discussion of tethytherian systematics. Smithsonian Contributions to Paleobiology 59:1–56.

Douady, C. J., and E. J. P. Douzery. 2003. Molecular estimation of eulipotyphlan divergence times and the evolution of "Insectivora." Molecular Phylogenetics and Evolution 28:285–296.

Douady, C. J., P. I. Chatelier, O. Madsen, W. W. de Jong, F. Catzeflis, M. S. Springer, and M. J. Stanhope. 2002. Molecular phylogenetic evidence confirming the Eulipotyphla concept and in support of hedgehogs as the sister group to shrews. Molecular Phylogenetics and Evolution 25:200–209.

Ducrocq, S. 1994. An Eocene peccary from Thailand and the biogeographical origins of the artiodactyl family Tayassuidae. Palaeontology 37:765–779.

———. 1997. The anthracotheriid genus *Bothriogenys* (Mammalia, Artiodactyla) in Africa and Asia during the Paleogene: Phylogenetical and paleogeographical relationships. Stuttgarter Beiträge zur Naturkunde, Serie B 250:1–44.

———. 1999. *Siamopithecus eocaenus*, a late Eocene anthropoid primate from Thailand: Its contribution to the evolution of anthropoids in southeast Asia. Journal of Human Evolution 36:613–635.

Ducrocq, S., E. Buffetaut, H. Buffetaut-Tong, J.-J. Jaeger, Y. Jongkanjanasoontorn, and V. Suteethorn. 1992. First fossil flying lemur: A dermopteran from the late Eocene of Thailand. Palaeontology 35:373–380.

Ducrocq, S., Y. Chaimanee, V. Suteethorn, and J.-J. Jaeger. 1998. The earliest known pig from the upper Eocene of Thailand. Palaeontology 41:147–156.

Ducrocq, S., J.-J. Jaeger, and B. Sigé. 1993. Un mégachiroptère dans l'Eocène supérieur de Thailande. Incidence dans la discussion phylogénique du groupe. Neues Jahrbuch für Geologie und Paläontologie, Monatshefte 9:561–575.

Dupuis, C., M.-P. Aubry, E. Steurbaut, W. A. Berggren, K. Ouda, R. Magioncalda, B. S. Cramer, D. V. Kent, R. P. Speijer, and C. Heilmann-Clausen. 2003. The Dababiya Quarry section: Lithostratigraphy, clay mineralogy, geochemistry and paleontology. Micropaleontology 49:41–59.

Easteal, S. 1999. Molecular evidence for the early divergence of placental mammals. BioEssays 21:1052–1058.

Eaton, J. G. 1993. Therian mammals from the Cenomanian (Upper Cretaceous) Dakota Formation, southwestern Utah. Journal of Vertebrate Paleontology 13:105–124.

Eberle, J. J. 1999. Bridging the transition between didelphodonts and taeniodonts. Journal of Paleontology 73:936–944.

Eberle, J. J., and J. A. Lillegraven. 1998. A new important record of earliest Cenozoic mammalian history: Eutheria and paleogeographic/biostratigraphic summaries. Rocky Mountain Geology 33:49–117.

Eberle, J. J., and M. C. McKenna. 2002. Early Eocene Leptictida, Pantolesta, Creodonta, Carnivora, and Mesonychidae (Mammalia) from the Eureka Sound Group, Ellesmere Island, Nunavut. Canadian Journal of Earth Sciences 39:899–910.

Effinger, J. A. 1998. Entelodontidae; pp. 375–380 in C. M. Janis, K. M. Scott, and L. L. Jacobs (eds.), Evolution of Tertiary Mammals of North America. Volume 1: Terrestrial Carnivores, Ungulates, and Ungulatelike Mammals. Cambridge University Press, Cambridge, UK.

Egi, N. 2001. Body mass estimates in extinct mammals from limb bone dimensions: The case of North American hyaenodontids. Palaeontology 44:497–528.

Eisenberg, J. F. 1981. The Mammalian Radiations. An Analysis of Trends in Evolution, Adaptation, and Behavior: University of Chicago Press, Chicago.

Eizirik, E., W. J. Murphy, and S. J. O'Brien. 2001. Molecular dating and biogeography of the early placental mammal radiation. Journal of Heredity 92:212–219.

Emry, R. J. 1970. A North American Oligocene pangolin and other additions to the Pholidota. Bulletin of the American Museum of Natural History 142:455–510.

———. 1972. A new species of *Agnotocastor* (Rodentia, Castoridae) from the early Oligocene of Wyoming. American Museum Novitates 2485:1–7.

———. 1981. New material of the Oligocene muroid rodent *Nonomys*, and its bearing on muroid origins. American Museum Novitates 2712:1–14.

———. 1989. A tiny new Eocene ceratomorph and comments on "tapiroid" systematics. Journal of Mammalogy 70:794–804.

———. 2004. The edentulous skull of the North American pangolin, *Patriomanis americanus*. Bulletin of the American Museum of Natural History 285:130–138.

Emry, R. J., and W. W. Korth. 1989. Rodents of the Bridgerian (middle Eocene) Elderberry Canyon local fauna of eastern Nevada. Smithsonian Contributions to Paleobiology 67:1–14.

———. 1996a. Cylindrodontidae; pp. 399–416 in D. R. Prothero and R. J. Emry (eds.), The Terrestrial Eocene-Oligocene Transition in North America. Cambridge University Press, New York.

———. 1996b. The Chadronian squirrel "*Sciurus*" *jeffersoni* Douglass, 1901: A new generic name, new material, and its bearing on the early evolution of Sciuridae (Rodentia). Journal of Vertebrate Paleontology 16:775–780.

Emry, R. J., and R. W. Thorington, Jr. 1982. Descriptive and comparative osteology of the oldest fossil squirrel, *Protosciurus* (Rodentia: Sciuridae). Smithsonian Contributions to Paleobiology 47:1–35.

Emry, R. J., L. A. Tyutkova, S. G. Lucas, and B. Wang. 1998. Rodents of the middle Eocene Shinzhaly fauna of eastern Kazakstan. Journal of Vertebrate Paleontology 18:218–227.

Emry, R. J., B. Wang, L. A. Tjutkova, and S. G. Lucas. 1997. A late Eocene eomyid rodent from the Zaysan Basin of Kazakhstan. Journal of Vertebrate Paleontology 17:229–234.

Engelmann, G. F. 1987. A new Deseadan sloth (Mammalia: Xenarthra) from Salla, Bolivia, and its implications for the primitive condition of the dentition in edentates. Journal of Vertebrate Paleontology 7:217–223.

Engesser, B. 1975. Revision der europäischen Heterosoricinae (Insectivora, Mammalia). Eclogae Geologicae Helvetiae 68:649–671.

Erfurt, J. 2000. Rekonstruktion des Skelettes und der Biologie von *Anthracobunodon weigelti* (Artiodactyla, Mammalia) aus dem Eozän des Geiseltales. Hallesches Jahrbuch für Geowissenschaften B 12:57–141.

Erfurt, J., and H. Altner. 2003. Habitus-Rekonstruktion von *Anthracobunodon weigelti* (Artiodactyla, Mammalia) aus dem Eozän des Geiseltales. Veröffentlichungen des Landesamtes für Archäologie 57:153–175, 353 (color plate).

Erfurt, J., and H. Haubold. 1989. Artiodactyla aus den Eozänen Braunkohlen des Geiseltales dei Halle (DDR). Palaeovertebrata 19:131–160.

Escarguel, G. 1999. Les rongeurs de l'Éocène inférieur et moyen d'Europe occidentale. Systematique, phylogénie, biochronologie et paléobiogéographie des niveaux-repères MP 7 à MP 14. Palaeovertebrata 28:89–351.

Escarguel, G., C. Seiffert, and G. Storch. 2001. Un loir dort depuis 50 millions d'années. Pour la Science 279:20.

Estravís, C, and D. E. Russell. 1992. The presence of Taeniodonta (Mammalia) in the early Eocene of Europe. Ciências da Terra (Universidade Nova de Lisboa) 11:191–201.

Estravís Fernández, C. 1992. Estudo dos mamíferos do Eocénico inferior de Silveirinha (Baixo Mondego). Ph.D. dissertation, Universidade Nova de Lisboa, Lisbon.

Evans, F. G. 1942. The osteology and relationships of the elephant shrews (Macroscelididae). Bulletin of the American Museum of Natural History 80:85–125.

Fischer, M. S. 1986. Die Stellung der Schliefer (Hyracoidea) im phylogenetischen System der Eutheria. Courier Forschungsinstitut Senckenberg 84:1–132.

———. 1989. Hyracoids, the sister-group of perissodactyls; pp. 37–56 in D. R. Prothero and R. M. Schoch (eds.), The Evolution of Perissodactyls. Oxford University Press, New York.

Fischer, M. S., and P. Tassy. 1993. The interrelation between Proboscidea, Sirenia, Hyracoidea, and Mesaxonia: The morphological evidence; pp. 217–234 in F. S. Szalay, M. J. Novacek, and M. C. McKenna (eds.), Mammal Phylogeny: Placentals. Springer-Verlag, New York.

Flannery, T. 1994. Possums of the World: A Monograph of the Phalangeroidea. Geo Productions and Australian Museum, Sydney.

Flannery, T. F., M. Archer, T. H. Rich, and R. Jones. 1995. A new family of monotremes from the Cretaceous of Australia. Nature 377:418–420.

Fleagle, J. G. 1999. Primate Adaptation and Evolution. Second edition. Academic Press, San Diego.

Fleagle, J. G., and R. F. Kay. 1987. The phyletic position of the Parapithecidae. Journal of Human Evolution 16:483–532.

Fleagle, J. G., and E. L. Simons. 1982a. Skeletal remains of *Propliopithecus chirobates* from the Egyptian Oligocene. Folia Primatologica 39:161–177.

———. 1982b. The humerus of *Aegyptopithecus zeuxis*: A primitive anthropoid. American Journal of Physical Anthropology 59:175–193.

———. 1995. Limb skeleton and locomotor adaptations of *Apidium phiomense*, an Oligocene anthropoid from Egypt. American Journal of Physical Anthropology 97:235–289.

Flerov, K. K. 1967. Dinocerata of Mongolia. Transactions of the Institute of Paleontology, Academy of Sciences of the USSR 67:1–84.

Flower, W. H. 1885. An Introduction to the Osteology of the Mammalia. Macmillan and Co., London.

Flynn, J. J. 1998. Early Cenozoic Carnivora ("Miacoidea"); pp. 110–123 in C. M. Janis, K. M. Scott, and L. L. Jacobs (eds.), Evolution of Tertiary Mammals of North America. Cambridge University Press, Cambridge, UK.

Flynn, J. J., and H. Galiano. 1982. Phylogeny of Early Tertiary Carnivora, with a description of a new species of *Protictis* from the middle Eocene of northwestern Wyoming. American Museum Novitates 2725:1–64.

Flynn, J. J., and C. C. Swisher III. 1995. Cenozoic South American land mammal ages: Correlation to global geochronologies; pp. 317–333 in W. A. Berggren, D. V. Kent, M.-P. Aubry, and J. Hardenbol (eds.), Geochronology, Time Scales and Global Stratigraphic Correlation. SEPM (Society for Sedimentary Geology) Special Publication 54, Tulsa, Okla.

Flynn, J. J., and G. D. Wesley-Hunt. 2005. Carnivora; pp. 175–198 in K. D. Rose and J. D. Archibald (eds.), The Rise of Placental Mammals: Origins and Relationships of the Major Extant Clades. The Johns Hopkins University Press, Baltimore.

Flynn, J. J., and A. R. Wyss. 1999. New marsupials from the Eocene-Oligocene transition of the Andean main range, Chile. Journal of Vertebrate Paleontology 19:533–549.

Flynn, J. J., J. A. Finarelli, S. Zehr, J. Hsu, and M. A. Nedbal. 2005. Molecular phylogeny of the Carnivora (Mammalia): Assessing the impact of increased sampling on resolving enigmatic relationships. Systematic Biology 54:317–337.

Flynn, J. J., N. A. Neff, and R. H. Tedford. 1988. Phylogeny of the Carnivora; pp. 73–116 in M. J. Benton (ed.), The Phylogeny and Classification of the Tetrapods. Clarendon Press, Oxford.

Flynn, J. J., J. M. Parrish, B. Rakotosamimanana, W. F. Simpson, and A. R. Wyss. 1999. A Middle Jurassic mammal from Madagascar. Nature 401:57–60.

Flynn, J. J., A. R. Wyss, D. A. Croft, and R. Charrier. 2003. The Tinguiririca Fauna, Chile: Biochronology, paleoecology, biogeography, and a new earliest Oligocene South American Land Mammal "Age." Palaeogeography, Palaeoclimatology, Palaeoecology 195:229–259.

Flynn, L. J. 1994. Roots of rodent radiation. Nature 370:97–98.

Flynn, L. J., L. L. Jacobs, and I. U. Cheema. 1986. Baluchimyinae, a new ctenodactyloid rodent subfamily from the Miocene of Baluchistan. American Museum Novitates 2841:1–58.

Foote, M., J. P. Hunter, C. M. Janis, and J. J. Sepkoski, Jr. 1999. Evolutionary and preservational constraints on origins of biologic groups: Divergence times of eutherian mammals. Science 283:1310–1314.

Fordyce, R. E. 1982. A review of Australian fossil Cetacea. Memoirs of the National Museum, Victoria 43:43–58.

———. 1989. Problematic early Oligocene toothed whale (Cetacea, ?Mysticeti) from Waikari, North Canterbury, New Zealand. New Zealand Journal of Geology and Geophysics 32:385–390.

———. 2002. Neoceti; pp. 787–791 in W. F. Perrin et al. (ed.), Encyclopedia of Marine Mammals. Academic Press, San Diego.

Fordyce, R. E., and L. G. Barnes. 1994. The evolutionary history of whales and dolphins. Annual Review of Earth and Planetary Sciences 22:419–455.

Fordyce, R. E., and C. de Muizon. 2001. Evolutionary history of cetaceans: A review; pp. 169–233 in J.-M. Mazin and V. de Buffrénil (eds.), Secondary Adaptation of Tetrapods to Life in Water. Verlag Dr. Friedrich Pfeil, Munich.

Fortelius, M. 1985. Ungulate cheek teeth: Developmental, functional, and evolutionary interrelations. Acta Zoologica Fennica 180:1–76.

Fortelius, M., and J. Kappelman. 1993. The largest land mammal ever imagined. Zoological Journal of the Linnean Society 107:85–101.

Fox, R. C. 1979a. Mammals from the Upper Cretaceous Oldman Formation, Alberta. I. *Alphadon* Simpson (Marsupialia). Canadian Journal of Earth Sciences 16:91–102.

———. 1979b. Mammals from the Upper Cretaceous Oldman Formation, Alberta. II. *Pediomys* Marsh (Marsupialia). Canadian Journal of Earth Sciences 16:103–113.

———. 1984a. *Paranyctoides maleficus* (new species), an early eutherian mammal from the Cretaceous of Alberta. Carnegie Museum of Natural History Special Publication 9:9–20.

———. 1984b. The dentition and relationships of the Paleocene primate *Micromomys* Szalay, with description of a new species. Canadian Journal of Earth Sciences 21:1262–1267.

———. 1987. Palaeontology and the early evolution of marsupials; pp. 161–169 in M. Archer (ed.), Possums and Opossums: Studies in Evolution. Surrey Beatty & Sons and the Royal Zoological Society of New South Wales, Sydney.

———. 1991. *Saxonella* (Plesiadapiformes: ?Primates) in North America: *S. naylori*, sp. nov., from the Late Paleocene of Alberta, Canada. Journal of Vertebrate Paleontology 11:334–349.

———. 1993. The primitive dental formula of the Carpolestidae (Plesiadapiformes, Mammalia) and its phylogenetic implications. Journal of Vertebrate Paleontology 13:516–524.

Fox, R. C., and B. G. Naylor. 1986. A new species of *Didelphodon* Marsh (Marsupialia) from the Upper Cretaceous of Alberta, Canada: Paleobiology and phylogeny. Neues Jahrbuch für Geologie und Paläontologie, Abhandlungen 172:357–380.

———. 1995. The relationships of the Stagodontidae, primitive North American Late Cretaceous mammals; pp. 247–250 in A. Sun and Y. Wang (eds.), Sixth Symposium on Mesozoic Terrestrial Ecosystems and Biotas, Short Papers. China Ocean Press, Beijing.

———. 2003. A Late Cretaceous taeniodont (Eutheria, Mammalia) from Alberta, Canada. Neues Jahrbuch für Geologie und Paläontologie Abhandlungen 229:393–420.

Fox, R. C., and C. S. Scott. 2005. First evidence of a venom delivery apparatus in extinct mammals. Nature 435:1091–1093.

Fox, R. C., and G. P. Youzwyshyn. 1994. New primitive carnivorans (Mammalia) from the Paleocene of western Canada, and their bearing on relationships of the order. Journal of Vertebrate Paleontology 14:382–404.

Frailey, C., and K. Campbell. 2004. Eocene rodents of South America and the evolution of the Caviidae. Journal of Vertebrate Paleontology 24(supplement to no. 3):60A.

Franzen, J. L. 1968. Revision der Gattung *Palaeotherium* Cuvier 1804 (Palaeotheriidae, Perissodactyla, Mammalia). Ph.D. thesis, Albert-Ludwigs-Universität, Freiburg, Germany.

———. 1981. Das erste Skelett eines Dichobuniden (Mammalia, Artiodactyla), geborgen aus mitteleozänen Ölschiefern der "Grube Messel" bei Darmstadt (Deutschland, S-Hessen). Senckenbergiana Lethaea 61:299–353.

———. 1983. Ein zweites Skelett von *Messelobunodon* (Mammalia, Artiodactyla, Dichobunidae) aus der "Grube Messel" bei Darmstadt (Deutschland, S-Hessen). Senckenbergiana Lethaea 64:403–445.

———. 1987. Ein neuer Primate aus dem Mitteleozän der Grube Messel (Deutschland, S-Hessen). Courier Forschungsinstitut Senckenberg 91:151–187.

———. 1988. Skeletons of *Aumelasia* (Mammalia, Artiodactyla, Dichobunidae) from Messel (M. Eocene, W. Germany). Courier Forschungsinstitut Senckenberg 107:309–321.

———. 1989. Origin and systematic position of the Palaeotheriidae; pp. 102–108 in D. R. Prothero and R. M. Schoch (eds.), The Evolution of Perissodactyls. Oxford University Press, New York.

———. 1990. *Hallensia* (Mammalia, Perissodactyla) aus Messel und dem Pariser Becken sowie Nachträge aus dem Geiseltal. Bulletin de l'Institut Royal des Sciences Naturelles de Belgique, Sciences de la Terre 60:175–201.

———. 1994. The Messel primates and anthropoid origins; pp. 99–122 in J. G. Fleagle and R. F. Kay (eds.), Anthropoid Origins. Plenum Press, New York.

———. 2000. Der sechste Messel-Primate (Mammalia, Primates, Notharctidae, Cercamoniinae). Senckenbergiana Lethaea 80:289–303.

———. 2003. Mammalian faunal turnover in the Eocene of central Europe; pp. 455–461 in S. L. Wing, P. D. Gingerich, B. Schmitz, and E. Thomas (eds.), Causes and Consequences of Globally Warm Climates in the Early Paleogene. Geological Society of America Special Paper 369, Boulder.

Franzen, J. L., and E. Frey. 1993. *Europolemur* completed. Kaupia—Darmstädter Beiträge zur Naturgeschichte 3:113–130.

Fraser, F. C., and P. E. Purves. 1960. Hearing in cetaceans: Evolution of the accessory air sacs and the structure of the outer and middle ear in recent cetaceans. Bulletin of the British Museum of Natural History (Zoology) 7:1–140.

Freudenthal, M. 1972. *Deinogalerix koenigswaldi* nov. gen., nov. spec., a giant insectivore from the Neogene of Italy. Scripta Geologica 14:1–19.

———. 2004. Gliridae (Rodentia, Mammalia) from the Eocene and Oligocene of the Sierra Palomera (Teruel, Spain). Treballs del Museu de Geologia de Barcelona 12:97–173.

Frey, E., B. Herkner, F. Schrenk, and C. Seiffert. 1993. Reconstructing organismic constructions and the problem of *Leptictidium*'s locomotion. Kaupia—Darmstädter Beiträge zur Naturgeschichte 3:89–95.

Frick, C. 1937. Horned ruminants of North America. Bulletin of the American Museum of Natural History 69:1–669.

Fricke, H. C., W. C. Clyde, J. R. O'Neil, and P. D. Gingerich. 1998. Evidence for rapid climate change in North America during the latest Paleocene thermal maximum: Oxygen isotope compositions of biogenic phosphate from the Bighorn Basin (Wyoming). Earth and Planetary Science Letters 160:193–208.

Froehlich, D. J. 1999. Phylogenetic systematics of basal perissodactyls. Journal of Vertebrate Paleontology 19:140–159.

———. 2002. Quo vadis eohippus? The systematics and taxonomy of the early Eocene equids (Perissodactyla). Zoological Journal of the Linnean Society 134:141–256.

Frye, M. S., and S. B. Hedges. 1995. Monophyly of the order Rodentia inferred from mitochondrial DNA sequences of the genes for 12S rRNA, 16S rRNA, and tRNA-valine. Molecular Biology and Evolution 12:168–176.

Fu, J.-F., J.-W. Wang, and Y.-S. Tong. 2002. The new discovery of the Plesiadapiformes from the early Eocene of Wutu Basin, Shandong Province. Vertebrate PalAsiatica 40:219–227.

Galis, F. 1999. Why do almost all mammals have seven cervical vertebrae? Developmental constraints, *Hox* genes, and cancer. Journal of Experimental Zoology (Molecular and Developmental Evolution) 285:19–26.

Gambaryan, P. P. 1974. How Mammals Run. John Wiley and Sons, New York.

Gambaryan, P. P., and Z. Kielan-Jaworowska. 1995. Masticatory musculature of Asian taeniolabidoid multituberculate mammals. Acta Palaeontologica Polonica 40:45–108.

———. 1997. Sprawling versus parasagittal stance in multituberculate mammals. Acta Palaeontologica Polonica 42:13–44.

Gatesy, J. 1997. More DNA support for a Cetacea/Hippopotamidae clade: The blood-clotting protein gene gamma-fibrinogen. Molecular Biology and Evolution 14:537–543.

Gatesy, J., and M. A. O'Leary. 2001. Deciphering whale origins with molecules and fossils. Trends in Ecology and Evolution 16:562–570.

Gatesy, J., C. Hayashi, M. A. Cronin, and P. Arctander. 1996. Evidence from milk casein genes that cetaceans are close relatives of hippopotamid artiodactyls. Molecular Biology and Evolution 13:954–963.

Gatesy, J., M. Milinkovitch, V. Waddell, and M. Stanhope. 1999. Stability of cladistic relationships between Cetacea and higher-level artiodactyl taxa. Systematic Biology 48:6–20.

Gaudin, T. J. 1995. The ear region of edentates and the phylogeny of the Tardigrada (Mammalia, Xenarthra). Journal of Vertebrate Paleontology 15:672–705.

———. 2003. Phylogeny of the Xenarthra. Senckenbergiana Biologica 83:27–40.

Gaudin, T. J., and A. A. Biewener. 1992. The functional morphology of xenarthrous vertebrae in the armadillo *Dasypus novemcinctus* (Mammalia, Xenarthra). Journal of Morphology 214:63–81.

Gaudin, T. J., and R. J. Emry. 2002. The late Eocene pangolin *Patriomanis* from North America, and a new genus of pangolin from the late Eocene of Nei Mongol, China (Mammalia, Pholidota). Journal of Vertebrate Paleontology 22(supplement to no. 3):57A.

Gaudin, T. J., J. R. Wible, J. A. Hopson, and W. D. Turnbull. 1996. Reexamination of the morphological evidence for the cohort Epitheria (Mammalia, Eutheria). Journal of Mammalian Evolution 3:31–79.

Gawne, C. E. 1978. Leporids (Lagomorpha, Mammalia) from the Chadronian (Oligocene) deposits of Flagstaff Rim, Wyoming. Journal of Paleontology 52:1103–1118.

Gazin, C. L. 1952. The Lower Eocene Knight Formation of western Wyoming and its mammalian faunas. Smithsonian Miscellaneous Collections 117(18):1–82.

———. 1953. The Tillodontia: An early Tertiary order of mammals. Smithsonian Miscellaneous Collections 121(10):1–110.

———. 1955. A review of the upper Eocene Artiodactyla of North America. Smithsonian Miscellaneous Collections 128(8):1–96.

———. 1957. A skull of the Bridger middle Eocene creodont, *Patriofelis ulta* Leidy. Smithsonian Miscellaneous Collections 134(8):1–20.

———. 1959. Early Tertiary *Apheliscus* and *Phenacodaptes* as pantolestid insectivores. Smithsonian Miscellaneous Collections 139(7):1–7.

———. 1962. A further study of the Lower Eocene mammalian faunas of southwestern Wyoming. Smithsonian Miscellaneous Collections 144(1):1–98.

———. 1965. A study of the early Tertiary condylarthran mammal *Meniscotherium*. Smithsonian Miscellaneous Collections 149(2):1–98.

———. 1968. A study of the Eocene condylarthran mammal *Hyopsodus*. Smithsonian Miscellaneous Collections 153(4):1–90.

———. 1969. A new occurrence of Paleocene mammals in the Evanston Formation, southwestern Wyoming. Smithsonian Contributions to Paleobiology 2:1–17.

Gebo, D. L. 2002. Adapiformes: Phylogeny and adaptation; pp. 21–43 in W. C. Hartwig (ed.), The Primate Fossil Record. Cambridge University Press, Cambridge, UK.

Gebo, D. L., and K. D. Rose. 1993. Skeletal morphology and locomotor adaptation in *Prolimnocyon atavus,* an early Eocene hyaenodontid creodont. Journal of Vertebrate Paleontology 13:125–144.

Gebo, D. L., M. Dagosto, K. C. Beard, T. Qi, and J. Wang. 2000. The oldest known anthropoid postcranial fossils and the early evolution of higher primates. Nature 404:276–278.

Gebo, D. L., G. F. Gunnell, R. L. Ciochon, M. Takai, T. Tsubamoto, and N. Egi. 2002. New eosimiid primate from Myanmar. Journal of Human Evolution 43:549–553.

Geisler, J. H. 2001. New morphological evidence for the phylogeny of Artiodactyla, Cetacea, and Mesonychidae. American Museum Novitates 3344:1–53.

Geisler, J. H., and M. D. Uhen. 2003. Morphological support for a close relationship between hippos and whales. Journal of Vertebrate Paleontology 23:991–996.

———. 2005. Phylogenetic relationships of extinct cetartiodactyls: Results of simultaneous analyses of molecular, morphological, and stratigraphic data. Journal of Mammalian Evolution 12:145–160.

Gelfo, J. N., and R. Pascual. 2001. *Peligrotherium tropicalis* (Mammalia, Dryolestida) from the early Paleocene of Patagonia, a survival from a Mesozoic Gondwanan radiation. Geodiversitas 23:369–379.

Gentry, A. W., and J. J. Hooker. 1988. The phylogeny of the Artiodactyla; pp. 235–272 in M. J. Benton (ed.), The Phylogeny and Classification of the Tetrapods, Volume 2: Mammals. Clarendon Press, Oxford.

Geraads, D., G. Bouvrain, and J. Sudre. 1987. Relations phylétiques de *Bachitherium* Filhol, ruminant de l'Oligocène d'Europe occidentale. Palaeovertebrata 17:43–73.

Getty, R. 1975. Sisson and Grossman's The Anatomy of the Domestic Animals. W. B. Saunders, Philadelphia.

Gheerbrant, E. 1992. Les mammifères Paléocènes du Bassin d'Ouarzazate (Maroc). I. Introduction générale et Palaeoryctidae. Palaeontographica Abteilung A 224:67–132.

———. 1994. Les mammifères Paléocènes du Bassin d'Ouarzazate (Maroc). II. Todralestidae (Proteutheria, Eutheria). Palaeontographica Abteilung A 231:133–188.

———. 1995. Les mammifères Paléocènes du Bassin d'Ouarzazate (Maroc). III. Adapisoriculidae et autres mammifères (Carnivora, ?Creodonta, Condylarthra, ?Ungulata et incertae sedis). Palaeontographica Abteilung A 237:39–132.

Gheerbrant, E., and J.-L. Hartenberger. 1999. Nouveau mammifère insectivore (?Lipotyphla, ?Erinaceomorpha) de l'Eocène inférieur de Chambi (Tunisie). Paläontologische Zeitschrift 73:143–156.

Gheerbrant, E., and D. E. Russell. 1991. *Bustylus cernaysi* nov. gen., nov. sp., nouvel adapisoriculidé (Mammalia, Eutheria) Paléocène d'Europe. Géobios 24:467–481.

Gheerbrant, E., V. Codrea, A. Hosu, S. Sen, C. Guernet, F. de Lapparent de Broin, and J. Riveline. 1999. Découverte de vertébrés dans les Calcaires de Rona (Thanétien ou Sparnacien), Transylvanie, Roumanie: Les plus anciens mammifères cénozoïques d'Europe Orientale. Eclogae Geologicae Helvetiae 92:517–535.

Gheerbrant, E., D. P. Domning, and P. Tassy. 2005. Paenungulata (Sirenia, Proboscidea, Hyracoidea, and Relatives); pp. 84–105 in K. D. Rose and J. D. Archibald (eds.), The Rise of Placental Mammals: Origins and Relationships of the Major Extant Clades. The Johns Hopkins University Press, Baltimore.

Gheerbrant, E., K. D. Rose, and M. Godinot. 2005. First palaeanodont (Mammalia, ?Pholidota) from the Eocene of Europe. Acta Palaeontologica Polonica 50:185–194.

Gheerbrant, E., J. Sudre, and H. Cappetta. 1996. A Palaeocene proboscidean from Morocco. Nature 383:68–70.

Gheerbrant, E., J. Sudre, H. Cappetta, and G. Bignot. 1998. *Phosphatherium escuilliei* du Thanétien du Bassin des Ouled Abdoun (Maroc), plus ancien proboscidien (Mammalia) d'Afrique. Géobios 30:247–269.

Gheerbrant, E., J. Sudre, H. Capetta, M. Iarochène, M. Amaghzaz, and B. Bouya. 2002. A new large mammal from the Ypresian of Morocco: Evidence of surprising diversity of early proboscideans. Acta Palaeontologica Polonica 47: 493–506.

Gheerbrant, E., J. Sudre, H. Cappetta, C. Mourer-Chauviré, E. Bourdon, M. Iarochène, M. Amaghzaz, and B. Bouya. 2003. Les localités à mammifères des carrières de Grand Daoui, bassin des Ouled Abdou, Maroc, Ypresian: Premier état des lieux. Bulletin de la Société Géologique de France 174:279–293.

Gheerbrant, E., J. Sudre, M. Iarochène, and A. Moumni. 2001. First ascertained African "condylarth" mammals (primitive ungulates: cf. Bulbulodentata and cf. Phenacodonta) from the earliest Ypresian of the Ouled Abdoun Basin, Morocco. Journal of Vertebrate Paleontology 21:107–118.

Gheerbrant, E., J. Sudre, P. Tassy, M. Amaghzaz, B. Bouya, and M. Iarochène. 2005. Nouvelles données sur *Phosphatherium escuilliei* (Mammalia, Proboscidea) de l'Éocène inférieur du Maroc, apports à la phylogénie des Proboscidea et des ongulés lophodontes. Geodiversitas 27:239–333.

Gill, P. 1974. Resorption of premolars in the early mammal *Kuehneotherium praecursoris*. Archives of Oral Biology 19:327–328.

Gingerich, P. D. 1975. Dentition of *Adapis parisiensis* and the origin of lemuriform primates; pp. 65–80 in I. Tattersall and R. Sussman (eds.), Lemur Biology. Plenum Press, New York.

———. 1976. Cranial anatomy and evolution of Early Tertiary Plesiadapidae (Mammalia, Primates). University of Michigan Papers on Paleontology 15:1–141.

———. 1980. *Tytthaena parrisi*, oldest known oxyaenid (Mammalia, Creodonta) from the late Paleocene of western North America. Journal of Paleontology 54:570–576.

———. 1981a. Early Cenozoic Omomyidae and the evolutionary history of tarsiiform primates. Journal of Human Evolution 10:345–374.

———. 1981b. Radiation of Early Cenozoic Didymoconidae (Condylarthra, Mesonychia) in Asia, with a new genus from the early Eocene of western North America. Journal of Mammalogy 62:526–538.

———. 1981c. Cranial morphology and adaptations in Eocene Adapidae. I. Sexual dimorphism in *Adapis magnus* and *Adapis parisiensis*. American Journal of Physical Anthropology 56:217–234.

———. 1981d. Variation, sexual dimorphism, and social structure in the early Eocene horse *Hyracotherium* (Mammalia, Perissodactyla). Paleobiology 7:443–455.

———. 1982. *Aaptoryctes* (Palaeoryctidae) and *Thelysia* (Palaeoryctidae?): New insectivorous mammals from the late Paleocene and early Eocene of western North America. Contributions from the Museum of Paleontology, the University of Michigan 26:37–47.

———. 1983a. Systematics of early Eocene Miacidae (Mammalia, Carnivora) in the Clark's Fork Basin, Wyoming. Contributions from the Museum of Paleontology, the University of Michigan 26:197–225.

———. 1983b. New Adapisoricidae, Pentacodontidae, and Hyopsodontidae (Mammalia, Insectivora and Condylarthra) from the late Paleocene of Wyoming and Colorado. Contributions from the Museum of Paleontology, the University of Michigan 26:227–255.

———. 1985. South American mammals in the Paleocene of North America; pp. 123–137 in F. G. Stehli and S. D. Webb (eds.), The Great American Biotic Interchange. Plenum Press, New York.

———. 1987. Early Eocene bats (Mammalia, Chiroptera) and other vertebrates in freshwater limestones of the Willwood Formation, Clark's Fork Basin, Wyoming. Contributions from the Museum of Paleontology, the University of Michigan 27:275–320.

———. 1989. New earliest Wasatchian mammalian fauna from the Eocene of northwestern Wyoming: Composition and diversity in a rarely sampled high-floodplain assemblage. University of Michigan Papers on Paleontology 28:1–97.

———. 1991. Systematics and evolution of Early Eocene Perissodactyla (Mammalia) in the Clark's Fork Basin, Wyoming. Contributions from the Museum of Paleontology, the University of Michigan 28:181–213.

———. 1995. Sexual dimorphism in earliest Eocene *Cantius torresi* (Mammalia, Primates, Adapoidea). Contributions from the Museum of Paleontology, the University of Michigan 29:185–199.

———. 1998. Paleobiological perspectives on Mesonychia, Archaeoceti, and the origin of whales; pp. 423–449 in J. G. M. Thewissen (ed.), The Emergence of Whales. Evolutionary Patterns in the Origin of Cetacea. Plenum Press, New York.

———. 2003. Mammalian responses to climate change at the Paleocene-Eocene boundary: Polecat Bench record in the northern Bighorn Basin, Wyoming; pp. 463–478 in S. L. Wing, P. D. Gingerich, B. Schmitz, and E. Thomas (eds.), Causes and Consequences of Globally Warm Climates in the Early Paleogene. Geological Society of America Special Paper 369, Boulder.

———. 2005. Cetacea; pp. 234–252 in K. D. Rose and J. D. Archibald (eds.), The Rise of Placental Mammals: Origins and Relationships of the Major Extant Clades. The Johns Hopkins University Press, Baltimore.

Gingerich, P. D., and C. G. Childress, Jr. 1983. *Barylambda churchilli*, a new species of Pantolambdidae (Mammalia, Pantodonta) from the late Paleocene of western North America. Contributions from the Museum of Paleontology, the University of Michigan 26:141–155.

Gingerich, P. D., and H. A. Deutsch. 1989. Systematics and evolution of early Eocene Hyaenodontidae (Mammalia, Creodonta) in the Clarks Fork Basin, Wyoming. Contributions from the Museum of Paleontology, the University of Michigan 27:327–391.

Gingerich, P. D., and G. F. Gunnell. 1979. Systematics and evolution of the genus *Esthonyx* (Mammalia, Tillodontia) in the early Eocene of North America. Contributions from the Museum of Paleontology, the University of Michigan 25:125–153.

———. 2005. Brain of *Plesiadapis cookei* (Mammalia, Proprimates): Surface morphology and encephalization compared to those of Primates and Dermoptera. Contributions from the Museum of Paleontology, the University of Michigan 31:185–195.

Gingerich, P. D., and R. D. Martin. 1981. Cranial morphology and adaptations in Eocene Adapidae. II. The Cambridge skull of *Adapis parisiensis*. American Journal of Physical Anthropology 56:235–257.

Gingerich, P. D., and K. D. Rose. 1979. Anterior dentition of the Eocene condylarth *Thryptacodon*: Convergence with the tooth comb of lemurs. Journal of Mammalogy 60:16–22.

———. 1982. Dentition of Clarkforkian *Labidolemur kayi*. Contributions from the Museum of Paleontology, the University of Michigan 26:49–55.

Gingerich, P. D., and A. Sahni. 1984. Dentition of *Sivaladapis nagrii* (Adapidae) from the late Miocene of India. International Journal of Primatology 5:63–79.

Gingerich, P. D., and E. L. Simons. 1977. Systematics, phylogeny, and evolution of early Eocene Adapidae (Mammalia, Primates) in North America. Contributions from the Museum of Paleontology, the University of Michigan 24:245–279.

Gingerich, P. D., and M. D. Uhen. 1994. Time of origin of primates. Journal of Human Evolution 27:443–445.

———. 1996. *Ancalecetus simonsi*, a new dorudontine archaeocete (Mammalia, Cetacea) from the early late Eocene of Wadi Hitan, Egypt. Contributions from the Museum of Paleontology, the University of Michigan 29:359–401.

Gingerich, P. D., and D. A. Winkler. 1985. Systematics of Paleocene Viverravidae (Mammalia, Carnivora) in the Bighorn Basin and Clark's Fork Basin, Wyoming. Contributions from the Museum of Paleontology, the University of Michigan 27:87–128.

Gingerich, P. D., S. G. Abbas, and M. Arif. 1997. Early Eocene *Quettacyon parachai* (Condylarthra) from the Ghazij Formation of Baluchistan (Pakistan): Oldest Cenozoic land mammal from South Asia. Journal of Vertebrate Paleontology 17:629–637.

Gingerich, P. D., M. Arif, M. A. Bhatti, and W. C. Clyde. 1998. Middle Eocene stratigraphy and marine mammals (Mammalia: Cetacea and Sirenia) of the Sulaiman Range, Pakistan; pp. 239–259 *in* K. C. Beard and M. R. Dawson (eds.), Dawn of the Age of Mammals in Asia. Bulletin of Carnegie Museum of Natural History 34.

Gingerich, P. D., M. Arif, M. A. Bhatti, H. A. Raza, and S. M. Raza. 1995. *Protosiren* and *Babiacetus* (Mammalia, Sirenia and Cetacea) from the middle Eocene Drazinda Formation, Sulaiman Range, Punjab (Pakistan). Contributions from the Museum of Paleontology, the University of Michigan 29:331–357.

Gingerich, P. D., M. Arif, and W. C. Clyde. 1995. New archaeocetes (Mammalia, Cetacea) from the middle Eocene Domanda Formation of the Sulaiman Range, Punjab (Pakistan). Contributions from the Museum of Paleontology, the University of Michigan 29:291–330.

Gingerich, P. D., M. Arif, I. H. Khan, W. C. Clyde, and J. I. Bloch. 1999. *Machocyon abbasi*, a new early Eocene quettacyonid (Mammalia, Condylarthra). Contributions from the Museum of Paleontology, the University of Michigan 30:233–250.

Gingerich, P. D., M. Arif, I. H. Khan, M. ul Haq, J. I. Bloch, W. C. Clyde, and G. F. Gunnell. 2001. Gandhera Quarry, a unique mammalian faunal assemblage from the early Eocene of Baluchistan (Pakistan); pp. 251–262 *in* G. F. Gunnell (ed.), Eocene Biodiversity: Unusual Occurrences and Rarely Sampled Habitats. Kluver Academic/Plenum Press, New York.

Gingerich, P. D., D. Dashzeveg, and D. E. Russell. 1991. Dentition and systematic relationships of *Altanius orlovi* (Mammalia, Primates) from the early Eocene of Mongolia. Géobios 24:637–646.

Gingerich, P. D., D. P. Domning, C. E. Blane, and M. D. Uhen. 1994. Cranial morphology of *Protosiren fraasi* (Mammalia, Sirenia) from the middle Eocene of Egypt: A new study using computed tomography. Contributions from the Museum of Paleontology, the University of Michigan 29:41–67.

Gingerich, P. D., P. Houde, and D. W. Krause. 1983. A new earliest Tiffanian (late Paleocene) mammalian fauna from Bangtail Plateau, western Crazy Mountain Basin, Montana. Journal of Paleontology 57:957–970.

Gingerich, P. D., S. M. Raza, M. Arif, M. Anwar, and X. Zhou. 1994. New whale from the Eocene of Pakistan and the origin of cetacean swimming. Nature 368:844–847.

Gingerich, P. D., D. E. Russell, and S. M. I. Shah. 1983. Origin of whales in epicontinental remnant seas: New evidence from the early Eocene of Pakistan. Science 220:403–406.

Gingerich, P. D., D. E. Russell, and N. A. Wells. 1990. Astragalus of *Anthracobune* (Mammalia, Proboscidea) from the early–middle Eocene of Kashmir. Contributions from the Museum of Paleontology, the University of Michigan 28:71–77.

Gingerich, P. D., B. H. Smith, and E. L. Simons. 1990. Hind limbs of Eocene *Basilosaurus*: Evidence of feet in whales. Science 249:154–157.

Gingerich, P. D., M. ul Haq, I. S. Zalmout, I. H. Khan, and M. S. Malkani. 2001. Origin of whales from early artiodactyls: Hands and feet of Eocene Protocetidae from Pakistan. Science 293:2239–2242.

Ginsburg, L. 1974. Les tayassuidés des Phosphorites du Quercy. Palaeovertebrata 6:55–85.

Ginsburg, L., and P. Mein. 1987. *Tarsius thailandica*, nov. sp., premier Tarsiidae (Primates, Mammalia) fossile d'Asie. Comptes Rendus de l'Académie des Sciences Paris, sér. II 304:1213–1215.

Ginsburg, L., K. H. Durrani, A. M. Kassi, and J.-L. Welcomme. 1999. Discovery of a new Anthracobunidae (Tethytheria, Mammalia) from the lower Eocene lignite of the Kach-Harnai area in Baluchistan (Pakistan). Comptes Rendus de l'Académie des Sciences de Paris. Sciences de la Terre et des Planètes 328:209–213.

Godfrey, L. R., and W. L. Jungers. 2002. Quaternary fossil lemurs; pp. 97–122 *in* W. C. Hartwig (ed.), The Primate Fossil Record. Cambridge University Press, Cambridge, UK.

Godinot, M. 1984. Un nouveau genre de Paromomyidae (Primates) de l'Eocène Inférieur d'Europe. Folia Primatologica 43:84–96.

———. 1988. Le gisement du Bretou (Phosphorites du Quercy, Tarn-et-Garonne, France) et sa faune de vertébrés de l'Eocène superieur: VI. Primates. Palaeontographica Abteilung A 205:113–127.

———. 1991. Toward the locomotion of two contemporaneous *Adapis* species. Zeitschrift für Morphologie und Anthropologie 78:387–405.

———. 1992a. Apport à la systématique de quatre genres d'Adapiformes (Primates, Eocène). Comptes Rendus de l' Académie des Sciences Paris, sér. II 314:237–242.

———. 1992b. Early euprimate hands in evolutionary perspective. Journal of Human Evolution 22:267–283.

———. 1994. Early North African primates and their significance for the origin of Simiiformes (=Anthropoidea); pp. 235–295 in J. G. Fleagle and R. F. Kay (eds.), Anthropoid Origins. Plenum Press, New York.

———. 1998. A summary of adapiform systematics and phylogeny. Folia primatologica 69(supplement 1):218–249.

Godinot, M., and F. de Lapparent de Broin. 2003. Arguments for a mammalian and reptilian dispersal from Asia to Europe during the Paleocene-Eocene boundary interval; pp. 255–275 in J. W. F. Reumer and W. Wessels (eds.), Distribution and Migration of Tertiary Mammals in Eurasia. A Volume in Honour of Hans de Bruijn. DEINSEA 10 (Annual of the Natural History Museum of Rotterdam).

Godinot, M., and M. Mahboubi. 1992. Earliest known simian primate found in Algeria. Nature 357:324–326.

———. 1994. Les petits primates simiiformes de Glib Zegdou (Éocène inférieur à moyen d'Algérie). Comptes Rendus de l'Académie des Sciences Paris, sér. II 319:357–364.

Godinot, M., J.-Y. Crochet, J.-L. Hartenberger, B. Lange-Badré, D. E. Russell, and B. Sigé. 1987. Nouvelles données sur les mammifères de Palette (Eocène inférieur, Provence). Münchner Geowissenschaftliche Abhandlungen, Reihe A, Geologie und Paläontologie 10:273–288.

Godinot, M., T. Smith, and R. Smith. 1996. Mode de vie et affinités de *Paschatherium* (Condylarthra, Hyopsodontidae) d'après ses os du tarse. Palaeovertebrata volume jubilaire de D. E. Russell 25:225–242.

Godthelp, H., M. Archer, R. Cifelli, S. J. Hand, and C. F. Gilkeson. 1992. Earliest known Australian Tertiary mammal fauna. Nature 356:514–516.

Godthelp, H., S. Wroe, and M. Archer. 1999. A new marsupial from the early Eocene Tingamarra Local Fauna of Murgon, southeastern Queensland: A prototypical Australian marsupial? Journal of Mammalian Evolution 6:289–313.

Goin, F. J., and A. A. Carlini. 1995. An Early Tertiary microbiotheriid marsupial from Antarctica. Journal of Vertebrate Paleontology 15:205–207.

Goin, F. J., A. M. Candela, and C. de Muizon. 2003. The affinities of *Roberthoffstetteria nationalgeographica* (Marsupialia) and the origin of the polydolopine molar pattern. Journal of Vertebrate Paleontology 23:869–876.

Goin, F. J., J. A. Case, M. O. Woodburne, S. F. Vizcaino, and M. A. Reguero. 1999. New discoveries of "opossum-like" marsupials from Antarctica (Seymour Island, medial Eocene). Journal of Mammalian Evolution 6:335–365.

Golz, D. J. 1976. Eocene Artiodactyla of southern California. Natural History Museum of Los Angeles County Science Bulletin 26:1–85.

Gould, G. C. 2001. The phylogenetic resolving power of discrete dental morphology among extant hedgehogs and the implications for their fossil record. American Museum Novitates 3340:1–52.

Gow, C. E. 1980. The dentitions of the Trithelodontidae (Therapsida: Cynodontia). Proceedings of the Royal Society of London B 208:461–481.

Gradstein, F. M., F. P. Agterberg, J. G. Ogg, J. Hardenbol, P. Van Veen, J. Thierry, and Z. Huang. 1995. A Triassic, Jurassic, and Cretaceous time scale; pp. 95–126 in W. A. Berggren, D. V. Kent, M.-P. Aubry, and J. Hardenbol (eds.), Geochronology, Time Scales and Global Stratigraphic Correlation. SEPM (Society for Sedimentary Geology) Special Publication 54, Tulsa, Okla.

Gradstein, F. M., J. G. Ogg, and A. G. Smith (eds.). 2004. A Geologic Time Scale 2004. Cambridge University Press, Cambridge, UK.

Granger, W., and W. K. Gregory. 1934. An apparently new family of amblypod mammals from Mongolia. American Museum Novitates 720:1–8.

———. 1935. A revised restoration of the skeleton of *Baluchitherium,* gigantic fossil rhinoceros of central Asia. American Museum Novitates 787:1–3.

Grassé, P.-P. 1955a. Ordre des hyracoïdes ou hyraciens; pp. 878–898 in P.-P. Grassé (ed.), Traité de Zoologie, 17(1), Mammifères. Masson et Cie, Paris.

———. 1955b. Ordre des pholidotes; pp. 1267–1282 in P.-P. Grassé (ed.), Traité de Zoologie, 17(2). Masson et Cie, Paris.

Graur, D., and D. G. Higgins. 1994. Molecular evidence for the inclusion of cetaceans within the order Artiodactyla. Molecular Biology and Evolution 11:357–364.

Graur, D., and W. Martin. 2004. Reading the entrails of chickens: Molecular timescales of evolution and the illusion of precision. Trends in Genetics 20:80–86.

Graur, D., W. A. Hide, and W.-H. Li. 1991. Is the guinea-pig a rodent? Nature 351:649–652.

Green, M. 1977. A new species of *Plesiosorex* (Mammalia, Insectivora) from the Miocene of South Dakota. Neues Jahrbuch für Geologie und Paläontologie, Monatshefte 4:189–198.

Greenwald, N. S. 1988. Patterns of tooth eruption and replacement in multituberculate mammals. Journal of Vertebarte Paleontology 8:265–277.

Gregory, W. K. 1910. The orders of mammals. Bulletin of the American Museum of Natural History 27:1–524.

———. 1920. On the structure and relations of *Notharctus,* an American Eocene primate. Memoirs of the American Museum of Natural History 3:49–243.

———. 1922. The Origin and Evolution of the Human Dentition. A Palaeontological Review. Williams & Wilkins, Baltimore.

———. 1951. Evolution Emerging. Two volumes. Macmillan, New York.

Gunnell, G. F. 1988. New species of *Unuchinia* (Mammalia: Insectivora) from the middle Paleocene of North America. Journal of Paleontology 62:139–141.

———. 1989. Evolutionary history of Microsyopoidea (Mammalia, ?Primates) and the relationship between Plesiadapiformes and Primates. University of Michigan Papers on Paleontology 27:1–157.

———. 1995. Omomyid primates (Tarsiiformes) from the Bridger Formation, middle Eocene, southern Green River Basin, Wyoming. Journal of Human Evolution 28:147–187.

———. 1998. Creodonta; pp. 91–109 in C. M. Janis, K. M. Scott, and L. L. Jacobs (eds.), Evolution of Tertiary Mammals of North America. Cambridge University Press, Cambridge, UK.

Gunnell, G. F., and P. D. Gingerich. 1991. Systematics and evolution of late Paleocene and early Eocene Oxyaenidae (Mammalia, Creodonta) in the Clarks Fork Basin, Wyoming. Contributions from the Museum of Paleontology, the University of Michigan 28:141–180.

———. 1993. Skeleton of *Brachianodon westorum*, a new middle Eocene metacheiromyid (Mammalia, Palaeanodonta) from the early Bridgerian (Bridger A) of the southern Green River Basin, Wyoming. Contributions from the Museum of Paleontology, the University of Michigan 28:365–392.

Gunnell, G. F., and K. D. Rose. 2002. Tarsiiformes: Evolutionary history and adaptation; pp. 45–82 in W. C. Hartwig (ed.), The Primate Fossil Record. Cambridge University Press, Cambridge, UK.

Gunnell, G. F., B. F. Jacobs, P. S. Herendeed, J. J. Head, E. Kowalski, C. P. Msuya, F. A. Mizambwa, T. Harrison, J. Habersetzer, and G. Storch. 2003. Oldest placental mammal from sub-Saharan Africa: Eocene microbat from Tanzania—Evidence for early evolution of sophisticated echolocation. Palaeontologia Electronica 5:1–10.

Guo, J.-W., M. R. Dawson, and K. C. Beard. 2000. *Zhailimeryx*, a new lophiomerycid artiodactyl (Mammalia) from the late middle Eocene of central China and the early evolution of ruminants. Journal of Mammalian Evolution 7:239–258.

Guo, J.-W., T. Qi, and H.-J. Sheng. 1999. A restudy of the Eocene ruminants from Baise and Yongle Basins, Guangxi, China, with a discussion of the systematic positions of *Indomeryx*, *Notomeryx*, *Gobiomeryx*, and *Prodremotherium*. Vertebrata PalAsiatica 37:18–39.

Habersetzer, J., and G. Storch. 1987. Klassifikation und funktionelle Flügelmorphologie paläogener Fledermäuse (Mammalia, Chiroptera). Courier Forschungsinstitut Senckenberg 91: 117–150.

———. 1989. Ecology and echolocation of the Eocene Messel bats; pp. 213–233 in V. Hanak, I. Horacek, and J. Gaisler (eds.), European Bat Research 1987. Charles University Press, Prague.

———. 1993. Radiographic studies of the cochlea in extant Chiroptera and microchiropterans from Messel. Kaupia—Darmstädter Beiträge zur Naturgeschichte 3:97–105.

Habersetzer, J., G. Richter, and G. Storch. 1992. Bats: Already highly specialized insect predators; pp. 181–191 in S. Schaal and W. Ziegler (eds.), Messel. An Insight into the History of Life and of the Earth. Clarendon Press, Oxford.

———. 1994. Paleoecology of early middle Eocene bats from Messel, FRG. Aspects of flight, feeding and echolocation. Historical Biology 8:235–260.

Hahn, G. 1977. Neue Schädel-Reste von Multituberculaten (Mamm.) aus dem Malm Portugals. Geologica et Palaeontologica 11:161–186.

———. 1978. Die Multituberculata, eine fosille Säugetier-Ordnung. Sonderband des Naturwissenschaftlichen Vereins, Hamburg 3:61–95.

———. 1993. The systematic arrangement of the Paulchoffatiidae (Multituberculata) revisited. Geologica et Palaeontologica 27:201–214.

Hahn, G., R. Hahn, and P. Godefroit. 1994. Zur Stellung der Dromatheriidae (Ober-Trias) zwischen den Cynodontia und den Mammalia. Geologica et Palaeontologica 28: 141–159.

Hahn, G., D. Sigogneau-Russell, and G. Wouters. 1989. New data on Theroteinidae—Their relations with Paulchoffatiidae and Haramiyidae. Geologica et Palaeontolologica 23:205–215.

Hamrick, M. W. 1996. Locomotor adaptations reflected in the wrist joints of Early Tertiary primates (Adapiformes). American Journal of Physical Anthropology 100:585–604.

———. 1998. Functional and adaptive significance of primate pads and claws: Evidence from New World anthropoids. American Journal of Physical Anthropology 106:113–127.

Hamrick, M. W., and J. P. Alexander. 1996. The hand skeleton of *Notharctus tenebrosus* (Primates, Notharctidae) and its significance for the origin of the primate hand. American Museum Novitates 3182:1–20.

Hamrick, M. W., B. A. Rosenman, and J. A. Brush. 1999. Phalangeal morphology of the Paromomyidae (?Primates, Plesiadapiformes): The evidence for gliding behavior reconsidered. American Journal of Physical Anthropology 109:397–413.

Hand, S., M. J. Novacek, H. Godthelp, and M. Archer. 1994. First Eocene bat from Australia. Journal of Vertebrate Paleontology 14:375–381.

Hanson, C. B. 1989. *Teletaceras radinskyi*, a new primitive rhinocerotid from the late Eocene Clarno Formation of Oregon; pp. 379–398 in D. R. Prothero and R. M. Schoch (eds.), The Evolution of Perissodactyls. Oxford University Press, New York.

Hartenberger, J.-L. 1969. Les Pseudosciuridae (Mammalia, Rodentia) de l'Eocène moyen de Bouxwiller, Egerkingen et Lissieu. Palaeovertebrata 3:27–61.

———. 1971. Contribution à l'étude des genres *Gliravus* et *Microparamys* (Rodentia) de l'Éocène d'Europe. Palaeovertebrata 4:97–135.

———. 1973. Étude systématique des Theridomyoidea (Rodentia) de l'Éocène supérieur. Mémoires de la Société Géologique de France, Nouvelle série 52:1–76.

———. 1985. The order Rodentia: Major questions on their evolutionary origin, relationships and suprafamilial systematics; pp. 1–33 in W. P. Luckett and J.-L. Hartenberger (eds.), Evolutionary Relationships among Rodents: A Multidisciplinary Analysis. Plenum Press, New York.

———. 1986. Hypothèse paléontologique sur l'origine des Macroscelidea (Mammalia). Comptes Rendus de l'Académie des Sciences Paris, sér. II 302:247–249.

———. 1990. L'origine des Theridomyoidea (Mammalia, Rodentia): Données nouvelles et hypothèses. Comptes Rendus de l'Académie des Sciences, sér. II 311:1017–1023.

———. 1994. The evolution of the Gliroidea; pp. 19–33 in Y. Tomida, C.-K. Li, and T. Setoguchi (eds.), Rodent and Lagomorph Families of Asian Origins and Diversification. National Science Museum, Tokyo.

———. 1995. Place des Ailuravinae dans la radiation initiale des rongeurs en Europe. Comptes Rendus de l'Académie des Sciences, sér. II 321:631–637.

———. 1998. Description de la radiation des Rodentia (Mammalia) du Paléocène supérieur au Miocène; incidences phylogénétiques. Comptes Rendus de l'Académie des Sciences, Paris, Sciences de la terre et des planètes 326:439–444.

Hartenberger, J.-L., and B. Marandat. 1992. A new genus and species of early Eocene primate from North Africa. Human Evolution 7:9–16.

Hedges, S. B., P. H. Parker, C. G. Sibley, and S. Kumar. 1996. Continental breakup and the ordinal diversification of birds and mammals. Nature 381:226–229.

Heinrich, R. E., and K. D. Rose. 1995. Partial skeleton of the primitive carnivoran *Miacis petilus* from the early Eocene of Wyoming. Journal of Mammalogy 76:148–162.

———. 1997. Postcranial morphology and locomotor behavior of two early Eocene miacoid carnivorans, *Vulpavus* and *Didymictis*. Palaeontology 40:279–305.

Heissig, K. 1993. The astragalus in anoplotheres and oreodonts, phylogenetical and paleogeographical implications. Kaupia—Darmstädter Beiträge zur Naturgeschichte 3:173–178.

———. 2003. Origin and early dispersal of the squirrels and their relatives; pp. 277–286 in J. W. F. Reumer and W. Wessels (eds.), Distribution and Migration of Tertiary Mammals in Eurasia. A Volume in Honour of Hans de Bruijn. DEINSEA 10 (Annual of the Natural History Museum of Rotterdam).

Hennig, W. 1966. Phylogenetic Systematics. University of Illinois Press, Urbana.

Heyning, J. E. 1989. Comparative facial anatomy of beaked whales (Ziphiidae) and a systematic revision among the families of extant Odontoceti. Contributions in Science, the Natural History Museum of Los Angeles County 405:1–64.

———. 1997. Sperm whale phylogeny revisited: Analysis of the morphological evidence. Marine Mammal Science 13:596–613.

Hickey, L. J. 1977. Stratigraphy and paleobotany of the Golden Valley Formation (Early Tertiary) of western North Dakota. Geological Society of America Memoir 150:1–183.

Hildebrand, M. 1995. Analysis of Vertebrate Structure. John Wiley and Sons, New York.

Hildebrand, M., D. M. Bramble, K. F. Liem, and D. B. Wake (eds.). 1985. Functional Vertebrate Morphology. Belknap Press, Cambridge, Mass.

Hillson, S. 1986. Teeth. Cambridge University Press, Cambridge, UK.

Hoffstetter, R. 1970. *Colombitherium tolimense*, pyrothérien nouveau de la formation Gualanday (Colombie). Annales de Paléontologie, Vertébrés 56:149–169.

———. 1972. Origine et dispersion des rongeurs hystricognathes. Comptes Rendus des séances de l'Académie des Sciences, sér. D 274:2867–2870.

———. 1975. El origen de los Caviomorpha y el problema de los Hystricognathi (Rodentia). Actas del Primer Congreso Argentino de Paleontologia y Bioestratigrafia, Tucumán 2:505–528.

———. 1982. Les édentés xénarthres, un group singulier de la fauna néotropicale; pp. 385–443 in E. M. Gallitelli (ed.), Palaeontology, Essential of Historical Geology. S.T.E.M. Mucchi, Venice.

Holbrook, L. T. 1999. The phylogeny and classification of tapiromorph perissodactyls (Mammalia). Cladistics 15:331–350.

———. 2001. Comparative osteology of early Tertiary tapiromorphs (Mammalia, Perissodactyla). Zoological Journal of the Linnean Society 132:1–54.

———. 2005. On the skull of *Radinskya* (Mammalia, ?Phenacolophidae) and its affinities. Journal of Vertebrate Paleontology 25(supplement to no. 3):70A–71A.

Holbrook, L. T., and S. G. Lucas. 1997. A new genus of rhinocerotoid from the Eocene of Utah and the status of North American *"Forstercooperia."* Journal of Vertebrate Paleontology 17:384–396.

Honey, J. G., J. A. Harrison, D. R. Prothero, and M. S. Stevens. 1998. Camelidae; pp. 439–462 in C. M. Janis, K. M. Scott, and L. L. Jacobs (eds.), Evolution of Tertiary Mammals of North America. Volume 1: Terrestrial Carnivores, Ungulates, and Ungulatelike Mammals. Cambridge University Press, Cambridge, UK.

Honeycutt, R. L., and R. M. Adkins. 1993. Higher level systematics of eutherian mammals: An assessment of molecular characters and phylogenetic hypotheses. Annual Review of Ecology and Systematics 24:279–305.

Hooker, J. J. 1979. Two new condylarths (Mammalia) from the early Eocene of southern England. Bulletin of the British Museum of Natural History (Geology) 32:43–56.

———. 1989. Character polarities in early perissodactyls and their significance for *Hyracotherium* and infraordinal relationships; pp. 79–101 in D. R. Prothero and R. M. Schoch (eds.), The Evolution of Perissodactyls. Oxford University Press, New York.

———. 1992a. British mammalian paleocommunities across the Eocene-Oligocene transition and their environmental implications; pp. 494–515 in D. R. Prothero and W. A. Berggren (eds.), Eocene-Oligocene Climatic and Biotic Evolution. Princeton University Press, Princeton.

———. 1992b. An additional record of a placental mammal (order Astrapotheria) from the Eocene of West Antarctica. Antarctic Science 4:107–108.

———. 1994. The beginning of the equoid radiation. Zoological Journal of the Linnean Society 112:29–63.

———. 1996. A primitive emballonurid bat (Chiroptera, Mammalia) from the earliest Eocene of England. Palaeovertebrata, Volume jubilaire D. E. Russell 25:287–300.

———. 1998. Mammalian faunal change across the Paleocene-Eocene transition in Europe; pp. 428–450 in M.-P. Aubry, S. G. Lucas, and W. A. Berggren (eds.), Late Paleocene–Early Eocene Climatic and Biotic Events in the Marine and Terrestrial Records. Columbia University Press, New York.

———. 2001. Tarsals of the extinct insectivoran family Nyctitheriidae (Mammalia): Evidence for archontan relationships. Zoological Journal of the Linnean Society 132:501–529.

———. 2005. Perissodactyla; pp. 199–214 in K. D. Rose and J. D. Archibald (eds.), The Rise of Placental Mammals: Origins and Relationships of the Major Extant Clades. The Johns Hopkins University Press, Baltimore.

Hooker, J. J., and D. Dashzeveg. 2003. Evidence for direct mammalian faunal interchange between Europe and Asia near the Paleocene-Eocene boundary; pp. 479–500 in S. L. Wing, P. D. Gingerich, B. Schmitz, and E. Thomas (eds.), Causes and Consequences of Globally Warm Climates in the Early Paleogene. Geological Society of America Special Paper 369, Boulder.

———. 2004. The origin of chalicotheres (Perissodactyla, Mammalia). Palaeontology 47:1363–1386.

Hooker, J. J., and K. M. Thomas. 2001. A new species of *Amphiragatherium* (Choeropotamidae, Artiodactyla, Mammalia) from the Late Eocene Headon Hill Formation of southern England and phylogeny of endemic European "anthracotherioids." Palaeontology 44:827–853.

Hooker, J. J., and M. Weidmann. 2000. The Eocene mammal faunas of Mormont, Switzerland. Systematic revision and resolution of dating problems. Schweizerische Paläontologische Abhandlungen 120:1–141.

Hooker, J. J., M. E. Collinson, and N. P. Sille. 2004. Eocene-Oligocene mammalian faunal turnover in the Hampshire Basin, UK: Calibration to the global time scale and the major cooling event. Journal of the Geological Society, London 161:161–172.

Hooker, J. J., D. E. Russell, and A. Phélizon. 1999. A new family of Plesiadapiformes (Mammalia) from the Old World Lower Paleogene. Palaeontology 42:377–407.

Hopson, J. A. 1994. Synapsid evolution and the radiation of non-eutherian mammals; pp. 190–219 in D. R. Prothero and R. M.

Schoch (eds.), Major Features of Vertebrate Evolution. Paleontological Society, Lawrence, Kansas.

Hopson, J. A., and H. R. Barghusen. 1986. An analysis of therapsid relationships; pp. 83–106 in N. Hotton III, P. D. MacLean, J. J. Roth, and E. C. Roth (eds.), The Ecology and Biology of Mammal-Like Reptiles. Smithsonian Institution Press, Washington, D. C.

Hopson, J. A., and A. W. Crompton. 1969. Origin of mammals. Evolutionary Biology 3:15–72.

Hopson, J. A., and J. W. Kitching. 2001. A probainognathian cynodont from South Africa and the phylogeny of non-mammalian cynodonts. Bulletin of the Museum of Comparative Zoology 156:5–35.

Hopson, J. A., and G. W. Rougier. 1993. Braincase structure in the oldest known skull of a therian mammal: Implications for mammalian systematics and cranial evolution. American Journal of Science 293A:268–299.

Hopson, J. A., Z. Kielan-Jaworowska, and E. F. Allin. 1989. The cryptic jugal of multituberculates. Journal of Vertebrate Paleontology 9:201–209.

Horovitz, I. 2000. The tarsus of *Ukhaatherium nessovi* (Eutheria, Mammalia) from the Late Cretaceous of Mongolia: An appraisal of the evolution of the ankle in basal therians. Journal of Vertebrate Paleontology 20:547–560.

———. 2003. Postcranial skeleton of *Ukhaatherium nessovi* (Eutheria, Mammalia) from the Late Cretaceous of Mongolia. Journal of Vertebrate Paleontology 23:857–868.

Hough, J. 1956. A new insectivore from the Oligocene of the Wind River Basin, Wyoming, with notes on the taxonomy of the Oligocene Tenrecoidea. Journal of Paleontology 30:531–541.

Howell, A. B. 1944. Speed in Animals. Their Specialization for Running and Leaping. University of Chicago Press, Chicago.

Hu, Y. 1993. Two new genera of Anagalidae (Anagalida, Mammalia) from the Paleocene of Qianshan, Anhui, and the phylogeny of anagalids. Vertebrata PalAsiatica 31:153–182.

Hu, Y., J. Meng, Y. Wang, and C.-K. Li. 2005. Large Mesozoic mammals fed on young dinosaurs. Nature 433:149–152.

Hu, Y., Y. Wang, Z.-X. Luo, and C. Li. 1997. A new symmetrodont mammal from China and its implications for mammalian evolution. Nature 390:137–142.

Huang, X.-S. 1977. *Archaeolambda* fossils from Anhui. Vertebrata PalAsiatica 15:249–260.

———. 1995. Classification of Pantolambdodontidae (Pantodonta, Mammalia). Vertebrata PalAsiatica 33:194–215.

Huang, X.-S., and J.-J. Zheng. 1987. A new pantodont-like mammal from the Paleocene of Chienshan Basin, Anhui. Vertebrata PalAsiatica 25:20–31.

———. 1999. A new tillodont from the Paleocene of Nanxiong Basin, Guangdong. Vertebrata PalAsiatica 37:96–104.

———. 2003. A tillodont-like mammal from the middle Paleocene of Qianshan Basin, Anhui, China. Vertebrata PalAsiatica 41:131–136.

Huchon, D., O. Madsen, M. J. J. B. Sibbald, K. Ament, M. J. Stanhope, F. Catzeflis, W. W. de Jong, and E. J. P. Douzery. 2002. Rodent phylogeny and a timescale for the evolution of Glires: Evidence from an extensive taxon sampling using three nuclear genes. Molecular Biology and Evolution 19:1053–1065.

Hulbert, R. C., Jr., R. M. Petkewich, G. A. Bishop, D. Bukry, and D. P. Aleshire. 1998. A new middle Eocene protocetid whale (Mammalia: Cetacea: Archaeoceti) and associated biota from Georgia. Journal of Paleontology 72:907–927.

Hunt, R. M., Jr. 1974a. *Daphoenictis,* a cat-like carnivore (Mammalia, Amphicyonidae) from the Oligocene of North America. Journal of Paleontology 48:1030–1047.

———. 1974b. The auditory bulla in Carnivora: An anatomical basis for reappraisal of carnivore evolution. Journal of Morphology 143:21–76.

———. 1977. Basicranial anatomy of *Cynelos* Jourdan (Mammalia: Carnivora), an Aquitanian amphicyonid from the Allier Basin, France. Journal of Paleontology 51:826–843.

———. 1987. Evolution of the aeluroid Carnivora: Significance of auditory structure in the nimravid cat *Dinictis*. American Museum Novitates 2886:1–74.

———. 1989. Evolution of the aeluroid Carnivora: Significance of the ventral promontorial process of the petrosal, and the origin of basicranial patterns in the living families. American Museum Novitates 2930:1–32.

———. 1991. Evolution of the aeluroid Carnivora: Viverrid affinities of the Miocene carnivoran *Herpestides*. American Museum Novitates 3023:1–34.

———. 1996. Amphicyonidae; pp. 476–485 in D. R. Prothero and R. J. Emry (eds.), The Terrestrial Eocene-Oligocene Transition in North America. Cambridge University Press, Cambridge, UK.

———. 1998a. Ursidae; pp. 174–195 in C. M. Janis, K. M. Scott, and L. L. Jacobs (eds.), Evolution of Tertiary Mammals of North America. Cambridge University Press, Cambridge, UK.

———. 1998b. Amphicyonidae; pp. 196–227 in C. M. Janis, K. M. Scott, and L. L. Jacobs (eds.), Evolution of Tertiary Mammals of North America. Cambridge University Press, Cambridge, UK.

———. 1998c. Evolution of the aeluroid Carnivora: Diversity of the earliest aeluroids from Eurasia (Quercy, Hsanda-Gol) and the origin of felids. American Museum Novitates 3252:1–65.

———. 2001. Basicranial anatomy of the living linsangs *Prionodon* and *Poiana* (Mammalia, Carnivora, Viverridae), with comments on the early evolution of aeluroid carnivorans. American Museum Novitates 3330:1–24.

Hunt, R. M., Jr., and L. G. Barnes. 1994. Basicranial evidence for ursid affinity of the oldest pinnipeds. Proceedings of the San Diego Society of Natural History 29:57–67.

Hunt, R. M., Jr., and W. W. Korth. 1980. The auditory region of Dermoptera: Morphology and function relative to other living mammals. Journal of Morphology 164:167–211.

Hunt, R. M., Jr., and R. H. Tedford. 1993. Phylogenetic relationships within the aeluroid Carnivora and implications of their temporal and geographic distribution; pp. 53–73 in F. S. Szalay, M. J. Novacek, and M. C. McKenna (eds.), Mammal Phylogeny: Placentals. Springer-Verlag, New York.

Hunter, J. P., and J. Jernvall. 1995. The hypocone as a key innovation in mammalian evolution. Proceedings of the National Academy of Sciences, USA 92:10718–10722.

Hurum, J. H. 1994. The snout and orbit of Mongolian multituberculates studied by serial sections. Acta Palaeontologica Polonica 39:181–221.

———. 1998. The braincase of two Late Cretaceous Asian multituberculates studied by serial sections. Acta Palaeontologica Polonica 43:21–52.

Hurum, J. H., R. Presley, and Z. Kielan-Jaworowska. 1996. The middle ear in multituberculate mammals. Acta Palaeontologica Polonica 41:253–275.

Hürzeler, J. 1936. Osteologie und Odontologie der Cainotheriden. Schweizerische Palaeontologische Gesellschaft Abhandlungen 58–59:1–89, 90–112.
———. 1944. Beiträge zur Kenntnis der Dimylidae. Schweizerische Palaeontologische Gesellschaft Abhandlungen 65:1–44.
Hussain, S. T., H. de Bruijn, and J. M. Leinders. 1978. Middle Eocene rodents from the Kala Chitta Range (Punjab, Pakistan). Proceedings of the Koninklijke Nederlandse Akademie van Wetenschappen, ser. B 81:74–112.
Hutcheon, J. M., J. A. W. Kirsch, and J. D. Pettigrew. 1998. Base-compositional biases and the bat problem. III. The question of microchiropteran monophyly. Philosophical Transactions of the Royal Society of London B 353:607–617.
Iakovleva, A. I., H. Brinkhuis, and C. Cavagnetto. 2001. Late Palaeocene–early Eocene dinoflagellate cysts from the Turgay Strait, Kazakhstan; Correlations across ancient seaways. Palaeogeography, Palaeoclimatology, Palaeoecology 172:243–268.
Ivy, L. D. 1990. Systematics of late Paleocene and early Eocene Rodentia (Mammalia) from the Clarks Fork Basin, Wyoming. Contributions from the Museum of Paleontology, the University of Michigan 28:21–70.
Jaeger, J.-J., Y. Chaimanee, P. Tafforeau, S. Ducrocq, A. N. Soe, L. Marivaux, J. Sudre, S. T. Tun, W. Htoon, and B. Marandat. 2004. Systematics and paleobiology of the anthropoid primate *Pondaungia* from the late middle Eocene of Myanmar. Comptes Rendus Palevol 3:243–255.
Jaeger, J.-J., V. Courtillot, and P. Tapponnier. 1989. Paleontological view of the ages of the Deccan Traps, the Cretaceous/Tertiary boundary, and the India-Asia collision. Geology 17:316–319.
Jaeger, J.-J., C. Denys, and B. Coiffait. 1985. New Phiomorpha and Anomaluridae from the late Eocene of north-west Africa: Phylogenetic implications; pp. 567–588 in W. P. Luckett and J.-L. Hartenberger (eds.), Evolutionary Relationships Among Rodents. A Multidisciplinary Analysis. Plenum Press, New York.
Jaeger, J.-J., A. N. Soe, A. K. Aung, M. Benammi, Y. Chaimanee, R.-M. Ducrocq, T. Tun, T. Thein, and S. Ducrocq. 1998. New Myanmar middle Eocene anthropoids. An Asian origin for catarrhines? Comptes Rendus de l'Académie des Sciences, Paris, Sciences de la vie 321:953–959.
Jaeger, J.-J., T. Thein, M. Benammi, Y. Chaimanee, A. N. Soe, T. Lwin, T. Tun, S. Wai, and S. Ducrocq. 1999. A new primate from the middle Eocene of Myanmar and the Asian early origin of anthropoids. Science 286:528–530.
Janis, C. M. 1976. The evolutionary strategy of the Equidae and the origins of rumen and cecal digestion. Evolution 30:757–774.
———. 1987. Grades and clades in hornless ruminant evolution: The reality of the Gelocidae and the systematic position of *Lophiomeryx* and *Bachitherium*. Journal of Vertebrate Paleontology 7:200–216.
———. 1989. A climatic explanation for patterns of evolutionary diversity in ungulate mammals. Palaeontology 32:463–481.
Janis, C. M., J. D. Archibald, R. L. Cifelli, S. G. Lucas, C. R. Schaff, R. M. Schoch, and T. E. Williamson. 1998. Archaic ungulates and ungulatelike mammals; pp. 247–259 in C. M. Janis, K. M. Scott, and L. L. Jacobs (eds.), Evolution of Tertiary Mammals of North America. Cambridge University Press, Cambridge, UK.
Janis, C. M., M. W. Colbert, M. C. Coombs, W. D. Lambert, B. J. MacFadden, B. J. Mader, D. R. Prothero, R. M. Schoch, J. Shoshani, and W. P. Wall. 1998. Perissodactyla and Proboscidea; pp. 511–524 in C. M. Janis, K. M. Scott, and L. L. Jacobs (eds.), Evolution of Tertiary Mammals of North America. Cambridge University Press, Cambridge, UK.
Janis, C. M., J. A. Effinger, J. A. Harrison, J. G. Honey, D. G. Kron, B. Lander, E. Manning, D. R. Prothero, M. S. Stevens, R. K. Stucky, S. D. Webb, and D. B. Wright. 1998. Artiodactyla; pp. 337–357 in C. M. Janis, K. M. Scott, and L. L. Jacobs (eds.), Evolution of Tertiary Mammals of North America. Cambridge University Press, Cambridge, UK.
Janis, C. M., J. M. Theodor, and B. Boisvert. 2002. Locomotor evolution in camels revisited: A quantitative analysis of pedal anatomy and the acquisition of the pacing gait. Journal of Vertebrate Paleontology 22:110–121.
Janke, A., O. Magnell, G. Wieczorek, M. Westerman, and U. Arnason. 2002. Phylogenetic analysis of 18S rRNA and the mitochondrial genomes of the wombat, *Vombatus ursinus*, and the spiny anteater, *Tachyglossus aculeatus*: Increased support for the Marsupionta hypothesis. Journal of Molecular Evolution 54:71–80.
Janke, A., X. Xu, and U. Arnason. 1997. The complete mitochondrial genome of the wallaroo (*Macropus robustus*) and the phylogenetic relationship among Monotremata, Marsupialia, and Eutheria. Proceedings of the National Academy of Sciences, USA 94:1276–1281.
Janovskaja, N. M. 1980. The Brontotheres of Mongolia [in Russian]. Joint Soviet-Mongolian Paleontological Expedition Transactions 12:1–219.
Jayne, H. 1898. Mammalian Anatomy. Part I. The Skeleton of the Cat. J. B. Lippincott, Philadelphia.
Jefferies, R. P. S. 1979. The origin of chordates—A methodological essay; pp. 443–477 in M. R. House (ed.), The Origin of Major Invertebrate Groups. Academic Press, London.
Jenkins, F. A., Jr. 1969a. The evolution and development of the dens of the mammalian axis. Anatomical Record 164:173–184.
———. 1969b. Occlusion in *Docodon* (Mammalia, Docodonta). Postilla 139:1–24.
———. 1971. The postcranial skeleton of African cynodonts. Bulletin of Peabody Museum of Natural History 36:1–216.
———. 1974. Tree shrew locomotion and the origins of primate arborealism; pp. 85–115 in F. A. Jenkins, Jr. (ed.), Primate Locomotion. Academic Press, New York.
———. 1984. A survey of mammalian origins; pp. 32–47 in P. D. Gingerich and C. E. Badgley (eds.), Mammals. Notes for a Short Course. University of Tennessee Department of Geological Sciences, Studies in Geology 8, Knoxville.
Jenkins, F. A., Jr., and A. W. Crompton. 1979. Triconodonta; pp. 74–90 in J. A. Lillegraven, Z. Kielan-Jaworowska, and W. A. Clemens (eds.), Mesozoic Mammals. The First Two-Thirds of Mammalian History. University of California Press, Berkeley.
Jenkins, F. A., Jr., and D. W. Krause. 1983. Adaptations for climbing in North American multituberculates (Mammalia). Science 220:712–715.
Jenkins, F. A., Jr., and D. McClearn. 1984. Mechanisms of hind foot reversal in climbing mammals. Journal of Morphology 182:197–219.
Jenkins, F. A., Jr., and F. R. Parrington. 1976. The postcranial skeleton of the Triassic mammals *Eozostrodon*, *Megazostrodon*

and *Erythrotherium*. Philosophical Transactions of the Royal Society of London B 273:387–431.
Jenkins, F. A., Jr., and C. R. Schaff. 1988. The Early Cretaceous mammal *Gobiconodon* (Mammalia, Triconodonta) from the Cloverly Formation in Montana. Journal of Vertebrate Paleontology 8:1–24.
Jenkins, F. A., Jr., A. W. Crompton, and W. R. Downs. 1983. Mesozoic Mammals from Arizona: New evidence on mammalian evolution. Science 222:1233–1235.
Jenkins, F. A., Jr., S. M. Gatesy, N. H. Shubin, and W. W. Amaral. 1997. Haramiyids and Triassic mammalian evolution. Nature 385:715–718.
Jepsen, G. L. 1930. Stratigraphy and paleontology of the Paleocene of northeastern Park County, Wyoming. Proceedings of the American Philosophical Society 69:463–528.
———. 1966. Early Eocene bat from Wyoming. Science 154:1333–1339.
———. 1970. Bat origins and evolution; pp. 1–64 *in* W. A. Wimsatt (ed.), Biology of Bats. Academic Press, New York.
Jernvall, J. 2000. Linking development with generation of novelty in mammalian teeth. Proceedings of the National Academy of Sciences, USA 97:2641–2645.
Ji, Q., Z.-X. Luo, and S. Ji. 1999. A Chinese triconodont mammal and mosaic evolution of the mammalian skeleton. Nature 398:326–330.
Ji, Q., Z.-X. Luo, C.-X. Yuan, and A. R. Tabrum. 2006. A swimming mammaliaform from the Middle Jurassic and ecomorphological diversification of early mammals. Science 311:1123–1127.
Ji, Q., Z.-X. Luo, C.-X. Yuan, J. R. Wible, J.-P. Zhang, and J. A. Georgi. 2002. The earliest known eutherian mammal. Nature 416:816–822.
Joeckel, R. M. 1990. A functional interpretation of the masticatory system and paleoecology of entelodonts. Paleobiology 16:459–482.
Joeckel, R. M., and J. M. Stavas. 1996. Basicranial anatomy of *Syndyoceras cooki* (Artiodactyla, Protoceratidae) and the need for a reappraisal of tylopod relationships. Journal of Vertebrate Paleontology 16:320–327.
Johannessen, C. L., and J. A. Harder. 1960. Sustained swimming speeds of dolphins. Science 132:1550–1551.
Johnson, K. R., and B. Ellis. 2002. A tropical rainforest in Colorado 1.4 million years after the Cretaceous-Tertiary boundary. Science 296:2379–2383.
Jones, C. 1984. Tubulidentates, proboscideans, and hyracoideans; pp. 523–535 *in* S. Anderson and J. K. Jones, Jr. (eds.), Orders and Families of Recent Mammals of the World. John Wiley and Sons, New York.
Kalthoff, D. C., W. von Koenigswald, and C. Kurz. 2004. A new specimen of *Heterohyus nanus* (Apatemyidae, Mammalia) from the Eocene of Messel (Germany) with unusual soft-part preservation. Courier Forschungsinstitut Senckenberg 252:1–12.
Kappelman, J., E. L. Simons, and C. C. Swisher III. 1992. New age determination for the Eocene-Oligocene boundary sediments in the Fayum Depression, northern Egypt. Journal of Geology 100:647–668.
Katz, M. E., D. K. Pak, G. R. Dickens, and K. G. Miller. 1999. The source and fate of massive carbon input during the latest Paleocene thermal maximum. Science 286:1531–1533.
Kay, R. F., and B. A. Williams. 1994. Dental evidence for anthropoid origins; pp. 361–445 *in* J. G. Fleagle and R. F. Kay (eds.), Anthropoid Origins. Plenum Press, New York.
Kay, R. F., R. H. Madden, M. G. Vucetich, A. A. Carlini, M. M. Mazzoni, G. H. Re, M. Heizler, and H. Sandeman. 1999. Revised geochronology of the Casamayoran South American Land Mammal Age: Climatic and biotic implications. Proceedings of the National Academy of Sciences, USA 96:13235–13240.
Kay, R. F., C. Ross, and B. A. Williams. 1997. Anthropoid origins. Science 275:797–804.
Kay, R. F., J. G. M. Thewissen, and A. D. Yoder. 1992. Cranial anatomy of *Ignacius graybullianus* and the affinities of the Plesiadapiformes. American Journal of Physical Anthropology 89:477–498.
Kay, R. F., R. W. Thorington, Jr., and P. Houde. 1990. Eocene plesiadapiform shows affinities with flying lemurs, not primates. Nature 345:342–344.
Kay, R. F., B. A. Williams, C. F. Ross, M. Takai, and N. Shigehara, 2004. Anthropoid origins: A phylogenetic analysis; pp. 91–135 *in* C. F. Ross and R. F. Kay (eds.), Anthropoid Origins: New Visions. Kluwer Academic/Plenum Press, New York.
Kemp, T. S. 1983. The relationships of mammals. Zoological Journal of the Linnean Society 77:353–384.
———. 2005. The Origin and Evolution of Mammals. Oxford University Press, Oxford, UK.
Kennett, J. P., and L. D. Stott. 1991. Abrupt deep-sea warming, palaeoceanographic changes and benthic extinctions at the end of the Palaeocene. Nature 353:225–229.
Kent, D. V., B. S. Cramer, L. Lanci, D. Wang, J. D. Wright, and R. Van der Voo. 2003. A case for a comet impact trigger for the Paleocene/Eocene thermal maximum and carbon isotope excursion. Earth and Planetary Science Letters 211:13–26.
Kermack, K. A., D. M. Kermack, P. M. Lees, and J. R. E. Mills. 1998. New multituberculate-like teeth from the Middle Jurassic of England. Acta Palaeontologica Polonica 43:581–606.
Kermack, K. A., A. J. Lee, P. M. Lees, and F. Mussett. 1987. A new docodont from the Forest Marble. Zoological Journal of the Linnean Society 89:1–39.
Kermack, K. A., F. Mussett, and H. W. Rigney. 1973. The lower jaw of *Morganucodon*. Zoological Journal of the Linnean Society 53:87–175.
———. 1981. The skull of *Morganucodon*. Zoological Journal of the Linnean Society 71:1–158.
Kielan-Jaworowska, Z. 1969. Preliminary data on the Upper Cretaceous eutherian mammals from Bayn Dzak, Gobi Desert. Palaeontologia Polonica 19:171–191.
———. 1975a. Preliminary description of two new eutherian genera from the Late Cretaceous of Mongolia. Palaeontologia Polonica 33:5–16.
———. 1975b. Evolution of the therian mammals in the Late Cretaceous of Asia. Part I. Deltatheridiidae. Palaeontologia Polonica 33:103–132.
———. 1978. Evolution of the therian mammals in the Late Cretaceous of Asia. Part III. Postcranial skeleton in Zalambdalestidae. Palaeontologia Polonica 38:3–41.
———. 1979. Pelvic structure and nature of reproduction in Multituberculata. Nature 277:402–403.
———. 1984. Evolution of the therian mammals in the Late Cretaceous of Asia. Part V. Skull structure in Zalambdalestidae. Palaeontologia Polonica 46:107–117.
———. 1989. Postcranial skeleton of a Cretaceous multituberculate mammal. Acta Palaeontologica Polonica 34:75–85.
———. 1992. Interrelationships of Mesozoic mammals. Historical Biology 6:185–202.

Kielan-Jaworowska, Z., and J. F. Bonaparte. 1996. Partial dentary of a multituberculate mammal from the Late Cretaceous of Argentina and its taxonomic implications. Museo Argentino de Ciencias Naturales "Bernardino Rivadavia" e Instituto Nacional de Investigacion de las Ciencias Naturales, Extra, Nueva Serie 145:1–9.

Kielan-Jaworowska, Z., and D. Dashzeveg. 1989. Eutherian mammals from the Early Cretaceous of Mongolia. Zoologica Scripta 18:347–355.

———. 1998. Early Cretaceous amphilestid ("triconodont") mammals from Mongolia. Acta Palaeontologica Polonica 43:413–438.

Kielan-Jaworowska, Z., and P. P. Gambaryan. 1994. Postcranial anatomy and habits of Asian multituberculate mammals. Fossils and Strata 36:1–92.

Kielan-Jaworowska, Z., and J. H. Hurum. 1997. Djadochtatheria—A new suborder of multituberculate mammals. Acta Palaeontologica Polonica 42:201–242.

———. 2001. Phylogeny and systematics of multituberculate mammals. Palaeontology 44:389–429.

Kielan-Jaworowska, Z., and T. E. Lancaster. 2004. A new reconstruction of multituberculate endocranial casts and encephalization quotient of *Kryptobaatar*. Acta Palaeontologica Polonica 49:177–188.

Kielan-Jaworowska, Z., and T. Qi. 1990. Fossorial adaptations of a taeniolabidoid multituberculate mammal from the Eocene of China. Vertebrata PalAsiatica 28:81–94.

Kielan-Jaworowska, Z., R. L. Cifelli, and Z.-X. Luo. 1998. Alleged Cretaceous placental from down under. Lethaia 31:267–268.

———. 2002. Dentition and relationships of the Jurassic mammal *Shuotherium*. Acta Palaeontologica Polonica 47:479–486.

———. 2004. Mammals from the Age of Dinosaurs: Origins, Evolution, and Structure. Columbia University Press, New York.

Kielan-Jaworowska, Z., A. W. Crompton, and F. A. Jenkins, Jr. 1987a. The origin of egg-laying mammals. Nature 326:871–873.

Kielan-Jaworowska, Z., D. Dashzeveg, and B. A. Trofimov. 1987b. Early Cretaceous multituberculates from Mongolia and a comparison with Late Jurassic forms. Acta Palaeontologica Polonica 32:3–47.

Kielan-Jaworowska, Z., R. Presley, and C. Poplin. 1986. The cranial vascular system in taeniolabidoid multituberculate mammals. Philosophical Transactions of the Royal Society of London B 313:525–602.

Kingdon, J. 1974. East African Mammals, Volume 1. University of Chicago Press, Chicago.

Kirsch, J. A. W., F.-J. Lapointe, and M. S. Springer. 1997. DNA-hybridisation studies of marsupials and their implications for metatherian classification. Australian Journal of Zoology 45:211–280.

Knox, R. W.O'B. 1998. The tectonic and volcanic history of the North Atlantic region during the Paleocene-Eocene transition: Implications for NW European and global biotic events; pp. 91–102 in M.-P. Aubry, S. G. Lucas, and W. A. Berggren (eds.), Late Paleocene–Early Eocene Climatic and Biotic Events in the Marine and Terrestrial Records. Columbia University Press, New York.

Koch, P. L., W. C. Clyde, R. P. Hepple, M. L. Fogel, S. L. Wing, and J. C. Zachos. 2003. Carbon and oxygen isotope records from paleosols spanning the Paleocene-Eocene boundary, Bighorn Basin, Wyoming; pp. 49–64 in S. L. Wing, P. D. Gingerich, B. Schmitz, and E. Thomas (eds.), Causes and Consequences of Globally Warm Climates in the Early Paleogene. Geological Society of America Special Paper 369, Boulder.

Koenigswald, W. von. 1979. Ein Lemurenrest aus dem eozänen Ölschiefer der Grube Messel bei Darmstadt. Paläontologische Zeitschrift 53:63–76.

———. 1980. Das Skelett eines Pantolestiden (Proteutheria, Mamm.) aus dem mittleren Eozän von Messel bei Darmstadt. Paläontologische Zeitschrift 54:267–287.

———. 1983. Skelettfunde von *Kopidodon* (Condylarthra, Mammalia) aus dem mitteleozänen Ölschiefer von Messel bei Darmstadt. Neues Jahrbuch für Geologie und Paläontologie Abhandlungen 167:1–39.

———. 1987. Ein zweites Skelett von *Buxolestes* (Pantolestidae, Proteutheria, Mammalia) aus dem Mitteleozän von Messel bei Darmstadt. Carolinea 45:36–42.

———. 1990. Die Paläobiologie der Apatemyiden (Insectivora s. l.) und die Ausdeutung der Skelettfunde von *Heterohyus nanus* aus dem Mitteleozän von Messel bei Darmstadt. Palaeontographica Abteilung A 210:41–77.

———. 1995. Lagomorpha versus Rodentia: The number of layers in incisor enamels. Neues Jahrbuch für Geologie und Paläontologie Monatshefte 10:605–613.

———. 1997a. Brief survey of enamel diversity at the Schmelzmuster level in Cenozoic placental mammals; pp. 137–161 in W. von Koenigswald and P. M. Sander (eds.), Tooth Enamel Microstructure. A. A. Balkema, Rotterdam.

———. 1997b. Evolutionary trends in the differentiation of mammalian enamel ultrastructure; pp. 203–235 in W. von Koenigswald and P. M. Sander (eds.), Tooth Enamel Microstructure. A. A. Balkema, Rotterdam.

———. 2004. The three basic types of schmelzmuster in fossil and extant rodent molars and their distribution among rodent clades. Palaeontographica Abteilung A 270:95–132.

Koenigswald, W. von, and W. A. Clemens. 1992. Levels of complexity in the microstructure of mammalian enamel and their application in studies of systematics. Scanning Microscopy 6:195–218.

Koenigswald, W. von, and F. Goin. 2000. Enamel differentiation in South American marsupials and a comparison of placental and marsupial enamel. Palaeontographica Abteilung A 255:129–168.

Koenigswald, W. von, and R. Pascual. 1990. The Schmelzmuster of the Paleogene South American rodentlike marsupials *Groeberia* and *Patagonia* compared to rodents and other Marsupialia. Paläontologische Zeitschrift 64:345–358.

Koenigswald, W. von, and K. D. Rose. (2005). The enamel microstructure of the early Eocene pantodont *Coryphodon* and the nature of the zigzag enamel. Journal of Mammalian Evolution 12:419–432.

Koenigswald, W. von, and H.-P. Schierning. 1987. The ecological niche of an extinct group of mammals, the early Tertiary apatemyids. Nature 326:595–597.

Koenigswald, W. von, and G. Storch. 1983. *Pholidocercus hassiacus*, ein Amphilemuride aus dem Eozän der "Grube Messel" bei Darmstadt (Mammalia, Lipotyphla). Senckenbergiana Lethaea 64:447–495.

———. 1992. The marsupials: Inconspicuous opossums; pp. 155–158 in S. Schaal and W. Ziegler (eds.), Messel. An Insight into the History of Life and of the Earth. Clarendon Press, Oxford.

Koenigswald, W. von, F. Goin, and R. Pascual. 1999. Hypsodonty and enamel microstructure in the Paleocene gondwanatherian mammal *Sudamerica ameghinoi*. Acta Palaeontologica Polonica 44:263–300.

Koenigswald, W. von, J. M. Rensberger, and H. U. Pfretzschner. 1987. Changes in the tooth enamel of early Paleocene mammals allowing increased diet diversity. Nature 328:150–152.

Koenigswald, W. von, G. Richter, and G. Storch. 1981. Nachweis von Hornschuppen bei *Eomanis waldi* aus der "Grube Messel" bei Darmstadt (Mammalia, Pholidota). Senckenbergiana Lethaea 61:291–298.

Koenigswald, W. von, K. D. Rose, L. Grande, and R. D. Martin. 2005. First apatemyid skeleton from the lower Eocene Fossil Butte Member, Wyoming, compared to the European apatemyid from Messel, Germany. Palaeontographica Abteilung A 272:149–169.

Koenigswald, W. von, G. Storch, and G. Richter. 1992a. Primitive insectivores, extraordinary hedgehogs, and long-fingers; pp. 159–177 *in* S. Schaal, and W. Ziegler (eds.), Messel. An Insight into the History of Life and of the Earth. Clarendon Press, Oxford.

———. 1992b. Rodents: At the start of a great career; pp. 217–222 *in* S. Schaal and W. Ziegler (eds.), Messel. An Insight into the History of Life and of the Earth. Clarendon Press, Oxford.

Köhler, R., and R. E. Fordyce. 1997. An archaeocete whale (Cetacea: Archaeoceti) from the Eocene Waihao Greensand, New Zealand. Journal of Vertebrate Paleontology 17:574–583.

Kondrashov, P. E., and A. Agadjanian. 2005. A nearly complete skeleton of *Ernanodon* (Mammalia, Ernanodonta) from Mongolia: Functional analysis. Journal of Vertebrate Paleontology 25(supplement to no. 3):79A.

Kondrashov, P. E., and S. G. Lucas. 2004. *Palaeostylops iturus* from the upper Paleocene of Mongolia and the status of Arctostylopida (Mammalia, Eutheria). New Mexico Museum of Natural History and Science Bulletin 26:195–204.

Kondrashov, P. E., A. V. Lopatin, and S. G. Lucas. 2004. The oldest known Asian artiodactyl (Mammalia). New Mexico Museum of Natural History and Science Bulletin 26: 205–208.

Koopman, K. F. 1984. Bats; pp. 145–186 *in* S. Anderson and J. K. Jones, Jr. (eds.), Orders and Families of Recent Mammals of the World. John Wiley and Sons, New York.

Korth, W. W. 1984. Earliest Tertiary evolution and radiation of rodents in North America. Bulletin of Carnegie Museum of Natural History 24:1–71.

———. 1987. Sciurid rodents (Mammalia) from the Chadronian and Orellan (Oligocene) of Nebraska. Journal of Paleontology 61:1247–1255.

———. 1994. The Tertiary Record of Rodents in North America. Plenum Press, New York.

———. 2001. Comments on the systematics and classification of the beavers (Rodentia, Castoridae). Journal of Mammalian Evolution 8:279–296.

Korth, W. W., and R. J. Emry. 1991. The skull of *Cedromus* and a review of the Cedromurinae (Rodentia, Sciuridae). Journal of Paleontology 65:984–994.

Korth, W. W., J. H. Wahlert, and R. J. Emry. 1991. A new species of *Heliscomys* and recognition of the family Heliscomyidae (Geomyoidea: Rodentia). Journal of Vertebrate Paleontology 11:247–256.

Kraus, M. J. 1979. Eupantotheria; pp. 162–171 *in* J. A. Lillegraven, Z. Kielan-Jaworowska, and W. A. Clemens (eds.), Mesozoic Mammals. The First Two-Thirds of Mammalian History. University of California Press, Berkeley.

Krause, D. W. 1982. Jaw movement, dental function, and diet in the Paleocene multituberculate *Ptilodus*. Paleobiology 8:265–281.

———. 1986. Competitive exclusion and taxonomic displacement in the fossil record: The case of rodents and multituberculates in North America; pp. 95–117 *in* K. M. Flanagan and J. A. Lillegraven (eds.), Vertebrates, Phylogeny, and Philosophy. Contributions to Geology, University of Wyoming, Special Paper 3, Laramie.

———. 2001. Fossil molar from a Madagascan marsupial. Nature 412:497–498.

Krause, D. W., and J. F. Bonaparte. 1993. Superfamily Gondwanatherioidea: A previously unrecognized radiation of multituberculate mammals in South America. Proceedings of the National Academy of Sciences, USA 90:9379–9383.

Krause, D. W., and F. A. Jenkins, Jr. 1983. The postcranial skeleton of North American multituberculates. Bulletin of the Museum of Comparative Zoology 150:199–246.

Krause, D. W., and Z. Kielan-Jaworowska. 1993. The endocranial cast and encephalization quotient of *Ptilodus* (Multituberculata, Mammalia). Palaeovertebrata 22:99–112.

Krause, D. W. and Maas, M. C. 1990. The biogeographic origins of late Paleocene–early Eocene mammalian immigrants to the Western Interior of North America; pp. 71–105 *in* T. M. Bown and K. D. Rose (eds.), Dawn of the Age of Mammals in the Northern Part of the Rocky Mountain Interior, North America. Geological Society of America Special Paper 243, Boulder.

Krause, D. W., M. D. Gottfried, P. M. O'Connor, and E. M. Roberts. 2003. A Cretaceous mammal from Tanzania. Acta Palaeontologica Polonica 48:321–330.

Krause, D. W., Z. Kielan-Jaworowska, and J. F. Bonaparte. 1992. *Ferugliotherium* Bonaparte, the first known multituberculate from South America. Journal of Vertebrate Paleontology 12:351–376.

Krause, D. W., G. V. R. Prasad, W. von Koenigswald, A. Sahni, and F. E. Grine. 1997. Cosmopolitanism among Gondwanan Late Cretaceous mammals. Nature 390:504–507.

Krebs, B. 1991. Das Skelett von *Henkelotherium guimarotae* gen. et sp. nov. (Eupantotheria, Mammalia) aus dem Oberen Jura von Portugal. Berliner Geowissenschaftliche Abhandlungen, Reihe A 133:1–121.

———. 1993. Das Gebiss von *Crusafontia* (Eupantotheria, Mammalia)—Funde aus der Unter-Kreide von Galve un Uña. Berliner Geowissenschaftliche Abhandlungen E 9:233–252.

Krishtalka, L. 1976a. Early Tertiary Adapisoricidae and Erinaceidae (Mammalia, Insectivora) of North America. Bulletin of Carnegie Museum of Natural History 1:1–40.

———. 1976b. North American Nyctitheriidae (Mammalia, Insectivora). Annals of Carnegie Museum 46:7–28.

Krishtalka, L., and T. Setoguchi. 1977. Paleontology and geology of the Badwater Creek area, central Wyoming. Part 13. The late Eocene Insectivora and Dermoptera. Annals of Carnegie Museum 46:71–99.

Krishtalka, L., and R. K. Stucky. 1985. Revision of the Wind River faunas, early Eocene of central Wyoming. Part 7. Revision of *Diacodexis* (Mammalia, Artiodactyla). Annals of Carnegie Museum 54:413–486.

Krishtalka, L., and R. M. West. 1979. Paleontology and geology of the Bridger Formation, southern Green River Basin, southwestern Wyoming. Part 4. The Geolabididae (Mammalia,

Insectivora). Milwaukee Public Museum Contributions in Biology and Geology 27:1–10.

Krishtalka, L., R. J. Emry, J. E. Storer, and J. F. Sutton. 1982. Oligocene multituberculates (Mammalia: Allotheria): Youngest known record. Journal of Paleontology 56: 791–794.

Krishtalka, L., R. K. Stucky, and K. C. Beard. 1990. The earliest fossil evidence for sexual dimorphism in primates. Proceedings of the National Academy of Sciences, USA 87:5223–5226.

Kron, D. G. 1979. Docodonta; pp. 91–98 in J. A. Lillegraven, Z. Kielan-Jaworowska, and W. A. Clemens (eds.), Mesozoic Mammals. The First Two-Thirds of Mammalian History. University of California Press, Berkeley.

Kron, D. G., and E. Manning. 1998. Anthracotheriidae; pp. 381–388 in C. M. Janis, K. M. Scott, and L. L. Jacobs (eds.), Evolution of Tertiary Mammals of North America. Volume 1: Terrestrial Carnivores, Ungulates, and Ungulatelike Mammals. Cambridge University Press, Cambridge, UK.

Krusat, G. 1991. Functional morphology of *Haldanodon exspectatus* (Mammalia, Docodonta) from the Upper Jurassic of Portugal. Fifth Symposium on Mesozoic Terrestrial Ecosystems and Biota, Contributions from the Paleontological Museum, University of Oslo 363:37–38.

Kühne, W. G. 1956. The Liassic Therapsid *Oligokyphus*. British Museum of Natural History, London.

Kumar, K. 1991. *Anthracobune aijiensis* nov. sp. (Mammalia: Proboscidea) from the Subathu Formation, Eocene from NW Himalaya, India. Géobios 24:221–239.

Kumar, K., and A. Sahni. 1985. Eocene mammals from the Upper Subathu Group, Kashmir Himalaya, India. Journal of Vertebrate Paleontology 5:153–168.

———. 1986. *Remingtonocetus harudiensis*, new combination, a middle Eocene archaeocete (Mammalia, Cetacea) from western Kutch, India. Journal of Vertebrate Paleontology 6:326–349.

Kumar, S., and S. B. Hedges. 1998. A molecular timescale for vertebrate evolution. Nature 392:917–920.

Kurz, C. 2001. Osteologie einer Beutelratte (Didelphimorphia, Marsupialia, Mammalia) aus dem Mitteleozän der Grube Messel bei Darmstadt. Kaupia—Darmstädter Beiträge zur Naturgeschichte 11:83–109.

Lander, B. 1998. Oreodontoidea; pp. 402–420 in C. M. Janis, K. M. Scott, and L. L. Jacobs (eds.), Evolution of Tertiary Mammals of North America. Volume 1: Terrestrial Carnivores, Ungulates, and Ungulatelike Mammals. Cambridge University Press, Cambridge, UK.

Lanèque, L. 1992. Analyse de matrice de distance euclidienne de la région du museau chez *Adapis* (Adapiforme, Éocène). Comptes Rendus de l'Académie des Sciences Paris, sér. II 314:1387–1393.

———. 1993. Variation of orbital features in adapine skulls. Journal of Human Evolution 25:287–317.

Lange-Badré, B. 1975. Données récentes sur les créodontes européens. Colloque international du Centre National de la Recherche Scientifique 218:675–682.

———. 1979. Les créodontes (Mammalia) d'Europe occidentale de l'Eocène supérieur à l'Oligocène supérieur. Mémoires du Muséum National d'Histoire Naturelle, sér. C, Sciences de la Terre 42:1–249.

Langer, P. 1974. Stomach evolution in the Artiodactyla. Mammalia 38:295–314.

———. 2001. Evidence from the digestive tract on phylogenetic relationships in ungulates and whales. Journal of Zoological Systematics and Evolutionary Research 39:77–90.

Lavergne, A., E. Douzery, T. Stichler, F. M. Catzeflis, and M. S. Springer. 1996. Interordinal mammalian relationships: Evidence for paenungulate monophyly is provided by complete mitochondrial 12S rRNA sequences. Molecular Phylogenetics and Evolution 6:245–258.

Lavocat, R. 1955. Ordres des Pantodonta Cope, 1873, et des Dinocerata Marsh, 1873; pp. 902–913 in P.-P. Grassé (ed.), Traité de Zoologie XVII(1). Masson et Cie, Paris.

———. 1969. La systématique des rongeurs hystricomorphes et la dérive des continents. Comptes Rendus de l'Académie des Sciences, Paris, sér. II 264:1496–1497.

———. 1973. Les rongeurs du Miocène d'Afrique orientale. I. Miocène inférieur. École Pratique des Hautes Études: Mémoires et Travaux de l'Institut de Montpellier 1:1–284.

———. 1974. The interrelationships between the African and South American rodents and their bearing on the problem of the origin of South American monkeys. Journal of Human Evolution 3:323–326.

Lavocat, R., and J.-P. Parent. 1985. Phylogenetic analysis of middle ear features in fossil and living rodents; pp. 333–354 in W. P. Luckett and J.-L. Hartenberger (eds.), Evolutionary Relationships among Rodents: A Multidisciplinary Analysis. Plenum Press, New York.

Legendre, S. 1985. Molossidés (Mammalia, Chiroptera) cénozoïques de l'Ancien et du Nouveau Monde; statut systématique; Intégration phylogénique des données. Neues Jahrbuch für Geologie und Paläontologie Abhandlungen 170:205–227.

———. 1987. Les immigrations de la "Grande Coupure" sont-elles contemporaines en Europe occidentale? Münchner Geowissenschaftliche Abhandlungen, Reihe A, Geologie und Paläontologie 10:141–148.

Legendre, S., and J.-L. Hartenberger. 1992. Evolution of mammalian faunas in Europe during the Eocene and Oligocene; pp. 516–528 in D. R. Prothero and W. A. Berggren (eds.), Eocene-Oligocene Climatic and Biotic Evolution. Princeton University Press, Princeton.

Legendre, S., J.-Y. Crochet, M. Godinot, J.-L. Hartenberger, B. Marandat, J. A. Remy, B. Sigé, J. Sudre, and M. Vianey-Liaud. 1991. Evolution de la diversité des faunes de mammifères d'Europe occidentale au Paléogène (MP 11 à MP 30). Bulletin de la Société Géologique de France 162:867–874.

Legendre, S., B. Sigé, J. G. Astruc, L. de Bonis, J.-Y. Crochet, C. Denys, M. Godinot, J.-L. Hartenberger, F. Lévêque, B. Marandat, C. Mourer-Chauviré, J.-C. Rage, J. A. Remy, J. Sudre, and M. Vianey-Liaud. 1997. Les Phosphorites du Quercy: 30 ans de recherché. Bilan et perspectives. Géobios Mémoire Spécial 20:331–345.

Lessertisseur, J., and R. Saban. 1967a. Squelette axial; pp. 584–708 in P.-P. Grassé (ed.), Traité de Zoologie: Anatomie, Systématique, Biologie, XVI(1). Masson et Cie, Paris.

———. 1967b. Squelette appendiculaire; pp. 709–1078 in P.-P. Grassé (ed.), Traité de Zoologie: Anatomie, Systématique, Biologie, XVI(1). Masson et Cie, Paris.

Li, C.-K., and S.-Y. Ting. 1983. The Paleogene mammals of China. Bulletin of Carnegie Museum of Natural History 21:1–93.

———. 1993. New cranial and postcranial evidence for the affinities of the eurymylids (Rodentia) and mimotonids (Lagomorpha); pp. 151–158 *in* F. S. Szalay, M. J. Novacek, and M. C. McKenna (eds.), Mammal Phylogeny: Placentals. Springer-Verlag, New York.

Li, C.-K., Y.-Q. Wang, Y.-M. Hu, and J. Meng. 2003. A new species of *Gobiconodon* (Triconodonta, Mammalia) and its implications for the age of Jehol Biota. Chinese Science Bulletin (English Edition) 48:1129–1134.

Li, C.-K., R. W. Wilson, M. R. Dawson, and L. Krishtalka. 1987. The origin of rodents and lagomorphs. Current Mammalogy 1:97–108.

Li, C.-K., J.-J. Zheng, and S.-Y. Ting. 1989. The skull of *Cocomys lingchaensis*, an early Eocene ctenodactyloid rodent of Asia; pp. 179–192 *in* C. C. Black and M. R. Dawson (eds.), Papers on Fossil Rodents in Honor of Albert Elmer Wood. Natural History Museum of Los Angeles County Science Series 33, Los Angeles.

Lillegraven, J. A. 1969. Latest Cretaceous mammals of upper part of Edmonton Formation of Alberta, Canada, and review of marsupial-placental dichotomy in mammalian evolution. University of Kansas Paleontological Contributions 50:1–122.

Lillegraven, J. A., and J. J. Eberle. 1999. Vertebrate faunal changes through Lancian and Puercan time in southern Wyoming. Journal of Paleontology 73:691–710.

Lillegraven, J. A., and G. Krusat. 1991. Cranio-mandibular anatomy of *Haldanodon exspectatus* (Docodonta; Mammalia) from the Late Jurassic of Portugal and its implications to the evolution of mammalian characters. Contributions to Geology, University of Wyoming 28:39–138.

Lillegraven, J. A., and M. C. McKenna. 1986. Fossil mammals from the "Mesaverde" Formation (Late Cretaceous, Judithian) of the Bighorn and Wind River basins, Wyoming, with definitions of Late Cretaceous North American Land-Mammal "Ages." American Museum Novitates 2840:1–68.

Lillegraven, J. A., and R. W. Wilson. 1975. Analysis of *Simimys simplex*, an Eocene rodent (?Zapodidae). Journal of Paleontology 49:856–874.

Lillegraven, J. A., Z. Kielan-Jaworowska, and W. A. Clemens (eds.). 1979. Mesozoic Mammals. The First Two-Thirds of Mammalian History. University of California Press, Berkeley.

Lillegraven, J. A., M. C. McKenna, and L. Krishtalka. 1981. Evolutionary relationships of middle Eocene and younger species of *Centetodon* (Mammalia, Insectivora, Geolabididae) with a description of the dentition of *Ankylodon* (Adapisoricidae). University of Wyoming Publications 45:1–115.

Lillegraven, J. A., S. D. Thompson, B. K. McNab, and J. L. Patton. 1987. The origin of eutherian mammals. Biological Journal of the Linnean Society 32:281–336.

Liu, F.-G. R., M. M. Miyamoto, N. P. Freire, P. Q. Ong, M. R. Tennant, T. S. Young, and K. F. Gugel. 2001. Molecular and morphological supertrees for eutherian (placental) mammals. Science 291:1786–1789.

Liu, L. 2001. Eocene suoids (Artiodactyla, Mammalia) from Bose and Yongle basins, China, and the classification and evolution of the Paleogene suoids. Vertebrata PalAsiatica 39:116–128.

Lofgren, D. L. 1995. The Bug Creek problem and the Cretaceous-Tertiary transition at McGuire Creek, Montana. University of California Publications in Geological Sciences 140:1–185.

Lofgren, D. L., J. A. Lillegraven, W. A. Clemens, P. D. Gingerich, and T. E. Williamson. 2004. Paleocene biochronology: The Puercan through Clarkforkian Land Mammal Ages; pp. 43–105 *in* M. O. Woodburne (ed.), Late Cretaceous and Cenozoic Mammals of North America. Columbia University Press, New York.

Lopatin, A. V. 2001. The skull structure of *Archaeoryctes euryalis* sp. nov. (Didymoconidae, Mammalia) from the Paleocene of Mongolia and the taxonomic position of the family. Paleontological Journal 35:320–329.

———. 2002. The earliest shrew (Soricidae, Mammalia) from the middle Eocene of Mongolia. Paleontological Journal 36:650–659.

———. 2003a. A zalambdodont insectivore of the family Apternodontidae (Insectivora, Mammalia) from the middle Eocene of Mongolia. Paleontological Journal 37:187–195.

———. 2003b. A new species of *Ardynictis* (Didymoconidae, Mammalia) from the middle Eocene of Mongolia. Paleontological Journal 37:303–311.

———. 2004. A new genus of the Galericinae (Erinaceidae, Insectivora, Mammalia) from the middle Eocene of Mongolia. Paleontological Journal 38:319–326.

Lourens, L. J., A. Sluijs, D. Kroon, J. C. Zachos, E. Thomas, U. Röhl, J. Bowles, and I. Raffi. 2005. Astronomical pacing of late Palaeocene to early Eocene global warming events. Nature 435:1083–1087.

Lucas, S. G. 1982. The phylogeny and composition of the order Pantodonta (Mammalia, Eutheria). Proceedings of the Third North American Paleontological Convention 2:337–342.

———. 1992. Extinction and the definition of the class Mammalia. Systematic Biology 41:370–371.

———. 1993. Pantodonts, tillodonts, uintatheres, and pyrotheres are not ungulates; pp. 182–194 *in* F. S. Szalay, M. J. Novacek, and M. C. McKenna (eds.), Mammal Phylogeny: Placentals. Springer-Verlag, New York.

———. 1998. Pantodonta; pp. 274–283 *in* C. M. Janis, K. M. Scott, and L. L. Jacobs (eds.), Evolution of Tertiary Mammals of North America. Volume 1: Terrestrial Carnivores, Ungulates, and Ungulatelike Mammals. Cambridge University Press, Cambridge, UK.

Lucas, S. G., and L. T. Holbrook. 2004. The skull of the Eocene perissodactyl *Lambdotherium* and its phylogenetic significance. New Mexico Museum of Natural History and Science Bulletin 26:81–85.

Lucas, S. G., and P. E. Kondrashov. 2004. Early Eocene (Bumbanian) perissodactyls from Mongolia and their biochronological significance. New Mexico Museum of Natural History and Science Bulletin 26:215–220.

Lucas, S. G., and Z.-X. Luo. 1993. *Adelobasileus* from the Upper Triassic of west Texas: The oldest mammal. Journal of Vertebrate Paleontology 13:309–334.

Lucas, S. G., and R. M. Schoch. 1989a. Taxonomy and biochronology of *Eomoropus* and *Grangeria*, Eocene chalicotheres from the western United States and China; pp. 422–437 *in* D. R. Prothero and R. M. Schoch (eds.), The Evolution of Perissodactyls. Oxford University Press, New York.

———. 1989b. Taxonomy of *Duchesneodus* (Brontotheriidae) from the late Eocene of North America; pp. 490–503 *in* D. R. Prothero and R. M. Schoch (eds.), The Evolution of Perissodactyls. Oxford University Press, New York.

———. 1998a. Tillodontia; pp. 268–273 *in* C. M. Janis, K. M. Scott, and L. L. Jacobs (eds.), Evolution of Tertiary Mammals of North America. Volume 1: Terrestrial Carnivores, Ungulates, and Ungulatelike Mammals. Cambridge University Press, Cambridge, UK.

———. 1998b. Dinocerata; pp. 284–291 in C. M. Janis, K. M. Scott, and L. L. Jacobs (eds.), Evolution of Tertiary Mammals of North America. Volume 1: Terrestrial Carnivores, Ungulates, and Ungulatelike Mammals. Cambridge University Press, Cambridge, UK.

Lucas, S. G., R. M. Schoch, and T. E. Williamson. 1998. Taeniodonta; pp. 260–267 in C. M. Janis, K. M. Scott, and L. L. Jacobs (eds.), Evolution of Tertiary Mammals of North America. Volume 1: Terrestrial Carnivores, Ungulates, and Ungulatelike Mammals. Cambridge University Press, Cambridge, UK.

Luckett, W. P. (ed.) 1980. Comparative Biology and Evolutionary Relationships of Tree Shrews. Plenum Press, New York.

———. 1985. Superordinal and intraordinal affinities of rodents: Developmental evidence from the dentition and placentation; pp. 227–276 in W. P. Luckett and J.-L. Hartenberger (eds.), Evolutionary Relationships among Rodents: A Multidisciplinary Analysis. Plenum Press, New York.

———. 1993. An ontogenetic assessment of dental homologies in therian mammals; pp. 182–204 in F. S. Szalay, M. J. Novacek, and M. C. McKenna (eds.), Mammal Phylogeny: Mesozoic Differentiation, Multituberculates, Monotremes, Early Therians, and Marsupials. Springer-Verlag, New York.

Luckett, W. P., and J.-L. Hartenberger. 1985. Evolutionary relationships among rodents: Comments and conclusions; pp. 685–712 in W. P. Luckett and J. L. Hartenberger (eds.), Evolutionary Relationships among Rodents. A Multidisciplinary Analysis. Plenum Press, New York.

———. 1993. Monophyly or polyphyly of the order Rodentia: Possible conflict between morphological and molecular interpretations. Journal of Mammalian Evolution 1:127–147.

Luckett, W. P., and N. Hong. 1998. Phylogenetic relationships between the orders Artiodactyla and Cetacea: A combined assessment of morphological and molecular evidence. Journal of Mammalian Evolution 5:127–182.

Luo, Z.-X. 1991. Variability of dental morphology and the relationships of the earliest arctocyonid species. Journal of Vertebrate Paleontology 11:452–471.

———. 1994. Sister-group relationships of mammals and transformations of diagnostic mammalian characters; pp. 98–128 in N. C. Fraser and H.-D. Sues (eds.), In the Shadow of the Dinosaurs—Early Mesozoic Tetrapods. Cambridge University Press, Cambridge, UK.

———. 1998. Homology and transformation of cetacean ectotympanic structures; pp. 269–301 in J. G. M. Thewissen (ed.), The Emergence of Whales. Evolutionary Patterns in the Origin of Cetacea. Plenum Press, New York.

Luo, Z.-X., and P. D. Gingerich. 1999. Terrestrial Mesonychia to aquatic Cetacea: Transformation of the basicranium and evolution of hearing in whales. University of Michigan Papers on Paleontology 31:1–98.

Luo, Z.-X., and J. R. Wible. 2005. A Late Jurassic digging mammal and early mammalian diversification. Science 308:103–107.

Luo, Z.-X., R. L. Cifelli, and Z. Kielan-Jaworowska. 2001. Dual origin of tribosphenic mammals. Nature 409:53–57.

Luo, Z.-X., A. W. Crompton, and S. G. Lucas. 1995. Evolutionary origins of the mammalian promontorium and cochlea. Journal of Vertebrate Paleontology 15:113–121.

Luo, Z.-X., A. W. Crompton, and A.-L. Sun. 2001. A new mammaliaform from the Early Jurassic and evolution of mammalian characteristics. Science 292:1535–1540.

Luo, Z.-X., Q. Ji, J. R. Wible, and C.-X. Yuan. 2003. An Early Cretaceous tribosphenic mammal and metatherian evolution. Science 302:1934–1940.

Luo, Z.-X., Z. Kielan-Jaworowska, and R. Cifelli. 2002. In quest for a phylogeny of Mesozoic mammals. Acta Palaeontologica Polonica 47:1–78.

Maas, M. C., and J. G. M. Thewissen. 1995. Enamel microstructure of *Pakicetus* (Mammalia: Archaeoceti). Journal of Paleontology 69:1154–1163.

Maas, M. C., D. W. Krause, and S. G. Strait. 1988. The decline and extinction of Plesiadapiformes (Mammalia: ?Primates) in North America: Displacement or replacement? Paleobiology 14:410–431.

Maas, M. C., J. G. M. Thewissen, and J. Kappelman. 1998. *Hypsamasia seni* (Mammalia: Embrithopoda) and other mammals from the Eocene Kartal Formation of Turkey; pp. 286–297 in K. C. Beard and M. R. Dawson (eds.), Dawn of the Age of Mammals in Asia. Bulletin of Carnegie Museum of Natural History 34.

MacFadden, B. J. 1992. Fossil Horses: Systematics, Paleobiology, and Evolution of the Family Equidae. Cambridge University Press, Cambridge, UK.

MacFadden, B. J., and C. D. Frailey. 1984. *Pyrotherium*, a large enigmatic ungulate (Mammalia, *incertae sedis*) from the Deseadan (Oligocene) of Salla, Bolivia. Palaeontology 27:867–874.

MacFadden, B. J., P. Higgins, M. T. Clementz, and D. S. Jones. 2004. Diets, habitat preferences, and niche differentiation of Cenozoic sirenians from Florida: Evidence from stable isotopes. Paleobiology 30:297–324.

Machlus, M., S. R. Hemming, P. E. Olsen, and N. Christie-Blick. 2004. Eocene calibration of geomagnetic polarity time scale reevaluated: Evidence from the Green River Formation of Wyoming. Geology 32:137–140.

MacLeod, N., and K. D. Rose. 1993. Inferring locomotor behavior in Paleogene mammals via eigenshape analysis. American Journal of Science 293A:300–355.

MacPhee, R. D. E. (ed.) 1993. Primates and Their Relatives in Phylogenetic Perspective. Plenum Press, New York.

MacPhee, R. D. E., and M. Cartmill. 1986. Basicranial structures and primate systematics; pp. 219–275 in D. R. Swindler and J. Erwin (eds.), Comparative Primate Biology. Volume 1: Systematics, Evolution, and Anatomy. Alan R. Liss, New York.

MacPhee, R. D. E., and M. J. Novacek. 1993. Definition and relationships of Lipotyphla; pp. 13–31 in F. S. Szalay, M. J. Novacek, and M. C. McKenna (eds.), Mammal Phylogeny: Placentals. Springer-Verlag, New York.

MacPhee, R. D. E., K. C. Beard, C. Flemming, and P. Houde. 1995. Petrosal morphology of *Tinimomys graybulliensis* is plesiadapoid, not microsyopoid. Journal of Vertebrate Paleontology 15(supplement to no. 3):42A.

MacPhee, R. D. E., M. Cartmill, and K. D. Rose. 1989. Craniodental morphology and relationships of the supposed Eocene dermopteran *Plagiomene* (Mammalia). Journal of Vertebrate Paleontology 9:329–349.

MacPhee, R. D. E., M. J. Novacek, and G. Storch. 1988. Basicranial morphology of Early Tertiary erinaceomorphs and the origin of Primates. American Museum Novitates 2921:1–42.

Madar, S. I., J. G. M. Thewissen, and S. T. Hussain. 2002. Additional holotype remains of *Ambulocetus natans* (Cetacea, Ambulocetidae), and their implications for locomotion in early whales. Journal of Vertebrate Paleontology 22:405–422.

Mader, B. J. 1998. Brontotheriidae; pp. 525–536 *in* C. M. Janis, K. M. Scott, and L. L. Jacobs (eds.), Evolution of Tertiary Mammals of North America. Volume 1: Terrestrial Carnivores, Ungulates, and Ungulatelike Mammals. Cambridge University Press, Cambridge, UK.

Madsen, O., M. Scally, C. J. Douady, D. J. Kao, R. W. DeBry, R. Adkins, H. M. Amrine, M. J. Stanhope, W. W. de Jong, and M. S. Springer. 2001. Parallel adaptive radiations in two major clades of placental mammals. Nature 409: 610–614.

Maglio, V. J. 1973. Origin and evolution of the Elephantidae. Transactions of the American Philosophical Society 63(3): 1–149.

Mahboubi, M., R. Ameur, J.-Y. Crochet, and J.-J. Jaeger. 1984. Earliest known proboscidean from early Eocene of northwest Africa. Nature 308:543–544.

———. 1986. El Kohol (Saharan Atlas, Algeria): A new Eocene mammal locality in northwestern Africa. Palaeontographica Abteilung A 192:15–49.

Maier, W. 1979. *Macrocranion tupaiodon,* an adapisoricid (?) insectivore from the Eocene of "Grube Messel" (western Germany). Paläontologische Zeitschrift 53:38–62.

Maier, W., G. Richter, and G. Storch. 1986. *Leptictidium nasutum*—ein archaisches Säugetier aus Messel mit aussergewöhnlichen biologischen Anpassungen. Natur und Museum (Frankfurt) 116:1–19.

Maisch, M. W., A. T. Matzke, F. Grossman, H. Stöhr, H.-U. Pfretzschner, and G. Sun. 2005. The first haramiyoid mammal from Asia. Naturwissenschaften 92:40–44.

Malia, M. J., Jr., R. M. Adkins, and M. W. Allard. 2002. Molecular support for Afrotheria and the polyphyly of Lipotyphla based on analyses of the growth hormone receptor gene. Molecular Phylogenetic and Evolution 24:91–101.

Manchester, S. R. 1999. Biogeographical relationships of North American Tertiary floras. Annals of the Missouri Botanical Garden 86:472–522.

Marandat, B. 1997. La disparité des faunes mammaliennes du niveau MP7 (Eocène inférieur) des domaines péri-mésogéen et nordiques. Investigation d'un provincialisme intra-européen. Newsletters on Stratigraphy 35:63–82.

Marenssi, S. A., M. A. Reguero, S. N. Santillana, and S. F. Vizcaíno. 1994. Eocene land mammals from Seymour Island, Antarctica: Palaeobiogeographical implications. Antarctic Science 6:3–15.

Marincovich, L., Jr., and A. Y. Gladenkov. 1999. Evidence for an early opening of the Bering Strait. Nature 397:149–151.

Marivaux, L., Y. Chaimanee, S. Ducrocq, B. Marandat, J. Sudre, A. N. Soe, S. T. Tun, W. Htoon, and J.-J. Jaeger. 2003. The anthropoid status of a primate from the late middle Eocene Pondaung Formation (central Myanmar): Tarsal evidence. Proceedings of the National Academy of Sciences, USA 100:13173–13178.

Marivaux, L., M. Vianey-Liaud, and J.-J. Jaeger. 2004. High-level phylogeny of early Tertiary rodents: Dental evidence. Zoological Journal of the Linnean Society 142:105–134.

Marivaux, L., J.-L. Welcomme, P.-O. Antoine, G. Métais, I. M. Baloch, M. Benammi, Y. Chaimanee, S. Ducrocq, and J.-J. Jaeger. 2001. A fossil lemur from the Oligocene of Pakistan. Science 294:587–591.

Marivaux, L., J.-L. Welcomme, S. Ducrocq, and J.-J. Jaeger. 2002. Oligocene sivaladapid primate from the Bugti Hills (Balochistan, Pakistan) bridges the gap between Eocene and Miocene adapiform communities in southern Asia. Journal of Human Evolution 42:379–388.

Marsh, O. C. 1886. Dinocerata, a monograph of an extinct order of gigantic mammals. Monographs of the U. S. Geological Survey 10:1–243.

Marshall, L. G. 1978. Evolution of the Borhyaenidae, extinct South American predaceous marsupials. University of California Publications in Geological Sciences 117:1–89.

———. 1980. Systematics of the South American marsupial family Caenolestidae. Fieldiana Geology, new series 5:1–145.

———. 1981. Review of the Hathlyacyninae, an extinct subfamily of South American "dog-like" marsupials. Fieldiana Geology, new series 7:1–120.

———. 1987. Systematics of Itaboraian (middle Paleocene) age "opossum-like" marsupials from the limestone quarry at São José de Itaboraí, Brazil; pp. 91–160 *in* M. Archer (ed.), Possums and Opossums: Studies in Evolution. Surrey Beatty & Sons and the Royal Zoological Society of New South Wales, Sydney.

Marshall, L. G., and C. de Muizon. 1988. The dawn of the age of mammals in South America. National Geographic Research 4:23–55.

Marshall, L. G., J. A. Case, and M. O. Woodburne. 1990. Phylogenetic relationships of the families of marsupials. Current Mammalogy 2:433–505.

Marshall, L. G., C. de Muizon, and B. Sigé. 1983. Late Cretaceous mammals (Marsupialia) from Bolivia. Géobios 16:739–745.

Marshall, L. G., C. de Muizon, and D. Sigogneau-Russell. 1995. *Pucadelphys andinus* (Marsupialia, Mammalia) from the early Paleocene of Bolivia. Mémoires du Muséum National d'Histoire Naturelle 165:1–164.

Marshall, L. G., T. Sempere, and R. F. Butler. 1997. Chronostratigraphy of the mammal-bearing Paleocene of South America. Journal of South American Earth Sciences 10:49–70.

Martin, L. D. 1998. Nimravidae; pp. 228–235 *in* C. M. Janis, K. M. Scott, and L. L. Jacobs (eds.), Evolution of Tertiary Mammals of North America. Cambridge University Press, Cambridge, UK.

Martin, R. D. 1990. Primate Origins and Evolution. A Phylogenetic Reconstruction. Princeton University Press, Princeton.

———. 1993. Primate origins: Plugging the gaps. Nature 363:223–234.

Martin, T. 1993. Early rodent incisor enamel evolution: Phylogenetic implications. Journal of Mammalian Evolution 1:227–254.

———. 1994. African origin of caviomorph rodents is indicated by incisor enamel microstructure. Paleobiology 20:5–13.

———. 1997a. Incisor enamel microstructure and systematics in rodents; pp. 163–175 *in* W. von Koenigswald and P. M. Sander (eds.), Tooth Enamel Microstructure. A. A. Balkema, Rotterdam.

———. 1997b. Tooth replacement in Late Jurassic Dryolestidae (Eupantotheria, Mammalia). Journal of Mammalian Evolution 4:1–18.

———. 1999a. Evolution of incisor enamel microstructure in Theridomyidae (Rodentia). Journal of Vertebrate Paleontology 19:550–565.

———. 1999b. Dryolestidae (Dryolestoidea, Mammalia) aus dem Oberen Jura von Portugal. Abhandlungen der Senckenbergischen Naturforschenden Gesellschaft 550:1–119.

———. 1999c. Phylogenetic implications of Glires (Eurymylidae, Mimotonidae, Rodentia, Lagomorpha) incisor enamel microstructure. Mitteilungen aus dem Museum für Naturkunde in Berlin, Zoologische Reihe 75:257–273.

———. 2002. New stem-lineage representatives of Zatheria (Mammalia) from the Late Jurassic of Portugal. Journal of Vertebrate Paleontology 22:332–348.

———. 2004. Evolution of incisor enamel microstructure in Lagomorpha. Journal of Vertebrate Paleontology 24:411–426.

———. 2005. Postcranial anatomy of *Haldanodon exspectatus* (Mammalia, Docodonta) from the Late Jurassic (Kimmeridgian) of Portugal and its bearing for mammalian evolution. Zoological Journal of the Linnean Society 145:219–248.

Martin, T., and B. Krebs (eds.) 2000. Guimarota: A Jurassic Ecosystem. Verlag Dr. Friedrich Pfeil, Munich.

Martin, T., and Z.-X. Luo. 2005. Homoplasy in the mammalian ear. Science 307:861–862.

Martin, T., and O. W. M. Rauhut. 2005. Mandible and dentition of *Asfaltomylus patagonicus* (Australosphenida, Mammalia) and the evolution of tribosphenic teeth. Journal of Vertebrate Paleontology 25:414–425.

Martinez, R. N., C. L. May, and C. A. Forster. 1996. The skull of *Probelesodon sanjuanensis*, sp. nov., from the Late Triassic Ischigualasto Formation of Argentina. Journal of Vertebrate Paleontology 16:271–284.

Matsumoto, H. 1926. Contribution to the knowledge of the fossil Hyracoidea of the Fayûm, Egypt, with description of several new species. Bulletin of the American Museum of Natural History 56:253–350.

Matthew, W. D. 1901. Fossil mammals of the Tertiary of northeastern Colorado. Memoirs of the American Museum of Natural History 1:353–447.

———. 1906. The osteology of *Sinopa*, a creodont mammal of the middle Eocene. Proceedings of the U. S. National Museum 30:203–233.

———. 1909. The Carnivora and Insectivora of the Bridger Basin, Middle Eocene. Memoirs of the American Museum of Natural History 9:291–567.

———. 1910a. On the osteology and relationships of *Paramys*, and the affinities of the Ischyromyidae. Bulletin of the American Museum of Natural History 28:43–72.

———. 1910b. The phylogeny of the Felidae. Bulletin of the American Museum of Natural History 28:289–316.

———. 1915a. A revision of the Lower Eocene Wasatch and Wind River faunas. Part I: Order Ferae (Carnivora). Suborder Creodonta. Bulletin of the American Museum of Natural History 34:1–103.

———. 1915b. A revision of the Lower Eocene Wasatch and Wind River faunas. Part II: Order Condylarthra, family Hyopsodontidae. Bulletin of the American Museum of Natural History 34:311–328.

———. 1915c. A revision of the Lower Eocene Wasatch and Wind River faunas. Part IV: Entelonychia, Primates, Insectivora (part). Bulletin of the American Museum of Natural History 34:429–483.

———. 1918. A revision of the Lower Eocene Wasatch and Wind River faunas. Part V: Insectivora (continued), Glires, Edentata. Bulletin of the American Museum of Natural History 38:565–657.

———. 1921. *Stehlinius*, a new Eocene insectivore. American Museum Novitates 14:1–5.

———. 1937. Paleocene faunas of the San Juan Basin, New Mexico. Transactions of the American Philosophical Society, new series 30:1–510.

Matthew, W. D., and W. Granger. 1925a. Fauna and correlation of the Gashato Formation of Mongolia. American Museum Novitates 189:1–12.

———. 1925b. New creodonts and rodents from the Ardyn Obo formation of Mongolia. American Museum Novitates 193:1–7.

Mayr, E. 1969. Principles of Systematic Zoology. McGraw-Hill, New York.

McDowell, S. B., Jr. 1958. The Greater Antillean insectivores. Bulletin of the American Museum of Natural History 115:113–214.

McGehee, S. G., and G. C. Gould. 1991. New evidence uniting pyrotheres to Notoungulata. Journal of Vertebrate Paleontology 11(supplement to no. 3):46A.

McKenna, M. C. 1960a. Fossil Mammalia from the early Wasatchian Four Mile Fauna, Eocene of northwest Colorado. University of California Publications in Geological Sciences 37:1–130.

———. 1960b. The Geolabidinae, a new subfamily of early Cenozoic erinaceoid insectivores. University of California Publications in Geological Sciences 37:131–164.

———. 1963. New evidence against tupaioid affinities of the mammalian family Anagalidae. American Museum Novitates 2158:1–16.

———. 1966. Paleontology and the origin of the Primates. Folia Primatologica 4:1–25.

———. 1968. *Leptacodon*, an American Paleocene nyctithere (Mammalia, Insectivora). American Museum Novitates 2317:1–12.

———. 1972. Was Europe connected directly to North America prior to the middle Eocene? pp. 179–188 *in* T. Dobzhansky, M. K. Hecht, and W. C. Steere (eds.), Evolutionary Biology. Appleton-Century-Crofts, New York.

———. 1973. Sweepstakes, filters, corridors, Noah's Arks, and beached Viking funeral ships in palaeogeography; pp. 295–308 *in* D. H. Tarling and S. K. Runcorn (eds.), Implications of Continental Drift to the Earth Sciences. Academic Press, New York.

———. 1975a. Toward a phylogenetic classification of the Mammalia; pp. 21–46 *in* W. P. Luckett, and F. S. Szalay (eds.), Phylogeny of the Primates. Plenum Press, New York.

———. 1975b. Fossil mammals and early Eocene North Atlantic land continuity. Annals of the Missouri Botanical Garden 62:335–353.

———. 1980a. Eocene paleolatitude, climate, and mammals of Ellesmere Island. Palaeogeography, Palaeoclimatology, Palaeoecology 30:349–362.

———. 1980b. Early history and biogeography of South America's extinct land mammals; pp. 43–77 *in* R. L. Ciochon and A. B. Chiarelli (eds.), Evolutionary Biology of the New World Monkeys and Continental Drift. Plenum Press, New York.

———. 1982. Lagomorph interrelationships. Géobios Mémoire Spécial 6:213–223.

———. 1983. Holarctic landmass rearrangement, cosmic events, and Cenozoic terrestrial organisms. Annals of the Missouri Botanical Garden 70:459–489.

———. 1990. Plagiomenids (Mammalia: ?Dermoptera) from the Oligocene of Oregon, Montana, and South Dakota, and middle Eocene of northwestern Wyoming; pp. 211–234 in T. M. Bown and K. D. Rose (eds.), Dawn of the Age of Mammals in the Northern Part of the Rocky Mountain Interior, North America. Geological Society of America Special Paper 243, Boulder.

———. 2003. Semi-isolation and lowered salinity of the Arctic Ocean in late Paleocene to earliest Eocene time. Journal of Vertebrate Paleontology 23(supplement to no. 3):77A.

McKenna, M. C., and S. K. Bell. 1997. Classification of Mammals above the Species Level. Columbia University Press, New York.

———. 2002. Unitaxon Data Base. ftp://ftp.amnh.org/pub/people/mckenna/.

McKenna, M. C., and D. L. Lofgren. 2003. *Mimotricentes tedfordi*, a new arctocyonid from the late Paleocene of California; pp. 632–643 in L. J. Flynn (ed.), Vertebrate Fossils and Their Context. Contributions in Honor of Richard H. Tedford. Bulletin of the American Museum of Natural History 279.

McKenna, M. C., and E. Manning. 1977. Affinities and palaeobiogeographic significance of the Mongolian Paleogene genus *Phenacolophus*. Géobios Mémoire Spécial 1:61–85.

McKenna, M. C., and J. Meng. 2001. A primitive relative of rodents from the Chinese Paleocene. Journal of Vertebrate Paleontology 21:565–572.

McKenna, M. C., M. Chow, S. Ting, and Z.-X. Luo. 1989. *Radinskya yupingae*, a perissodactyl-like mammal from the late Paleocene of China; pp. 24–36 in D. R. Prothero and R. M. Schoch (eds.), The Evolution of Perissodactyls. Oxford University Press, New York.

McKenna, M. C., Z. Kielan-Jaworowska, and J. Meng. 2000. Earliest eutherian mammal skull, from the Late Cretaceous (Coniacian) of Uzbekistan. Acta Palaeontologica Polonica 45:1–54.

McKenna, M. C., X. Xue, and M. Zhou. 1984. *Prosarcodon lonanensis*, a new Paleocene micropternodontid palaeoryctoid insectivore from Asia. American Museum Novitates 2780:1–17.

Mead, J. G. 1975. Anatomy of the external nasal passages and facial complex in the Delphinidae (Mammalia: Cetacea). Smithsonian Contributions to Zoology 207:1–72.

Mellett, J. S. 1977. Paleobiology of North American *Hyaenodon* (Mammalia, Creodonta). Contributions to Vertebrate Evolution 1:1–134.

Meng, J. 1990. The auditory region of *Reithroparamys delicatissimus* (Mammalia, Rodentia) and its systematic implications. American Museum Novitates 2972:1–35.

Meng, J., and R. C. Fox. 1995. Osseous inner ear structures and hearing in early marsupials and placentals. Zoological Journal of the Linnean Society 115:47–71.

Meng, J., and Y.-M. Hu. 2004. Lagomorphs from the Yihesubu Late Eocene of Nei Mongol (Inner Mongolia). Vertebrate PalAsiatica 42:261–275.

Meng, J., and M. C. McKenna. 1998. Faunal turnovers of Palaeogene mammals from the Mongolian Plateau. Nature 394:364–367.

Meng, J., and A. R. Wyss. 1994. Enamel microstructure of *Tribosphenomys* (Mammalia, Glires): Character analysis and systematic implications. Journal of Mammalian Evolution 2:185–203.

———. 1995. Monotreme affinities and low-frequency hearing suggested by multituberculate ear. Nature 377:141–144.

———. 1997. Multituberculate and other mammal hair recovered from Palaeogene excreta. Nature 385:712–714.

———. 2001. The morphology of *Tribosphenomys* (Rodentiaformes, Mammalia): Phylogenetic implications for basal Glires. Journal of Mammalian Evolution 8:1–71.

Meng, J., G. J. Bowen, J. Ye, P. L. Koch, S.-Y. Ting, Q. Li, and X. Jin. 2004. *Gomphos elkema* (Glires, Mammalia) from the Erlian Basin: Evidence for the Early Tertiary Bumbanian Land Mammal Age in Nei-Mongol, China. American Museum Novitates 3425:1–24.

Meng, J., Y. Hu, and C. Li. 2003. The osteology of *Rhombomylus* (Mammalia: Glires): Implications for phylogeny and evolution of Glires. Bulletin of the American Museum of Natural History 275:1–247.

Meng, J., S. Ting, and J. A. Schiebout. 1994. The cranial morphology of an early Eocene didymoconid (Mammalia, Insectivora). Journal of Vertebrate Paleontology 14:534–551.

Meng, J., A. R. Wyss, M. R. Dawson, and R. Zhai. 1994. Primitive fossil rodent from Inner Mongolia and its implications for mammalian phylogeny. Nature 370:134–136.

Meng, J., R. Zhai, and A. Wyss. 1998. The late Paleocene Bayan Ulan fauna of Inner Mongolia, China; pp. 148–185 in K. C. Beard and M. R. Dawson (eds.), Dawn of the Age of Mammals in Asia. Bulletin of Carnegie Museum of Natural History 34.

Meyer, G. E. 1978. Hyracoidea; pp. 284–314 in V. J. Maglio and H. B. S. Cooke (eds.), Evolution of African Mammals. Harvard University Press, Cambridge, Mass.

Miao, D. 1988. Skull morphology of *Lambdopsalis bulla* (Mammalia, Multituberculata) and its implications to mammalian evolution. Contributions to Geology, University of Wyoming, Special Paper 4:1–104.

———. 1991. On the origins of mammals; pp. 579–597 in H.-P. Schultze and L. Trueb (eds.), Origins of Major Groups of Tetrapods: Controversies and Consensus. Cornell University Press, Ithaca, N. Y.

———. 1993. Cranial morphology and multituberculate relationships; pp. 63–74 in F. S. Szalay, M. J. Novacek, and M. C. McKenna (eds.), Mammal Phylogeny: Mesozoic Differentiation, Multituberculates, Monotremes, Early Therians, and Marsupials. Springer-Verlag, New York.

Miao, D., and J. A. Lillegraven. 1986. Discovery of three ear ossicles in a multituberculate mammal. National Geographic Research 2:500–507.

Mihlbachler, M. C., S. G. Lucas, and R. J. Emry. 2004. The holotype specimen of *Menodus giganteus*, and the "insoluble" problem of Chadronian brontothere taxonomy. New Mexico Museum of Natural History and Science Bulletin 26:129–135.

Milinkovitch, M. C., M. Bérubé, and P. J. Palsbøll. 1998. Cetaceans are highly derived artiodactyls; pp. 113–131 in J. G. M. Thewissen (ed.), The Emergence of Whales. Evolutionary Patterns in the Origin of Cetacea. Plenum Press, New York.

Milinkovitch, M. C., G. Ortí, and A. Meyer. 1993. Revised phylogeny of whales suggested by mitochondrial ribosomal DNA sequences. Nature 361:346–348.

Miller, E. R., and E. L. Simons. 1997. Dentition of *Proteopithecus sylviae*, an archaic anthropoid from the Fayum, Egypt. Proceedings of the National Academy of Sciences, USA 94:13760–13764.

Misawa, K., and A. Janke. 2003. Revisiting the Glires concept— Phylogenetic analysis of nuclear sequences. Molecular Phylogenetics and Evolution 28:320–327.

Mitchell, E. D. 1989. A new cetacean from the late Eocene La Meseta Formation, Seymour Island, Antarctic Peninsula. Canadian Journal of Fisheries and Aquatic Sciences 46:2219–2235.

Miyamoto, M. M. 1996. A congruence study of molecular and morphological data for eutherian mammals. Molecular Phylogeny and Evolution 6:373–390.

Miyata, K., and Y. Tomida. 1998. A new tillodont from the early middle Eocene of Japan and its implication to the subfamily Trogosinae (Tillodontia: Mammalia). Paleontological Research 2:53–66.

Moeller, H. 1990. Marsupials: Introduction; pp. 212–219 in S. B. Parker (ed.), Grzimek's Encyclopedia of Mammals, vol. 1, first edition. McGraw-Hill, New York.

Mones, A. 1982. An equivocal nomenclature: What means hypsodonty? Paläontologische Zeitschrift 56:107–111.

Morlo, M., and J. Habersetzer. 1999. The Hyaenodontidae (Creodonta, Mammalia) from the lower middle Eocene (MP11) of Messel (Germany) with special remarks on new x-ray methods. Courier Forschungsinstitut Senckenberg 216:31–73.

Mouchaty, S. K., F. Catzeflis, A. Janke, and U. Arnason. 2001. Molecular evidence of an African Phiomorpha–South American Caviomorpha clade and support for Hystricognathi based on the complete mitochondrial genome of the cane rat (*Thryonomys swinderianus*). Molecular Phylogenetic and Evolution 18:127–135.

Muizon, C. de. 1994. A new carnivorous marsupial from the Palaeocene of Bolivia and the problem of marsupial monophyly. Nature 370:208–211.

———. 1998. *Mayulestes ferox*, a borhyaenoid (Metatheria, Mammalia) from the early Palaeocene of Bolivia. Phylogenetic and palaeobiologic implications. Geodiversitas 20:19–142.

Muizon, C. de, and C. Argot. 2003. Comparative anatomy of the Tiupampa didelphimorphs; An approach to locomotory habits of early marsupials; pp. 43–62 in M. Jones, C. Dickman, and M. Archer (eds.), Predators with Pouches: The Biology of Carnivorous Marsupials. CSIRO, Collingwood, Victoria.

Muizon, C. de, and R. L. Cifelli. 2000. The "condylarths" (archaic Ungulata, Mammalia) from the early Palaeocene of Tiupampa (Bolivia): Implications on the origin of the South American ungulates. Geodiversitas 22:47–150.

———. 2001. A new basal "didelphoid" (Marsupialia, Mammalia) from the early Paleocene of Tiupampa (Bolivia). Journal of Vertebrate Paleontology 21:87–97.

Muizon, C. de, and B. Lange-Badré. 1997. Carnivorous dental adaptations in tribosphenic mammals and phylogenetic reconstruction. Lethaia 30:353–366.

Muizon, C. de, and L. G. Marshall. 1992. *Alcidedorbignya inopinata* (Mammalia: Pantodonta) from the early Paleocene of Bolivia: Phylogenetic and paleobiogeographic implications. Journal of Paleontology 66:499–520.

Muizon, C. de, R. L. Cifelli, and L. P. Bergqvist. 1998. Eutherian tarsals from the early Paleocene of Bolivia. Journal of Vertebrate Paleontology 18:655–663.

Muizon, C. de, R. L. Cifelli, and R. Céspedes Paz. 1997. The origin of the dog-like borhyaenoid marsupials of South America. Nature 389:486–489.

Munthe, K. 1998. Canidae; pp. 124–143 in C. M. Janis, K. M. Scott, and L. L. Jacobs (eds.), Evolution of Tertiary Mammals of North America. Cambridge University Press, Cambridge, UK.

Murphy, W. J., E. Eizirik, W. E. Johnson, Y. P. Zhang, O. A. Ryder, and S. J. O'Brien. 2001. Molecular phylogenetics and the origins of placental mammals. Nature 409:614–618.

Mussell, J. C. 2005. A reexamination of Lipotyphla and Afrotheria based on both molecular and morphological analysis. Ph.D. dissertation, The Johns Hopkins University, Baltimore.

Nessov, L. A. 1987. Research on Cretaceous and Paleocene mammals of the territory of the USSR [in Russian]. Ezhegodnik Vsesoyuznogo Paleontologicheskogo Obshchestva, Akademiya Nauk SSSR 30:199–218.

Nessov, L. A., J. D. Archibald, and Z. Kielan-Jaworowska. 1998. Ungulate-like mammals from the Late Cretaceous of Uzbekistan and a phylogenetic analysis of Ungulatomorpha; pp. 40–88 in K. C. Beard and M. R. Dawson (eds.), Dawn of the Age of Mammals in Asia. Bulletin of Carnegie Museum of Natural History 34.

Nessov, L. A., D. Sigogneau-Russell, and D. E. Russell. 1994. A survey of Cretaceous tribosphenic mammals from middle Asia (Uzbekistan, Kazakhstan and Tajikistan), of their geological setting, age and faunal environment. Palaeovertebrata 23:51–92.

Ni, X., Y. Wang, Y. Hu, and C. Li. 2004. A euprimate skull from the early Eocene of China. Nature 427:65–68.

Nikaido, M., F. Matsuno, H. Hamilton, R. L. Brownell, Jr., Y. Cao, W. Ding, Z. Zuoyan, A. M. Shedlock, R. E. Fordyce, M. Hasegawa, and N. Okada. 2001. Retroposon analysis of major cetacean lineages: The monophyly of toothed whales and the paraphyly of river dolphins. Proceedings of the National Academy of Sciences, USA 98:7384–7389.

Nikaido, M., H. Nishihara, Y. Hukumoto, and N. Okada. 2003. Ancient SINEs from African endemic mammals. Molecular Biology and Evolution 20:522–527.

Nikaido, M. A., P. Rooney, and N. Okada. 1999. Phylogenetic relationships among cetartiodactyls based on insertions of short and long interspersed elements: Hippopotamuses are the closest extant relatives of whales. Proceedings of the National Academy of Sciences, USA 96:10261–10266.

Norberg, U. M. 1989. Ecological determinants of bat wing shape and echolocation call structure with implications for some fossil bats; pp. 197–211 in V. Hanák, I. Horáček, and J. Gaisler (eds.), European Bat Research 1987. Charles University Press, Prague.

Norris, C. A. 1999. The cranium of *Bunomeryx* (Artiodactyla: Homacodontidae) from the Upper Eocene Uinta deposits of Utah and its implications for tylopod systematics. Journal of Vertebrate Paleontology 19:742–751.

———. 2000. The cranium of *Leptotragulus*, a hornless protoceratid (Artiodactyla: Protoceratidae) from the middle Eocene of North America. Journal of Vertebrate Paleontology 20:341–348.

Norris, K. S. 1968. The evolution of acoustic mechanisms in odontocete cetaceans; pp. 297–324 in E. T. Drake (ed.), Evolution and Environment: A Symposium Presented on the Occasion of the 100th Anniversary of the Foundation of the Peabody Museum of Natural History at Yale University. Yale University Press, New Haven, Conn.

Norris, R. D., and U. Röhl. 1999. Carbon cycling and chronology of climate warming during the Palaeocene/Eocene transition. Nature 401:775–778.

Novacek, M. J. 1977. A review of Paleocene and Eocene Leptictidae (Eutheria: Mammalia) from North America. PaleoBios 24:1–42.

———. 1986a. The skull of leptictid insectivorans and the higher-level classification of eutherian mammals. Bulletin of the American Museum of Natural History 183:1–112.

———. 1986b. The primitive eutherian dental formula. Journal of Vertebrate Paleontology 6:191–196.

———. 1987. Auditory features and affinities of the Eocene bats *Icaronycteris* and *Palaeochiropteryx* (Microchiroptera, incertae sedis). American Museum Novitates 2877:1–18.

———. 1990. Morphology, paleontology, and the higher clades of mammals. Current Mammalogy 2:507–543.

———. 1992a. Mammalian phylogeny: Shaking the tree. Nature 356:121–125.

———. 1992b. Fossils, topologies, missing data, and the higher level phylogeny of eutherian mammals. Systematic Biology 41:58–73.

———. 1993. Patterns of diversity in the mammalian skull; pp. 438–545 in J. Hanken and B. K. Hall (eds.), The Skull. Patterns of Structural and Systematic Diversity. University of Chicago Press, Chicago.

———. 1999. 100 million years of land vertebrate evolution: The Cretaceous–Early Tertiary transition. Annals of the Missouri Botanical Garden 86:230–258.

Novacek, M. J., and A. Wyss. 1986. Higher-level relationships of the recent eutherian orders: Morphological evidence. Cladistics 2:257–287.

Novacek, M. J., T. M. Bown, and D. M. Schankler. 1985. On the classification of the early Tertiary Erinaceomorpha (Insectivora, Mammalia). American Museum Novitates 2813:1–22.

Novacek, M. J., I. Ferrusquia-Villafranca, J. J. Flynn, A. R. Wyss, and M. Norell. 1991. Wasatchian (early Eocene) mammals and other vertebrates from Baja California, Mexico: The Lomas las Tetas de Cabra Fauna. Bulletin of the American Museum of Natural History 208:1–88.

Novacek, M. J., M. C. McKenna, N. A. Neff, and R. L. Cifelli. 1983. Evidence from earliest known erinaceomorph basicranium that insectivorans and primates are not closely related. Nature 306:683–684.

Novacek, M. J., G. W. Rougier, J. R. Wible, M. C. McKenna, D. Dashzeveg, and I. Horovitz. 1997. Epipubic bone in eutherian mammals from the Cretaceous of Mongolia. Nature 389:483–486.

Novacek, M. J., A. R. Wyss, and M. C. McKenna. 1988. The major groups of eutherian mammals; pp. 31–71 in M. J. Benton (ed.), The Phylogeny and Classification of the Tetrapods. Volume 2: Mammals. Clarendon Press, Oxford.

Nowak, R. M. (ed.). 1999. Walker's Mammals of the World. The Johns Hopkins University Press, Baltimore.

O'Leary, M. A. 1996. Dental evolution in the early Eocene Notharctinae (Primates, Adapiformes) from the Bighorn Basin, Wyoming: Documentation of gradual evolution in the oldest true primates. Ph.D. dissertation, The Johns Hopkins University, Baltimore.

———. 1998. Phylogenetic and morphometric reassessment of the dental evidence for a mesonychian and cetacean clade; pp. 133–161 in J. G. M. Thewissen (ed.), The Emergence of Whales. Evolutionary Patterns in the Origin of Cetacea. Plenum Press, New York.

———. 1999. Parsimony analysis of total evidence from extinct and extant taxa, and the cetacean-artiodactyl question. Cladistics 15:315–330.

O'Leary, M. A., and J. H. Geisler. 1999. The position of Cetacea within Mammalia: Phylogenetic analysis of morphological data from extinct and extant taxa. Systematic Biology 48:455–490.

O'Leary, M. A., and K. D. Rose. 1995. Postcranial skeleton of the early Eocene mesonychid *Pachyaena* (Mammalia: Mesonychia). Journal of Vertebrate Paleontology 15:401–430.

O'Leary, M. A., and M. D. Uhen. 1999. The time of origin of whales and the role of behavioral changes in the terrestrial-aquatic transition. Paleobiology 25:534–556.

O'Leary, M. A., S. G. Lucas, and T. E. Williamson. 2000. A new specimen of *Ankalagon* (Mammalia, Mesonychia) and evidence of sexual dimorphism in mesonychians. Journal of Vertebrate Paleontology 20:387–393.

Oliveira, E. V., and L. P. Bergqvist. 1998. A new Paleocene armadillo (Mammalia, Dasypodoidea) from the Itaboraí Basin, Brazil. Asociacion Paleontologica Argentina Publicacion Especial 5:35–40.

———. 1999. A new Paleocene armadillo (Mammalia, Xenarthra, Astegotheriini) from Itaboraí, Brazil, and phylogeny of early Tertiary astegotheriines. Anais da Academia Brasileira de Ciências 71:814–815.

Osborn, H. F. 1898a. Remounted skeleton of *Phenacodus primaevus*. Comparison with *Euprotogonia*. Bulletin of the American Museum of Natural History 10:159–164.

———. 1898b. Evolution of the Amblypoda. Part I. Taligrada and Pantodonta. Bulletin of the American Museum of Natural History 10:169–218.

———. 1904. An armadillo from the middle Eocene (Bridger) of North America. Bulletin of the American Museum of Natural History 20:163–165.

———. 1918. Equidae of the Oligocene, Miocene, and Pliocene of North America, iconographic type revision. Memoirs of the American Museum of Natural History, new series 2:1–330.

———. 1924. *Andrewsarchus*, giant mesonychid of Mongolia. American Museum Novitates 146:1–5.

———. 1929. The titanotheres of ancient Wyoming, Dakota, and Nebraska. Two volumes. U. S. Geological Survey Monograph 55(1–2):1–953.

Osborn, H. F., and W. Granger. 1931. Coryphodonts of Mongolia, *Eudinoceras mongoliensis* Osborn, *E. kholobolchiensis* sp. nov. American Museum Novitates 459:1–13.

———. 1932. Coryphodonts and uintatheres from the Mongolian expedition of 1930. American Museum Novitates 552:1–16.

Pascual, R. 1980a. Nuevos y singulares tipos ecologicos de marsupiales extinguidos de America del Sur (Paleoceno tardio o Eoceno temprano) del noroeste Argentino. Actas II Congreso Argentino de Paleontología y Bioestratigrafia y I Congreso Latinoamericano de Paleontología, Buenos Aires, 1978 2:151–173.

———. 1980b. Prepidolopidae, nueva familia de Marsupialia Didelphoidea del Eoceno Sudamericano. Ameghiniana 17:216–242.

Pascual, R., M. Archer, E. Ortiz Jaureguizar, J. L. Prado, H. Godthelp, and S. J. Hand. 1992. The first non-Australian monotreme: An early Paleocene South American platypus (Monotremata, Ornithorhynchidae); pp. 2–15 in M. L. Augee (ed.), Platypus and Echidnas. Royal Zoological Society of New South Wales, Sydney.

Pascual, R., F. J. Goin, L. Balarino, and D. E. Udrizar. 2002. New data on the Paleocene monotreme *Monotrematum sudamericanum* and the convergent evolution of triangulate molars. Acta Palaeontologica Polonica 47:487–492.

Pascual, R., F. J. Goin, and A. A. Carlini. 1994. New data on the Groeberiidae: Unique late Eocene–early Oligocene South American marsupials. Journal of Vertebrate Paleontology 14:247–259.

Pascual, R., F. J. Goin, P. Gonzalez, A. Ardolino, and P. F. Puerta. 2000. A highly derived docodont from the Patagonian Late Cretaceous: Evolutionary implications for Gondwanan mammals. Geodiversitas 22:395–414.

Pascual, R., F. J. Goin, D. W. Krause, E. Ortiz-Jaureguizar, and A. A. Carlini. 1999. The first gnathic remains of *Sudamerica*: Implications for gondwanathere relationships. Journal of Vertebrate Paleontology 19:373–382.

Pascual, R., M. G. Vucetich, and J. Fernandez. 1978. Los primeros mamíferos (Notoungulata, Henricosborniidae) de la Formación Mealla (Grupo Salta, Subgrupo Santa Barbara). Sus implicancias filogenéticas, taxonómicas y cronológicas. Ameghiniana 15:366–390.

Patterson, B. 1936. The internal structure of the ear in some notoungulates. Field Museum of Natural History, Geology series 6:199–227.

———. 1949. Rates of evolution in taeniodonts; pp. 243–278 *in* G. L. Jepsen, E. Mayr, and G. G. Simpson (eds.), Genetics, Paleontology, and Evolution. Princeton University Press, Princeton.

———. 1956. Early Cretaceous mammals and the evolution of mammalian molar teeth. Fieldiana Geology 13:1–105.

———. 1965. The fossil elephant shrews (Family Macroscelididae). Bulletin of the Museum of Comparative Zoology 133:295–335.

———. 1977. A primitive pyrothere (Mammalia, Notoungulata) from the Early Tertiary of northwestern Venezuela. Fieldiana Geology 33:397–422.

Patterson, B., and R. Pascual. 1968. Evolution of mammals on southern continents. V. The fossil mammal fauna of South America. Quaterly Review of Biology 43:409–451.

Patterson, B., and A. E. Wood. 1982. Rodents from the Deseadan Oligocene of Bolivia and the relationships of the Caviomorpha. Bulletin of the Museum of Comparative Zoology 149:371–543.

Patterson, B., W. Segall, W. Turnbull, and T. Gaudin. 1992. The ear region in xenarthrans (=Edentata: Mammalia). Part II. Pilosa (sloths, anteaters), palaeanodonts, and a miscellany. Fieldiana Geology, new series 24:1–79.

Patton, T. H., and B. E. Taylor. 1973. The Protoceratinae (Mammalia, Tylopoda, Protoceratidae) and the systematics of the Protoceratidae. Bulletin of the American Museum of Natural History 150:347–414.

Paula Couto, C. de. 1952a. Fossil mammals from the beginning of the Cenozoic in Brazil. Marsupialia: Polydolopidae and Borhyaenidae. American Museum Novitates 1559:1–27.

———. 1952b. Fossil mammals from the beginning of the Cenozoic in Brazil. Marsupialia: Didelphidae. American Museum Novitates 1567:1–26.

———. 1952c. Fossil mammals from the beginning of the Cenozoic in Brazil. Condylarthra, Litopterna, Xenungulata, and Astrapotheria. Bulletin of the American Museum of Natural History 99:355–394.

———. 1962. Didelfideos fósiles del Paleoceno de Brazil. Revista del Museo Argentino de Ciencias Naturales "Bernardino Rivadavia," Ciencias Zoologia 112:135–166.

———. 1970a. News on the fossil marsupials from the Riochican of Brazil. Anais da Academia Brasileira de Ciências 42:19–34.

———. 1970b. Novo notoungulado no Riochiquense de Itaboraí. Iheringia Geologia 3:77–86.

———. 1978. Ungulados fósseis do Riochiquense de Itaboraí, RJ, Brasil. I—Xenungulata. Anais da Academia Brasileira de Ciências 50:203–207.

———. 1979. Tratado de Paleomastozoologia. Academia Brasiliera de Ciências, Rio de Janeiro.

Pecon Slattery, J., W. J. Murphy, and S. J. O'Brien. 2000. Patterns of diversity among SINE elements isolated from three Y-chromosome genes in carnivores. Molecular Biology and Evolution 17:825–829.

Peigné, S., and L. de Bonis. 1999. The genus *Stenoplesictis* Filhol (Mammalia, Carnivora) from the Oligocene deposits of the Phosphorites of Quercy, France. Journal of Vertebrate Paleontology 19:566–575.

Peláez-Campomanes, P., N. López-Martínez, M. A. Álvarez-Sierra, and R. Daams. 2000. The earliest mammal of the European Paleocene: The multituberculate *Hainina*. Journal of Paleontology 74:701–711.

Peters, D. S., and G. Storch. 1993. South American relationships of Messel birds and mammals. Kaupia—Darmstädter Beiträge zur Naturgeschichte 3:263–269.

Petit, G. 1955. Ordre des siréniens; pp. 918–1001 *in* P. P. Grassé (ed.), Traité de Zoologie XVII(1). Masson et Cie, Paris.

Pettigrew, J. D. 1986. Flying primates? Megabats have the advanced pathway from eye to midbrain. Science 231:1304–1306.

Pettigrew, J. D., B. G. M. Jamieson, S. K. Robson, L. S. Hall, K. I. McAnally, and H. M. Cooper. 1989. Phylogenetic relations between microbats, megabats, and primates (Mammalia: Chiroptera and Primates). Philosophical Transactions of the Royal Society of London B 325:489–559.

Peyer, B. 1968. Comparative Odontology. University of Chicago Press, Chicago.

Pfretzschner, H.-U. 1993. Muscle reconstruction and aquatic locomotion in the middle Eocene *Buxolestes piscator* from Messel near Darmstadt. Kaupia—Darmstädter Beiträge zur Naturgeschichte 3:75–87.

Phillips, E. M., and A. Walker. 2002. Fossil lorisoids; pp. 83–96 *in* W. C. Hartwig (ed.), The Primate Fossil Record. Cambridge University Press, Cambridge, UK.

Pickford, M. 1983. On the origins of Hippopotamidae together with descriptions of two new species, a new genus, and a new subfamily from the Miocene of Kenya. Géobios 16:193–217.

———. 1989. Update on hippo origins. Comptes Rendus de l'Académie des Sciences, Paris, sér. II 309:163–168.

Pickford, M., S. Moyà Solà, and P. Mein. 1997. A revised phylogeny of Hyracoidea (Mammalia) based on new specimens of Pliohyracidae from Africa and Europe. Neues Jahrbuch für Geologie und Paläontologie Abhandlungen 205:265–288.

Polly, P. D. 1996. The skeleton of *Gazinocyon vulpeculus* gen. et comb. nov. and the cladistic relationships of Hyaenodontidae (Eutheria, Mammalia). Journal of Vertebrate Paleontology 16:303–319.

Prasad, G. V. R., and M. Godinot. 1994. Eutherian tarsal bones from the Late Cretaceous of India. Journal of Paleontology 68:892–902.

Prasad, G. V. R., J.-J. Jaeger, A. Sahni, E. Gheerbrant, and C. K. Khajuria. 1994. Eutherian mammals from the Upper Cretaceous (Maastrichtian) Intertrappean Beds of Naskal, Andhra Pradesh, India. Journal of Vertebrate Paleontology 14:260–277.

Prasad, V., C. A. E. Strömberg, H. Alimohammadian, and A. Sahni. 2005. Dinosaur coprolites and the early evolution of grasses and grazers. Science 310:1177–1180.

Prothero, D. R. 1996a. Magnetostratigraphy of the Eocene-Oligocene transition in Trans-Pecos Texas; pp. 189–198 in D. R. Prothero and R. J. Emry (eds.), The Terrestrial Eocene-Oligocene Transition in North America. Cambridge University Press, Cambridge, UK.

———. 1996b. Camelidae; pp. 609–651 in D. R. Prothero and R. J. Emry (eds.), The Terrestrial Eocene-Oligocene Transition in North America. Cambridge University Press, Cambridge, UK.

———. 1998a. Oromerycidae; pp. 426–430 in C. M. Janis, K. M. Scott, and L. L. Jacobs (eds.), Evolution of Tertiary Mammals of North America. Volume 1: Terrestrial Carnivores, Ungulates, and Ungulatelike Mammals. Cambridge University Press, Cambridge, UK.

———. 1998b. Protoceratidae; pp. 431–438 in C. M. Janis, K. M. Scott, and L. L. Jacobs (eds.), Evolution of Tertiary Mammals of North America. Volume 1: Terrestrial Carnivores, Ungulates, and Ungulatelike Mammals. Cambridge University Press, Cambridge, UK.

———. 1998c. Rhinocerotidae; pp. 595–605 in C. M. Janis, K. M. Scott, and L. L. Jacobs (eds.), Evolution of Tertiary Mammals of North America. Volume 1: Terrestrial Carnivores, Ungulates, and Ungulatelike Mammals. Cambridge University Press, Cambridge, UK.

———. 2005. The Evolution of North American Rhinoceroses. Cambridge University Press, Cambridge, UK.

Prothero, D. R., and R. J. Emry. 2004. The Chadronian, Orellan, and Whitneyan North American Land Mammal Ages; pp. 156–168 in M. O. Woodburne (ed.), Late Cretaceous and Cenozoic Mammals of North America. Biostratigraphy and Geochronology. Columbia University Press, New York.

Prothero, D. R., and S. G. Lucas. 1996. Magnetic stratigraphy of the Duchesnean part of the Galisteo Formation, New Mexico; pp. 199–205 in D. R. Prothero and R. J. Emry (eds.), The Terrestrial Eocene-Oligocene Transition in North America. Cambridge University Press, Cambridge, UK.

Prothero, D. R., and R. M. Schoch. 1989. Origin and evolution of the Perissodactyla: Summary and synthesis; pp. 504–529 in D. R. Prothero and R. M. Schoch (eds.), The Evolution of Perissodactyls. Oxford University Press, New York.

Prothero, D. R., and C. C. Swisher III. 1992. Magnetostratigraphy and geochronology of the terrestrial Eocene-Oligocene transition in North America; pp. 46–73 in D. R. Prothero and W. A. Berggren (eds.), Eocene-Oligocene Climatic and Biotic Evolution. Princeton University Press, Princeton.

Prothero, D. R., C. Guérin, and E. Manning. 1989. The history of the Rhinocerotoidea; pp. 321–340 in D. R. Prothero and R. M. Schoch (eds.), The Evolution of Perissodactyls. Oxford University Press, New York.

Prothero, D. R., E. M. Manning, and M. Fischer. 1988. The phylogeny of the ungulates; pp. 201–234 in M. J. Benton (ed.), The Phylogeny and Classification of the Tetrapods. Volume 2: Mammals. Clarendon Press, Oxford.

Qi, T., and K. C. Beard. 1998. Late Eocene sivaladapid primate from Guangxi Zhuang Autonomous Region, People's Republic of China. Journal of Human Evolution 35:211–220.

Radinsky, L. B. 1963. Origin and early evolution of North American Tapiroidea. Peabody Museum of Natural History Bulletin 17:1–115.

———. 1964. *Paleomoropus*, a new early Eocene chalicothere (Mammalia, Perissodactyla), and a revision of Eocene chalicotheres. American Museum Novitates 2179:1–28.

———. 1965a. Evolution of the tapiroid skeleton from *Heptodon* to *Tapirus*. Bulletin of the Museum of Comparative Zoology 134:69–106.

———. 1965b. Early Tertiary Tapiroidea of Asia. Bulletin of the American Museum of Natural History 129:181–264.

———. 1966a. The adaptive radiation of the phenacodontid condylarths and the origin of the Perissodactyla. Evolution 20:408–417.

———. 1966b. The families of the Rhinocerotoidea (Mammalia, Perissodactyla). Journal of Mammalogy 47:631–639.

———. 1967a. A review of the rhinocerotoid family Hyracodontidae (Perissodactyla). Bulletin of the American Museum of Natural History 136:1–46.

———. 1967b. *Hyrachyus, Chasmotherium,* and the early evolution of helaletid tapiroids. American Museum Novitates 2313:1–23.

———. 1969. The early evolution of the Perissodactyla. Evolution 23:308–328.

———. 1977. Brains of early carnivores. Paleobiology 3:333–349.

———. 1981. Brain evolution in extinct South American ungulates. Brain, Behavior and Evolution 18:169–187.

Radulesco, C., and J. Sudre. 1985. *Crivadiatherium iliescui* n. sp., nouvel embrithopode (Mammalia) dans le Paléogène ancien de la Dépression de Hateg (Roumanie). Palaeovertebrata 15:139–157.

Rana, R. S., and G. P. Wilson. 2003. New Late Cretaceous mammals from the Intertrappean beds of Rangapur, India and paleobiogeographic framework. Acta Palaeontologica Polonica 48:331–348.

Rana, R. S., H. Singh, A. Sahni, K. D. Rose, and P. K. Saraswati. 2005. Early Eocene chiropterans from a new mammalian assemblage (Vastan lignite mine, Gujarat, western peninsular margin): Oldest known bats from Asia. Journal of the Palaeontological Society of India 50:93–100.

Rasmussen, D. T. 1989. The evolution of the Hyracoidea: A review of the fossil evidence; pp. 57–78 in D. R. Prothero and R. M. Schoch (eds.), The Evolution of Perissodactyls. Oxford University Press, New York.

———. 1990. The phylogenetic position of *Mahgarita stevensi*: Protoanthropoid or lemuroid? International Journal of Primatology 11:439–469.

———. 2002. Early catarrhines of the African Eocene and Oligocene; pp. 203–220 in W. C. Hartwig (ed.), The Primate Fossil Record. Cambridge University Press, Cambridge, UK.

Rasmussen, D. T., and K. A. Nekaris. 1998. Evolutionary history of lorisiform primates. Folia Primatologica 69(supplement 1):250–285.

Rasmussen, D. T., and E. L. Simons. 1988. New Oligocene hyracoids from Egypt. Journal of Vertebrate Paleontology 8:67–83.

———. 1991. The oldest hyracoids (Mammalia: Pliohyracidae): New species of *Saghatherium* and *Thyrohyrax* from the Fayum. Neues Jahrbuch für Geologie und Paläontologie Abhandlungen 182:187–209.

———. 2000. Ecomorphological diversity among Paleogene hyracoids (Mammalia): A new cursorial browser from the Fayum, Egypt. Journal of Vertebrate Paleontology 20:167–176.

Rasmussen, D. T., G. C. Conroy, and E. L. Simons. 1998. Tarsier-like locomotor specializations in the Oligocene primate

Afrotarsius. Proceedings of the National Academy of Sciences, USA 95:14848–14850.

Rasmussen, D. T., M. Gagnon, and E. L. Simons. 1990. Taxeopody in the carpus and tarsus of Oligocene Pliohyracidae (Mammalia: Hyracoidea) and the phylogenetic position of hyraxes. Proceedings of the National Academy of Sciences, USA 87:4688–4691.

Rathbun, G. B. 1984. Elephant-shrews; pp. 730–735 in D. MacDonald (ed.), The Encyclopedia of Mammals. Facts on File, New York.

Rauhut, O. W. M., T. Martin, E. Ortiz-Jaureguizar, and P. Puerta. 2002. A Jurassic mammal from South America. Nature 416:165–168.

Ray, C. E., D. P. Domning, and M. C. McKenna. 1994. A new specimen of *Behemotops proteus* (Order Desmostylia) from the marine Oligocene of Washington. Proceedings of the San Diego Society of Natural History 29:205–222.

Reed, C. A. 1951. Locomotion and appendicular anatomy in three soricoid insectivores. American Midland Naturalist 45:513–671.

———. 1954. Some fossorial mammals from the Tertiary of western North America. Journal of Paleontology 28:102–111.

Reed, C. A., and W. D. Turnbull. 1965. The mammalian genera *Arctoryctes* and *Cryptoryctes* from the Oligocene and Miocene of North America. Fieldiana Geology 15:99–170.

Reguero, M. A., S. A. Marenssi, and S. N. Santillana. 2002. Antarctic Peninsula and South America (Patagonia) Paleogene terrestrial faunas and environments: Biogeographic relationships. Palaeogeography, Palaeoclimatology, Palaeoecology 179:189–210.

Reig, O. A., J. A. W. Kirsch, and L. G. Marshall. 1987. Systematic relationships of the living and Neocenozoic American "opossum-like" marsupials (suborder Didelphimorphia), with comments on the classification of these and of the Cretaceous and Paleogene New World and European metatherians; pp. 1–89 in M. Archer (ed.), Possums and Opossums: Studies in Evolution. Surrey Beatty & Sons and the Royal Zoological Society of New South Wales, Sydney.

Reilly, S. M., and T. D. White. 2003. Hypaxial motor patterns and the function of epipubic bones in primitive mammals. Science 299:400–402.

Remy, J. A. 1985. Nouveaux gisements de mammifères et reptiles dans les Grès de Célas (Eocène sup. du Gard). Etude des Palaeothériidés (Perissodactyla, Mammalia). Palaeontographica Abteilung A 189:171–225.

———. 2001. Sur le crâne de *Propalaeotherium isselanum* (Mammalia, Perissodactyla, Palaeotheriidae) de Pépieux (Minervois, Sud de la France). Geodiversitas 23:105–127.

Remy, J. A., J.-Y. Crochet, B. Sigé, J. Sudre, L. de Bonis, M. Vianey-Liaud, M. Godinot, J.-L. Hartenberger, B. Lange-Badré, and B. Comte. 1987. Biochronologie des phosphorites du Quercy: Mise à jour des listes fauniques et nouveaux gisements de mammifères fossiles. Münchner Geowissenschaftliche Abhandlungen A 10:169–188.

Rensberger, J. M. 1975. *Haplomys* and its bearing on the origin of the aplodontoid rodents. Journal of Mammalogy 56:1–14.

Rensberger, J. M., and W. von Koenigswald. 1980. Functional and phylogenetic interpretation of enamel microstructure in rhinoceroses. Paleobiology 6:477–495.

Rensberger, J. M., and H.-U. Pfretzschner. 1992. Enamel structure in astrapotheres and its functional implications. Scanning Microscopy 6:495–510.

Repenning, C. A. 1967. Subfamilies and genera of the Soricidae. U. S. Geological Survey Professional Paper 565:1–74.

Reshetov, V. Y. 1979. Early Tertiary Tapiroidea of Mongolia and the USSR. Joint Soviet-Mongolian Paleontological Expedition Transactions 11:1–143.

Rich, T. H., T. F. Flannery, P. Trusler, L. Kool, N. A. van Klaveren, and P. Vickers-Rich. 2001. A second tribosphenic mammal from the Mesozoic of Australia. Records of the Queen Victoria Museum 110:1–9.

———. 2002. Evidence that monotremes and ausktribosphenids are not sister groups. Journal of Vertebrate Paleontology 22:466–469.

Rich, T. H., J. A. Hopson, A. M. Musser, T. F. Flannery, and P. Vickers-Rich. 2005. Independent origins of middle ear bones in monotremes and therians. Science 307:910–914.

Rich, T. H., P. Vickers-Rich, A. Constantine, T. F. Flannery, L. Kool, and N. von Klaveren. 1997. A tribosphenic mammal from the Mesozoic of Australia. Science 278:1438–1442.

———. 1999. Early Cretaceous mammals from Flat Rocks, Victoria, Australia. Records of the Queen Victoria Museum 106:1–29.

Rich, T. H., P. Vickers-Rich, P. Trusler, T. F. Flannery, R. Cifelli, A. Constantine, L. Kool, and N. van Klaveren. 2001. Monotreme nature of the Australian Early Cretaceous mammal *Teinolophos*. Acta Palaeontologica Polonica 46:113–118.

Rigby, J. K., Jr. 1980. Swain Quarry of the Fort Union Formation, middle Paleocene (Torrejonian), Carbon County, Wyoming: Geologic setting and mammalian fauna. Evolutionary Monographs 3:1–179.

———. 1981. A skeleton of *Gillisonchus gillianus* (Mammalia; Condylarthra) from the early Paleocene (Puercan) Ojo Alamo Sandstone, San Juan Basin, New Mexico, with comments on the local stratigraphy of Betonnie Tsosie Wash; pp. 89–126 in S. G. Lucas, J. K. Rigby, Jr., and B. S. Kues (eds.), Advances in San Juan Basin Paleontology. University of New Mexico Press, Albuquerque.

Robertson, D. S., M. C. McKenna, O. B. Toon, S. Hope, and J. A. Lillegraven. 2004. Survival in the first hours of the Cenozoic. Geological Society of America Bulletin 116:760–768.

Robinson, P. 1968. Nyctitheriidae (Mammalia, Insectivora) from the Bridger Formation of Wyoming. Contributions to Geology, University of Wyoming 7:129–138.

Robinson, P., and D. G. Kron. 1998. *Koniaryctes*, a new genus of apternodontid insectivore from lower Eocene rocks of the Powder River Basin, Wyoming. Contributions to Geology, University of Wyoming 32:187–190.

Robinson, P., G. F. Gunnell, S. L. Walsh, W. C. Clyde, J. E. Storer, R. K. Stucky, D. J. Froehlich, I. Ferrusquia-Villafranca, and M. C. McKenna. 2004. Wasatchian through Duchesnean biochronology; pp. 106–155 in M. O. Woodburne (ed.), Late Cretaceous and Cenozoic Mammals of North America. Biostratigraphy and Geochronology. Columbia University Press, New York.

Rodriguez-Trelles, F., R. Tarrio, and F. J. Ayala. 2002. A methodological bias toward overestimation of molecular evolutionary time scales. Proceedings of the National Academy of Sciences, USA 99:8112–8115.

Röhl, U., R. D. Norris, and J. G. Ogg. 2003. Cyclostratigraphy of upper Paleocene and lower Eocene sediments at Blake Nose Site 1051 (western North Atlantic); pp. 567–588 in S. L. Wing, P. D. Gingerich, B. Schmitz, and E. Thomas (eds.), Causes and Consequences of Globally Warm Climates in the Early

Paleogene. Geological Society of America Special Paper 369, Boulder.

Rohr, J. J., F. E. Fish, and J. W. Gilpatrick. 2002. Maximum swim speeds of captive and free-ranging delphinids: Critical analysis of extraordinary performance. Marine Mammal Science 18:1–19.

Romer, A. S. 1966. Vertebrate Paleontology. University of Chicago Press, Chicago.

———. 1970. The Chanares (Argentina) Triassic reptile fauna. VI. A chiniquodontid cynodont with an incipient squamosal-dentary jaw articulation. Breviora 344:1–18.

Romer, A. S., and A. Lewis. 1973. The Chanares (Argentina) Triassic reptile fauna: XIX. Postcranial materials of the cynodonts *Probelesodon* and *Probainognathus*. Breviora 407:1–26.

Romer, A. S., and T. S. Parsons. 1977. The Vertebrate Body. W. B. Saunders, Philadelphia.

Rose, K. D. 1972. A new tillodont from the Eocene upper Willwood Formation of Wyoming. Postilla 155:1–13.

———. 1973. The mandibular dentition of *Plagiomene* (Dermoptera, Plagiomenidae). Breviora 411:1–17.

———. 1975a. The Carpolestidae, Early Tertiary primates from North America. Bulletin of the Museum of Comparative Zoology 147:1–74.

———. 1975b. *Elpidophorus*, the earliest dermopteran (Dermoptera, Plagiomenidae). Journal of Mammalogy 56:676–679.

———. 1978. A new Paleocene epoicotheriid (Mammalia), with comments on the Palaeanodonta. Journal of Paleontology 52:658–674.

———. 1979. A new Paleocene palaeanodont and the origin of the Metacheiromyidae. Breviora 455:1–14.

———. 1981. The Clarkforkian Land-Mammal Age and mammalian faunal composition across the Paleocene-Eocene boundary. University of Michigan Papers on Paleontology 26:1–197.

———. 1982a. Anterior dentition of the early Eocene plagiomenid dermopteran *Worlandia*. Journal of Mammalogy 63:179–183.

———. 1982b. Skeleton of *Diacodexis*, oldest known artiodactyl. Science 216:621–623.

———. 1985. Comparative osteology of North American dichobunid artiodactyls. Journal of Paleontology 59:1203–1226.

———. 1987. Climbing adaptations in the early Eocene mammal *Chriacus* and the origin of Artiodactyla. Science 236:314–316.

———. 1988. Early Eocene mammal skeletons from the Bighorn Basin, Wyoming: Significance to the Messel fauna. Courier Forschungsinstitut Senckenberg 107:435–450.

———. 1990. Postcranial skeletal remains and adaptations in early Eocene mammals from the Willwood Formation, Bighorn Basin, Wyoming; pp. 107–133 in T. M. Bown and K. D. Rose (eds.), Dawn of the Age of Mammals in the Northern Part of the Rocky Mountain Interior, North America. Geological Society of America Special Paper 243, Boulder.

———. 1995. The earliest primates. Evolutionary Anthropology 3:159–173.

———. 1996a. On the origin of the order Artiodactyla. Proceedings of the National Academy of Sciences, USA 93:1705–1709.

———. 1996b. Skeleton of early Eocene *Homogalax* and the origin of Perissodactyla. Palaeovertebrata, Volume jubilaire D. E. Russell, 25:243–260.

———. 1999a. Postcranial skeleton of Eocene Leptictidae (Mammalia), and its implications for behavior and relationships. Journal of Vertebrate Paleontology 19:355–372.

———. 1999b. *Eurotamandua* and Palaeanodonta: Convergent or related? Paläontologische Zeitschrift 73:395–401.

———. 2000. Land mammals from the late Paleocene Aquia Formation: The first early Cenozoic mammals from Maryland. Proceedings of the Biological Society of Washington 113:855–863.

———. 2001a. Compendium of Wasatchian mammal postcrania from the Willwood Formation; pp. 157–183 in P. D. Gingerich (ed.), Paleocene-Eocene Stratigraphy and Biotic Change in the Bighorn and Clarks Fork Basins of Northwestern Wyoming. University of Michigan Papers on Paleontology 33, Ann Arbor.

———. 2001b. The ancestry of whales. Science 293:2216–2217.

———. In press. The postcranial skeleton of early Oligocene *Leptictis* (Mammalia: Leptictida), with a comparison to *Leptictidium* from the middle Eocene of Messel. Palaeontographica Abteilung A.

Rose, K. D., and J. D. Archibald (eds.). 2005. The Rise of Placental Mammals: Origins and Relationships of the Major Extant Clades. The Johns Hopkins University Press, Baltimore.

Rose, K. D., and T. M. Bown. 1982. New plesiadapiform primates from the Eocene of Wyoming and Montana. Journal of Vertebrate Paleontology 2:63–69.

———. 1984. Gradual phyletic evolution at the generic level in early Eocene omomyid primates. Nature 309:250–252.

———. 1991. Additional fossil evidence on the differentiation of the earliest euprimates. Proceedings of the National Academy of Sciences, USA 88:98–101.

———. 1996. A new plesiadapiform (Mammalia: Plesiadapiformes) from the early Eocene of the Bighorn Basin, Wyoming. Annals of Carnegie Museum 65:305–321.

Rose, K. D., and B. J. Chinnery. 2004. The postcranial skeleton of early Eocene rodents. Bulletin of Carnegie Museum of Natural History 36:211–244.

Rose, K. D., and R. J. Emry. 1983. Extraordinary fossorial adaptations in the Oligocene palaeanodonts *Epoicotherium* and *Xenocranium* (Mammalia). Journal of Morphology 175:33–56.

———. 1993. Relationships of Xenarthra, Pholidota, and fossil "edentates": The morphological evidence; pp. 81–102 in F. S. Szalay, M. J. Novacek, and M. C. McKenna (eds.), Mammal Phylogeny: Placentals. Springer-Verlag, New York.

Rose, K. D., and P. D. Gingerich. 1987. A new insectivore from the Clarkforkian (earliest Eocene) of Wyoming. Journal of Mammalogy 68:17–27.

Rose, K. D., and W. von Koenigswald. 2005. An exceptionally complete skeleton of *Palaeosinopa* (Mammalia, Cimolesta, Pantolestidae) from the Green River Formation, and other postcranial elements of the Pantolestidae from the Eocene of Wyoming (USA). Palaeontographica Abteilung A 273:55–96.

Rose, K. D., and D. W. Krause. 1982. Cyriacotheriidae, a new family of early Tertiary pantodonts from western North America. Proceedings of the American Philosophical Society 126:26–50.

———. 1984. Affinities of the primate *Altanius* from the early Tertiary of Mongolia. Journal of Mammalogy 65:721–726.

Rose, K. D., and S. G. Lucas. 2000. An early Paleocene palaeanodont (Mammalia, ?Pholidota) from New Mexico, and the origin of Palaeanodonta. Journal of Vertebrate Paleontology 20:139–156.

Rose, K. D., and M. A. O'Leary. 1995. The manus of *Pachyaena gigantea* (Mammalia: Mesonychia). Journal of Vertebrate Paleontology 15:855–859.

Rose, K. D., and E. L. Simons. 1977. Dental function in the Plagiomenidae: Origin and relationships of the mammalian order Dermoptera. Contributions from the Museum of Paleontology, University of Michigan 24:221–236.

Rose, K. D., and A. Walker. 1985. The skeleton of early Eocene *Cantius,* oldest lemuriform primate. American Journal of Physical Anthropology 66:73–89.

Rose, K. D., K. C. Beard, and P. Houde. 1993. Exceptional new dentitions of the diminutive plesiadapiforms *Tinimomys* and *Niptomomys* (Mammalia), with comments on the upper incisors of Plesiadapiformes. Annals of Carnegie Museum 62:351–361.

Rose, K. D., J. J. Eberle, and M. C. McKenna. 2004. *Arcticanodon dawsonae,* a primitive new palaeanodont from the lower Eocene of Ellesmere Island, Canadian High Arctic. Canadian Journal of Earth Sciences 41:757–763.

Rose, K. D., R. J. Emry, T. J. Gaudin, and G. Storch. 2005. Xenarthra and Pholidota; pp. 106–126 *in* K. D. Rose and J. D. Archibald (eds.), The Rise of Placental Mammals: Origins and Relationships of the Major Extant Clades. The Johns Hopkins University Press, Baltimore.

Rose, K. D., M. Godinot, and T. M. Bown. 1994. The early radiation of euprimates and the initial diversification of Omomyidae; pp. 1–28 *in* J. G. Fleagle and R. F. Kay (eds.), Anthropoid Origins. Plenum Press, New York.

Rose, K. D., L. Krishtalka, and R. K. Stucky. 1991. Revision of the Wind River faunas, early Eocene of central Wyoming. Part 11. Palaeanodonta (Mammalia). Annals of Carnegie Museum 60:63–82.

Rose, K. D., R. D. E. MacPhee, and J. P. Alexander. 1999. Cranium of early Eocene *Cantius abditus* (Primates: Adapiformes) and its phylogenetic implications, with a re-evaluation of *"Hesperolemur" actius.* American Journal of Physical Anthropology 109:523–539.

Rose, K. D., A. Walker, and L. L. Jacobs. 1981. Function of the mandibular tooth comb in living and extinct mammals. Nature 289:583–585.

Rosenberger, A. L., E. Strasser, and E. Delson. 1985. Anterior dentition of *Notharctus* and the adapid-anthropoid hypothesis. Folia Primatologica 44:15–39.

Ross, C., B. Williams, and R. F. Kay. 1998. Phylogenetic analysis of anthropoid relationships. Journal of Human Evolution 35:221–306.

Roth, V. L., and J. Shoshani. 1988. Dental identification and age determination in *Elephas maximus.* Journal of Zoology, London 214: 567–588.

Rougier, G. W. 1993. *Vincelestes neuquenianus* Bonaparte (Mammalia, Theria) un primitivo mamífero del Cretácico Inferior de la Cuenca Neuquina. Ph.D. dissertation, University of Buenos Aires, Buenos Aires.

Rougier, G. W., Q. Ji, and M. J. Novacek. 2003. A new symmetrodont mammal with fur impressions from the Mesozoic of China. Acta Geologica Sinica 77:7–14.

Rougier, G. W., M. J. Novacek, E. Ortiz-Jaureguizar, D. Pol, and P. Purerta. 2003. Reinterpretation of *Reigitherium bunodontum* as a Reigitheriidae dryolestoid and the interrelationships of the South American dryolestoids. Journal of Vertebrate Paleontology 23(supplement to no. 3):90A–91A.

Rougier, G. W., J. R. Wible, and J. A. Hopson. 1996a. Basicranial anatomy of *Priacodon fruitaensis* (Triconodontidae, Mammalia) from the Late Jurassic of Colorado, and a reappraisal of mammaliaform interrelationships. American Museum Novitates 3183:1–38.

Rougier, G. W., J. R. Wible, and M. J. Novacek. 1996b. Middle-ear ossicles of the multituberculate *Kryptobaatar* from the Mongolian Late Cretaceous: Implications for mammaliamorph relationships and the evolution of the auditory apparatus. American Museum Novitates 3187:1–43.

———. 1998. Implications of *Deltatheridium* specimens for early marsupial history. Nature 396:459–463.

Rowe, T. 1988. Definition, diagnosis and origin of Mammalia. Journal of Vertebrate Paleontology 8:241–264.

———. 1993. Phylogenetic systematics and the early history of mammals; pp. 129–145 *in* F. S. Szalay, M. J. Novacek, and M. C. McKenna (eds.), Mammal Phylogeny: Mesozoic Differentiation, Multituberculates, Monotremes, Early Therians, and Marsupials. Springer-Verlag, New York.

Runestad, J. A., and C. B. Ruff. 1995. Structural adaptations for gliding in mammals with implications for locomotor behavior in paromomyids. American Journal of Physical Anthropology 98:101–119.

Russell, D. A. 1960. A review of the Oligocene insectivore *Micropternodus borealis.* Journal of Paleontology 34: 940–949.

Russell, D. E. 1964. Les mammifères Paléocènes d'Europe. Mémoires du Muséum National d'Histoire Naturelle Sér. C, Sciences de la Terre 13:1–324.

———. 1980. Sur les condylarthres Cernaysiens *Tricuspiodon* et *Landenodon* (Paléocène supérieur de France). Palaeovertebrata, Mémoire Jubilaire en Hommage à René Lavocat: 127–166.

Russell, D. E., and P. D. Gingerich. 1981. Lipotyphla, Proteutheria(?), and Chiroptera (Mammalia) from the early-middle Eocene Kuldana Formation of Kohat (Pakistan). Contributions from the Museum of Paleontology, the University of Michigan 25:277–287.

Russell, D. E., and M. Godinot. 1988. The Paroxyclaenidae (Mammalia) and a new form from the early Eocene of Palette, France. Paläontologische Zeitschrift 62:319–331.

Russell, D. E., and B. Sigé. 1970. Révision des chiroptères lutétiens de Messel (Hesse, Allemagne). Palaeovertebrata 3:83–182.

Russell, D. E., and H. Tobien. 1986. Mammalian evidence concerning the Eocene-Oligocene transition in Europe, North America and Asia; pp. 299–307 *in* C. Pomerol and I. Premoli-Silva (eds.), Terminal Eocene Events. Elsevier Science, Amsterdam.

Russell, D. E., and R.-J. Zhai. 1987. The Paleogene of Asia: Mammals and stratigraphy. Mémoires du Muséum National d'Histoire Naturelle, Sér. C, Sciences de la Terre 52:1–488.

Russell, L. S. 1958. The dentition of rabbits and the origin of the Lagomorpha. National Museum of Canada Bulletin 166:41–45.

———. 1980. Tertiary mammals of Saskatchewan. Part V: The Oligocene entelodonts. Life Sciences Contributions Royal Ontario Museum 122:1–42.

Rzebik-Kowalska, B. 2003. Distribution of shrews (Insectivora, Mammalia) in time and space; pp. 499–508 *in* J. W. F. Reumer and W. Wessels (eds.), Distribution and Migration of Tertiary Mammals in Eurasia. A Volume in Honour of Hans de Bruijn. DEINSEA 10 (Annual of the Natural History Museum of Rotterdam).

Sagne, C., 2001. La diversification des siréniens à l'Eocène (Sirenia, Mammalia): Etude morphologique et analyse phylogénetique du sirénien de Taulanne, *Halitherium taulannense*. Ph.D. dissertation, Muséum National d'Histoire Naturelle, Paris.

Sahni, A. 1984. Cretaceous-Paleocene terrestrial faunas of India: Lack of endemism during drifting of the Indian Plate. Science 226:441–443.

———. 1985. Enamel structure of early mammals and its role in evaluating relationships among rodents; pp. 133–150 in W. P. Luckett and J.-L. Hartenberger (eds.), Evolutionary Relationships among Rodents. A Multidisciplinary Analysis. Plenum Press, New York.

Sahni, A., and S. Bajpai. 1991. Eurasiatic elements in the Upper Cretaceous nonmarine biotas of peninsular India. Cretaceous Research 12:177–183.

Samonds, K., I. Zalmout, and D. Krause. 2005. New sirenian fossils from the late Eocene of Madagascar. Journal of Vertebrate Paleontology 25(supplement to no. 3):108A.

Sánchez-Villagra, M. R., and R. F. Kay. 1997. A skull of *Proargyrolagus*, the oldest argyrolagid (late Oligocene Salla beds, Bolivia), with brief comments concerning its paleobiology. Journal of Vertebrate Paleontology 17:717–724.

Sánchez-Villagra, M. R., and B. A. Williams. 1998. Levels of homoplasy in the evolution of the mammalian skeleton. Journal of Mammalian Evolution 5:113–126.

Sánchez-Villagra, M. R., R. J. Burnham, D. C. Campbell, R. M. Feldmann, E. S. Gaffney, R. F. Kay, R. Lozsán, R. Purdy, and J. G. M. Thewissen. 2000. A new near-shore marine fauna and flora from the early Neogene of northwestern Venezuela. Journal of Paleontology 74:957–968.

Sanderson, M. J., and H. B. Shaffer. 2002. Troubleshooting molecular phylogenetic analyses. Annual Review of Ecology and Systematics 33:49–72.

Sanderson, M. J., A. Purvis, and C. Henze. 1998. Phylogenetic supertrees: Assembling the trees of life. Trends in Ecology and Evolution 13:105–109.

Sargis, E. J. 2002. The postcranial morphology of *Ptilocercus lowii* (Scandentia, Tupaiidae): An analysis of primatomorphan and volitantian characters. Journal of Mammalian Evolution 9:137–160.

Savage, D. E. 1971. The Sparnacian-Wasatchian mammalian fauna, early Eocene, of Europe and North America. Abhandlungen des Hessischen Landesamtes für Bodenforschung 60:154–158.

Savage, D. E., and D. E. Russell. 1983. Mammalian Paleofaunas of the World. Addison-Wesley, New York.

Savage, D. E., D. E. Russell, and P. Louis. 1965. European Eocene Equidae (Perissodactyla). University of California Publications in Geological Sciences 56:1–94.

Savage, R. J. G. 1969. Early Tertiary mammal locality in southern Libya. Proceedings of the Geological Society of London 1648:98–101.

———. 1973. *Megistotherium*, gigantic hyaenodont from Miocene of Gebel Zelten, Libya. Bulletin of the British Museum of Natural History (Geology) 22:483–511.

———. 1976. Review of early Sirenia. Systematic Zoology 25:344–351.

———. 1977. Evolution in carnivorous mammals. Palaeontology 20:237–271.

Savage, R. J. G., and M. R. Long. 1986. Mammal Evolution—An Illustrated Guide. Facts on File, New York.

Savage, R. J. G., D. P. Domning, and J. G. M. Thewissen. 1994. Fossil Sirenia of the west Atlantic and Caribbean region. V. The most primitive known sirenian, *Prorastomus sirenoides* Owen, 1855. Journal of Vertebrate Paleontology 14:427–449.

Scally, M., O. Madsen, C. J. Douady, W. W. de Jong, M. J. Stanhope, and M. S. Springer. 2001. Molecular evidence for the major clades of placental mammals. Journal of Mammalian Evolution 8:239–277.

Schaeffer, B. 1947. Notes on the origin and function of the artiodactyl tarsus. American Museum Novitates 1356:1–24.

Schmidt-Kittler, N. (ed.). 1987. International Symposium on Mammalian Biostratigraphy and Paleoecology of the European Paleogene, Mainz, February 18–21, 1987. Münchner Geowissenschaftliche Abhandlungen A 10:1–312.

Schoch, R. M. 1985. Preliminary description of a new late Paleocene land-mammal fauna from South Carolina, U.S.A. Postilla 196:1–13.

———. 1986. Systematics, functional morphology and macroevolution of the extinct mammalian order Taeniodonta. Bulletin of the Peabody Museum of Natural History, Yale University 42:1–307.

———. 1989. A review of the tapiroids; pp. 298–320 in D. R. Prothero and R. M. Schoch (eds.), The Evolution of Perissodactyls. Oxford University Press, New York.

Schoch, R. M., and S. G. Lucas. 1985. The phylogeny and classification of the Dinocerata (Mammalia, Eutheria). Bulletin of the Geological Institutions of the University of Uppsala, new series 11:31–58.

Schultz, C. B., and C. H. Falkenbach. 1968. The phylogeny of the oreodonts. Parts 1 and 2. Bulletin of the American Museum of Natural History 139:1–498.

Schwartz, G. T., D. T. Rasmussen, and R. J. Smith. 1995. Body-size diversity and community structure of fossil hyracoids. Journal of Mammalogy 76:1088–1099.

Schwartz, J. H., and L. Krishtalka. 1976. The lower antemolar teeth of *Litolestes ignotus*, a late Paleocene erinaceid (Mammalia, Insectivora). Annals of Carnegie Museum 46:1–6.

Schwartz, J. H., and I. Tattersall. 1987. Tarsiers, adapids and the integrity of Strepsirhini. Journal of Human Evolution 16:23–40.

Scillato-Yané, G. 1976. Sobre un Dasypodidae (Mammalia, Xenarthra) de Edad Riochiquense (Paleoceno Superior) de Itaboraí, Brasil. Anais da Academia Brasileira de Ciências 48:527–530.

Scott, K. M., and C. M. Janis. 1993. Relationships of the Ruminantia (Artiodactyla) and an analysis of the characters used in ruminant taxonomy; pp. 282–302 in F. S. Szalay, M. J. Novacek, and M. C. McKenna (eds.), Mammal Phylogeny: Placentals. Springer-Verlag, New York.

Scott, W. B. 1888. On some new and little known creodonts. Journal of the Academy of Natural Sciences, Philadelphia 9:155–185.

———. 1894. The structure and relationships of *Ancodus*. Journal of the Academy of Natural Sciences, Philadelphia 9:461–497.

———. 1903–1904. Mammalia of the Santa Cruz Beds. Part I. Edentata. Reports of the Princeton University Expeditions to Patagonia, 1896–1899 V:1–364.

———. 1928. Part IV. Astrapotheria of the Santa Cruz Beds. Reports of the Princeton University Expeditions to Patagonia, 1896–1899 VI:301–342.

———. 1937. A History of Land Mammals in the Western Hemisphere. Second edition. Macmillan, New York.

———. 1938. A problematical cat-like mandible from the Uinta Eocene, *Apataelurus kayi*, Scott. Annals of the Carnegie Museum 27:113–120.

———. 1940. The mammalian fauna of the White River Oligocene. Part IV. Artiodactyla. Transactions of the American Philosophical Society, new series 28:363–746.

Scott, W. B., and G. L. Jepsen. 1936. The mammalian fauna of the White River Oligocene. Part I. Insectivora and Carnivora. Transactions of the American Philosophical Society, new series 28:1–153.

Seiffert, E. R., and E. L. Simons. 2000. *Widanelfarasia,* a diminutive placental from the late Eocene of Egypt. Proceedings of the National Academy of Sciences, USA 97:2646–2651.

Seiffert, E. R., E. L. Simons, and Y. Attia. 2003. Fossil evidence for an ancient divergence of lorises and galagos. Nature 422:421–424.

Seiffert, E. R., E. L. Simons, and C. V. M. Simons. 2004. Phylogenetic, biogeographic, and adaptive implications of new fossil evidence bearing on crown anthropoid origins and early stem catarrhine evolution; pp. 157–181 *in* C. F. Ross and R. F. Kay (eds.), Anthropoid Origins: New Visions. Kluwer Academic/Plenum Press, New York.

Seiffert, E. R., E. L. Simons, T. M. Ryan, and Y. Attia. 2005. Additional remains of *Wadilemur elegans*, a primitive stem galagid from the late Eocene of Egypt. Proceedings of the National Academy of Sciences, USA 102:11396–11401.

Sen, S., and E. Heintz. 1979. *Palaeoamasia kansui* Ozansoy 1966, embrithopode (Mammalia) de l'Eocène d'Anatolie. Annales de Paléontologie (Vertébrés) 65:73–91.

Sereno, P. C. 1982. An early Eocene sirenian from Patagonia (Mammalia, Sirenia). American Museum Novitates 2729:1–10.

Sereno, P. C., and M. C. McKenna. 1995. Cretaceous multituberculate skeleton and the early evolution of the mammalian shoulder girdle. Nature 377:144–147.

Shevyreva, N. S. 1989. New rodents (Ctenodactyloidea, Rodentia, Mammalia) from the lower Eocene of Mongolia. Paleontologicheskiy Zhurnal 1989:60–72.

———. 1995. The oldest lagomorphs (Lagomorpha, Mammalia) of the Eastern Hemisphere. Doklady Akademii Nauk 345:377–379.

Shigehara, N., M. Takai, R. F. Kay, A. K. Aung, A. N. Soe, S. T. Tun, T. Tsubamato, and T. Thein. 2002. The upper dentition and face of *Pondaungia cotteri* from central Myanmar. Journal of Human Evolution 43:143–166.

Shimamura, M., H. Yasue, K. Ohshima, H. Abe, H. Kato, T. Kishiro, M. Goto, I. Munechika, and N. Okada. 1997. Molecular evidence from retroposons that whales form a clade within even-toed ungulates. Nature 388:666–670.

Shockey, B. J. 1997. Two new notoungulates (family Notohippidae) from the Salla Beds of Bolivia (Deseadan: late Oligocene): Systematics and functional morphology. Journal of Vertebrate Paleontology 17:584–599.

Shockey, B. J., and F. Anaya Daza. 2004. *Pyrotherium macfaddeni,* sp. nov. (late Oligocene, Bolivia) and the pedal morphology of pyrotheres. Journal of Vertebrate Paleontology 24:481–488.

Shoshani, J. 1986. Mammalian phylogeny: Comparison of morphological and molecular results. Molecular Biology and Evolution 3:222–242.

———. 1993. Hyracoidea-Tethytheria affinity based on myological data; pp. 235–256 *in* F. S. Szalay, M. J. Novacek, and M. C. McKenna (eds.), Mammal Phylogeny: Placentals. Springer-Verlag, New York.

Shoshani, J., and M. C. McKenna. 1998. Higher taxonomic relationships among extant mammals based on morphology, with selected comparisons of results from molecular data. Molecular Phylogenetics and Evolution 9:572–584.

Shoshani, J., C. P. Groves, E. L. Simons, and G. F. Gunnell. 1996. Primate phylogeny: Morphological vs. molecular results. Molecular Phylogenetics and Evolution 5:102–154.

Shoshani, J., M. C. McKenna, K. D. Rose, and R. J. Emry. 1997. *Eurotamandua* is a pholidotan, not a xenarthran. Journal of Vertebrate Paleontology 17(supplement to no. 3):76A.

Shoshani, J., R. M. West, N. Court, R. J. G. Savage, and J. M. Harris. 1996. The earliest proboscideans: General plan, taxonomy, and palaeoecology; pp. 57–75 *in* J. Shoshani and P. Tassy (eds.), The Proboscidea: Evolution and Palaeoecology of Elephants and Their Relatives. Oxford University Press, Oxford, UK.

Shubin, N. H., A. W. Crompton, H.-D. Sues, and P. E. Olsen. 1991. New fossil evidence on the sister-group of mammals and early Mesozoic faunal distributions. Science 251:1063–1065.

Sigé, B. 1974. Données nouvelles sur le genre *Stehlinia* (vespertilionoidea, Chiroptera) du Paléogène d'Europe. Palaeovertebrata 6:253–272.

———. 1976. Insectivores primitifs de l'Eocène supérieur et Oligocène inférieur d'Europe occidentale. Nyctithériidés. Mémoires du Muséum National d'Histoire Naturelle, Sér. C, Sciences de la Terre 34:1–140.

———. 1988. Le gisement du Bretou (phosphorites du Quercy, Tarn-et-Garonne, France) et sa faune de vertébrés de l'Eocène supérieur. IV. Insectivores et chiroptères. Palaeontographica Abteilung A 205:69–102.

———. 1991. Rhinolophoidea et Vespertilionoidea (Chiroptera) du Chambi (Eocène inférieur de Tunisie). Aspects biostratigraphiques, biogéographiques et paléoécologiques de l'origine des chiroptères modernes. Neues Jahrbuch für Geologie und Paläontologie Abhandlungen 182:355–376.

Sigé, B., and B. Marandat. 1997. Apport à la faune du Paléocène inférieur d'Europe: Un plésiadapiforme du Montien de Hainin (Belgique). Mémoires et Travaux de l'Ecole Pratique des Hautes Etudes, Institut de Montpellier 21:679–686.

Sigé, B., and D. E. Russell. 1980. Compléments sur les chiroptères de l'Eocène moyen d'Europe. Les genres *Palaeochiropteryx* et *Cecilionycteris*. Palaeovertebrata Mémoire Jubilaire en Hommage à René Lavocat:91–126.

Sigé, B., J.-Y. Crochet, and A. Insole. 1977. Les plus vieilles taupes. Géobios Mémoire Spécial 1:141–157.

Sigé, B., J.-J. Jaeger, J. Sudre, and M. Vianey-Liaud. 1990. *Altiatlasius koulchii* n. gen. et sp., primate omomyidé du Paléocène supérieur du Maroc, et les origines des Euprimates. Palaeontographica Abteilung A 214:31–56.

Sigé, B., T. Sempere, R. F. Butler, L. G. Marshall, and J.-Y. Crochet. 2004. Age and stratigraphic reassessment of the fossil-bearing Laguna Umayo red mudstone unit, SE Peru, from regional stratigraphy, fossil record, and paleomagnetism. Géobios 37:771–794.

Sigogneau-Russell, D. 1989. Haramyidae (Mammalia, Allotheria) en Provenance du Trias Supérieur de Lorraine (France). Palaeontographica Abteilung A 206:137–198.

———. 1991. Découverte du premier mammifère tribosphénique du Mesozoique africain. Comptes Rendus de l'Académie des Sciences Paris, sér. II 313:1635–1640.

———. 1998. Discovery of a Late Jurassic Chinese mammal in the Upper Bathonian of England. Comptes Rendus de l'Académie des Sciences Paris, Sciences de la Terre et des Planètes 327:571–576.

———. 1999. Réévaluation des Peramura (Mammalia, Cladotheria) sur la base de nouveaux spécimens du Crétacé inférieur d'Angleterre et du Maroc. Geodiversitas 21:93–127.

———. 2003a. Holotherian mammals from the Forest Marble (Middle Jurassic of England). Geodiversitas 25:501–537.

———. 2003b. Diversity of triconodont mammals from the Early Cretaceous of North Africa—Affinities of the amphilestids. Palaeovertebrata 32:27–55.

Sigogneau-Russell, D., and P. Ensom. 1998. *Thereuodon* (Theria, Symmetrodonta) from the Lower Cretaceous of North Africa and Europe, and a brief review of symmetrodonts. Cretaceous Research 19:445–470.

Sigogneau-Russell, D., and R. Hahn. 1995. Reassessment of the Late Triassic symmetrodont mammal *Woutersia*. Acta Palaeontologica Polonica 40:245–260.

Sigogneau-Russell, D., R. M. Frank, and J. Hemmerlé. 1986. A new family of mammals from the lower part of the French Rhaetic; pp. 99–108 *in* K. Padian (ed.), The Beginning of the Age of Dinosaurs. Cambridge University Press, Cambridge, UK.

Sigogneau-Russell, D., J. J. Hooker, and P. C. Ensom. 2001. The oldest tribosphenic mammal from Laurasia (Purbeck Limestone Group, Berriasian, Cretaceous, UK) and its bearing on the "dual origin" of Tribosphenida. Comptes Rendus de l'Académie des Sciences Paris, Sciences de la Terre et des Planètes 333:141–147.

Silcox, M. T. 2001. A phylogenetic analysis of Plesiadapiformes and their relationship to Euprimates and other archontans. Ph.D. dissertation, The Johns Hopkins University, Baltimore.

———. 2003. New discoveries on the middle ear anatomy of *Ignacius graybullianus* (Paromomyidae, Primates) from ultra high resolution X-ray computed tomography. Journal of Human Evolution 44:73–86.

Silcox, M. T., and K. D. Rose. 2001. Unusual vertebrate microfaunas from the Willwood Formation, early Eocene of the Bighorn Basin, Wyoming; pp. 131–164 *in* G. F. Gunnell (ed.), Eocene Biodiversity: Unusual Occurrences and Rarely Sampled Habitats. Kluwer Academic/Plenum Press, New York.

Silcox, M. T., J. I. Bloch, E. J. Sargis, and D. M. Boyer. 2005. Euarchonta (Dermoptera, Scandentia, Primates); pp. 127–144 *in* K. D. Rose, and J. D. Archibald (eds.), The Rise of Placental Mammals: Origins and Relationships of the Major Extant Clades. The Johns Hopkins University Press, Baltimore.

Silcox, M. T., D. W. Krause, M. C. Maas, and R. C. Fox. 2001. New specimens of *Elphidotarsius russelli* (Mammalia, ?Primates, Carpolestidae) and a revision of plesiadapoid relationships. Journal of Vertebrate Paleontology 21:132–152.

Simmons, N. B. 1993. Phylogeny of Multituberculata; pp. 146–164 *in* F. S. Szalay, M. J. Novacek, and M. C. McKenna (eds.), Mammal Phylogeny: Mesozoic Differentiation, Multituberculates, Monotremes, Early Therians, and Marsupials. Springer-Verlag, New York.

———. 1994. The case for chiropteran monophyly. American Museum Novitates 3103:1–54.

———. 1995. Bat relationships and the origin of flight. Symposium of the Zoological Society of London 67:27–43.

———. 2005. Chiroptera; pp. 159–174 *in* K. D. Rose and J. D. Archibald (eds.), The Rise of Placental Mammals: Origins and Relationships of the Major Extant Clades. The Johns Hopkins University Press, Baltimore.

Simmons, N. B., and J. H. Geisler. 1998. Phylogenetic relationships of *Icaronycteris*, *Archaeonycteris*, *Hassianycteris*, and *Palaeochiropteryx* to extant bat lineages, with comments on the evolution of echolocation and foraging strategies in Microchiroptera. Bulletin of the American Museum of Natural History 235:1–182.

Simons, E. L. 1960. The Paleocene Pantodonta. Transactions of the American Philosophical Society 50:1–99.

———. 1962. A new Eocene primate genus, *Cantius*, and a revision of some allied European lemuroids. Bulletin of the British Museum of Natural History (Geology) 7:1–36.

———. 1972. Primate Evolution. Macmillan, New York.

———. 1992. Diversity in the early Tertiary anthropoidean radiation in Africa. Proceedings of the National Academy of Sciences, USA 89:10743–10747.

———. 1995a. Crania of *Apidium*: Primitive anthropoidean (Primates, Parapithecidae) from the Egyptian Oligocene. American Museum Novitates 3124:1–10.

———. 1995b. Skulls and anterior teeth of *Catopithecus* (Primates: Anthropoidea) from the Eocene and anthropoid origins. Science 268:1885–1888.

———. 1995c. Egyptian Oligocene Primates: A review. Yearbook of Physical Anthropology 38:199–238.

———. 1997a. Discovery of the smallest Fayum Egyptian primates (Anchomomyini, Adapidae). Proceedings of the National Academy of Sciences, USA 94:180–184.

———. 1997b. Preliminary description of the cranium of *Proteopithecus sylviae,* an Egyptian late Eocene anthropoidean primate. Proceedings of the National Academy of Sciences, USA 94:14970–14975.

———. 2003. The fossil record of tarsier evolution; pp. 9–34 *in* P. C. Wright, E. L. Simons, and S. Gursky (eds.), Tarsiers. Past, Present, and Future. Rutgers University Press, New Brunswick, N. J.

Simons, E. L., and T. M. Bown. 1985. *Afrotarsius chatrathi,* first tarsiiform primate (?Tarsiidae) from Africa. Nature 313:475–477.

———. 1995. Ptolemaiida, a new order of Mammalia—With description of the cranium of *Ptolemaia grangeri*. Proceedings of the National Academy of Sciences, USA 92:3269–3273.

Simons, E. L., and D. T. Rasmussen. 1994. A remarkable cranium of *Plesiopithecus teras* (Primates, Prosimii) from the Eocene of Egypt. Proceedings of the National Academy of Sciences, USA 91:9946–9950.

———. 1996. Skull of *Catopithecus browni,* an early Tertiary catarrhine. American Journal of Physical Anthropology 100:261–292.

Simons, E. L., and D. E. Russell. 1960. Notes on the cranial anatomy of *Necrolemur*. Breviora 127:1–14.

Simons, E. L., and E. R. Seiffert. 1999. A partial skeleton of *Proteopithecus sylviae* (Primates, Anthropoidea): First associated dental and postcranial remains of an Eocene anthropoidean. Comptes Rendus de l'Académie des Sciences Paris, Sciences de la Terre et des Planètes 329:921–927.

Simons, E. L., P. A. Holroyd, and T. M. Bown. 1991. Early Tertiary elephant-shrews from Egypt and the origin of the Macroscelidea. Proceedings of the National Academy of Sciences, USA 88:9734–9737.

Simons, E. L., E. R. Seiffert, P. S. Chatrath, and Y. Attia. 2001. Earliest record of a parapithecid anthropoid from the Jebel

Qatrani Formation, northern Egypt. Folia Primatologica 72:316–331.

Simpson, G. G. 1928. A Catalogue of the Mesozoic Mammalia in the Geological Department of the British Museum. Oxford University Press, London.

———. 1931a. *Metacheiromys* and the Edentata. Bulletin of the American Museum of Natural History 59:295–381.

———. 1931b. A new insectivore from the Oligocene, Ulan Gochu Horizon, of Mongolia. American Museum Novitates 505:1–22.

———. 1932. A new Paleocene mammal from a deep well in Louisiana. Proceedings of the U. S. National Museum 82(2):1–4.

———. 1933. The "plagiaulacoid" type of mammalian dentition. Journal of Mammalogy 14:97–107.

———. 1936. A new fauna from the Fort Union of Montana. American Museum Novitates 873:1–27.

———. 1937a. The Fort Union of the Crazy Mountain Field, Montana, and its mammalian faunas. Bulletin of the U. S. National Museum 169:1–287.

———. 1937b. Notes on the Clark Fork, Upper Paleocene, fauna. American Museum Novitates 954:1–24.

———. 1937c. The beginning of the Age of Mammals. Biological Reviews 12:1–47.

———. 1945. The principles of classification and a classification of mammals. Bulletin of the American Museum of Natural History 85:1–350.

———. 1947. Holarctic mammalian faunas and continental relationships during the Cenozoic. Bulletin of the Geological Society of America 58:613–688.

———. 1948. The beginning of the Age of Mammals in South America. Part 1. Introduction. Systematics: Marsupialia, Edentata, Condylarthra, Litopterna and Notioprogonia. Bulletin of the American Museum of Natural History 91:1–232.

———. 1953. Evolution and Geography. Oregon State System of Higher Education, Eugene.

———. 1955. The Phenacolemuridae, new family of early primates. Bulletin of the American Museum of Natural History 105:411–441.

———. 1961. Principles of Animal Taxonomy. Columbia University Press, New York.

———. 1965. Attending Marvels. A Patagonian Journal. Time Reading Program Special Edition, New York. (Reprint of 1934 edition published by Macmillan).

———. 1967. The beginning of the Age of Mammals in South America. Part 2. Systematics: Notoungulata, concluded (Typotheria, Hegetotheria, Toxodonta, Notoungulata incertae sedis); Astrapotheria; Trigonostylopoidea; Pyrotheria; Xenungulata; Mammalia incertae sedis. Bulletin of the American Museum of Natural History 137:1–260.

———. 1970a. The Argyrolagidae, extinct South American marsupials. Bulletin of the Museum of Comparative Zoology, Harvard 139:1–86.

———. 1970b. Mammals from the early Cenozoic of Chubut, Argentina. Breviora 360:1–13.

———. 1970c. Additions to knowledge of Groeberia (Mammalia, Marsupialia) from the mid-Cenozoic of Argentina. Breviora 362:1–17.

———. 1980. Splendid Isolation. Yale University Press, New Haven, Conn.

———. 1984. Discoverers of the Lost World. An Account of Some of Those Who Brought Back to Life South American Mammals Long Buried in the Abyss of Time. Yale University Press, New Haven, Conn.

Simpson, G. G., J. L. Minoprio, and B. Patterson. 1962. The mammalian fauna of the Divisadero Largo Formation, Mendoza, Argentina. Bulletin of the Museum of Comparative Zoology 127:239–293.

Sinclair, W. J. 1914. A revision of the bunodont Artiodactyla of the middle and lower Eocene of North America. Bulletin of the American Museum of Natural History 33:267–295.

Slaughter, B. H. 1971. Mid-Cretaceous (Albian) therians of the Butler Farm local fauna, Texas. Zoological Journal of the Linnean Society 50:131–143.

Sloan, L. C., and E. Thomas. 1998. Global climate of the late Paleocene epoch: Modeling the circumstances associated with a climatic "event"; pp. 138–157 in M.-P. Aubry, S. G. Lucas, and W. A. Berggren (eds.), Late Paleocene–Early Eocene Climatic and Biotic Events in the Marine and Terrestrial Records. Columbia University Press, New York.

Smith, A. B., and K. J. Peterson. 2002. Dating the time of origin of major clades: Molecular clocks and the fossil record. Annual Review of Earth and Planetary Sciences 30:65–88.

Smith, A. G., D. G. Smith, and B. M. Funnell. 1994. Atlas of Mesozoic and Cenozoic Coastlines. Cambridge University Press, Cambridge, UK.

Smith, J. D., and G. Storch. 1981. New middle Eocene bats from "Grube Messel" near Darmstadt, W-Germany (Mammalia: Chiroptera). Senckenbergiana Biologica 61:153–167.

Smith, M. E., B. S. Singer, and A. R. Carroll. 2003. $^{40}Ar/^{39}Ar$ geochronology of the Eocene Green River Formation, Wyoming. Geological Society of America Bulletin 115:549–565.

———. 2004. Discussion and reply: $^{40}Ar/^{39}Ar$ geochronology of the Eocene Green River Formation, Wyoming. Reply. Geological Society of America Bulletin 116:253–256.

Smith, T. 2000. Mammals from the Paleocene-Eocene transition in Belgium (Tienen Formation, MP7): Palaeobiogeographical and biostratigraphical implications. GFF 122:148–149.

Smith, T., and R. Smith. 2001. The creodonts (Mammalia, Ferae) from the Paleocene-Eocene transition in Belgium (Tienen Formation, MP7). Belgian Journal of Zoology 131:117–135.

Smith, T., K. D. Rose, and P. D. Gingerich. In press. Rapid Asia–Europe–North America geographic dispersal of earliest Eocene primate *Teilhardina* during the Paleocene-Eocene thermal maximum. Proceedings of the National Academy of Sciences, USA.

Smith, T., J. Van Itterbeeck, and P. Missiaen. 2004. Oldest plesiadapiform (Mammalia, Proprimates) from Asia and its paleobiogeographical implications for faunal interchange with North America. Comptes Rendus Palevol 3:43–52.

Soria, M. F. 1982. *Tetragonostylops apthomasi* (Price y Paula Couto, 1950): Su asignacion a Astrapotheriidae (Mammalia: Astrapotheria). Ameghiniana 19:234–238.

———. 1987. Estudios sobre los Astrapotheria (Mammalia) del Paleoceno y Eoceno. Parte I: Descripcion de *Eoastrapostylops riolorense* Soria y Powell, 1982. Ameghiniana 24:21–34.

———. 1989a. Notopterna: Un nuevo orden de Mamiferos ungulados Eogenos de America del Sur. Parte I. Los Amilnedwardsidae. Ameghiniana 25:245–258.

———. 1989b. Notopterna: Un nuevo orden de Mamiferos ungulados Eogenos de America del Sur. Parte II. *Notonychops powelli* gen. et sp. nov. (Notonychopidae nov.) de la Formacion Rio Loro (Paleoceno Medio), Provincia de Tucaman, Argentina. Ameghiniana 25:259–272.

———. 2001. Los Proterotheriidae (Mammalia, Litopterna): Sistemática, origen y filogenia. Monografias del Museo Argentino de Ciencias Naturales 1:1–167.

Soria, M. F., and M. Bond. 1984. Adiciones al conocimiento de *Trigonostylops* Ameghino, 1897 (Mammalia, Astrapotheria, Trigonostylopidae). Ameghiniana 21:43–51.

Soria, M. F., and J. E. Powell. 1981. Un primitivo Astrapotheria (Mammalia) y la edad de la Formacion Rio Loro, Provincia de Tucuman, Republica Argentina. Ameghiniana 18:155–168.

Springer, M. S. 1997. Molecular clocks and the timing of the placental and marsupial radiations in relation to the Cretaceous-Tertiary boundary. Journal of Mammalian Evolution 4:285–302.

Springer, M. S., and W. W. de Jong. 2001. Which mammalian supertree to bark up? Science 291:1709–1711.

Springer, M. S., and J. A. W. Kirsch. 1993. A molecular perspective on the phylogeny of placental mammals based on mitochondrial 12S rDNA sequences, with special reference to the problem of the Paenungulata. Journal of Mammalian Evolution 1:149–166.

Springer, M. S., W. J. Murphy, E. Eizirik, and S. J. O'Brien. 2003. Placental mammal diversification and the Cretaceous-Tertiary boundary. Proceedings of the National Academy of Sciences, USA 100:1056–1061.

———. 2005. Molecular evidence for major placental clades; pp. 37–49 in K. D. Rose, and J. D. Archibald (eds.), The Rise of Placental Mammals: Origins and Relationships of the Major Extant Clades. The Johns Hopkins University Press, Baltimore.

Stafford, B. J., and R. W. Thorington, Jr. 1998. Carpal development and morphology in archontan mammals. Journal of Morphology 235:135–155.

Stanhope, M. J., M. R. Smith, V. G. Waddell, C. A. Porter, M. S. Shivji, and M. Goodman. 1996. Mammalian evolution and the interphotoreceptor retinoid binding protein (IRBP) gene: Convincing evidence for several superordinal clades. Journal of Molecular Evolution 43:83–92.

Stanhope, M. J., V. G. Waddell, O. Madsen, W. de Jong, S. B. Hedges, G. C. Cleven, D. Kao, and M. S. Springer. 1998. Molecular evidence for multiple origins of Insectivora and for a new order of endemic African insectivore mammals. Proceedings of the National Academy of Sciences, USA 95:9967–9972.

Stanley, S. M. 1974. Relative growth of the titanothere horn: A new approach to an old problem. Evolution 28:447–457.

Stefen, C. 1997. Differentiations in Hunter-Schreger bands of carnivores; pp. 123–136 in W. von Koenigswald and P. M. Sander (eds.), Tooth Enamel Microstructure. A. A. Balkema, Rotterdam.

———. 1999. Evolution of enamel microstructure of archaic ungulates ("Condylarthra") and comments on some other early Tertiary mammals. PaleoBios 19:15–36.

Stehlin, H. G. 1909. Remarques sur les faunules des mammifères des couches Éocènes et Oligocènes de Bassin de Paris. Bulletin de la Société Géologique de France 9:488–520.

———. 1912. Die Säugetiere des schweizerischen Eocaens. Critischer Catalog der Materialien. Siebenter Teil, erste Hälfte: *Adapis*. Abhandlungen der Schweizerischen Paläontologischen Gesellschaft 38:1165–1298.

Steurbaut, E., R. Magioncalda, C. Dupuis, S. Van Simaeys, E. Roche, and M. Roche. 2003. Palynology, paleoenvironments, and organic carbon isotope evolution in lagoonal Paleocene-Eocene boundary settings in North Belgium; pp. 291–317 in S. L. Wing, P. D. Gingerich, B. Schmitz, and E. Thomas (eds.), Causes and Consequences of Globally Warm Climates in the Early Paleogene. Geological Society of America Special Paper 369, Boulder.

Stevens, M. S., and J. B. Stevens. 1996. Merycoidodontinae and Miniochoerinae; pp. 498–573 in D. R. Prothero and R. J. Emry (eds.), The Terrestrial Eocene-Oligocene Transition in North America. Cambridge University Press, Cambridge, UK.

Stevens, N., P. O'Connor, M. Gottfried, E. Roberts, S. Ngasala, and S. Kapilima. 2005. New Paleogene mammals and other vertebrates from the Rukwa Rift Basin, southwestern Tanzania. Journal of Vertebrate Paleontology 25(supplement to no. 3):118A.

Stirton, R. A., and J. M. Rensberger. 1964. Occurrence of the insectivore genus *Micropternodus* in the John Day Formation of central Oregon. Bulletin of the Southern California Academy of Sciences 63:57–80.

Storch, G. 1978. *Eomanis waldi*, ein Schuppentier aus dem Mittel-Eozän der "Grube Messel" bei Darmstadt (Mammalia: Pholidota). Senckenbergiana Lethaea 59:503–529.

———. 1981. *Eurotamandua joresi*, ein Myrmecophagide aus dem Eozän der "Grube Messel" bei Darmstadt (Mammalia, Xenarthra). Senckenbergiana Lethaea 61:247–289.

———. 1993a. Morphologie und Paläobiologie von *Macrocranion tenerum*, einem Erinaceomorphen aus dem Mittel-Eozän von Messel bei Darmstadt (Mammalia, Lipotyphla). Senckenbergiana Lethaea 73:61–81.

———. 1993b. "Grube Messel" and African–South American faunal connections; pp. 76–86 in W. George and R. Lavocat (eds.), The Africa–South America Connection. Clarendon Press, Oxford.

———. 1996. Paleobiology of Messel erinaceomorphs. Palaeovertebrata, Volume jubilaire D. E. Russell 25:215–224.

———. 2003. Fossil Old World "edentates." Senckenbergiana Biologica 83:51–60.

Storch, G., and A. M. Lister. 1985. *Leptictidium nasutum*, ein Pseudorhyncocyonide aus dem Eozän der "Grube Messel" bei Darmstadt (Mammalia, Proteutheria). Senckenbergiana Lethaea 66:1–37.

Storch, G., and G. Richter. 1992. Pangolins: Almost unchanged for 50 million years; pp. 203–207 in S. Schaal and W. Ziegler (eds.), Messel. An Insight into the History of Life and of the Earth. Clarendon Press, Oxford.

Storch, G., and M. Rummel. 1999. *Molaetherium heissigi* n. gen., n. sp., an unusual mammal from the early Oligocene of Germany (Mammalia: Palaeanodonta). Paläontologische Zeitschrift 73:179–185.

Storch, G., B. Sigé, and J. Habersetzer. 2002. *Tachypteron franzeni* n. gen., n. sp., earliest emballonurid bat from the middle Eocene of Messel (Mammalia, Chiroptera). Paläontologische Zeitschrift 76:189–199.

Strait, S. G. 2001. Dietary reconstruction of small-bodied omomyoid primates. Journal of Vertebrate Paleontology 21:322–334.

Strömberg, C. A. E. 2005. Decoupled taxonomic radiation and ecological expansion of open-habitat grasses in the Cenozoic of North America. Proceedings of the National Academy of Sciences, USA 102:11980–11984.

Stucky, R. K. 1998. Eocene bunodont and bunoselenodont Artiodactyla ("dichobunids"); pp. 358–374 in C. M. Janis, K. M. Scott, and L. L. Jacobs (eds.), Evolution of Tertiary Mammals of North America. Volume 1: Terrestrial Carnivores, Ungulates, and Ungulatelike Mammals. Cambridge University Press, Cambridge, UK.

Sudre, J. 1972. Révision des artiodactyles de l'Eocène moyen de Lissieu (Rhône). Palaeovertebrata 5:111–156.

———. 1977. L'évolution du genre *Robiacina* Sudre 1969, et l'origine des Cainotheriidae; Implications systématiques. Géobios Mémoire Spécial 1:213–231.

———. 1978a. La poche à phosphate de Ste-Neboule (Lot) et sa faune de vertébrés du Ludien supérieur. 9.—Primates et artiodactyles. Palaeovertebrata 8:269–290.

———. 1978b. Les artiodactyles de l'Éocène moyen et supérieur d'Europe occidentale (systématique et évolution). Mémoires et Travaux de l'Institut de Montpellier 7:1–229.

———. 1979. Nouveaux mammifères Éocènes du Sahara occidental. Palaeovertebrata 9:83–115.

———. 1983. Interprétation de la denture et description des éléments du squelette appendiculaire de l'espèce *Diplobune minor* (Filhol 1877): Apports à la connaissance de l'anatomie des Anoplotheriinae Bonaparte 1850; pp. 439–458 in E. Buffetaut, J. M. Mazin, and E. Salmon (eds.), Actes du Symposium Paléontologique G. Cuvier, Montbeliard, France.

———. 1984. *Cryptomeryx* Schlosser, 1886, tragulidé de l'Oligocène d'Europe: Relations du genre et considérations sur l'origine des ruminants. Palaeovertebrata 14:1–31.

———. 1988. Apport à la connaissance de *Dichobune robertiana* Gervais, 1848–1852 (Mammalia, Artiodactyla) du Lutétien: Considérations sur l'évolution des dichobunidés. Courier Forschungsinstitut Senckenberg 107:409–418.

Sudre, J., and G. Lecomte. 2000. Relations et position systématique du genre *Cuisitherium* Sudre et al., 1983, le plus dérivé des artiodactyles de l'Eocène inférieur d'Europe. Geodiversitas 22:415–432.

Sudre, J., and D. E. Russell. 1982. Les mammifères Montiens de Hainin (Paléocène moyen de Belgique). Part II. Les condylarthres. Palaeovertebrata 12:173–184.

Sues, H.-D. 1983. Advanced mammal-like reptiles from the Early Jurassic of Arizona. Ph.D. dissertation, Harvard University, Cambridge, Mass.

———. 1985. The relationships of the Tritylodontidae (Synapsida). Zoological Journal of the Linnean Society 85:205–217.

———. 1986. The skull and dentition of two tritylodontid synapsids from the Lower Jurassic of western North America. Bulletin of the Museum of Comparative Zoology 151:217–268.

———. 2001. On *Microconodon,* a Late Triassic cynodont from the Newark Supergroup of eastern North America. Bulletin of the Museum of Comparative Zoology 156:37–48.

Sulimski, A. 1968. Paleocene genus *Pseudictops* Matthew, Granger & Simpson, 1929 (Mammalia) and its revision. Palaeontologia Polonica 19:101–129.

Svensen, H., S. Planke, A. Malthe-Sørenssen, B. Jamtveit, R. Myklebust, T. R. Eldem, and S. S. Rey. 2004. Release of methane from a volcanic basin as a mechanism for initial Eocene global warming. Nature 429:542–545.

Swisher, C. C., III, J. M. Grajales-Nishimura, A. Montanari, S. V. Margolis, P. Claeyes, W. Alvarez, P. Renne, E. Cedillo-Pardo, F. Muarassee, G. H. Curtis, J. Smit, and M. O. McWilliams. 1992. Coeval ^{40}Ar/^{39}Ar ages of 65.0 million years ago from Chicxulub Crater melt rock and Cretaceous-Tertiary boundary tektites. Science 257:954–958.

Swisher, C. C., III, L. Dingus, and R. F. Butler. 1993. ^{40}Ar/^{39}Ar dating and magnetostratigraphic correlation of the terrestrial Cretaceous-Paleogene boundary and Puercan mammal age. Canadian Journal of Earth Sciences 30:1981–1996.

Sych, L. 1971. Mixodontia, a new order of mammals from the Paleocene of Mongolia. Palaeontologia Polonica 25:147–158.

Szalay, F. S. 1968. The Picrodontidae, a family of early primates. American Museum Novitates 2329:1–55.

———. 1969a. Mixodectidae, Microsyopidae, and the insectivore-primate transition. Bulletin of the American Museum of Natural History 140:193–330.

———. 1969b. The Hapalodectinae and a phylogeny of the Mesonychidae (Mammalia, Condylarthra). American Museum Novitates 2361:1–26.

———. 1971. The European adapid primates *Agerina* and *Pronycticebus.* American Museum Novitates 2466:1–19.

———. 1974. New genera of European adapid primates. Folia Primatologica 22:116–133.

———. 1976. Systematics of the Omomyidae (Tarsiiformes, Primates), taxonomy, phylogeny, and adaptations. Bulletin of the American Museum of Natural History 156:157–450.

———. 1977. Phylogenetic relationships and a classification of the eutherian Mammalia; pp. 315–374 in M. K. Hecht, P. C. Goody, and B. M. Hecht (eds.), Major Patterns in Vertebrate Evolution. Plenum Press, New York.

———. 1982. A new appraisal of marsupial phylogeny and classification; pp. 621–640 in M. Archer (ed.), Carnivorous Marsupials. Royal Zoological Society of New South Wales, Sydney.

———. 1985. Rodent and lagomorph morphotype adaptations, origins, and relationships: Some postcranial attributes analyzed; pp. 83–132 in W. P. Luckett and J.-L. Hartenberger (eds.), Evolutionary Relationships among Rodents. A Multidisciplinary Analysis. Plenum Press, New York.

———. 1994. Evolutionary History of the Marsupials and an Analysis of Osteological Characters. Cambridge University Press, Cambridge, UK.

Szalay, F. S., and R. L. Decker. 1974. Origin, evolution, and function of the tarsus in Late Cretaceous Eutheria and Paleocene Primates; pp. 223–259 in F. A. Jenkins, Jr. (ed.), Primate Locomotion. Academic Press, New York.

Szalay, F. S., and E. Delson. 1979. Evolutionary History of the Primates. Academic Press, New York.

Szalay, F. S., and G. Drawhorn. 1980. Evolution and diversification of the Archonta in an arboreal milieu; pp. 133–169 in W. P. Luckett (ed.), Comparative Biology and Evolutionary Relationships of Tree Shrews. Plenum Press, New York.

Szalay, F. S., and C.-K. Li. 1986. Middle Paleocene euprimate from southern China and the distribution of primates in the Paleogene. Journal of Human Evolution 15:387–397.

Szalay, F. S., and S. G. Lucas. 1993. Cranioskeletal morphology of archontans, and diagnoses of Chiroptera, Volitantia, and Archonta; pp. 187–226 in R. D. E. MacPhee (ed.), Primates and Their Relatives in Phylogenetic Perspective. Plenum Press, New York.

———. 1996. The postcranial morphology of Paleocene *Chriacus* and *Mixodectes* and the phylogenetic relationships of archontan mammals. New Mexico Museum of Natural History and Science Bulletin 7:1–47.

Szalay, F. S., and E. J. Sargis. 2001. Model-based analysis of postcranial osteology of marsupials from the Palaeocene of

Itaboraí (Brazil) and the phylogenetics and biogeography of Metatheria. Geodiversitas 23:139–302.

Szalay, F. S., and F. Schrenk. 1998. The middle Eocene *Eurotamandua* and a Darwinian phylogenetic analysis of "edentates." Kaupia—Darmstädter Beiträge zur Naturgeschichte 7:97–186.

———. 2001. An enigmatic new mammal (Dermoptera?) from the Messel middle Eocene, Germany. Kaupia—Darmstädter Beiträge zur Naturgeschichte 11:153–164.

Szalay, F. S., and B. A. Trofimov. 1996. The Mongolian Late Cretaceous *Asiatherium*, and the early phylogeny and paleobiogeography of Metatheria. Journal of Vertebrate Paleontology 16:474–509.

Szalay, F. S., A. L. Rosenberger, and M. Dagosto. 1987. Diagnosis and differentiation of the order Primates. Yearbook of Physical Anthropology 30:75–105.

Tabuce, R., B. Coiffait, P.-E. Coiffait, M. Mahboubi, and J.-J. Jaeger. 2001. A new genus of Macroscelidea (Mammalia) from the Eocene of Algeria: A possible origin for elephant-shrews. Journal of Vertebrate Paleontology 21:535–546.

Tabuce, R., M. Mahboubi, and J. Sudre. 2001. Reassessment of the Algerian Eocene hyracoid *Microhyrax*. Consequences on the early diversity and basal phylogeny of the order Hyracoidea (Mammalia). Eclogae Geologicae Helvetiae 94:537–545.

Tabuce, R., M. Mahboubi, P. Tafforeau, and J. Sudre. 2004. Discovery of a highly specialized plesiadapiform primate in the early-middle Eocene of northwestern Africa. Journal of Human Evolution 47:305–321.

Takai, M., N. Shigehara, A. K. Aung, S. T. Tun, A. N. Soe, T. Tsubamoto, and T. Thein. 2001. A new anthropoid from the latest middle Eocene of Pondaung, central Myanmar. Journal of Human Evolution 40:393–409.

Tassy, P. 1981. Le crane de *Moeritherium* (Proboscidea, Mammalia) de l'Eocène de Dor el Talha (Libye) et le problème de la classification phylogénétique du genre dans les Tethytheria McKenna, 1975. Bulletin Muséum National d'Histoire Naturelle, Paris 3:87–147.

———. 1988. The classification of Proboscidea: How many cladistic classifications? Cladistics 4:43–57.

———. 1996. Who is who among the Proboscidea? pp. 39–48 *in* J. Shoshani and P. Tassy (eds.), The Proboscidea. Evolution and Palaeoecology of Elephants and Their Relatives. Oxford University Press, Oxford.

Tassy, P., and J. Shoshani. 1988. The Tethytheria: Elephants and their relatives; pp. 283–315 in M. J. Benton (ed.), The Phylogeny and Classification of the Tetrapods. Volume 2: Mammals. Clarendon Press, Oxford.

Tavaré, S., C. R. Marshall, O. Will, C. Soligo, and R. D. Martin. 2002. Using the fossil record to estimate the age of the last common ancestor of extant primates. Nature 416:726–729.

Teeling, E. C., O. Madsen, R. A. Van Den Bussche, W. W. de Jong, M. J. Stanhope, and M. S. Springer. 2002. Microbat paraphyly and the convergent evolution of a key innovation in Old World rhinolophoid microbats. Proceedings of the National Academy of Sciences, USA 99:1431–1436.

Teeling, E. C., M. S. Springer, O. Madsen, P. Bates, S. J. O'Brien, and W. J. Murphy. 2005. A molecular phylogeny for bats illuminates biogeography and the fossil record. Science 307:580–584.

Thalmann, U. 1994. Die Primaten aus dem eozänen Geiseltal bei Halle/Saale (Deutschland). Courier Forschungsinstitut Senckenberg 175:1–161.

Thalmann, U., H. Haubold, and R. D. Martin. 1989. *Pronycticebus neglectus*—An almost complete adapid primate specimen from the Geiseltal (GDR). Palaeovertebrata 19:115–130.

Theodor, J. M. 1999. *Protoreodon walshi*, a new species of agriochoerid (Oreodonta, Artiodactyla, Mammalia) from the late Uintan of San Diego County, California. Journal of Paleontology 73:1179–1190.

Thewissen, J. G. M. 1990. Evolution of Paleocene and Eocene Phenacodontidae (Mammalia, Condylarthra). University of Michigan Papers on Paleontology 29:1–107.

———. 1991. Limb osteology and function of the primitive Paleocene ungulate *Pleuraspidotherium* with notes on *Tricuspiodon* and *Dissacus* (Mammalia). Géobios 24:483–495.

———. 1994. Phylogenetic aspects of cetacean origins: A morphological perspective. Journal of Mammalian Evolution 2:157–184.

———. 1998. Cetacean origins. Evolutionary turmoil during the invasion of the oceans; pp. 451–462 *in* J. G. M. Thewissen (ed.), The Emergence of Whales. Evolutionary Patterns in the Origin of Cetacea. Plenum Press, New York.

Thewissen, J. G. M., and S. K. Babcock. 1992. The origin of flight in bats. BioScience 42:340–345.

Thewissen, J. G. M., and D. P. Domning. 1992. The role of phenacodontids in the origin of the modern orders of ungulate mammals. Journal of Vertebrate Paleontology 12:494–504.

Thewissen, J. G. M., and P. D. Gingerich. 1987. Systematics and evolution of *Probathyopsis* (Mammalia, Dinocerata) from the late Paleocene and early Eocene of western North America. Contributions from the Museum of Paleontology, the University of Michigan 27:195–219.

———. 1989. Skull and endocranial cast of *Eoryctes melanus*, a new palaeoryctid (Mammalia: Insectivora) from the early Eocene of western North America. Journal of Vertebrate Paleontology 9:459–470.

Thewissen, J. G. M., and S. T. Hussain. 1990. Postcranial osteology of the most primitive artiodactyl *Diacodexis pakistanensis* (Dichobunidae). Anatomia, Histologia, Embryologia 19:37–48.

———. 1993. Origin of underwater hearing in whales. Nature 361:444–445.

Thewissen, J. G. M., and S. T. Hussain. 1998. Systematic review of the Pakicetidae, early and middle Eocene Cetacea (Mammalia) from Pakistan and India; pp. 220–238 in K. C. Beard and M. R. Dawson (eds.), Dawn of the Age of Mammals in Asia. Bulletin of Carnegie Museum of Natural History 34.

Thewissen, J. G. M., and E. L. Simons. 2001. Skull of *Megalohyrax eocaenus* (Hyracoidea, Mammalia) from the Oligocene of Egypt. Journal of Vertebrate Paleontology 21:98–106.

Thewissen, J. G. M., and E. M. Williams. 2002. The early radiations of Cetacea (Mammalia): Evolutionary pattern and developmental correlations. Annual Review of Ecology and Systematics 33:73–90.

Thewissen, J. G. M., P. D. Gingerich, and D. E. Russell. 1987. Artiodactyla and Perissodactyla (Mammalia) from the early-middle Eocene Kuldana Formation of Kohat (Pakistan). Contributions from the Museum of Paleontology, the University of Michigan 27:247–274.

Thewissen, J. G. M., S. T. Hussain, and M. Arif. 1994. Fossil evidence for the origin of aquatic locomotion in archaeocete whales. Science 263:210–212.

Thewissen, J. G. M., S. I. Madar, and S. T. Hussain. 1996. *Ambulocetus natans,* an Eocene cetacean (Mammalia) from Pakistan. Courier Forschungsinstitut Senckenberg 191:1–86.

Thewissen, J. G. M., S. I. Madar, and S. T. Hussain. 1998. Whale ankles and evolutionary relationships. Nature 395:452.

Thewissen, J. G. M., L. J. Roe, J. R. O'Neil, S. T. Hussain, A. Sahni, and S. Bajpai. 1996. Evolution of cetacean osmoregulation. Nature 381:379–380.

Thewissen, J. G. M., E. M. Williams, and S. T. Hussain. 2001a. Eocene mammal faunas from northern Indo-Pakistan. Journal of Vertebrate Paleontology 21:347–366.

Thewissen, J. G. M., E. M. Williams, L. J. Roe, and S. T. Hussain. 2001b. Skeletons of terrestrial cetaceans and the relationship of whales to artiodactyls. Nature 413:277–281.

Thomas, H., E. Gheerbrant, and J.-M. Pacaud. 2004. Découverte de squelettes subcomplets de mammifères (Hyracoidea) dans le Paléogène d'Afrique (Libye). Comptes Rendus Palevol 3:209–217.

Tiffney, B. H. 2000. Geographic and climatic influences on the Cretaceous and Tertiary history of Euramerican floristic similarity. Acta Universitatis Carolinae—Geologica 44:5–16.

Ting, S.-Y. 1998. Paleocene and early Eocene land mammal ages of Asia; pp. 124–147 in K. C. Beard and M. R. Dawson (eds.), Dawn of the Age of Mammals in Asia. Bulletin of Carnegie Museum of Natural History 34.

Ting, S.-Y., and J. Zheng. 1989. The affinities of *Interogale* and *Anchilestes* and the origin of Tillodontia. Vertebrata PalAsiatica 27:77–86.

Ting, S.-Y., J. Meng, M. C. McKenna, and C.-K. Li. 2002. The osteology of *Matutinia* (Simplicidentata, Mammalia) and its relationship to *Rhombomylus*. American Museum Novitates 3371:1–33.

Ting, S.-Y., J. A. Schiebout, and M.-C. Chow. 1982. Morphological diversity of early Tertiary pantodonts: A new tapir-like pantodont from China. Third North American Paleontological Convention Proceedings 2:547–550.

Tobien, H. 1962. Insectivoran (Mamm.) aus dem Mitteleozän (Lutetium) von Messel bei Darmstadt. Notizblatt des Hessischen Landesamtes für Bodenforschung zu Wiesbaden 90:7–47.

———. 1978. The structure of the mastodont molar (Proboscidea, Mammalia). Part 3. The Oligocene mastodont genera *Palaeomastodon, Phiomia,* and the Eo/Oligocene paenungulate *Moeritherium*. Mainzer Geowissenschaftliche Mitteilungen 6:177–208.

———. 1980. Ein anthracotherioider Paarhufer (Artiodactyla, Mammalia) aus dem Eozän von Messel bei Darmstadt (Hessen). Geologisches Jahrbuch Hessen 108:11–22.

Tong, Y.-S. 1979. A late Paleocene primate from S. China. Vertebrata PalAsiatica 17:65–70.

———. 1988. Fossil tree shrews from the Eocene Hetaoyuan Formation of Xichuan, Henan. Vertebrata PalAsiatica 26:214–220.

———. 1992. *Pappocricetodon,* a pre-Oligocene cricetid genus (Rodentia) from central China. Vertebrata PalAsiatica 30:1–16.

———. 1997. Middle Eocene small mammals from Liguanqiao Basin of Henan Province and Yuanqu Basin of Shanxi Province, Central China. Palaeontologia Sinica, new series C 18:1–256.

Tong, Y.-S., and M. R. Dawson. 1995. Early Eocene rodents (Mammalia) from Shandong Province, People's Republic of China. Annals of Carnegie Museum 64:51–63.

Tong, Y.-S, and S. G. Lucas. 1982. A review of Chinese uintatheres and the origin of the Dinocerata (Mammalia, Eutheria). Third North American Paleontological Convention Proceedings 2:551–556.

Tong, Y.-S., and J. Wang. 1997. A new palaeanodont (Mammalia) from the early Eocene of Wutu Basin, Shandong Province. Vertebrata PalAsiatica 35:110–120.

———. 1998. A preliminary report on the early Eocene mammals of the Wutu fauna, Shandong Province, China; pp. 186–193 in K. C. Beard and M. R. Dawson (eds.), Dawn of the Age of Mammals in Asia. Bulletin of Carnegie Museum of Natural History 34.

Tong, Y.-S., J.-W. Wang, and J.-F. Fu. 2003. *Yuesthonyx,* a new tillodont (Mammalia) from the Paleocene of Henan. Vertebrata PalAsiatica 41:55–65.

Tong, Y.-S., J.-W. Wang, and J. Meng. 2004. *Olbitherium millenariusum,* a new perissodactyl-like archaic ungulate (Mammalia) from the early Eocene Wutu Formation, Shandong. Vertebrata PalAsiatica 42:27–38.

Tong, Y.-S., S. Zheng, and Z. Qiu. 1995. Cenozoic mammal ages of China. Vertebrata PalAsiatica 33:290–314.

Turnbull, W. D. 1991. *Protoptychus hatcheri* Scott, 1895. The Mammalian faunas of the Washakie Formation, Eocene Age, of southern Wyoming. Part II. The Adobetown Member, middle division (=Washakie B), Twka/2 (in part). Fieldiana Geology, new series 21:1–33.

———. 2002. The mammalian faunas of the Washakie Formation, Eocene age, of southern Wyoming. Part IV. The Uintatheres. Fieldiana Geology, new series 47:1–189.

———. 2004. Taeniodonta of the Washakie Formation, southwestern Wyoming. Bulletin of Carnegie Museum of Natural History 36:302–333.

Uhen, M. D. 1998. Middle to late Eocene basilosaurines and dorudontines; pp. 29–61 in J. G. M. Thewissen (ed.), The Emergence of Whales. Evolutionary Patterns in the Origin of Cetacea. Plenum Press, New York.

———. 1999. New species of protocetid archaeocete whale, *Eocetus wardii* (Mammalia: Cetacea) from the middle Eocene of North Carolina. Journal of Paleontology 73:512–528.

———. 2004. Form, function, and anatomy of *Dorudon atrox* (Mammali, Cetacea): An archaeocete from the middle to late Eocene of Egypt. University of Michigan Papers on Paleontology 34:1–222.

Uhen, M. D., and P. D. Gingerich. 1995. Evolution of *Coryphodon* (Mammalia, Pantodonta) in the late Paleocene and early Eocene of northwestern Wyoming. Contributions from the Museum of Paleontology, the University of Michigan 29:259–289.

Upchurch, G. R., Jr., and J. A. Wolfe. 1987. Mid-Cretaceous to Early Tertiary vegetation and climate: Evidence from fossil leaves and woods; pp. 75–105 in E. M. Friis, W. G. Chaloner, and P. R. Crane (eds.), The Origins of Angiosperms and Their Biological Consequences. Cambridge University Press, Cambridge, UK.

Ursing, B. M., and U. Arnason. 1998. Analyses of mitochondrial genomes strongly support a hippopotamus-whale clade. Proceedings of the Royal Society of London B 265:2251–2255.

Van Dijk, M. A. M., O. Madsen, F. Catzeflis, M. J. Stanhope, W. W. de Jong, and M. Pagel. 2001. Protein sequence signatures support the African clade of mammals. Proceedings of the National Academy of Sciences, USA 98:188–193.

Van Valen, L. 1963. The origin and status of the mammalian order Tillodontia. Journal of Mammalogy 44:364–373.
———. 1964. A possible origin for rabbits. Evolution 18:484–491.
———. 1965. Treeshrews, primates and fossils. Evolution 19:137–151.
———. 1966. Deltatheridia, a new order of mammals. Bulletin of the American Museum of Natural History 132:1–126.
———. 1967. New Paleocene insectivores and insectivore classification. Bulletin of the American Museum of Natural History 135:217–284.
———. 1968. Monophyly or diphyly in the origin of whales. Evolution 22:37–41.
———. 1971. Toward the origin of artiodactyls. Evolution 25:523–529.
———. 1978. The beginning of the Age of Mammals. Evolutionary Theory 4:45–80.
———. 1979. The evolution of bats. Evolutionary Theory 4:103–121.
———. 1988. Paleocene dinosaurs or Cretaceous ungulates in South America. Evolutionary Monographs 10:1–79.
———. 1994a. Serial homology: The crests and cusps of mammalian teeth. Acta Palaeontologica Polonica 38:145–158.
———. 1994b. The origin of the plesiadapid primates and the nature of *Purgatorius*. Evolutionary Monographs 15:1–79.
Van Valen, L., and R. E. Sloan. 1966. The extinction of the multituberculates. Systematic Zoology 15:261–278.
Van Valkenburgh, B. 1987. Skeletal indicators of locomotor behavior in living and extinct carnivores. Journal of Vertebrate Paleontology 7:162–182.
———. 1999. Major patterns in the history of carnivorous mammals. Annual Review of Earth and Planetary Sciences 27:463–493.
Vaughan, T. A. 1978. Mammalogy. Second edition. W. B. Saunders, Philadelphia.
Vaughan, T. A., J. M. Ryan, and N. J. Czaplewski. 2000. Mammalogy. Fourth edition. W. B. Saunders College Publishing, Philadelphia.
Vianey-Liaud, M. 1974. *Palaeosciurus goti* nov. sp., écureuil terrestre de l'Oligocène moyen du Quercy. Données nouvelles sur l'apparition des sciuridés en Europe. Annales de Paléontologie 60:103–122.
———. 1976. Les Issiodoromyinae (Rodentia, Theridomyidae) de l'Éocène supérieur à l'Oligocène supérieur en Europe occidentale. Palaeovertebrata 7:1–115.
———. 1979a. Les mammifères Montiens de Hainin (Paléocène moyen de Belgique). Part I: Multituberculés. Palaeovertebrata 9:117–131.
———. 1979b. Evolution des rongeurs à l'Oligocène en Europe occidentale. Palaeontographica Abteilung A 166:136–236.
———. 1985. Possible evolutionary relationships among Eocene and lower Oligocene rodents of Asia, Europe, and North America; pp. 277–309 in W. P. Luckett and J.-L. Hartenberger (eds.), Evolutionary Relationships among Rodents. A Multidisciplinary Analysis. Plenum Press, New York.
———. 1994. La radiation des Gliridae (Rodentia) à l'Éocène supérieur en Europe Occidentale, et sa descendance Oligocène. Münchner Geowissenschaftliche Abhandlungen A 26:117–160.
Vianey-Liaud, M., and J.-J. Jaeger. 1996. A new hypothesis for the origin of African Anomaluridae and Graphiuridae (Rodentia). Palaeovertebrata 25:349–358.

Vianey-Liaud, M., J.-J. Jaeger, J.-L. Hartenberger, and M. Mahboubi. 1994. Les rongeurs de l'Éocène d'Afrique nord-occidentale [Glib Zegdou (Algérie) et Chambi (Tunisie)] et l'origine des Anomaluridae. Palaeovertebrata 23:93–118.
Villarroel, C. 1987. Caracteristicas y afinidades de *Etayoa* n. gen., tipo de una nueva familia de Xenungulata (Mammalia) del Paleoceno Medio (?) de Colombia. Comunicaciones Paleontologicas del Museo de Historia Natural de Montevideo 1:241–253.
Vizcaíno, S. F. 1994. Sistematica y anatomia de los Astegotheriini Ameghino, 1906 (nuevo rango) (Xenarthra, Dasypodidae, Dasypodinae). Ameghiniana 31:3–13.
Vizcaíno, S. F., and G. J. Scillato-Yané. 1995. An Eocene tardigrade (Mammalia, Xenarthra) from Seymour Island, West Antarctica. Antarctic Science 7:407–408.
Wahlert, J. H. 1968. Variability of rodent incisor enamel as viewed in thin section, and the microstructure of the enamel in fossil and recent rodent groups. Breviora 309:1–18.
———. 1973. *Protoptychus*, a hystricomorphous rodent from the late Eocene of North America. Breviora 419:1–14.
———. 1974. The cranial foramina of protrogomorphous rodents; An anatomical and phylogenetic study. Bulletin of the Museum of Comparative Zoology 146:363–410.
———. 1977. Cranial foramina and relationships of *Eutypomys* (Rodentia, Eutypomyidae). American Museum Novitates 2626:1–8.
———. 1978. Cranial foramina and relationships of the Eomyoidea (Rodentia, Geomorpha). Skull and upper teeth of *Kansasimys*. American Museum Novitates 2645:1–16.
———. 1989. The three types of incisor enamel in rodents; pp. 6–17 in C. C. Black and M. R. Dawson (eds.), Papers on Fossil Rodents in Honor of Albert Elmer Wood. Natural History Museum of Los Angeles County Science Series 33, Los Angeles.
Wahlert, J. H., and W. von Koenigswald. 1985. Specialized enamel in incisors of eomyid rodents. American Museum Novitates 2832:1–12.
Wahlert, J. H., S. L. Sawitzke, and M. E. Holden. 1993. Cranial anatomy and relationships of dormice (Rodentia, Myoxidae). American Museum Novitates 3061:1–32.
Wake, D. B. 1979. The endoskeleton: The comparative anatomy of the vertebral column and ribs; pp. 192–237 in M. H. Wake (ed.), Hyman's Comparative Vertebrate Anatomy. University of Chicago Press, Chicago.
Wall, C., and D. W. Krause. 1992. A biomechanical analysis of the masticatory apparatus of *Ptilodus* (Multituberculata). Journal of Vertebrate Paleontology 12:172–187.
Wall, W. P. 1980. Cranial evidence for a proboscis in *Cadurcodon* and a review of snout structure in the family Amynodontidae (Perissodactyla, Rhinocerotoidea). Journal of Paleontology 54:968–977.
———. 1989. The phylogenetic history and adaptive radiation of the Amynodontidae; pp. 341–354 in D. R. Prothero and R. M. Schoch (eds.), The Evolution of Perissodactyls. Oxford University Press, New York.
———. 1998. Amynodontidae; pp. 583–588 in C. M. Janis, K. M. Scott, and L. L. Jacobs (eds.), Evolution of Tertiary Mammals of North America. Volume 1: Terrestrial Carnivores, Ungulates, and Ungulatelike Mammals. Cambridge University Press, Cambridge, UK.
Walsh, S. L. 1997. New specimens of *Metanoiamys*, *Pauromys*, and *Simimys* (Rodentia: Myomorpha) from the Uintan (middle

Eocene) of San Diego County, California, and comments on the relationships of selected Paleogene Myomorpha. Proceedings of the San Diego Society of Natural History 32:1–20.

———. 1998. Fossil datum and paleobiological event terms, paleontostratigraphy, chronostratigraphy, and the definition of land mammal "age" boundaries. Journal of Vertebrate Paleontology 18:150–179.

Wang, B.-Y. 1979. A new species of *Harpyodus* and its taxonomic position; pp. 366–372 *in* The Mesozoic and Cenozoic Red Beds of South China. Selected Papers from "The Field Conference on the South China Cretaceous–Early Tertiary Red Beds" Held at Nanxiong, Guangdong Province, 24 November–6 December 1976. IVPP Academia Sinica and Nanjing Institute of Geology and Paleontology, Academia Sinica, Kexue Chubanshe, Beijing.

———. 1994. The Ctenodactyloidea of Asia; pp. 35–47 *in* Y. Tomida, C. Li, and T. Setoguchi (eds.), Rodent and Lagomorph Families of Asian Origins and Diversification. National Science Museum Monographs 8, Tokyo.

———. 1997. The mid-Tertiary Ctenodactylidae (Rodentia, Mammalia) of eastern and central Asia. Bulletin of the American Museum of Natural History 234:1–88.

———. 2001a. Late Eocene ctenodactyloids (Rodentia, Mammalia) from Qujing, Yunnan, China. Vertebrata PalAsiatica 39:24–42.

———. 2001b. Eocene ctenodactyloids (Rodentia, Mammalia) from Nei Mongol, China. Vertebrata PalAsiatica 39:102–114.

Wang, B.-Y., and M. R. Dawson. 1994. A primitive cricetid (Mammalia: Rodentia) from the middle Eocene of Jiangsu Province, China. Annals of Carnegie Museum 63:239–256.

Wang, B.-Y., and C. Li. 1990. First Paleogene mammalian fauna from northeast China. Vertebrate PalAsiatica 28:165–205.

Wang, X. 1993. Transformation from plantigrady to digitigrady: Functional morphology of locomotion in *Hesperocyon* (Canidae: Carnivora). American Museum Novitates 3069:1–23.

———. 1994. Phylogenetic systematics of the Hesperocyoninae (Carnivora: Canidae). Bulletin of the American Museum of Natural History 221:1–207.

Wang, X., and R. H. Tedford. 1994. Basicranial anatomy and phylogeny of primitive canids and closely related miacids (Carnivora: Mammalia). American Museum Novitates 3092:1–34.

Wang, X., and R. Zhai. 1995. *Carnilestes*, a new primitive lipotyphlan (Insectivora: Mammalia) from the early and middle Paleocene, Nanxiong Basin, China. Journal of Vertebrate Paleontology 15:131–145.

Wang, X., W. Downs, J. Xie, and G. Xie. 2001. *Didymoconus* (Mammalia: Didymoconidae) from Lanzhou Basin, China and its stratigraphic and ecological significance. Journal of Vertebrate Paleontology 21:555–564.

Wang, Y. 1995. A new primitive chalicothere (Perissodactyla, Mammalia) from the early Eocene of Hubei, China. Vertebrata PalAsiatica 33:138–159.

Wang, Y., and X. Jin. 2004. A new Paleocene tillodont (Tillodontia, Mammalia) from Qianshan, Anhui, with a review of Paleocene tillodonts from China. Vertebrata PalAsiatica 42:14–26.

Wang, Y., W. A. Clemens, Y. Hu, and C. Li. 1998. A probable pseudo-tribosphenic upper molar from the Late Jurassic of China and the early radiation of the Holotheria. Journal of Vertebrate Paleontology 18:777–787.

Wang, Y., Y. Hu, J. Meng, and C. Li. 2001. An ossified Meckel's Cartilage in two Cretaceous mammals and origin of the mammalian middle ear. Science 294:357–361.

Webb, S. D. 1985. The interrelationships of tree sloths and ground sloths; pp. 105–112 *in* G. G. Montgomery (ed.), The Evolution and Ecology of Armadillos, Sloths, and Vermilinguas. Smithsonian Institution Press, Washington, D.C.

———. 1998. Hornless ruminants; pp. 463–476 *in* C. M. Janis, K. M. Scott, and L. L. Jacobs (eds.), Evolution of Tertiary Mammals of North America. Volume 1: Terrestrial Carnivores, Ungulates, and Ungulatelike Mammals. Cambridge University Press, Cambridge, UK.

Webb, S. D., and B. E. Taylor. 1980. The phylogeny of hornless ruminants and a description of the cranium of *Archaeomeryx*. Bulletin of the American Museum of Natural History 167:121–157.

Weigelt, J. 1960. Die Arctocyoniden von Walbeck. Freiberger Forschungshefte C77:1–241.

Wells, N. A., and P. D. Gingerich. 1983. Review of Eocene Anthracobunidae (Mammalia, Proboscidea) with a new genus and species, *Jozaria palustris*, from the Kuldana Formation of Kohat (Pakistan). Contributions from the Museum of Paleontology, the University of Michigan 26:117–139.

Wesley-Hunt, G. D. 2005. The morphological diversification of carnivores in North America. Paleobiology 31:35–55.

Wesley-Hunt, G. D., and J. J. Flynn. 2005. Phylogeny of the Carnivora: Basal relationships among the carnivoramorphans, and assessment of the position of "Miacoidea" relative to Carnivora. Journal of Systematic Palaeontology 3:1–28.

West, R. M. 1973a. Antemolar dentitions of the Paleocene apatemyid insectivorans *Jepsenella* and *Labidolemur*. Journal of Mammalogy 54:33–40.

———. 1973b. Review of the North American Eocene and Oligocene Apatemyidae (Mammalia: Insectivora). Special Publications, The Museum, Texas Tech University 3:1–42.

———. 1976. The North American Phenacodontidae. Contributions in Biology and Geology, Milwaukee Public Museum 6:1–78.

———. 1984. A review of the South Asian Middle Eocene Moeritheriidae. Mémoires de la Société Géologique de France, N. S. 147:183–190.

West, R. M., M. R. Dawson, and J. H. Hutchison. 1977. Fossils from the Paleogene Eureka Sound Formation, N.W.T., Canada: Occurrence, climatic and paleogeographic implications. Milwaukee Public Museum Special Publications in Biology and Geology 2:77–93.

Wheeler, W. H. 1961. Revision of the uintatheres. Bulletin of the Peabody Museum of Natural History, Yale University 14:1–93.

Whidden, H. P. 2002. Extrinsic snout musculature in Afrotheria and Lipotyphla. Journal of Mammalian Evolution 9:161–184.

Wible, J. R. 1991. Origin of Mammalia: The craniodental evidence reexamined. Journal of Vertebrate Paleontology 11:1–28.

———. 2003. On the cranial osteology of the short-tailed opossum *Monodelphis brevicaudata* (Didelphidae, Marsupialia). Annals of Carnegie Museum 72:137–202.

Wible, J. R., and J. A. Hopson. 1993. Basicranial evidence for early mammal phylogeny; pp. 45–62 *in* F. S. Szalay, M. J. Novacek, and M. C. McKenna (eds.), Mammal Phylogeny:

Mesozoic Differentiation, Multituberculates, Monotremes, Early Therians, and Marsupials. Springer-Verlag, New York.

Wible, J. R., and M. J. Novacek. 1988. Cranial evidence for the monophyletic origin of bats. American Museum Novitates 2911:1–19.

Wible, J. R., M. J. Novacek, and G. W. Rougier. 2004. New data on the skull and dentition in the Mongolian Late Cretaceous eutherian mammal *Zalambdalestes*. Bulletin of the American Museum of Natural History 281:1–144.

Wible, J. R., G. W. Rougier, and M. J. Novacek. 2005. Anatomical evidence for superordinal/ordinal eutherian taxa in the Cretaceous; pp. 15–36 in K. D. Rose and J. D. Archibald (eds.), The Rise of Placental Mammals: Origins and Relationships of the Major Extant Clades. The Johns Hopkins University Press, Baltimore.

Wible, J. R., G. W. Rougier, M. J. Novacek, and M. C. McKenna. 2001. Earliest eutherian ear region: A petrosal referred to *Prokennalestes* from the Early Cretaceous of Mongolia. American Museum Novitates 3322:1–44.

Wilf, P. 1997. When are leaves good thermometers? A new case for Leaf Margin Analysis. Paleobiology 23:373–390.

Wilf, P., K. C. Beard, K. S. Davies-Vollum, and J. W. Norejko. 1998. Portrait of a late Paleocene (early Clarkforkian) terrestrial ecosystem: Big Multi Quarry and associated strata, Washakie Basin, southwestern Wyoming. Palaios 13: 514–532.

Wilf, P., K. R. Johnson, and B. T. Huber. 2003. Correlated terrestrial and marine evidence for global climate changes before mass extinction at the Cretaceous-Paleogene boundary. Proceedings of the National Academy of Sciences, USA 100: 599–604.

Wilhelm, P. B., 1993. Morphometric analysis of the limb skeleton of generalized mammals in relation to locomotor behavior, with applications to fossil mammals. Ph.D. dissertation, Brown University, Providence, R. I.

Williamson, T. E. 1996. The beginning of the age of mammals in the San Juan Basin, New Mexico: Biostratigraphy and evolution of Paleocene mammals of the Načimiento Formation. Bulletin of the New Mexico Museum of Natural History and Science 8:1–141.

Williamson, T. E., and S. G. Lucas. 1992. *Meniscotherium* (Mammalia, "Condylarthra") from the Paleocene-Eocene of western North America. Bulletin of the New Mexico Museum of Natural History and Science 1:1–75.

Wilson, D. E., and D. M. Reeder. 1993. Mammal Species of the World: A Taxonomic and Geographic Reference. Second edition. Smithsonian Institution Press, Washington, D. C.

———. 2005. Mammal Species of the World: A Taxonomic and Geographic Reference. Third edition. The Johns Hopkins University Press, Baltimore.

Wilson, G. P., and N. C. Arens. 2001. The evolutionary impact of an epeiric seaway on Late Cretaceous and Paleocene palynofloras of South America. Asociación Paleontológica Argentina Publicación Especial 7:185–189.

Wilson, J. A. 1971a. Early Tertiary Vertebrate faunas, Vieja Group, Trans-Pecos Texas: Agriochoeridae and Merycoidodontidae. Bulletin of the Texas Memorial Museum 18:1–83.

———. 1971b. Early Tertiary vertebrate faunas, Vieja Group, Trans-Pecos Texas: Entelodontidae. The Pearce-Sellards Series, Texas Memorial Museum 17:1–17.

———. 1974. Early Tertiary vertebrate faunas, Vieja Group and Buck Hill Group, Trans-Pecos Texas; Protoceratidae, Camelidae, Hypertragulidae. Bulletin of the Texas Memorial Museum 23:1–34.

Wilson, J. A., and F. S. Szalay. 1976. New adapid primate of European affinities from Texas. Folia Primatologica 25:294–312.

Wilson, J. A., and J. W. Westgate. 1991. A lophodont rodent from the middle Eocene of the Gulf Coastal Plain, Texas. Journal of Vertebrate Paleontology 11:257–260.

Wing, S. L. 1998a. Tertiary vegetation of North America as a context for mammalian evolution; pp. 37–65 in C. M. Janis, K. M. Scott, and L. L. Jacobs (eds.), Evolution of Tertiary Mammals of North America. Cambridge University Press, Cambridge, UK.

———. 1998b. Late Paleocene–early Eocene floral and climatic change in the Bighorn Basin, Wyoming; pp. 380–400 in M.-P. Aubry, S. G. Lucas, and W. A. Berggren (eds.), Late Paleocene–Early Eocene Climatic and Biotic Events in the Marine and Terrestrial Records. Columbia University Press, New York.

———. 2001. Hot times in the Bighorn Basin. Natural History 110(3):48–54.

Wing, S. L., and D. R. Greenwood. 1993. Fossils and fossil climate: The case for equable continental interiors in the Eocene. Philosophical Transactions of the Royal Society of London B 341:243–252.

Wing, S. L., and G. J. Harrington. 2001. Floral response to rapid warming in the earliest Eocene and implications for concurrent faunal change. Paleobiology 27:539–563.

Wing, S. L., and B. H. Tiffney. 1987. The reciprocal interaction of angiosperm evolution and tetrapod herbivory. Review of Palaeobotany and Palynology 50:179–210.

Wing, S. L., H. Bao, and P. L. Koch. 1999. An early Eocene cool period? Evidence for continental cooling during the warmest part of the Cenozoic; pp. 197–237 in B. T. Huber, K. G. MacLeod, and S. L. Wing (eds.), Warm Climates in Earth History. Cambridge University Press, Cambridge, UK.

Wing, S. L., T. M. Bown, and J. D. Obradovich. 1991. Early Eocene biotic and climatic change in interior western North America. Geology 19:1189–1192.

Wolfe, J. A. 1979. Temperature parameters of humid to mesic forests of eastern Asia and relation to forests of other regions of the Northern Hemisphere and Australasia. U. S. Geological Survey Professional Paper 1106:1–37.

———. 1985. Distribution of major vegetational types during the Tertiary. Geophysical Monograph 32:357–375.

———. 1990. Palaeobotanical evidence for a marked temperature increase following the Cretaceous/Tertiary boundary. Nature 343:153–156.

———. 1993. A method of obtaining climatic parameters from leaf assemblages. U. S. Geological Survey Bulletin 2040:1–71.

———. 1994. Tertiary climatic changes at middle latitudes of western North America. Palaeogeography, Palaeoclimatology, Palaeoecology 108:195–205.

Wolfe, J. A., and G. R. Upchurch, Jr. 1986. Vegetation, climatic and floral changes at the Cretaceous-Tertiary boundary. Nature 324:148–152.

Wolsan, M. 1993. Phylogeny and classification of early European Mustelida (Mammalia: Carnivora). Acta Theriologica 38:345–384.

Wolsan, M., and B. Lange-Badré. 1996. An arctomorph carnivoran skull from the Phosphorites du Quercy and the origin of procyonids. Acta Palaeontologica Polonica 41:277–298.

Wood, A. E. 1937. The Mammalian Fauna of the White River Oligocene. Part II. Rodentia. Transactions of the American Philosophical Society, new series 28:155–269.
———. 1940. The Mammalian Fauna of the White River Oligocene. Part III. Lagomorpha. Transactions of the American Philosophical Society, new series 28:271–362.
———. 1957. What, if anything, is a rabbit? Evolution 11: 417–425.
———. 1962. The Early Tertiary rodents of the family Paramyidae. Transactions of the American Philosophical Society, new series 52:1–261.
———. 1965. Grades and clades among rodents. Evolution 19: 115–130.
———. 1968. Early Cenozoic Mammalian faunas, Fayum Province, Egypt. Part II. The African Oligocene Rodentia. Peabody Museum of Natural History Bulletin 28:23–105.
———. 1974a. Early Tertiary vertebrate faunas Vieja Group Trans-Pecos Texas: Rodentia. Texas Memorial Museum Bulletin 21:1–112.
———. 1974b. The evolution of the Old World and New World hystricomorphs. Symposium Zoological Society of London 34:21–60.
———. 1975. The problem of the hystricognathous rodents. University of Michigan Papers on Paleontology 12:75–80.
———. 1976. The paramyid rodent *Ailuravus* from the middle and late Eocene of Europe, and its relationships. Palaeovertebrata 7:117–149.
———. 1980. The Oligocene rodents of North America. Transactions of the American Philosophical Society 70:1–68.
———. 1984. Hystricognathy in the North American Oligocene rodent *Cylindrodon* and the origin of the Caviomorpha. Carnegie Museum of Natural History Special Publication 9:151–160.
———. 1985. The relationships, origin and dispersal of the hystricognathous rodents; pp. 475–513 in W. P. Luckett and J.-L. Hartenberger (eds.), Evolutionary Relationships among Rodents. A Multidisciplinary Analysis. Plenum Press, New York.
Wood, C. B., and W. A. Clemens, Jr. 2001. A new specimen and a functional reassociation of the molar dentition of *Batodon tenuis* (Placentalia, incertae sedis), latest Cretaceous (Lancian), North America. Bulletin of the Museum of Comparative Zoology 156:99–118.
Wood, H. E. II, R. W. Chaney, J. Clark, E. H. Colbert, G. L. Jepsen, J. B. Reeside, Jr., and C. Stock. 1941. Nomenclature and correlation of the North American continental Tertiary. Bulletin of the Geological Society of America 52:1–48.
Woodburne, M. O. (ed.) 1987. Cenozoic Mammals of North America. Geochronology and Biostratigraphy. University of California Press, Berkeley.
———. 2003. Monotremes as pretribosphenic mammals. Journal of Mammalian Evolution 10:195–248.
———. (ed.) 2004. Late Cretaceous and Cenozoic Mammals of North America. Biostratigraphy and Geochronology. Columbia University Press, New York.
Woodburne, M. O., and J. A. Case. 1996. Dispersal, vicariance, and the Late Cretaceous to Early Tertiary land mammal biogeography from South America to Australia. Journal of Mammalian Evolution 3:121–162.
Woodburne, M. O., and C. C. Swisher III. 1995. Land mammal high-resolution geochronology, intercontinental overland dispersals, sea level, climate, and vicariance; pp. 335–364 in W. A. Berggren, D. V. Kent, M.-P. Aubry, and J. Hardenbol (eds.), Geochronology, Time Scales and Global Stratigraphic Correlation. SEPM (Society for Sedimentary Geology) Special Publication 54, Tulsa, Okla.
Woodburne, M. O., and R. H. Tedford. 1975. The first Tertiary monotreme from Australia. American Museum Novitates 2588:1–11.
Woodburne, M. O., T. H. Rich, and M. S. Springer. 2003. The evolution of tribosphony and the antiquity of mammalian clades. Molecular Phylogenetics and Evolution 28:360–385.
Wright, D. B. 1998. Tayassuidae; pp. 389–401 in C. M. Janis, K. M. Scott, and L. L. Jacobs (eds.), Evolution of Tertiary Mammals of North America. Volume 1: Terrestrial Carnivores, Ungulates, and Ungulatelike Mammals. Cambridge University Press, Cambridge, UK.
Wyss, A. R., and J. J. Flynn. 1993. A phylogenetic analysis and definition of the Carnivora; pp. 32–52 in F. S. Szalay, M. J. Novacek, and M. C. McKenna (eds.), Mammal Phylogeny: Placentals. Springer-Verlag, New York.
Wyss, A. R., and J. Meng. 1996. Application of phylogenetic taxonomy to poorly resolved crown clades: A stem-modified node-based definition of Rodentia. Systematic Biology 45:559–568.
Wyss, A. R., J. J. Flynn, M. A. Norell, C. C. Swisher III, R. Charrier, M. J. Novacek, and M. C. McKenna. 1993. South American's earliest rodent and recognition of a new interval of mammalian evolution. Nature 365:434–437.
Wyss, A. R., J. J. Flynn, M. A. Norell, C. C. Swisher III, M. J. Novacek, M. C. McKenna, and R. Charrier. 1994. Paleogene mammals from the Andes of Central Chile: A preliminary taxonomic, biostratigraphic, and geochronologic assessment. American Museum Novitates 3098:1–31.
Yalden, D. W. 1966. The anatomy of mole locomotion. Journal of Zoology, London 149:55–64.
Yoder, A. D., and Z. Yang. 2004. Divergence dates for Malagasy lemurs estimated from multiple gene loci: Geological and evolutionary context. Molecular Ecology 13:757–773.
Yoder, A. D., M. Cartmill, M. Ruvolo, K. Smith, and R. Vilgalys. 1996. Ancient single origin for Malagasy primates. Proceedings of the National Academy of Sciences, USA 93: 5122–5126.
Youlatos, D. 2003. Osteological correlates of tail prehensility in carnivorans. Journal of Zoology, London 259:423–430.
Zachos, J., M. Pagani, L. Sloan, E. Thomas, and K. Billups. 2001. Trends, rhythms, and aberrations in global climate 65 Ma to present. Science 292:686–693.
Zack, S. P. 2004. An early Eocene arctostylopid (Mammalia: Arctostylopida) from the Green River Basin, Wyoming. Journal of Vertebrate Paleontology 24:498–501.
Zack, S. P., T. A. Penkrot, J. I. Bloch, and K. D. Rose. 2005. Affinities of "hyopsodontids" to elephant shrews and a Holarctic origin of Afrotheria. Nature 434:497–501.
Zapfe, H. 1979. *Chalicotherium grande* (Blainv.) aus der miozänen Spaltenfüllung von Neudorf an der March (Devinska Nova Ves), Tschechoslowakei. Neue Denkschriften des Naturhistorischen Museums in Wien 2:1–282.
Zeller, U., J. R. Wible, and M. Elsner. 1993. New ontogenetic evidence on the septomaxilla of *Tamandua* and *Choloepus* (Mammalia, Xenarthra), with a reevaluation of the homology of the mammalian septomaxilla. Journal of Mammalian Evolution 1:31–46.

Zhang, F., A. W. Crompton, Z.-X. Luo, and C. R. Schaff. 1998. Pattern of dental replacement of *Sinoconodon* and its implication for evolution of mammals. Vertebrata PalAsiatica 36:197–217.

Zhang, Y.-P. 1978. Two new genera of condylarthran phenacolophids from the Paleocene of Nanxiong Basin, Guangdong. Vertebrata PalAsiatica 16:267–274.

Zhou, M., Y. Zhang, B. Wang, and S. Ding. 1977. Mammalian fauna from the Paleocene of Nanxiong Basin, Guangdong. Palaeontologia Sinica, new series C 20:1–100.

Zhou, X., W. J. Sanders, and P. D. Gingerich. 1992. Functional and behavioral implications of vertebral structure in *Pachyaena ossifraga* (Mammalia, Mesonychia). Contributions from the Museum of Paleontology, the University of Michigan 28:289–319.

Zhou, X., R. Zhai, P. D. Gingerich, and L. Chen. 1995. Skull of a new mesonychid (Mammalia, Mesonychia) from the late Paleocene of China. Journal of Vertebrate Paleontology 15:387–400.

INDEX

Information presented in figures and tables is denoted by *f* and *t*, respectively. Major discussions are indicated by **boldface** page numbers.

Aaptoryctes, 96, 96f
aardvarks, 10t, 35, 198, 211–212, 285
aardwolf, 35
Abderitidae, 74t
Abdounodus, 217–218
abducent nerve, 24
Absarokius, 188–189
Abuqatrania, 194
accessory nasal organ, 26
accessory nerve, 26
acetabulum, 33, 36f
Achaenodon, 290–291, 296
Acidomomys, 176, 177
Acotherulum, 298f
acoustic meatus, internal, 25–26
Acreodi, **274–275**
Acritoparamys, 319, 321–322
Acrobates, 175
acromion process, 32, 32f
Acropithecus, 232
Adapidae, 166t, 182, **185–186**
Adapidium, 113
Adapiformes, 179
Adapinae, 182
Adapis, 168f, 180, 184f, 185–186
Adapis parisiensis, 182
Adapisorex, 145
Adapisoricidae, 140t, 145
Adapisoriculidae, 140t, **144**, 144f, 197, 338, 342

Adapisoriculus, 144
adapoid primate, Plate 6
Adapoidea, 166, 166t, 167f, 168f, 172f–173f, 178–179, **179–186,** 181f–184f, 191, 192, 339, 343, Plate 6
Adapoides, 186
adaptations, skeletal, **34–40,** 39f
Adelobasileus, 9t, 49t, 50–51, 51f
Adianthidae, 213t, 227f, 233, 235, 235f
Adiantoides, 235, 235f
Adunator, 145–146, 342
Aegialodon, 67f, 69–70, 78f
Aegialodontia, 9t, 49t, **69–70**
Aegialodontidae, 49t, 51f
Aegyptopithecus, 194, 196–197, 196f
Aeluroidea, 126
Aenigmadelphys, 77
Aetiocetidae, 273t, 284
Aetiocetus, 284
Aframonius, 185
Afredentata, 203
Africa, 52, 56, 58, 62, 64, 65, 79, 80
  adapids, 186
  adapisoriculids, 144
  adapoids, 185
  altungulates, 345
  Anagalida, 346
  anomaluromorphs, 347
  anthracotheres, 293

Africa (*continued*)
  anthropoids, 192, 195f, 196
  and Arabian Peninsula, faunal exchange between, 340
  archaic ungulates, 344
  Archonta, 343
  Artiodactyla, 346
  carnivorans, 342
  chiropterans, 159, 161
  cimolestids, 96
  Early Cenozoic mammal record for, 336–337
  elephant shrews, 311, 312
  embrithopods, 265
  equids, 247
  erinaceomorphs, 146
  euprimates, 179
  hedgehogs, 147
  hyaenodontids, 123
  hyopsodontids, 222
  hyracoids, 257, 260
  lorisoids, 186
  mammalian dispersal to/from, 20
  mioclaenid, 219
  notharctids, 183
  omomyids, 188
  paleogeography, 18, 19f
  pangolins, 204
  pantolestids, 100
  plesiadapiforms, 178
  primates, 166, 169
  proboscideans, 260, 263
  rhinocerotoids, 255
  tayassuids, 296
  tribosphenic mammals, 68
  tribotheres, 70
  ungulates, 211
  xenarthrans, 198
Afrodon, 144, 144f
Afrosoricida, 143
Afrotarsius, 166t, 192
Afrotheria, 8, 12f, 139f, 143, 199, 211–212, 212f, 223–225, 242, 311, 312, 336, 342–344, 346
Ageina, 159
Ages. *See also* land-mammal ages
  standard, 10, 11, 13f, 14f, 15
Agnotocastor, 328
Agoutidae, 307t
agoutis, 318
Agriochoeridae, 286f, 289t, 300
Agriochoerus, 299f, 300
Ailuravinae, 321
Ailuravus, 320f, 321, 324f, 326
Ailuridae, 120f
Aksyiromys, 330
Alabama, 337
Alag Tsab (Mongolia), 132
Alagomyidae, 307t, 308f, 317, 319, 321, 347
Alagomys, 319–320, 321f
Alberta, Canada, 108, 164, 337
Albertatherium, 79f
Albertogaudrya, 236
Albian Age, 13f, 74, 76, 89
Alcidedorbignya, 110f, 114–115, 115f
Aleutian area, 18
Algeria, 192, 259, 260, 262, 312
Algeripithecus, 192, 195f
alisphenoid, 24, 25f, 57, 73, 81, 84, 85
  archaeocete, 281f
  elephant shrew, 311
  leptictidan, 142f
  lipotyphlan, 143
  sirenian, 269f
Allodontidae, 57f, 58
Allomyidae, 307t, 326
Allotheria, 9t, 49t, 51
Allqokirus, 85
ALMA. *See* Asian Land Mammal Ages
Alocodontulum, 36f, 207
Alostera, 89f, 213, 214
Alouatta, 197

Alphadelphia, 73, 75f, 78
Alphadon, 77–78, 78f, 79f
Altanius, 166t, 167f, 179, 180f, 339
alternating carpus. *See* carpus, alternating
Altiatlasius, 166t, 167f, 170f, 178, 179, 180f, 192, 336, 339
altricial young, 61, 73, 92
Altungulata, 10t, 22, 213t, 224–225, **241–270**, 242t, 344, **345**
alveoli (jaw), 27
Alveugena, 105, 107f, 109, 110f
Amaramnis, 101
amastoidy, 291, 293, 297
  artiodactyl, 285, 289
  Cetacea, 276
Amblocotoninae, 122
Ambloctonus, 122
Amblypoda, 114, 240
Ambondro, 51f, 67, 67f, 68
Ambulocetidae, 272f, 273t, 278–285
Ambulocetus, 280–282, 282f, 284, Plate 8.1
ambush predation, 282
Ameghino, Florentino, 202, 230, 232, 237
ameloblasts, 319
Amelotabes, 206, 207, 208f
Ameridelphia, 9t, 73–74, 74t, 75f, **79–86**
Amilnedwardsia, 213t, 234
Amphechinus, 147
amphibious mammals, 236, 262, 269, 280, 282, 283
Amphicynodon, 135, 136f
Amphicyonidae, 120f, 121t, 126, 135–136, 136f, 137f, 342
Amphidontidae, 49t, 64
Amphidozotherium, 150, 150f
Amphilemuridae, 140t, 145, 146
Amphilestes, 62
Amphilestidae, 49t, 61–63
Amphimerycidae, 286f, 289t, 300, **303**, 346
Amphimeryx, 303
Amphiperatherium, 80f, 81
Amphipithecidae, 166t, 193
Amphipithecinae, 185
Amphipithecus, 185, 193, 193f
Amphirhagatherium, 292
Amphitheriida, 9t, 49t
Amphitheriidae, 49t, 64
Amphitherium, 51f, 64–65, 65f
Amynodontidae, 242t, 255
Anacodon, 36f, 216, 220
Anagale, 139f, 309, 309f, 346
Anagalida, 4, 9t, 22, 113, 114, 216, **306–334**, 307t, 308f, 339, **346–347**
  primitive Asian, 307–310
Anagalidae, 9t, 306, 307t, 308f, **309**, 309f, 346–347
analyses
  Bayesian, 7
  maximum likelihood, 7
  supertree, 7
  total evidence, 7
anapophysis, 30, 31
Anaptomorphinae, 188–189, 191
Anasazia, 171
Ancalecetus, 284
Anchilestes, 113
Anchilophidae, 242t
Anchilophus, 248, 249
Anchistodelphys, 77, 78f
Anchomomys, 180, 183f, 185
Ancylopoda, 242t, 244, 245–246, 246f, **255–257**, 256f
Andinodelphys, 75f, 85
Andrewsarchus, 212f, 217, 219f, 274, 275, 286f
Anemorhysis, 189
Angelocabrerus, 85, 86f
angiosperms, 21, 340
angular (bone), 25f, 42f, 45f, 47f
Anhui Province (China), 339
Anisolambda, 234, 235f

Anisolambdidae, 233
Anisolambdinae, 234–235
Ankalagon, 274, 275
Ankylodon, 145f
Anomaluridae, 307t, 331–332, 332f
Anomaluroidea, 307t
Anomaluromorpha, 307t, 308f, **331–332**, 347
Anoplotheriidae, 289t, **297–298**, 298f
Anoplotherioidea, 289t, **297–298**, 346
Anoplotherium, 297–298
Antarctica, 20, 83, 236, 341. *See also* La Meseta Formation
  carnivorans, 126
  Early Cenozoic mammal record for, 336–337
  glaciation, 20
  gondwanatheres, 70–71
  marsupials, 83, 87
  paleogeography, 18–20, 19f
  polydolopoids, 83
  xenarthrans, 202, 343
anteater(s), 9t, 35, 198, 200–204, 343, Plate 7.1. *See also* Vermilingua
  scaly. *See* Pholidota
ant-eating. *See* myrmecophagy
antebrachium, dermopteran, 163
antecrochet, 227–228, 231
antelope, 10t, 285
antemolar teeth, 26–27
anterior lacerate foramen, 24
anteroconid, 316
Anthracobune, 261
Anthracobunidae, 8f, 242t, 243, 261–262, 261f, 262f, 265, 339, 345
Anthracobunodon, 292–293, 292f, Plate 8.4
anthracotheres. *See* Anthracotheriidae
Anthracotheriidae, 273, 274, 286f, 289t, **293–294**, 336, 338–340, 345–346
Anthracotherioidea, 289t, 291, 293, 294
Anthropoidea, 166, 166t, 167f, 179, 185, 186, 191–192, **192–197**, 339, 340, 343
  earliest North African, 192
  early Asian, 192–194
  Fayum, 194–197
anticlinal vertebra, 31
Antidorcas, 260
Antilocapridae, 289t
Antilohyrax, 260
antlers, 285, 303
Apataelurus, 122, 123f
Apatemyidae, 10t, 95t, 103–105, 138, Plate 3.1, Plate 3.2
Apatemys, 104, 104f, Plate 3.2
Apatotheria, 10t, 22, 94, 95t, **103–105**, 104f, 341
apes, 10t, 36, 166
apheliscines, 222, 344, 346
Apheliscus, 222–223, 312, 312f
Aphronorus, 100f, 101
Apidium, 195f, 196
Aplodontidae, 307t, 317, 320, 326
Aplodontoidea, 307t, 322, 322f, **326**, 347
apomorphic, definition of, 6
appendicular skeleton, definition of, 30
Apternodontidae, 140t, 147, **151–153**, 153, 342
Apternodus, 152–153, 152f, 153f
Aptian Age, 13f, 89
aquatic adaptations, 39f, 40, 101, 271, 275, 276, 279, 283, 284, 346
Aquilan Age, 77
Arabia
  cercamoniines, 185
  notharctids, 183
Arabian Peninsula, and Africa, faunal exchange between, 340
Arapahovius, 188
arboreal mammals, 37f, 58, 60, 65, 74, 79, 81, 85, 86, 89, 100, 101, 121, 126, 131, 163, 164, 166–167, 169–171, 177, 179, 180, 183, 186, 188, 191, 192, 194, 196–197, 201, 204, 215, 216, 327, 339–341
  skeletal adaptations, 32f, 37f, 39, 39f

Archaeoceti, 272f, 273t, 275, **278–284,** 278f, 279f, 281f, 282f, 336, 339, 345–346, Plate 8.1, Plate 8.2
Archaeohyracidae, 213t, 228f, 233
Archaeohyrax, 230
Archaeolambda, 117
archaeolambdids, 339
Archaeomeryx, 303, 305f
Archaeonycteridae, 158t
Archaeonycteris, 159–161, 160f, Plate 5.2
Archaeopithecidae, 213t, 228f, 231–232
Archaeopithecus, 230, 232
Archaeopteropus, 158t, 162
Archaeoryctes, 97
Archaeotherium, 295–296, 295f
Archimetatheria, 78
Archonta, 10t, 22, **156–197,** 157f, 158t. *See also* bat(s); Dermoptera; Primates; tree shrews
  Paleocene-Eocene, synopsis, **343**
archosaurs, 2, 44
Arcius, 176
Arctic Canada. *See* Ellesmere Island
Arcticanodon, 209
Arctictis, 32f, 33f, 36f, 101
Arctocyon, 215, 216, 216f
Arctocyonia, 113, 215, 238, 337, 338
Arctocyonidae, 113, 114, 120, 212f, 213t, 214, 214f, **215–217,** 216f–218f, 240, 245f, 272f, 275, 286f, 287, 288f, 341, 344–346, Plate 6, Plate 7.2
Arctodontomys, 178
Arctoidea, 120f, 121t, 126, 133, 135, 136f, 342
Arctostylopida, 10t, 213t, **225–226,** 225f
Arctostylopidae, 18, 22, 225–226, 229–230, 337, 339, 344–345
Arctostylops, 226
Ardynictis, 97, 98f
Ardynomys, 323, 325f
Arenahippus, 247
Arfia, 125
Argentina, 16, 64, 65, 68, 69, 71, 84, 86, 199, 201, 222, 230, 234, 236, 269, 336, 340. *See also* Patagonia
Arginbaatar, 57f, 58
Arginbaataridae, 57f
Arguimuridae, 49t, 66
Arguimus, 65f, 66
Arguitheriidae, 49t, 66
Arguitherium, 66
Argyrolagidae, 9t, 74t, 75f
Argyrolagoidea, 74t, 81–84, 341
Arkansas, 337
armadillos, 9t, 35, 40, 71, 198, 200, 201, 343. *See also* Dasypodidae
Arminiheringia, 85, 86f
Armintomyidae, 307t, 329f
Armintomys, 330, 331, 347
Arshantan ALMA, 14f, 17
Arsinoea, 194
Arsinoitheres. *See* Arsinoitheriidae
Arsinoitheriidae, 22, 211–212, 242t, 265, 336
Arsinoitherium, 237, 262, 265–266, 266f, 267f
artery(ies)
  external carotid, 128–129
  internal carotid, 24, 26, 79, 96, 128–129, 134, 169, 171, 174–176, 178, 180, 188, 194
  stapedial, 303, 328
  vertebral, 30, 301
articular (bone), 42f, 44, 45f–47f, 52
Artiocetus, 273, 283, 283f
Artiodactyla, 7, 8, 10t, 11f, 12, 22, 36, 178, 211–212, 212f, 213t, 214–217, 222, 241, 260, 262, 271–274, 272f, **285–305,** 286f, 289t, 298f, 337–339, 344, **345–346,** Plate 6, Plate 8.3, Plate 8.4
Asfaltomylos, 68
Asia, 52, 55, 58, 59, 62, 64, 65, 74, 79, 80, 271, 342
  adapids, 186
  amphipithecids, 193

Anagalida, 306, 309, 346
anthracotheres, 293
anthropoids, 192–194, 193f
apternodontids, 152
archaic ungulates, 344–345
arctocyonids, 216
arctostylopids, 225, 225f, 226
Artiodactyla, 346
artiodactyls, 288
brontotheres, 250, 252
carpolestids, 172
chalicotheres, 257
chiropterans, 159, 161, 162
cimolestans, 341
cimolestids, 96
coryphodontids, 118
dermopterans, 163
didymoconids, 97
embrithopods, 265
entelodonts, 295
equids, 247, 248
Erinaceidae, 342
erinaceids, 146
euprimates, 178, 179
  and Europe, faunal exchange between, 338
eutherians, 89, 90, 90f, 92
feliforms, 132
Glires, 313
hedgehogs, 147
helohyids, 290
heterosoricines, 153–154
hyaenodontids, 123, 125
hyopsodontids, 222
hyracoids, 257
lagomorphs, 314, 347
leptomerycids, 303
mammals, 338–340
mesonychians, 274, 275
metatherians, 340–341
miacoids, 130
micropternodontids, 150
mimotonids, 315
  and North America, faunal exchange between, 337, 339, 340
notharctids, 183
nyctitheres, 149
omomyids, 188, 190
oxyaenids, 122
palaeanodonts, 207
palaeotheres, 248
pangolins, 204
pantodonts, 114, 117, 118
pantolestids, 100
paroxyclaenids, 101
periptychids, 219
phenacodontids, 223
pholidotan, 205
plesiosoricids, 153
primates, 166, 169, 188
proboscideans, 260
rhinocerotoids, 255
rodents, 319, 321, 324f, 326, 328
Scandentia, 343
Sciuravidae, 347
shrews, 154
sivaladapids, 186
soricomorphs, 148
suids, 297
tillodonts, 110, 113, 118
trogosines, 113
uintatheres, 238
ursids, 135
xenarthrans, 198
zhelestids, 213
Asiabradypus, 202
Asiadelphia, 9t, 72, 74t, **76,** 340
Asian Land Mammal Ages, 11, 14f, **17**
Asiatherium, 72, 75f, 76, 77f
Asiavorator, 132, 133f
Asiomomys, 190

Asioplesiadapis, 172, 339
Asioryctes, 90, 92f, 95
Asioryctitheria, 9t, 11f, 89f, 90, 95, 140
Asiostylops, 226
Asmithwoodwardia, 221, 227f, 233–234, 234f
Aspanlestes, 213, 213f, 214
Astigale, 215
astragalar canal, 45
astragalar foramen, 53
astragalus (pl. astragali), 30f, 33, 34, 37f, 38f, 75f, 308, 312
  adapoid, 168f
  alagomyid, 320
  anoplothere, 298
  archaeocete, 273, 283f
  archontan, 159f
  arctocyonid, 218f
  artiodactyl, 283f, 285, 287f
  astrapothere, 233f, 236
  Bunophorus, 38f, 283f
  Chriacus, 38f, 218f
  cimolestid, 98f
  condylarth, 223f
  Dasypus, 38f
  Deccanolestes, 159f
  didolodontid, 233f
  double-pulley (trochleated), 273, 285, 290
  euprimate, 168f
  eutherian, 90, 92
  glyptodont, 202
  hyaenodontid, 126
  hyopsodontid, 223, 223f
  hypertragulid, 303
  hyracoid, 260
  Hyracotherium, 38f, 246f
  ischyromyid, 325f
  litoptern, 233f, 234
  Manis, 38f
  mesonychian, 274, 277f, 283f
  miacid, 131, 131f
  mimotonid, 315
  notoungulate, 228, 233f
  omomyid, 168f
  oreodont, 300
  oxyaenid, 121
  Palaeanodon, 38f
  palaeoryctid, 97
  pangolin, 204
  pecoran, 303
  periptychid, 220
  perissodactyl, 244, 246f
  phenacodontid, 224
  Phenacodus, 38f, 246f
  pholidotan, 205
  plesiadapiform, 168f
  proboscidean, 260
  Procerberus, 98f
  Protungulatum, 98f
  pseudictopid, 310
  rodent, 325f
  uintathere, 240
  ungulate, 98f, 211
  viverravid, 131, 131f
  xenarthran, 201
Astraponotus, 235
astrapotheres. *See* Astrapotheria
Astrapotheria, 10t, 22, 213t, 215, 226, **235–236,** 236f, 337, 344
Astrapotheriidae, 213t, 233f, 236
Astrapotherium, 235, 236
atlas (vertebra C1), 30, 30f, 31f, 53, 270
auditory bulla, 24, 73, 85, 96, 121, 203, 309
  adapoid, 180
  amphicyonid, 135
  anthropoid, 194
  apternodont, 152
  archaeocete, 280
  canid, 134
  caniform, 133
  carnivoran, 128–129, 129f, 132

auditory bulla (*continued*)
  cetacean, 277
  chiropteran, 158
  creodont, 121
  cylindrodontid, 323
  dermopteran, 163
  didymoconid, 97
  elephant shrew, 311
  Ernanodon, 210
  euprimate, 179
  feliform, 132
  lagomorph, 313
  leptictidan, 140, 141
  lipotyphlan, 143
  marsupial, 73, 81, 85
  microsyopid, 178
  nimravid, 133
  notoungulate, 228
  omomyid, 188
  oromerycid, 301
  palaeoryctid, 96
  paromomyid, 176
  peccary, 297
  pentacodontid, 101
  phenacodontid, 223
  plagiomenid, 164
  plesiadapid, 171
  protoceratid, 302
  rodent, 321
  sciuravid, 322
  squirrel, 328
  talpid, 154–155
  ungulate, 211
  xiphodont, 301
auditory region (of basicranium), 24
auditory (eustachian) tube, 26, 259
Aumelasia, 290, Plate 8.3
Auroratherium, 207
Ausktribosphenida, 9t, 49t
Ausktribosphenidae, 51f
Ausktribosphenos, 67–68, 67f
Australia, 4, 68, 341
  carnivorans, 126
  chiropterans, 159, 343
  Early Cenozoic mammal record for, 336–337
  mammalian dispersal to, 20
  marsupials, 9t, 72–74, 79, 85–87, 86f, 341
  monotremes, 68–69
  paleogeography, 18, 19f
  Tingamarra, 86f, 87, 225
Australidelphia, 9t, 73, 74, 74t, 75f, 78, 85, **86–87**
Australonycteris, 158t, 159
Australosphenida, 9t, 49t, **67–69**, 67f, 70
Austria, 338
Austrotriconodontidae, 49t, 61, 62
Autoceta, 273t, 275
Avenius, 178
Avitotherium, 89f, 213, 214
Axel Heiberg Island, 164, 337
axial skeleton, definition of, 30
axis (vertebra C2), 30, 30f, 31f
  morganucodont, 53
  pantolestan, 101
  tritylodontid, 45
aye-aye, 341. See also *Daubentonia*
Azibiidae, 169, 178
Azibius, 178
Azygonyx, 111–113

Bachitheriidae, 289t, 304
Bachitherium, 304
baculum, 33
badgers, 9t, 40, 144
Bahinia, 193, 193f
Baioconodon, 212f, 214, 216–217
Baja California, 337
Balaenidae, 273t, 284–285
balaenids, 278
Balaenoptera, 1
Balaenopteridae, 273t, 278, 284–285

balance, organs of, 24, 26
baleen, 35, 276–277, 279f, 284, 285
baleen whales. See Mysticeti
Balochistan, 186
Baluchitherium, 248f, 255
bandicoots, 9t
Banyuesminthus, 330
Barrancan subage, 17
Bartonian Stage/Age, 13f, 14f, 20, 338
Barunlestes, 90–92, 307
Barylambda, 116, 116f–118f, 117
Barylambdidae, 95t, 110f, 116, 116f–118f
barytheres. See Barytheriidae
Barytheriidae, 8f, 242t, 260–261, 265
Barytherium, 261f, 265
Basalina, 110
base-pairs, 7
basicranium, 24–26. See also auditory region (of basicranium); foramen/foramina
  air sinuses, archaeocete, 284
  archaeocete, 280
  carnivoran, 119, 127, 129
  creodont, 119
  dichobunid, 289
  fossil whale, 273
  paromomyid, 176
  plagiomenid, 164
  plesiadapiform, 178
  protoceratid, 302
Basilosauridae, 272f, 273t, 278–285, 282f, 284, 284f
Basilosaurinae, 272f, 284
Basilosaurus, 282f, 284
basioccipital, 24, 25f
basisphenoid, 24, 25f
  anagalid, 310f
  lipotyphlan, 143
bat(s), 6, 22, 159–162, 175, 343, Plate 5.1–Plate 5.4. See also Chiroptera
  dentition, 35, 157–159, 160f
  Early Cenozoic
    of Asia, 339
    of Australia, 337
    of North America, 337
    of Southern Hemisphere, 336–337
  molossid, 162
  natalid, 162
  nectivorous, 36
  niche partitioning by, 161, 162f
  phylogeny and classification, 8, 10t
  primitive, 160f
  skeletal adaptations, 39
  vespertilionid, 162
Bathonian Age, 56
Bathyergidae, 334, 347
Bathyopsis, 238, 239
Batodon, 89f, 140t, 148, 149f, 342
Batodonoides, 29f, 149, 149f
Bayan Ulan (Mongolia), 339
Bayesian analysis, 7
bear-dogs. See Amphicyonidae
bears, 35, 39. See also Ursidae
beavers, 9t, 40, 101, 318, 320, 328
Belgium, 338. See also Dormaal
Bemalambda, 114, 115f, 118f
Bemalambdidae, 95t, 110f, 114–115, 115f, 339
Benaius, 110, 113
Bering land bridge, 18, 19f
Beringia, 18, 342. See also Bering land bridge
Berruvius, 170f, 178
Betulaceae, 21
bicipital tuberosity, 33
Bighorn Basin (Wyoming), 16, 59, 189, 191f, 222, Plate 3.3, Plate 6
bilophodonty/bilophodont dentition, 36, 69, 107, 237–238, 241, 242, 253, 254, 262–266, 270, 291, 297, 322
binturong, 101
biochronology, of Early Cenozoic, 8, 10–12, 13f, 14f, 15–17

biochrons, 16
bipedalism, 142, 142f, 257
Bishops, 68
Bisonalveus, 101
blowhole, cetacean, 276
blubber, 276
Bobolestes, 148
bolide impact, 4–5, 21, 340
Bolivia, 202, 227f, 336. See also Tiupampa
Bolodon, 57f
Bonapartheriidae, 74t, 83
Bonapartherium, 83
bone(s), 41–42
  articular ends, 23
  growth, 23
  of skull, 24. See also skull; specific bone
Boreastylops, 229
Boreoeutheria, 8
Boreosphenida, 9t, 49t, 51f, 67f, 68, **69–70**
Borhyaenidae, 9t, 74, 74t, 85
Borhyaenoidea, 75f, 78–79, **84–86,** 86f
borophagines, 135
Bothriodon, 293, 294f
Bothriostylops, 226
Bouxwiller (France), 338
Bovidae, 289t
Bovoidea, 289t
Brachianodon, 206
brachium, dermopteran, 163
brachydonty/brachydont dentition, 27, 29f, 35, 36, 71, 83, 92, 231, 234, 289, 294, 297, 300, 301, 322, 325
Brachyhyops, 295–296, 295f
Bradypodidae, 199t
Bradypus, 202
brain, 41, 49. See also encephalization quotient (EQ)
  anthropoid, 194, 197
  archaeocete, 279
  carnivoran, 121, 129
  Coryphodon, 118
  creodont, 121
  euprimate, 179
  mesonychian, 275
  miacoid, 129
  multituberculate, 61
  notoungulate, 229
braincase, 24, 49
  bemalambdid, 114
  carnivoran, 129
  entelodont, 295
  euprimate, 167
  monotreme, 57
  multituberculate, 57
  peccary, 297
Bransatoglis, 329
Brasilitherium, 46
Brasilodon, 46
Brazil, 221, 236, 336
  xenarthrans, 201
Bridger Formation, 321
Bridgerian NALMA, 14f, 111, 112f, 122, 125, 126, 129, 152, 178, 206, 207, 223, 252, 326
Bridgerian/Uintan boundary, 14f, 16
Brontops, 251, 252f
Brontotheriidae, 242t, 244–246, 248f, 249f, 250, 250f, **251–252,** 251f, 252f, 257, 339, 345
Brontotheriinae, 250
Brontotherioidea, 242t, 246f
Brontotherium, 235, 251f
browsing mammals, 116, 244, 247, 250, 255, 256, 260, 300, 301, 303
Bryanpattersonia, 230, 233f
buccal, definition of, 27
Bug Creek (Montana), 214
Buginbaatar, 57f
Bugti Hills (Balochistan), 186
Bugtilemur, 186–187, 187f
Bulganbaatar, 58
Bulgaria, 338

Bumbanian ALMA, 14f, 244, 257, 333, 339, 347
bumblebee bat, 1, 153
bunodonty/bunodont dentition, 28, 29f, 35–36,
    69, 79, 83, 105, 135, 144–146, 155, 158,
    164, 167, 171, 174, 176, 183–185, 192,
    194, 196, 211, 214, 215, 217, 220, 222,
    223, 225, 235–237, 250, 259, 260, 264,
    286, 288–291, 292f, 293, 296, 297, 310,
    311, 311f, 346
Bunohyrax, 259f, 260
bunolophodonty, 233, 252, 260, 262, 264, 291, 326
Bunomeryx, 286f, 289
Bunophorus, 38f, 283f, 289
bunoselenodonty/bunoselenodont dentition,
    29f, 83, 250, 260, 288, 292–294, 297, 300,
    326
Burma. *See* Myanmar
Bustylus, 144, 144f
Butler, Percy, 145
Butselia, 153
Buxolestes, 101, 102f

Cadurcodon, 255
Caenolambda, 116f
Caenolestes, 81
Caenolestidae, 74, 74t, 81, 81f
Caenolestoidea, 75f, 83–84
Caenopithecus, 185
Cainotheriidae, 286f, 289t, 297, 298f, **299**, 346
Cainotherium, 305f
Calcaires de Rona (Romania), 338
calcaneal canal, mimotonid, 315
calcaneal tuber, eutherian, 90
calcaneofibular articulation, pseudictopid, 310
calcaneus, 30f, 33, 34, 37f, 75f
    archaeocete, 273, 283f
    archontan, 159f
    artiodactyl, 283f, 287f
    astrapothere, 233f, 236
    cimolestid, 98f
    Deccanolestes, 159f
    didolodontid, 233f
    eosimiid, 193
    fossil whale, 273
    hyopsodontid, 223
    ischyromyid, 325f
    lipotyphlan, 143
    litoptern, 233f
    mesonychian, 277f, 283f
    miacid, 131, 131f
    miacoid, 130
    mimotonid, 315
    notoungulate, 233f
    palaeoryctid, 97
    periptychid, 220
    perissodactyl, 244, 245f, 246f
    Procerberus, 98f
    Protungulatum, 98f
    pseudictopid, 310
    ungulate, 98f
calcar, chiropteran, 157
California, 337
Callistoe, 85, 86f
Camelidae, 10t, 12f, 272f, 285, 286–287, 286f, 289t,
    301, **302**
Cameloidea, 289t
camels. *See* Camelidae
Campanian Age, 13f, 77, 90, 92
Campanorcidae, 213t, 231
Canada
    Early Cenozoic mammal record for, 337
    Saxonella, 173
cane rats, 334
canid(s). *See* Canidae
Canidae, 120f, 121t, 126, 131, 133, **134–135**,
    342
Caniformia, 127, **133–137**, 342
    classification, 120f, 121t, 126
    origin, 131, 134, 135f
caniforms, 131. *See also* Caniformia

canine teeth, 25f, 26, 27, 85
    adapoid, 182
    anagalid, 309
    anaptomorphine, 188
    anoplothere, 298
    anthracothere, 294
    anthropoid, 192, 194
    apternodont, 152
    arctocyonid, 216
    artiodactyl, 286
    cebochoerid, 291
    choeropotamid, 292
    condylarth, 215
    conoryctid, 105–106
    coryphodontid, 116f, 118
    creodont, 120
    dichobunid, 288–289
    didymoconid, 97
    Dinocerata, 239
    entelodont, 295
    eosimiid, 193
    Ernanodon, 210
    eupantothere, 64
    euprimate, 168
    helohyid, 290
    hyopsodontid, 222
    lipotyphlan, 144
    mesonychian, 274
    nimravid, 133
    omomyid, 188
    oreodont, 300
    oxyaenid, 122
    palaeanodont, 206
    pantodont, 114, 117
    pantolambdodontid, 117
    peccary, 297
    phenacodontid, 223
    plagiaulacoid, 58
    plesiosoricid, 153
    primate, 186
    procumbent, 182
    pseudictopid, 310
    Purgatoriidae, 171
    ruminant, 302
    Sinoconodon, 53, 53f
    strepsirrhine, 187
    stylinodontid, 107
    suid, 297
    symmetrodont, 63
    taeniodont, 105, 109
    venomous mammals, 101
    xenarthran, 200
Canis, 32f, 33f, 101, 144
cannonbone, 285, 297
    leptomerycid, 303
    peccary, 297
Cantius, 172–173, 172f–173f, 180, 182–185, 182f,
    Plate 6
capitate, 33
captorhinomorphs, 44
capybaras, 9t, 40, 320f
Carbon Isotope Excursion (CIE), 12, 16, 19
Cardiolophus, 243f, 245f, 246, 246f, 253
Caribbean, 338
    altungulates, 345
carnassial notch, 85, 128
carnassials, 28, 35, 85, 119–120, 122f, 126,
    127–128, 130, 134, 342
    amphicyonid, 135
    creodont, 120
    hyaenodontid, 126
    ursid, 135
Carnian Age, 13f, 50
Carnilestes, 150, 151f
Carnivora, 11f, 12f, 22, 94, 120, **126–137**, 156, 199,
    337–338, **341–342**. *See also* carnivores
    age of divergence, 5t
    auditory structures, 128–129, 129f
    classification and phylogeny, 9t, 119, 120f, 121t,
        126, 139f, 272f, 286f

crown-group, 120f, 127, 342
dental formula, 127
dentition, 119, 122f, 127, 129, 130f
diet, 35, 126
origin, 96, 129
Carnivoramorpha, 127
carnivoran(s), 337, 341. *See also* Carnivora
    crown, 342
    Early Cenozoic, of North America, 337–338
    feliform, 338
    musteloid, 338
    stem, 342
carnivores, 8, 9t, 35, 39, 84, 119–120, 129, 274.
    See also Carnivora
    dentition, 78, 85
Carodnia, 237, 238, 238f, 240
Carodniidae, 213t
Caroloameghinia, 83f
Caroloameghiniidae, 79, 83
Caroloameghinioidea, 74t, 81–84
Carolodarwinia, 230
Carolopaulacoutoia, 81f, 83–84
Carolozittelia, 237
carotid artery
    external, 128–129
    internal, 24, 26, 79, 96, 128–129, 134
carotid canal, 26
carpals, 40
    hyracoid, 258
    perissodactyl, 244
Carpocristes, 172, 339
Carpodaptes, 172
Carpolestes, 172, 173, 174f, 176f, Plate 4.3
Carpolestidae, 18, 166t, 169, 170, 170f, **172–173**,
    343
Carpomegadon, 172
carpus, 30f, 33, 35f, 111, 129
    alternating, 223, 225
    archaeocete, 284
    carnivoran, 129
    creodont, 121
    dermopteran, 163
    hypertragulid, 303
    hyracoid, 259
    leptomerycid, 303
    metatherian, 73
    miacoid, 130
    Numidotherium, 263
    paromomyid, 176
    phenacodontid, 223, 224
    serial, 223, 225, 260, 263
    tillodont, 111
cartilage, 23
Casamayoran SALMA, 14f, 17, 20, 83, 85, 201,
    202, 221, 227, 228, 229f, 230–233,
    235–237
Castor, 101
Castoridae, 307t, 317, 320, 322, 322f, 326, 328
Castorocauda, 56
Castoroidea, 307t, 328, 347
cat(s), 126. *See also* Felidae
    auditory structures, 129f
    classification, 9t
    saber-toothed, 6, 122
Catarrhini, 7, 166t, 167f, 194
Catopithecus, 194, 195f
Catopsbaatar, 57f
cattle, 10t, 285
caudal vertebrae, 29–30, 30f, 31, 31f
    archaeocete, 284
    arctocyonid, 218f
    cetacean, 276
    xenarthran, 201
Caviomorpha, 307t, 316, 334, 347
cavum epiptericum, 43, 45, 50
Cebochoeridae, 286f, 288, 289t, **291**, 293, 297,
    298f
Cebochoerus, 291, 298f
Ceboidea, 166t
Cecilionycteris, 161

cecum, lipotyphlan, 143
Cedrocherus, 146, 147*f*
cedromurines, 327–328
cementum, 27, 261
Cenomanian Age, 13*f*, 78, 148
Cenozoic Era, geochronology and
    biochronology, 8, 10–12, 13*f*, 14*f*, 15–17
Centetodon, 146, 148–149, 149*f*
Central America
    paleogeography, 18, 19*f*
    xenarthrans, 203
centrale (bone), 33, 224, 225
    hyracoid, 259
    Numidotherium, 263
centrocrista, 28, 28*f*
Cephalogale, 135, 342
Ceratomorpha, 242*t*, 244, 245, 246*f*, **252–255,**
    254*f*
Cercamoniinae, 183–185
Cercamonius, 184
Cercopithecoidea, 166*t*
cerebrum, 41
Cernay (France), 15
Cernaysian Age, 15
cervical nerves, suiform, 293
cervical vertebrae, 30, 30*f*–31*f*
    of aquatic mammals, 40
    cetacean, 276
    embrithopod, 266
    of fossorial mammals, 40
    fused, 40, 60, 207, 276
    manatee, 30
    palaeanodont, 207
    palaeothere, 249
    of saltatorial mammals, 40
    sloth, 30, 201
    suiform, 293
    taeniolabidoid, 60
Cervidae, 289*t*
Cervoidea, 289*t*
Cetacea, 5, 7, 8, 12*f*, 22, 211–212, 212*f*, 213*t*, 215,
    241, 271, 273*t*, **275–285,** 286*f*, 291, 339,
    344–346. *See also* Archaeoceti; Mysticeti;
    Odontoceti
    classification and phylogeny, 10*t*, 11*f*, 211–212,
        212*f*, 213*t*, 272*f*, 273–274, 273*t*, 286*f*
Cetartiodactyla, 12*f*, 274, 344, 346
    age of divergence, 5*t*
    classification, 139*f*, 272*f*
Cete, 215, **271–285, 345–346**
    classification and phylogeny, 10*t*, 213*t*, 272*f*,
        273*t*, 286*f*
Cetotheriidae, 273*t*, 284–285
Ceutholestes, 150, 150*f*
Chadrolagus, 314
Chadronian NALMA, 15, 17, 59, 135, 136, 140,
    155, 164, 204, 207, 251, 252, 255, 293,
    301, 302, 314, 323, 327
Chadronian/Orellan boundary, 14*f*, 15
Chalicomomys, 174
Chalicothere(s), 246. *See also* Ancylopoda;
    Chalicotheriidae
Chalicotheriidae, 10*t*, 242*t*, 244–245, 249*f*, 255,
    257
Chalicotherium, 257
Chambilestes, 146
Chambilestidae, 140*t*, 145
Chambius, 311–312, 311*f*
Chapattimyidae, 307*t*, 332, 334
Chapattimys, 332
character polarity. *See* polarity
cheek teeth, 26, 36, 44, 45, 87. *See also* molars
    adapid, 185
    adapoid, 183
    alagomyid, 319
    anagalid, 309
    anaptomorphine, 189
    anthropoid, 196
    archaeocete, 278, 279, 284
    arctostylopid, 225

artiodactyl, 286, 287
astrapothere, 235
camel, 302
cimolestid, 95
cingulate, 201*f*
cylindrodont, 325
elephant shrew, 311
fossil whale, 273
geolabidid, 149
gliran, 313
Hadroconium, 55
ischyromyid, 321
lagomorph, 313
mesonychian, 274
mimotonid, 315
notharctid, 183
notoungulate, 227, 228*f*
ochotonid, 314
oreodont, 300
pantodont, 114, 117
pentacodontid, 101
periptychid, 220
phenacodontid, 223
pseudictopid, 310
Pseudoglyptodon, 202
rodent, 316
squirrel, 328
taeniodont, 108
talpid, 154–155
tethythere, 242
theridomyid, 326
tillodont, 110
trogosine, 111
xenarthran, 200
Cheirogaleidae, 166*t*
Cheirogaleus, 188
chevron bones, 31
chevrotains, 285. *See also* Tragulidae
chewing. *See also* mastication
    haramiyid, 51, 52
    lagomorphs, 314
    multituberculate, 56
    muscles, 26
    orthal, 51, 52, 56, 128
    palinal, 51, 52, 56, 71
    propalinal, 56, 316, 318
    in rodents, 56, 316
Chile, 84, 86
China, 51, 55, 56, 63, 64, 70, 339, 340, Plate 4.4
    adapids, 186
    Anagalida, 346
    anagalids, 309
    anthropoids, 192
    apternodontids, 152
    carpolestids, 172, 173
    chalicotheres, 257
    chiropterans, 161
    didymoconids, 97
    embrithopods, 266
    entelodonts, 295
    eosimiids, 193
    erinaceids, 147
    Ernanodon, 209
    euprimates, 179
    eutherians, 90*f*
    heterosoricines, 153–154
    metatherians, 340–341
    miacoids, 130
    palaeanodonts, 205, 207
    pantodonts, 114
    plesiosoricids, 153
    Scandentia, 343
    Simplicidentata, 315, 316*f*
    sivaladapids, 186
    suids, 297
    tarsiids, 191–192
    tayassuids, 296
    tillodonts, 113, 114
    triisodontines, 217
    tupaiids, 197

chinchillas, 9*t*
Chiromyoides, 172
Chiroptera, 11*f*, 12*f*, 119, 156, **157–162,** 163, 343
    age of divergence, 5*t*
    classification, 10*t*, 139*f*, 157, 157*f*, 158*t*
Chlamyphorus, 207
choanae, 26
Choeropotamidae, 288, 289*t*, **291–293,** 292*f*, 294
Choeropotamus, 291–293
Choloepus, 38*f*, 202
Chonecetus, 284
Chriacus, 34*f*, 36*f*–38*f*, 215, 216, 218*f*, 245*f*, 287,
    Plate 6, Plate 7.2
Chronolestes, 166*t*, 170*f*, 173, 339
chronostratigraphy, units for, 10, 13*f*, 14*f*
Chrysochloridae, 8, 12*f*, 139*f*, 140*t*, 143, 148, 152,
    212, 212*f*, 311
Chrysochloromorpha, 143
Chulpas, Peru, 83
Chulsanbaatar, 57*f*
Chungchienia, 113
CIE. *See* Carbon Isotope Excursion (CIE)
Cimolesta, 22, **94–118,** 99, 107*f*, 113, 199, 214,
    336, 337, **341,** 344
    classification, 10*t*, 94, 95*t*, 140, 199*t*
Cimolestes, 95–97, 95*f*, 108, 110*f*, 120, 129, 130*f*,
    342
Cimolestidae, 89*f*, 94–95, 95*t*, 109, 113, 114, 129,
    337–338
Cimolodonta, 49*t*, 57*f*, 58
Cimolomyidae, 57*f*
Cingulata, 199*t*, 200–202, 201*f*, 343
cingulid/cingulum (pl. cingula), 28, 28*f*, 67*f*, 68,
    99, 109, 152, 171, 185–187, 193, 194, 220,
    222, 238, 291, 311, 319, 321, 322, 328
civet(s), 9*t*, 39, 126, 132
clade(s)
    definition, 6
    "modern," of extant orders, 4
cladograms, definition of, 7
Claenodon, 111, 215, 216
Clark, W. E. Le Gros, 197
Clarkforkian NALMA, 12, 14*f*, 59, 104, 113, 116,
    117, 122, 130, 159, 160, 172, 177, 206,
    223, 226, 319, 321, 338, 339
Clarkforkian/Wasatchian boundary, 12, 14*f*
classification, 7–8, 9*t*–10*t*, 22
clavicle, 32, 40, 65
    dichobunid, 290
    fossil whale, 273
    lagomorph, 314
    perissodactyl, 244
    pholidotan, 204
    talpid, 155
claws, 33, 39, 40, 65, 215
    ancylopod, 256
    arctocyonid, 216
    chalicothere, 257
    chiropteran, 157, 159, 161
    dermopteran, 163
    erinaceomorph, 146
    grooming, adapoid, 181
    hyopsodontid, 222
    hyracoid, 258
    ischyromyid, 321
    nimravid, 133
    oreodont, 300
    pantodont, 116–117
    paroxyclaenid, 101
    retractile, 132–133
    stylinodontid, 107
    taeniodont, 106
    tillodont, 111
    xenarthran, 201
Climacoceratidae, 289*t*
climate, Paleocene-Eocene, 20–21
climbing, skeletal adaptations for, 39, 216, 341.
    See also arboreal mammals; scansorial
    mammals
coatimundi, 136, 216

cochlea, 24, 43, 50, 60, 79
  chiropteran, 158, 161
  eutherian, 89, 90
  eutriconodont, 61
  monotreme, 68
  multituberculate, 57
  Sinoconodon, 53
  symmetrodont, 64
  Vincelestes, 66
cochlear canal, 24
Cocomyidae, 307t, 332
Cocomys, 332–333, 333f
Colbertia, 230, 233f
colobus monkeys, 36
Colombia, 237, 238
Colombitheriidae, 213t, 237
Colombitherium, 236–237, 237f
Colorado, 55, 59, 71, 116
colugo(s), 10t, 162–164, 260
Comanchea, 70
Comoro Islands, 186
Conacodon, 220
condylarth(s). *See* Condylarthra
Condylarthra, 2, 7, 10t, 17, 22, 61, 109–110, 113, 114, 214, **215–240**, 245f, 246f, 287, 311, 311f, 312, 337, 344–345
  classification, 211–212, 212f, 213t, 215
  dental formula, 215
  Early Cenozoic
    of Europe, 338
    of North America, 337–338
    of Southern Hemisphere, 336
Coniacian Age, 13f, 90
conies, 257
Conoryctes, 105–106, 108f
Conoryctidae, 105, 109, 110f
continental drift, 8, 18–20, 19f
conules, 77, 78f, 96, 105, 110, 114, 120, 148, 149, 164, 171, 172, 196, 214, 215, 217, 250, 289, 295, 298, 299, 303, 311, 319, 321, 322, 325
  definition, 28, 28f
convergence, 6, 34, 43
coracoid, 32, 32f, 53, 63, 68
coronoid (process), 25f, 47f, 56, 58, 64
  artiodactyl, 295, 298
  lagomorph, 314
  mesonychian, 277f
  proboscidean, 263
  soricomorph, 148
coronoid crest, 25f
Coryphodon, 115f, 116–118, 116f, 315, 337–339, 341, Plate 3.3
Coryphodontidae, 95t, 110f, 114, 117–118
cotylar fossa, 260
coyote, 101, 134–135, 144
cranial nerves, 24–26
  I. *See* olfactory nerve
  II. *See* optic nerve
  III. *See* oculomotor nerve
  IV. *See* trochlear nerve
  V. *See* trigeminal nerve
  VI. *See* abducent nerve
  VII. *See* facial nerve
  VIII. *See* vestibulocochlear nerve
  IX. *See* glossopharyngeal nerve
  X. *See* vagus nerve
  XI. *See* accessory nerve
  XII. *See* hypoglossal nerve
  multituberculate, 58
Craseonycteridae, 158t
Craseonycteris, 1, 153. *See also* bumblebee bat
Craseops, 178
Creodonta, 6, 12, 22, 94, 99, **119–126**, 341–342
  classification and relationships, 9t, 11f, 119, 120f, 121t
  dental formula, 120
  dentition, 35, 119, 120, 122f
  Early Cenozoic, of North America, 337
  origin, 96

Creotarsidae, 140t, 145
crest(s) (tooth), 27
Cretaceous Period, 13f
  bolide impact, 4–5
  fossil record quality, 5
  mammalian diversification, 3f, 4, 335
  mammalian geochronology and biochronology, 13f, 14f
  mammals, 50f
  mass extinctions, 2
Cretaceous/Tertiary boundary. *See* K/T boundary
Cretasorex, 154
cribriform plate (ethmoid bone), 24, 61
Cricetidae, 307t, 330–331, 338, 339
cristid obliqua, 28, 28f, 73f, 78f, 171, 186, 196, 225, 238, 239f, 321, 326
Crivadiatherium, 266
crochet, 227, 228f, 231
Crocidurinae, 153
crocodilians, 18, 19
crown, of tooth, 27
crown-group taxon, 8, 8f, 43
Crown-Therian radiation, timing of, 3–5
crural index
  anthropoid, 196
  ischyromyid, 322
Crusafontia, 64
Cryptadapis, 185–186
Cryptoryctes, 150–151
Ctenacodon, 57f
Ctenodactylidae, 307t, 332
Ctenodactyloidea, 307t, 308f, 317, 318, 325f, **332–334**, 332f, 333f, 339, 347
cuboid, 30f, 33, 37f, 75f
  artiodactyl, 285
  astrapothere, 236
  dichobunid, 290
  hyopsodontid, 223
  mesonychian, 274, 277f
  oreodont, 299
  oxyaenid, 121
  protoceratid, 301
cubonavicular
  amphimerycid, 303
  artiodactyl, 285, 287f
  ruminant, 302
Cuisitherium, 292, 292f, 299
cuneiforms, 30f, 33, 35f, 37f
  palaeanodont, 208f
cursorial mammals, 6, 37f, 121, 125, 126, 131, 134–135, 211, 219, 223, 226, 233, 233f, 234, 244, 247, 253, 255, 257, 258f, 260, 273, 274, 280, 281, 283, 285, 289, 290, 292, 293, 302, 303, 310, 312, 314, 315, 342
  bipedal, 142, 142f
  skeletal adaptations, 6, 32f, 39f, 40, 308
cusp(s), of teeth, 64, 83, 96, 99, 101, 104, 144, 152, 164, 170–172, 174, 185, 188, 216, 217, 219, 220, 223, 224, 273–275, 278, 280, 289–292, 297. *See also* stylar cusps
  alagomyid, 319
  docodont, 55
  eutricodont, 54
  eutriconodont, 62
  hypoconulid-entoconid, 76–77
  Kuehneotherium, 54
  marsupial, 72
  monotreme, 69
  morganucodont, 55
  multituberculate, 56
  nomenclature for, 27–28, 28f
  plagiaulacoid, 58
  symmetrodont, 54, 63
  therian, 54
  triconodont, 61
  Woutersia, 54
cuspules, 78
Cuvierimops, 162
Cyclopedidae, 199t

Cylindrodon, 323, 326, 327f
Cylindrodontidae, 307t, 308f, 322f, **323–326**, 325f, 327f, 332, 332f, 347
Cymbalophus, 253
Cynocephalidae, 163
Cynocephalus, 38f, 163, 163f, 164, 174f, 176–177, 260
Cynodictis, 135
cynodonts, 23, 26, 42–46, 42f, 43f, 44–46, 44f–47f, 52, 53
Cynoidea, 121t, 133
Cynomys, 222
Cyriacotheriidae, 95t, 110f, 116, 117
Cyriacotherium, 115f, 117

Dacrytheriidae, 289t, 297, 298f, **299**
Dacrytherium, 298f, 299
Dactylopsila, 104–105, 106f
dagger symbol (†), 22
Danian Stage/Age, 13f, 14f, 338
Danjiangia, 257
Dano-Montian Age, 15
Daouitherium, 8f, 261f, 262f, 263
Daphoenictis, 135
Daphoenus, 135, 136f, 137f
Darbonetus, 149
dassies, 257. *See also* Hyracoidea
Dasypodidae, 199t, 200f, 201–202, 201f, 343
Dasypodoidea, 199t
Dasypus, 32f, 38f
Dasyuromorphia, 9t, 74t, 75f
dating methods, 10
Daubentonia, 104, 106f, 313, 341
Daubentoniidae, 166t
Daulestes, 89f, 90, 91f, 92f
dawn horse(s), Plate 6. *See also Hyracotherium*
Dawsonomys, 323, 326
De Geer Route, 18, 19f
Deccanolestes, 19–20, 157, 158t, 159f, 343
deciduous teeth, 26–27
Decipomys, 316
Decoredon, 179
deer, 10t, 285
Deinogalerix, 144
Deinotheriidae, 242t, 260, 261f
Deltatheridiidae, 74
Deltatheridium, 51f, 74–76, 75f, 76f, 78f
Deltatherium, 110f, 113
Deltatheroida, 9t, 11f, 72, **74–76**, 74t, 340
Deltatheroididae, 74
deltopectoral crest, 32, 55
  didymoconid, 97
  Ernanodon, 210
  Gobiconodon, 62, 62f
  hyaenodontid, 126
  miacoids, 131
  micropternodontid, 151
  palaeanodont, 206–208
  pangolin, 204
  pantolestid, 101
  perissodactyl, 244
  phenacodontid, 223
  taeniodont, 106, 107
Demidoff's bushbaby, 188
Dendrogale, 197
dens, 30, 45
dental formula, definition, 27
dental occlusion, 42f, 44, 53–56, 53f, 54f
  amphilestid, 61
  eutriconodont, 61
  haramiyid, 51, 52
  megazostrodontid, 61
  morganucodont, 61
  multituberculate, 56
dentary(ies), 26, 41, 44, 45f, 46f, 49, 56, 61, 63, 68, 83, 201f
  adapoid, 183f
  anagalid, 309, 310f
  angle, 47f
  angular process, 64, 73

dentary(ies) (continued)
    anoplothere, 298
    anthracothere, 294
    anthropoid, 193f
    condylar process, 24
    conoryctid, 105
    cylindrodontid, 323
    epoicotheriid, 209
    eupantothere, 64
    Kuehneotherium, 54
    leptictidan, 142f
    morganucodont, 52
    oxyaenid, 121
    palaeanodont, 206, 208f
    pantolestid, 101
    pholidotan, 204
    rodent, 316
    saxonellid, 176f
    taeniodont, 108
    tarsiid, 187f
    Wyolestes, 99f
dentary-squamosal joint, 41–43, 42f, 52, 64
dentine, 27, 261
dentition, **26–30,** 28f, 29f, 34–36, 38f, 45–46. See also canine teeth; carnassials; cheek teeth; dental formula; incisors; milk teeth; molars; premolars
    adapid, 185
    adapisoriculid, 144, 144f
    adapoid, 180, 181f, 182–183, 183f
    adianthid, 235, 235f
    aegialodont, 69–70
    alagomyid, 319, 321f
    amphilestid, 61, 62, 62f
    amphimerycid, 303
    amphipithecid, 185
    anaptomorphine, 188
    anoplothere, 298
    anthracobunid, 262, 262f
    anthracothere, 294, 294f
    anthropoid, 192–194, 193f, 195f, 196, 196f
    apatemyid, 103, 104f
    apternodont, 152, 152f
    archaeocete, 278–279, 282, 284
    archontan, 157, 159f
    arctocyonid, 214, 214f, 215f, 216–217, 217f, 288f
    arctoid, 136f
    arctostylopid, 225, 225f
    artiodactyl, 285–287, 288f, 292f, 298f
    Asiatherium, 76, 77f
    astrapothere, 235, 236f
    ausktribosphenid, 68
    australosphenidan, 67f
    basal metatherian, 76–77, 77f
    bat. See dentition, chiropteran
    bemalambdid, 114–115, 115f
    bilophodont, 36, 237
    boreosphenidan, 67f
    brachydont, 27, 29f, 35, 36
    brontothere, 249f, 250, 250f, 252
    bunodont. See bunodonty/bunodont dentition
    bunoselenodont. See bunoselendonty/bunoselenodont dentition
    camel, 302, 302f
    canid, 135, 135f
    carnivoran, 119, 122f, 127–129, 130f
    carpolestid, 172, 176f
    cebochoerid, 291
    ceratomorph, 253–254, 254f
    cercamoniine, 185
    cetacean, 276
    chalicothere, 256–257, 256f
    chiropteran, 157–160, 160f
    choeropotamid, 291, 292, 292f
    Cimolestes, 95f, 129
    cimolestid, 95, 95f, 96f, 107f, 130f
    cingulate, 201, 201f
    condylarth, 215
    conoryctid, 105–106

coryphodontid, 116f, 118
creodont, 119–121, 122f, 129. See also dentition, hyaenodontid; dentition, oxyaenid
ctenodactyloid, 325f, 332–333
cylindrical teeth, 36
cylindrodontid, 325f
dasypodid, 201, 201f
dermopteran, 163, 163f, 164
Diacodexis, 288f
dichobunid, 288, 292f
didelphodont, 95, 95f
didelphoid, 77, 81, 83, 337
didolodontid, 220f, 221
didymoconid, 97, 98f
and diet, 29, 34, 45, 54, 67, 71, 78, 83, 96, 101, 108, 110, 144, 145, 163, 170, 180, 196, 213, 215, 216, 220, 221, 274, 276–277, 279, 284, 289, 291, 321
dilambdodont. See dilambdodonty/dilambdodont dentition
Dinictis, 38f
dinoceratan, 238–239
diphyodont, 26, 41, 43, 51, 56
djadochtathere, 58
docodont, 55, 55f
dromatheriid, 45–46, 47f
dryolestid, 65, 66f
edentate, 199
elephant, 261
elephant shrew, 311, 311f
embrithopod, 266, 267f
entelodont, 294
eomoropid, 257
eosimiid, 193, 193f
epoicotheriid, 206–207, 208f
equid, 247–248
equoid, 249f
erinaceid, 144–147, 147f
erinaceomorph, 144–146, 145f–147f
eupantothere, 64, 67f
euprimate, 167, 172f–173f, 179, 180f
eurymylid, 316, 321f
eutherian, 88, 89, 91f, 92f
eutriconodont, 61–62, 62f
ever-growing. See hypselodonty
feliform, 132, 133f
fossil whale, 273
galericine, 147
geolabidid, 148–149, 149f
Glires, 313
glyptodont, 201f, 202
gomphodont, 45
gondwanathere, 71, 71f
haramiyid, 51–52, 52f
helohyid, 290, 292f
herbivores, 35–36
heterodont, 25f, 26, 41, 44
heterosoricine, 153
homodont, 35
homologies, 27
hyaenodontid, 124, 125, 125f, 126, 126f
hyopsodontid, 220f, 222
hypertragulid, 303, 303f
hypselodont. See hypselodonty
hypsodont, 27, 29f, 35, 36, 38f, 60, 113, 228
hyracoid, 259, 259f, 260
ischyromyid, 321
Jeholodens, 62f, 63
Kuehneotherium, 53f, 54
lagomorph, 313–314, 315f
lemuroid, 186, 187f
leptictid, 140
leptictidan, 140, 141, 141f, 142f
lipotyphlan, 144
litopteran, 233–235, 234f, 235f
lophodont, 28, 29f, 35, 36, 225, 228
macraucheniid, 235, 235f
marsupial, 36, 72, 73, 78, 78f–79f, 81f–82f, 83–87, 83f–86f

megazostrodontid, 61
mesonychian, 274, 275, 275f, 277f
metacheiromyid, 206–207, 208f
metatherian, 76, 76f
miacoid, 130, 130f
micromomyid, 174
micropternodontid, 150
microsyopid, 178
mimotonid, 314f
mioclaenid, 217, 218, 220f, 227f
mixodectid, 164, 165f
monotreme, 69
morganucodont, 52, 53f
multituberculate, 56, 59, 59f
musteloid, 136, 136f
myrmecophagous, 38f
notharctid, 183, 184–185
notoungulate, 227–228, 228f–233f
nyctithere, 149–150, 150f
omomyid, 172f–173f, 188, 189–190, 190f, 191, 191f, 192
oreodont, 299, 299f, 300
oromerycid, 301
oxyaenid, 121, 122, 122f, 123f
palaeanodont, 205–206, 208f
palaeoryctid, 96, 96f, 97f
palaeothere, 248, 249
pantodont, 114, 115f
pantolestan, 99, 100f
paromomyid, 176
periptychid, 220, 220f
perissodactyl, 244, 249f
phenacodontid, 220f, 223, 224
picrodontid, 175, 177f
picromomyid, 175, 177f
pigmented, 153
pilosan, 201f, 202
plagiaulacoid, 58, 83, 172
plagiomenid, 164, 165f
plesiadapid, 171
plesiadapiform, 168f, 169–170, 172–173, 172f–173f, 175f, 176, 176f–177f, 178
plesiosoricid, 153
proboscidean, 260, 261f, 262–264, 262f, 265f
prosciurine, 326
proterotheriid, 234, 235f
pseudictopid, 310, 310f
ptilodontid, 59, 59f
Ptolemaia, 103, 103f
Purgatoriidae, 171, 172f, 173f
pyrothere, 236–237, 237f
raoellid, 291, 292f
Reigitherium, 55
rhinocerotoid, 249f, 254f, 255
rodent, 84, 107, 316, 321, 321f, 325f, 328
saxonellid, 173, 176f
sciuravid, 322
secodont, 28, 34, 35, 167
selenodont, 28, 29f, 35, 36, 110, 167, 286–287, 289, 294, 297, 298f, 346
shrew. See dentition, soricid
simplicidentate, 315, 316f
Sinoconodon, 53, 53f
sirenian, 268–269, 269f
sivaladapid, 186
sloths, 201f, 202
soricid, 153, 154f, 154–155
soricoid, 151f
soricomorph, 148, 149f
sparnotheriodontid, 235, 235f
squirrel, 328
strepsirrhine, 187
stylinodontid, 107, 108f
suoid, 296f
symmetrodont, 63, 63f
taeniodont, 105, 107f, 109
taeniolabidoid, 60
talpid, 154f, 154–155
tapiroid, 249f, 253, 254f
tarsiid, 187f, 191

tethythere, 242
theridomyid, 325f, 326
Theroteinidae, 51
tillodont, 110, 111f, 112f, 113, 114
titanothere, 250, 250f. *See also* dentition, brontothere
tooth cusps, nomenclature for, 27–28, 28f
toxodont, 230–231, 230f
tree shrews, 197, 197f
tribothere, 70, 70f
triconodont, 61
triconodontid, 62, 62f
trithelodont, 45
tritylodont, 45, 47f
trogosine, 111, 113
typothere, 231–232, 231f, 232f
uintathere, 238–240, 239f
ungulate, 211, 214
ursid, 135, 136f
wear patterns, 42f, 51, 53, 53f, 54f, 55, 56, 58, 68, 69, 84, 94, 101, 103, 105, 107, 108, 110, 194, 223, 274, 295
wyolestid, 99, 99f
xenarthran, 200, 201f
xenungulate, 238, 238f
xiphodont, 301
zalambdodont. *See* zalambdodonty/zalambdodont dentition
zambdalestid, 307–308, 308f
zhelestid, 213–214, 213f
Deperetellidae, 242t, 254, 339
derived features or taxa, definition, 6
dermal ossicles, 201
dermal scutes, 201–202
Dermoptera, 10t, 11f, 12f, 22, 39, 139f, 156, 157f, 158, 158t, **162–166**, 165f, 170f, 171, 176, 335, 337, 343. *See also* Archonta
plagiomenid, 191
and Primatomorpha, 165–166
Dermotherium, 163, 163f
Deseadan SALMA, 83, 202, 221, 231, 234, 235, 237
Desmatoclaenus, 217, 224
Desmatolagus, 314–315
Desmostylia, 215, 242, 265, 270
classification and relationships, 8f, 11f, 211–212, 213t, 242f, 243f
Desmostylidae, 242t
Diacocherus, 145f, 146
Diacodexeidae, 288
Diacodexeinae, 286f
Diacodexis, 34f, 37f, 273, 286f, 288–290, 288f, 290f, 299, 303, Plate 6
ungual phalanges, 36f
Diademodontidae, 45
Diamantomyidae, 307t, 334
diaphragm, 41
diaphysis, 23, 42
Diarthrognathus, 44, 45, 47f
diastema/diastemata, 35, 45, 51, 56, 83, 92, 110, 141, 149, 163, 170, 171, 176, 229, 233, 236, 244, 259, 262, 264, 265, 278, 280, 288, 290, 292, 293, 298, 301, 307, 310, 313, 315, 316, 319
palaeanodont, 206
Diatryma, Plate 6
Dichobune, 292f
Dichobunidae, **288–290**, 288f, 289f, 290f–292f
Dichobunoidea, 288, 289t, 293, 296, 297, 346
Didelphidae, 74t, 75f, 78, **79–81**. *See also* opossums
Didelphimorphia, 4, 9t, 73, 74t, 78, 83, 341
Didelphinae, 80, 81f
Didelphis, 27, 42f, 73, 106
Didelphodon, 77f, 78, 108
Didelphodonta, 10t, **94–97**, 95f, 95t, 110f, 341
Didelphodus, 95f, 96, 114
Didolodontidae, 213t, 215, **220–222**, 220f, 227f, 234, 344–345
Didolodontoidea, 235

Didolodus, 221
Didomyconidae, 10t
Didymictis, 34f, 131, 131f
Didymoconidae, 10t, 22, 95t, **97–99**, 98f, 140
Didymoconus, 97, 98f
diet, 34–36. *See also* durophagous diet
abrasive, 105
carnivoran, 126
dentition and, 29, 34–36, 45, 54, 67, 71, 78, 83, 96, 101, 103, 108, 110, 144, 145, 163, 170, 180, 196, 213, 215, 216, 220, 221, 274, 276–277, 279, 284, 289, 291, 321
faunivorous, 35
omomyid, 191
digging mammals. *See* fossorial mammals
digit(s), 33, 71
adapoid, 180
anoplothere, 298
anthracothere, 294
apatemyid, 104, 106f
aquatic mammals, 40
arboreal mammals, 39
archaeocete, 281, 284
artiodactyl, 285
cetacean, 276
chiropteran, 157, 159, 161
dermopteran, 163
dichobunid, 290
elephant shrew, 311
equid, 247, 248
euprimate, 168, 179
flying mammals, 39
lagomorph, 314
mesonychian, 274
notoungulate, 228
oxyaenid, 121
pangolin, 204
paroxyclaenid, 101
peccary, 297
pedal, 34
periptychid, 219
perissodactyl, 244
phenacodontid, 223
plesiadapiform, 174f
proboscidean, 260
rhinocerotoid, 255
stylinodontid, 107
uintathere, 240
webbed, of semi-aquatic mammals, 40
xenarthran, 201
xiphodont, 301
digitigrade posture/stance, 39–40, 125, 126, 129, 134, 135, 240, 260, 290, 290f, 293, 302
dilambdodonty/dilambdodont dentition, 29f, 34, 38f, 113–115, 115f, 117, 144, 147–148, 150, 155, 157–159, 197, 249, 257, 263, 266, 297
Dimetrodon, 44
dimorphism. *See* sexual dimorphism
Dimylidae, 140t, 154–155
Dinictis, 38f, 133, 134f
Dinnetherium, 61
Dinocerata, 10t, 211–212, 213t, 215, 226, **238–240**, 239f, 242, 306, 339, 345
Dinohyus, 295
dinosaurs, extinction of, 11–12, 21
Dipassalus, 207
diphyodonty/diphyodont dentition, 26, 41, 43, 51, 56
Diplobune, 297–298
Dipodidae, 307t
Dipodoidea, 307t, 329f, 330, **331**
Diprotodontia, 9t, 74t, 75f
Dipsalidictis, 122
Dipsalodon, 122
dispersal routes, 18, 19f, 20. *See also* Bering land bridge; De Geer Route
Dissacus, 272f, 275, 275f, 337
distal, definition, 27
diurnal mammals, 180, 188, 194, 197

divergence
genetic, timing, 5
morphological, timing, 5
of placental clades, ages, 5t
superordinal, 8
diversity, across epochs, 3f
Divisaderan SALMA, 14f, 17, 84, 201, 231–233
Dizzya, 159
djadochtatheres. *See* Djadochtatherioidea
Djadochtatheria, 58
Djadochtatherioidea, 49t, 56, 57f, **58**, 59f, 61
Djarthia, 74t, 86f, 87
Djebelemur, 183f, 185
Docodon, 55, 55f
Docodonta, 9t, 49, 49t, 50f, 51f, **55–56**, 55f, 70
dogs, 9t, 126, 129f, 133–135. *See also* Canidae
Doliochoerinae, 293
Doliochoerus, 296
dolphins, 10t, 35, 40, 271, 275, 276
Domnina, 153, 154f
Donrussellia, 172–173, 172f–173f, 182–184, 188
Dormaal (Belgium), 338
Dormaaliidae, 145, 146
dormice. *See* Gliridae
Dorudon, 281f, 282f, 284, 284f
Dorudontinae, 272f, 278, 284
Douglassciurus, 323f, 327–328
Douglassia, 327
Dralestes, 178
Dremotherium, 305
Dromatheriidae, 45–46, 47f
Dromatherium, 47f
Dromiciops australis, 86
Dryolestida, 9t, 49t, 336
Dryolestidae, 49t, 51f, **64–66**, 66f
Dryolestoidea, 9t, 17, 49t, 55, 64–66, 222, **340**
Duchesnean ALMA, 14f, 15, 134–135, 152, 255, 293, 297, 300, 301, 342
Duchesnean/Chadronian boundary, 16
Dugong, 268
Dugongidae (dugongs), 10t, 242t, 243f, 267–268
Duplicidentata, 9t, 307t, 308f, **313–315**, 346–347
durophagous diet, 99, 101, 103, 122, 125
dwarf lemurs, 188

eardrum, 24
archaeocete, 280, 282
cetacean, 277
Early Eocene Climatic Optimum, 20
echidna, 35, 43f, 68, 69
echolocation, 157, 158, 161, 276–278, 285
ectocingulid/ectocingulum, 78, 113, 228
Ectocion, 220f, 223, 224, 244
Ectoconus, 219, 221f
ectocuneiform, 33
dichobunid, 290
mesonychian, 277f
ectoflexus, 28f, 73f, 108–109, 114, 115
Ectoganus, 36f, 107, 107f, 108f
ectoloph, 28, 28f, 34, 114, 115f, 117, 150, 163–164, 223–225, 227, 228, 233, 249, 250, 252, 255–257, 259, 260, 294, 297, 299–303, 326
ectolophid, 225, 321, 326
ecto-mesocuneiform
archaeocete, 281
artiodactyl, 285
ectopterygoid bones, 57
ectotympanic bone, 24, 25f, 128, 129f, 194, 309
amphicyonid, 135
archaeocete, 282
canid, 134
caniform, 133
carnivoran, 128–129, 129f
cetacean, 277
chiropteran, 158
dermopteran, 163
elephant shrew, 311
lagomorph, 313
leptictidan, 142f
lipotyphlan, 143

ectotympanic bone (*continued*)
  nimravid, 133
  omomyid, 191
  palaeoryctid, 96
  phenacodontid, 223
  ungulate, 211
ectotympanic bulla, 96, 309, 313, 346
ectotympanic ring
  adapoid, 180
  omomyid, 188
Ectypodus, 59
Edentata, 6, 9t, 11f, 22, 198, 199, 199t, 344
edentates, 9t, **198–210**, 199t
Edvardocopeia, 230
Edvardotrouessartia, 230
Egatochoerus, 296
Egerkingen (Switzerland), 338
Egypt, 148, 185, 192, 259, 263, 269, 284, 311, 336. See also Fayum Depression
  anthropoids, 194
  archaeocetes, 278, 282
  embrithopods, 265
  hyracoids, 259–260
  macroscelideans, 311
  proboscideans, 263
  ptolemaiids, 101
  rodents, 334
  sirenians, 270
Ekgmowechashala, 191
El Molina Formation, 16
elbow, 33, 64
  adapoid, 180
  archaeocete, 281, 284
  arctocyonid, 216
  cetacean, 276
  chiropteran, 157
  hyopsodontid, 222
  lagomorph, 314
  paroxyclaenid, 101
elephant shrews. See Macroscelidea
Elephantidae, 8f, 242t, 260
elephants, 8, 8f, 10t, 40, 211–212, 260–261, 265
Elephas, 8f
Eleutherodon, 51
Eleutherodontidae, 49t, 51
Elfomys, 326
Ellesmene, 164
Ellesmere Island, 18, 21, 337, 343
  palaeanodonts, 209
  plagiomenids, 164
Ellipsodon, 29f, 220f
ELMA. See European Land-Mammal Ages
Elomeryx, 294, 294f
Elphidotarsius, 172
Elpidophorus, 164, 165f
Elwynella, 176
Elymys, 330
Emballonuridae, 158t, 159, 161
Embolotheriinae, 250
embrasure pits, 275
Embrithopoda, 215, 242, 262, **265–267**, 266f, 270, 345
  classification and phylogeny, 8f, 10t, 11f, 211–212, 213t, 242t, 243f
Enaliarctos, 121t
enamel, 26, 27, 29–30, 45, 71, 201
  alagomyid, 319–320
  amphipithecid, 185
  anagalid, 309
  anaptomorphine, 189
  aprismatic, 29–30
  archaeocete, 278, 280, 283–284
  astrapothere, 235
  cimolodont, 58
  crenulated, 164, 175, 185, 188–191, 194, 216, 220, 290, 297, 321, 326
  creodont, 121
  cylindrodont, 323
  double-layered prisms, 319
  eurymylid, 313, 316

folivores, 36
Glires, 313, 318, 319
grazers, 36
herbivores, 36, 36f
lagomorph, 313, 319
Lambdopsalis, 60
marsupial, 83, 84
microscopic structure, 29–30, 318–320, 320f
mimotonid, 313, 315
multiserial, 319, 320f, 347
multituberculate, 56, 58, 60
nonprismatic, 29–30
omomyid, 188
palaeanodont, 207
pauciserial, 316, 319, 320, 320f, 321–323, 326
peccary, 297
periptychid, 220
perissodactyl, 244
perptychid, 220
picrodontid, 175
pigmented, 60, 153
pikas, 319
proboscidean, 260
pseudo-multiserial, 326
radial, 29–30, 319, 320
raoellid, 291
rodents, 84, 313, 316–317, **318–319**, 320, 320f
Simplicidentata, 315
Sinomylus, 315
soricid, 153
stylinodontid, 107
suid, 297
taeniodont, 105
tangential, 29–30
theridomyid, 326
tillodont, 110, 113
uniserial, 319, 320f, 321, 323, 326, 328
wrinkled. See enamel, crenulated
zalambdalestid, 307
encephalization quotient (EQ)
  multituberculate, 61
  Plesiadapis, 171
endocranial cast, 57
endothermy, 41
England, 21, 51, 56, 66, 69, 70, 155, 338
Entelodontidae, 286f, 289t, 293, **294–296**, 295f
Entelodontoidea, 289t
Entelopidae, 199t
Entelops, 202
entepicondyle, palaeanodont, 206, 208, 208f
entoconid, 28, 28f, 73, 73f, 74, 76, 77, 85, 89, 91f, 153, 155, 158, 164, 180f, 186, 192, 194, 217, 218, 228, 311f, 316, 328
entoconulid, 28, 28f
entocristid, 28, 28f, 69
entocuneiform, 33
  euprimate, 179
  mesonychian, 277f
entolophid, 225, 228, 328
entotympanics, 24, 128–129, 129f, 143, 309
  canid, 134
  caniform, 133
  carnivoran, 129f
  chiropteran, 158
  didymoconid, 97
  elephant shrew, 311
  leptictidan, 140, 141, 142f
  palaeoryctid, 96
  paromomyid, 176
  plesiadapiform, 169
Eoastrapostylopidae, 213t, 236
Eoastrapostylops, 213t, 236, 236f
Eobaatar, 57f
Eobaataridae, 57f
Eobasileus, 238, 239
eobrasiliine, 81f
Eocenchoerus, 297
Eocene Epoch, 13f
  boundary with Paleocene, 12, 15
  climate, 20–21

flora, 20–21
geochronology and biochronology, 8, 10, 14f
mammalian diversity in, 3, 3f, 335
mammals, synopsis, 340–346
Eocene/Oligocene boundary, 12, 15–17, 19, 20
Eochenus, 147, 147f
Eochiroptera, 161
Eoconodon, 217
Eodendrogale, 197, 197f
Eoentelodon, 295
Eogalericius, 147
Eogliravus, 324f, 328–329
Eohaplomys, 321
Eohippus, 247. See also *Hyracotherium*
Eohyrax, 233
Eomaia, 9t, 75f, 89, 89f–91f
Eomanidae, 95t, 199t, 200f, 204–205
Eomanis, 199, 203, 204, 205f, 344
Eometatheria, 9t, 74t
Eomoropidae, 242t, 255, 257
Eomoropus, 257
Eomorphippus, 231
Eomyidae, 307t, 329f, 331, 338
Eoryctes, 96, 97f
Eosigale, 309, 310f
Eosimias, 192–193, 193f
Eosimiidae, 166t, 167f, 192–193, 339
Eosiren, 243f, 270
Eotalpa, 154f, 155
Eotheroides, 243f, 270
Eotitanops, 29f, 249f–251f, 252
Eotylopus, 301, 301f
Eoungulatum, 213
Eozostrodon, 52
Ephelcomenus, 297–298
Epidolops, 83, 84f
Epihippus, 247
epiphyses, 23, 42
epipubic bones, 33, 57, 65, 68, 73, 76, 89, 90, 92
Epitheria, 199
epitympanic recess, apternodont, 152
epitympanic sinus, notoungulate, 228
Epoicotheriidae, 95t, 199t, 200f, 205–209, 208f, 209f
Epoicotherium, 207–209, 208f
Eppsinycteris, 158t, 159, 160f
EQ. See encephalization quotient (EQ)
Equidae, 242t, 246f, **247–248**, 248f, 249f, 345
Equoidea, 242t, **247**, 248, 248f, 249f, 253, 345
Equus, 38f, 113
erector spinae muscles, djadochtathere, 58
Erethizontidae, 334, 347
Ergilian ALMA, 14f, 17
Erinaceidae, 139f, 140t, 143, 144, **146–147**, 147f, 342
Erinaceomorpha, 143, **144–147**, 342
  classification and relationships, 140t, 145
  dentition, 144–146, 145f–147f
Erinaceus, 37f, 101
Ernanodon, 198, 200f, 207f, 209–210, 339
Ernanodonta, 95t, 199t, 209–210
Ernanodontidae, 95t, 199t
Ernestohaeckelia, 230
Ernestokokenia, 221
Ernosorex, 153–154, 154f
Escavadodon, 206, 207f
Escavadodontidae, 95t, 199t, 200f
Escribania, 217
Esthonychidae, 95t, 110
Esthonychinae, 110–111, 110f
Esthonyx, 111, 111f, 113, 234
Etayoa, 238, 240
Ethegotherium, 233, 233f
ethmoid bone, 24, 26
Euarchonta, 10t, 156, 158t, 343
  classification, 10t, 158t
Euarchontoglires, 8, 12f, 156
  age of divergence, 5t
Eucosmodon, 57f, 60
Eucosmodontidae, 57f, 59–60
Eucricetodon, 330

Eudaemonema, 164, 165*f*
euhypsodont, 27
Eulipotyphla, 5*t*, 8, 12*f*, 139*f*, 143, 342–343
Eumys, 327*f*, 330–331
eupantotheres, 49, 50*f*, 63, **64–66**, 65*f*, 67*f*, 340
Eupecora, 304
Euphractus, 32*f*, 36*f*, 201
Euprimates, 12, 61, 166, 168, 168*f*, 169, 171, 173, 173*f*, 176–178, **178–179**, 180*f*, 343, Plate 4.4
    classification, 157*f*, 166, 166*t*, 170*f*
    Early Cenozoic
        of Africa, 336
        of Asia, 339
        of Europe, 338
        of North America, 337
Eurodon, 109
Euromys, 321
Europe, 52, 55, 58, 59, 62, 65, 79, 80, 341–342
    adapids, 185, 186
    adapisoriculids, 144
    adapoids, 180, 183–185
    altungulates, 345
    amphicyonids, 135
    amphimerycids, 303
    anoplotherioids, 297
    anthracotheres, 293, 294
    apatemyids, 103
    archaic ungulates, 344
    arctocyonids, 215
    and Asia, faunal exchange between, 18–19, 19*f*, 338
    brontotheres, 250
    cainotheres, 299
    cercamoniines, 185
    chiropterans, 159, 161, 162
    cimolestids, 96
    coryphodontids, 118
    dimylids, 154–155
    embrithopods, 265
    entelodonts, 295
    epoicotheriid, 208
    equids, 247
    erinaceomorphs, 145–146
    euprimates, 178
    feliforms, 132
    hedgehogs, 147
    heterosoricines, 153
    hyaenodontids, 123, 125
    hyopsodontids, 222
    hyracoids, 257
    lagomorphs, 314
    lorisoids, 186
    miacoids, 130
    mixtotheriids, 297
    musteloids, 136
    and North America, faunal exchange between, 18, 19*f*, 337
    notharctids, 183–184
    nyctitheres, 149, 150
    omomyids, 188, 191
    oxyaenids, 122
    palaeanodonts, 205
    palaeotheres, 248
    paleogeography, 18, 19*f*
    pantolestids, 100
    paromomyids, 343
    paroxyclaenids, 101
    periptychids, 219
    phenacodontids, 223
    plesiadapiforms, 178
    plesiosoricids, 153
    primates, 185, 188
    rhinocerotoids, 254, 255
    rodents, 321, 324*f*, 326, 328
    Saxonella, 173
    sciurine clades, 328
    selenodont artiodactyls, 346
    sirenians, 270
    suids, 297
    taeniodonts, 105, 109
    talpids, 154–155
    tayassuids, 296
    tillodonts, 113
    toliapinids, 178
    ursids, 135
    xenarthrans, 203
    xiphodonts, 301
    zhelestids, 213
European Land-Mammal Ages, 11, **15**
European reference levels, 15
Europolemur, 180, 184–185
Eurotamandua, 198, 200*f*, 202–204, 203*f*, 343–344, Plate 7.1
Eurygenium, 231
Eurymylidae, 307*t*, 308*f*, 313, **315–316**, 317*f*, 320*f*, 339, 346–347
Eurymylus, 316
Eutheria, 2–4, 22, 48, 68, 69–70, 72, 73, 79, 92*f*, 99, 148, 335, 338
    characteristics, 73, 88
    classification and relationships, 9*t*, 11*f*, 49*t*, 50*f*, 51*f*, 75*f*, 88–89, 89*f*
    earliest, **88–93**, 90*f*–92*f*, Plate 1.3, Plate 1.4
    primitive, 141
Eutriconodonta, 52, **61–63**, 62*f*
    classification, 9*t*, 49*t*, 50*f*, 51*f*
Eutypomyidae, 307*t*, 322*f*, 328
Eutypomys, 327*f*, 328
exoccipital, 24, 25*f*, 45*f*
    artiodactyl, 285
    dichobunid, 289
    proboscidean, 264*f*
    sirenian, 269*f*
exodaenodonty, 155
Exodaenodus, 155
explosive model, of therian radiation, 3, 4*f*
external auditory canal/tube
    archaeocete, 280
    cetacean, 277
    notoungulate, 228
    omomyid, 188
    oreodont, 300
external auditory meatus
    omomyid, 191
    proboscidean, 264*f*
    suiform, 293
external carotid artery. *See* artery, external carotid
external nares
    archaeocete, 281*f*, 283
    cetacean, 276
    proboscidean, 262, 264*f*
    pyrothere, 237
extinctions
    dinosaur, 11–12, 21
    K/T boundary, 2, 21
    multituberculate, 61
    palaeanodont, 209
    plant, 21
    uintathere, 240

facial nerve, 25
facial skeleton, 26
fairy armadillo, 207
false sabertooths, 126, 132
false vampire bats, 161
Faroe Islands, paleogeography, 18
faunivory/faunivorous mammals, 191
    definition, 35, 191
Fayum Depression, 186–187, 192, 260, 263–265
    anthropoids, 194–197
    faunas, 311, 312, 336
feet. *See* pes (foot)
Felidae, 120*f*, 121*t*, 126, 131–132, 342
Feliformia, 120*f*, 121*t*, 126, 127, **131–133**, 133*f*, 134*f*, 342
femur, 30*f*, 33, 37*f*, 53, 64
    adapid, 186
    adapoid, 180
    anoplothere, 298
    anthropoid, 194, 196
    archaeocete, 273, 283, 284
    arctocyonid, 218*f*
    artiodactyl, 285, 287*f*, 290*f*
    carnivoran, 129
    cebochoerid, 291
    chalicothere, 257
    choeropotamid, 292
    dichobunid, 290, 290*f*
    elephant shrew, 311, 312*f*
    entelodont, 295
    hyaenodontid, 125, 126
    hyopsodontid, 223, 312, 312*f*
    hyracoid, 257
    ischyromyid, 325*f*
    lagomorph, 314
    leptictid, 140
    mesonychian, 277*f*
    miacoids, 131, 131*f*
    omomyid, 188
    palaeothere, 248
    pangolin, 204
    periptychid, 220
    perissodactyl, 244, 246*f*
    pholidotan, 204, 205
    proboscidean, 260
    pseudictopid, 310
    tillodont, 111
    uintathere, 240
fenestra cochleae, 25*f*
fenestra ovalis, 24
Ferae, 9*t*, 94, 95, 95*t*, 119–120, 121*t*
fermentation
    foregut, 286–287
    hindgut, 345
ferns, 21
Ferugliotherium, 71, 71*f*
fibula, 30*f*, 33, 40, 308
    anthracothere, 294
    anthropoid, 196
    archaeocete, 284
    arctocyonid, 218*f*
    artiodactyl, 285
    cainothere, 299
    camel, 302
    chiropteran, 157, 161
    choeropotamid, 292
    coryphodontid, 118
    dichobunid, 290
    elephant shrew, 311
    entelodont, 295
    equid, 247
    erinaceid, 144
    erinaceomorph, 146
    eutherian, 90, 92
    Heterohyus, 164–165
    hyopsodontid, 223
    hypertragulid, 303
    hyracoid, 257, 260
    lagomorph, 314
    leptictid, 140
    leptictidan, 142
    leptomerycid, 303
    miacoid, 130
    mimotonid, 315
    omomyid, 188, 191
    oreodont, 300
    proboscidean, 260
    pseudictopid, 310
    tarsiid, 192
fins, cetacean, 276
flight
    gliding, 6, 39, 163, 175–177, 327, 331–332
    powered, 157, 158, 161, 162
    skeletal adaptations for, 39, 162
flora
    and fauna
        coevolution, 340
        relationships, 21
    Paleocene-Eocene, 20–21, 340

Florentiamyidae, 307t
Florentinoameghinia, 237
Florida, 337
flukes
    archaeocete, 284
    cetacean, 276
flying foxes, 157
flying lemurs, 8, 10t, 156, 162–163. *See also* Dermoptera
flying squirrels, 39
folivory/folivorous dentition, 21, 36, 38f, 178, 185, 200, 253, 260, 336
Fontllonga-3 (Tremp Basin, Spain), 338
foot. *See* hind foot; pes
foramen magnum, 24, 26
foramen ovale, 25, 247
    lipotyphlan, 143
foramen rotundum, 24–25
foramen/foramina
    anterior condyloid, 26
    anterior lacerate, 24
    antorbital, 264f
        archaeocete, 281f
    astragalar, 53
        fossil whale, 273
    basicranial, 24–26, 133, 328
    cranial, 50
    entepicondylar, 245f, 263
        hyopsodontid, 312f
    incisive, 25f, 26, 316f
    infraorbital, 25, 25f, 91f, 100, 264f, 316f, 318, 321
        archaeocete, 281f
    jugular, 26
    lacerate, 26
    lacrimal, 25f, 82f
    mandibular, 25, 47f, 68, 91f
        archaeocete, 280, 284
        mysticete, 278
        sirenian, 269
    mental, 25, 25f
    middle lacerate, 26
    obturator, sirenian, 270
    optic, 24, 247
        archaeocete, 281f
    palatine, 26
        major, 25f
        minor, 25f
    posterior lacerate, 26
    post-temporal, 82f
    postzygomatic, 82f
    premolar, 314
    sphenorbital, 24
    stylomastoid, 26
    supratrochlear, 277f
        hyopsodontid, 312f
    vertebral, 30, 31, 31f
forearm, 33, 39
    artiodactyl, 287f
    dermopteran, 163
    supination, 101
forefoot, 35f
    anagalid, 309f
    anoplothere, 298
    anthracothere, 294f
    apatemyid, 106f
    arctocyonid, 218f
    artiodactyls, 287f
    chalicothere, 256f
    early whale, 283f
    Eomaia, 75f
    equid, 247
    hypertragulid, 303
    lagomorph, 314
    oreodont, 299f, 300
    oxyaenid, 124f
    pantodonts, 118f
    periptychid, 221f
    protoceratid, 301
    ruminant, 304f

Sinodelphys, 75f
taeniodont, 109f
tylopod, 304f
foregut fermentation, 300
forelimb skeleton, 32, 308
    adapoid, 180, 185
    anagalid, 309, 309f
    anthracothere, 294
    anthropoid, 196
    aquatic mammals, 40
    archaeocete, 280, 284
    artiodactyl, 287f
    cainothere, 299
    cetacean, 276
    chalicothere, 257
    chiropteran, 157
    choeropotamid, 292
    dichobunid, 290
    didymoconid, 97
    edentate, 199
    elephant shrew, 311
    entelodont, 295
    erinaceomorph, 146
    flying mammals, 39
    fossorial mammals, 40
    Gobiconodon, 62, 62f
    helohyid, 291
    hypertragulid, 303
    ischyromyid, 321, 322
    lagomorph, 314
    leptictid, 140
    leptictidan, 142
    miacid, 131
    multituberculate, 58
    omomyid, 188
    Pachyaena, 277f
    palaeanodont, 208, 208f
    pangolin, 204
    pholidotan, 204
    saltatorial mammals, 40
    talpid, 155
    xenarthran, 200–201
    xenungulate, 238f
fossetids, 326
fossettes, 231, 326
fossil record, xi–xii, 4, 41, 42–43, 335
    Cretaceous, quality, 5
    informativeness, 347
    skeletal features preserved in, 41–42
    vs. molecular evidence, 4
fossorial mammals, 35, 37f, 40, 56, 58, 60, 62, 71, 97, 100, 101, 126, 144, 146, 151, 152, 155, 198, 200, 204–208, 222, 300, 309, 311, 312, 326, 327, 340, 341
    skeletal adaptations, 32f, 39f, 40
Fouchia, 255
Four Mile (Colorado), 59
France, 54, 144, 183, 207, 338. *See also* Paris Basin; Quercy (France)
    palaeanodonts, 209
Franchaius, 113
Franimorpha, 322, 334
Franimys, 322
free-tailed bats, 161
freshwater environment, 126, 271, 279, 284
Friasian, 235
frontal bones, 24, 25f, 45f, 58, 82f
    anagalid, 310f
    anthropoid, 194
    archaeocete, 281f
    entelodont, 295
    euprimate, 167
    leptictidan, 142f
    proboscidean, 264f
    sirenian, 269f
frugivores/frugivorous mammals, 21, 35, 36, 38f, 126, 158, 175, 180, 184, 185, 191, 194, 196
fruit bats, 36, 157
Fruitafossor, 71, 200

fur, 90f, 146, 216, 321
    oldest evidence of, 56, 64
Furipteridae, 158t

Galagidae, 166t, 187
Galago, 188
Galagoides demidoff, 188
galagos, 40, 186
Galeopithecidae, 157f, 158t, 163, 163f, 343
Galericinae, 147, 342
Galerix, 147
Gallolestes, 89f, 92, 92f, 93, 213
Garatherium, 144, 144f
"Garden of Eden" hypothesis, 5
Gashatan ALMA, 14f, 17, 172, 209, 319, 339
Gashato (Mongolia), 339
Gashatostylops, 225f, 226
Gaudeamus, 334
Gaviacetus, 280f
Gaylordia, 81f
Gazinocyon, 125, 125f
Geiseltal (Germany), 161, 185, 291, 292, 338, Plate 8.4
Gelocidae, 289t, 304–305
genetic distance, 3–4
geochronology, 14f
    of Early Cenozoic, 8, 10–12, 13f, 14f, 15–17
Geolabididae, 140t, **148–149**, 149f
geologic time scale, 13f
geomagnetic polarity time scale, 13f, 14f
Geomorpha, 307t, 329, 329f, **331**, 347
Geomyidae, 307t, 329f
Geomyoidea, 307t, 331
Georgia, 337
Georgiacetus, 281f, 283
gerbils, 9t
Germany, 338, Plate 2.2, Plate 3.1, Plate 4.2, Plate 5.2–Plate 5.4, Plate 7.1, Plate 7.3, Plate 8.3, Plate 8.4. *See also* Geiseltal (Germany); Messel (Germany)
Gervachoerus, 291, 292f
ghost bats, 161
Gillisonchus, 219
giraffes, 10t, 285
Giraffidae, 10t, 289t
Giraffoidea, 289t
glaciation, 20
Glamys, 329
Glasbiidae, 74t
Glasbius, 79, 83, 83f
glenoid fossa (of scapula), 24, 32, 53
Glibemys, 331
gliders, 6, 39, 163, 175–177, 327, 331–332
glirans. *See* Glires
Gliravus, 329
Glires, 92, 141, 156, 306, **312–334**, 339, 342, 346–347
    classification and relationships, 9t, 89f, 307t, 308f, 311
Gliridae, 307t, 322, 322f, 324f, 328–329, 329f, 332f, 347
Glirimetatheria, 83
glissant mammals, skeletal adaptations, 39. *See also* gliders
global warming, 12, 18, 20. *See also* Initial Eocene Thermal Maximum
glossopharyngeal nerve, 26
gluteobiceps muscle, artiodactyl, 285
Glyptatelus, 202
Glyptodon, 201f
Glyptodontidae, 198, 199t, 200–202, 343
Glyptodontoidea, 199t, 200f
Glyptostrobus, 21
gnawing
    rodents, 316
    tillodonts, 113
Gobi Desert, 76, 90, 310
Gobiatherium, 239–240
Gobiconodon, 62, 62f
Gobiconodontidae, 62

Gobiohyus, 290
Gobiolagus, 314
Gobiomyidae, 307t, 332
Godinotia, 180, 185
golden moles, 9t, 34, 40, 138, 143, 148, 152, 212, 342. *See also* Chrysochloridae
Goler Formation (California), 337
gomphodonts, 45
Gomphos, 315
Gomphotheriidae, 8, 8f, 242t, 260
Gondwana, 68, 69
Gondwanatheria, 9t, 17, 49t, 70–71, 198, 336
Gondwanatherium, 71, 71f
GPTS. *See* geomagnetic polarity time scale
Grande Coupure, 15, 19, 132, 147, 155, 186, 326, 328, 330, 331, 338, 340, 342, 346
Grangeria, 256, 257
graviportal mammals, 40, 116, 117f, 118, 236, 237, 240, 251, 260, 261, 263, 265, 266
grazers, 36, 38f, 71, 200, 286
Greater Green River Basin (Wyoming), 16
Green River Formation (Wyoming), 101, 159, 343, Plate 2.1, Plate 3.2
Greenland, 51
Groeberia, 84, 85f
Groeberiidae, 74t, 75f, 84
ground sloths, 116, 198, 200, 202
Guanajuato (Mexico), 337
Guangdong Province (China), 339
Guangxilemur, 186
Guggenheimia, 81f
Guilielmofloweria, 230
Guilielmoscottia, 230
Guimarota lignites (Portugal), 55, 64, 65
Guimarotodon, 57f
guinea pigs, 9t, 316
gymnures, 147
Gypsonictopidae, 89, 90, 140
Gypsonictops, 11f, 89f, 92–93, 92f, 93, 140, 140t, 141f, 214, 342

Hadrocodium, 9t, 49t, 51f, 54f, **55**
haemal arches, 31
Hahnotherium, 56
Hainin (Belgium), 338
hair, 41, 61
Haldanodon, 55–56, 55f
Halitherium, 270
Hallensia, 249
hallux, 34, 308
  adapoid, 180
  in arboreal mammals, 39
  carpolestid, 173
  dichobunid, 290
  euprimate, 179
  opposable, 81, 173, 179, 180
  oreodont, 300
  plesiadapid, 171
  plesiadapiform, 169
  ptilodontid, 59, 60f
hamate, 33, 74, 75f
hamulus, 25f
Hanna Basin (Wyoming), 2, 215
Hantkeninidae, 15
Hapalemur, 185
Hapalodectes, 274, 275, 275f
Hapalodectidae, 273t, 275
Haplaletes, 222
Haplobunodon, 292, 292f
Haplobunodontidae, 289t, 292–294
Haplohippus, 247
Haplolambda, 116f
Haplomylus, 222–223, 311f, 312, 312f
Haplorhini, 166t, 167, 167f, 178, 191
Haramiyavia, 51–52, 51f, 52f
Haramiyaviidae, 51
Haramiyida, 49, 49t, 51
Haramiyidae, 9t, 49t, 50f, **51–52**, 51f, 60–61
Harbicht Hill (Montana), 214
hares, 9t, 313. *See also* Leporidae

Harpagolestes, 274
Harpyodidae, 95t
Harpyodus, 110f, 114, 115f
Hartenbergeromys, 326
Hassianycteridae, 158t
Hassianycteris, 159, 160–161, Plate 5.3
Hathliacynidae, 74t, 85
Headon Hill (England), 338
hearing
  archaeocete, 280, 282–284
  cetacean, 276, 277
  organs of, 24, 26, 79
heart, 41
hedgehogs, 9t, 39, 101, 138, 143, 144, 147, 342. *See also* Erinaceidae
Hegetotheria, 213t, 228f, 229, **232–233**, 233f
Hegetotheriidae, 213t, 228f
Helaletes, 254f
Helaletidae, 242t, 253, 254
Heliscomyidae, 307t
Heliscomys, 331
Hell Creek (Montana), 2
Helohyidae, 286f, 288, 289t, **290–291**, 339
Helohyus, 290–291, 292f, 296
Hemiacodon, 168f, 188
Hemimastodontidae, 242t
Hemipsalodon, 126
Henkelotherium, 51f, 64, 65, 65f, 67f
Henricosbornia, 228f, 229, 230
Henricosborniidae, 213t, 228f, 229
Heomys, 315–316, 319, 321f, 333, 347
Heptacodon, 293–294, 294f
Heptodon, 245f, 248f, 249f, 253–254, 253f, 254, 254f
herbivores, 2, 21, 35–39, 45, 67, 110, 114, 163, 166, 170, 200, 222, 239, 244, 259, 267, 285, 297, 313, 314, 340, 341
Heroditiinae, 311
Herodotius, 311, 311f
Herpestidae, 120f, 121t, 126, 132
herpetotheriines, 80–81
Hesperocyon, 134–135, 135f
hesperocyonines, 134
heterodonty/heterodont dentition, 25f, 26, 41, 44, 285
Heterohyus, 103, 104, 106f, 164–165, Plate 3.1
Heteromyidae, 307t
Heterosoricinae, 153, 343
Heteroxerus, 328
Hexacodus, 289
Higotherium, 111f, 113
Himalayacetus, 279, 346
hind foot, 86
  abduction, 59, 60f
  amphicyonid, 137f
  anagalid, 309, 309f
  anoplothere, 298
  archaeocete, 281
  arctocyonid, 218f
  artiodactyl, 287f
  dichobunid, 290
  early whale, 283f
  entelodont, 295
  Eomaia, 75f
  equid, 247
  fossil whale, 273
  hyaenodontid, 127f
  Hyopsodus, 223f
  hyperinversion ("reversal"), 39, 59, 60f, 125, 216
  hypertragulid, 303
  Leptictidium, 142f
  multituberculate, 60f
  oreodont, 300
  oxyaenid, 124f
  Pachyaena, 277f
  pantodont, 118f
  periptychid, 221f
  perissodactyl, 246f
  Phenacodus, 224f
  protoceratid, 301

ruminant, 304f
Sinodelphys, 75f
tylopod, 304f
hind limb skeleton, 33, 308
  adapoid, 180
  anthracothere, 294
  anthropoid, 194
  aquatic mammals, 40
  archaeocete, 279, 280, 282–284
  arctocyonid, 217, 218f
  artiodactyl, 287, 287f
  astrapothere, 236
  cainothere, 299
  cetacean, 276
  chalicothere, 257
  choeropotamid, 292
  Diacodexis, 290
  elephant shrew, 311
  erinaceomorph, 146
  euprimate, 179
  flying mammals, 39
  helohyid, 291
  hyopsodontid, 312f
  hypertragulid, 303
  hyracoid, 260
  ischyromyid, 321
  lagomorph, 314
  leptictid, 140
  leptictidan, 142
  miacid, 131
  mimotonid, 315
  omomyid, 188
  palaeanodont, 208
  palaeothere, 249
  pangolin, 204
  pantodont, 116
  pantolestid, 101
  perissodactyl, 246f
  rodent, 325f
  saltatorial mammals, 40
  semi-aquatic mammals, 40
  sirenian, 270
hindgut fermenters, 240, 345
hip bone, 33
Hippomorpha, 242t, 244–246
hippopotami, 10t, 40, 273–274, 285, 291, 293, 345–346. *See also* Hippopotamidae
Hippopotamidae, 272f, 274, 286f, 289t, 296, 345–346
Hippopotamus, 12f
Hipposideridae, 158t, 161
Hipposideros, 161
Hoanghonius, 186
Holarctidelphia, 76
Holocene Epoch, 13f
Holoclemensia, 11f, 70, 70f, 74, 74t, 75f, 78f, 341
holophyly, 6
Holotheria, 63
Homacodon, 286f, 288
Homalodotheriidae, 213t, 228f
Hominoidea, 166t
homodonty, 200, 276
Homogalax, 245f, 246, 246f, 249f, 253, 257, Plate 3.3
homoplasy, 6, 34
Hondadelphidae, 74t
honey possum, 36
Honrovits, 159, 160f
hoofed mammals, 10t, 33, 40, 114, 116, 215, 224, 230, 240, 244, 260–261, 266, 274, 285, 287f, 290, 293, 298, 341. *See also* ungulates
hoofs, 40, 211
  anoplothere, 298
  choeropotamid, 293
  condylarth, 215
  dichobunid, 290
  notoungulate, 230
  oreodont, 300
  pantodont, 114
  perissodactyl, 244

hoofs (continued)
  phenacodontid, 224
  uintathere, 240
Hoplitomerycidae, 289t
Hoplophoneus, 133
hopping. *See* saltatorial mammals
horn cores, embrithopod, 265–266
horns, 285
  brontothere, 250–251, 251f, 252
  protoceratid, 301
  rhinocerotoid, 255
  ruminant, 302–304
  uintathere, 238, 239
horses, 10t, 113, 233, 244, 247. *See also* Equidae
horsetails, 21
howler monkey, 197
HSB. *See* Hunter-Schreger bands
Hsiuannania, 309
Huananius, 113
humans, 10t, 24, 166
humerus, 30f, 32, 32f, 55, 60, 71, 85
  anoplothere, 298
  archaeocete, 283
  arctocyonid, 218f, 245f
  arctostylopid, 226
  artiodactyl, 287f, 290f
  bemalambdid, 114
  chiropteran, 157
  condylarth, 223f
  Dasypus, 32f
  Diacodexis, 290f
  didymoconid, 97
  epoicotheriid, 209
  Ernanodon, 210
  eutriconodont, 62
  Gobiconodon, 62, 62f
  hyaenodontid, 125, 126
  hyopsodontid, 222, 312, 312f
  ischyromyid, 325f
  Jeholodens, 63
  mesonychian, 274, 277f
  miacoid, 131, 131f
  micropternodontid, 150–151, 151f
  morganucodont, 53
  oxyaenid, 121
  Pachyaena, 277f
  palaeanodont, 206, 208, 208f
  palaeoryctid, 97
  pantolestid, 101
  paroxyclaenid, 101
  perissodactyl, 244, 245f
  phenacodontid, 223, 224f
  pholidotan, 204
  proboscidean, 260
  rodent, 325f
  semi-aquatic mammals, 40
  stylinodontid, 107
  taeniodont, 106
  talpid, 155
  xenungulate, 238, 238f
Hunanictis, 97
Hungary, 269, 338
Hunter-Schreger bands, 30, 84, 121, 220, 307–308, 313, 315, 316, 318, 319–320, 320f
Hyaenidae, 120f, 121t, 126, 132
Hyaenodon, 125–126, 126f, 127f
Hyaenodontidae, 120, 120f, 121t, **122–126**, 122f, 125f–128f, 336–339, 341–342
Hyaenodontinae, 123, 124
hyaenoid dogs, 135
Hydrochaeris, 320f
hyenas, 9t, 126
hyoid bone, notoungulate, 228
Hyopsodontidae, 109, 110, 212f, 213t, 215, 217, 220f, **222–223**, 223f, 286f, 287, 311–312, 312f, 344–346
  apheliscine, 222, 312f, 344, 346
  louisinine, 312, 344, 346
Hyopsodus, 220f, 222, 223f, 286f, 289, 337–339
hypercarnivores, 35, 38f, 122

Hypertragulidae, 289t, **303,** 303f
Hypertragulus, 287f, 303, 303f, 304f
Hypisodus, 303
hypocone, 28, 28f, 72, 99, 104, 110, 113–115, 140, 148–150, 153, 164, 168–169, 174, 176, 182–185, 187, 188, 191, 194, 196, 217, 220, 224, 247, 250, 259, 286, 289–291, 295, 296, 299, 300, 311f, 315, 316, 319, 321, 322, 325, 326, 328
hypocone shelf, 76, 169
hypoconid, 28, 28f, 73f, 91f, 149, 155, 180f, 185, 238, 311f
hypoconulid, 28, 28f, 73, 73f, 74, 76, 77, 78f, 89, 91f, 110, 117, 158, 159, 180f, 184, 192, 194, 196, 217, 250, 289, 291, 294, 295, 311, 311f, 319
hypoflexid, 28f, 35, 73f, 231
hypoglossal canal, 26
hypoglossal nerve, 26
hypolophid, 28, 28f, 228, 247, 253, 316
hypostria, 313
hypotympanic sinus, notoungulate, 228
Hypsamasia, 266
hypselodonty, 27, 71, 85, 105, 107, 108, 110, 113, 201, 228, 232–233, 313, 314
hypsodonty/hypsodont mammals, 27, 29f, 35, 36, 38f, 60, 71, 101, 105, 107, 111, 113, 228, 230, 231, 233, 235, 248, 249, 255, 259, 268, 286, 300, 302, 303, 309–311, 313, 315, 322, 325, 326, 328, 346
Hyrachyidae, 242t
Hyrachyus, 249f, 254–255, 338, Plate 7.3
Hyracodon, 254f
Hyracodontidae, 242t, 255
Hyracoidea, 12f, 22, 39, 212f, 215, 224–225, 241–242, **257–260,** 258f, 259f, 336, 345, Plate 7.4
  classification and relationships, 8, 8f, 10t, 11f, 211–212, 213t, 242t, 243f
Hyracotherium, 34f, 37f, 38f, 244, 245f, 247–249, 248f, 249f, 253, 257, Plate 6
Hyracotherium leporinum, 247
Hyracotherium sandrae, 255
hyraxes, 10t, 225, 257, 258f. *See also* Hyracoidea
Hystricidae, 334, 347
Hystricognatha, 307f, 308f, 318, 325f, 332, 333, **334**
Hystricognathi, 317
hystricognathy/hystricognathous rodents, 318, 319f, 322, 326, 347
Hystricomorpha, 334
hystricomorphy/hystricomorphous rodents, 318, 318f, 326, 347

ICA (internal carotid artery). *See* artery, internal carotid
Icaronycteridae, 158t
Icaronycteris, 159–161, 161f, Plate 5.1
ice caps, polar, 20
ice sheets, 20
Iceland, paleogeography, 18
Ichthyolestes, 273, 280
Ictidopappus, 130
ictidosaurs, 45
Ictops. *See Leptictis*
Ignacius, 169, 176, 177f
iliosacral articulation, archaeocete, 283
ilium, 30f, 33, 36f, 41–42, 53
  leptictidan, 143
  palaeothere, 248
  perissodactyl, 244
  proboscidean, 260
incisive foramen, 25f, 26
incisors, 25f, 26, 27, 34, 35–36
  adapoid, 180, 182
  anagalid, 309
  anaptomorphine, 189
  anoplothere, 298
  anthropoid, 192, 194, 196
  apatemyid, 103, 104

apternodont, 152, 152f
artiodactyl, 286
carpolestid, 172
conoryctid, 105–106
creodont, 120
cylindrodont, 323–325
dermopteran, 163, 164
dichobunid, 288
dinoceratan, 238–239
duplicidentate, 313
elephant shrew, 311
enteledont, 295
erinaceid, 144, 147
Ernanodon, 210
eupantothere, 64
euprimate, 168
eutherian, 88, 90, 92
ever-growing, 71, 84, 313, 315, 316, 319. *See also* hypselodonty
geolabidid, 149
gliriform, 110
Hadrocodium, 55
herbivores, 35–36
homologies, 27
hyopsodontid, 222
lanceolate, 178
lipotyphlan, 144
marsupial, 73, 74, 83, 84
metatherian, 76
microsyopid, 178
mimotonid, 315
mixodectid, 164
multituberculate, 56
notharctid, 183
nyctithere, 149
omomyid, 188
oreodont, 300
pantodont, 114, 117
paromomyid, 176
picromomyid, 175
plagiaulacoid, 58
plagiomenid, 164
plesiadapid, 171
plesiadapiform, 169–170, 175f
plesiosoricid, 153
procumbent, 36, 51, 147, 152, 154, 170, 182, 307
pseudictopid, 310, 310f
ptilodontid, 59, 59f
Purgatoriidae, 171
rodent, 313, 319
shrew, 153, 155
Sinoconodon, 53, 53f
strepsirrhine, 187
symmetrodont, 63
taeniodont, 109
talpid, 155
tillodont, 110, 111
tree shrew, 197
triconodontid, 62
trogosine, 111, 113
tylopod, 300
ungulate, 215
xenarthran, 200
zambdalestid, 307
incus, 24, 42f, 44
  archaeocete, 280
Indalecia, 234, 235
India, 269, 279, 284, 339
  altungulates, 345
  archaeocetes, 278, 282, 283
  Archonta, 343
  archontans, 157, 159f
  arctocyonids, 216
  artiodactyls, 291
  Cetacea, 346
  collision with Asia, 19–20, 339
  gondwanatheres, 70–71
  paleogeography, 18
  primates, 169

Indian Plate, 19–20
Indomeryx, 305
Indonesia, 339
Indo-Pakistan. *See* India; Pakistan
Indricotherium, 255
Indriidae, 166t
infraorbital foramen. *See* foramen/foramina, infraorbital
infraorbital process, cetacean, 276
infraspinous fossa, 32, 32f
in-group taxa, 6
Initial Eocene Thermal Maximum, 12, 18, 20, 178, 247, 337, 341–343
inner ear, 24
innominate bone, 33, 36f
Insectivora, 9t, 22, 94, **138–155**, 140t, 178, 311, **342–343**
insectivores, 22, 34–35, 38f, 71, 93, 96, 104, 148, 149, 153, 156, 159, 166, 170, 174, 178, 180, 185, 194. *See also* Insectivora
  archaic, 138
  definition, 138
Interatheriidae, 213t, 228f, 231, 232f
interclavicle, 32, 68
intermembral index, 40, 308
  adapoid, 180
  anthropoid, 196
  elephant shrew, 311
  ischyromyid, 322
internal acoustic meatus, 25–26
internal carotid artery. *See* artery, internal carotid
internal nares, 26
Interogale, 111, 113
interparietal bone, 24, 25f
involucrum, 277
  archaeocete, 280, 282
Iqualadelphis, 77, 78f
Irdinmanhan ALMA, 14f, 17
ischiosacral synostosis, 203
ischium, 30f, 33, 36f, 42, 201
  omomyid, 188
  xenarthran, 201
Ischyromyidae, 307t, 317–319, 321–322, 322f, 324f, 325f, 326, 347
Ischyromyoidea, 326
Ischyromys, 321
Isectolophidae, 242t, 246f
Isectolophus, 246, 246f
Isotemnidae, 213t, 230
Isotemnus, 230
Itaboraí, Brazil, 81, 199, 201, 228, 336
Itaboraian SALMA, 14f, 83, 85, 87, 201, 202, 221, 231, 233–236, 240, 336, 343
Itaboratherium, 231f
Iugomortiferum, 77
ivory, 261

Jacobson's organ, 26, 182
Jamaica, 270f, 338, 345
Japan, 339
jaw(s), 26, 41, 42, 42f, 44, 46f, 47f. *See also* dental occlusion; dentary(ies); maxilla; postdentary bones; *specific taxon*
  adapoid, 179, 183
  alagomyid, 319
  amphipithecid, 185
  anagalid, 309
  apatemyid, 103
  apternodont, 152, 152f
  archaeocete, 280
  arctostylopid, 225, 225f
  ausktribosphenid, 68
  carnivoran, 128
  entelodont, 295
  equid, 248
  erinaceid, 147f
  geolabidid, 149
  hyaenodontid, 124
  hyopsodontid, 222
  Llanocetus, 284
  lower, 26
  mesonychian, 274, 275f
  nyctithere, 150f
  oxyaenid, 122
  pangolin, 204
  pantolambdodontid, 117
  Peligotherium, 222
  picromomyid, 175
  plesiadapid, 171
  plesiadapiform, 178
  Simpsonictis, 130f
  sloth, 202
jaw joint, 42f, 44, 49
Jebel Qatrani Formation, 260, 265
Jeholodens, 61, 62f, 63, 74
Jemezius, 190
Jepsenella, 103
jerboas, 40
Josepholeidya, 230
Jozaria, 262, 262f
Judithian Age, 77
jugal, 25f, 26, 45f, 57, 58, 69, 82f, 310f
  archaeocete, 281f
  didymoconid, 97
  euprimate, 167
  leptictidan, 142f
  lipotyphlan, 143
  proboscidean, 264f
  sirenian, 269f
jugular foramen, 26
jumping. *See also* saltatorial mammals
  skeletal adaptations for, 308
Jurassic Period, 13f, 45, 50f, 51, 52, 55, 56, 58, 61–66, 68, 70, 71

kangaroo(s), 9t, 40
kangaroo rats, 40
Karakia, 216
Karanisia, 187, 187f
Kayentatherium, 46f, 47f
Kazakhstan, 202, 331, 339
Kekenodontidae, 273t
Kellogg, Remington, 283
Kennalestes, 89f, 90, 91f, 92f, 140
Kenyapotamus, 293
Kermackia, 70, 70f
Kermackodon, 56
Kerodon, 300
Khashanagale, 215
Khasia, 87
Khirtharia, 291, 292f
Khovboor (Mongolia), 89
Kielantherium, 65f, 67f, 69–70
Kimmeridgian Age, 13f
kinkajou, 36
Klohnia, 84
knee cap. *See* patella
Knightomys, 322
koala, 9t, 36
Kokopellia, 75f, 76–77, 77f
Kollikodon, 69
Kollpaniinae, 226
Koniaryctes, 152
Kopidodon, 100f, 101, 103f, Plate 2.2
Korea, 339
Krebsotherium, 65, 66f
Kryptobaatar, 57f, 61
K/T boundary, 335
  climate around, 20
  dating, 12
  mammalian diversification after, 3, 4, 335, 340
  survivals beyond, 2
Kuehneodon, 56
Kuehneotheriidae, 9t, 49t, 50f, 51f, 52, **54–55**
Kuehneotherium, 49, 53f, **54–55**, 63
Kulbeckia, 92, 307
Kulbeckiidae, 140
Kumsuperus, 213
Kyrgyzstan, 314, 339

La Meseta Formation, 71, 87, 202, 337
Labes, 213
labial, definition, 27
Labidolemur, 103, 104, 104f
lacerate foramen. *See* foramen/foramina, lacerate
lacrimal (bone), 24, 25f, 26, 45f, 58, 82f, 90, 203
  anagalid, 310f
  archaeocete, 281f
  leptictidan, 140, 142f
  proboscidean, 264f
lacrimal canal, 26
lacrimal foramen. *See* foramen/foramina, lacrimal
Lagomorpha, 4, 5, 12f, 22, 156, 240, 306, **313–315**, 314f, 339, 340, 346–347
  age of divergence, 5t
  classification and relationships, 8, 9t, 11f, 139f, 307t, 308f, 312–315
Laguna Umayo, Peru, 81, 226, 336
Lainodon, 213
lambdoid plates, 152
lambdoidal crest, 24, 82f, 121
Lambdopsalis, 57f, 60
Lambdotherium, 249f, 252
Lamegoia, 221
Lancian NALMA, 14f, 108, 148
land bridges, 18, 19, 19f, 337, 340, 341, 343. *See also* Bering land bridge; De Geer Route
land-mammal ages, 10, 11, 14f, 15
langurs, 36
Lantianius, 216
Lapichiropteryx, 160f, 161
Laredomyidae, 307t, 329
Laredomys, 329
Laurasia, 68, 69, 123, 179
Laurasiatheria, 8, 12f, 156, 214, 242
Lavanify, 71
leaf eating, 180, 184
leaf-margin analysis, 20
leaf-nosed bats, 161
Leipsanolestes, 146
Lemur, 185
lemur(s), 10t, 40, 166, 178–181, 183, 186
Lemuridae, 166t
Lemuriformes (Lemuroidea), 166t, 167f, 179
Leonardus, 65
Leontiniidae, 213t, 228f, 231
Lepilemuridae, 166t
Leporidae, 307t, 313, 314, 319
Leptacodon, 149, 150f, 342
Leptadapis, 180, 184f, 185–186
Leptauchenia, 300
Leptictida, 9t, 93, 96, 139, **140–143**, 140t, 142f, 214, 335, 337, 341, 342, 344, Plate 4.2
  classification and relationships of, 9t, 97, 139–141, 140t
  dentition, 140, 141f, 142f
  skull, 140, 142f
Leptictidae, 22, 90, 138, 139f, 140, 140t, 206, 337–338, 342–344, Plate 4.1
  classification of, 139f, 140t
  Early Cenozoic, of North America, 337–338
Leptictidium, 141–142, 142f, Plate 4.2
Leptictis, 140, 141f, 142f, Plate 4.1
leptochoerines, 288–289
Leptochoerus, 289, 290
Leptolambda, 118f
Leptomerycidae, 289t, **303**
Leptomeryx, 303, 304f, 305f
Leptoreodon, 301f, 302
Leptotragulus, 302
Lesmesodon, 125, 128f
Lessnessina, 109–110, 219
Liassic, 52
Libya, 260, Plate 7.4
Libycosaurus, 293
Lightning Ridge (Australia), 68, 69
lignite deposits, 55–56

limb(s). *See also* forelimb skeleton; hind limb skeleton
  amphimerycid, 303
  anteater, 203
  anthropoid, 196
  archaeocete, 279, 280
  artiodactyl, 285, 287f
  canid, 134
  cetacean, 276
  creodont, 121
  dermopteran, 163
  dichobunid, 290
  diversification, 31–32, 40
  edentate, 199
  enteledont, 295
  equid, 247, 248
  euprimate, 168
  eutherian, 90
  fossil whale, 273
  hyaenodontid, 125
  hyopsodontid, 223
  hypertragulid, 303
  ischyromyid, 325f
  leptomerycid, 303
  mesonychian, 274
  miacid, 131, 131f
  micromomyid, 175
  Onychodectes, 106, 109f
  paromomyid, 176, 177
  perissodactyl, 244
  phenacodontid, 223
  proboscidean, 260
  pseudictopid, 310
  stylinodontid, 107, 109f
  taeniodont, 106, 109f
  tillodont, 111
  uintathere, 240
  ungulate, 214
limb girdle, 32, 44
limb joints, 23, 44
limb posture, 49
Limnocyon, 126
Limnocyoninae, 123, 124
Lingcha Formation (China), 17, 339
lingual, definition, 27
Lipotyphla, 4, 8, 22, 96–97, 119, 141, **143–155**, 156, 159, 212, 214, 335–336, 338, 342–343
  classification and relationships, 9t, 11f, 138, 139, 139f, 140t
Litaletes, 217
Litocherus, 145f, 146
Litolestes, 146–147, 147f
Litolophus, 256f, 257
Litomylus, 222
Litopterna, 22, 215, 221–222, 226, **233–235**, 233f–235f, 336–337, 344–345
  classification, 10t, 213t, 227f
Llanocetidae, 273t
Llanocetus, 284–285
locomotion, skeletal adaptations for, 37f, 39–40, 39f
Lofochaius, 110, 113
London Basin, 338
London Clay, 247
long-fuse model, of therian radiation, 3, 4f
Loomis, Frederick, 237
Lophialetes, 253f, 254
Lophialetidae, 242t, 254, 339
Lophiaspis, 257
Lophiodontidae, 242t, 249f, 255–257
Lophiohyus, 290
Lophiomerycidae, 289t, 304
Lophocion, 244
lophodonty, 116f, 118, 223, 225, 227–228, 234, 238, 239, 247–249, 252–254, 257, 259, 260, 262, 266, 315, 316, 319, 321, 322, 325, 328, 336
lophodonty/lophodont dentition, 28, 29f, 35, 36

lophoselenodonty, 236
Loridae, 166t, 187
lorises, 10t, 166, 178–181, 186
Lorisidae. *See* Loridae
lorisids, 187
Lorisoidea, 166t, 167f, 186–187, 187f
Loroidea, 186. *See also* Lorisoidea
Los Alamitos Formation (Patagonia), 65, 340
Louisiana, 337
Louisina, 222
Louisininae, 222–223, 312
Loxodonta, 8f
Loxolophus, 214f, 215, 216f
lumbar vertebrae, 30, 30f, 31, 31f
  archaeocete, 284
  didelphoid, 81
  djadochtathere, 58
  Fruitafossor, 71
  leptictidan, 143
  peradectid, 80
  pholidotan, 204
  xenarthrous articulations, in Xenarthra, 200, 200f
Lumbrera Formation, 230
lunate, 33, 35f
  creodont, 121
  miacoid, 130
  palaeanodont, 208f
  phenacodontid, 223
  pholidotan, 204
Lushilagus, 314
Lutetian Stage/Age, 13f, 14f, 160, 280, 338
Lutra, 101

Maastrichtian Stage/Age, 13f, 14f
Machaeroides, 122
Machaeroidinae, 122
machairodontines, 122
Machlydotherium, 201f, 202
Macrauchenia, 235
Macraucheniidae, 213t, 227f, 233, 235, 235f
Macrocranion, 145, 145f, 146, 146f
Macroscelidea (elephant shrews), 4, 5, 8, 12f, 22, 138, 141, 211–212, 223, 306, **310–312**, 342, 344, 346–347
  classification and relationships, 9t, 11f, 139f, 212f, 307t, 308f
Macroscelididae, 138, 307t, 310–312, 336, 342
Macrotarsius, 190
Madagascar, 20, 68, 336
  carnivorans, 342
  gondwanatheres, 70–71
  marsupials, 79
  primates, 186
  sirenian, 336
magnetostratigraphy, 10
magnum, 33
  leptomerycid, 303
  phenacodontid, 223
Mahgarita, 183f, 185
Maiorana, 220
malar bones, 26
malleolus
  dichobunid, 290
  leptomerycid, 303
malleus, 24, 42f, 44
  apternodont, 152
  cetacean, 277
  chiropteran, 158, 161
Mammal Paleogene (MP), 15, 338
Mammalia (Class). *See also* mammals
  crown-group definition, 43, 49
  monophyletic, 44–45, 50f, 56, 61
  polyphyletic, 44, 50f, 56
  stem-based definition, 43
  synoptic, 8, 9t–10t, 49t
Mammaliaformes, 43
Mammaliamorpha, 43
mammalian boundary, 42–43
Mammalodontidae, 273t

mammals. *See also* Mammalia (Class)
  basal, 52–53, 53f
  characteristics, 41–43, 42f, 43f
    evolution of, 41–43, 42f, 43f, 49
  classification, 7–8, 9t–10t
  classification and relationships, 5–6, 11f, 12f
  definition, 41–44
  diversity, 2–3, 3f, 335
  evolutionary transition to, 44–46, 44f–47f, 49
  extant
    genera, 1
    number, 1, 335
    orders, 1, 4, 335
  genera, numbers, 2
  Mesozoic, 48–71, 50f, 74–79, 83f, 88–93, 335–336
    classification and relationships, 8, 9t–10t, 48, 49, 49t, 50f, 51f
  nontherian, 49–50
  oldest, 11, 50–55
  orders
    extant, 1, 4, 335
    "modern," 4
  smallest known, 96
mammoths, 260
Mammutidae, 242t
manatees, 40, 267–268
  cervical vertebrae, 30
mandible, 26, 65
  adapoid, 182f
  amphipithecid, 185
  anagalid, 309
  anthropoid, 192
  apatemyid, 104f, 105f
  apternodont, 152
  archaeocete, 282
  Arsinoitherium, 267f
  cebochoerid, 291
  chalicothere, 257
  didymoconid, 97
  entelodont, 295
  eutherian, 89
  geolabidid, 148
  haramiyid, 52f
  herbivore, 36–39
  hyopsodontid, 222
  hyracoid, 259
  hystricognathous, 318, 319f, 347
  marsupial, 85–86, 85f, 86f
  mesonychian, 277f
  metatherian, 76f
  mixtotherid, 297
  oreodont, 300
  pangolin, 204
  pholidotan, 204
  Pseudoglyptodon, 201f, 202
  rodent, 318, 319f
  sciurognathus, 318, 319f, 321, 328
  strepsirrhine, 187
  stylinodontid, 107
  taeniodont, 105
  tillodont, 112f
  xenungulate, 238, 238f
mandibular angle
  inflected, 73, 84
  Asiatherium, 76
  deltatheroidian, 74
  pantodont, 117
mandibular condyle, 25f
mandibular foramen. *See* foramen/foramina, mandibular
mandibular nerve, 247
mandibular symphysis, 26, 85, 328
  adapoid, 180
  anthracothere, 293, 294
  anthropoid, 192, 194, 197
  archaeocete, 278
  cercamoniine, 185
  entelodont, 295
  Ernanodon, 210

notharctid, 184
omomyid, 188
palaeanodont, 205
pholidotan, 204
Plesiopithecus, 187
proboscideans, 263, 265
Pseudoglyptodon, 202
sirenian, 268–270
taeniodont, 107
tillodont, 111
Tricuspiodon, 222
uintathere, 239, 240
Manidae, 95t, 199t, 200f, 204
Manis, 38f
Manteoceras, 251f
manubrium, 31, 42f
manubrium sterni, 60
manus, 32, 33, 35f, 71. *See also* forefoot
  adapoid, 180
  anthracothere, 294, 294f
  apatemyid, 106f
  arboreal mammals, 39
  archaeocete, 281
  arctocyonid, 218f
  astrapothere, 236
  brontothere, 251
  choeropotamid, 292
  Chriacus, 218f
  dichobunid, 290
  elephant shrew, 311
  Eomaia, 75f
  Ernanodon, 210
  eutherian, 75f
  fossorial mammals, 40
  graviportal mammals, 40
  lemuroid, 106f
  marsupial, 106f
  mesonychian, 274
  notharctid, 183
  oreodont, 299, 299f
  oxyaenid, 124f
  palaeanodont, 208, 208f
  pangolin, 204
  pantodont, 118f
  paromomyid, 176
  periptychid, 221f
  pholidotan, 204
  plesiadapiform, 174f
  rhinocerotoid, 255
  semi-aquatic mammals, 40
  Sinodelphys, 75f
  taeniodont, 106, 107, 109f
  tillodont, 111
  uintathere, 240
  xenarthran, 200–201
  xenungulate, 238, 238f
Maotherium, 64
marine environment, 126, 284
marmoset, 194
Marmosopsis, 81f
marsupial bones. *See* epipubic bones
marsupial cats, 9t
marsupial mice, 9t
marsupial moles, 9t, 40
Marsupialia, 2, 12f, 42f, 61, 67, 72–87, 90, 121, 169, 225, 335, 336, 341, Plate 1.2
  auditory bulla, 24
  auditory ossicles, 42f
  Australian, 72–74, 79, 85–87, 86f, 341
    classification of, 9t
  borhyaenoid, 6, 84–86
  classification and relationships, 9t, 11f, 51f, 72, 73, 75f
  dental formula, 27
  dentition, 36, 73, 78, 78f, 79f, 81f, 82f, 83–87, 83f–86f
  didelphoid, 77, 79–81, 337
  Early Cenozoic
    of Europe, 338
    of Southern Hemisphere, 336–337

epipubic bones. *See* epipubic bones
in Madagascar, 79
microbiotheriid, 337
New World, 73–74
origin, 49, 67, 79
palatal vacuities, 26
Paleocene-Eocene, dispersal, 341
petaurid, 176
phalangeroid, 106f
polydolopoid, 83, 84f, 337
saber-toothed, 6
skeletal adaptations, 36, 39, 80f
skull, 25f
Marsupionta, 73
marsupium, 73
Martinmiguelia, 231
Maryland, 337
Masillabune, 292–293
Masillamys, 326
masseter muscle, 26, 36
  lagomorph, 314
  rodent, 318, 318f
masseteric fossa, 25f
mastication. *See* chewing
mastiff bats, 161
mastodons, 260
mastoid (of petrosal), 57
  artiodactyl, 285, 289, 293, 298, 299, 301, 303
Mattimys, 328
Matutinia, 316
maxilla, 25f, 26, 45f, 57, 69, 82f
  anagalid, 310f
  archaeocete, 281f
  cetacean, 276, 277
  haramiyid, 52f
  interathere, 231
  leptictidan, 142f
  lipotyphlan, 143
  proboscidean, 264f
  sirenian, 269f
  uintathere, 239
maxillary nerve, 100
maxilloturbinal, 26
maximum likelihood analysis, 7
Maxschlosseria, 230
Mayulestes, 82f, 85
Mayulestidae, 74t
Mckennatherium. *See Adunator*
Meckelian groove, 68, 89
Meckel's cartilage, 42f, 62
Mediterranean Sea, 18
Megacerops, 250f, 251, 251f
Megachiroptera, 157, 158f, 161, 162
Megadelphus, 178
Megadermatidae, 158t, 161, 162
Megadolodus, 221
Megalagus, 314
Megalesthonyx, 112f, 113
Megalohyrax, 259f, 260
Megalonychidae, 199t
Megatheria, 199t
Megatheriidae, 199t
Megazostrodon, 53f, 55
Megazostrodontidae, 49t, 52, 61–63
Megistotherium, 126
Meiostylodon, 113
melon (organ), of whales, 278
Meniscoessus, 57f
Meniscotheriidae, 215
Meniscotheriinae, 219
Meniscotherium, 223, 224
Menotyphla, 138, 311
mental foramen. *See* foramen/foramina, mental
Meridiungulata, 218, 226–238, 227f, 344
  classification, 10t, 211–212, 213t
Merychippus, 248
Merycoidodon, 299f
Merycoidodontidae, 289t, 300
Merycoidodontoidea, 299–300
Merycopotamus, 293

mesaxonic symmetry, 228, 231, 242, 244, 257, 290, 294
mesial, definition, 27
mesoconid, 28, 28f, 259, 316, 322, 325
mesocuneiform, 33
  dichobunid, 290
  mesonychian, 277f
Mesodmops, 339
Mesohippus, 247, 249f
mesoloph, 326
mesolophid, 316, 322, 326
Mesonychia, 7, 22, 99, 120, 215, 217, 272f, **274–275**, 275f, 337–338, 344–346
  classification and relationships, 10t, 212f, 213t, 272f, 273–274, 273t, 286f
Mesonychidae, 273t, 339
Mesonyx, 274–275, 276f, 279
mesostyle, 28, 73, 73f, 110, 113, 144, 164, 184, 186, 188, 190, 223, 224, 244, 249, 292, 297, 303, 326
Mesotheriidae, 213t, 228f, 231, 232
Mesozoic, 13f
  mammalian radiation in. *See* mammals, Mesozoic
Messel (Germany), 79, 80f, 81, 101, 102f, 103f, 104, 106f, 125, 128f, 141, 142f, 146, 146f, 147f, 160, 161, 164, 185, 203, 203f, 204, 205f, 249, 250f, 290, 291f, 292, 321, 324f, 326, 338, 343, Plate 2.2, Plate 3.1, Plate 4.2, Plate 5.2–Plate 5.4, Plate 7.1, Plate 7.3, Plate 8.3
Messelobunodon, 288–290, 291f
Mesungulatum, 64, 340
metacarpals, 30f, 33, 35f, 37f, 75f. *See also* forefoot; manus
  anthracothere, 294f
  artiodactyl, 287f
  cainothere, 299
  chalicothere, 257
  didymoconid, 97
  early whale, 283f
  leptomerycid, 303
  notharctid, 183
  oreodont, 299f
  palaeothere, 249
  protoceratid, 302
  ruminant, 303, 304f
Metacheiromyidae, 95t, 199, 199t, 200f, 205–209
Metacheiromys, 32f, 206, 207f
metacone, 28, 28f, 34, 35, 54, 65, 65f, 66, 67, 70, 74, 77, 78f, 85, 96, 97, 99, 105, 114, 115, 115f, 120, 149, 150, 153, 180f, 238, 254, 255, 280, 298, 310, 311f
metaconid(s), 28, 28f, 65, 68, 73f, 76, 77, 78f, 89–90, 91f, 110, 114, 120, 132, 149, 153, 164, 180f, 196, 217, 238, 274, 298, 311f, 319
metaconules, 28, 28f, 72, 73f, 83, 136, 180f, 219, 224, 247, 256, 286, 289–292, 294, 297, 298, 300, 311f, 328
metacristid, 28, 28f, 145
metacromion, 32f
metaloph, 28, 227, 235, 238, 239f, 243, 247, 255–257, 322
metalophid(s), 28, 28f, 114, 228, 238, 239f
Metamynodon, 248f, 255
Metanoiamys, 331
Metaphiomys, 325f, 334
metapodials, 34, 40. *See also* metacarpals; metatarsals
  amphimerycid, 303
  anthracothere, 294
  artiodactyls, 292
  camelid, 302
  chalicothere, 257
  choeropotamid, 292
  dichobunid, 290
  entelodont, 295
  Ernanodon, 210
  helohyid, 291

metapodials (continued)
  palaeanodont, 206, 207
  ruminant, 303, 304
  uintathere, 240
metapophysis (mamillary process), 30, 31, 31f
Metasequoia, 21
metastyle, 28, 73, 73f, 90, 120, 255
metastylid, 28, 110, 185, 190, 238, 239f
metatarsals, 30f, 33–34, 75f, 308. *See also* hind foot; pes
  artiodactyl, 287f, 290–291, 290f
  camelid, 302
  chalicothere, 257
  choeropotamid, 292
  Diacodexis, 290f
  dichobunid, 290
  early whale, 283f
  equid, 247
  erinaceomorph, 146
  hyaenodontid, 126
  ischyromyid, 322
  lagomorph, 314
  leptictid, 140
  leptomerycid, 303
  mesonychian, 277f
  palaeothere, 249
  protoceratid, 301
  pseudictopid, 310
  ruminants, 303–304, 304f
  zalambdalestid, 308
Metatheria, 3–4, 22, 48, 70, **72–87, 340–341**
  basal, 74–76
  characteristics, 73
  classification and relationships, 9t, 49t, 50f, 51f, 72–73, 74t, 75f
  definition, 72
  dental formula, 88
  dentition, 73, 73f
  origin, 79
Metoldobotes, 312
Meurthodon, 47f
Mexico, 337
Miacidae, 120f, 121, 121t, 122f, 127, 129–131, 131f
Miacis, 131, 134, 135f
Miacoidea, 7, 126–127, 129, **130–131**, 130f, 342
mice, 9t, 318, 329–331. *See also* Muroidea
Microadapis, 183f, 185–186
Microbiotheria, 9t, 73, 74, 74t, 78, 81f, 341
Microbiotheriidae, 74f, 75f, 86
Microchiroptera, 157–158, 158t, 161, 162
Microchoerinae, 188, 191
Microchoerus, 190f, 191
Microconodon, 47f
Microcosmodon, 59
Microhyrax, 260
Microhyus, 223, 312
Micromomyidae, 157f, 165–166, 166t, 169, 170, 170f, **173–175**
Micromomys, 174
Microparamys, 321–322, 328, 329
Micropternodontidae, 97, 140t, 147, **150–151,** 151f
Micropternodus, 150–151, 151f
Microsyopidae, 164, 166t, 169, 170f, 175, **177–178,** 197
Microsyopoidea, 170f, 178
Microsyops, 178
middle ear, 24, 43, 44, 49, 62
  bones, sirenian, 268
  canid, 134
  carnivoran, 128
  Lambdopsalis, 60
  multituberculate, 60, 61
middle lacerate foramen. *See* foramen/foramina, middle lacerate
middle-ear ossicles. *See* ossicles, middle-ear
Miguelsoria, 227f, 230, 233, 233f, 234
milk, 41
milk teeth, 26–27, 88
Mimatuta, 214, 220, 227f

Mimolagus, 315
Mimotona, 314f, 315, 347
Mimotonida, 9t, 307t, 313
Mimotonidae, 307t, 308f, 313, **315,** 346–347
Minchenella, 265, 266
Mingotherium, 310, 337
Minippus, 247
mink, 126
Miocene Epoch, 3, 3f, 13f
Mioclaenidae, 212f, 213t, 215, **217–219,** 220f, 226, 227, 227f, 287, 336, 344–346
Miohippus, 247
Mirandatherium, 81f, 87
Mississippi, 337
Mithrandir, 219
mitochondrial genes, 6
Mixodectes, 164
Mixodectidae, 138, 158t, **164–165,** 165f, 197
mixodectoids, 117
Mixodontia, 9t, 307t, 313, 315
Mixtotheriidae, 286f, 288, 289t, **297,** 299, 346
Mixtotherium, 297, 298f
Moeripithecus, 196
Moeritheriidae, 8, 8f, 242t, 261
Moeritherium, 261, 261f, 262–264, 264f, 265f, 336
Molaetherium, 209
molars, 25f, 26–28, 28f, 71, 71f, 73, 73f, 96, 140, 144, 148. *See also* cusp(s), of teeth; dentition; *individual taxon*
  adapid, 185
  adapisoriculid, 144, 144f
  adapoid, 180, 182f, 183f
  adaptations, 34
  aegialodont, 69
  alagomyid, 319, 321f
  amphicyonid, 135, 136f
  amphimerycid, 298f, 303
  amphipithecid, 185
  anagalid, 309, 310f
  anaptomorphine, 189
  anoplothere, 297, 298f
  anthracothere, 294, 294f
  anthropoid, 192, 193f, 194, 195f, 196, 196f
  apatemyid, 103, 104, 104f
  apternodont, 152, 152f
  archaeocete, 278–280, 281f, 284
  arctocyonid, 214f, 216, 288f
  arctostylopid, 225
  artiodactyl, 286
  astrapothere, 235
  ausktribosphenid, 67f, 68
  bilophodont, 36, 237
  brachydont, 83
  brontothere, 250, 250f
  bunodont, 79, 144, 145, 214, 215, 217. *See also* bunodonty
  bunoselenodont, 83. *See also* bunoselenodonty/bunoselenodont dentition
  cainothere, 298f, 299
  camel, 302, 302f
  canid, 134, 135f
  caniform, 133
  carnivoran, 128
  carpolestid, 176f
  cebochoerid, 291, 298f
  cercamoniine, 185
  chalicothere, 256–257, 256f
  chiropteran, 159, 160f
  choeropotamid, 291–293, 292f
  cimolestid, 95, 95f, 96
  condylarth, 214, 215, 220f
  creodont, 120
  crown morphology, 27, 28f, 73f, 78f, 172f–173f, 228f
  ctenodactyloid, 332, 333f
  Deccanolestes, 159f
  deltatheroidian, 74
  dermopteran, 163, 163f, 164
  dichobunid, 288f, 289, 292f

didelphimorph, 77
didelphodont, 95, 95f
didelphoid, 77, 81, 83, 337
didymoconid, 97, 98f, 99
dilambdodont. *See* dilambdodonty/dilambdodont dentition
docodont, 55, 55f
dryolestid, 65, 66f
elephant, 261
elephant shrew, 311, 311f
embrithopod, 266, 267f
entelodont, 295, 295f
equid, 247
equoid, 249f
erinaceid, 144–145, 147f
erinaceomorph, 145–146, 145f, 147f
eupantothere, 64, 65f, 67f
euprimate, 167, 168, 172f–173f, 180f
eurymylid, 316, 321f
eutherian, 90, 91f, 92, 92f
ever-growing, 71, 85. *See also* hypselodonty
feliform, 133, 133f
geolabidid, 149
gondwanathere, 71, 71f
Hadrocodium, 54f, 55
haramiyid, 51, 52f
helohyid, 290
herbivores, 35
hyaenodontid, 122f, 124, 125f, 126
hyopsodontid, 222
hyracoids, 259–260, 259f
hystricognath, 325f, 334
ischyromyid, 321
lagomorph, 313, 315f
leptictid, 140, 142f
leptictidan, 141, 141f
lipotyphlan, 144
litoptern, 234, 234f, 235, 235f
lophodont, 28, 29f, 35, 36, 225, 228
lorisoid, 187f
macroscelidean, 311, 311f
marsupial, 72–74, 78, 79f, 81f, 82f, 83, 83f, 84, 85, 86f, 87
mesonychian, 274, 275, 275f
metatherian, 73f, 76, 76f, 77, 77f, 78f
miacoid, 130, 130f
micromomyid, 174
micropternodontid, 150
microsyopid, 178
mioclaenid, 217, 219, 227f
mixodectid, 164, 165f
monotremes, 67f, 69
morganucodont, 52, 55
multituberculate, 56, 59f
musteloid, 136, 136f
notharctid, 183–185
notoungulate, 227–228, 228f, 229f, 231, 231f–233f
nyctithere, 149, 150, 150f
omomyid, 188, 190f, 191
oreodont, 299f, 300
oxyaenid, 121, 122, 122f
palaeanodont, 206
palaeoryctid, 96, 96f
pantodont, 114, 115, 115f
pantolestan, 99, 100f
paroxyclaenid, 101
peccary, 296f, 297
pentacodontid, 101
peramurid, 66
periptychid, 220
perissodactyl, 249f
phenacodontid, 223
picrodontid, 175
picromomyid, 175
plagiomenid, 164, 165f
plesiadapiform, 169, 170, 172f–173f, 175f, 177f
primate, 186
proboscidean, 262f, 265, 265f
pseudictopid, 310, 310f

Ptolemaia, 103, 103f
Pyrothere, 237f
raoellid, 291, 292f
reversed-triangle pattern, 63
rhinocerotoid, 254f, 255
rodent, 325f, 326, 328, 331, 332, 333f
sciuravid, 322
secodont, 28, 34, 35
sectorial, 28, 29f
shrew, 153, 154f, 154–155
Sinoconodon, 53, 53f
Sinomylus, 315, 316f
sirenian, 268, 269f, 270
sivaladapid, 186
soricomorph, 148, 149f, 151f
strepsirrhine, 187, 187f
suoid, 296f, 297
symmetrodont, 63, 63f, 64
taeniodont, 105, 107, 107f
taeniolabidoid, 60
talpid, 154f, 155
tapiroid, 253–254, 254f
tarsiid, 187f, 192
terminology for, 65f, 73f, 172f–173f, 180f, 228f
tillodont, 110, 111f, 113
tree shrew, 197, 197f
tribosphenic, 27–28, 28f, 34, 49, 63–67, 73, 120, 138, 144, 149, 206, 345
"tribothere," 70f
triconodont, 61, 62, 62f
triisodontine, 217
uintathere, 238, 239f
ursoid, 135, 136f
xenungulate, 238, 238f
xiphodont, 301
zalambdalestid, 308, 308f
zalambdodont. *See* zalambdodonty/ zalambdodont dentition
mole(s), 9t, 40, 138, 143, 147, 148, 150, 153–155, 154f, 342
desman, 56
mole rats, 318, 334
molecular evidence, 3–4, 6, 8, 12f
molecular evolution, rates of, 4
molecular phylogeny. *See individual taxa*
Molinodus, 217, 218
Molossidae, 158t, 161
Mongolia, 58, 66, 69, 76, 89, 90, 97, 132, 147, 152, 154, 172, 179, 274, 307, 310, 315, 316, 323, 339, 342, Plate 1.3, Plate 1.4. *See also* Bayan Ulan (Mongolia); Gashato (Mongolia); Naran Bulak (Mongolia); Tsagan Khushu (Mongolia)
Anagalida, 346
Ernanodon, 210
pholidotan, 205
rhinocerotoids, 255
triisodontines, 217
Mongolian Remodeling, 340
Mongolotherium, 238
mongooses, 9t, 126
monito del monte, 86
monkeys, 10t, 36, 166
Monodelphis, 25f
monophyly, **6–7,** 22
Monotremata, 9t, 11f, 49, 49t, 51f, 336, **340.** *See also* monotremes
epipubic bones, 33
sternal ribs, 31
Monotrematum, 69, 340
monotremes, 4, 17, 32, 43, 43f, 48, 57, 60–61, 63, 67, **68–69,** 72. *See also* Monotremata
classification and relationships, 9t, 49, 50f, 72
Monshyus, 223
Montana, 59, 62, 89, 146, 155, 172, 214, 321
Montanalestes, 9t, 67f, 89–90, 89f
moon rats, 147
Morganucodon, 43f, 46
dental formula, 52
jaw joint, 42f, 52–53, 53f

occlusal relationships, 42f, 53f
skull and dentition, 53f
Morganucodonta, 9t, 45–46, 49–51, 49t, **52–55,** 53f
Morganucodontidae, 45–46, 49t, 50f, 51f, 52, 61–63
Mormoopidae, 158t
Morocco, 20, 70, 101, 144, 179, 219, 225, 259, 263, 336
Moropomorpha, 253
Moropus, 256f
Morrison Formation, 55, 71
Moschidae, 289t, 304–305
mountain beavers, 320, 326
Multituberculata, 22, 48–52, **56–61,** 57f, 71, 335, 337, **340,** Plate 1.1
arboreal, 58, 60f
classification and relationships, 9t, 49t, 50f, 51f, 58, **60–61**
definition, 2
dispersal, 61
Early Cenozoic
of Asia, 339
of Europe, 338
of North America, 337–338
extinction, 61
fossorial, 58
terrestrial, 58
multituberculates. *See* Multituberculata
Murgon (Queensland, Australia), 337
Muridae, 307t
Muroidea, 307t, 325f, 327f, 329f, **330–331**
Murtoilestes, 9t, 89, 89f
musk deer, 304–305
muskrats, 40
Mustelavus, 136–137
Mustelictis, 136–137, 136f
Mustelida, 121t, 136–137
Mustelidae, 120f, 121t, 126, 136–137
Musteloidea, 120f, **136–137,** 136f, 342
Mustersan SALMA, 14f, 17, 83, 84, 202, 230, 231, 235, 343
Myanmar, 185, 193, 339
Myanmarpithecus, 185
Mygatalpa, 155
Mylagaulidae, 307t, 326
Mylodonta, 199t, 201
Mylodontidae, 199t
Myodonta, 307t, 329, 329f
Myohyracinae, 311
Myomorpha, 307t, 308f, 318, 322f, **329–331,** 329f, 332f, 347
myomorphous rodents, 318, 318f, 328
Myophiomyidae, 307t, 334
Myoxidae, 328–329
Myrmecophagidae, 199t, 203
myrmecophagy/myrmecophagous mammals, 35, 38f, 71, 126, 200, 204, 206, 207
Mysops, 323, 326
Mystacinidae, 158t
Mysticeti, 271, 275–277, 278f, 279f, **284–285,** 346
classification, 272f, 273t
oldest, 273t, 284–285
Mytonolagus, 314
Mytonomys, 321
Myxomygale, 154f, 155
Myzopodidae, 158t

nails, 33, 39, 81, 260
adapoid, 180
carpolestid, 173
euprimate, 168, 179
hallucal, 173
hyracoid, 258
phenacodontid, 224
Nalacetus, 279
NALMAs. *See* North American Land-Mammal Ages
Nandinia, 120f, 129, 132
Nandiniidae, 121t

Nannodectes, 171–172, 174f
Nannopithex, 188, 191
nannopithex fold, 169, 172f, 178, 180f, 184, 188, 191
Nanolestes, 66
Nanxiong Basin (China), 339
Naran Bulak (Mongolia), 257, 339
nasal bone, 25f, 26, 45f, 82f
anagalid, 310f
archaeocete, 281f
astrapotheres, 236
embrithopod, 266
lepticitidan, 142f
phenacodontid, 223
rhinocerotoid, 255
sirenian, 268
tapiroid, 253
uintathere, 239, 240
nasal cavities, 26
nasal incisure, 249, 253, 254
nasal opening, 41
archaeocete, 283
litoptern, 235
pantodont, 117
phenacodontid, 223
proboscidean, 261, 263–265
sirenian, 268–270
nasofacial vacuities, oreodont, 300
nasolacrimal canal, 26
Nasua, 36f, 216
Natalidae, 158t, 159, 161
natatorial, definition, 40
Navajovius, 177f, 178
navicular, 30f, 33, 37f, 74, 75f
artiodactyl, 285
astrapothere, 236
dichobunid, 290
mesonychian, 274, 277f
oreodont, 299
oxyaenid, 121
protoceratid, 301
neck, of teeth, 27
Necrolemur, 188, 189f, 191, 192
Necromanis, 204
Necromantis, 162
nectivores/nectivorous mammals, 36, 158, 175
Nei Mongol, 339
Nemegbaatar, 59f
Nementchamys, 332
Nementchatherium, 311–312
Neoceti, 273t, 275, 276, **284–285**
Neocomian Epoch, 13f, 66
Neogene Period, 13f, 84
Neoliotomus, 59
neoplagiaulacids, 59, 339
nerve(s)
abducent, 24
accessory, 26
cranial, 24–26, 58
facial, 25
hypoglossal, 26
mandibular, 25
oculomotor, 24
olfactory, 24
ophthalmic, 24
optic, 24, 247
trigeminal, 24–25, 43, 45
trochlear, 24
vagus, 26
vestibulocochlear, 25–26
Nesophontes, 38f
Nesophontidae, 140t
Neurogymnurus, 147
Neustrian ELMA, 15
New Guinea, monotremes, 68–69
New Mexico, 16, 21, 206, 216–217, 274, 337
New Zealand, 284
niche partitioning, 161
Nigeria, 278, 336
archaeocetes, 282

Nimravidae, 120f, 121t, 126, 129, 132–133, 134f, 342
Niptomomys, 177f, 178
Noctilionidae, 158t
nocturnal mammals, 180, 187, 188
node-based taxon, 8
Nongshanian ALMA, 14f, 17, 209, 226, 339
Nonomys, 330
Norian Age, 13f, 51
North America, 52, 58, 59, 65, 74, 78–81. *See also* Western Interior (North America)
    adapoids, 180
    amphicyonids, 135
    anaptomorphines, 189
    anthracotheres, 293, 294
    apatemyids, 103
    apheliscines, 222
    apternodontids, 151
    archaeocetes, 278, 282–284
    archaic ungulates, 344–345
    arctocyonids, 215
    arctostylopids, 225, 226
    and Asia, faunal exchange between, 18, 19f, 337, 339, 340
    brontotheres, 250, 252
    camels, 302
    canids, 133–134
    caniforms, 133
    carpolestids, 172–173
    cercamoniines, 185
    chalicotheres, 257
    chiropterans, 159
    cimolestans, 341
    cimolestids, 95
    condylarths, 215, 226
    creodonts, 119–121
    didymoconids, 99
    entelodonts, 295
    equids, 247
    erinaceids, 146
    erinaceomorphs, 145–146
    euprimates, 178
    and Europe, faunal exchange between, 18, 19f, 337
    eutherians, 89, 92
    feliforms, 132
    geolabidids, 148
    hedgehogs, 147
    helohyids, 290
    heterosoricines, 153
    homacodontines, 289
    hyaenodontids, 122–123, 125, 126
    hyopsodontids, 222
    hypertragulids, 303
    lagomorphs, 314, 347
    leptictidans, 140
    leptomerycids, 303
    mesonychians, 274
    metatherians, 340–341
    miacoids, 130
    micromomyids, 174
    mioclaenids, 218
    musteloids, 136
    myomorph, 347
    nimravids, 133
    notharctids, 183, 184
    nyctitheres, 149, 150
    omomyids, 188, 190
    oreodonts, 346
    oromerycids, 301
    oxyaenids, 122
    palaeanodonts, 205
    Paleocene-Eocene climate and flora, 20
    paleogeography, 18, 19f
    pantodonts, 114–116, 116f
    pantolestids, 100
    pentacodontids, 101
    periptychids, 219
    phenacodontids, 223
    plesiadapids, 172
    plesiadapiforms, 169
    plesiosoricids, 153
    primates, 169, 188
    Purgatoriidae, 171
    rhinocerotoids, 254
    rodents, 319, 321, 324f, 326, 328
    Sciuravidae, 347
    shrews, 154
    sirenians, 270
    soricomorphs, 148
    and South America, faunal exchange between, 18, 19f, 337
    taeniodonts, 105, 108, 109
    talpids, 154–155
    tapiroids, 253
    tayassuids, 296
    tillodonts, 113
    tribotheres, 70
    trogosines, 113
    tylopods, 301
    uintatheres, 238
    ungulates, 212
    ursids, 135
    xenarthrans, 198
    zhelestids, 213
North American Land-Mammal Ages, 11, 14f, **15–16**
Notharctidae, 166t, 181f, 182, **183–185**
Notharctinae, 183
Notharctus, 168f, 180, 181f, 183–185
Nothrotheriidae, 199t
Notioprogonia, 213t, 228f, **229–230**, 229f
Notohippidae, 213t, 228f, 230, 231
Notohippus, 230, 231
Notonychopidae, 213t, 233, 234
Notonychops, 234
Notopithecus, 230, 231, 232f
Notopterna, 234, 235
Notoryctemorphia, 9t, 74t, 75f
Notostylopidae, 213t, 227, 228f, 229, 229f
Notostylops, 29f, 229, 229f
Notoungulata, 22, 215, 225–226, **227–233**, 229f, 237, 336, 337, 344
    classification and relationships, 10t, 213t, 226, 227f, 228f
    dental formula, 227
numbats, 9t, 35
Numidotheriidae, 8, 8f, 242t, 261
Numidotherium, 259, 261f, 262–263, 262f, 264f
Nycterididae, 158t
Nyctitheriidae, 93, 140t, 147, **149–150**, 150f, 154, 155, 158, 342, 343
Nyctitherium, 158

Obdurodon, 69
Obik Sea, 18, 19f, 340
occipital condyles, 24, 41, 44, 50
occipitals, 25f
occiput, 24
    apternodont, 152
    brontothere, 250
    cetacean, 276
    stylinodontid, 107
occlusion, dental. *See* dental occlusion
Ocepeia, 225
Ochotona, 315
Ochotonidae, 9t, 307t, 313–315, 319
Octodontotherium, 202
oculomotor nerve, 24
Odobenidae, 121t
Odontoceti, 271, 275–277, 278f, 279f, **284–285**, 346
    classification, 272f, 273t
    oldest, 273t, 285
odontoid process, 30, 31f, 45
Olbitherium, 242t, 244, 253, 345
Oldfieldthomasia, 231, 231f
Oldfieldthomasiidae, 213t, 228f, 231, 231f, 233f
olfactory bulb, multituberculate, 57
olfactory nerve, 24

Oligocene Epoch, 3, 3f, 12, 13f, 14f
Oligokyphus, 43f, 46f, 47f
Oligopithecidae, 166t, 167f, 194–197
Oligopithecus, 194
Oligoryctes, 152, 153
Oligoryctidae, 153
Oligospermophilus, 327–328
Olson, Everett C., 44
Omanodon, 185
omnivores, 34, 35, 67, 101, 105, 110, 114, 122, 126, 135, 144, 145, 170, 197, 286, 295
    multituberculates as, 56
Omomyidae, 18, 166, 166t, 167f, 168f, 173f, 178–179, 187, **188–191**, 189f–191f, 192, 343
Omomyiformes, 179
Omomyinae, 188, 189
Omomys, 188, 190, 191
Onychodectes, 105–106, 107f, 108f, 109
ophthalmic nerve, 24
opossums, 9t, 73, 79, 106. *See also* Didelphidae
optic foramen. *See* foramen/foramina, optic
optic nerve, 24, 247
orbit, 26
    adapoid, 180, 185
    anthropoid, 194, 197
    archaeocete, 282, 283
    artiodactyl, 285
    cetacean, 276
    dermopteran, 163
    didymoconid, 97
    elephant shrew, 311
    entelodont, 295
    euprimate, 167
    interathere, 231
    lagomorph, 313
    leptictidan, 141
    omomyid, 188, 189f, 191, 192
    oreodont, 300
    proboscidean, 260, 264
    rhinocerotoid, 255
    sirenian, 270
    strepsirrhine, 187
    suiform, 293
    tethythere, 242
    uintathere, 239
orbitosphenoid, 24
    leptictidan, 142f
Oregon, 191, 337
Orellan NALMA, 14f, 15, 17, 136, 140, 152, 293, 314
Oreodontidae, 289t, 297, 300, 346
Oreodontoidea, 286f, 289t, **299–300**, 299f
oreodonts. *See* Oreodontoidea
Orientalophus, 253
Ornithorhynchidae, 68
Ornithorhynchus, 69
Orohippus, 247
Oromerycidae, 286f, 289t, **301**, 301f
Orophodon, 202
Orthaspidotherium, 219
os penis, 33
Osborn, H.F., 265
ossicles, middle-ear, 24, 41, 42f, 43, 49, 268
    cetacean, 277
    Didelphis, 42f
    Hadrocodium, 55
    Lambdopsalis, 60
    multituberculate, 57, 60, 61
ossicones, 285, 303, 304
ossification, 23, 42
osteoderms, 201, 201f, 202
    xenarthran, 201–202
Otariidae, 121t
Othnielmarshia, 229
otic capsule, 24
Otlestes, 89f, 140t, 148, 149f, 342
Otlestidae, 148
otters, 9t, 35, 40, 101, 136
Ottoryctes, 96

Ouarzazate Basin (Morocco), 101, 336
Ouled Abdoun (Morocco), 259, 263
out-group taxon, 6
oval window. *See* fenestra ovalis
Owen, Richard, 117, 247
Oxacron, 299
Oxalidaceae, 21
Oxetocyon, 135
Oxyaena, 34f, 36f, 37f, 122, 123f, 124f, 339, 342
Oxyaenidae, 120, 120f, **121–122**, 121t, 122f, 123f, 124, 124f, 341–342
Oxyclaeninae, 215, 287
Oxyclaenus, 212f, 214f, 217
oxygen isotope analysis, 20
Oxyprimus, 214–215, 217, 227f

Pachyaena, 35f, 272f, 274, 275, 275f–277f, 283f
Pachygenelus, 45, 47f
Pachynolophidae, 248f
Pachynolophus, 248, 249, 255
Paenungulata, 114, 240, 242, **257–270**, 336, 344, 345
  age of divergence, 5t
  classification and relationships, 10t, 139f, 211–212, 212f, 213t, 242t, 243f
Pakicetidae, 272f, 273, 273t, 278–285, 280f
Pakicetus, 273, 279–282, 280f, 281f, 284
Pakilestes, 153
Pakistan, 173, 186, 269, 273, 279, 280, 284, 339, Plate 8.1, Plate 8.2
  altungulates, 345
  archaeocetes, 278, 282–283
  arctocyonids, 216
  artiodactyls, 291
  Cetacea, 346
  chiropterans, 161
  lemuroid, 186
  Parvocristes, 173
  plesiosoricids, 153
  quettacyonids, 216
  sirenians, 269
Palaeanodon, 34f, 36f–38f, 206–207, 208f, 209
Palaeanodonta, 7, 22, 198, 199t, 204, 205–209, 337, 338, 341–344, Plate 7.1
  classification and relationships, 95t, 198–199, 200f
Palaechthon, 168f, 171–173, 172f–173f
Palaechthonidae, 166t, 169, 170f, **171**, 178
Palaeictops, 140
Palaeoamasia, 266, 267f
Palaeocastor, 328
Palaeochiropterygidae, 158t
Palaeochiropteryx, 159–161, 160f
Palaeochoerus, 297
Palaeodonta, 288
Palaeogale, 133, 134f
Palaeolagus, 314, 315f
Palaeolemur, 180, 185–186
Palaeomastodon, 261f, 262, 264–265, 264f, 265f
Palaeomastodontidae, 8f, 242t, 261
Palaeomerycidae, 289t
Palaeomoropus, 249f
Palaeonictis, 122, 123f
Palaeopeltidae, 199t
Palaeopeltis, 202
Palaeoprionodon, 132
Palaeoryctes, 96, 96f
Palaeoryctidae, 94–95, 95t, 96, 138, 139–140, 150, 336
palaeoryctoids, 95
Palaeosciurus, 328
Palaeosinopa, 101, Plate 2.1
Palaeostylops, 225f, 226
Palaeosyops, 248f, 252
Palaeothentidae, 74t
Palaeotheriidae, 242t, 246f, 247, **248–249**, 250f, 345
Palaeotherium, 249, 249f, 253f
Palaeoxonodon, 64–65
Palasiomys, 330

palatal vacuities, 26, 73, 78, 84
palate
  altungulate, 243f
  carpolestid, 176f
  condylarth, 243f
  hard (=bony), 26, 41, 44
    pholidotan, 204
  mysticete, 284
  perissodactyl, 243f
palatine, 25f, 26
  anagalid, 310f
  lepticidan, 142f
  lipotyphlan, 143
palatine foramen. *See* foramen/foramina, palatine
Palenochtha, 170f, 177f
Paleocene Epoch, 13f
  climate, 20–21
  flora, 20–21
  geochronology and biochronology in, 8, 10, 13f, 14f
  mammalian diversity in, 3f, 335
  mammals, synopsis, 340–346
Paleocene/Eocene boundary, 12, 13f, 15, 19
Paleocene-Eocene Thermal Maximum. *See* Initial Eocene Thermal Maximum
paleofelids, 132
Paleogene Period, 13f
paleogeography, 17–20, 19f
paleogeography of Early Cenozoic, 17–20, 19f
Paleomoropus, 249f, 257
Palette (France), 338
palm civet, 35, 132, 216
palms, 21
Pampa Grande, Argentina, 230, 336
Pampahippus, 230, 230f
Pampatemnus, 230
Pampatheriidae, 199t, 343
Panameriungulata, 218, 219, 221, 226, 227f
Pandemonium, 170f, 172
pangolins. *See* Manidae; Pholidota
Pantodonta, 22, 94, 113, **114–118**, 115f, 116f, 118f, 215, 240, 242, 336, 337, 341, 345, Plate 3.3
  classification and relationships, 10t, 95t, 110f, 114, 211–212
  dental formula, 114
  Early Cenozoic
    of Asia, 339, 340
    of North America, 337–338
Pantolambda, 110f, 115–116, 115f, 116f, 118f
Pantolambdidae, 95t, 115–116
Pantolambdodon, 117
Pantolambdodontidae, 95t, 110f, 114, 117
Pantolesta, 22, 94, **99–101**, 102f, 206, 222, 336, 341
  classification, 10t, 95t
  dental formula, 99
  dentition, 99, 100f, 101
Pantolestes, 101
Pantolestidae, 95t, 99, 101, 138, 200f, 342, 344, Plate 2.1
Pantomesaxonia, 241, 266
Pappictidops, 130
Pappocricetodon, 320f, 325f, 330–331
Pappotherium, 51f, 70, 70f, 78f
Paraceratherium, 248f, 255
paracone, 28, 28f, 34, 35, 54, 65, 66–67, 70, 73, 73f, 74, 77, 85, 89, 96, 97, 99, 105, 114, 115, 115f, 120, 128, 149, 150, 180f, 196, 238, 280, 297, 310, 311f
paraconid, 28, 28f, 68, 73, 73f, 74, 76, 77, 78f, 89, 91f, 114, 117, 128, 132, 140, 144, 145, 149, 167, 171, 180, 180f, 183, 186, 187, 192, 194, 217, 223, 228, 238, 247, 274, 279, 289–291, 295, 298, 303, 308, 311f, 319
paraconule, 28, 28f, 73f, 180f, 247, 256, 257, 290–292, 294, 297, 298, 300, 302, 311f
paracrista, 73, 73f
paracristid, 28, 28f, 73f, 85, 114, 120, 122, 126, 145, 187

Paradelomys, 326
Paradjidaumo, 331
Paradoxurus, 36f, 37f
parallelism, 6, 34
paraloph, 238
paralophid, 28, 28f, 238
Paramyidae, 307t, 320, 321, 326
Paramys, 315–316, 319, 321–322, 323f–325f, 327f
Paranisolambda, 234
Paranyctoides, 89f, 92–93, 92f, 93, 140t, 148, 342
Paraphiomys, 334
paraphyly, **6–7**
Parapithecidae, 166t, 167f, **194–197**
Parapithecus, 196
Parapternodontidae, 153
Parapternodus, 152, 153
parastyle, 28, 65, 73, 73f, 90, 96, 104, 130, 257, 259
Paratriisodon, 217
Paratylopus, 302
paraxonic symmetry, 228, 231, 273, 274, 281, 285, 287f, 290, 292, 294
Parazhelestes, 213
Pariadens, 78
Parictis, 135, 136f
parietal bones, 24, 25f, 45f, 82f
  anagalid, 310f
  archaeocete, 281f
  entelodont, 295
  lepticidan, 142f
  proboscidean, 264f
  protoceratid, 301
  sirenian, 268, 269f
  uintathere, 239
Paris Basin, 222, 338
Paromomyidae, 157f, 165–166, 166t, 169–171, 170f, 173, **175–177**, 343
Paromomyoidea, 166t, 170f
Paromomys, 176
Paroxacron, 298f, 299
Paroxyclaenidae, 95t, 99, 100f, 101, 103, 103f, Plate 2.2
parsimony, 7
Parvocristes, 173
Paschatherium, 222
Pastoralodontidae, 95t, 117, 339
patagium, 39, 157, 163, 171, 177
Patagonia, 20, 55, 71, 201, 221, 226, 336, 340
Patagoniidae, 74t
patella, 30f, 34
Patene, 85
Patriofelis, 122, 124f
Patriomanidae, 95t, 199t, 200f, 204–205
Patriomanis, 36f, 199, 204–205, 206f
Patterson, Bryan, 44
Paucituberculata, 4, 9t, 73–74, 74t, 78, **81–84**
Paulacoutoia, 220f, 221, 233f
Paulchoffatiidae, 56, 57f, 58
Paurodontidae, 49f, 64, 65
peccaries, 273, 285, 293, 296–297, 346. *See also* Tayassuidae
Pecora, 289t, 302, **304–305**
pectoral girdle. *See* shoulder girdle
Pedetidae, 307t, 331–332
pedetoids, 347
Pediomyidae, 74t, 78, 79
Pediomys, 77f, 79f
Peligran SALMA, 14f, 226
Peligrotheriidae, 222
Peligrotherium, 65, 222, 226, 340
Peltephilidae, 199t
pelvic girdle, 33, 41–42, 43f
  monotreme, 68
pelvis, 30f, 36f, 64
  archaeocete, 283, 284
  cetacean, 276
  embrithopod, 266
  morganucodont, 53
  sirenian, 268, 270
  uintathere, 240

Pelycomys, 326
Pelycosauria, 44
Pentacemylus, 289
Pentacodontidae, 95t, 99, 101, 103, 138
Pentapassalus, 207
Peradectes, 79–80, 80f
Peradectidae, 74t, 75f, 78, **79–81**, 83
Peradectinae, 79
Peramelia, 9t, 74t, 75f
　classification, 9t, 74t
Peramelina, classification of, 75f
Peramura, 9t, 49t, 64–66
Peramuridae, 49t, 64, 66
Peramus, 51f, 65f, 66, 67f
Perchoerus, 286f, 296–297, 296f
pericone, 185, 186, 194, 220, 313
Periconodon, 185
periotic, 85, 328. *See* petrosal bone
　cetacean, 277
Periptychidae, 212f, 213t, 214, 215, **219–220**, 220f, 221f, 227f, 344–345
Periptychus, 216f, 219–220, 220f
Perissodactyla, 6–8, 12, 12f, 22, 178, 211–212, 214–215, 224–225, 227, 241–242, **244–257**, 245f, 246f, 248f, 249f, 258, 260, 262, 267, 274, 285, 286f, 344–345
　age of divergence, 5t
　classification and relationships, 8, 10t, 11f, 12f, 139f, 211–212, 212f, 213t, 242t, 246f, 272f, 286f
　Early Cenozoic
　　of Asia, 339, 340
　　of Europe, 338
　　of North America, 337
Peru, 83, 336. *See also* Laguna Umayo, Peru
Perutherium, 226
pes (foot), 33, 37f, 137f. *See also* feet
　adapoid, 180
　arboreal mammals, 36, 39–40
　archaeocete, 281, 283f
　arctocyonid, 218f
　artiodactyl, 285
　astrapothere, 236
　brontothere, 251
　cainothere, 299
　canid, 134
　carnivoran, 129
　choeropotamid, 292
　Coryphodon, 118f
　creodont, 121
　cursorial mammals, 36
　Diacodexis, 37f
　elephant shrew, 311
　erinaceid, 144
　eutherian, 75f
　graviportal mammals, 40
　mesonychian, 274, 277f
　pantodont, 114
　periptychid, 221f
　perissodactyl, 246f
　phenacodontid, 224f
　saltatorial mammals, 36
　semi-aquatic mammals, 40
　Sinodelphys, 75f
　stylinodontid, 107
　taeniodont, 106
　terrestrial mammals, 36
　uintathere, 240
PETM (Paleocene-Eocene Thermal Maximum). *See* Initial Eocene Thermal Maximum
Petrolemur, 179, 215
petromastoid
　apternodont, 152
　leptictidan, 142f
Petromuridae, 334
Petromus, 334
petrosal bone, 24, 25f, 43, 60, 82f, 242
　anoplothere, 298
　archaeocete, 280
　artiodactyl, 285, 289, 293, 298

　carnivoran, 128
　cetacean, 277, 280
　dichobunid, 289
　euprimate, 167, 179
　eutherian, 89
　marsupial, 79
　palaeoryctid, 96
　plesiadapiform, 169, 171
　sirenian, 268
　tethythere, 242
petrotympanic complex
　archaeocete, 282
　cetacean, 277
Pezosiren, 243f, 269, 270f
phalangeal formula, 33
phalangers, 9t, 39
phalanges, 30f, 33, 37f. *See also* forefoot; hind foot; manus; pes
　anagalid, 309, 309f
　ancyclopod, 256, 257
　apatemyid, 104
　archaeocete, 273, 281
　arctocyonid, 216, 218f
　camel, 302
　canid, 134
　cetacean, 276
　chalicothere, 257
　chiropteran, 157
　creodont, 121
　didymoconid, 97
　Eomaia, 75f
　erinaceomorph, 146
　Ernanodon, 210
　euprimate, 168, 179, 183
　eutherian, 89
　Gobiconodon, 62, 62f
　hyaenodontid, 125
　hyracoid, 260
　marsupial, 81
　mesonychian, 273, 274
　miacid, 131
　notharctid, 183
　palaeanodont, 206–207
　pantodont, 116
　pantolestid, 101
　paromomyid, 176, 177
　paroxyclaenid, 101
　phenacodontid, 224
　pholidotan, 204
　plagiomenid, 164
　plesiadapiform, 169, 174f, 176
　pseudictopid, 310
　Sinodelphys, 75f
　tillodont, 111
　uintathere, 240
　ungual. *See* ungual phalanges
　viverravid, 131
Phascolotherium, 62
Phenacodaptes, 222
Phenacodontidae, 212f, 213t, 215, 217, 220f, **223–225**, 224f, 243, 244, 246f, 286f, 344, 345
Phenacodus, 30f, 34f, 36f, 37f, 223, 224f, 228, 243f, 245f
Phenacolemur, 174f, 175f, 176
Phenacolophidae, 242t, 243, 265, 266, 339, 345
Phenacolophus, 266, 267f
Phenacopithecus, 193
Philisidae, 158t
Phiocricetomys, 334
Phiomia, 262, 265
Phiomiidae, 242t, 261
　classification of, 242t
Phiomyidae, 307t, 334, 347
　classification of, 307t
Phiomys, 334
Phocidae, 121t
Phocoidea, 121t
Pholidocercus, 145, 146, 147f

Pholidota, 119, 198, 204–210, 341, 344
　classification and relationships, 8, 10t, 11f, 12f, 95t, 139f, 198–199, 199t, 200f, 272f
　Paleocene-Eocene, synopsis, 337, **343–344**
Phosphatheriidae, 242t, 261, 263
Phosphatherium, 8f, 261f, 262, 262f, 263, 263f
Phosphorites (Quercy, France), 338. *See also* Quercy (France)
Phyllophaga, 199t, 200, 200f, 202
Phyllostomidae, 158t
phylogeny, 5–8
Physeteridae, 285
phytoliths, 71
Picopsis, 70
Picrodontidae, 166t, 169, 170f, **175**
Picrodus, 175, 177f
Picromomyidae, 166t, 169, 170f, **175**
Picromomys, 169, 175, 177f
pigs. *See* Suidae
pikas. *See* Ochotonidae
Pilgrimella, 262, 262f
Pilosa, 199t, 200, 200f, 202–204
Pinnipedia, 120f, 121t, 126, 128, 129, 135
piriform fenestra, 97, 150
piscivores, 35, 126
pisiform, 33, 35f
　palaeanodont, 208f
placental mammals, 61, 67, 70, 89, 199
　auditory bulla, 24
　clades, age of divergence, 5t
　classification and relationships, 9t–10t, 11f, 12f
　dental formula, 27
　diversification, timing, 3–4, 5t
　extant order, 4
　origin, 3–4, 5t, 49, 67
Placentalia, 5t, 9t, 11f, 12f. *See also* placental mammals
Plagiaulacida, 56, 57f
Plagiaulacidae, 57f, 58
Plagiaulacoidea, 49t, 57f, **58**
Plagiolophus, 249
Plagiomene, 164, 165f
Plagiomenidae, 117, 158t, 163, **164–165**, 165f, 337, 343
plantigrade stance, 39–40, 101, 121, 129, 134, 144, 219, 240, 257, 263
plants. *See* angiosperms; flora
Platanus, 21
Platychoerops, 172
Platypoda, 4, 10t
platypus, 68–69
Platyrrhini, 7, 166t, 167f, 180, 194
Plesiadapidae, 166t, 168f, 169–170, 170f, **171–172**, 173, 174f, 175f, 337
Plesiadapiformes, 7, 61, 103, 156, 163, 165, 168, 168f, **169–178**, 173f, 177f, 337, 341, 343, Plate 4.3
　classification and relationships, 166t, 167f, 169, 170f
　Early Cenozoic
　　of Europe, 338
　　of North America, 337–338
Plesiadapis, 168f, 169, 170–172, 174f
Plesiadapoidea, 157f, 166t, 339
Plesiesthonyx, 113
Plesiofelis, 85
Plesiolestes, 171
plesiomorphic, 6
Plesiopithecidae, 166t
Plesiopithecus, 187, 187f
Plesiosminthus, 330
Plesiosorex, 151f, 153
Plesiosoricidae, 140t, **153**
Plethorodon, 110f, 111f, 113–114
Pleuraspidotherium, 219
Pleurostylodon, 230, 230f
Plexotemnus, 230
Pliohyracidae, 242t, 257, 259, 260
Pliolophus, 247

Pliopithecidae, 166t
pneumatization
　of skeleton, 260
　of skull, 26, 164, 262, 264, 265
Poabromylus, 29f
pocket gophers, 40, 329–331
Poebrodon, 302
Poebrotherium, 302, 302f, 304f
polar ice caps, 20
polarity, 6
pollex, 33
　in arboreal mammals, 39
　chiropteran, 157, 159
　choeropotamid, 292
　hyracoid, 258
　oreodont, 300
　suiform, 293
Polydolopidae, 9t, 74t, 341
Polydolopimorphia, 74, 74t, 83
Polydolopoidea, 74t, 75f, 81–84, 84f
Polymorphis, 235, 235f
polyphyletic groups, 6, 44
polytomy(ies), 7
Pondaung Formation (Myanmar), 185, 193
Pondaungia, 185, 193
Pondaungimys, 331–332
Pontifactor, 150, 159
porcupines, 9t, 318, 334
porpoises, 275
Portugal, 55, 66, 338
postcanine teeth, 26, 27, 64. See also cheek teeth; molars
　Ernanodon, 210
　palaeanodont, 207
　paurodontid, 65
　soricomorph, 148
　symmetrodont, 63
postcingula, 76, 78f, 114, 115, 140, 148, 149, 188, 214, 289, 309
postcranial skeleton, 30–34, 30f–33f. See also specific taxon
postcristid, 28, 28f, 73f
postdentary bones, 45, 61
　Ausktribosphenos, 68
　docodonts, 56
　eupantothere, 64
　fossa for, 45, 47f
　haramiyid, 51–52
　Kuehneotherium, 54
　monotreme, 69
　Sinoconodon, 53
　symmetrodonts, 63
postdentary trough (groove), 55, 61, 68, 69
posterior lacerate foramen. See foramen/foramina, posterior lacerate
postmetacone crista, 85, 183
postmetacrista, 78f
postorbital bar, 167
　anthracothere, 293
　euprimate, 167, 168f, 179
　hypertragulid, 303
　hyracoid, 259
　oreodont, 300
　plesiadapiform, absence of, 169
postorbital bone, 45f
postorbital closure, 167, 185, 187
　anthropoid, 194, 197
postpalatine torus, 25f
postparietal, 82f
postprotocingulum, 169. See also nannopithex fold
postprotocrista, 28, 28f, 73f, 136, 172f, 225, 310
post-tympanic process, 25f
posture, 44, 44f
　digitigrade, 39–40
　djadochtathere, 58
　graviportal, 40, 261
　plantigrade, 39–40
　unguligrade, 40
postvallum, 78

postzygapophyses, 30, 31f
Potamotelses, 70, 78f
Potos, 36f
potto, 186
prearticular, 47f
precingula, 76, 114, 148, 149, 309
prefrontal bone, 45f
premaxilla, 25f, 26, 45f, 82f, 310f
　archaeocete, 281f
　cetacean, 276
　chiropteran, 158
　leptictidan, 142f
　metatherian, 76f
　palaeothere, 249
　proboscidean, 260, 264f
　rhinocerotoid, 255
　sirenian, 269f, 270
premolars, 25f, 26, 27, 35, 36
　adapid, 185
　adapoid, 180, 183
　aegialodont, 69
　Altungulata, 241
　amphicyonid, 135
　anagalid, 309
　anaptomorphine, 188, 189
　anthracothere, 294
　anthropoid, 194, 196
　apatemyid, 104
　apternodont, 152
　archaeocete, 279, 280
　arctocyonid, 216
　arctostylopid, 225
　artiodactyl, 286, 288–290, 294, 295, 297, 299, 301, 304
　Asiatherium, 76
　astrapothere, 236
　ausktribosphenid, 68
　camel, 302
　carnivoran, 122f, 128, 135, 136
　carpolestid, 172
　choeropotamid, 292
　cimolestid, 96
　condylarth, 215
　creodont, 120, 122, 122f, 125, 126
　cylindrodont, 323
　deltatheroidian, 74
　dermopteran, 163, 164
　dichobunid, 288, 289
　didymoconid, 97, 99
　dryolestid, 65
　elephant shrew, 311
　embrithopod, 266
　entelodont, 295
　eosimiid, 193
　erinaceid, 144, 147f
　erinaceomorph, 145–146
　euprimate, 167
　eutherian, 88, 89–90, 92–93
　Glires, 313
　gondwanathere, 71
　Hadrocodium, 55
　helohyid, 290
　hyaenodontid, 125, 126
　hyopsodontid, 222
　hyracoid, 259, 260
　ischyromyid, 321
　leptictid, 140
　leptictidan, 141
　litoptern, 231
　macroscelidid, 311, 311f
　marsupial, 72, 73, 77, 78, 83, 85
　metatherian, 77
　micromomyid, 174
　mioclaenid, 217, 219
　mixodectid, 164
　multituberculate, 56, 58, 59, 59f
　musteloid, 136
　notharctid, 185
　notoungulate, 229, 231
　nyctithere, 149, 150

　omomyid, 188
　oxyaenid, 122
　palaeanodont, 206
　palaeoryctid, 96
　pantodont, 114, 115, 117
　pantolestan, 99
　paromomyid, 176
　paroxyclaenid, 101
　peccary, 297
　peramurid, 66
　periptychid, 220
　perissodactyl, 244, 247, 249, 250, 253–257
　phenacodontid, 223
　plagiaulacoid, 58
　plagiomenid, 164
　plesiadapiform, 170
　plesiosoricid, 153
　primates, 185, 187–189, 194, 196
　proboscidean, 261–263
　pseudictopid, 240, 310, 310f
　ptilodontid, 59, 59f
　Ptolemaia, 103
　Purgatoriidae, 171
　pyrothere, 236
　rodents, 316, 321, 322, 326, 328–334
　sciuravid, 322
　shrew, 153
　sirenian, 88, 268
　sivaladapid, 186
　soricomorph, 148, 149, 152, 153
　spalacotheriid, 64
　taeniodont, 106, 107
　taeniolabidoid, 60
　tillodont, 110, 111
　tree shrew, 197
　triconodontid, 62
　trogosine, 111
　uintathere, 238
　ungulate, 214, 215
　xiphodont, 301
　zalambdalestid, 307
　zhelestid, 214
preparacrista, 35
Prepidolopidae, 74t, 83
Prepidolops, 83
preprotocrista, 28, 28f, 73f, 78f, 225, 259, 310
Preptotheria, 141
Presbymys, 331
preselenodonts, 289
presphenoid, 24, 25f
prevallid, 78
prezygapophyses, 30, 31f
Priabonian Stage/Age, 13f, 14f, 15, 17, 338
Primates, 22, 39, 119, 141, 156, 165, **166–197,** 168f, 216, 336, 340, 343
　age of divergence, 5t
　anthropoid, 192–197, 336
　archaic, 169. See also Plesiadapiformes
　classification and relationships, 7–8, 10t, 11f, 12f, 139f, 157f, 158, 158t, 166t, 167f, 170f
　Early Cenozoic
　　of Asia, 339
　　of Southern Hemisphere, 336
　lemuroid, 106f, 186
　lorisoid, 186–187, 336
　omomyid, 83, 188–191
　origin, 168–169
Primatomorpha, 156, 157f, 165–166
Primisminthus, 330
Priodontes, 200
prism sheath, 29–30
prisms, of tooth enamel, 29–30, 45, 58, 220, 319, 320f
Proailurus, 132, 133f
Proardynomys, 323
Probainognathus, 44, 45f, 51f
Probathyopsis, 238
Probelesodon, 44f
Proborhyaenidae, 74t, 85

Proboscidea, 5, 22, 215, 241–243, 258, 259, **260–265**, 345
　　classification and relationships, 8, 8f, 11f, 12f, 211–212, 212f, 213t, 242t, 243f, 261f, 270
　　Early Cenozoic, of Africa, 336
proboscis, 260
　　astrapothere, 236
　　lipotyphlan, 143
　　palaeothere, 249
　　pantodont, 117
　　proboscidean, 260, 262, 263
　　pyrothere, 237
　　rhinocerotoid, 255
　　sirenian, 270
　　tapiroid, 253, 254
Procaprolagus, 314
Procavia, 258f
Procaviidae, 242t, 257
Procerberus, 95f, 96, 97, 98f, 110f
Procreodi, 215
Procynodictis, 134, 135
Procyonidae, 39, 120f, 121t, 126, 128, 136–137
Prodiacodon, 140, 141f
Prodinoceras, 238–240, 239f
Prodinoceratidae, 213t
Prodremotheriidae, 305
Prodremotherium, 305
Proectocion, 235
Progaleopithecus, 230
Prohesperocyon, 135, 135f
Prokennalestes, 9t, 11f, 89, 89f, 140, 148
Prolimnocyon, 123, 125, 125f, 126, 127f, 339, 342
Promioclaenus, 218, 220f
promontorium, 50
pronation, forearm, 33
Pronothodectes, 171–173, 175f
Pronycticebus, 180, 184f, 185
Propachynolophus, 248, 249
Propalaeanodon, 206, 208f
Propalaeocastor, 328
Propalaeosinopa, 100, 100f
Propalaeotherium, 248, 249, 249f, 250f
Propithecus, 183
Propliopithecidae, 166t, 167f, 194–197
Propliopithecoidea, 166t
Propliopithecus, 196–197
Prorastomidae, 242t, 268–270
Prorastomus, 243f, 268, 269, 269f
Prosarcodon, 150
Proscalopidae, 140t, 154
Prosciurinae, 326
Prosciurus, 326
prosimians, 178–179, 185, 187
Prosimii, 166, 167f
Protadelomys, 326
Protadjidaumo, 331
Protalphadon, 77, 79f
Protamandua, 203
Protapirus, 253
Proteopithecidae, 166t, 167f, 194–197
Proteopithecus, 194
Proterix, 147
Proterotheriidae, 213t, 227f, 233, 234, 235f
Proteutheria, 94, 103, 138
Proticia, 236–237, 237f, 238
Protictis, 130
Protitanotherium, 251f
Protoadapinae, 183
Protoadapis, 180, 184
protoanthropoid, 179
Protoceras, 301f, 304f
Protoceratidae, 286f, 289t, 300, **301–302**, 301f
Protoceratoidea, 289t
Protocetidae, 272t, 273t, 278–285, 280f, 282
Protocetus, 282–283
protocone, 28, 28f, 34, 65–67, 73f, 78f, 85, 96, 105, 120, 128, 169, 171, 180f, 184–185, 188, 196, 217, 220, 250, 292, 298, 299, 308, 309, 311f

protoconid, 28, 28f, 54, 73f, 74, 77, 91f, 132, 148, 149, 180f, 217, 238, 274, 311f
protocristid, 28, 28f, 35, 73f, 187, 235
Protodidelphinae, 74t, 81f, 83
Protodidelphys, 81f
Protolipterna, 227f, 233–234
Protolipternidae, 213t, 227f, 233, 234
protoloph, 28, 227, 235, 238, 239f, 247, 255, 256
protolophid, 238, 240, 247, 253
Protomoropus, 257
Protophiomys, 334
Protoptychidae, 307t, 331
Protoptychus, 323f, 331
Protoreodon, 299f, 300
Protorohippus, 247
Protosciurus jeffersoni, 327
Protoselene, 219
protosimiiform, 179
Protosiren, 243f, 269–270, 269f, 270
Protosirenidae, 242t, 268, 269
protostylid, 311f
Prototheria, 49–50
Prototherium, 243f, 270
protothyrids, 44
Prototomus, 34f, 125
Protrogomorpha, 322
protrogomorphous rodents, 318, 318f, 322, 327
Protungulatum, 89f, 98f, 114, 212f, 214–215, 214f, 216, 220, 227f, 344
Protylopus, 301
Proviverrinae, 123–125
Prozeuglodon, 281f
Prozostrodon, 45, 47f
Pseudamphimeryx, 298f, 303
pseudangular process, 52
Pseudictopidae, 9t, 306, 307t, 308f, **310**, 310f, 339, 345–347
Pseudictops, 226, 240, 310, 310f
Pseudobassaris, 137
Pseudocreodi, 119
Pseudoglyptodon, 199t, 200f, 201f, 202
pseudohypocone, 182, 184, 185, 225
Pseudoloris, 190f, 191
Pseudoltinomys, 325f, 326
pseudoprotocone, 70
Pseudorhyncocyonidae, 140t, 141, 342
Pseudorophodon, 202
pseudosacrals, 201
pseudotalonid, 70
Pseudotetonius, 191f
Pseudotriconodon, 47f
Psittacosaurus, 62
Psittacotherium, 107, 108f
Pterodon, 126
Pterodontinae, 123, 124
Pteropodidae, 157, 158t
pterygoid bone, 25f
　　leptictidan, 142f
pterygoid fossa, 61
pterygoid muscles, 26, 36
　　lagomorph, 314
Ptilocercus, 197f
Ptilodontoidea, 57f, **58–59**, 340
　　classification, 49t
Ptilodus, 57f, 59, 59f–60f, Plate 1.1
Ptolemaia, 103, 103f
Ptolemaiidae, 95t, 99, 101–103
pubic symphysis, 33, 143
pubis, 33, 36f, 42, 73
Pucadelphys, 81, 82f, Plate 1.2
Puercan NALMA, 14f, 100, 105, 106, 172, 214–216, 218, 219, 275
Punta Peligro, Argentina, 218, 226, 234, 336
Purgatoriidae, 166t, 169, **171**
Purgatorius, 170f, 171–174, 172f–173f, 178, 343
pygmy gliding possum, 175
Pyrocyon, 125
Pyrotheria, 10t, 22, 213t, 215, 226, 227f, **236–237**, 237f, 240, 242, 306, 344

Pyrotheriidae, 213t, 237
Pyrotherium, 237, 237f, 238

Qatrania, 196
Qianshan Basin (China), 339, 347
Qipania, 309
quadrate, 42f, 43, 44, 45f, 46f, 52, 83
quadratojugal, 45f
Quercitherium, 125, 125f
Quercy (France), 131–133, 135, 136, 147, 161, 186, 296, 297, 338, 342
Quercysorex, 153, 154f
Quettacyonidae, 213t, 216, 339

rabbits, 9t, 313–314, 319
raccoons, 9t, 35, 126, 136
Radinskya, 242t, 243, 243f, 253, 266–267, 339, 345
radiometric dating, 10, 11
radioulnar joint, 33f
radius, 30f, 32, 33, 34f, 39, 40, 75f, 85
　　anoplothere, 298
　　anthracothere, 294
　　archaeocete, 281, 283
　　arctocyonid, 218f
　　artiodactyl, 285, 287f, 290f, 294, 295, 297, 298, 301, 302, 304
　　camel, 302
　　chiropteran, 157, 159
　　condylarth, 220, 222, 224
　　coryphodontid, 118
　　dermopteran, 163
　　entelodont, 295
　　euprimate, 166, 168
　　eutherian, 90
　　hyaenodontid, 126
　　hypertragulid, 303
　　hyracoid, 257, 260
　　ischyromyid, 325f
　　mesonychian, 274, 277f
　　miacid, 131, 131f
　　oromerycid, 301
　　palaeanodont, 206
　　pantodont, 118
　　peccary, 297
　　perissodactyl, 244
　　phenacodontid, 224
　　proboscidean, 260, 261
　　sirenian, 268
　　stylinodontid, 107
　　taeniodont, 107
　　tillodont, 111
　　xenungulate, 238, 238f
Ragnorak, 214
rainforest
　　paratropical, 21
　　tropical, 20–21
Raoellidae, 274, 288, 289t, **291**, 292f, 339, 345–346
Raricricetodon, 330
rat(s), 9t, 318, 329–331. See also Muroidea
rat opossums, 9t, 81
Ratufa, 32f
Ravenictis, 130, 130f
Reigitherium, 55, 340
Reithroparamyinae, 321–322, 322f
Reithroparamys, 321–322, 326, 329
Remingtonocetidae, 272t, 273t, 278–285, 280f, 283
Remingtonocetus, 280f
Rencunius, 186
Repenomamus, 62
reproductive tract, marsupial, 73
reptiles, 44
　　growth in, 23
　　lower jaw, 26
　　skull, 24
Requisia, 234
Rhaetian Age, 13f, 51, 52, 54
Rhaeto-Liassic taxa, 52, 54

rhinarium, 182
rhinoceros, 10t, 40, 244, 246
Rhinocerotidae, 242t
Rhinocerotoidea, 242t, 246f, 248f, 249f, 252–253, **254–255**, 254f, 338, 339, 345, Plate 7.3
Rhinolophidae, 158t, 161
Rhinolophus, 161
Rhinopomatidae, 158t
Rhombomylus, 316, 317f
Rhynchippus, 231
Rhynchocyon, 37f, 141, 311f, 312f
Rians (France), 338
rib(s), 30–31, 41, 44
    cervical, 30, 58, 68
    dermopteran, 163
    sirenian, 268, 270
    sternal, 199
        Ernanodon, 210
Ricardolydekkeria, 230
Ricardowenia, 230
ricochetal gait, 40
ricochetal mammals, 322
right whales, 285
Rio Loro Formation, 234, 236
Riochican SALMA, 14f, 17, 199, 201, 221, 230, 231, 233, 236, 238f, 240, 336, 343
Riostegotherium, 201, 201f
Roberthoffstetteria, 83, 83f
Robiac (France), 338
Robiacina, 297–298, 298f, 299
rock cavy, 300
rock rat, 334
Rocky Mountains, 16, 164, 337
Rodentia, 4, 5, 22, 61, 107, 156, 240, 306, 315, **316–334**, 324f, 327f, 336–337, 346–347
    age of divergence, 5t, 316–317
    classification and relationships, 8, 9t, 11f, 12f, 139f, 286f, 307t, 308f, 312–313, 316–318
    dental formula, 316
    Early Cenozoic
        of Asia, 339, 340
        of Europe, 338
        of North America, 337
    fossorial, 40, 326–328
    hystricognath radiation, 334, 347
    hystricomorphous, 318, 318f, 326
    monophyly, 316
    myomorphous, 318, 318f, 328
    oldest known, 319–320
    phiomorph, 334, 336
    protrogomorphous, 318, 318f, 322, 327
    sciurognathous, 318, 319f, 321, 326, 328, 332
    sciuromorphous, 318, 318f, 326–329, 331
    tooth enamel, 318–319, 320f
    zygomasseteric anatomy, 318, 318f, 319f
Rodentiaformes, 317
Rodhocetus, 273, 280f, 282f, 283, 283f, Plate 8.2
Romania, 338
Rooneyia, 188, 189f, 190f, 191
root(s), of teeth, 27
Rowe, 43
Ruminantia, 12f, 273, 285, 293, 298f, 300–302, **302–305**, 304f, 338, 346
    classification, 272f, 286f, 289t, 302
    stomach, 286–287
runners. *See* cursorial mammals
Rupelian Stage/Age, 13f, 14f, 15, 17
Russia, eutherians, 89

saber-toothed mammals, 6, 122, 132–133, 134f, 216, 238–239
Sabiaceae, 21
sacral vertebrae, 30, 30f, 31, 31f
    archaeocete, 283
    cetacean, 283
    embrithopod, 266
    sirenian, 270
sacroiliac joint
    archaeocete, 283
    sirenian, 270

sacrum, 36f
    archaeocete, 282, 284
    xenarthran, 201
sac-winged bats, 161
Saghacetus, 284–285
Saghatherium, 258f, 260, Plate 7.4
sagittal crest, 24, 35, 121
Sahara Desert, 20, 336
Saharagalago, 187, 187f
Saimiri, 194, 196
Salladolodus, 221
SALMA. *See* South American Land-Mammal Ages
saltatorial mammals, 37f, 40, 142–143, 142f, 223, 234, 257, 289, 303, 310, 314, 315, 322
San Juan Basin (New Mexico), 21, 219
Sanitheriidae, 289t
Santa Lucía Formation, 16
Santacrucian, 202, 231
Santiagorothia, 230
Santonian Age, 13f, 148
Sarcodon, 150
Sarkastodon, 122, 123f, 342
Saskatchewan, 130, 162, 337
Saturninia, 149, 150f
Saxonella, 172, 173, 176f
Saxonellidae, 166t, 169, 170f, **173**, 343
Scaglia, 236, 236f
scaly-anteaters, 10t. *See also* Pholidota
Scandentia, 138, 156, 176, **197**, 197f, 335, 343
    classification, 10t, 157f, 158t, 167f, 170f
    relationships, 11f, 12f, 157f
scansorial mammals, 37f, 74, 89, 111, 121, 122, 125–126, 131, 134, 150, 197, 215, 216, 219, 222
    skeletal adaptations, 39, 39f
scaphoid, 33, 35f, 74, 75f
    creodont, 121
    miacoid, 130
    palaeanodont, 208f
    phenacodontid, 223
    pholidotan, 204
scapholunate, 129
    carnivore, 129
    pholidotan, 204
scapula, 30f, 32, 32f, 53, 55, 65, 68, 201
    Ernanodon, 210
    Euphractus, 32f
    eutriconodont, 62
    Jeholodens, 63
    palaeanodont, 206
    pholidotan, 204
    proboscidean, 260
    talpid, 155
Scarrittia, 231
Scelidotheriidae, 199t
Scenopagidae, 140t, 145
Schizotherium, 257
Schowalteria, 108–109, 108f
Sciuravida, 326, **332**
Sciuravidae, **322**, 329, 332, 347
    classification and relationships, 307t, 308f, 322f, 329f, 332f
Sciuravus, 322
Sciuridae, 317, 318, 320, 322, **327–328**, 338, 347
    classification, 9t, 307t, 322f
    skeletal adaptations, 39–40
Sciurognathi, 317
sciurognathy/sciurognathous rodents, 318, 319f, 321, 326, 328, 332
Sciuromorpha, 317, **320–329**, 347
    classification and relationships, 307t, 308f, 322f
sciuromorphous rodents, 318, 318f, 326–329, 331
scutes, 201–202
sea cows, 10t, 267. *See also* Sirenia
sea levels, Paleocene, 18
seals, 9t, 35, 40, 126
seawater. *See* marine environment

secodonty/secodont dentition, 28, 34, 35
sectorial teeth, 28, 29f
Seggeurius, 259–260, 259f
Selandian Stage/Age, 13f, 14f
selenodonty/selenodont dentition, 28, 29f, 35, 36, 110, 167, 184, 219, 222–224, 233–235, 250, 259, 260, 286–287, 289, 291, 293–294, 297, 299–303, 311, 346
semiaquatic mammals, 40, 56, 71, 100, 101, 103, 118, 144, 240, 264, 266, 294, 298, 300, 341
semi-fossorial mammals, 40
Senegal, 284, 336
septomaxilla, 45f, 52, 53, 56, 68, 199
Serapia, 194
serial carpus. *See* carpus, serial
sesamoid bones, 34
    palaeanodont, 208, 208f
    stylinodontid, 107
Sespedectes, 145
Sespedectidae, 140t, 145
sexual dimorphism, 118, 168, 180, 186, 193, 194, 197, 223, 239, 247, 250, 255, 259, 274, 295, 297
Seymour Island (Antarctica), 71, 87, 202, 284, 337
Shamolagus, 314
Shandgolian ALMA, 14f, 17
Shandong Province (China), 339
Shanghuan ALMA, 14f, 339
Shanxi Province (China), 191–192
Sharamurunian ALMA, 14f, 17
Shizarodon, 185
short-fuse model, of therian radiation, 3, 4f
Shoshonius, 188, 189f, 190, 190f, 191–192
shoulder girdle, 32, 41, 43f, 53, 65, 68
    symmetrodont, 64
shrews, 9t, 93, 138, 143, 147, 148, 150, 153, 342, 343. *See also* Soricidae
    primitive, 153–155, 154f
Shuotheridia, 9t, 49t
Shuotherium, 51f, 70
Siamochoerus, 296f
Siamopithecus, 185, 193, 193f
Sifrhippus, 247
Sillustania, 83
Sillustaniidae, 74t
Silveirinha (Spain), 338
Simiacritomys, 330
Simimeryx, 303
Simimyidae, 307t, 329f
Simimys, 330
Simplicidentata, 9t, 307t, 308f, 313, **315–334**, 316f, 346–347
Simplodon, 110, 111f, 114
Simpson, George Gaylord, xii, 44
Simpsonictis, 129, 130f
Simpsonodon, 70
Simpsonotus, 229, 230
Sinclairella, 103, 104, 105f
Sinemurian Age, 13f, 52, 55
Sinoconodon, 45f, 51f, **52–53**, 53, 53f
Sinoconodontidae, 9t, 49t
Sinodelphys, 74, 74t, 75f, 340
Sinomylus, 9t, 307t, 308f, 315–316, 316f, 319
Sinonyx, 272f, 275f
Sinopa, 125, 126f, 127f
Sinosinopa, 150
Sinostylops, 226
sinuses, 26
    basicranial, archaeocete, 284
    cetacean, 277
    inferior petrosal, 135
    nasal, 26
    paranasal
        archaeocete, 282
        phenacodontid, 223
        sirenian, 269
Sirenia, 5, 22, 88, 215, 237, 241–242, 260, **267–270**, 269f, 270f, 345
    classification and relationships, 8, 8f, 10t, 11f, 12f, 211–212, 212f, 213t, 242t, 243f, 270

Sirenia (continued)
  Early Cenozoic
    of Africa, 336
    of North America, 337–338
Sivaladapidae, 166t, 182, **186**
Sivaladapis, 186
skeleton, 23, 310. *See also* dentition; skull
  adapoid, 179–182, 181f, 185
  adaptations, 34–40, 39f
  amphicyonid, 135, 137f
  anagalid, 309, 309f
  anoplothere, 298
  anteater, 203f
  anthracothere, 294, 294f
  anthropoid, 195f, 196, 196f
  apatemyid, 104, 106f, Plate 3.1, Plate 3.2
  appendicular, 30
  archaeocete, 278f, 280, 282f
  arctocyonid, 216, 218f, Plate 8.1
  arctostylopid, 226
  artiodactyl, 285, 287f, 290f, 292f, 305f, Plate 8.3
  astrapothere, 235–236
  axial, 30
  bemalambdid, 114
  brontothere, 251, 252f
  cainothere, 299, 305f
  camel, 302, 302f
  canid, 135f
  carnivoran, 129, 132f, 134f, 135f, 137f
  carpolestid, 173, 174f
  cebochoerid, 292f
  cetacean, 276, 278f. *See also* skeleton, archaeocete
  chalicothere, 256f
  chiropteran, 157, 161f, Plate 5
  choeropotamid, 292, 292f
  condylarth, 215, 223f
  creodont, 120, 121. *See also* skeleton, hyaenodontid; skeleton, oxyaenid
  dermopteran, 163f
  dichobunid, 289–290, 290f, 291f
  djadochtathere, 58, 59f
  docodont, 55–56
  edentate, 199
  elephant shrew, 311
  embrithopod, 266, 266f
  entelodont, 294, 295f
  Eomaia, 90f
  epoicotheriid, 207
  erinaceid, 144
  erinaceomorph, 145, 146, 146f, 147f
  Ernanodon, 207f, 209
  euprimate, 179
  Eurotomandua, 203–204, 203f, Plate 7.1
  eutherian, 89, 90f, Plate 1.3, Plate 1.4
  feliform, 132
  glirid, 324f
  glyptodont, 202
  Gobiconodon, 62, 62f
  Haldanodon, 55–56
  hyaenodontid, 125, 127f, 128f
  hypertragulid, 287f, 303
  hyracoid, 257, 258f, 260, Plate 7.4
  ischyromyid, 321, 322, 323f–325f
  Jeholodens, 62f, 63
  lagomorph, 314
  leptictid, 140, Plate 4.1
  leptictidan, Plate 4.2
  lipotyphlan, 144. *See also* skeleton, erinaceomorph
  marsupial, 80f, 82f, 85, Plate 1.2
  mesonychian, 274, 276f
  metatherian, 75f, 77f
  miacid, 131, 131f, f132
  micropternodontid, 150–151
  mimotonid, 315
  mioclaenid, 219
  mixodectid, 164
  morganucodont, 52–53, 53f

  multituberculate, 57–58, 59f, 60f
  mysticete, 278f
  nimravid, 134f
  notharctid, 181f, 183, 185
  notoungulate, 228, 229f
  odontocete, 278f
  omomyid, 192
  oreodont, 299, 299f, 300
  oxyaenid, 121, 124f
  palaeanodont, 100, 205, 206, 207f
  pangolin, 204
  pantodont, 114, 115, 117, 117f, 118
  pantolestan, 100
  pantolestid, 101, 102f, Plate 2.1
  paroxyclaenid, 101, 103f, Plate 2.2
  periptychid, 219, 220, 221f
  perissodactyl, 244, 245f, 246f, 248f, 250, 252f, 256f, Plate 7.3
  phenacodontid, 223, 224, 224f
  pholidotan, 204, 205f, 206f
  plesiadapiform, 170, 174f
  pneumatized, 260
  postcranial, **30–34**, 30f–33f, 42
  proboscidean, 260, 264
  Protoptychus, 323f
  ptilodontid, 59, 60f
  rhinocerotoid, 255, Plate 7.3
  rodent, 323f, 324f
  ruminant, 305f
  sciurid, 323f
  sirenian, 269, 270f
  taeniodont, 105, 107–108, 109f
  tillodont, 111
  tree shrew, 197, 197f
  tritylodontid, 45, 47f
  uintathere, 239f, 240
  ungulate, 211, 215
  viverravid, 131
  xenarthran, 200–201
  xenungulate, 238, 238f
  xiphodont, 301
  zalambdalestid, 308, 308f, Plate 1.4
skull, **24–26**, 25f, 42
  adapid, 184f
  adapoid, 168f, 179–180, 182f
  altungulate, 243f
  amphicyonid, 137f
  anagalid, 309, 309f, 310f
  Andrewsarchus, 219f
  anteater, 203
  anthracothere, 293, 294f
  anthropoid, 194, 195f, 196f
  apatemyid, 104, 104f, 105f
  apternodont, 152, 152f
  archaeocete, 278, 279f–281f, 282
  arctocyonid, 216, 216f
  Arsinoitherium, 267f
  artiodactyl, 285–286
  Asioryctes, 92f
  astrapothere, 236f
  basal mammal, 45f
  bemalambdid, 114
  brontothere, 251, 251f, 252
  cainothere, 299
  camel, 302, 302f
  carnivoran, 35
  castorimorph, 327f
  cetacean, 276, 279f
  chalicothere, 257
  chiropteran, 160f
  condylarth, 215, 243f
  creodont, 121
  ctenodactyloid, 333, 333f
  cylindrodontid, 323, 327f
  cynodont, 44–46, 45f–47f
  Daulestes, 92f
  dermopteran, 163
  dichobunid, 289
  didymoconid, 97, 98f
  djadochtathere, 57f, 58

  docodont, 55f
  edentate, 199
  elephant shrew, 311
  entelodont, 294, 295, 295f
  equid, 248
  Ernanodon, 210
  euprimate, 179
  eurymylid, 317f
  eutherian, 90–92, 92f
  feliform, 134f
  fossorial mammals, 40
  geolabidid, 149
  Gobiconodon, 62
  Hadrocodium, 55
  Haldanodon, 55, 55f
  hegetothere, 233, 233f
  helohyid, 291
  herbivores, 36–39
  human, 24
  hyaenodontid, 124, 125f, 126, 126f
  hyopsodontid, 222
  hypertragulid, 303, 303f
  hyracoid, 258f, 259, 259f
  ischyromyid, 321, 327f
  Kennalestes, 92f
  lagomorph, 313, 315f
  Lambdopsalis, 60
  lemuroid, 186
  leptictidan, 140, 142f
  litopteran, 234f
  Llanocetus, 284
  marsupial, 73, 79, 82f, 84f, 85
  mesonychian, 275f
  metatherian, 76f
  monotreme, 68, 69
  morganucodont, 52, 53f
  multituberculate, 57, 57f
  muroid, 327f
  myrmecophagous mammals, 35, 38f
  mysticete, 279f
  nimravid, 134f
  notoungulate, 228, 229f
  odontocete, 279f
  omomyid, 168f, 188, 189f, 191
  oreodont, 299f, 300
  oromerycid, 301f
  oxyaenid, 121, 124f
  palaeanodont, 205, 207–208, 209f
  palaeoryctid, 96–97, 96f, 97f
  palaeothere, 253f
  pangolin, 204
  pantodont, 114, 115f–116f, 118
  pantolestan, 99
  pantolestid, 101
  paromomyid, 176
  paroxyclaenid, 101
  pentacodontid, 101
  perissodactyl, 243f, 244
  phenacodontid, 223, 243f
  pholidotan, 204
  plagiomenid, 164
  plesiadapid, 171
  plesiadapiform, 168f, 169
  Plesiopithecus, 187f
  pneumatization, 26, 164, 262, 264–265
  proboscidean, 263, 263f–264f
  procyonid, 136
  protoceratid, 301f
  protrogomorphous, 321–323, 326, 327, 329
  pseudomyomorphous, 329
  Ptolemaia, 103
  pyrothere, 237f
  Radinskya, 243f
  Repenomamus, 62
  reptile, 24
  rhinocerotoid, 255
  rodent, 327f, 333f
  ruminant, 303f
  sciuravid, 322
  sciuromorphous, 326, 328, 329

simplicidentate, 316f
Sinoconodon, 53, 53f
sirenian, 268, 269f
soricomorph, 147
strepsirrhine, 187
suoid, 296f
taeniodont, 105, 107, 108f
tapiroid, 253f
tillodont, 111, 112f
triisodontine, 217, 219f
trithelodont, 45
tritylodontid, 45, 47f
typothere, 232f
uintathere, 239
ungulate, 36
whale. *See* skull, cetacean
xiphodont, 301
Zalambdalestes, 92f
Slaughteria, 70f
sloths, 9t, 198, 200, 201, 201f, 202, 343. *See also* Phyllophaga
   cervical vertebrae, 30, 201
   megalonychid, 202
   megatheriid, 201, 202
   three-toed, 202
   two-toed, 202
Smilodectes, 180, 184
Smilodon, 6
snout, 26, 143. *See also* skull
   adapoid, 180, 185
   amphicyonid, 135
   anagalid, 310f
   anthracothere, 293
   anthropoid, 194
   bemalambdid, 114
   camel, 302
   condylarth, 215
   cylindrodontid, 323
   elephant shrew, 310
   eutherian, 92
   geolabidid, 149
   leptictidan, 141
   lipotyphlan, 143
   marsupial, 85f
   mimotonid, 314f
   Numidotherium, 263
   oromerycid, 301
   palaeoryctid, 96
   pantolestid, 100, 101
   paromomyid, 176
   perissodactyl, 244
   plesiadapiform, 169
   proboscidean, 263
   squirrel, 328
   taeniodont, 108, 109
   tillodont, 111, 113
   xenarthran, 202
   zalambdalestid, 307
Solenodon, 34, 38f, 101, 143, 147, 148, 152–153
Solenodontidae, 138, 140t, 143, 147
Somalia, 336
Soricidae, 139f, 140t, 143, 147, **153–154**, 154f, 342–343
Soricinae, 153
Soricoidea, 140t
Soricolestes, 154f, 154
Soricomorpha, 93, 96–97, 139, 140t, 143, 144, **147–154**, 149f, 150, 153, 342–343
Sorlestes, 213
South America, 8, 56, 61, 65, 74, 78, 80, 81, 83, 84, 271. *See also* Meridiungulata
   australosphenidan, 68
   carnivorans, 342
   didolodontids, 220
   Early Cenozoic mammal record for, 336–337
   equids, 247
   gondwanatheres, 70–71
   mammalian dispersal to, 20
   metatherians, 72, 341
   monotreme, 69
   and North America, faunal exchange between, 337
   paleogeography, 18, 19f
   pantodonts, 114
   periptychids, 219
   primates, 166
   proboscideans, 260
   pyrotheres, 236, 240
   tayassuids, 296
   triisodontine, 217, 219f
   ungulates, 226
      archaic, 344
      endemic. *See* Meridiungulata
   xenarthrans, 202, 343
South American Land-Mammal Ages, 11, 14f, **16–17**, 237, 336
South Carolina, 310, 337
South Dakota, 191
Spain, 20, 338
Spalacolestes, 63f
Spalacotheriidae, 49t, 63
Sparassocynidae, 74t, 75f
Sparassodonta, 78, 81, **84–86**, 341
   classification, 9t, 74t
Sparnacian Age, 12, 15
Sparnotheriodontidae, 221–222, 227f, 235, 235f
sperm whales, 275–276, 285
sphenacodontid pelycosaurs, 44
sphenoid bone, 24
sphenorbital fissure, 58
sphenorbital foramen. *See* foramen/foramina, sphenorbital
splenial (bone), 47f
springhares, 331–332
Spurimus, 326
squamosal (bone), 24, 25f, 43, 44, 45f, 46f, 49, 52, 56, 57, 82f
   apternodont, 152
   archaeocete, 281f
   artiodactyl, 285
   dichobunid, 289
   leptictidan, 142f
   notoungulate, 228
   palaeoryctid, 96
   proboscidean, 264f
   sirenian, 269f
squirrel monkeys, 194
squirrels. *See also* Sciuridae
   gliding scaly-tailed (Anomaluridae), 331–332
stages, 10
   standard, 11, 14f
Stagodontidae, 74t, 75f, 77f, 78
stance. *See* posture
stapes, 24, 42f, 44, 56
Stegotherium, 38f
Stehlinia, 162
Steinius, 173f, 188, 189
stem taxa, 4
stem-based taxon, 7–8, 8f
Steneofiber, 328
Stenogale, 132, 133f
Stenoplesictis, 132, 133f, 342
sternal ribs, 31
Sternbergiidae, 74t, 83
sternebra, 31
sternum, 30f, 31
   chiropteran, 157
   dermopteran, 163
Steropodon, 67, 67f, 68, 69
Stibarus, 289
stomach, ruminant, 302
Strenulagus, 314
Strepsirrhini, 166t, 167f, 168f, 178–179, 181–182, 183f, 185, 187f, 343
   earliest, 186–187
strepsirrhinism, 182
stylar cusps, 28, 66, 73, 73f, 74–76, 148, 153, 164, 321
   marsupial, 77, 78, 83, 87
   metatherian, 78f
      designations, 73

stylar shelf, 28, 70, 89, 92, 95–97, 99, 101, 104, 105, 108, 110, 113, 115, 140, 148–150, 152, 168, 169, 186, 188, 214, 215, 217
   marsupial, 72–74, 76, 78, 87
   zalambdalestid, 308
Stylinodon, 107, 107f, 108, 108f
Stylinodontidae, 95t, 105, 106–108, 110f
stylocone, 28, 54, 65, 65f, 73, 73f, 153
stylohyal bone, chiropteran, 158, 161
stylohyoid bone, notoungulate, 228
stylomastoid foramen. *See* foramen/foramina, stylomastoid
subages, 16
Subengius, 172
Subparictidae, 135
subterranean mammals, 40, 208–209
Sudamerica, 17, 71, 71f
Sudameridelphia, 83
Suevosciurus, 326
sugar gliders, 9t
Suidae, 12f, 272f, 286f, 289t, 296, 296f, **297**
Suiformes, 285, 286f, 288, 289, 289t, **293**, 297, 300, 346
Suina. *See* Suiformes
Suoidea, 274, 289t, 296, 296f
superior orbital fissure, 24
supernumerary teeth, 200
supertree analysis, 7
supination, forearm, 33
suprameatal fossa, 133
supraoccipital bones, 24, 25f
   proboscidean, 264f
   sirenian, 268, 269f
supraorbital process, cetacean, 276
supraorbital shield, archaeocete, 282
supraspinous fossa, 32, 32f, 58
   Jeholodens, 63
surangular, 43, 45f–47f
Sus, 32f, 33f
suspensory posture, 197
Swain Quarry (Wyoming), 59
sweepstakes dispersal, 18, 19f, 20
swimming mammals, 40, 276, 281, 283. *See also* aquatic adaptations
Switzerland, 338
Symmetrodonta, 9t, 49, 49t
symmetrodonts, 49, 50f, **63–64**, 63f, 70
   basal, 52
   as paraphyletic assemblage, 63
symplesiomorphies, 6
Symplokeomys, 331
synapomorphies, 6
Synapsida, 44
synapsids, 2
Syndactyli, 9t, 74t
Synoplotherium, 274–275

Tabelia, 179, 192, 195f
tabular bones, 52, 57
Tachyglossa, 10t
Tachyglossidae, 68
Tachypteron, 161–162, Plate 5.4
Tachytheriinae, 232
Taeniodonta, 22, 94, **105–110**, 107f, 109f, 198, 335, 337–338, 341
   classification and relationships, 10t, 95t, 105, 109, 110f
Taeniolabididae, 59–60
Taeniolabidoidea, 49t, 56, 57f, 58, **59–60**, 340
Taeniolabis, 57f, 60
tail
   apatemyid, 104
   arboreal mammals, 39
   archaeocete, 281, 283
   Arctictis, 101
   cebochoerid, 291
   cetacean, 276
   chiropteran, 159
   choeropotamid, 293
   erinaceomorph, 146

tail (continued)
  hyaenodontid, 125
  hyracoid, 257
  lagomorph, 314
  leptictidan, 142
  marsupial, 80–81
  pantodont, 116
  pantolestid, 100, 101
  paurodontid, 65
  prehensile, 39, 59, 79, 85
    arctocyonid, 216
    semi-aquatic mammals, 40
  taeniodont, 106
  uintathere, 240
  vertebrae, 31, 31f
Takracetus, 280f
talonid(s), 28, 34, 35, 63–70, 73f, 76, 77, 78f, 85, 87, 90, 92, 95–97, 99, 103–105, 109, 110, 117, 120, 134, 140, 144, 145, 148–150, 152, 154, 164, 167, 169, 175, 176, 180, 180f, 186, 188, 194, 215, 217, 220, 222, 223, 231, 274, 275, 280, 289, 308, 309, 316, 321
talonid basin, 28, 28f
Talpidae, 139f, 140t, 143, 154–155
Talpoidea, 140t, 147, 154f, **154–155,** 343
talus, 33. See also astragalus
  amphipithecid, 185
Tamandua, 344
Tamquammyidae, 307t, 332
Tamquammys, 325f
Tanzania, 20, 71, 161, 336
Tanzanycteris, 161
Taphozous, 162
tapir(s), 10t, 36, 244, 246
Tapiridae, 242t, 253
Tapiroidea, 246, 248f, 249f, 252, **253–254,** 253f, 254f, 256, 345
  classification, 242t, 246f
tapiroids, 244–245, 248f
Tapiromorpha, 246, 249f, 252–253, Plate 3.3
  classification, 242t
Tapirulus, 299
Tapirus, 33f, 253f
Tarka, 164
Tarkadectes, 164
tarsal bones, 33, 34, 40. See also tarsus
  adapid, 186
  adapoid, 168f
  anthropoid, 192
  archaeocete, 284
  archontan, 157
  arctocyonid, 216
  arctostylopid, 226
  artiodactyl, 285, 345
  cainothere, 299
  carnivoran, 119, 131
  cetacean, 283, 283f, 345
  creodont, 119
  dermopteran, 163
  didolodontid, 221
  elephant shrew, 311
  fossil whale, 197, 273, 274
  hyopsodontid, 222
  hyracoid, 257, 258
  lagomorph, 314
  leptictid, 140
  litopteran, 233, 233f, 234, 235
  marsupial, 86
  notoungulate, 227, 228
  nyctithere, 150
  omomyid, 168f, 188, 192
  palaeanodont, 206
  palaeothere, 249
  pantodont, 114
  plesiadapoid, 168f
  ptilodontid, 59, 60f
  pyrothere, 237
  serial, 260
  taeniodont, 105

toxodont, 230
uintathere, 240
ungulate, 233, 233f
tarsiers, 10t, 35, 40, 166, 179, 188, 191–192
Tarsiidae, 166t, 167f, 179, 187, **191–192,** 339, 343
Tarsiiformes, 166, 166t, 167f, 179, **187–192,** 187f, 191
Tarsioidea, 179
Tarsipes, 36
Tarsius, 161, 187, 188, 191–192
tarsus, 30f
  archaeocete, 284
  arctocyonid, 216
  arctostylopid, 226
  artiodactyl, 290f
  astrapothere, 236
  carnivoran, 119
  creodont, 119
  didolodontid, 221–222
  elephant shrew, 312
  hyaenodontid, 126
  hyracoid, 260
  lagomorph, 314
  leptictid, 140
  metatherian, 73
  pantodont, 114
  paromomyid, 176
  plesiadapid, 171
  pyrothere, 237
  toxodont, 231
Tasmanian devil, 9t
Tasmanian wolf, 9t
taxeopody, 258–259. See also carpus, serial
Taxidea, 32f
Taxodiaceae, 21
taxon (pl., taxa)
  apomorphic, 6
  crown-group, 8, 8f, 43
  derived, 6
  in-group, 6
  node-based, 8
  out-group, 6
  plesiomorphic, 6
  primitive, 6
  stem, 4
  stem-based, 7–8, 8f
taxonomy, 7
Tayassuidae, 273, 286f, 289t, **296–297,** 296f
teeth. See canine teeth; dental formula; dentition; incisors; molars; premolars
Teilhardina, 168f, 172–173, 172f–173f, 188–189, Plate 4.4
Teinolophos, 69
Teletaceras, 255
temperature
  and mammalian diversity, 340
  ocean, 20
  in Paleocene-Eocene, 20–21
temporal bones, 24, 69
temporal fossa, cetacean, 276
temporal line, 25f
temporalis muscle, 24, 35
  lagomorph, 314
  rodent, 318f
  stylinodontid, 107
temporomandibular joint, 128
  entelodont, 295
  lagomorph, 314
  shrew, 153
Tenrecidae, 8, 12f, 34, 39, 138, 143, 147, 148, 212, 311, 342
  classification, 9t, 139f, 140t, 212f
Tenrecoidea, 139, 140t, 143, 148
tenrecs, 152. See Tenrecidae
tentorium, 119
teres major tuberosity, mesonychian, 277f
termites, diet based on, 200. See also myrmecophagy/myrmecophagous mammals

terrestrial mammals, 37f, 58, 76, 81, 85, 97, 100, 111, 121, 122, 125, 131, 140, 144, 147, 204, 216, 219, 220, 222, 225, 228, 244, 274, 279–281, 283, 309, 310, 314, 316, 321, 327, 339–341
  skeletal adaptations, 39–40, 39f
Tethys Sea, 18, 260, 269, 278, 336, 338, 346
Tethytheria, 8f, 22, 212, 224–225, 241–243, **260–270,** 339, 340, 344, 345
  classification, 10t, 213t, 242t, 243f
Tetonius, 168f, 188–189, 190f, 191f
Tetraclaenodon, 223, 224, 244
Tetracus, 147
Tetragonostylops, 233f, 236
Tetrapassalus, 207
Texas, 16, 50, 74, 185, 191, 337, 341
Thailand, 163, 186, 296, 297, 339, 343
Thanetian Age, 13f, 14f, 145, 216, 338
Thanetian/Ypresian Stage/Age boundary, 12
Therapsida, 42, 44
Theria, 49–50, 61, 70
  definition, 50
therian mammals, 4, 49–50, 63
Theridomyidae, 307t, 308f, 325f, **326,** 332f, 347
Theridomys, 326
Therioherpeton, 45, 47f
Theroteinidae, 49t, 51, 61
Thinocyon, 126
Thinohyus, 296–297
Thoatherium, 233
Thomashuxleya, 228, 229f, 230
thoracic vertebrae, 30–31, 30f–31f
  altungulate, 241
  archaeocete, 284
  brontothere, 251
  sirenian, 269
Thrinaxodon, 42f, 44, 44f, 45f, 47f, 53f
  middle-ear ossicles, 42f
  occlusal relationships, 53f
Thryonomyidae, 307t, 334
Thryonomys, 334
Thryptacodon, 216, 217f
Thulean Route, 18, 19f
thumb, 33
Thylacosmilus, 6
Thylacotinga, 86f, 87
Thyrohyrax, 259, 260
Thyropteridae, 158t
tibia, 30f, 33, 37f, 40
  adapid, 186
  adapoid, 180
  alagomyid, 320
  anoplothere, 298
  anthropoid, 194, 196
  archaeocete, 283, 284
  arctocyonid, 218f
  artiodactyl, 285, 287f, 290f
  camel, 302
  cebochoerid, 291
  chalicothere, 257
  choeropotamid, 292
  coryphodontid, 118
  dichobunid, 290
  elephant shrew, 311, 312, 312f
  entelodont, 295
  epoicotheriid, 208
  equid, 248
  erinaceid, 144
  erinaceomorph, 146
  eurymylid, 315
  eutherian, 90, 92
  hyopsodontid, 223
  hyracoid, 257, 260
  ischyromyid, 322, 325f
  lagomorph, 314
  leptictid, 140
  leptictidan, 142
  leptomerycid, 303
  mimotonid, 315
  omomyid, 188, 191

perissodactyl, 244, 246f
proboscidean, 260, 261, 263
pseudictopid, 310
rodent, 328, 331
tarsiid, 192
tillodont, 111
uintathere, 240
zalambdalestid, 308
tibia-fibula, hyopsodontid, 312, 312f
tibiofibula, omomyid, 188
tibiotalar joint, adapoid, 181
Tiffanian NALMA, 14f, 16, 96, 104, 122, 124, 172, 176, 178, 206, 216, 226, 238, 337, 339, 342
Tillodon, 110, 113
Tillodontia, 22, 94, **110–114,** 111f, 112f, 118, 215, 337, 341
    classification and relationships, 10t, 95t, 110f, 113, 211–212
    dental formula, 110
    Early Cenozoic
        of Asia, 339
        of Europe, 338
        of North America, 337
    origin, 113–114
Tillotheriidae, 95t, 110
time scale, geologic, 13f
Tingamarra, 86f, 87, **225**
Tingamarra Local Fauna (Australia), 87, 336
Tinguiririean SALMA, 14f, 17–18, 84, 202, 343
Tinimomys, 174, 175f, 177f
Tinodon, 51f, 63f
Tinodontidae, 49t, 54, 64
Titanohyrax, 259–260
Titanoideidae, 95t, 116
Titanoides, 110f, 116–117, 116f–118f
titanotheres, 10t, 250. *See also* Brontotheriidae
Titanotheriomorpha, 242t, 244
Tithonian Age, 13f
Tiuclaenus, 217, 226, 227f
Tiupampa (Bolivia), 81, 82f, 83, 85, 114, 201, 218–219, 226, 227, 336, 337, 341, Plate 1.2
Tiupampan SALMA, 14f, 87
Todralestes, 100f, 101
Todralestidae, 101
Toliapinidae, 166t, 169, 170f, **178,** 179
tongue, pholidotan, 204
tooth combs, 166, 179, 186, 197
    adapoid, 182
    arctocyonid, 216, 217f
toothed whales. *See* Odontoceti
Torrejonian NALMA, 14f, 59, 100, 104, 105, 113, 115, 129, 130, 140, 145, 155, 164, 171, 172, 175, 176, 206, 215–219, 223, 224, 244, 274, 275, 338, 342, 344
total evidence analyses, 7
Toxodonta, 213t, 228f
Toxodontia, 229, 229f, **230–231,** 230f
Toxodontidae, 213t, 228f
Tragulidae, 289t, 303
Tragulina, 289t, 302, 303
Tragulus, 290f
trapezium, 33, 35f, 75f
    phenacodontid, 223
trapezoid (bone), 33, 35f
    leptomerycid, 303
    palaeanodont, 208f
    phenacodontid, 223
Traversodontidae, 45
Trechnotheria, 9t, 49t
tree shrews, 8, 22, 43f, 138, 156, 197. *See also* Scandentia
tree sloths, 36, 200, 201
Tremp Basin (Spain), 338
Triassic Period, 13f
    mammalian evolution in, 44
    mammals, 50–52, 50f
Tribactonodon, 69–70
tribosphenic mammals, **66–70,** 74, 96, 120, 138, 144, 149, 161, 167, 169, 197, 247, 316

tribosphenic molar. *See* molars, tribosphenic
Tribosphenomys, 319–320, 321f, 339
tribotheres, **70,** 70f, 74
Tribotheria, 70
Tribotherium, 70
Tricentes, 287
Trichechidae, 242t, 243f, 268
Trichechus, 268
Triconodontidae, 49, 49t, 61, 62
triconodonts, 52. *See also* Eutriconodonta; Triconodontidae
Tricuspes, 47f
Tricuspiodon, 222
trigeminal nerve, 24–25, 43, 45
trigon, 66
Trigonias, 255
trigonid(s), 28, 34, 35, 54, 64, 65, 67, 68, 70, 73f, 78f, 87, 90, 92, 95–97, 103–105, 110, 113, 120, 126, 128, 140, 144–145, 148–150, 152, 153, 154, 164, 169, 171, 175, 176, 180, 180f, 183, 186, 188, 192, 211, 215, 218, 222, 231, 240, 280, 289, 308, 309, 319, 321
Trigonostylopidae, 213t, 236
Trigonostylops, 236, 236f
Triisodon, 214f, 217
Triisodontidae, 273t
Triisodontinae, 212f, 217, 219f, 272f, 274, 275, 286f, 346
trilophodonty, 264
Trinititherium, 70f
Trinity Formation (Texas), 74
Trioracodon, 62f
Triplopus, 29f
triquetrum, 33, 74, 75f
Tritemnodon, 125, 125f, 127f
Tritheledontidae, 43, 45, 51f
Tritylodontidae, 43, 43f, 45, 47f, 51f, 53
trochiter, chiropteran, 157
trochlear nerve, 24
Trogosinae, 110–111, 110f, 113
Trogosus, 111, 111f, 112f
trophoblast, 88
trunk, elephant, 260. *See also* proboscis
Tsagan Khushu (Mongolia), 339
Tsinlingomys, 325f
Tubulidentata, 8, 198, 335
    classification and relationships, 10t, 11f, 12f, 139f, 211–212, 212f, 213t
Tubulodon, 207, 208f, 209f
Tunisia, 146, 185, 192, 259
    elephant shrews, 311
Tupaia, 37f, 43f, 197, 197f
Tupaiidae, 138, 139f, 158t, 197, 197f
turbinal bones, 26
Turgai Strait, 18, 19f, 340
Turgidodon, 77
Turkey, 266, 339, 345
Turonian, 148
Turpan Basin, 339
tusks
    anthracobunid, 262
    hyracoid, 260
    proboscidean, 260, 264
    sirenian, 270
    suid, 297
Tylopoda, 285, 289, 293, 297, 298f, 300, **300–302,** 301f, 304f, 346
    classification, 289t
tympanic bones, 328
tympanic cavity, 24
tympanic ligament, archaeocete, 282
tympanic membrane, 24
    lipotyphlan, 143
tympanic ring, 194
tympanum, 42f
Typotheria, 213t, 228, 228f, 229, **231–232,** 231f, 232f
Tytthaena, 122, 123f

Uintaceras, 255
Uintacyon, 130–131, 130f, 131f
Uintan NALMA, 14f, 153, 164, 175, 178, 252, 253, 255, 289, 301–303, 347
Uintasorex, 178
uintatheres, 10t, 18, 22, 114, 238–240, 239f, 337, 345. *See also* Dinocerata
    Early Cenozoic
        of Asia, 339
        of North America, 337
Uintatheriamorpha, 240, 306
Uintatheriidae, 213t, 226, 238
Uintatherium, 239–240, 239f
Ukhaatherium, 90, 91f, Plate 1.3
Ulanbulakian ALMA, 17
Ulangochuian ALMA, 17
ulna, 30f, 32–33, 33f, 40
    anagalid, 309
    anthracothere, 294
    arctocyonid, 218f
    artiodactyl, 285, 287f
    borhyaenoid, 85
    camel, 302
    chiropteran, 157
    cingulate, 201
    coryphodontid, 118
    dermopteran, 163
    docodont, 55–56
    entelodont, 295
    equid, 247, 248
    Ernanodon, 210
    eutherian, 90
    eutriconodont, 62, 63
    hyopsodontid, 222
    hypertragulid, 303
    hyracoid, 257, 260
    ischyromyid, 325f
    mesonychian, 274, 277f
    metatherian, 75f
    miacid, 131, 131f
    morganucodont, 53
    oreodont, 300
    oromerycid, 301
    oxyaenid, 121
    palaeanodont, 208, 208f
    peccary, 297
    perissodactyl, 244
    proboscidean, 260, 261
    pseudictopid, 310
    sirenian, 268
    symmetrodont, 64
    taeniodont, 106, 107
    xenungulate, 238, 238f
unciform, 33, 35f
    phenacodontid, 223
underwater hearing. *See* hearing
ungual phalanges, 33, 36f, 37f, 39, 39f, 202, 309. *See also* phalanges
    claw-bearing, 207, 256, 257
    hoof-bearing, 274, 281, 287f
    hooflike, 219, 223, 224, 238
    nail-bearing, 81
    palaeanodont, 208
    pangolin, 204
    phenacodontid, 223, 224
    pholidotan, 204
    tillodont, 111
Ungulata, 92, 114, 148, 156, **211–305,** 341. *See also* ungulates
    characteristics, 211–212
    classification, 10t, 139f, 211–212, 213t
    Paleocene-Eocene, synopsis, **344–346**
ungulates, 8, 35, 36, 40, 119, 273, 274, 341. *See also* Ungulata
    archaic, 211–240, **344–345**
    classification and relationships, 10t, 212f, 213t
    cloven-hoofed, 285
    endemic South American. *See* Meridiungulata
    even-toed, 10t, 285
    oldest relatives, 213–215, 213f, 214f

Ungulatomorpha, 10t, 140, 214
unguligrade stance, 40, 244, 290, 290f, 302
Unuchinia, 104
Uranotheria, 242, 242t
Ursida, 121t
Ursidae, 9t, 120f, 121t, 126, 128, 129f, 135–136, 136f, 342
Ursoidea, 121t, **135–136**
Ursus, 32f, 33f, 219f
Utaetus, 201, 201f
Utah, 76, 78, 255
Uzbekistan, 90, 92, 148, 154, 307, 339
　　zhelestids, 213

Vacan subage, 17
vagus nerve, 26
Valerilagus, 314
Venezuela, 237
venomous mammals, 101
Vermilingua, 199t, 200, 200f, 202–204
vertebrae, 30–31, 31f, 41, 44
　　anticlinal, 31
　　archaeocete, 281
　　astrapothere, 236
　　body (centrum), 30
　　caudal. *See* caudal vertebrae
　　cervical. *See* cervical vertebrae
　　hyrax, 257
　　lumbar. *See* lumbar vertebrae
　　morganucodont, 52–53
　　sacral. *See* sacral vertebrae
　　thoracic. *See* thoracic vertebrae
　　xenarthrous, 71, 200, 200f, 201, 203, 210
　　zygapophyses, 30, 31f
vertebral foramen. *See* foramen/foramina, vertebral
vertebralarterial canal, 301–302
Vespertiliavus, 161–162
Vespertilionidae, 158t, 161
vestibulocochlear nerve, 25–26
Victorlemoinea, 221–222, 235, 235f
Vincelestes, 51f, 64–66, 65f
Vincelestidae, 49t
Virginia, 337
Viverra, 37f, 131
Viverravidae, 120f, 121t, 122f, 127, 129, 130, 130f, 339, 342
Viverravus, 130f
Viverridae, 120f, 121t, 126, 131–132
volant mammals, skeletal adaptations, 39
volar pads
　　hyracoid, 257–258
　　proboscidean, 260
Volitantia, 157f, 158, 166

vomer, 26
vomeronasal organ, 26, 182
Vulpavus, 34f, 36f, 131, 131f, 132f

Wadilemur, 187
Wailekia, 186
Walbeck (Germany), 338
Wales, 52
Walia, 162
walruses, 9t, 35, 126
Wasatchian NALMA, 12, 14f, 16, 112f, 113, 117, 122, 126, 159, 176, 178, 206, 207, 244, 252, 288, 322, 326, 339
Wasatchian/Bridgerian boundary, 16
Washakius, 191
weasels, 9t, 126, 136
Western Interior (North America), 3, 337–338
　　Paleocene-Eocene climate and flora, 21
whale(s), 35, 40, 268, 271, 275, 276, 336, 345–346.
　　　See also Cetacea
　　classification, 8, 10t, 211–212
　　communication by, 278
　　ear regions, 277, 280
White River Group, 134, 135
Whitneyan, 294
Widanelfarasia, 148
wings, of bats, 158, 161, 161f, 162f
Worlandia, 164, 165f
Worlandiinae, 164
Wortmania, 107, 107f, 108f
Woutersia, 54–55
Wutu (Shandong Province, China), 172, 207, 244, 339
Wyolestes, 99, 99f
Wyolestidae, 10t, 94–95, 95t, 97–99
Wyoming, 16, 55, 59, 101, 104, 105, 113, 116, 122, 146, 159, 160, 164, 189, 206, 206f, 215, 222, 226, 255, 321, 331, 337, Plate 3.3, Plate 6
　　Paleocene-Eocene climate and flora, 20
Wyonycteris, 159, 160f

Xanthorhysis, 187f, 191–192
Xenarthra, 8, 12f, 22, 31, 39, 71, 198–204, 336–337
　　age of divergence, 5t
　　classification and relationships, 9t, 139f, 198–199, 199t, 200f
　　Early Cenozoic, of Southern Hemisphere, 336–337
　　origin, 201
　　Paleocene-Eocene, synopsis, **343–344**
xenarthrous articulations. *See* vertebrae, xenarthrous
Xenicohippus, 247

Xenocranium, 207–208, 208f, 209f
Xenungulata, 22, 215, 226, **238,** 238f, 240, 306, 336, 337, 344, 345
　　classification and relationships, 10t, 213t, 226
xiphisternum, 31
Xiphodon, 298f, 301
Xiphodontidae, 286f, 289t, 297, **301,** 346

Yangochiroptera, 158t
Yinochiroptera, 158t
Yixian Formation (China), 74, 89
Yoderimys, 331
Ypresian Stage/Age, 12, 13f–14f, 244, 259, 263, 288, 292, 321, 338
Yuesthonyx, 110, 113
Yuomyidae, 307t, 332

Zaglossus, 69
Zaisaneomys, 331
Zalambdalestes, 90, 92f, 306, 307, 308f, Plate 1.4
Zalambdalestidae, 4, 9t, 89f, 114, 140, 306, **307–309,** 307t, 308f, 342, 346–347
Zalambdalestoidea, 148
zalambdodonty/zalambdodont dentition, 34–35, 38f, 95, 96, 114, 144, 147–150, 152, 153
Zapodidae, 307t, 339
Zatheria, 9t, 49t, 66
Zegdoumyidae, 307t, 331, 347
Zeuglodon, 284
Zhailimeryx, 304
Zhangheotherium, 51f, 63f, 64, 74
Zhelestes, 140, 213
Zhelestidae, 4, 79, 92, 93, 140, 213, 213f, 214–215, 344
　　classification, 10t, 89f, 212f, 227f
Zhujegale, 215
zonation, floristic, 21
zygapophyses, 30, 31, 31f, 200, 200f, 204
zygodonty, 264
zygomasseteric anatomy (rodents), 318, 318f
zygomatic arch, 24, 26, 57
　　apternodont, 153
　　entelodont, 295
　　erinaceid, 144
　　geolabidid, 149
　　lipotyphlan, 143
　　notoungulate, 228
　　palaeoryctid, 96
　　plesiosoricid, 153
　　rodent, 318
　　taeniodont, 109
　　talpid, 155
zygomatic bones, 26
Zygorhiza, 284

CREDITS

Permission to reproduce the following figures was generously provided by the individuals and institutions listed below.

1.2, J. D. Archibald; 1.4, J. R. Wible.

2.1, J. R. Wible; 2.2, T. M. Bown; 2.3A, American Philosophical Society; 2.3B,D, Department of Library Services, American Museum of Natural History; 2.3C, J. I. Bloch; 2.3E, J. A. Wilson; 2.6–2.15, 2.16D, illustrations by Elaine Kasmer.

3.1, A. W. Crompton; 3.2, from F. A. Jenkins, Jr., and F. R. Parrington, 1976, The postcranial skeleton of the Triassic mammals *Eozostrodon*, Megazostrodon and *Erythrotherium*. Philosophical Transactions of the Royal Society of London 273:387–431. With permission of the Royal Society and F. A. Jenkins, Jr.; 3.3A, Museum of Comparative Zoology, Harvard University; 3.3B, F. A. Jenkins, Jr., and the Paleontological Society; 3.4A,B, J. A. Hopson and the Museum of Comparative Zoology, Harvard University; 3.4C, A. W. Crompton and the Linnean Society of London; 3.5, Museum of Comparative Zoology, Harvard University; 3.6A,C, A. W. Crompton; 3.6B, J. F. Bonaparte and the Museum of Comparative Zoology, Harvard University; 3.6D, fig. 4.14 from A. W. Crompton and Z.-X. Luo, Relationships of the Liassic mammals *Sinoconodon, Morganucodon oehleri,* and *Dinnetherium;* pp. 30–44 in Placental Mammals: Mesozoic Differentiation, Multituberculates, Monotremes, Early Therians, and Marsupials (F. S. Szalay, M. J. Novacek, and M. C. McKenna, eds.). Copyright 1993 Springer-Verlag. By permission of Springer-Verlag and A. W. Crompton; 3.7A, H.-D. Sues; 3.7B, from Kühne, 1956, with permission, © The Natural History Museum, London; 3.8, G. Hahn.

4.1, R. L. Cifelli and the Paleontological Society; 4.3, F. A. Jenkins, Jr.; 4.4A, A. W. Crompton and the Linnean Society of London; 4.4B,C, Linnean Society of London; 4.4D, Figure 18 from F. A. Jenkins, Jr., and F. R. Parrington, 1976, The postcranial skeleton of the Triassic mammals *Eozostrodon*, Megazostrodon and *Erythrotherium*. Philosophical Transactions of the Royal Society of London 273:387–431. With permission of the Royal Society and F. A. Jenkins, Jr.; 4.5, J. A. Hopson, A. W. Crompton, and the Paleontological Society; 4.6, illustrations by Mark Klingler, Carnegie Museum of Natural History, from Z.-X. Luo, A. W. Crompton, and A.-L. Sun, A new mammaliaform from the Early Jurassic and evolution of mammalian characteristics. Science 292:1535–1540. Copyright 2001 AAAS. Reprinted with permission from Z.-X. Luo, M. Klingler, and AAAS; 4.7A, J. A. Lillegraven; 4.7B, F. A. Jenkins, Jr., and Peabody Museum of Natural History, Yale University; 4.8, Z. Kielan-Jaworowska and Acta Palaeontologica Polonica; 4.9, Z. Kielan-Jaworowska and the Palaeontological Association; 4.10, reproduced from Kielan-Jaworowska and Gambaryan, 1994, Postcranial anatomy and habits of Asian multituberculate mammals. Fossils and Strata 36:1–92, www.tandf.no/fossils, by permission of Taylor & Francis AS; 4.11A–C, D. W. Krause and the Paleontological Society; 4.11D, Society of Vertebrate Paleontology; 4.12, from F. A. Jenkins, Jr., and D. W. Krause, Adaptations for climbing in North American multituberculates (Mammalia). Science 220:712–715. Copyright 1983 AAAS. Reprinted with permission from F. A. Jenkins, Jr., D. W. Krause, and AAAS; 4.13A, from Simpson, 1928, with permission, © The Natural History Museum, London; 4.13B, F. A. Jenkins, Jr., and the Society of Vertebrate Paleontology; 4.13C,D, Z.-X. Luo, illustration in 4.13D by Mark Klinger, Carnegie Museum of Natural History; 4.14A, reprinted with permission from A. W. Crompton and F. A. Jenkins, Jr., American Jurassic symmetrodonts and Rhaetic pantotheres. Science 155: 1006–1008. Copyright 1967 AAAS; 4.14B, from R. L. Cifelli and S. K. Madsen, 1999, Spalacotheriid symmetrodonts (Mammalia) from the medial Cretaceous (upper Albian or lower Cenomanian) Mussentuchit local fauna, Cedar Mountain Formation, Utah, USA. Geodiversitas 21:167–214. © Publications Scientifiques du Muséum National d'Histoire Naturelle, Paris; with permission of the Muséum National d'Histoire Naturelle, Paris, and R. L. Cifelli; 4.14C, Z.-X. Luo; 4.15B, G. Rougier; 4.16C, reprinted from A. W. Crompton and Z. Kielan-Jaworowska, Molar structure and occlusion in Cretaceous therian mammals, pp. 249–287 in Development, Function and Evolution of Teeth (P. M. Butler and K. A. Joysey, eds.), copyright 1978, with permission from A. W. Crompton and Elsevier; 4.16D, D. Dashzeveg; 4.17, T. Martin; 4.18, Z.-X. Luo and Acta Palaeontologica Polonica; 4.19, P. M. Butler and the Museum of Comparative Zoology, Harvard University; 4.20A,B, J. F. Bonaparte; 4.20C, R. Pascual and the Society of Vertebrate Paleontology.

5.1, Surrey Beatty & Sons; 5.3, illustrations by Mark Klingler, Carnegie Museum of Natural History, from Z.-X. Luo, Q. Ji, J. R. Wible, and C.-X. Yuan, 2003, An Early Cretaceous tribosphenic mammal and metatherian evolution. Science 302:1934–1940. Copyright 2003 AAAS. Reprinted with permission from M. Klingler, Z.-X. Luo, and AAAS; 5.4A, G. Rougier; 5.4B, Z. Kielan-Jaworowska; 5.5, F. S. Szalay and the Society of Vertebrate Paleontology; 5.6, fig. 1 from R. L. Cifelli, 1993a, Early Cretaceous mammal from North America and the evolution of marsupial dental characters, Proceedings of the National Academy of Sciences, USA 90:9413–9416. Copyright 1993 National Academy of Sciences, USA. With permission of R. L. Cifelli and NAS; 5.6B, R. C. Fox; 5.6C, 5.8A,C,D, from W. A. Clemens, Jr., Fossil mammals of the type Lance Formation, Wyoming. Part II. Marsupialia, © 1966 The Regents of the University of California; 5.7, fig. 14.2 from R. L. Cifelli, Theria of metatherian-eutherian grade and the origin of marsupials; pp. 205–215 in Placental Mammals: Mesozoic Differentiation, Multituberculates, Monotremes, Early Therians, and Marsupials (F. S. Szalay, M. J. Novacek, and M. C. McKenna, eds.). Copyright 1993 Springer-Verlag. By permission of Springer-Verlag and R. L. Cifelli; 5.8B, R. C. Fox and Surrey Beatty & Sons; 5.9A,B, C. Kurz and the Hessisches Landesmuseum Darmstadt, Germany; 5.9C,D, W. von Koenigswald and G. Storch; 5.10, Surrey Beatty & Sons; 5.11, 5.12D, C. de Muizon; 5.12A, from W. A. Clemens, Jr., Fossil mammals of the type Lance Formation, Wyoming. Part II. Marsupialia, © 1966 The Regents of the University of California; 5.12B,C, Surrey Beatty & Sons; 5.13, Department of Library Services, American Museum of Natural History; illustrations by John LeGrand; 5.14, R. Pascual and the Society of Vertebrate Paleontology; 5.15B, Department of Library Services, American Museum of Natural History; 5.15C, C. de Muizon; 5.15D, Museum of Comparative Zoology, Harvard University; 5.16, S. Hand.

6.2, 6.3A, illustrations by Mark Klingler, Carnegie Museum of Natural History, reprinted with permission of Z.-X. Luo and M. Klingler; 6.3B, R. L. Cifelli; 6.3C, 6.4A, M. C. McKenna and Acta Palaeontologica Polonica; 6.3D, 6.4B–D, Z. Kielan-Jaworowska; 6.3E, M. J. Novacek; 6.5A, P. M. Butler; 6.5B, P. M. Butler and Acta Palaeontologica Polonica; 6.5C, PaleoBios and the University of California Museum of Paleontology.

7.1A, J. A. Lillegraven; 7.1B, from W. A. Clemens, Jr., Fossil mammals of the type Lance Formation, Wyoming. Part III. Eutheria and summary, © 1973 The Regents of the University of California; 7.2A, P. D. Gingerich and the University of Michigan Museum of Paleontology; 7.2C, J. I. Bloch, P. D. Gingerich, and the University of Michigan Museum of Paleontology; 7.3, P. D. Gingerich and the Society of Vertebrate Paleontology; 7.4, reprinted with permission from F. S. Szalay and G. Drawhorn, Evolution and diversification of the Archonta in an arboreal milieu; pp. 133–169 in W. P. Luckett (ed.), Comparative Biology and Evolutionary Relationships of Tree Shrews. Plenum. Copyright 1980, Kluwer Academic/Plenum Publishers; 7.5A, Department of Library Services, American Museum of Natural History; 7.5B, X. Wang and the Society of Vertebrate Paleontology; 7.6, M. J. Novacek and the Department of Library Services, American Museum of Natural History; photographs by Chester S. Tarka; 7.7A, Department of Library Services, American Museum of Natural History. Illustrations by Mildred Klemens; 7.7B, E. Gheerbrant and Palaeontographica; 7.7C, Smithsonian Institution; 7.7D, W. von Koenigswald; 7.8, H.-U. Pfretzschner. Restoration by D. Kranz; 7.9, W. von Koenigswald. Restoration by D. Kranz; 7.10, figs. 1 and 2 from E. L. Simons and T. M. Bown, 1995, Ptolemaiida, a new order of Mammalia—With description of the cranium of *Ptolemaia grangeri*. Proceedings of the National Academy of Sciences, USA 92:3269–3273. Copyright 1995 National Academy of Sciences, USA. Reprinted with permission of E. L. Simons and NAS; 7.11B P. D. Gingerich and the University of Michigan Museum of Paleontology; 7.12, American Philosophical Society; 7.13, W. von Koenigswald and Palaeontographica. Drawings and restoration by D. Kranz; 7.14, W. von Koenigswald and Palaeontographica; 7.15A, Paleontological Society; 7.15B–F, R. M. Schoch and the Peabody Museum of Natural History, Yale University; 7.16A–C, American Philosophical Society; 7.16D, R. C. Fox; 7.16E, from G. L. Jepsen, E. Mayr, and G. G. Simpson (eds.), Genetics, Paleontology, and Evolution. Copyright 1949, Princeton University Press, 1977 renewed PUP. Reprinted by permission of Princeton University Press; 7.17, 7.18A, R. M. Schoch and the Peabody Museum of Natural History, Yale University; 7.18B, American Philosophical Society; 7.20C, Smithsonian Institution; 7.20D, Department of Library Services, American Museum of Natural History. Illustrations by L. M. Sterling; 7.20E,F, Y. Tomida and Paleontological Research; 7.21A, Smithsonian Institution; 7.21B, Peabody Museum of Natural History, Yale University; 7.22A,B, 7.23A, B.-Y. Wang; 7.22C, 7.23B, C. de Muizon and the Paleontological Society; 7.22D,E, 7.23C,E, American Philosophical Society; 7.22F, 7.23D,F, Department of Library Services, American Museum of Natural History; 7.24, 7.25A,B, 7.26L, E. L. Simons and the American Philosophical Society.

8.1A, J. J. Flynn; 8.3A, P. D. Gingerich and the Paleontological Society; 8.3C, Carnegie Museum of Natural History; 8.3D, 8.4A–C, from R. H. Denison, 1938, The broad-skulled Pseudocreodi. Annals of the New York Academy of Sciences 37:163–256. © 1938 New York Academy of Sciences, USA; 8.3E, University of Michigan Museum of Paleontology; 8.5A, P. D. Gingerich and the University of Michigan Museum of Paleontology; 8.5B, Society of Vertebrate Paleontology. Illustration by Elaine Kasmer; 8.5C, B. Lange-Badré; 8.6B, J. S. Mellett; 8.7A, Society of Vertebrate Paleontology. Illustration by Elaine Kasmer; 8.8, M. Morlo and the Hessisches Landesmuseum Darmstadt, Germany; 8.9, fig. 5.5 from R. M. Hunt, Jr., and R. H. Tedford, Phylogenetic relationships within the aeluroid Carnivora and implications of their temporal and geographic distribution; pp. 53–73 in Mammal Phylogeny: Placentals (F. S. Szalay, M. J. Novacek, and M. C. McKenna, eds.). Copyright © 1993 Springer-Verlag. By permission of R. M. Hunt, Jr., and Springer-Verlag; 8.10A, from W. A. Clemens, Jr., Fossil mammals of the type Lance Formation, Wyoming. Part III. Eutheria and summary, © 1973 The Regents of the University of California; 8.10B–D,F, P. D. Gingerich and the University of Michigan Museum of Paleontology; 8.10E, R. C. Fox and the Society of Vertebrate Paleontology; 8.12, Jay Matternes, illustrations © Jay Matternes; 8.13A, L. de Bonis and the Society of Vertebrate Paleontology; 8.13B, R. M. Hunt, Jr., and Department of Library Services, American Museum of Natural History; 8.13C,D, L. de Bonis; 8.14 (skull), 8.15: American Philosophical Society; 8.16B–E, X. Wang and Department of Library Services, American Museum of Natural History; 8.17B, L. de Bonis and Palaeontographica; 8.17C, L. de Bonis; 8.17D, 8.18, American Philosophical Society.

9.2A,C, J. A. Lillegraven; 9.2B, M. J. Novacek and PaleoBios and the University of California Museum of Paleontology; 9.3, M. J. Novacek and the Department of Library Services, American Museum of Natural History; drawing by S. B. McDowell, photographs by Chester S. Tarka; 9.4A,B,D, G. Storch; 9.4C, E. Frey; 9.5A, E. Gheerbrant and Palaeontographica; 9.5B, J.-Y. Crochet; 9.5C, E. Gheerbrant; 9.6A,B, P. D. Gingerich and the University of Michigan Museum of Paleontology; 9.6C, J. A. Lillegraven; 9.6D, Geological Survey of Hesse, Germany; 9.7, G. Storch and the Forschungsinstitut Senckenberg, Frankfurt, Germany; 9.8, W. von Koenigswald, G. Storch, and the Hessisches Landesmuseum Darmstadt, Germany; 9.9A, M. J. Novacek and the Department of Library Services, American Museum of Natural History. Photography by Chester S. Tarka; 9.9B, P. D. Gingerich and the University of Michigan Museum of Paleontology; 9.9C, B.-Y. Wang; 9.10A, D. Sigogneau-Russell and Palaeovertebrata; 9.10B, J. A. Lillegraven; 9.10C, J. I. Bloch; 9.10D, from M. C. McKenna, The Geolabidinae, a new subfamily of early Cenozoic erinaceoid insectivores, © 1960 The Regents of the University of California; with permission of M. C. McKenna and University of California Press; 9.11A, upper teeth, M. C. McKenna and the Department of Library Services, American Museum of Natural History. Drawing by Chester S. Tarka. Lower teeth, University of Michigan Museum of Paleontology. Drawing by Karen Klitz; 9.11B,C, from B. Sigé, 1976, Insectivores primitifs de l'Eocène supérieur et Oligocène inférieur d'Europe occidentale. Nyctithériidés. Mémoires du Muséum National d'Histoire Naturelle, Série C, Sciences de la Terre 34:1–140. © Publications Scientifiques du Muséum National d'Histoire Naturelle, Paris, with permission of B. Sigé and the Muséum National d'Histoire Naturelle, Paris; 9.11D, illustration by Karen Klitz; 9.12A, X. Wang and the Society of Vertebrate Paleontology; 9.12B, upper teeth, The Paleontological Society; lower teeth, J. M. Rensberger and the Southern California Academy of Sciences; 9.12C, fig. 24C from C. A. Reed and W. D. Turnbull, 1965, The mammalian genera *Arctoryctes* and *Cryptoryctes* from the Oligocene and Miocene of North America, Fieldiana Geology 15:99–170. Reprinted with permission of W. D. Turnbull and Field Museum of Natural History; 9.13, 9.14, photographs and life restoration with permission of R. Asher and the Department of Library Services, American Museum of Natural History. Illustrations by Chester S. Tarka; 9.15A, B.-Y. Wang; 9.15B, J.-Y. Crochet, B. Engesser, and Palaeovertebrata; 9.15C, B. Engesser; 9.15D, A. Lopatin; 9.15E, from Crochet, 1974, with permission of J.-Y. Crochet and Palaeovertebrata; 9.15F, B. Sigé.

10.2A, R. S. Rana and Acta Palaeontologica Polonica; 10.2B, M. Godinot and the Paleontological Society; 10.3A, D. E. Russell and Palaeovertebrata; 10.3B, Y. Tong; 10.3C, P. D. Gingerich and the University of Michigan Museum of Paleontology; 10.4A, G. Storch; 10.4B, J. J. Hooker and Palaeovertebrata; 10.4C; K. C. Beard; 10.5, reprinted from G. L. Jepsen, Bat origins and evolution, pp. 1–64 in Biology of Bats (W. A. Wimsatt, ed.). Copyright 1970, with permission from Elsevier; 10.6, J. Habersetzer; 10.7B, M. C. McKenna. Illustrations by Owen Poe; 10.7C, S. Ducrocq and the Paleontological Association; 10.8A, Smithsonian Institution; 10.8B, Department of Library Services, American Museum of Natural History. Illustrations by Mildred Klemens; 10.8C, upper teeth, Society of Vertebrate Paleontology. Drawing by Elaine Kasmer; lower teeth, drawing by R. B. Horsfall; 10.8D, upper teeth, University of Michigan Museum of Paleontology. Drawing by Karen Klitz; lower teeth, drawing by Jennifer Emry; 10.12, illustrations by Elaine Kasmer; 10.13A, P. D. Gingerich; 10.13B, J. I. Bloch and the University of Michigan Museum of Paleontology; 10.13C,D, K. C. Beard; 10.14A,C, illustrations by Elaine Kasmer; 10.14B,D, Department of Library Services, American Museum of Natural History. Illustrations by Chester S. Tarka; 10.15A, R. C. Fox and the Society of Vertebrate Paleontology; 10.15B, J. I. Bloch and the University of Michigan Museum of Paleontology; 10.16, 10.17, illustrations by Elaine Kasmer; 10.19, photographs by Lorraine Meeker; 10.20A–C, F. S. Szalay; 10.20D, J.-L. Hartenberger; 10.21, left, P. D. Gingerich; right, F. S. Szalay; 10.22A, fig. 4 from E. L. Simons and D. T. Rasmussen, 1994, A remarkable cranium of *Plesiopithecus teras* (Primates, Prosimii) from the Eocene of Egypt. Proceedings of the National Academy of Sciences, USA 91:9946–9950. Copyright 1994 National Academy of Sciences, USA. With permission of E. L. Simons and NAS; 10.22B, from L. Marivaux, J.-L. Welcomme, P.-O. Antoine, G. Métais, I. M. Baloch, M. Benammi, Y. Chaimanee, S. Ducrocq, and J.-J. Jaeger. A fossil lemur from the Oligocene of Pakistan. Science 294:587–591. Copyright 2001 AAAS. Reprinted with permission of J.-J. Jaeger and AAAS; 10.22C, E. L. Simons; 10.22D, K. C. Beard and the Carnegie Museum of Natural History; 10.23A,B, F. S. Szalay; 10.23C, E. L. Simons; 10.23D, K. C. Beard and Carnegie Museum of Natural History; 10.24A, The Paleontological Society. Illustrations by Elaine Kasmer; 10.24B,C, F. S. Szalay; 10.24D, Schweizerische Paläontologische Abhandlungen; 10.24E, M. Godinot; 10.25, The Paleontological Society. Illustration by Elaine Kasmer; 10.26A, from K. C. Beard, Y. Tong, M. R. Dawson, J. Wang, and X. Huang. Earliest complete dentition of an anthropoid primate from the late middle Eocene of Shanxi Province, China. Science 272:82–85. Copyright 1996 AAAS. Reprinted with permission of K. C. Beard and AAAS; 10.26B, from J.-J. Jaeger, T. Thein, M. Benammi, Y. Chaimanee, A. N. Soe, T. Lwin, T. Tun, S. Wai, and S. Ducrocq. A new primate from the middle Eocene of Myanmar and the Asian early origin of anthropoids. Science 286:528–530. Copyright 1999 AAAS. Drawings by L. Meslin, reprinted with permission of J.-J. Jaeger and AAAS; 10.26C, R. L. Ciochon; 10.26D, S. Ducrocq; 10.27A,B,E,F, 10.28A,B, E. L. Simons; 10.27C, M. Godinot; 10.27D, 10.28C, reprinted from J. G. Fleagle, Primate Adaptation and Evolution, second edition, copyright 1999, with permission from J. G. Fleagle and Elsevier; 10.29C, Y. Tong.

11.2, fig. 7.3 from K. D. Rose and R. J. Emry, Relationships of Xenarthra, Pholidota, and fossil edentates: The morphological evidence; pp. 81–102 in Mammal Phylogeny: Placentals (F. S. Szalay, M. J. Novacek, and M. C. McKenna, eds.). Copyright 1993 Springer-Verlag. By permission of Springer-Verlag. Illustration by Elaine Kasmer; 11.3A, L. Bergqvist; 11.3B,C, Department of Library Services, American Museum of Natural History; 11.3E, Society of Vertebrate Paleontology; 11.4, G. Storch; 11.5, G. Storch and the Forschungsinstitut Senckenberg, Frankfurt, Germany; 11.6, R. J. Emry. Illustration by Mary Parrish; 11.7A, Society of Vertebrate Paleontology. Illustration by Elaine Kasmer; 11.7B, Department of Library Services, American Museum of Natural History. Illustration by John Germann; 11.7C, S.-Y. Ting; 11.8, illustrations by Karen Klitz; 11.9, illustrations by D. P. Bichell; 11.10A, Smithsonian Institution; 11.10B, B. Dalzell. Illustration by Bonnie Dalzell; 11.10C, illustration by L. B. Isham.

12.2, J. D. Archibald; 12.3A, Z.-X. Luo and the Society of Vertebrate Paleontology; 12.3B–D, American Philosophical Society; 12.4A, from D. E. Russell, 1964, Les mammifères Paléocène d'Europe. Mémoires du Muséum National d'Histoire Naturelle, Série C, Sciences de la Terre 13:1–324. © Publications Scientifiques du Muséum National d'Histoire Naturelle, Paris, with permission of D. E. Russell and the Muséum National d'Histoire Naturelle, Paris; 12.4B,C, American Philosophical Society; 12.5A, P. D. Gingerich; 12.6, illustrations by Elaine Kasmer. With permission from Science; © 1987 AAAS; 12.7, Department of Library Services, American Museum of Natural History; 12.8A, Smithsonian Institution; 12.8B, American Philosophical Society; 12.8C, R. L. Cifelli; 12.8E, University of Michigan Museum of Paleontology; 12.9B,C, American Philosophical Society; 12.10A,B, Smithsonian Institution; 12.11B, illustrations by Elaine Kasmer; 12.12A, from R. L. Cifelli and C. R. Schaff, 1998, Arctostylopida; pp. 332–336 in Evolution of Tertiary Mammals of North America, volume 1, edited by C. M. Janis, K. M. Scott, and L. L. Jacobs. Reprinted with the permission of R. L. Cifelli and Cambridge University Press; 12.12B Department of Library Services, American Museum of Natural History; 12.14, from C. de Muizon and R. L. Cifelli, 2000, The condylarths (archaic Ungulata, Mammalia) from the early Paleocene of Tiupampa (Bolivia): Implications on the origin of the South American ungulates. Geodiversitas 22:47–150. © Publications Scientifiques du Muséum National d'Histoire Naturelle, Paris, with permission of C. de Muizon and the Muséum National d'Histoire Naturelle, Paris; 12.16–12.19A, 12.20A,B, Department of Library Services, American Museum of Natural History. Illustration in 12.17 by Chester S. Tarka; 12.20C, fig. 1 from Paula Couto, 1970, in Iheringia Geologia 3, by permission; 12.21, Department of Library Services, American Museum of Natural History. Illustration by Chester S. Tarka; 12.22A, Department of Library Services, American Museum of Natural History. Illustration by M. T. Cabrera; 12.22B, Museum of Comparative Zoology, Harvard University; 12.23, R. L. Cifelli; 12.24, from Paula Couto, 1979, in Tratado de Paleomastozoologia. Reprinted by permission of Academia Brasileira de Ciências; 12.25A–C, Department of Library Services, American Museum of Natural History; 12.25D, Museum of Comparative Zoology, Harvard University; 12.26B,C, from Simpson, 1967, Department of Library Services, American Museum of Natural History; 12.27B, C. de Muizon; 12.27C, B. J. MacFadden and the Palaeontological Association; 12.28, from Paula Couto, 1978, in Anais da Academia Brasileira de Ciências 50:203–207. Reprinted by permission of Academia Brasileira de Ciências.

13.1, D. P. Domning; 13.2A, University of Michigan Museum of Paleontology; 13.2B, M. C. McKenna; 13.2C, P. D. Gingerich and the University of Michigan Museum of Paleontology; 13.3, 13.4, illustrations by Elaine Kasmer; 13.6A, P. D. Gingerich and the University of Michigan Museum of Paleontology; 13.6B, Museum of Comparative Zoology, Harvard University; 13.6E, Department of Library Services, American Museum of Natural History. Illustration by Helen Ziska; 13.8A, Smithsonian Institution; 13.8B, from D. E. Savage, D. E. Russell, and P. Louis, European Eocene Equidae (Perissodactyla). © 1965 The Regents of the University of California, with permission of D. E. Russell and University of California Press; 13.8D, J. L. Franzen; 13.9, J. L. Franzen and the Forschungsinstitut Senckenberg, Frankfurt, Germany; 13.13A–C, Museum of Comparative Zoology, Harvard University; 13.13D, J. L. Franzen; 13.14A,B, Peabody Museum of Natural History, Yale University; 13.15, teeth, Department of Library Services, American Museum of Natural History; 13.16B, *Saghatherium*, E. Gheerbrant; 13.17A, E. L. Simons and the Society of Vertebrate Paleontology; 13.17C, E. Gheerbrant and J.-J. Jaeger; 13.19A, E. Gheerbrant; 13.19B–D, P. D. Gingerich and the University of Michigan Museum of Paleontology; 13.19E,F, E. Gheerbrant and Acta Palaeontologica Polonica; 13.20, E. Gheerbrant; 13.21A, J.-J. Jaeger; drawing by D. Visset; 13.25B, M. C. McKenna; 13.26A, D. P. Domning and the Society of Vertebrate Paleontology; 13.26B, P. D. Gingerich and the University of Michigan Museum of Paleontology; 13.27, D. P. Domning.

14.2A, P. D. Gingerich and the Society of Vertebrate Paleontology; 14.2B, American Philosophical Society; 14.2C, Society of Vertebrate Paleontology. Illustration by Elaine Kasmer; 14.3A, P. D. Gingerich and the University of Michigan Museum of Paleontology; 14.4A–C,E,F, Society of Vertebrate Paleontology. Illustrations by E. Kasmer; 14.5, 14.6, R. E. Fordyce; 14.7, P. D. Gingerich; 14.8A, from P. D. Gingerich, D. E. Russell, and S. M. I. Shah. Origin of whales in epicontinental remnant seas: New evidence from the early Eocene of Pakistan. Science 220:403–406. Copyright 1983 AAAS. Reprinted with permission of P. D. Gingerich and AAAS; 14.8C, Paleontological Society; 14.9A, J. G. M. Thewissen; 14.9B, from P. D. Gingerich, M. ul Haq, I. S. Zalmout, I. H. Khan, and M. S. Malkani. Origin of whales from early artiodactyls: Hands and feet of Eocene Protocetidae from Pakistan. Science 293:2239–2242. Copyright 2001 AAAS. Illustration by D. M. Boyer. Reprinted with permission of P. D. Gingerich and AAAS; 14.9C, P. D. Gingerich, and the University of Michigan Museum of Paleontology; 14.10, from P. D. Gingerich, M. ul Haq, I. S. Zalmout, I. H. Khan, and M. S. Malkani. Origin of whales from early artiodactyls: Hands and feet of Eocene Protocetidae from Pakistan. Science 293:2239–2242. Copyright 2001 AAAS. Illustration by B. Miljour. Reprinted with permission of P. D. Gingerich and AAAS; 14.11, reprinted, with permission, from K. D. Rose. The ancestry of whales. Science 293:2216–2217. Copyright 2001 AAAS. Drawings by Elaine Kasmer; 14.12, M. D. Uhen. Restoration by Darryl Leja; 14.14, modified from Scott, 1940, with permission of the American Philosophical Society; 14.15, illustrations by Elaine Kasmer; 14.16, reprinted, with permission, from K. D. Rose. Skeleton of *Diacodexis*, oldest known artiodactyl. Science 216:621–623. Copyright 1982 AAAS. Drawings by D. Bichell; 14.17, J. L. Franzen and the Forschungsinstitut Senckenberg, Frankfurt, Germany; 14.18, J. Erfurt and Palaeovertebrata; 14.19B,E, J. Sudre; 14.19C, University of Michigan Museum of Paleontology; 14.19D, from J. Sudre and G. Lecomte, 2000, Relations et position systématique du genre *Cuisitherium* Sudre et al., 1983, le plus dérivé des artiodactyles de l'Eocène inférieur d'Europe. Geodiversitas 22:415–432. © Publications Scientifiques du Muséum National d'Histoire Naturelle, Paris; with permission of J. Sudre and the Muséum National d'Histoire Naturelle, Paris; 14.20A,B, 14.21B,C, 14.22A,B, American Philosophical Society; 14.21A, Carnegie Museum of Natural History; 14.22C, S. Ducrocq and the Palaeontological Association; 14.23, J. J. Hooker and Schweizerische Paläontologische Abhandlungen; 14.24A,B,D, J. Sudre; 14.24C, J. Sudre and Palaeovertebrata; 14.25A,D, 14.26A,B, 14.27A, American Philosophical Society; 14.25B,C, 14.26C, J. A. Wilson; 14.27B, from D. R. Prothero, 1996, Camelidae; pp. 609–651 in The Terrestrial Eocene-Oligocene Transition in North America, edited by D. R. Prothero and R. J. Emry. Reprinted with permission of D. R. Prothero and Cambridge University Press; 14.28, 14.29, 14.30 (*Leptomeryx*), American Philosophical Society; 14.30 (*Cainotherium*), B. Engesser; 14.30 (*Archaeomeryx*), Department of Library Services, American Museum of Natural History. Illustrated by John C. Germann.

15.2, Z. Kielan-Jaworowska; 15.3, Department of Library Services, American Museum of Natural History. Illustrations by John and Louise Germann; 15.4, Y. Hu; 15.5, fig. 14.7 from S. G. Lucas, Pantodonts, tillodonts, uintatheres, and pyrotheres are not ungulates; pp. 182–194 in Mammal Phylogeny: Placentals (F. S. Szalay, M. J. Novacek, and M. C. McKenna, eds.). Copyright 1993 Springer-Verlag. By permission of S. G. Lucas and Springer-Verlag; 15.6, P. M. Butler; 15.7, S. P. Zack; 15.8, figs. 11.1, 11.2, 11.3 from C.-K. Li and S.-Y. Ting, New cranial and postcranial evidence for the affinities of the eurymylids (Rodentia) and mimotonids (Lagomorpha); pp. 151–158 in Mammal Phylogeny: Placentals (F. S. Szalay, M. J. Novacek, and M. C. McKenna, eds.). Copyright 1993 Springer-Verlag. By permission of C.-K. Li, S.-Y. Ting and Springer-Verlag; 15.9, American Philosophical Society; 15.10, M. C. McKenna and the Society of Vertebrate Paleontology; 15.11, J. Meng and the Department of Library Services, American Museum of Natural History. Illustrations by Jin Meng; 15.12, 15.13, from Mammalogy, fourth edition, by Vaughan et al. Copyright 2000. Reprinted with permission of Brooks/Cole, a division of Thomson Learning: www.thomsonrights.com. Fax 800 730-2215; 15.14, D. Kalthoff; 15.15A, D. Dashzeveg and the Palaeontological Association; 15.15B, J. Meng; 15.15C, M. R. Dawson and Carnegie Museum of Natural History; 15.17A, American Philosophical Society; 15.17B,C, reprinted with permission from W. W. Korth, The Tertiary Record of Rodents in North America. Plenum. Copyright 1994 Kluwer Academic/Plenum Publishers; 15.18A, Jay Matternes, illustration © Jay Matternes; 15.18B, Forschungsinstitut Senckenberg, Frankfurt, Germany; 15.18C, G. Storch and the Forschungsinstitut Senckenberg, Frankfurt, Germany; 15.20, illustrations by Elaine Kasmer; 15.19A, 15.21 American Philosophical Society; 15.19B, J.-L. Hartenberger; 15.19C,D, Y.-S. Tong; 15.19F, Peabody Museum of Natural History, Yale University; 15.24, C.-K. Li.

Plate 1.1, L. L. Sadler, F. A. Jenkins, Jr., and D. W. Krause. Illustration by L. L. Sadler; Plate 1.2, C. de Muizon; Plates 1.3, I. Horovitz, image courtesy of M. J. Novacek; Plate 1.4, M. J. Novacek; Plate 2.1, photograph by G. Oleschinski; Plate 2.2, W. von Koenigswald and the Forschungsinstitut Senckenberg, Frankfurt, Germany; Plate 3.1, W. von Koenigswald and the Hessisches Landesmuseum Darmstadt, Germany; Plate 3.2, W. von Koenigswald. Photograph by J. Weinstock; Plate 3.3, Utako Kikutani. Illustration by Utako Kikutani; Plate 4.1, image by T. Smith, skeleton courtesy of J.-P. Cavigelli; Plate 4.2, G. Storch and the Forschungsinstitut Senckenberg, Frankfurt, Germany; Plate 4.3, D. M. Boyer. Restoration by D. M. Boyer; Plate 4.4, X.-J. Ni and C.-K. Li; Plate 5.1, N. Simmons; Plates 5.2–5.4, G. Storch and the Forschungsinstitut Senckenberg, Frankfurt, Germany; Plate 6, Utako Kikutani. Illustration by Utako Kikutani; Plate 7.1, G. Storch and the Forschungsinstitut Senckenberg, Frankfurt, Germany; Plate 7.2, illustration by Elaine Kasmer; Plate 7.3, Hessisches Landesmuseum Darmstadt, Germany; Plate 7.4, E. Gheerbrant; Plate 8.1, from http://darla.neoucom.edu/DEPTS/ANAT/Hans/AmbulocetusPhoto.jpg; Plate 8.2, J. Klausmeyer. Illustration by J. Klausmeyer; Plate 8.3, J. L. Franzen; Plate 8.4, J. Erfurt and the Institute of Geological Sciences and Geiseltalmuseum, Martin-Luther-University, Halle, Germany. Restoration by Pawel Major (Prag) under supervision of J. Erfurt and O. Fejfar.